Structure and Performance of Cements

Structure and Performance of Cements

Second edition

Edited by

J. Bensted and P. Barnes

CRC Press
Taylor & Francis Group
Boca Raton London New York

CRC Press is an imprint of the
Taylor & Francis Group, an **informa** business

A TAYLOR & FRANCIS BOOK

CRC Press
Taylor & Francis Group
6000 Broken Sound Parkway NW, Suite 300
Boca Raton, FL 33487-2742

First issued in paperback 2019

© 2002 by Taylor & Francis Group, LLC
CRC Press is an imprint of Taylor & Francis Group, an Informa business

Typeset in 10/12, Sabon by Newgen Imaging Systems (P) Ltd.

No claim to original U.S. Government works

ISBN-13: 978-0-419-23330-5 (hbk)
ISBN-13: 978-0-367-86539-9 (pbk)

British Library Cataloguing in Publication Data
A catalogue record for this book is available from the British Library

Library of Congress Cataloging in Publication Data

Structure and performance of cements/J. Bensted and P. Barnes [editors].–2nd ed.
   p. cm.
Includes bibliographical references and index.
1. Cement. I. Bensted, J. II. Barnes, P. (John)

TA434. S94 2001
624.1′833–dc21   2001020916

0-419-23330-X

Publisher's Note
The publisher has gone to great lengths to ensure the quality of this
reprint but points out that some imperfections in the original may be apparent.

Visit the Taylor & Francis Web site at
http://www.taylorandfrancis.com

and the CRC Press Web site at
http://www.crcpress.com

Contents

Contents

Contents

Contents

Contents

Contributors

Dr L. P. Aldridge
Australian Nuclear Science and Technology
  Organisation
Lucas Heights Research Laboratories
PMB7 1, Menai
Bangor
New South Wales 2234
Australia

M. Attfield
Industrial Materials Group
Department of Crystallography
Birkbeck College
University of London
Malet Street
London, WC1E 7HX
UK

Stephen P. Bailey
Industrial Materials Group
Department of Crystallography
Birkbeck College
University of London
Malet Street
London, WC1E 7HX
UK

Prof Paul Barnes
Industrial Materials Group
Department of Crystallography
Birkbeck College
University of London
Malet Street
London, WC1E 7HX
UK

Prof John Bensted
Industrial Materials Group
Department of Crystallography
Birkbeck College
University of London
Malet Street
London, WC1E 7HX
UK

Prof S. Chandra
Division of Applied Concrete Chemistry
Chalmers University of Technology
Sven Hulting Gata 8
S-412 96 Gothenburg
Sweden

Dr A. K. Chatterjee
The Associated Cement Companies Ltd
Research & Consultancy Directorate
CRS Complex
L.B.S. Marg
Thane 400 604
India

Dr T. M. Chrisp
Department of Civil and Offshore Engineering
Heriot-Watt University
Riccarton
Edinburgh, EH14 4AS
UK

Dr Sally L. Colston
Industrial Materials Group
Department of Crystallography
Birkbeck College

University of London
Malet Street
London, WC1E 7HX
UK

Dr D. Damidot
Lafarge Central Laboratory
95 rue du Montmurier
BO 15, 38291
Saint Quentin Fallavier
France

Dr Wolfgang Ehrfeld
IMM Institute of Microtechnology
GmbH
Carl-Zeiss-Strasse 18–20
D-55129 Mainz-Hechsteim
Germany

Dr C. Famy
Lafarge
Laboratoire Central de Recherche
95 Rue du Montmurier – B.P. 15 38291
St Quentin Fallavier Cedex
France

Dr Herbert Freimuth
IMM Institute of Microtechnology
GmbH
Carl-Zeiss-Strasse 18–20
D-55129 Mainz-Hechsteim
Germany

Dr E. M. Gartner
Lafarge
Laboratoire Central de Recherche
95 Rue du Montmurier – B.P. 15 38291
St Quentin Fallavier Cedex
France

Prof Christopher Hall
Division of Engineering and Centre for
Materials Science & Engineering
The University of Edinburgh
King's Buildings
Edinburgh EH9 3JL
UK

Dr I. Hinczak
Cementech Pty. Ltd.
PO Box 362
Liverpool
New South Wales 2170
Australia

Dr D. Hobbs
British Cement Association
Century House
Telford Avenue
Crowthorne
Berkshire, RG11 6YS
UK

Dr S. D. M. Jacques
Industrial Materials Group
Department of Crystallography
Birkbeck College
University of London
Malet Street
London, WC1E 7HX
UK

Dr Hans J. Jakobsen
Instrument Centre for Solid-State
 NMR Spectroscopy
Department of Chemistry
University of Aarhus
DK-8000 Aarhus C
Denmark

Dr I. Jawed
Transportation Research Board
National Research Council
Washington D.C. 20418
USA

Dr Tom R. Jones
New Materials Group
Imerys Minerals Ltd
John Keay House
St Austell
Cornwall, PL25 4DJ
UK

Dr A. C. Jupe
Industrial Materials Group
Department of Crystallography
Birkbeck College
University of London
Malet Street
London, WC1E 7HX
UK

Prof H. Justnes
SINTEF
Civil & Environmental Engineering Group
Cement & Concrete Group
N-7034 Trondheim
Norway

Prof Wiesław Kurdowski
Institute of Building Materials
Academy of Mining and Metallurgy
Al. Mickiewicza
Pawilion A3
Krakow
Poland

Dr E. Lang
Abteilungsleiter
Forschungsgemeinschaft
 Eisenhuettenschlacken EV
Forschungsinstitut
Bliersheimer Str 62
D-47229 Duisberg
Germany

P. Livesey
Castle Cement Ribblesdale Ltd
Clitheroe
Lancashire BB7 4QF
UK

Dr Karen Luke
Now at:
Halliburton Energy Services
Duncan Technology Center
2600 South 2nd Street
Duncan, Oklahoma 73536
USA

Dr S. Lunt
Thermo VG Scientific
Imberhorne Lane
West Sussex, RH19 1UB
UK

Prof F. Massazza
Laboratorio Centrale
Italcementi-SCF S.p.A.
Via Gabriele Camozzi 124
I-24100 Bergamo
Italy

Dr C. E. Matulis
CSIRO Coal and Energy Technology Division
Lucas Heights Research Laboratories
PMB7
Bangor
New South Wales 2234
Australia

Prof W. J. McCarter
Department of Civil and Offshore Engineering
Heriot-Watt University
Riccarton
Edinburgh, EH14 4AS
UK

S. Morgan
Industrial Materials Group
Department of Crystallography
Birkbeck College
University of London
Malet Street
London, WC1E 7HX
UK

R. Pisula
Industrial Materials Group
Department of Crystallography
Birkbeck College
University of London
Malet Street
London, WC1E 7HX
UK

Dr David O'Connor
Industrial Materials Group
Department of Crystallography
Birkbeck College
University of London
Malet Street
London, WC1E 7HX
UK

Prof Herbert Pöllman
Institut für Geologische Wissenschaften und
 Geiseltalmuseum
Martin Luther Universität
D-06099 Halle (Saale)
Germany

Dr I. Richardson
Civil Engineering Materials Unit
School of Civil Engineering
University of Leeds
Leeds, LS2 9JT
UK

Prof K. L. Scrivener
Lafarge
Laboratoire Central de Recherche
95 Rue du Montmurier – B.P. 15 38291
 St Quentin Fallavier Cedex
France

Dr Jørgen Skibsted
Instrument Centre for Solid-State
NMR Spectroscopy
Department of Chemistry
University of Aarhus

DK-8000 Aarhus C
Denmark

Dr G. Starrs
Department of Civil and Offshore Engineering
Heriot-Watt University
Riccarton
Edinburgh, EH14 4AS
UK

Prof H. F. W. Taylor
Maundry Bank
Lake Road
Coniston, Cumbria LA21 8EW
UK

Dr J. C. Taylor
CSIRO Coal and Energy Technology Division
Lucas Heights Research Laboratories
PMB7
Bangor
New South Wales 2234
Australia

Prof J. F. Young
Center for Cement Composite Materials
205 Ceramics Building
105 S. Goodwin Avenue
University of Illinois at Champaign-Urbana
Urbana, Illinois 61801
USA
Now at:
R.D.2 Katikati
New Zealand

Preface

Structure and Performance of Cements was first published in 1983. Much has changed over the last 18 years, but still, a surprisingly large number of the reasons given then for the need for such a book are equally valid today; these include:

- Cements constitute the second largest manufactured commodity (by weight) in the world
- Most peoples' lives are continually dependent upon the properties of cements
- Only a small fraction of income derived from the manufacture and marketing of cements is spent on research into the properties and applications of cements, including new developments

However, although the basic starting materials for the manufacture of most cements have not changed very much, the subject has in fact moved on appreciably. Cements today have a greater number and range of applications than ever before. The various types of cements now available (with and without inclusions of additives/admixtures) have increased significantly in numbers. New techniques of examination have arrived and our understanding of the performance requirements and utilization of many different types of cements has improved.

In particular, wet process manufacture for Portland cements is declining rapidly because of the high energy costs involved in driving off the water from the raw material slurries. Modern developments now permit more environmentally friendly dry process manufacture from soft raw materials like chalk and clay with their significant moisture contents using flash calciners. Also important environmentally friendly uses for materials, previously regarded as industrial wastes in the manufacture of cements, save substantial amounts of traditional fuels like coal, gas and oil. Such pyrotechnic processes now allow wastes that used to be dumped, to be disposed off safely by burning in cement kilns on a routine basis.

This is complemented by cement extension, which means that industrial by-products such as pulverized fuel ash (pfa), ground granulated blast-furnace slag (ggbs) and condensed silica fume (csf or 'microsilica') can partially replace Portland cement clinker in cements. Such replacement permits substantial benefits in the range of applications of the finished cements in grouts, mortars and concretes.

The second edition of the book has been produced, like the first edition, for those scientists and engineers working in both the cement and general construction industry and for research and development specialists in universities and colleges, who need an up-to-date knowledge in key areas of cement technology that have moved on since 1983. It is not intended to be a standard textbook dealing with the whole range of cement types and their applications. Instead, a focused tome has been produced with contributions by a multinational consortium of authors to indicate the global developments that have arisen over the last seventeen years. Modern cement manufacturing methods, key types of cement extenders and important examining techniques, both new, and developments in existing important methods, have been highlighted for addressing a global audience. This updated book is intended to reflect both current production and research in the cement arena, but, as indicated above, is not intended to be an exhaustive treatise, which would have been unmanageably large.

The order of presentation reflects the evolutionary developments that have arisen, and is divided roughly into three broad categories:

- Basic materials and methods – cement manufacture, cement phase composition, Portland cement hydration, calcium aluminate cements, properties of concrete with admixtures, special cements, various reaction/corrosion mechanisms.
- Cement extenders – ggbs, natural pozzolans, pfa, metakaolin and csf – where there have been considerable developments in terms of quality and application; many other extenders have not been included where there is still much basic work needed.
- Techniques of examination highlighted include the well established X-ray diffraction and electron microscopy, where there have been numerous developments in the last seventeen years, together with more recently introduced methods (electrical impedance, NMR, synchrotron radiation-based techniques) which, with cement-based composite microstructures, need to be drawn to the attention of a wider audience. Some of the more traditional techniques have been excluded since relatively few novel applications have appeared in recent years.

As the editors, we have endeavoured to obtain a widespread balance of authors from a number of different countries around the world. Many of these authors are well known internationally. They are joined by others who are newer to the field and who help to establish a broader viewpoint, both technically and through wide international coverage. We feel privileged to have succeeded in bringing such a distinguished worldwide group of authors together in one book under the umbrella title of *Structure and Performance of Cements*.

Notation

Standard cement chemistry notation is assumed throughout this book:

$A=Al_2O_3$	$C=CaO$	$\overline{C}=CO_2$	$F=Fe_2O_3$	$f=FeO$
$H=H_2O$	$K=K_2O$	$M=MgO$	$N=Na_2O$	$S=SiO_2$
$\overline{S}=SO_3$	$T=TiO_2$	C-S-H denotes a variable composition		

AFm denotes a solid solution range within the monosulphate-type structure (i.e. calcium monosulpho-aluminate hydrate to calcium monosulpho-ferrite hydrate).

AFt denotes a solid solution range within the ettringite-type structure (i.e. calcium trisulpho-aluminate hydrate to calcium trisulpho-ferrite hydrate).

Chapter one

Cement manufacture

Wiesław Kurdowski

1.1 Introduction

Cement technology is very traditional and the basic principles have remained unchanged for a long time. However, very profound changes in the production techniques were introduced leading to the diminishing of energy consumption and decrease in employment as well as fundamental pollution limitations.

Especially quick development was achieved in the last thirty years, which embraced burning and grinding techniques. First, the introduction of precalciners gave a considerable increase in kiln capacity, and the specific capacity increased from $1.5\,t/m^3 \times 24\,h$ to 3.5, and even to $5\,t/m^3 \times 24\,h$ for extra short kilns.

Second, in the field, the application of the roller press became a general practice and the development of roller mills is noted, which are more and more frequently used for cement grinding. The grinding efficiency was further improved by the introduction of a new generation of dynamic classifiers.

The introduction of automatic control in all production processes, including kilns, resulted in a drastic fall of employment and during weekends plants are run by two or three people.

Significant progress has been achieved in pollution control. The dust emission decreased to about 0.2–0.3 g/kg of cement. NO_x, SO_2 and CO are of the order 1200, 400 and 500 ppm, respectively, per cubic metre of kiln gases. For the special precalciner NO_x can be as low as 400 ppm. The problem is CO_2 emission, which originates mainly from $CaCO_3$ decomposition during clinker burning. A partial solution to this problem is the increasing production of cement with additives, chiefly slag, fly ash and limestone. Also, alternative fuels are used to partially replace completely classic fuels giving a decrease of total CO_2 emission from the combustion process.

In the case of alternative fuels utilization, emission of dioxins and furans can be monitored continuously. The industrial experience has shown no increase in these emissions. Long-term measurements have proved that, at least in the case of some alternative fuels, the emission of hazardous gases is lower in comparison with classical fuels. Old tyres must be recognized as being number-one alternative fuels.

Cement kilns are also used for the destruction of hazardous wastes instead of incinerators and the neutralization and destruction of these substances is much more effective.

All these achievements make the cement industry friendly to the environment.

1.2 Raw material preparation

The properties of Portland cement is determined by the mineralogical composition of clinker. The classic Portland clinker has the following mineralogical composition:

$3CaO \cdot SiO_2$	alite	55–65%
$2CaO \cdot SiO_2$	belite	15–25%
$3CaO \cdot Al_2O_3$	aluminate	8–14%
$4CaO \cdot Al_2O_3 \cdot Fe_2O_3$	brownmillerite	8–12%

For special cements the content of C_3A and brownmillerite can differ significantly.

Indispensable conditions for obtaining clinker with this mineralogical composition is the appropriate chemical composition of the bath. The share of main oxides is equal to about 95 per cent and their content is as follows:

CaO	60–70%
SiO_2	18–22%
Al_2O_3	4–6%
Fe_2O_3	2–4%

The entire content of minor components such as MgO, K_2O, TiO_2, Mn_2O_3 and SO_3 is normally under 5 per cent.

For the raw mix preparation limestone is used and to a lesser degree marl or clay, or even shale or fly ash. As a rule, a small quantity of the iron-bearing component is also added. Sometimes, a small fraction of sand must be used to regulate the silica ratio. It is not a satisfactory solution because quartz, which is the main sand component, is very hard and remains usually in the coarse fraction and is difficult to burn.

The cement plant is located near the limestone deposit which very often contains, in some deposits, rock of higher clay minerals content called 'marl'. In cases where limestone is rich in calcium carbonate such as silica and alumina bearing components, clay is used which deposits can be found, as a rule, in the proximity of the limestone deposits.

The raw mix composition can be calculated on the basis of the chosen mineralogical composition of clinker. However, to simplify the control of the chemical composition of raw meal and kiln feed, several ratios were introduced. The most important

TABLE 1.1 LSF, S_R and A_R of clinker produced in Germany

1988	Highest value	Mean value	Lowest value
LSF*	101	96	90
S_R	4.2	2.5	1.4
A_R	4.2	2.3	0.6

* German LSF = Lime standard I = C/2.8 S + 1.1 A + 0.7 F.

is the lime saturation factor (LSF):

$$LSF = \frac{CaO}{2.8\,SiO_2 + 1.2\,Al_2O_3 + 0.65\,Fe_2O_3}$$

where CaO, SiO_2 etc. are the oxide contents in the raw mix on the ignition-free basis.

The detailed foundation of LSF is given by Lea [1].

Other ratios are as follows:

$$\text{Silica ratio:} \quad S_R = \frac{SiO_2}{Al_2O_3 + Fe_2O_3}$$

$$\text{Alumina ratio:} \quad A_R = \frac{Al_2O_3}{Fe_2O_3}$$

For A_R lower than 0.64 the clinker does not contain the C_3A phase but only C_4AF and C_2F. The content of the latter increases with the lowering of A_R.

In Table 1.1 the ratios occurring in the German cement industry are shown [2].

1.3 Raw material crushing

There have been no major changes in the technology of primary crushing of raw materials in recent years. As a rule, mainly the single- and twin-rotor hammer crushers and impact crushers are used for the primary size reduction. Jaw crushers, also in combination with roll crushers and gyratory crushers, are applied, especially for hard as well as for softer materials. For the latter, a crushing department equipped with roll crushers, arranged in two stages, is frequently used.

The most recent improvements in the construction of hammer and impact crushers consist in increasing the maximum feed size and raising the maximum feed moisture content of the raw material. For the latest construction of the twin-rotor hammer crusher, the permissible feed moisture content has been doubled from 10 to 20 per cent and the mass of the individual lumps of feed material has been increased to about 7 tonnes [3]. Simultaneously, the crusher itself has a lower weight and for a unit of throughput of 1100 t/h the mass is about 160 t.

In the limestone quarry, very often a mobile installation is used which provides the most economical solution. Orenstein and Koppel have furnished the mobile crusher with a very high capacity, namely 4500 t/h [3]. Depending on the terrain conditions and the required mobility, these crushers can travel with the aid of a hydraulically powered walking mechanism or can be mounted on rubber tyres or rails. The development of mobile crushing installations have been associated with a drastic weight reduction of the crusher which has become lighter by 30 per cent and more [3].

1.4. Raw material grinding

For wet and dry grinding of the raw materials, ball mills are used which operate either as open-circuit mills or in closed circuit, in case of the dry process with an air classifier. By wet grinding the process is closed through the vibrating screen or arc-screen. The quick elimination of the fine fraction in closed circuit increases the grinding efficiency by about 20 per cent.

The granulometric composition of the feed greatly influences the energy consumption. To diminish the coarse fraction, the preliminary grinding is applied. For this purpose, Aerofall autogenic mills are installed which can be used for dry or wet grinding. These are one of the methods of modernization of the grinding department.

The energy consumption in ball mills is also heavily influenced by the grinding media filling ratio, by the design of the mill liners and by grinding media classification in the last chamber. For the solution of the grinding media classification the so-called autoclassifying liners are used.

Because of the predominance of the dry method of clinker production, in the dry grinding

technology more new achievements have been introduced. In the first place, the development of roller presses for preliminary grinding has been very often used recently. For grinding installations containing the roller press, a good solution is the application of the so-called V-separator (Figure 1.1) [4,5]. This static separator is extremely well suited for operations in grinding systems equipped with roller presses. This technology has been applied for several years in Hundai Cement Company's Young Wol plant for cement grinding and for raw material grinding. The operation data for raw material grinding are given in Table 1.2.

There are efficient installations using hammer or impact crushers for pregrinding of the material. These solutions are also convenient for raw materials drying. The hot gases from the kiln passing through the crusher transport the material to the classifier, where 70–85 per cent of the crushed material returns to the ball mill, the rest having suitable granulometry joins the product. These installations cannot be used for a very hard material because the short life of elements of the crusher can negatively influence the efficiency of the ball mill.

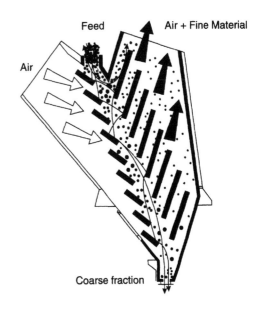

FIGURE 1.1 V-separator-principle scheme [5].

In the last decade, the application of roller mills increased because of the lower energy consumption of about 10 kWh/t in comparison with ball mills. Large quantities of gas conveying the material to the classifier enable drying of the feed up to 20 per cent of humidity.

In the past, there were some problems with the service life of the grinding elements, especially in the case of raw materials containing quartz [6]. In recent years, because of improvements in wear-resistant engineering materials, the service life has been substantially increased and, depending on the abrasiveness of the raw material, values ranging from 6000 to 16,000 h are reported [3].

A special high-drying capacity characterizes Loesche roller mills [7]. These mills can work with a very high gas temperature in the range of 600°C. Practical experience gathered during the exploitation of raw materials and coal grinding can be summarized as follows:

- for dry grinding of coal slurry with more than 30 per cent water content, operation temperatures of 700°C were used;
- for dry grinding of raw materials with 23 per cent water content, the hot gas temperature of 560°C has been used.

Among the various advantages of roller mills, the smaller dimensions and the lower capital cost (by about 10 per cent) in comparison with ball mills are worth mentioning.

The main goal of the grinding process is to ensure the suitable raw-meal granulometric composition, which usually corresponds to the residue of 1 per cent on 200 μm sieve and 12 per cent on 90 μm sieve. The large influence on raw-meal burnability is exerted by the coarse grains of quartz. Fundal [8] found that each per cent of silica grains larger than 44 μm increases the free lime content of the clinker by about 0.9 per cent. So the content of quartz greater than 44 μm should be lower than 2 per cent. Also the coarse grains of calcite, larger than 150 μm, should be under 2 per cent. Additionally, the raw-meal moisture must be very low, as a rule below 1 per cent, commonly 0.5–0.8 per cent. The higher moisture content negatively influences the flowing properties of raw meal.

1.5 Classifiers

A classifier exerts significant influence on mill efficiency and consequently on energy consumption. It is well documented by the curves showed in Figure 1.23 [47]. For the product containing 90 per cent of the fraction under 30 μm, the energy consumption of the ball mill is 65 kWh/t when the classifier efficiency is 25 per cent, but falls to 35 kWh/t for 100 per cent classifier efficiency.

Traditionally, two kinds of separators were used: the so-called grit separators and mechanical separators [9]. Air-swept mills are normally equipped with a grit separator and cyclone (Figure 1.2).

TABLE 1.2 Characteristic operation data: raw mill no. 3, Young Wol plant [5]

Operation mode of roller press	Finish grinding	Semi-finish grinding
Production rate tph	175	300
Raw mix fineness		
90 (μm% R)	15–17	15–17
212 (μm% R)	2–3	2–3
Specific energy consumption (kWh/t)		
Roller press	9.4	5.9
Ball mill	—	7.0
Separator SKS-L	0.7	0.4
Separator fan	4.6	2.7
Bucket elevators	0.9	1.0
Total	15.6	17.0

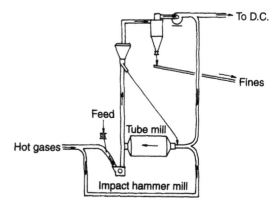

FIGURE 1.2 Scheme of the tandem drying-grinding plant.

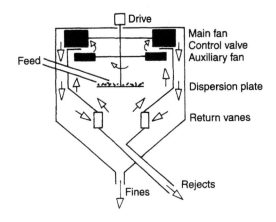

FIGURE 1.3 Scheme of mechanical separator.

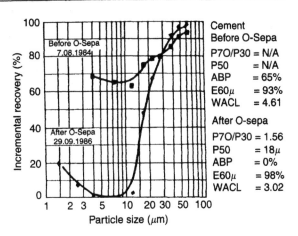

FIGURE 1.4 Tromp curves for classic and O-Sepa classifier [10].

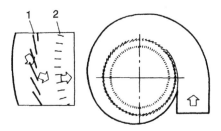

FIGURE 1.5 Principle of new classifier: (1) vertical guide vanes, (2) rotor blade.

Grit separators have no moving parts and the separation effect is due to diminishing of dust-entrained air and its tangential flow induced by guide vanes.

The main part of the mechanical separator is a rotating dispersion plate and two fans: the main fan and auxiliary fan. Coarse particles either fall directly from the dispersion plate or are rejected between the auxiliary fan blades and the control valve (Figure 1.3) The fines fall into the separation volume between two vessels and drop down under inertia forces. The air stream changes the direction forcibly and returns through the vanes and auxiliary fan, forming a closed circuit.

The mechanical separators have low efficiency in fine recovery, which is clearly seen on Tromp curve (Figure 1.4) [10]. To improve this efficiency a new generation of classifiers was introduced in the 1970s. The main part of this classifier is the rotor and tangential vanes (Figure 1.5). They ensure the even air velocity distribution along the full height of the rotor, which is a prerequisite for sharp separation classification. The new classifiers have a very good performance, and in comparison with classic separators produce an almost perfect Tromp curve (Figure 1.4). These separators significantly increase the mill capacity and decrease energy consumption. For example, Burgan [11] stated that the installation of a new classifier O-Sepa increased the cement mill capacity

by 30–32 per cent and a reduction of energy consumption by 20–30 per cent.

1.6 Homogenization

Raw material homogenization is a very important technological operation because the kiln feed should have a very stable chemical composition. The variations of chemical composition of raw mix influence unfavourably the kiln exploitation and clinker quality. For a stable kiln operation, the LSF factor fluctuations should be lower than one point. Each technological operation should increase the stability of raw materials (Figure 1.6) [12].

When raw materials of a very changeable composition are used, the prehomogenization is a very effective method. However, this technology is expensive with the capital cost being about 10 per cent of total expenditure for a new plant.

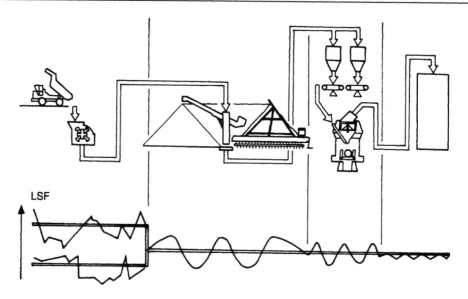

FIGURE 1.6 Homogeneity of raw material [12].

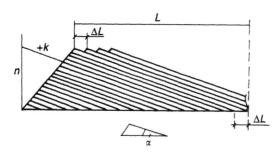

FIGURE 1.7 Chevcon method of stockpiling.

There are two methods of building blending beds: longitudinal or circular. The second is better because it overcomes the so-called 'end-cone problems'. It is difficult to keep the rate of reclaiming constant at the ends of the longitudinal pile and, especially when the reclaimer starts on the stockpile, the homogenizing effect is at first liable to be very unsatisfactory, because there will be a considerable segregation at the time of stacking. To avoid these problems, especially in longitudinal stacking, the so-called chevcon method is applied (Figure 1.7); the throw-off point of the stacker is varied by a radial distance ΔL in the course of each to and from cycle [13].

The effect of homogenization can be calculated using the following equation:

$$H = \frac{S_\alpha}{S_\beta}$$

where S_α is the standard deviation of input variations and S_β is the standard deviation of output variations.

For estimating the variations in output, the following equation is frequently used:

$$S_x = \frac{S_\alpha}{N^{1/2}}$$

where S_x is the standard deviation of the averages of slices and N is the number of layers.

In practice, there is no reason to use less than 50 or more than 500 layers (Figure 1.8)

Experience shows that with the current bed blending technology, the blending effects of up to 10 are attainable by qualitatively controlled build-up of the blending bed. For monitoring of the operation of blending beds, prompt gamma neutron activation analysis can be used which was elaborated and introduced in the cement industry in the 1980s by Gamma Metrics [14,15]. This method is designated for continuous analysis

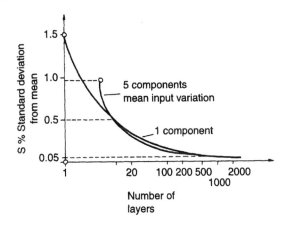

FIGURE 1.8 Blending effect as a function of the number of stacked layers of material.

TABLE 1.3 Comparative data of various homogenizing systems [3]

Blending installation	Homogenization			
	Batchwise	Cascade	Continuous	
			One silo	Two silos
Blending factor	>20	>20	5–7	7–12
Capital cost (%)	180	185	100	115
Specific power, consumption (kWh/t)	0.5–0.7	1.2–1.5	0.2–0.6	0.3–1.0

There are two major types of blending silo designs: turbulence and controlled flow. In a turbulence silo, the material is tumbled about by the injection of high volume air through air-pads on the silo floor. In the controlled flow system, the silo is divided into a large number of flow streams which run parallel at different flow rates, and the meal is finally blended in a tank (Figure 1.9). The fluidizing air aerates only a small part of the silo bottom. It results in low air and power consumption. The homogenizing factor for the CF FL Smidth silo is larger than seven. Modern blending silos are generally of the continuous, controlled flow type. The data for different types of blending silos after Malzig and Thier [3] are presented in Table 1.3.

1.7 Clinker burning

The dry method became predominant in the cement industry because of the much lower heat consumption, a typical value being 3000 kJ/kg of clinker for dry and 5500 kJ/kg for wet method. This heat expenditure for the wet method can be lowered when the water content in a slurry is under 30 per cent and the well chosen chain zone with special lifters ensures a low gas outlet temperature, not higher than 200°C. In such conditions, heat consumption can be as low as 4800–5000 kJ/kg of clinker [16]. The fuel cost can be lowered further by the use of alternative fuels. More details about this problem will be given in the point dealing with the pollution control.

When the moisture of limestone in the deposit is high, for example about 24 per cent for chalk,

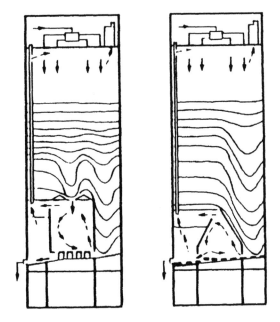

FIGURE 1.9 Two types of continuous homogenizing silos.

of the material stream of grain size up to 100 mm, for example on the belt conveyor.

There are several methods of raw meal homogenization [3]. Significant progress in pneumatic homogenization of raw meal has been made with continuous blending systems.

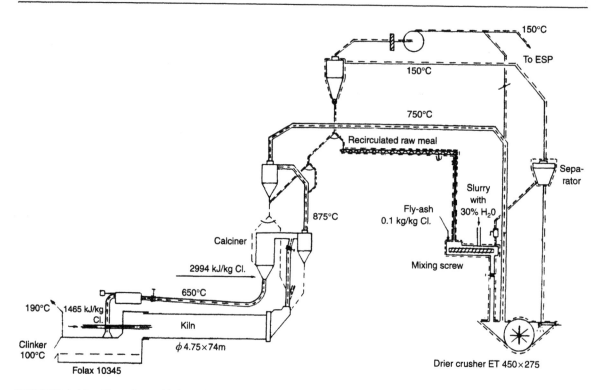

FIGURE 1.10   Flow sheet of kiln and drying departments. Only one of two identical preheater strings is shown [20].

the wet method become competitive with the dry process.

When the high humidity raw material is used, frequently the combined method of clinker production is applied. This technology consists in preparing the raw slurry in the same way as for the wet method, but then the slurry is injected into the drier-crusher and burned in the short kiln with a cyclone preheater [17] (Figure 1.10). The gas temperature is selected for the water content of the slurry through the adoption of appropriate cyclone stages (Figure 1.11). This technology was developed first in the Aalborg plant [17] and introduced successively in the Lagersdorf [18] and Rugby [19] plants. However, in the last two plants the slurry filtration in filter presses is applied and only the filter cake containing around 19 per cent of water is introduced to the drier-crusher.

A very original technology was developed by F. L. Smidth in the Chelm plant in Poland [18] the chalk and marl used in this plant are very homogenous and of stable composition, being

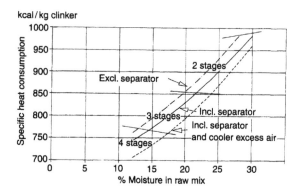

FIGURE 1.11   Drying capacity of compact kiln system with calciner [20].

simultaneously very soft [21]. The raw materials grinding process is totally eliminated and drier-crusher are fed directly by crushed raw materials (Figure 1.12). In this technology the splendid mixing behaviour of the cyclone preheater is utilized. This behaviour was firstly proved in the Rorhbach plant in Dotterhausen where limestone

8

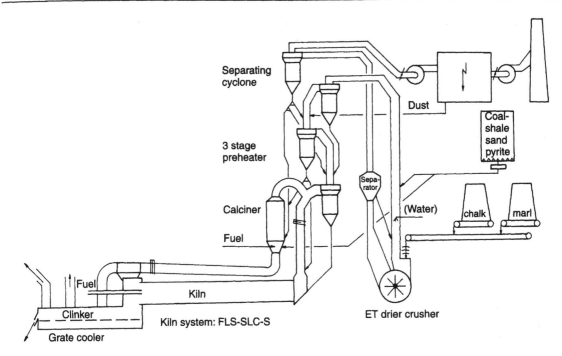

FIGURE 1.12   Semi-dry technology without raw material grinding, Chelm case.

from the preheater was mixed with shale ash in the last preheater stage. Thermal balances of different combined technology are compared in Table 1.4 [20,21,22]

The data in Table 1.4 comprise the heat consumption for the technology of filtration cake burning in the kiln with the Lepol grate preheater. This technology, that is, slurry filtration and cake nodulizing and burning in the Lepol kiln was very popular in France in the 1960s and 1970s, but totally disappeared in the 1980s. However, the dry method with granulation of raw mix and its burning in Lepol kilns does exist in France and the heat consumption is no higher than in the cyclone preheater kiln [23].

The dominating dry technology uses short kilns with cyclone preheaters. The kiln gases temperature is usually about 1100°C and the raw mix temperature about 800°C, with about 35 per cent calcination. The exit gas temperature is about 350°C for a four-stages preheater, but it can be lowered to 250°C for a six-stage preheater. The number of preheater stages is now only the question

TABLE 1.4   Thermal balances of combined methods [20–22]

Amount of heat (kJ/kg)	Semi-dry processes[a]		
	1[b]	2[c]	3[d]
Heat of reaction	1671	1654	1750
Smoke gas loss	419	482	452
Evaporation of water	921	1570	1072
Surface heat loss	507	419	289
Cooler loss	586	419	176
Heat in filter-cake, air and fuel	84	84	–
Net fuel consumption	4020	4460	3739[e]

[a] 1: Filtration, drier-crusher for cake treatment, kiln with two-stage cyclone preheater; 2: drier-crusher feeding with slurry/cake (Aalborg); 3: cake granulated and burnt in kiln grate preheater.
[b] Water content in kiln feed 18 per cent.
[c] Water content in kiln feed 16 per cent.
[d] Water content in kiln feed 21.7 per cent.
[e] Total heat consumption [20].

of moisture content of the raw materials, because the gases from the kiln are used for raw meal drying. To avoid the necessity of an additional furnace for the raw mill, as a rule the four stages preheater is adopted, which ensures raw material drying of about 6 per cent humidity. In the past, the four-stage preheater was the most economical because of a high pressure drop in the cyclones. Now, new constructions of cyclones (Figure 1.13) give a lower pressure drop for six stages than before for four [24]. Simultaneously, the separation efficiency of these cyclones is also higher. The examples of typical heat balances of dry kilns are given in Table 1.5.

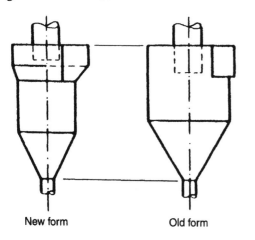

New form   Old form

FIGURE 1.13 Comparison of new FLS cyclone with the old form.

A great development of dry kilns was the introduction of precalcination in the 1970s. This technology results in an important increase of kiln throughput by a factor of about 2.5. Two types of precalcining technology are used; with partial and with total calcination. In the first one the air for fuel combustion in the precalciner is introduced through the kiln, increasing the coefficient of air excess, and in the other a special pipe for tertiary air must be installed (Figure 1.14). To facilitate the correct choice of precalcining technology one can consult some relations found by Herchenbach [26] and presented in Figure 1.15.

The degree of calcination in the precalciner is not higher than 90–95 per cent. The unfinished calcining process constitutes a very convenient regulator preventing an excessive increase of the temperature in the precalciner. This protects against a too high temperature for the exit gases from the precalciner, and thereby, against the increase of the heat loss with the exit gases. At the same time, it protects from the possibility of a clinkering process starting in the calciner and leading to the formation of build-ups in the precalciner, which can block the inlet of the gases supplied from the kiln.

The combustion of the fuel in the precalciner is carried out flamelessly and the heat exchange, which is very intensive, takes place with 90 per cent by convection. These favourable conditions of heat exchange are the reason for a very short

TABLE 1.5 Heat balances of dry kilns with oil or coal fired calciners in kJ/kg of clinker [16,20,25]

Kiln size (m)	—	5.4×85	4.75×85[a]	Last construction[b]
Preheater type	PASEC	RSP	FLS	FLS
Kiln output tpd	2440	3850	4200	—
Spec. cap. (t/m$^3 \times$ 24 h)	—	1.98	3.82	up to 5
Heat of clinkering	1792	1696	1796	1680
Dust in exit gases	—	38	40	—
Evaporation of water	12	29	12	—[c]
Exit gas	395[d]	812	608	609
Cooler vent air	549[e]	243	472	417
Clinker leaving, cooler	—	75	74	66
Radiation loss	212	210	216	273
Total	2960	3103	3218	3045

[a] Oil firing; [b] 6 stage preheater, calciner, modern grate cooler; [c] in exit gas losses; [d] 5 stage preheater; [e] total losses of cooler and clinker.

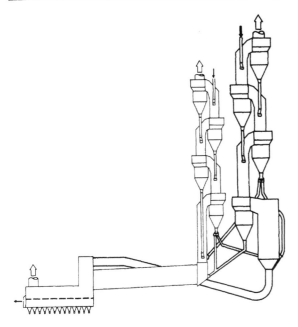

FIGURE 1.14 Conversion of SP kiln to an SLC system by FLS at Adelaide Brighton, Australia, the increase of production from 1800 to 4100 tpd [20].

time of calcination which is controlled by the rate of the chemical reactions.

The latest achievement in burning technology is constituted by very short kilns with L/D ratio of about 10 [27–29]. These kilns ensure a very rapid clinker formation, which gives a very quick reactions run without recrystallization phenomena and, as a result, a higher hydraulic activity of the clinker is obtained.

These kilns are characterized by two supports, selfadjusting roller suspensions, a tangential tyre suspension system (Figure 1.16) and a kiln drive via supporting rollers. FL Smidth also introduced a new sealing system of kiln inlet and outlet ends, a so-called double-lamella seal, which allows a higher degree of vibration at the kiln ends. The volumetric load of these kilns is of the order of $5\,t/m^3 \times 24\,h$. FL Smidth mention further important advantages of these kilns, namely, the kiln alignment is not affected by foundation settlement [29]; also, the kiln shell ovality remains very low and the risk of kiln shell constriction is eliminated.

High volumetric loading of these kilns causes the drop in heat consumption due to a reduction of the heat losses from radiation from the kiln shell and preheater, but the main problem remains the heat losses with the exit gases.

The construction of precalciners also underwent a substantial development, especially in the view of limiting pollution problems. First of all, combustion of the fuel must be very good with very low CO level in the exit gases. For this aim, the retention time of fuel in the precalciner is increased through the elongation of this furnace. It found its expression in a large volume of calciners which operate with a volume rating of $7 \pm 2\,t$ of clinker/ $m^3 \times day$ which corresponds to the gas retention time of 2–3 s and a meal retention time of 6–12 s [30].

The lowering of NO_x emission is another important feature. For this aim, two zones in the calciner are created: a reducing zone in the lower part, and an oxidation zone in the upper part of the precalciner [30–32]. In the first zone, the following reactions occur:

$$CO + NO \rightarrow CO_2 + \tfrac{1}{2}N_2$$
$$H_2 + NO \rightarrow \tfrac{1}{2}N_2 + H_2O$$
$$1\tfrac{1}{2}H_2 + NO \rightarrow NH_3 + H_2O$$
$$NH_3 \rightarrow \tfrac{1}{2}N_2 + 1\tfrac{1}{2}H_2$$

These reactions are favoured by a high temperature and catalysed by the raw meal.

In the upper part, the remaining CO is oxidized to CO_2, but because of the low temperature in the precalciner no additional oxidation of nitrogen will take place.

A very interesting example of a low CO and NO_x calciner is the ILC low-NO_x calciner of FL Smidth [31]. In the inlet of the calciner, the reduction zone is created where coal is introduced and the only oxygen available is the amount present in the rotary kiln gases. This favours NO_x reduction, according to the previously mentioned reactions. The high temperature that improves the efficiency of the first reaction is provided by splitting the raw meal between the oxidizing zone and the reduction zone. The temperature is controlled to 925–1050°C, or as high as possible without getting any encrustations in the kiln riser

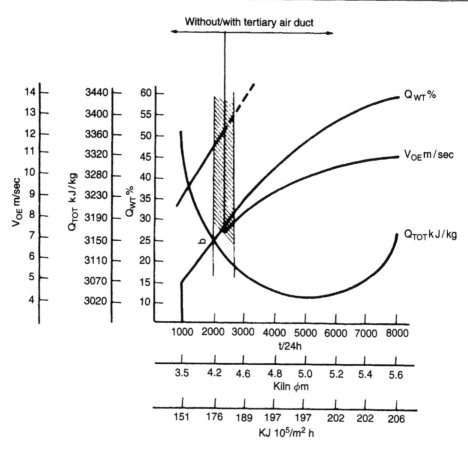

FIGURE 1.15 Relation between clinker production, kiln diameter, heat input in preheater, burning zone rating, gas velocity in kiln inlet and tertiary air control [26].

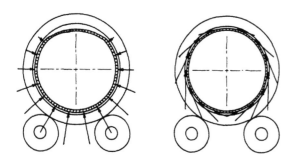

FIGURE 1.16 Plan view of FLS tangential tyre suspension system [29].

duct and reduction zone. Typically, 15–30 per cent of the raw meal should be fed to the reduction zone inlet (Figure 1.17). It is very important to assure the strongest possible reducing conditions

in order to favour the reactions of NO_x decomposition. In the ILC low-NO_x calciner, all the fuel is introduced to this zone before the tertiary air. The result of joint effects of all the decomposition reactions is the removal of approximately 70 per cent of the rotary kiln NO_x. The restriction in the middle of the calciner and the bend in the top are applied in order to improve the mixing, thus minimizing the CO emission and at the same time improving the fuel burnout. Even though after the reduction zone more than 3 per cent CO is in the gas, the concentration of the CO in the calciner exit gas will only be around 100–200 ppm, due to these facilities.

A very similar principle was adopted in the Pyrotop calciner of KHD Humboldt [30] (Figure 1.18). The kiln exit gases and tertiary air are

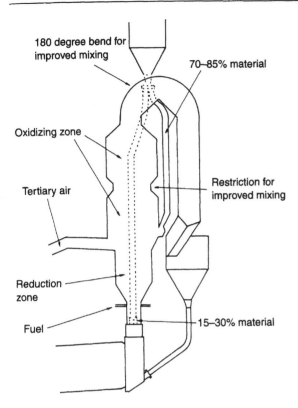

FIGURE 1.17 ILC low-NO$_x$ calciner of FLS [31].

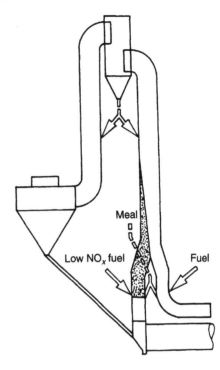

FIGURE 1.18 Pyrotop calciner of KHD Humboldt [30].

introduced into the calciner at an acute angle, which results in the slow mixing of the two flows. The fuel is burnt at two different locations and also the meal is admitted at two different points. The calciner is equipped with a new reaction chamber mounted at the location of the 180° elbow. It ensures a reliable mixing of the flow portions originating from the kiln waste-gas duct and from the tertiary air duct. Furthermore, a portion of the meal is discharged from the chamber and may either be recycled, that is, be routed back to the ascending calciner section, or directed to the calciner section running downward. A repeated deflection of gas/meal flow especially achieves an intensive mixing and thereby enhances combustion. The turbulent flow at the calciner outlet is a decisive prerequisite for good burning out of the fuel [30].

The construction of burners also underwent substantial changes and instead of very simple burners, the multi-channel burners were introduced [32–34]. Pillard was among the first people to develop this type of burner, which was adopted for a simultaneous use of several fuels [34]. The new constructions eliminate a very low fuel/air ratio, which was typical for old coal burners, and result in the higher combustion efficiency and a greater possibility for flame regulation. The volume and the shape of the flame may be changed by adjusting the ratio between the radial and axial air amount and velocity by adjusting the burner nozzle areas (Figure 1.19). Special constructions which diminish a NO$_x$ formation in the flame, have been very popular recently [35].

1.8 Coolers

Clinker cooling is very important for heat consumption in the kiln and for cement quality. This process is conducted in the coolers. There are three types of coolers which are generally used: rotary coolers, planetary coolers and grate coolers (Table 1.6).

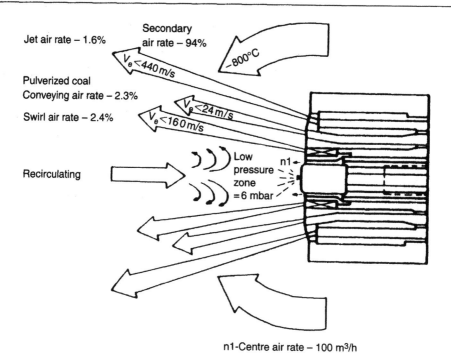

FIGURE 1.19 Multi-channel burner.

TABLE 1.6 Data of clinker coolers for dry kilns [40–43]

Cooler type	Rotary	Satellite[a]	Grate		
			SF	Coolax	Combined
Capacity (tpd)	up to 4500	up to 5000	Up to 5000	Up to 10,000	Up to 10,000
Power consumption (kWh/t)	4	2.5	5	7	6
Load/m² × d	150[b]	2[c]	<50	<60	50
Clinker (°C)	160–210	120–200	65[e]	65[e]	80–110
Cooling air (nm³/kg)	0.85	0.85	1.78	1.97	2.0–2.2
Second air (°C)	800–870	700–750	1040	1000	840–900
Exit air (°C)	—	—	290	280	240–280
Exit air (nm³/kg)	—	—	0.97	1.13	1.1–1.3
Heat expenditure (kJ/kg)					
Secondary air	1006	922	1285[f]	1246[f]	1006
Clinker	168	126	66	66	67
Radiation	210	293	25	25	25
Exit air	–	—	368	423	377
Water chute	25	—[d]	—	—	—
Thermal efficiency	71	69	75	71	70

[a] FLS Unax-cooler.

[b] Cross-sectional rating, inclination 4.5–5 per cent.

[c] Specific area rating.

[d] In hot climatic conditions, about 20 kJ/kg of clinker.

[e] Above ambient temperature.

[f] With tertiary air.

Grate coolers are most common because they ensure the lowest clinker temperature and highest kiln capacity [16,36]. Planetary coolers take second place – they are mounted on the kiln, and the air for cooling is provided by the kiln fan. Rotary coolers are seldom used. The superiority of grate coolers is finally caused by the convenience of tertiary air supply to the calciner.

The effiency of rotary and planetary coolers is closely connected with the heat transfer, which is enhanced through a geometric configuration of the cooler and arrangement of internal fittings. A special system of elbows and different lifters, both for planetary and rotary coolers, has been developed by Magotteaux and Estanda [37,38].

Planetary and rotary coolers have a limited airflow, usually 0.85–0.95 nm^3/kg of clinker, depending on the fuel consumption of the kiln and regulated by the kiln ID fan. For these reasons, the clinker temperature exceeds 150°C and water injection is applied. On the contrary, the quantity of air in grate coolers, which is supplied by special fans, is much higher, up to 2.5 nm^3/kg of clinker and the clinker temperature is low, usually 50°C + ambient temperature. However, the well-known problems with exit air must be resolved. The best solution is to use the exit air for drying raw materials, slag and coal. In some cases electrical energy is produced [39]. For dedusting of exit air, the electrofilter or gravel bed filter is used.

To diminish the quantity of cooling air different, variants of air recirculation have been employed and the exit air of low temperature, that is, 250°C, is reintroduced to the hot part of the cooler [39].

The grate cooler is frequently equipped with the roll crusher, which can work at a high temperature [41]. It is very important because the clinker cooling rate depends strongly on the grain size of this material.

The grate coolers regularly underwent developments in construction. The construction of the plates, and air supply directly to the plates in the so-called air-beam technology (Figure 1.20) have been recent achievements. The new plates have narrow air outlet slots, which ensure a sharp air

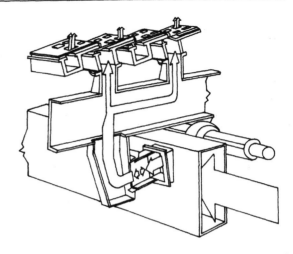

FIGURE 1.20 Aeration system for directly aerated troughed plates [41].

jet of about 40 m/s in the direction of clinker transport.

This solution permits the specific loading of the cooler to be raised, even the level to 100 t/m^2 × d has been tested [41].

The Coolax of F. L. Smidth [40] is an example of a new cooler. In the heat recuperation zone, cooling air is supplied directly to the grate plates through a system of ducts and hollow beams instead of via an undergrate compartment as in conventional coolers. This system also eliminates the so-called 'red river' phenomena in the cooler. The principal Coolax features are:

- fuel savings of 130–170 kJ/kg of clinker;
- 30 per cent less cooling air and 40 per cent less air to be dedusted;
- smaller overall dimensions due to the high specific grate load-up to 60 t/m^2 × 24 h.

A very original construction was developed by Smidth and Fuller [42,44]. It is a so-called SF Cross-Bar Cooler in which the cooling process is separated from the conveying action. A large drag chain pulls the clinker over a stationary bed through which cooling air is blown. Each grate plate is equipped with a rectangular tower containing a weighed mechanical flow regulator, which acts as a barometric damper in a chimney (Figure 1.21). All these solutions enable a decrease of the cooling air input to only 1.5–1.7 nm³/kg of clinker. The heat losses are also diminished to 400 kJ/kg of clinker.

Environmental protection regulations limiting dust emission have led to developments in construction of the second degree coolers working in conjunction with a short grate cooler [45]. The latter supplies hot air to the kiln, and the clinker at approximately 500°C, after being crushed in a breaker, is fed to a static membrane g-cooler (Figure 1.22). In this cooler the clinker, descending slowly at the speed of 2–3 cm/min, is cooled

down to 100°C, heating the air flowing in the tubes up to 60–100°C. The specific capacity of this cooler amounts to 10 t/m³ × 24 h. The specific power demand of the static cooler is about 1.2 kWh/t of clinker and of the whole installation is about 5 kWh/t. The g-cooler has been applied at the Robilante plant [46], for example.

1.9 Cement grinding

For cement grinding, ball mills are more frequently used. They have a rather high-energy consumption, typically 34 kWh/t for cement having a specific surface of 300 m²/kg. For good operational results, the level of ventilation, typically 0.3–0.5 nm³/kg clinker and volumetric charge loading which is 25–30 per cent for the first and 26–28 per cent for the second compartment, are very important. The modern mills are two chamber mills with lifting liners in the first and classifying liners in the second compartment. The diaphragms play also a very important role. The new construction of Slegten–Magotteaux or Combidan of F. L. Smidth provide a good ratio of the mass of material to the mass of the balls and thus improve the effectiveness of grinding [43].

Much better results are achieved by the mills working in closed circuits which have lower energy consumption and higher capacity of 20 per cent in comparison with the same mill working in an open circuit. The efficiency of classifiers is also important. This is presented in Figure 1.23 [47].

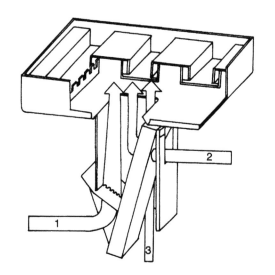

FIGURE 1.21 Aerating system for plates in SF Cross-Bar Cooler [42]. (1) air flow through holes in tongue; (2) air flow through idle hole; (3) air flow through gaps.

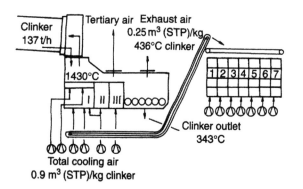

FIGURE 1.22 Short grate cooler used in combination with a gravitation cooler [45].

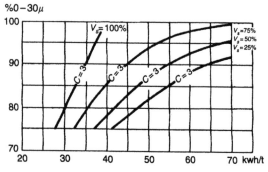

FIGURE 1.23 Influence of classifier performance on mill power consumption [47].

Ball mills have low efficiency, the majority of energy being transformed into heat. Significant progress in grinding technology has been due to the application of roller press for pregrinding or finish grinding [48]. The material between the rolls is submitted to a very high pressure ranging from 100 to 200 MPa. During the grinding process, partial breakdown of the grains occurs and there is a partial formation of incipient cracks in the interior of the grains. The material leaves the press in the form of agglomerates. When the roller press works in the closed circuit for finish grinding, a disintegrator must be used or a special separator must be installed to deglomerate the product. Frequently the roller press is applied for pregrinding with the ball mill as a second stage, thus the rise in capacity is of 30–60 per cent and 15–25 per cent for economy of energy [49]. Patzelt [49] gives, as an example, the installation of the roller press (dimensions 1.7×0.63 m) with a motor 2×575 kW which raised the raw mill throughput from 190 to 280 t/h that is, by about 47 per cent.

Regardless of the fact that special hard materials are used for the lining of the rollers, the shortcoming of this technology is relatively low reliability. It is due to the fact that the rollers are operating at a very high pressure. The maximum local pressures are near 350 MPa for clinker grinding [50]. It was the main reason for further development in the construction of special mills, and horizontal roller mills were produced. The so-called Horo-mill (Figure 1.24) was the first with a grinding pressure evidently lower, that is, 40–70 MPa. These mills are working in several cement plants in Europe, Turkey and Mexico [51,52], having a capacity up to 100 tph. The energy consumption of the mill alone is about 16 kWh/t. Another horizontal mill, the so-called Cemax, was developed by F. L. Smidth [53]. The Cemax grinds the material by compressing it between a ring and a roller. The energy consumption of the installation is about 25 kWh/t and is similar, both in the Horo-mill and in the Cemax.

The quality of cement from the Horo-mill, Cemax and roller press is also quite similar and

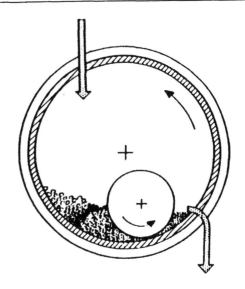

FIGURE 1.24 Principle of the Horo-mill.

comparable to the product of ball mills. A little higher water demand for the product of the roller press is noted.

Experience with roller mills used for cement grinding grows constantly. Ten years' experience in exploitation of a Pfeiffer roller mill in the Teutonia Zementwerk AG near Hanover has proved that these mills can be used for cement grinding [54]. The slag cement can be ground without preliminary slag drying. A specific power consumption by the grinding plant in the range 21–33 kWh/t can be attained. Thus the economy in energy, in comparison with ball mills, is between 25 and 45 per cent.

At the end of the 1970s, Loesche [55,56] started to develop new technologies in vertical roller mills, the so called $2+2$ and $3+3$ technology, which were applied with great success for cement and slag grinding. The main difference between the $2+2$ mills and the conventional four-roller-mills for raw material grinding is that two differently sized pairs of rollers are used (Figure 1.25). The so-called s-rollers are used to prepare the material bed on the grinding table. These s-rollers are of light design and have a smaller diameter compared to the large grinding rollers, the so-called m-rollers. By means of

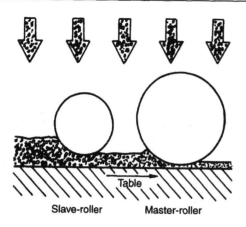

FIGURE 1.25 Diagramatic representation of grinding bed preparation [55].

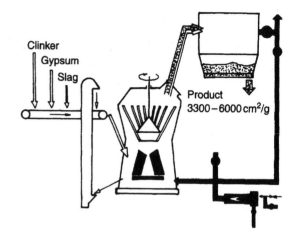

FIGURE 1.26 Cement grinding plant with roller mills [55].

TABLE 1.7 Power consumption for grinding of 60 tph blast furnace slag with 4300 cm²/g [55]

	Ball mill	Loesche mill
Mill (kW)	4400	1560
Separator (kW)	100	70
Fan (kW)	180	550
Total (kW)	4680	2180
Specific consumption (kWh/t)	78.7	36.3

hydraulic attachments the small, light s-rollers are positioned at a certain height above the grinding table, where they slightly consolidate and deaerate the material bed.

After this preparation the material is properly nipped under the large and heavy grinding rollers of conventional design. The forces required for comminution underneath the m-rollers are created by means of hydraulic cylinders attached to the rocker arms of the m-rollers. All m-rollers are supported on individual stands, which are relocated on the mill foundation.

Furthermore, individual hydropneumatic spring systems for each roller provide a very smooth and almost vibration-free operation for the mill. The 2 + 2 system is available for capacities of up to 170 tph OPC. In case of higher throughput the 3 + 3 system is used. Service lives of the roller shells, without any intermediate reworking, of more than 8000 h, have been achieved in cement mills in Japan and Korea and up to 15,000 h have been achieved for grinding track plates. The specific energy consumption for a complete cement grinding plant shown on Figure 1.26 is between 26 and 29 kWh/t for a finished product fineness, corresponding to Blaine's value of 3,300 cm²/g. The comparison of energy consumption for grinding slag in a ball mill and Loesche mill is given in Table 1.7.

There is the Hon Chon of the Morning Star Co. cement plant in Vietnam, in which only roller mills are used and no ball mill is installed [57].

1.10 Automation

The process technology and mechanical equipment for cement production has became increasingly complex in recent years. Product quality and diversity, environmental protection, secondary fuels, and energy management require more sensitive plant management. These are the main reasons for the introduction of computers for process control and for information systems.

The classical structure of the process control and information systems is distinguished by a relatively clear pyramid hierarchy (Figure 1.27) [58]. There is a constant increase in information density which affects every level in the computer integrated manufacture pyramid. Hardware and software with appropriate functions are installed at different levels of the computer integrated manufacture pyramid and linked with one another

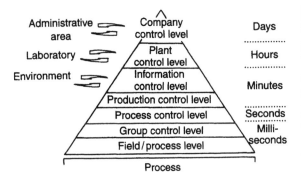

FIGURE 1.27 Structure of the process control and information systems (CIM pyramid) [58].

through communication buses to an integral process control and information system [58].

Nowadays, process control and information systems are not just a means of operating production plants; they are integrated parts of the plant and are essential for efficient, low-cost manufacture of quality products and for effective plant control and monitoring.

Many well-functioning computer-based systems for supervision, dialog and reporting have been introduced by different cement equipment producers as well as cement producers. Among others the HLC of Holderbank [59], TOPEXPERT of Ciments Français [60], SDR of FL Smidth [61], PRODUX of KHD and POLCID of Krupp–Polysius are the most popular ones [58].

The task of the higher-ranking process computer is to determine control inputs for the process and make them available for process control, to provide long-term storage of the process data and put them out as listings or diagrams, and also to make possible the dialogue between the operating personnel and decentralized processors with maximum convenience [62].

For many years, the automation concepts for raw meal preparation have been successfully introduced into cement manufacture. It was a very important step towards achieving a uniform kiln operation. But the problem of identifying a suitable microprocessor, based on the supervisory kiln control technique, took longer to resolve. Therefore, several methods were developed which were based on modelling of the operator actions

[62–64]. Basically, the technique involves expressing ill-defined or 'fuzzy' operator actions or commands in terms of precise mathematical logic. The fuzzy logic kiln control was applied to hundreds of kilns and the result fully proved its usefulness for automatic control of the burning process [65–67]. The introduction of automatic control of kilns saves thermal energy by 3 per cent, electrical energy by 2 per cent, results in production increase by 5 per cent, consumption of refractory lining diminution by 6–10 per cent and variance of clinker quality diminution of 4 per cent [60]. In case of mills the production increase is of up to 10 per cent, the decrease of energy consumption is 7 per cent, and, simultaneously, the quality variations are eliminated.

The actual breakthrough in the automation of rotary kilns and also the mills appeared with the introduction of new heuristic control methods, such as fuzzy logic, rule-based logic and online expert systems, coupled with the availability of powerful computers. In Europe alone there are now more than six suppliers of expert systems for cement kilns and ball mills [53].

In the last twenty years, new and very important measurement techniques were introduced. Perhaps the adoption of NO_x measurements to follow the temperature level in the sintering zone is most important [68]. Another very important tool in cement kiln control is the automatic supervision of protective coating thickness and hot spots on the kiln surface through kiln shell temperature measurements by the infrared technique. Several solutions of this problem are proposed by cement equipment producers [69]. Also the laser technique for cement fineness control or free lime control in clinker, which was introduced by Lafarge, is worthy of mention.

Laboratory automation has also become extremely important recently. Its particular function is to ensure that the data from raw materials and the products relevant to quality are continuously available to ensure a quality-driven process control. Full automation of laboratory functions is supplied by different producers. As an example, the Polysius system POLAB 3 can be mentioned which uses modular technology and includes two completely separate sample preparation lines with five robots, including even the automatic fusion equipment [70]. The entire system normally comprises a representative sampling system, a plant for rapid sample transport, a sample processing system and analysis technology suited to the special features of the material samples, and reliable open- and closed-loop control equipment for ensuring and monitoring the quality [71].

For the raw materials, chemical composition control, very great progress can be achieved through prompt gamma neutron activation analysis [14,15] which was mentioned in point 3.

1.11 Environmental protection

The manufacture of cement necessarily involves interference with the environment, although the cement industry is doing its best to minimize the negative effects of this. The problem comprises the water discharge, solid waste, noise and air pollution. Water discharge should not be a concern beyond handling normal domestic waste and storm water run-off with its potential for leaching from stockpiles and spillage. Solid waste practically concerns kiln bricks. When chromium bricks are eliminated, the alumina bricks can be added to the raw material and magnesium bricks sent to the landfill. Air pollution has been the main problem for the cement industry and will be discussed in more detail.

For twenty years, a significant progress has been made within the cement industry in the field of air pollution protection. The dust emission problem is now practically solved and the new construction of electrostatic precipitators and fabric filters, operating with a compressed air (reverse-pulse) cleaning system [72], have ensured a very low emission level, typically under $50 \, mg/nm^3$ of gases. However, for good electrostatic precipitator efficiency, the dew point of the gases is very important and for the dry method is always too low. It affects negatively the dust resistivity (Figure 1.28) Thus the conditioning towers in the gas circuit are installed and are in operation when the grinding/drying unit is not working.

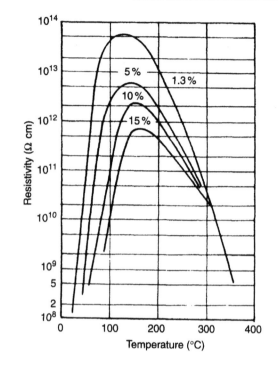

FIGURE 1.28 Resistivity of kiln dust as a function of temperature for different moisture levels of kiln gases (1.3, 5, 10 and 15 per cent).

Electrostatic precipitators underwent substantial modernization in 1990s and all modifications are reviewed in Peterson's paper [73]. Maybe the most important is the pulse energization system. By superimposing short-duration high voltage pulses on DC voltage, the charging of particles as well as the collecting efficiency of the precipitator are improved. The improvement factor of the precipitator performance is defined as a ratio between the dust migration velocity value w_k for pulse and for DC energization ascertained when both are adjusted to the optimum performance [74].

The w_k concept is useful since the required precipitator volume is inversely proportional to this value (Figure 1.29). Precipitator efficiency and the w_k migration velocity can be determined from the inlet and outlet dust concentration and gas volume flow rate:

$$w_k = \frac{Q}{A} \left\{ \ln \frac{100}{100 - \eta} \right\}^{1/k},$$

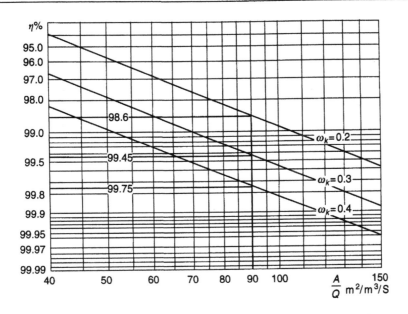

FIGURE 1.29   Electrofilter efficiency as a function of A/Q for three different values of w_k.

where w_k = particle migration velocity, m/s; Q = actual gas volume flow rate, m^3/s; A = total collecting area, m^2; k = constant (0.5); η = collecting efficiency, %.

The efficiency as a function of A/Q is shown for three different values of w_k in Figure 1.28.

In pulse energization the discharge current can be regulated independently of the precipitator voltage by variation of the pulse repetition frequency and the pulse height (Figure 1.30). This makes it possible to reduce the discharge current to the threshold limit for the back corona with high resistivity dust without reducing the precipitator voltage.

The gaseous emission comprises NO$_x$ and SO$_2$, and in the case of alternative fuels, dioxins and furans. The nitrogen oxides are formed mainly in the burning zone and are mostly thermal NO$_x$ [75]. The typical level of NO$_x$ for a cement rotary kiln is 1000–1500 ppm in which 90 per cent is thermal NO$_x$. There is also a fairly stable 200–250 ppm background contribution of NO$_x$ associated with fuel nitrogen and this is generated at relatively low temperatures within the flame itself. In the kilns with calciners, the NO$_x$ content becomes considerably lower because of the low burning temperature in the calciner. Also, the spe-

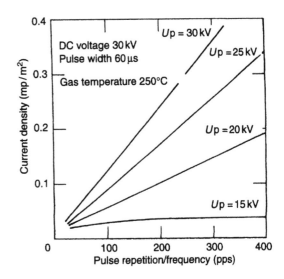

FIGURE 1.30   Current density as a function of repetition frequency for different Up values.

cial deNO$_x$ calciners are used which were described earlier. In the case of these calciners only 0.35 g/kg of clinker comes from the kiln and around 0.7 g NO$_2$ from the calciner, so that only around 1.05 g NO$_2$/kg of clinker is emitted, which is a little lower than 500 mg NO$_2$/nm^3 (10 per cent

21

of oxygen level). Because there is no ascertainable elimination of NO_x in conditioning towers or grinding/drying plants, when the decrease of NO_x in exit gases is indispensable, a very promising method is a non-catalytic decomposition of NO with NH_3 [76]. The selective reaction of NH_3 with NO takes place in the presence of oxygen, preferably at temperatures of about 950°C. Kreft [77] determined the best location of NH_3 injection to the cyclone preheater (Figure 1.31).

The emission of sulphur usually presents no problems. Additionally, the emission of SO_2 can be reduced by passing the gas through a grinding mill or conditioning tower. The cement burning process and grinding process thus function as ideal desulphurizing systems in which more than 90% of SO_2 is retained [72]. However, in some cases, when sulphides are present in raw material, the problem must be specially solved, because sulphur is combusted in the preheater and SO_2 leaves the kiln with the exit gases [78]. Bonn and Hasler [78] present the opinion that only absorption in a circulating fluidized bed can promise sufficient lowering of the SO_2. In the fluidized bed the exhaust gas is brought into contact with water to lower the temperature down to the dew-point range, then with a mixture of raw meal and slaked lime, the SO_2 is combined as sulphite and the product is fed back to the kiln, though in limited quantities. A very efficient desulphurization system is also in use at the Aalborg cement plant [79].

Polysius implemented an innovative system, called POLVITEC, which solved the problem of high SO_2 and NO_x emission and simultaneously provided the preconditions for environmentally acceptable utilization of sewage slurry as a fuel [80]. The system comprises an activated coke-packed bed filter in combination with an ammonia liquid injection into a kiln inlet. The pollutants contained in the kiln exhaust gases, including sulphur dioxide, ammonium compounds, heavy metals, hydrocarbons and residual dust from the upstream electrostatic precipitator are collected in the activated coke-packed bed filter. The gas passes the filter in the cross-current mode. The activated coke moves vertically through the housing. Collecting the replacement of activated coke takes place discontinuously in small steps. The exhausted activated coke is used as a fuel in the kiln.

Cement kilns are very promising reactors which effectively burn even hazardous waste. Because of the high flame temperature, the air excess cement kiln presents excellent conditions for burning of the most difficult organic compounds, that is PCB and PAH, so all hazardous wastes are destroyed and neutralized [81]. Simultaneously, a strongly basic medium in cement kilns absorbs the heavy metals and the immobilization is of the order of 99.99 per cent [82]. As long practice has shown, the emission of toxic substances and heavy metals from the cement kiln is as a rule lower than in the case of classic fuels [82]. However, the quantity of noxious components in the alternative fuels must not exceed the threshold quantity, which is shown in Table 1.8 [82].

The use of alternative fuels has been developed especially in France where the share of these fuels reached 52.4 per cent in 1994 [83]. The different

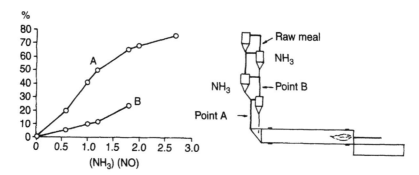

FIGURE 1.31   Reduction of NO_x for two location of feeding points in the suspension preheater.

TABLE 1.8 Pitch acceptability limits [82] in mg/kg of fuel

Chlorinated organic substances	<100
PAH	<50
PCB	<50
Metals	
Hg	<2.5
Hg + T1 + Cd	<20
Total	<1000

combustible wastes which are used as fuels are: petroleum coke, used lubricants, oils, paints, organic solvents, tar, waste-containing cyanides, pesticides, sundry organic and inorganic and others. The most popular alternative fuel for a long time has been old tyres.

1.12 Acknowledgements

The author wishes to express his gratitude to F. L. Smidth for sending very useful technical papers and especially to Mr. Richard Eimert and Bo Bentsen for their kind help. The author would like to thank Professor Bensted for his careful correction of the manuscript. Special thanks are due to Mr. Roland Nizecki for technical help.

1.13 References

1. Lea, F. M., *The Chemistry of Cement and Concrete*, 3rd edn., Chemical Publishing Company, New York (1971).
2. Sprung, S. and Rechenberg, W., *VDZ Kongress 1993*, Bauverlag GmbH, Wiesbaden, Berlin, p. 121 (1994).
3. Malzig, G. and Their, B., *VDZ Kongress 1985*, Bauverlag GmbH, Wiesbaden, Berlin, p. 178 (1987).
4. Strasser, S., *Cement-Wapno-Beton (in Polish)*, 63, 171 (1996).
5. Schultz, A. and Lube, B., *World Cem.*, June, p. 28 (1997).
6. Duda, W., *Cement Data Book*, Bauverlag GmbH, Wiesbaden, Berlin (1985).
7. Brundiek, H. and Poeschl, H. J., *Cement-Wapno-Beton (in Polish)*, 65, 56 (1998).
8. Fundal, E., *World Cem. Technol.*, 10, 604 (1979).
9. Alsop, P. A. and Post, J. W., Cement Plant Operations Handbook, 1st edn, Tradeship Publications Ltd (1995).
10. Knoflicek, *VDZ Kongress 1985*, Bauverlag GmbH, Wiesbaden, Berlin, p. 222 (1987).
11. Burgan, J. M., *Zem.-Kalk-Gips*, 41, 350 (1988).
12. Jung, C., *Int. Cem. Rev.*, June, p. 71 (1999).
13. Frommholtz, W., *Aufbereitung-Techn.*, 22, 119 (1981).
14. Poleszak, J., *Cement-Wapno-Gips (in Polish)*, 60, 189 (1993).
15. Tschudin, M., *VDZ Kongress 1993*, Bauverlag GmbH, Wiesbaden, Berlin, p. 259 (1994).
16. Kurdowski, W., 'Clinker burning technologies', in *Advances in Cement Technology*, S. N. Ghosh (ed.), p. 68, Pergamon Press, London (1983).
17. Borgholm, H. E. and Nielsen, P. B., *World Cem.*, 20, 74 (1989).
18. Hargreaves, D., *Int. Cem. Rev.*, June, p. 28 (1999).
19. Jackson, P., *Int. Cem. Rev.*, June, p. 30 (1996).
20. Palle, Grydgaard, 'Kiln conversions', FL Smidth (1998).
21. Kurdowski, W., *3rd NCB Int. Sem. Cem. Build. Mater.*, New Delhi, Vol. 2, p. III-29 January (1991).
22. 'Energy saving in cement manufacturing', FL Smidth Technical Literature, LST/RS/H 87-3A-4177.
23. Erhard, H. S. and Scheuer, A., *VDZ Kongress 1993*, Bauverlag GmbH , Wiesbaden, Berlin, p. 278 (1994).
24. Igawa, T. and Hanato, H., *VDZ Kongress 1985*, Bauverlag GmbH, Wiesbaden, Berlin, p. 420 (1987).
25. Kwech, L., *VDZ Kongress 1985*, Bauverlag GmbH, Wiesbaden, Berlin p. 430 (1987).
26. Herchenbach, H., *Zem.-Kalk-Gips*, 34, 395 (1981).
27. Mrowald, U. and Hartmann, R., *VDZ Kongress 1993*, Bauverlag GmbH, Wiesbaden, Berlin p. 416 (1994).
28. Weber, H., *World Cem.*, 13, 255 (1982).
29. ROTAX-2, FL Smidth Technical Data.
30. Bauer, C., *KHD Humboldt Symp*, p. 67 (1990).
31. Smidth, F. L., 'ILC low NO_X calciner', FL Smidth Technical Literature.
32. DUOFLEX burner, 'Swirlax kiln burner', FLS Technical Literature.
33. Steinbiss, E., Bauer, C. and Breidenstein, W., *VDZ Kongress 1993*, Bauverlag GmbH, Wiesbaden, Berlin, p. 454 (1994).
34. Pillard Technical Literature.
35. Thomsen, K., *VDZ Kongress 1993*, Bauverlag GmbH, Wiesbaden, Berlin, p. 624, (1994).
36. Herchenbach, H., *Zem.-Kalk-Gips*, 31, 42 (1978).
37. Fonderies Magotteaux Technical Literature.
38. Estanda Technical Literature.
39. Nakamura, N., *7th Int. Con. Chem. Cem.*, Paris, Vol. IV, p. 725 (1980).
40. Coolax grate cooler, FL Smidth Technical Literature.
41. Steffen, E., Koeberer, G. and Meyer, H., *VDZ Kongress 1993*, Bauverlag GmbH, Wiesbaden, Berlin, p. 449 (1994).
42. Bensten, B. and Keefe, B. P., *Zem.-Kalk-Gips*, 52, November, p. 608 (1999).

43. Kurdowski, W., 'Cement manufacture', in *Cement and Concrete Science and Technology*, S. N. Ghosh (ed.), Vol. 1, Part I, ABI Books, New Delhi (1991).

44. Keefe, B. P. and Puschock, E. L., *World Cem.* p. 52, January (1999).

45. Buzzi, S. and Sassone, G., *VDZ Kongress 1993*, Bauverlag GmbH, Wiesbaden, Berlin, p. 296 (1994).

46. Wedel, K., *Zem.-Kalk-Gips*, 38, 87 (1985).

47. Cleeman, J. and Modweg-Hansen, C., *Zem.-Kalk-Gips*, 27, 337 (1974).

48. Wustner, H., *VDZ Kongress 1985*, Bauverlag GmbH, Wiesbaden, Berlin, p. 279, (1987).

49. Patzelt, N., *Tonind.-Ztg.*, 113, 383, 662 (1989).

50. Cordonnier, A., *Zem.-Kalk-Gips*, 47, 643 (1994).

51. Buzzi, S., *Zem.-Kalk-Gips*, 50, 127 (1997).

52. *Int. Cem. Rev.*, February, p. 30 (1995).

53. CEMAX, FL Smidth Technical Data.

54. Leyser, W., Hill, P. and Sillem, H., *Ciment, Bétons Plâtres Chaux*, 761, 200 (1986).

55. Brundiek, H., *Cement-Wapno-Beton (in Polish)*, 65, 42 (1998).

56. Schafer, H. U., *Cement-Wapno-Beton (in Polish)*, 65, 48 (1998).

57. Schafer, H. U., Personal Communication.

58. Sauberli, R., Herzog, U. and Rosemann, H., *VDZ Kongress 1993*, Bauverlag GmbH, Wiesbaden, Berlin, p. 148 (1994).

59. Hasler, R. and Dekkiche, E. A., *VDZ Kongress 1993*, Bauverlag GmbH, Wiesbaden, Berlin, p. 216 (1994).

60. Kaminski, D. *VDZ Kongress 1993*, Bauverlag GmbH, Wiesbaden, Berlin, p. 225 (1994).

61. Holmblad, L. P., *VDZ Kongress 1985*, Bauverlag GmbH, Wiesbaden, Berlin, p. 539 (1987).

62. Holmblad, L. P., *Zem.-Kalk-Gips*, 39, 192 (1986).

63. Munk, R. and Ruhland, W., *VDZ Kongress 1985*, Bauverlag GmbH, Wiesbaden, Berlin, p. 504 (1987).

64. Haspel, D. W., Taylor, R. A., *VDZ Kongress 1985*, Bauverlag GmbH, Wiesbaden, Berlin, p. 530 (1987).

65. Holmblad, L. P. and Oestergaard, J. J., *Control of a Cement Kiln by Fuzzy Logic*, North-Holland Publishing Company, Amsterdam (1982).

66. Oestergaard, J. J., 'Fuzzy control of cement kilns', FL Smidth Technical Literature.

67. Bentrup, K., *VDZ Kongress 1985*, Bauverlag GmbH, Wiesbaden, Berlin, p. 600 (1987).

68. Lowes, T. M. and Evans, L. P., *VDZ Kongress 1993*, Bauverlag GmbH, Wiesbaden, Berlin, p. 305 (1994).

69. Pisters, H., Becke, M., Spielhagen, U. and Jager, G., *VDZ Kongress 1985*, Bauverlag GmbH, Wiesbaden, Berlin, p. 460 (1987).

70. Triebel, W., *VDZ Kongress 1993*, Bauverlag GmbH, Wiesbaden, Berlin, p. 255 (1994).

71. Teutenberg, J., *VDZ Kongress 1993*, Bauverlag GmbH, Wiesbaden, Berlin, p. 252 (1994).

72. Kroboth, K., Xeller, H., *VDZ Kongress 1985*, Bauverlag GmbH, Wiesbaden, Berlin, p. 600 (1987).

73. Petersen, H. H., 'New trends in electrostatic precipitation', FLSmidth-Review no. 35, 1980

74. Coromax, upse energisation. FLSmidth-Newsfront, 1983

75. Haspel, D.W., *World Cem.*, 14, 176 (1989).

76. Scheuer, A., *Zem.-Kalk-Gips*, 43, 1 (1990).

77. Kreft, W., *Zem.-Kalk-Gips*, 43, 153 (1990).

78. Bonn, W. and Hasler, R., *Zem.-Kalk-Gips*, 43, 139 (1990).

79. *World Cem.*, Anon., 54 (1998).

80. Rother, W., *Int. Cem. Rev.*, January, p. 39 (1996).

81. Xeller, H., *Cement-Wapno-Beton (in Polish)*, 64, 223 (1997).

82. Salvarode, D. and Zappa, F., 'Improved technologies for rational use of energy in the cement industry', *Proceedings*, Berlin, October 26–28, p. 409 (1992).

83. Cembureau, *Eur. Annu. Rev.*, 17, XIII (1994).

Chapter two

Composition of cement phases

Herbert Pöllmann

2.1 Introduction

Cement components are usually described by their phase relations out of phase diagram determinations. Basically, the most important cement phases can be shown in the phase diagram CaO–Al$_2$O$_3$–SiO$_2$ (Figure 2.1) with additional MgO and Fe$_2$O$_3$ and some minor, but also important components. Different hydraulic and latent hydraulic materials are shown in Figure 2.2. The stability fields of the phases can be given by these diagrams, but due to the multicomponent system, it is often difficult to describe solid solutions,

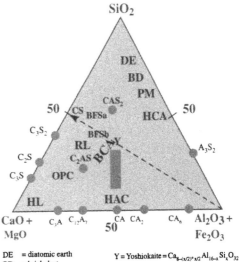

DE = diatomic earth
BD = brick dust
PM = pozzolanic material
HCA = hard coal ash
RL = roman lime
HL = hydraulic lime
BFSa = blast furnace slag (acid)
BFSb = blast furnace slag (basic)
OPC = ordinary portland cement
HAC = high alumina cement
BCA = brown coal ash

Y = Yoshiokaite = Ca$_{8-(x/2)^\circ x/2}$Al$_{16-x}$Si$_x$O$_{32}$

C$_2$AS = Gehlenite = Ca$_2$Al$_2$SiO$_7$

CAS$_2$ = Anorthite = CaAl$_2$Si$_2$O$_8$

FIGURE 2.1 Hydraulic and latent hydraulic materials in the system (CaO + MgO)–(Al$_2$O$_3$ + Fe$_2$O$_3$)–SiO$_2$.

phase transitions, phase stabilization and true element contents. For that, often isoplethic or isothermal sections are used to show the details. For these already quite well-known phase diagrams, the phase boundaries, melting points,

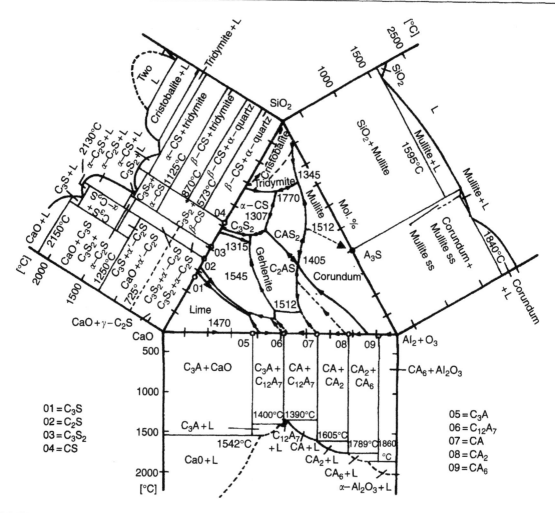

FIGURE 2.2 The ternary phase diagram $CaO–Al_2O_3–SiO_2$ showing the three binary phase diagrams $CaO–SiO_2$, $CaO–Al_2O_3$ and $Al_2O_3–SiO_2$ (Bowen, N. L. and Greig, J. W. 1924; Greig, J. W. 1927; Phillips, B. and Muan, A. 1959; Rankin, G. A. and Wright 1915; Roy, D. M. 1958; Welch, J. H. and Gutt, W. 1959).

transition points and stability regions can be given in detail. In a "cement system" characterization of compositional changes, solid solutions and the additional knowledge of the influence of minor contents of elements as a function of temperature can lead to differences in non-equilibrium systems. The existence of non-equilibrium conditions can often be seen by zoned crystals in cement clinkers. The complex phase composition of cement minerals is also now increasingly influenced by the use of alternative fuels, waste compounds and fluxing agents which can cause phase

changes or formation of new phases. Besides, for phase diagrams which are based on pure components, the *in situ* determination of compositions of phases is possible by using micro-techniques such as microdiffractometry, EPMA or EDX analysis. The varying chemical contents and impurities, the crystallographic changes of clinker minerals coming from different sources, kilns and production techniques can be measured and shown directly.

Due to these varying conditions, the equilibrium mineralogical calculations such as Bogue or modified Bogue are increasingly replaced by direct quantification methods to describe the real phase assemblages quantitatively. As microscopic techniques are mainly used to determine quality and quantity of clinker minerals at moderate times, the X-ray Rietveld technique can be developed for different clinkers even as a completely automized method for quality control. The quantification results can be obtained in at least 10–20 min after a proper installation of the control files. Phase compositions, crystal chemistry and crystallographic parameters of all phases must therefore be known.

The next steps will therefore provide direct determination methods qualitatively and quantitatively. A comparison of the advantages and disadvantages of microscopy, X-ray methods and Bogue calculations are summarized in Table 2.1. Small amounts of additional phases which can occur due to increasingly used impure raw and secondary materials cannot be found directly by X-ray methods, but must also be identified by different additional methods. Despite the knowledge of phase formation and crystallization conditions derived from phase diagrams, a description and investigation of clinkers and phase formation requires different methods:

1. Bogue calculation of theoretical phase contents including modified Bogue.
2. Direct X-ray analysis.
3. Rietveld analysis.
4. Chemical methods like free-lime determination.
5. X-ray fluorescence.
6. Electron microprobe analysis for small-sized samples and individual chemical information.
7. Optical methods like reflected-light microscopy, powder mounts and thin sections.
8. Scanning electron microscopy with fluorescence, secondary and back-scattered images.
9. Selective dissolution methods for enrichment of phases.
10. Thermoanalytical methods for alkali phases and carbonization effects.
11. Spectroscopic methods.
12. EPMA, ESR, ICP, AAS.

2.2 Simple oxides, elements and their potential role in cement chemistry

The role of simple oxides in cement chemistry is dominated by the free lime content which is measured in cement plants for quality control. High free lime contents can be an indication that burning conditions or homogenization are not successfully adapted. Low-burnt clinkers can have increased content of free lime. With grinding and transport of fine ground material, often carbonization can take place. This must be taken into account when free lime determinations are made routinely after some time, because lower free lime contents are measured. Free lime and free magnesia contents of clinkers are restricted due to the hydration of phase on-expansion reactions which can occur in the final product.

Other oxides like SiO_2 can be due to coarse quartz in the raw meal or due to incomplete homogenization. Corrective additives like iron-containing compounds can lead to the formation of iron spinels like magnetite or magnesium containing spinels. Spinels may occur due to reactions of $metals^{2+}$ and $metals^{3+}$.

Other iron oxides like wüstite were identified when reducing conditions occured in the clinker. The incorporation of Fe^{2+} in alite can lead to a destabilization of alite. With increasing reducing conditions, even native iron was found in clinkers.

Free titanium dioxide is normally not found in clinkers, but can be found when increased amounts of painting residues are added and incomplete homogenization takes place.

In Figures 2.3 and 2.4, free lime and an iron-rich spinel coming from reducing conditions are shown. Some details for these phases are summarized in Table 2.2.

2.3 Calcium silicates (C_xS_y)

The system of CaO–SiO_2 was investigated by Welch and Gutt (1959), Philips and Muan (1958) and Roy (1958). Four binary phases exist within the system which are increasingly hydraulic with increasing C/S-ratio. The different modifications

TABLE 2.1 Comparison of Bogue, microscopy and X-ray techniques for phase determinations (Pöllmann et al. 1997)

	Preparation	Measurement	Calculation	Precision	Advantage	Disadvantage
Microscopy	crushing; embedding in epoxy resin; polishing; surface etching (high personnel expense; no automatization; time consumption 2 days)	visual investigation; point counting of > 10,000 points; description of the microstructure (well educated and trained personnel necessary; high time consumption (1 days); no automatization	calculation of vol% into mass% (table calculator)	Standard deviation of about 2%	quantification and microstructure simulatenously; information on clinker genesis and on crystal shape and size	high time consumption; high preparation effort; well trained worker necessary; no automatization possible
Rietveld refinement	grinding; sample preparation (low personnel and equipment effort; automatization; time consumption ca. 20 min)	PC-controlled XRD (high equipment effort; automatization possible; time consumption ca. 1 h)	quantification with a PC program; automatic calculation within 1 min;	absolute error of 1 mass%	exact results on phase composition no texture effects; standardless; additional information on minor phases (like MgO);	rel. high measuring time; only quantification data
Bogue calculation	grinding; powder/glas pellets preparation (low personnel and equipment effort; automatization possible; time consumption ca. 20 min)	XRF analysis; wet analysis of volatile compounds and of free lime (XRF automatically; measurement time ca. 5 min; wet analysis needs high personnel and time effort)	normative phase quantification from chemical analysis (automatic calculation)	depends on the phase composition	fast and automatic analysis; within defined phase composition good quantification results	only normative phase; quantification; wet analysis necessary for exact quantification

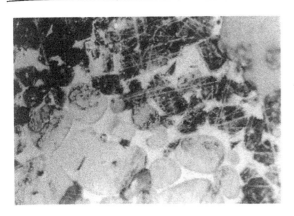

FIGURE 2.3 Reflected light microscopy nest of free lime (magnification 800×).

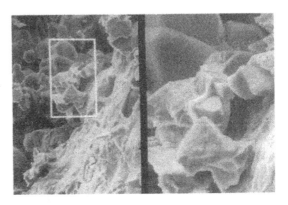

FIGURE 2.4 Iron-rich octahedra of spinel closely related to magnetite surrounded by interstitial phase (Octahedra 10 γm).

of CS are not hydraulic. C_3S_2 (Rankinite) is also not hydraulic and does not occur in Portland cements. C_3S and C_2S are the main hydraulic phases, undergoing several phase transformations. The complex solid solutions of C_3S and C_2S are named 'alite' and 'belite'. Important data on calcium silicates are summarized in Table 2.3.

2.3.1 C_3S – alite

C_3S, a highly hydraulic compound occurs in amounts of 50–90 per cent in Portland cements. C_3S is formed above 1250°C by a reaction of C_2S and C and can be metastably obtained by rapid cooling of the mixtures. The formation of C_3S and its polymorphs is highly influenced by other ions present, so high SO_3 contents even can suppress alite-formation which otherwise can be partly avoided by the presence of fluorine-ions. C_3S undergoes several phase transformations which are summarized in Figure 2.5. The high temperature polymorph R melts congruently at 2070°C. The polymorphs are all structurally similar with only small distortions, therefore they can be described by pseudohexagonal unit cells. By Rietveld refinement, mainly monoclinic polymorphs can be identified in clinkers. A stabilization of polymorphs can be obtained by incorporation of foreign ions in alite-lattice. Due to foreign ions and stabilization of C_2S, only small amounts of alkalies are incorporated in C_3S; instead alkalies are incorpo-

rated in belite. At 1500°C up to 1.4 per cent K_2O, 1.4 per cent Na_2O and 1.2 per cent Li_2O can be taken up, but in commercial clinkers, normally less alkalis are incorporated. The incorporation of 2.2 per cent MgO and 1 per cent Al_2O_3 and 1.1 per cent Fe_2O_3 at 1550°C was described. ZnO can be incorporated up to 5 per cent, Cr_2O_3 1.4 per cent and Ga_2O_3 up to 1.4 per cent. A combined incorporation stabilizes triclinic and monoclinic polymorphs. A combined addition of Al_2O_3 and Fe_2O_3 leads to an influence on each other on incorporation. A fluorine-stabilized C_3S was described by a formula $Ca_{6-0.5x}Si_2O_{10-x}F_x$. The incorporation of small amounts of Fe^{2+} in C_3S under reducing conditions leads to a destabilization and formation of belite and calcium-ferrite, but can also lead to increased free lime contents.

A diagram of varying contents of MgO and SO_3 showing the stabilization of different M_3 and M_1 modifications in industrial clinkers was determined by Maki. A zonal structure of alite is obtained due to lower cooling (less foreign ions), and the size of alite crystals is influenced by residence time in the furnace.

Due to rapid cooling, often a zonation of alite crystals can be found caused by M_1 and M_3 modifications or varying chemical compositions. The chemical variations and varying reactivities are not well established yet, but it seems to be clear that the inner cores of the crystals contain increased contents of foreign ions. The determination and

TABLE 2.2 Simple oxides and elements occurring in cements

Chemistry	Mineral name	Melting Point* in °C	Space group	Lattice parameters (Å)						JCPDS-No	Occurrence and properties
				a_o	b_o	c_o	α	β	γ		
CaO	Lime	2570	Fm3m	4,811	—	—	—	—	—	37–1497	free lime
MgO	Periclase	2800	Fm3m	4,211	—	—	—	—	—	45–946	free periclase
SiO₂	α-Quartz*	573	P3₂2	4,913	—	5,405	—	—	—	46–1045	coarse quartz in raw meal, insufficient homogenization
SiO₂	Tridymite*₂H	870	P6₃/mm2	5,046	—	8,236	—	—	—	18–1169 and others	HAC From rotary kiln production
SiO₂	Cristobalite*	1470 1723	P4₁4₁2	4,973 —	—	6,924 —	—	—	—	39–1425	high alumina cements
TiO₂	Rutile	1830	P4₂/mnm	4,593	—	2,959	—	—	—	21–1276	
Al₂O₃	Corundum	2072	R3c	4,759	—	12,99	—	—	—	46–1212	Al-rich HAC
Fe₂O₃	Hematite	1565	R3c	5,036	—	13,75	—	—	—	33–664	
FeO	Wustite	1369	Fm3m	4,293	—	—	—	—	—	46–1312	HAC from rotary kiln production
Fe₃O₄	Magnetite	1594	Fd3m	8,396	—	—	—	—	—	19–629	reducing kiln conditions
MgAl₂O₄	Spinel_ss	2105	Fd3m	8,083	—	—	—	—	—	21–1152	
MgFe₂O₄	Magnesio-ferrite_ss	1750	Fd3m	8,375	—	—	—	—	—	17–464	
Fe	Native iron	1535	Im3m	8,866	—	—	—	—	—	6–696	reactions with refractories strong reducing condition

*Transformation point.
SS = Solid solution; 2H = Ramsdell symbols.

TABLE 2.3 CS in cements and their properties

Chemistry	Mineral name	Melting Point	Space group	Lattice parameters (Å)						JCPDS-No	Occurrence and proper
				a_o	b_o	c_o	α	β	γ		
Para $CaO \cdot SiO_2$	Wollastonite-1A*	1126	C1	10,10	11,05	7,30	99,53	100,56	83,44	42-547	not occurring in cementitious systems
α-$CaO \cdot SiO_2$	Wollastonite-2M	1545	$P2_1/a$	15,43	7,07	—	95,38	—	—	43-1460	mainly known from slays
$3CaO \cdot 2SiO_2$	Rankinite	1465	$P2_1/a$	10,61	8,91	7,85	—	120,00	—	22-539	
α_L-$2CaO \cdot SiO_2$	Bredigite $(Ca_{14}Mg(SiO_4)_8)$	2130 / 650	Pmnn	10,904	18,38	6,75	—	—	—	36-399	
β-$2CaO \cdot SiO_2$	Larnite*		$P2_1/n$	9,31	6,76	5,51	—	94,46	—	33-302	belite-phase
γ-$2CaO \cdot SiO_2$	Skannonite		Pcmn	5,08	6,76	11,22	—	—	—	31-297	belite-phase
-$3CaO \cdot SiO_2$	R Hartrurite	2070	R3m	7,15	—	25,56	—	—	—	16-406	dusting, not hydraulic alite phase in cement
-$3CaO \cdot SiO_2$	*M_{III}	1060									
-$3CaO \cdot SiO_2$	*M_{II}	1050	P	12,23	7,03	24,96	—	90,10	—	42-551	
-$3CaO \cdot SiO_2$	*M_{I}	990									
-$3CaO \cdot SiO_2$	*T_{III}	980	P1	14,08	14,21	25,10	90,10	90,22	120,00	31-301	
-$3CaO \cdot SiO_2$	*T_{II}	920									
-$3CaO \cdot SiO_2$	*T_{I}	600									

*Phase transition.

quantification of the C_3S modifications by X-ray diffractometry is increasingly simplified by using the structural models and calculations of Rietveld methods.

A comparison of X-ray diagrams of high temperature R-phase to monoclinic phase is shown in Figure 2.6. The crystalline structural model is given in Figure 2.9 and the SEM micrographs of single pseudohexagonal shaped C_3S is given in Figures 2.7 and 2.8.

The crystalline structure was determined by Maki. A triclinic structure determination was done by Golovastikov (1975). Monoclinic alites show pseudotrigonal symmetry, in clinker often M_1 or M_3 and mixtures of both in zoned crystals are quite common (Kato 1978; Maki 1982).

The structure is composed of SiO_4-tetrahedra which are linked by calcium ions. Calcium is coordinated by eight oxygens (1978, 1982).

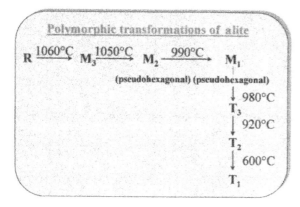

FIGURE 2.5 Polymorphic phase transformations of C_3S.

FIGURE 2.7 SEM micrograph of pseudohexagonal alite crystals showing slight surface coatings from starting hydration.

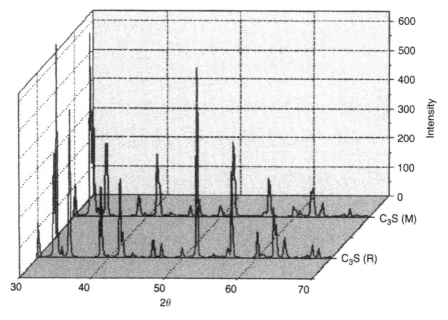

FIGURE 2.6 Comparison of X-ray powder data (CuKα) diagrams for rhombohedric high temperature C_3S and monoclinic C_3S.

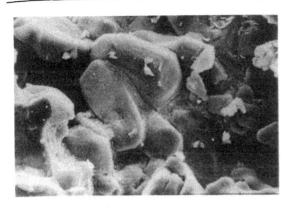

FIGURE 2.8 SEM micrograph of alite crystals separated by interstitial phase.

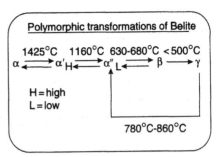

FIGURE 2.10 Polymorphic transformations of Belite.

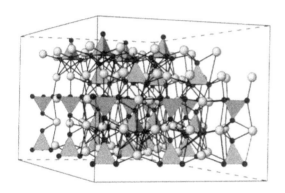

FIGURE 2.9 Crystal structure of C_3S.

2.3.2 C_2S – belite

C_2S mainly occurs in an amount of 10–40 per cent in OPCs. In special cements like belitic cements, low energy cements and slag materials, different modifications of C_2S may occur.

C_2S undergoes several phase transformations in the temperature range (Figure 2.10). The phase transformation to γ-C_2S is known as 'dusting' due to an enormous volume change. γ-C_2S is barely hydraulic. β-C_2S can be obtained by quenching or crystallochemical doping. In cements, often α- or β-polymorphs occur. These phase transformations cause the twin lamellae which can be seen in the optical microscope. The different C_2S-modifications can be stabilized at room temperature with doping by B, P, Mg, alkalis, Ba, Sr preferen-

tially. Mainly, rather large or small ions have stabilizing effects. It is quite obvious that belite is capable of incorporating larger amounts of foreign ions than alite. The stabilization of highly reactive α-polymorphs by foreign ions using alkalis is used in technical applications. Optically, four different types can be distinguished (F. P. Sorrentino 1998). Activated belite can be obtained by rapid cooling, lattice distortions by solid solutions, crystal size, crystallochemical changes and chemical shift of bonding energies. A comparison of X-ray powder diffractograms of three different polymorphs is shown in Figure 2.11.

In OPC clinkers, different belite origins coming from different sources can occur:

1. Primary belite formed by reaction of lime and silicon sources.
2. Secondary belite formed by the decomposition reaction of Alite according $C_3S \Rightarrow C_2S + C$ and forming small crystals on the rims of alite crystals.
3. Tertiary belite due to the recrystallization of the interstitial phase coming from the decomposition and decreased solid solution of SiO_2 in C_3A-phases.

Some compositions of belites in different clinkers were given by (F. P. Sorrentino) (1998):

OPC	$(Ca_{1.97}Na_{0.01}K_{0.02})(Mg_{0.02}Fe_{0.02})(Al_{0.07}Si_{0.92}S_{0.01})O_4$
white cement	$(Ca_{1.97}K_{0.02})(Mg_{0.01}Fe_{0.01})(Al_{0.08}Si_{0.90}S_{0.02})O_4$

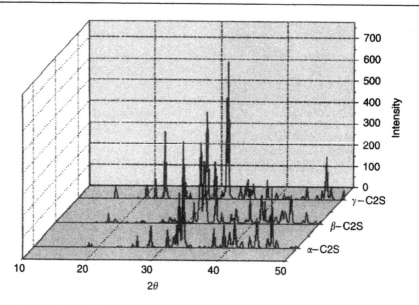

FIGURE 2.11 Comparison of X-ray powder data (CuKα) diagrams for different polymorphic phases of C_2S.

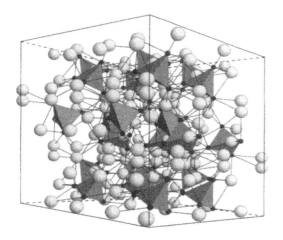

FIGURE 2.12 Crystal structure of C_2S.

iron-rich cement	$(Ca_{1.96}K_{0.01}Mg_{0.01}Fe_{0.02})$
	$(Fe_{0.02}Al_{0.06}Si_{0.91}S_{0.01})O_4$
oil-well cement	$(Ca_{1.95}K_{0.02}Mg_{0.02} Fe_{0.01})$
	$(Fe_{0.01}Al_{0.06}Si_{0.90} S_{0.03})O_4$

The attempts to form active belitic cements by doping C_2S or rapid cooling are increasingly interesting because of lowered temperature requirements and also because of the possibilities of using other lower-quality raw materials for cement production. The stabilization of β-C_2S by excess CaO and of γ-C_2S by excess SiO_2 was reported. Bredigite is now known as an one phase with a composition of $Ca_{14}Mg(SiO_4)_8$. Bredigite was formerly believed to be similar to α-C_2S but contrary to belite modifications, it is not hydraulic at room temperatures.

The structural model of these phases and their polyhedral links of SiO_4-tetrahedra linked by calcium is shown in Figure 2.12

1. α-C_2S, β-C_2S, γ-C_2S

A comparison of selected X-ray diagrams is summarized in Figure 2.11. The crystal structure models of α-C_2S is given in Figure 2.12. Optical and SEM micrographs are given on Figures 2.13–2.16.

2.4 Aluminum silicates, calcium aluminium silicates, their properties and occurrence in cements

Aluminum silicates and calcium aluminum silicates do not play an important role in OPC production, but metakaolinite and mullite can occur as decomposition products of clay materials in raw meal composition. Calcium aluminum silicates as gehlenite and anorthite typically occur in

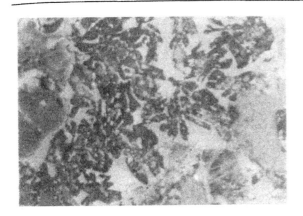

FIGURE 2.13 Optical reflected microscopical ragged belite crystals (magnification 700 ×).

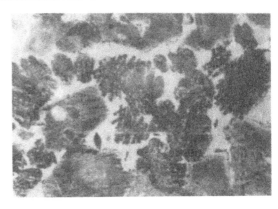

FIGURE 2.14 Optical reflected microscopical ragged belite crystals (magnification 700 ×).

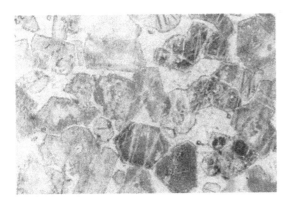

FIGURE 2.15 Reflected-light microscopy, belite aggregates (magnification 700 ×).

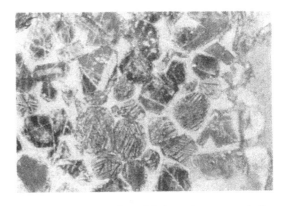

FIGURE 2.16 Reflected-light microscopy, belites with typical lamellae structure (magnification 700 ×).

slags and ashes from different sources. Gehlenite is also a typical mineral for high-alumina cements containing increased contents of silicon. Yoshiokaite is a metastable mineral showing some compositional variation within a nephelite–similar structure. Compositional variations within the calcium aluminum silicates with glassy structure occur in glass cements. Table 2.4 summarizes the most important data on some of the phases.

2.5 Calcium aluminates

In the system $CaO–Al_2O_3$ (Rankin and Wright 1915) four binary and one pseudobinary compounds exist (see Table 2.5).

C_3A is the most abundant Al-containing phase in OPC, but CA is the most important for high-alumina cements. Due to the Ca/Al-ratio, also CA_2 and CA_6 can occur in special HACs. $C_{12}A_7$ is not a pure compound in the system $CaO–Al_2O_3$, but instead, must be given by the formula $C_{11}A_7 \cdot Ca(OH)_2$. $Ca(OH)_2$ can be replaced by CaF_2, $CaCl_2$ and others.

2.5.1 Tricalcium aluminate – C_3A

Several polymorphs as a function of Na, Fe and S-content were synthezised by Lee, F. C., Banda, H. M. and Glasser, F. P. (1982). The cubic pure C_3A crystallizes in Pa3, but cubic C_3A doped with alkalis seems to crystallize in $P2_13$.

TABLE 2.4 Aluminum silicates and calcium aluminium silicates, their properties and occurrence in cements

Chemistry	Mineral name	Melting point	Space group	Lattice parameters (Å)						JCPDS-No	Occurrence and properties
				a_o	b_o	c_o	α	β	γ		
A_3S_2	Mullite	1810	Pbam	7.55	7.69	2.88	—	—	—	15–776	slags, ceramics
C_2AS	Gehlenite	1593	$P42_1m$	7.69	—	5.07	—	—	—	35–755	HAC's rich in SiO_2
CAS_2	Anorthite	1553	$P1$	8.18	12.87	14.18	93.17	115.91	91.20	41–1486	slags
$Ca(Al,Si)_2O_4$	Yoshiokaite		$P3$	9.94	—	8.25	—	—	—	46–1336	glass cements and glass slags
AS_2	Metakaolinite		amorphous	—	—	—	—	—	—	—	dehydrated China clay

TABLE 2.5 Calcium aluminates in cements and their properties

Chemistry	Mineral name	Melting point	Space group	Lattice parameters (Å)						JCPDS-No	Occurrence and properties
				a_o	b_o	c_o	α	β	γ		
$3CaO \cdot Al_2O_3$		1540	$Pa3$	15.26	—	—	—	—	—	38–1429	OPC
$CaO \cdot Al_2O_3$		1600	$P2_1/2$	8.70	8.09	15.21	—	90.14	—	23–1036	HAC
$CaO \cdot 2Al_2O_3$	Grossite	1790	$C2/c$	12.44	8.91	5.45	—	107.0	—	46–1475	HAC
$CaO \cdot 6Al_2O_3$	Hibonite-5A	1860	$P6_3/mmc$	5.56	—	21.91	—	—	—	38–470	HAC
$(C_{11}A_7 \cdot Ca(OH)_2)$	Mayenite	(1400)	$I43d$	11.98	—	—	—	—	—	9–413	HAC
$3CaO \cdot Al_2O_3$ (3.8% Na_2O)			Pbca	10.875	10.859	15.105	—	—	—		OPC
$3CaO \cdot Al_2O_3$ (5.7% Na_2O)			$P2_1/a$	10.877	10.854	15.135	—	90.1	—		OPC
$3CaO \cdot Al_2O_3$ (4% Na_2O)				10.851	—	15.109	—	—	(T>450°C)		high-temperature phase

Different stabilized cubic, orthorhombic and monoclinic lattice parameters are given by Moranville-Regourd and Boikova. The crystal structure is composed of $(AlO_4)^{5-}$-tetrahedra linked to $(Al_6O_{18})^{18-}$-rings which are connected by the Ca^{2+}-ions. For solid solutions with alkalies 2 Na^+ replace Ca^{2+}. Therefore a Na^+-ion takes the place of the Ca^{2+} and the second is located in the centre of the $(Al_6O_{18})^{18-}$-rings. Metastable forms of C_3A (proto-C_3A) were synthesized by crystallization of highly undercooled melts with increased contents of Fe_2O_3 and SiO_2 (Lee, Bandd and Glasser 1982). The crystal structure of an orthorhombic complex solid solution with the composition $Ca_{0.8375}Na_{0.875}[Al_{5.175}Fe_{0.45}Si_{0.375}O_{18}]$ was determined by Takeuchi, Nishi and Maki (1980).

The orthorhombic and monoclinic solid solutions of C_3A can be transformed at high temperatures ($T > 450°C$) into tetragonal (pseudoorthorhombic) polymorphs (Meissner 1986).

The Na-doped tricalcium aluminates seem to show decreased reactivity. The ranges of solid solutions and the symmetry changes can be given by:

0–1.9%	Na_2O	cubic
1.9–3.7%	Na_2O	cubic + orthorhombic
3.7–4.6%	Na_2O	orthorhombic
4.6–5.9%	Na_2O	monoclinic

These contents of sodium can be increased significantly in the presence of other foreign ions.

The existence of a NC_8A_3 can probably only be confirmed in the case when Si is present and rapid cooling of melts is occuring. A similar compound containing potassium can also be obtained preferentially when additionally iron and silicon-ions are present. A tetragonal KC_8A_3 was synthesized by Adams.

C_3A crystallizes in a cubic lattice, but due to incorporation of foreign ions, mainly alkalies and SiO_2, the symmetry changes to orthorhombic and monoclinic point groups. A detailed structure schemata is given in Figure 2.17. SEM micrographs of tricalciumaluminate are shown in Figures 2.18 and 2.19. X-ray data examples of tricalciumaluminate and sodium containing phases are given in Figure 20.

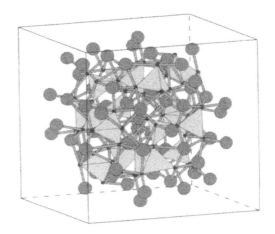

FIGURE 2.17 Crystal structure of tricalciumaluminate.

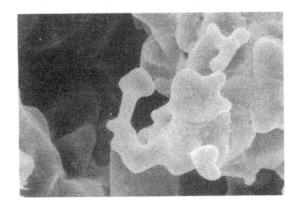

FIGURE 2.18 Molten calcium aluminate phase – SEM micrograph of molten interstitial phase, mainly aluminate.

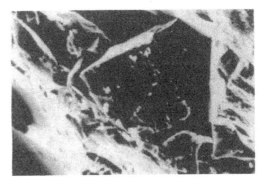

FIGURE 2.19 Cubic tricalcium aluminate – SEM micrograph of cubic C_3A-crystal edge.

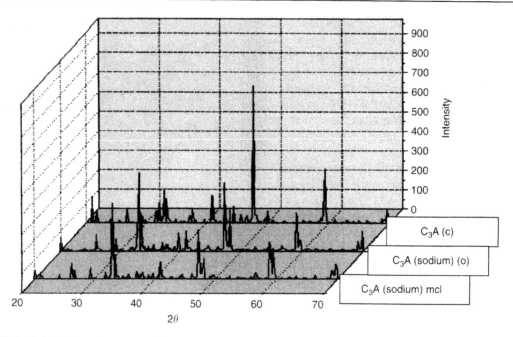

FIGURE 2.20 X-ray diagrams of typical cubic, orthorhombic and monoclinic tricalciumaluminate with different sodium contents.

2.5.2 Other calcium aluminates

Monocalcium aluminate is the main phase of HAC and crystallizes in a monoclinic point group. The hydroxide in $C_{11}A_7 \cdot Ca(OH)_2$. can be replaced by F_2, Cl_2 and others. These phases can play quite an important role in special cements like early hydrating cements. The fixation of halogenide ions is also very important in the combination of burning raw meals with chlorine or fluorine-bearing waste materials. In OPCs, the calcium content is too high, so that these phases cannot be found. Mayenite can also be formed as an additional constituent in high-alumina cements. CA_2 and CA_6 can occur in aluminum-rich high-alumina cements.

2.6 Calcium aluminium ferrites

Iron is mainly fixed under oxidizing conditions in the calcium aluminum ferrite phase. The composition of the ferrite phase can be described by a limited solid solution between C_2F and C_6A_2F with $C_4A_xF_{(1-x)}$, $0 < x < 0.7$ (Table 2.6). According to Büssem, the crystal structure of C_4AF is composed of layers of $(Al,Fe)O_6$-octahedra and $(Al,Fe)O_4$-tetrahedra linked along joint edges. In the open spaces in-between, the calcium atoms are located. A natural mineral was named 'brownmillerite'. Due to the occupation of iron and aluminum in tetrahedral and octahedral sites within the solid solution, a change of space group from Pcmn to Ibm2 takes place at a composition $Ca_4Fe_{4-2x} Al_{2x}O_{10}$ $(x < 1.14)$ (Colville and Geller). Preferentially, aluminum occupies the tetrahedral sites. Mn^{3+} and Cr^{3+} also can be incorporated in brownmillerite. C_4AMn was synthesized by Puertas, F.; Blanco Varela, M. and Dominguez, R. (1990). The crystal structure is shown in Figure 2.21. According to Mayerhofer, brownmillerite shows preferential orientation in $(0k0)$-direction. With decreasing Al-content of the solid solution, lattice parameter b_o mainly decreases. But, due to the increased uptake of silicon and magnesium at elevated temperatures and varying cooling rates, lattice parameters of calcium aluminum ferrites are also

TABLE 2.6 Calcium ferrites and calcium aluminum ferrates occurring in cements

Chemistry	Mineral name	Melting Point* (°C)	Space group	Lattice parameters (Å)						JCPDS-No	Occurrence and properties
				a_o	b_o	c_o	α	β	γ		
C_4AF	brownmillerite	1410	Pcmn	5.57	14.52	5.35	—	—	—	30–226	members of ss.-series
$Ca_2Al_{1.38}F_{0.62}O_5$			Ibm2	5.51	14.48	5.31	—	—	—	42–1469	ferrite phase in OPC
C_2F	srebrodolskite	1450	Pnma	5.42	14.75	5.60	—	—	—	38–408	end member of ss series
CF		1210	tetr.	5.33	—	7.58	—	—	—	41–753	minor phase
CF_2		1205	C2	10.41	6.01	31.65	—	96.5	—	39–1033	in HAC
C_4AMA				5.465	14.904	5.244	—	—	—	42–105	synthetic laboratory product

affected. In quenched laboratory samples, peak-broadening of X-ray peaks is rather common. The intensities of peaks within the solid solution is influenced by the site occupancies of aluminum and iron (Mayerhofer 1996; Walenta 1997) (Figure 2.22).

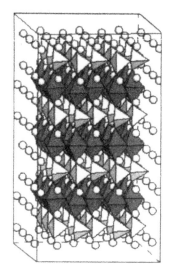

FIGURE 2.21 Crystal structure of C_4AF showing tetrahedral and octahedral coordination of $(Al,Fe)O_6$ and $(Al,Fe)O_4$ with Ca-ions.

In typical clinkers, the uptake of magnesium, titanium and silicon takes place, and can be given by the following equation:

$$2Fe^{3+} \rightarrow (Si^{4+} + Ti^{4+}) + Mg^{2+}$$

with a Si/Ti ratio $= 3:1$ (Harrison *et al.*) The composition of these ferrite phases typically lies within the range: $Ca_4Al_2Fe_{1.2}Mg_{0.4}Si_{0.3}Ti_{0.1}O_{10}$.

The composition in iron-rich cements can vary from the composition given above. The influence of varying burning conditions and reducing atmosphere on the aluminum/iron ratio was shown by Sieber *et al.* Due to increased silicon solution at high temperatures, some exsolution can occur at lower temperatures forming the tertiary belite in the interstitial phase. Titanium forms in high-alumina cements phases like perovskite or C-f-T with similar structure to perovskite but with Fe^{2+} (Motzet and Pöllmann 1998). OPCs normally have only low titanium contents which may increase when addition of wastes like painting dusts are added in fabrication. Then, TiO_2-phases or ilmenite may additionally occur.

Reducing conditions can lead to the formation of Fe^{2+}. This leads to increased formation of

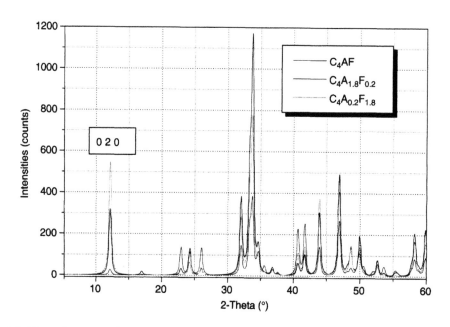

FIGURE 2.22 X-ray diagrams of ferrate phase showing different aluminum/iron contents.

calcium aluminate and free lime. Fe^{2+} cannot be incorporated in the ferrate phase. In some cases wüstite or spinels can be identified.

2.7 Influence of minor components (alone or in combinations) – alkalies, MgO, SO_3, phosphates, fluorides, borates, chlorine, titanium, heavy metals

2.7.1 Alkalis in cement

The amount of alkalis and their fixation in cement phases is very important due to their incorporation in cement phases, and also due to crystallization of new alkali phases. Alkalis can also cause alkali-aggregate reaction in concretes. Alkalis in cement can form in the cooler part of the cement kiln by reaction with SO_3 own alkalisulphates and contribute mainly to the vapourisation of SO_3. The formation of syngenite was also described in lower freshness of cements and was the cause of increased setting times and lumpiness due to aeration of cements (J. Bensted). No Na-analogue of syngenite is known. Other alkali phases are given in Table 2.7 as arcanite, aphtitalite, Ca-langbeinite. Ca-langbeinite improves the paste viscosity. Alkalis can also be incorporated in cement phases. The change of symmetry and the lowered hydration behaviour of alkali-containing orthorhombic C_3A was described. Alkalis are mainly incorporated in C_2S and stabilize the α and α' polymorphs. But also the increased alite size, the decrease of viscosity of melt and the increase in belite contents of cements are described.

2.8 Increased MgO content

This becomes more and more important due to the fact that less good-quality primary materials are available. Normally MgO forms free periclase or will participate in some solid solutions with tricalcium silicate and the calciumaluminumferrate phases. MgO can also be a constituent of spinel-structure types which occur scarcely. MgO also reduces the viscosity of the interstitial melt.

Bredigite and merwinite which were reported also contain magnesium as a constituent. Taylor summarized some other magnesium-containing phases as diopside, akermanite, etc., in combination with SO_3 stabilization of M_I and M_{III} polymorphs of C_3S polymorphs.

2.9 Sulphate-containing phases

Sulphate bound to calcium is normally bound to anhydrite which can hydrate to bassanite or gypsum in ground clinkers. Anhydrite only occurs for high SO_3/alkali-ratios. Yéelimite is the main phase in sulphoaluminate clinkers and shows a quite rapid hydrating behaviour. Yéelimite can also be an important phase in shrinkage-compensating mixtures. Silicosulphate or sulphatespurrite formation is mentioned from reaction zones of refractories and clinkers. Ellestadite is a typical non-hydraulic mineral with apatite structure capable of incorporating different toxic cat- and anions.

$C_2S \cdot CaSO_4 \cdot CaCl_2$ is a mineral which is stable up to about 750°C, which is quite hydraulic and can be used in special cement applications. Low burnt sulphate rich waste-mixtures can contain this phase. In Table 2.8, some phases and their properties are summarized.

2.10 Chlorine-containing phases

Chlorine contents can be accomodated by raw materials, alternative fuels, combustion of chlorine-containing wastes or production of low energy cements due to the addition of $CaCl_2$. The alkali chlorides can be vapourized and enriched in kiln dusts. In combustion products high in alkali and aluminum, the phase $C_{11}A_7 \cdot CaCl_2$ can often be detected. In presence of sulphate chloro-ellestadites can crystallize. Chloro-ellestadites can incorporate different toxic elements, but are not hydraulic. At lower temperatures, $C_2S \cdot CaCl_2 \cdot CaSO_4$, a hydraulic phase form (Kammerers 1993).

The use of different kinds of wastes as garbage combustion ash, lime and sulphate can result in the production of alinite-cements with the main component alinite. The composition and diadochic substitutions are summarized in Table 2.9a. Belinite

TABLE 2.7 Major alkali[a] phases in cements

Chemistry	Mineral name	Space group	Lattice parameters in Å						JCPDS-No	Occurrence and properties
			a_o	b_o	c_o	α	β	γ		
K_2SO_4	Arcanite	Pcmn	5.77	10.07	7.48	—	—	—	5–613	minor phase in OPC
Na_2SO_4	Thenardite	Fddd	9.82	12.31	5.86	—	—	—	37–1465	minor phase in OPC
$2CaO \cdot K_2O \cdot 3SO_3$	Ca-Langbeinite	$P2_12_12_1$	10.33	10.50	10.19	—	—	—	20–867	minor phase in OPC
NaCl	Halite	Fm3m	5.64	—	—	—	—	—	5–628	alkali phase at high Cl^- concentrations
KCl	Sylvite	Fm3m	6.29	—	—	—	—	—	41–1476	alkali phase at high Cl^- concentrations
$Na_{2x}Ca_{3-x}Al_2O_6$ ($Na_2O=5.3\%$) Orth. $C_3A(Na)$		Pba*	15.33	15.40	15.12	—	—	—	26–959	sodium containing C_3A
KC_8A_3									in prep.	potassium containing C_3A
$CaSO_4 \cdot K_2SO_4 \cdot H_2O$	Syngenite	$P2_1/m$	9.78	7.15	6.25	—	104.01	—	28–739	minor phase in OPC
$3K_2SO_4 \cdot Na_2SO_4$	Aphtitalite	P3m	5.68	—	7.33	—	—	—	20–928	minor phase in OPC

[a]Two types exist: alkalis incorporated in clinker minerals and those forming own minerals.

TABLE 2.8 Sulphate containing phases (without alkalis) in cements[a]

Chemistry	Mineral name	Space group	Lattice parameters in Å						JCPDS-No	Occurrence and properties
			a_o	b_o	c_o	α	β	γ		
$CaSO_4$	Anhydrite	Bmmb	6.99	7.00	6.24	—	—	—	37–1496	clinker mineral
$CaSO_4 \cdot 1/2H_2O$	Bassanite	I2	12.03	6.93	12.69	—	90.18	—	41–224	not in fresh clinker
$CaSO_4 \cdot 2H_2O$	Gypsum	C2/c	6.28	15.21	5.68	—	114.09	—	33–311	not in fresh clinker
$3CaO \cdot 3Al_2O3 \cdot CaSO_4$	Yéelimite		13.03	—	9.16	—	—	—	42–1418	sulphoaluminate cements
$4CaO \cdot 2SiO_2 \cdot CaSO_4$	Sulphate-Spurrite	Pcmn	10.18	15.41	6.85	—	—	—	26–1071	kiln-rings in OPC production
$3Ca_2SiO_4 \cdot 3CaSO_4 \cdot CaCl_2$	Ellestadite	P6$_3$/m	9.67	—	6.85	—	—	—	41–479	kiln rings in OPC production
$C_2S \cdot CaSO_4 \cdot CaCl_2$	—	—	7.01	10.91	6.29	—	—	—	41–247	SO4/Cl-Containing hydraulic phase

[a]Other suphates containing additionally alkali – see alkali-suphates.

TABLE 2.9a Alinite compositions with several solid solutions and incorporation of different ions (Von Lampe 1986; Neubauer; Pöllmann 1992)

Composition	a_o (Å)	c_o (Å)	Symmetry
$Ca_{10}Mg_{1-x/2}\square_{x/2}[(SiO_4)_{3+x}(AlO_4)_{1-x}/O_2/Cl]$ mit $0.35 < x < 0.45$	10.472–10451	8.583–8.615	tetragonal
$3(CaO_{0.875}MgO_{0.07}CaBr_{2\ 0.055})SiO_{2\ 0.885}Al_2O_{3\ 0.115}$	—	—	—
$Ca_{22}[SiO_4]_8O_4S_2$	10.452	8.6977	tetragonal
$19CaO \cdot 7SiO_2 \cdot 2CaF_2$	—	—	—
$Ca_{10}Mn_{1-z}\square_z[(SiO_4)_{3+x}(AlO_4)_{1-x}/O_{2-y}/Cl]$ $(y = z + x/2)$	—	—	—
$Ca_{10}Fe_{1-z}\square_z[(SiO_4)_{3+x}(AlO_4)_{1-x}/O_{2-y}/Cl]$ $(y = z + x/2)$	—	—	—
$Ca_{10}CO_{1-z}\square_z[(SiO_4)_{3+x}(AlO_4)_{1-x}/O_{2-y}/Cl]$ $(y = z + x/2)$	10.510	8.653	tetragonal
$Ca_{10}Ni_{1-z}\square_z[(SiO_4)_{3+x}(AlO_4)_{1-x}/O_{2-y}/Cl]$ $(y = z + x/2)$	10.481	8.591	tetragonal
$Ca_{10}Cu_{1-z}\square_z[(SiO_4)_{3+x}(AlO_4)_{1-x}/O_{2-y}/Cl]$ $(y = z + x/2)$	10.492	8.618	tetragonal
$Ca_{10}Zn_{1-z}\square_z[(SiO_4)_{3+x}(AlO_4)_{1-x}/O_{2-y}/Cl]$ $(y = z + x/2)$	10.529	8.665	tetragonal
$Ca_{10}Mg_{1-z}\square_z[(GeO_4)_{3+x}(AlO_4)_{1-x}/O_{2-y}/Cl]$ $(y = z + x/2)$	10.494	8.582	tetragonal
$Ca_{10}Mg_{1-z}\square_z[(SiO_4)_{3+x}(GaO_4)_{1-x}/O_{2-y}/Cl]$ $(y = z + x/2)$	10.703	8.692	tetragonal

$\square$ Empty lattice space.

was described by Von Lampe, F. *et al.* Chlorine is also able to form with alkali-elements and heavy metals low-temperature volatile compounds which can lead to special enrichment cycles.

Stemmermann and Pöllmann described several other calcium-silico-chlorides phase equilibria and different polymorphs. A natural mineral jasmundite (No. 33–0296) has a alinite-like structure. Different alinites are summarized in Table 2.9a. Various chlorine-containing phases and their properties are listed in Table 2.9b. Figures 2.23 and 2.24 show two typical chlorine containing phases.

2.11 Phosphate in cement

In raw materials, normally there are only low P_2O_5 concentrations, but due to water-clarification muds as a secondary raw material increased phosphate contents occur. The systems $CaO–Al_2O_3–P_2O_5$ and $CaO–SiO_2–P_2O_5$ were investigated by Hill, D. S., Faust, G. T. and Reynolds, D. S. (1994); St. Pierre, P. D. S. (1956) and Welch and Gutt (1967); Trömel, G (1943).

Phosphate only enters tricalcium silicate in very limited concentrations but is reported to form some limited solid solution with C_3P. At higher temperatures, some complete solid solution may occur. The C_4P does not occur in cementitious systems. The reported stabilization of dicalcium silicate tends to stabilize α'-polymorph at lower concentrations, and α-C_2S at increased concentrations. In the presence of fluorine, the formation of fluoro-apatite was reported. Increased contents of phosphate lead to decreased formation of C_3S and increased amounts of C_2S-polymorphs (Götz-Neunhöffer and Neubauer 1998). Silicocarnotite and nagelschmidtite are other reported phases. Others with varying stochiometry exist, but have not yet been identified in cementitious systems. The most important phosphorous-containing phases are listed in Table 2.10.

TABLE 2.9b Cl^--containing phases in cements

Chemistry	Mineral name	Space group	Lattice parameters in Å						JCPDS-No	Occurrence and properties
			a_o	b_o	c_o	α	β	γ		
$9CaO \cdot 6SiO_2 \cdot CaCl_2$		—	18.66	14.11	18.14	—	111.65	—	42–1453	Cl-rich wastes
$C_2S \cdot CaSO_4 \cdot CaCl_2$		—	7.01	10.91	6.29	—	—	—	41–247	SO$_4$/Cl-containing hydraulic phase
$C_{11}A_7 \cdot CaCl_2$	Chloro-Mayenite	I43d	12.01	—	—	—	—	—	45–568	Cl-bonding phase in HAC–
$Ca_{10}Mg_{0,8}[(SiO_4)_{3,4}(AlO_4)_{0,6}O_2Cl]$	Alinite	I42m	10.46	—	8.60	—	—	—	43–84	Alinite cements
$CaCl_2$ free	Hydrophilite	Pnnm	6.26	6.43	4.17	—	—	—	24–223	Cl-rich wastes
β-$CaO \cdot SiO_2 \cdot CaCl_2$	—		12.56	15.66	7.72	—	—	—	42–1455	Cl-rich wastes
α-$CaO \cdot SiO_2 \cdot CaCl_2$	—		10.71	—	9.35	—	—	—	43–0086	Cl-rich wastes
$4CaO \cdot 6SiO_2 \cdot CaCl_2$										
$3Ca_2SiO_4 \cdot 3CaSO_4 \cdot CaCl_2$	Ellestadite-Cl	P6$_3$/m	9.67	—	6.85	—	—	—	41–479	Reservoir mineral for immobilization, not hydraulic
$Ca_8Mg(SiO_4)Cl_2$	Belinite					—				low hydraulic mineral

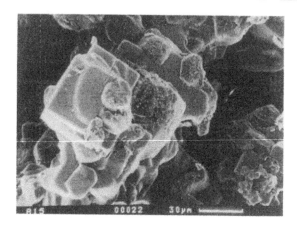

FIGURE 2.23 SEM micrograph of Sylvite on top of dust material.

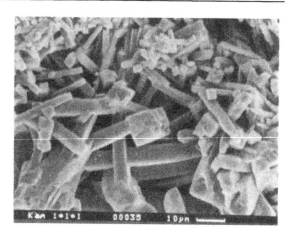

FIGURE 2.24 SEM micrograph of crystals of $C_2S \cdot CaSO_4 \cdot CaCl_2$.

2.12 Heavy metals in cement

Heavy metals normally occur only in minor quantities in cements and do not cause any problems. Only small amounts of ZnO normally come with the raw meal. An enrichment can take place when car tires or slags from metallurgical processes are added. But, often, artificial addition is made because ZnO improves the burning behaviour and lowers the temperature. The addition of ZnO enlowers the clinkerization temperature (1.5 per cent = 50°C and 2–4 per cent = 100–150°C). The phase formation with CaO is developed by addition of ZnO. In combination with sulphate and fluorides, burning temperature can be lowered to about 1050°C. According to Odler up to 4.7 per cent can stabilize different polytypes of C_3S. Other phases with Zn were described – Ca_2ZnSiO_5 and different calcium aluminum zinc oxides.

Chromium especially can be oxidized to Cr^{6+}, forming calcium chromate. Due to their causing skin eczema, the contents of chromium[6+] are strongly limited and reducing agents such as Fe^{2+}-salts are sometimes added. Despite this, the contents are very low and chromium will be incorporated in ettringite on hydrating. Other heavy metals only occur in small quantities as we know from the raw materials, carbonates and clays. Therefore, we normally do not detect single phases of these elements. The formation of low volatile compounds with sulphate or halogenides can cause an enrichment of these phases as was demonstrated with Tl(Cl,Br)-crystals by Bambauer. Addition of different kinds of wastes or alternative fuels can lead to an increase of heavy metals and should be controlled carefully. The use of CuO and Co_2O_3 were described as fluxes by Odler and Abdul-Maula (1980). Cadmium can replace calcium in different phases. It can replace Ca in tricalcium aluminate, but low volatile cadmium halogenides can be formed. Barium preferentially enters aluminates forming $BaO \cdot Al_2O_3$ (Jovanovic 1992). Some selected heavy metal containing phases are given in Table 2.11.

2.13 Elements used as fluxes (boron and fluorine)

Fluoride can be present in cements as a fluxing agent or occurs as minor quantities from natural sources where F^- – can replace OH^- – in micas. Other sources can be some secondary raw materials used in cement manufacturing.

Fluorides accelerate the formation of silicates in the burning of the raw meals. Besides $2C_2S \cdot CaF_2$, $Ca_{6-0.5x}Si_2O_{10-x}F_x$ is also described. Fluoride can also react with the aluminate phase to form calcium aluminium fluoride compounds. The compound $C_{11}A_7Ca(OH)_2$ forms a complete solid solution series with $C_{11}A_7CaF_2$, a sodalitic compound $C_3A_3CaF_2$ also exists (Table 2.12). These

TABLE 2.10 Phosphate-containing phases in cements

Chemistry[a]	Mineral name	Space group	Lattice parameters in Å						JCPDS-No	Occurrence and properties
			a_o	b_o	c_o	α	β	γ		
$3C_3P \cdot CaF_2$	F-Apatite	$P6_3/m$	9.3786	—	6.883	—	—	—	34–11	raw materials
$Ca_5(PO_4)_2SiO_4$	Silicocarnotite	Pmcn	15.468	10.086	6.72	—	—	—	21–0157	slags
		Pmcn	15.489	10.095	6.703	—	—	—	40–0393	
$2Ca_2SiO_4 \cdot Ca_3(PO_4)_2$	Nagelschmidtite		5.38	—	7.10	—	—	—	3–706	
$Ca_7(PO_4)_2(SiO_4)_2$	Nagelschmidtite		5.33	—	6.84	—	—	—	5–646	
$Ca_7Si_2P_2O_{16}$			21.8	—	21.57	—	—	—	11–0676	
$Ca_2SiO_4 \cdot 0.05Ca_3(PO_4)_2$		Pnma	6.7673	5.5191	9.303	—	—	—	49–1674	combustion products
$Ca_3(PO_4)_2$	Withlockite	R3c	10.429	—	37.38	—	—	—	9–169	
$Ca_4(PO_4)_2O$	Hilgenstockite	$P2_1$	7.023	11.986	9.473	—	90.9	—	70–1379	
$7CaO \cdot 5P_2O_5$	Trömelite									

[a]Many other stochiometries from different phases exist.

TABLE 2.11 Selected phases from cements containing Zn, Cd, Pb, Ba, Mn[a]

Chemistry	Mineral name	Melting point	Space group	Lattice parameters (Å)						JCPDS-No.	Occurrence and properties
				a_o	b_o	c_o	α	β	γ		
Ca_2ZnSiO_5	—	—	—	16.72	10.71	5.14	—	—	—	—	OPC
$Ca_3ZnAl_4O_{10}$	—	1350°C	—	14.91	—	—	—	—	—	—	interstitial phase in OPC
$Ca_6Zn_3Al_4O_{15}$	—	1260°C	—	—	—	—	—	—	—	—	interstitial phase in OPC
$CdO \cdot 6Al_2O_3$	—	—	—	5.63	—	22.58	—	—	—	22–1060	
$CdO \cdot Al_2O_3$	—	—	R3	14.21	9.57	—	—	—	—	34–0071	
$CdO \cdot 2Al_2O_3$	—	—	P21/c	12.67	8.86	5.40	105.93	—	—	22–1061	
$3BaO \cdot Al_2O_3$	—	—	Pa3	16.48	—	—	—	—	—	15–0331	barium cements
$BaO \cdot Al_2O_3$	—	—	P6322	10.44	8.79	—	—	—	—	17–0306	
$BaO \cdot 4.6Al_2O_3$	—	—	?	5.60	22.90	—	—	—	—	—	
C_4AMn	—	—	—	5.465	14.90	5.24	—	—	—	42–0105	
$PbOAl_2O_3$	—	—	P21	5.27	8.46	5.07	—	118.6	—	19–0613	laboratory product
$PbO \cdot 2Al_2O_3$	—	—	?	13.24	—	—	—	—	—	19–0672	

[a]Different slightly volatile halogenides of heavy metals not included.

TABLE 2.12 Phases occuring in presence of fluxes (B^{3+}, F^-)

Chemistry	Mineral name	Space group	Lattice parameters (Å)						JCPDS-No.	Occurrence and properties
			a_o	b_o	c_o	α	β	γ		
$C_{11}S_4B_2$		C?	10.65	55.43	6.89	—	—	—	33–304	jet-Cement
$C_{11}A \cdot CaF_2$	Fluoro-mayenite	I43d	17.29	11.97	—	—	—	—	25–394	
$C_3A_3CaF_2$	Fluoro-sodalite				7.09					
$2C_2S \cdot CaF_2$		P21/c		5.05	8.78	—	109.09	—	29–0324	
$3CsCaF_2$			11.49							
$C_4S_4CaF_2$										
$Ca_4Si_2O_7F_2$	Cuspidine	P21/c	10.90	10.54	7.54	—	109.60	—	41–1474	reservoir mineral, not hydraulic
$3Ca_2SiO_4 \cdot 3CaSO_4 \cdot CaF_2$	Ellestadite-F	P63/m	9.4417	—	6.9393	—	—	—	45–0009	

phases mainly play some important role in special cements, slags and metallurgical problems. F^- can also be incorporated instead of O^{2-} in calcium silicates. Some Al_2O_3 is probably necessary.

Cuspidine $Ca(Si_2O_7)F_2$ does not ocur in cementitious systems. The use of boron in cements is mainly due to the effect of a fluxing agent. To some extent, there is some incorporation in C_3S. Due to the amount of addition, different polymorphs of C_2S can be stabilized. A calciumsilicoborate is mentioned in Table 2.9. Basic investigations were undertaken by Fletcher.

Ellestadite crystallizes as a reservoir mineral capable of fixing different toxic cations and anions (Figure 2.25).

2.14 Quantification of cementitious materials by Rietveld method

The important aspect of studying the quantitative phase compositions of mineral phases by Rietveld refinement is possible nowadays. This helps to study the phase compositions of cements not only

by Bogue calculations but also by direct and fast analysis. By knowing the crystal structures of all phases present in the clinker mixture, it is possible to calculate a theoretical X-ray diagram which will be compared to the measured X-ray diagram of the sample, and a least-squares refinement will take place to adapt both diagrams. By means of

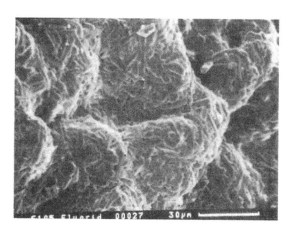

FIGURE 2.25 Crystals of fluoro-ellestadite.

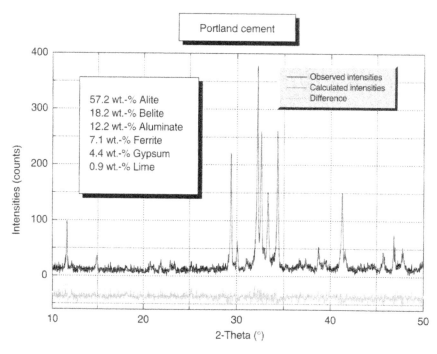

FIGURE 2.26 Rietveld-plot with quantification result of a typical Portland cement.

this technique, a good and reliable quantification can be installed. Good quantification results were already obtained for Portland cements, high-alumina cements and other construction materials.

A diagram illustrating the Rietveld refinement of measured, calculated and difference plot of an OPC is shown in Figure 2.26.

2.15 Acknowledgements

All crystal drawings were made using the program Atoms. Many thanks to Prof. Dr. R. Wenda for critical reading of the manuscript.

2.16 References

Adams, L. D. (year). 'Manufacture of Portland cement C_3A-compounds, including C_3A', *Proc. 14th Int. Conf. Cem. Micr.*, Costa Mesa, California, 134–44.

Abdul-Maula, S. and Odler I. (1992) 'SO_3-rich Portland cements: synthesis and strength development', *Mater. Res. Soc. Symp. Proc.*, 245, 315–20.

Aggarwal, P. S., Gard, J. A. and Glasser, F. P. (1972). 'Synthesis and properties of dicalcium aluminate, $2CaO \cdot Al_2O_3$', *Cem. Concr. Res.*, 2, 291–7.

Auer, S. T. (1992). 'Bindung umweltrelevanter Ionen in Ettringit und in Schichtstrukturen vom Typus TCAH', Dissertation, Erlangen.

Bambauer, H. U. and Schäfer, H. (1984). 'Der Mineralbestand eines thalliumhaltigen Reingasstaubes aus der Zementproduktion', *Fortschr. Min.*, 62/1, 33–50.

Bakdock, P. J., Parker, A. and Sladdin, I. (1970). 'X-ray powder diffraction data for calcium monoaluminate and calcium dialuminate', *J. Appl. Crystallogr.*, 3, 188–91.

Berggren, J., (1971). 'Refinement of the crystal structure of dicalcium ferrite, $Ca_2Fe_2O_5$', *Acta Chem. Scand.*, 25, 3616–24.

Bertaut, E. F., Blum, P. and Sagnieres, A. (1959). 'Structure du ferrite bicalcique et de la brownmillerite', *Acta Crystallogr.*, 12, 149–59.

Bigare, M., Guinier, A., Mazieres, C., Regourd, M., Yannaquis, N. and Eysel, W. (1967). 'Polymorphism of tricalcium silicate and its solid solutions', *J. Am. Ceram. Soc.*, 50, 609–19.

Black, P. M. (1969). 'Rankinite and kilchoanite from Tokatoka, New Zealand', *Miner. Mag.*, 37, 517–9.

Boikova, A. I. (1986). 'Chemical composition of raw materials as the main factor responsible for the composition, structure and properties of clinker phases', *8th ICCC*, Rio de Janeiro, Vol. 1, 19–33.

Bolio-Arceo, H. and Glasser, F. P. (1998). 'Zinc oxide in cement clinkering: part 1. Systems CaO-ZnO-Al_2O_3

and CaO-ZnO-Fe_2O_3', *Adv. Cem. Res.*, 10, (1), 25–32.

Bowen, N. L. and Greig, J. W. (1924). *J. Am. Ceram. Soc.*, 7, 238–54.

Bredig, M. A. (1950). 'Polymorphism of calcium orthosilicate', *J. Am. Ceram. Soc.*, 33, 188–92.

Bucchi, R. (1981). 'Role of minor elements in clinker', *World Technol. Part 1*, 12(5), 210–31; *World Technol. PartII*, 12(6), 258–73.

Büssem, W. (1937). 'Die Struktur des Tetracalciumaluminatferrits', *Fortschr. Min.*, 22, 31.

Cervantes-Lee, F. and Glasser, F. P. (1979). 'Powder diffraction data for compounds in the series $Na_x(Ca_{3-x}Na_x)Al_2O_6$', *J. Appl. Crystallogr.*, 12, 407–10.

Chatterjee, A. K. and Zhmoidin, G. I. (1972). 'The phase equilibrium diagram of the system CaO-Al_2O_3-CaF_2', *J. Mater. Sci.*, 7 93–7.

Chopra, S., Ghose, A. and Young, J. F. 'Electron-optical studies of stabilized α'- and β-dicalcium silicate', Abstract.

Colville, A. A. and Geller, S. (1971) 'The crystal structure of brownmillerite, Ca_2FeAlO_5', *Acta Cryst allogr.*, 27, 2311–15.

Colville, A. A. and Geller, S. (1972). 'Crystal structure of $Ca_2Fe_{1.43}Al_{0.57}O_5$ and $Ca_2Fe_{1.28}Al_{0.72}O_5$', *Acta Crystallogr.*, 28, 3196–200.

Colville, A. A. (1970). 'The crystal structure of $Ca_2Fe_2O_5$ and its relation to the nuclear electric field gradient at the iron sites', *Acta Crystallogr.*, 26, 1469–73.

Cruickshank, D. W. J. (1967). 'Refinements of structures containing bonds between Si, P, S or Cl and O or N. X. beta-Ca_2SiO_4', *Acta Crystallogr.*, 1, 1948–23; (1964). *ACCRA*, 17, 685–6.

Curien, H., Guillemin, C., Orcel, J. and Sternberg, M. (1956). 'Hibonite, a new mineral species', *Comptes Rendus hebdomadaires des Séances de l'Académie des Sciences*, 242, 2845–7.

Czaya, R. (1982). 'Refinement of the structure of gamma – Ca_2SiO_4', *Acta Crystallogr. B*, 24, 1968–38. (1971). *ACBCA*, 27, 848–9.

Della Giusta, A., Ottonellog, Seccol (1983). 'Precision estimates of interatomic distances using site occupancies, ionization potentials and polarizability in Pbnm silicate olivines', *Acta Crystallogr. B*, 39; (1990). *ASBSD* 46, 160–5.

Della, M. Roy and Oyefesobi, S. O. (1977). 'Preparation of very reactive Ca_2SiO_4 powder', *J. Am. Ceram. Soc.*, 60, 3–4.

Dougill, M. W. (1957). 'Crystal structure of calcium monoaluminate', *Nature (London)*, 180, 292–3.

Eysel, W. and Hahn, T. H. (1970). 'Polymorphism and solid solution of Ca_3GeO_5 and Ca_3SiO_5', *Z. Kristallogr.*, 131, S.40–59.

Felsche, J. (1968). 'The alkali problem in the crystal structure of β-alumina', *Z. Kristallogr.*, 127, 94–100.

Fierens, P. and Tirlocq, J. (1983). 'Effect of synthesis temperature and cooling conditions of beta-dicalcium silicate on its hydration rate', *Cem. Concr. Res.*, 13, 41–8.

Fischer, R. X. (1995). 'Die Mehrphasenanalyse in Industrie und Forschung', 6. *Philips-Symposium Röntgenbeugung*, Bad Hersfeld.

Fletcher, J. G. and Glasser, F. P. (1993) 'Phase relations in the system $CaO-B_2O_3-SiO_2$', *J. Mater. Sci.*, 28, 2677–86.

Fletcher, J. G., Borate fluxes in ordinary Portland cement production: a feasibility study', Ph.D. Thesis, Aberdeen University (1991).

Flint, E. P. and Wells, L. S. (1936). 'The system lime-boric oxide-silica', *J. Res. Natl. Bur. Stand.*, 17, 727–2.

Füllmann, T. 'Quantitative Phasenanalyse von Tonerdezementen', Dissertation, University of Erlangen 1997.

Fumito Nishi and Yoshio Takeuchi (1985). 'Tricalcium silicate $Ca_3O[SiO_4]$: the monoclinic superstructure', *Z. Kristallogr.*, 172, 297–314.

Geller, S., Grant, R. W. and Fullmer, L. D. (1970). 'LD – magnetic structures in $Ca_2Fe_{2-x}Al_xO_5$ system', *J. Phas. Chem. Solids*, 31, 793.

Gebhardt, R. F. (1987). 'KC_8A_3: X-ray diffraction characterization of the compound and comparison with NC8A3 and C3A', *Cem., Concr. Aggr.*, 9(2), 117–28.

Ghosh, S. N., Mathur, V. K. and Chopra, S. K. (1984). 'Low temperature OPC-type cement containing super-hydraulic C_2S phase', *Cem. Concr. Res.*, 14, 437–8.

Glasser, F. P. (1998). 'A discussion of the papers 'Influence of ZnO on clinkerization and properties of VSK Cement' by D. Bordoli, A. C. Bakuah, P. Barkakati, and P. C. Borthakur, and 'Hydration of ordinary Portland cements made from raw mix containing transition metal oxides' by G. Kakali, S. Tsivills and A. Tsialtas', *Cem. Concr. Res.*, 28(10), 1511–12.

Götz-Neunhöffer, F. and Neubauer, J. (1998). 'Effects of raw meal substitution by sewage sludge on OPC clinker studied by Rietveld analysis', *Porc. ICMA*, Guadalajara, 130–8.

Golovastikov, N. I., Matveeva, R. G. and Belov, N. V. (1975). 'Crystal structure of the tricalcium silicate $(CaSiO_2)_3 = C_3S$', *Kristallografiya, KRISA*, 20, 721–9.

Grant, R. W. (1969). 'Nuclear electric field gradient at iron sites in $Ca_2Fe_2O_5$ and Ca_2FeAlO_5', *J. Chem. Phys.*, 51, 1156.

Grant, R. W., Geller, S., Wiedersich, H., Gonser, U. and Fullmer, L. D. (1968). 'Spin orientation and magnetic properties of Ca_2FeAlO_5', *J. Appl. Phys.*, 39, 1122.

Greig, J. W. (1927). *Am. J. Sc.*, 13, 1–14.

Guinier, A. and Reyourd, M. 'Structure of Portland cement minerals', *Proc. 5th Int. Symp. Chem. Cem.*, Tokio, Vol. 1 pp. 1–32 (1968).

Gutt, W., Chatterjee, A. K. and Zhmoidin, G. I. (1970). 'The join calcium monoaluminate-calcium fluoride', *J. Mater. Sci.*, 5, 960–3.

Gutt, W. (1963). 'High temperature phase equilibria in the system $2CaO$ SiO_2-$3CaO$ P_2O_5-CaO', *Nature (London)*, 197, 142–3.

Gutt, W. and Osborne, G. J. (1970). 'The system $CaO-2CaO \cdot SiO_2-CaF_2$', *Trans. Br. Ceram. Soc.*, 69, 125–9.

Hansen, L. T., Brownmiller, L. T. and Bogue, R. H. (1927). 'Studies on the system calcium oxide – alumina-ferric Oxide', *J. Am. Chem. Soc.*, 50, 396–406.

Harrison, A. M., Taylor, H. F. W. and Winter, N. B. (1985). 'Electron-optical analysis of the phases in a Portland cement clinker, with some observations on the calculation of quantitative phase composition', *Cem. Concr. Res.*, 15, 775–80.

Hazen, R. M. (1976). 'Effect of temperature and pressure on the cell dimension and X-ray temperature factors of periclase', *Am. Miner.*, 61, 266–71.

Hentschel, G. (1964). 'Mayenit, $12CaO \cdot 7Al_2O_3$, und Brownmillerit, $2CaO \cdot (Al,Fe)_2O_3$, zwei neue Minerale in den Kalksteineinschlüssen der Lava des Ettringer Bellerberges', *Neues Jb. Min.*, 1, 22–9.

Hill, W. L., Faust, G. T. and Reynolds, D. S. (1994). *Am. J. Sci.*, 242, 470.

Ichikawa, M. and Komukai, Y. (1993). 'Effect of burning conditions and minor components on the color of Portland cement clinker', *Cem. Concr. Res.*, 23, 933–8.

Ichikawa, M., Ikeda, S. and Komukai, Y. (1994). 'Effect of cooling rate and Na_2O content on the character of the interstitial materials in Portland cement clinker', *Cem. Concr. Res.*, 24, 1092–96.

ICSD: Inorganic Structure Data Base, FIZ Karlsruhe.

Ilinetz, A. M., Melinowskiiya and Newskiinn (1985). 'The crystal structure of the rhombohedral modifikation of tricalcium silicate', *Doklady Akademii Nauk SSSR, DANKA*, 281, 332–6.

Imlach, J. A. and Glasser, F. P. (1973). 'Phase relations in the system $CaO-Al_2O_3-FeO-Fe_2O_3$ at liquidus temperatures', *Br. Ceram. Trans. J.*, 72, 221–8.

JCPDS: International Centre for Diffraction Data.

Jeeveratnam, J., Glasser, F. P. Dent-Glasser, L. S. (1964). 'Anion substitution and structure of $12CaO \cdot 7Al_2O_3$. *J. Am. Ceram. Soc.*, 47, 105–6.

Jeffrey, J. W. (1967). 'The crystal structure of tricalsium silicate', *Acta Crystallogr.*, 1, 1948–23 (1952). *ACCRA*, 5, 26–35.

Jost, K. H., Ziemer, B. and Seydel, R. (1977). 'Redetermination of the structure of β-dicalcium silicate', *Acta Crystallogr.*, 33, 1696–700.

Jost, K. H. and Ziemer, B. (1984). *Cem. Concr. Res.*, 14, 177.

Jovanovic, I. (1992). 'Untersuchungen in den ternären Systemen $CaO-CdO-Al_2O_3$ und $CaO-BaO-Al_2O_3$', *Dipl. Arb. Erlangen.*

Kakali, G. and Parissakis, G. (1995). 'Investigation of the effect of Zn oxide on the formation of Portland cement clinker', *Cem. Concr. Res.*, 25(1), 79–85.

Kakali, G. (1998). A reply to the discussion of the paper, 'Hydration of ordinary Portland cements made from raw mix containing transition element oxides' by F. P. Glasser', *Cem. Concr. Res.*, 28(10), 1513.

Kammerer, F.: Gleichgenichts reaktionen in System CaO-SiO₂-CaSO₄-CaCl₂-Dipl.-Arbeit (1993).

Kapràlik, I., Hanic, F. Havlica, J. and Ambrùz, V. (1986). 'Subsolidus phase relations in the system CaO-Al₂O₃-SiO₂-Fe₂O₃-MgO-CaSO₄-K₂SO₄ at 950°C in air referred to sulphoaluminate cement clinker', *Br. Ceram. Trans. J.*, 85, 107–10.

Klemm, W. A. and Skalny, J. (1977). 'Selective dissolution of clinker minerals and its applications', *XII Cont. Silikate Industry and Silikate Science*, Budapest.

Knöfel, D. (1978). 'Beeinflussung einiger Eigenschaften des Portlandzementklinkers und des Portlandzementes durch ZnO und ZnS', *Zem.-Kalk-Gips*, 3, 157–61.

Leary, J. K. (1962). 'New compound in the system CaO-Al₂O₃-CaF₂', *Nature*, 194, 79–80.

Lee, F. C., Banda, H. M. and Glasser, F. P. (1982). 'Substitution of Na, Fe and Si in tricalcium aluminate and the polymorphism of solid solution', *CCR*, 12, 237–46.

Lieber, W. (1967). 'Einfluβ von Zinkoxyd auf das Erstarren und Erhärten von Portlandzementen', *Zem.-Kalk-Gips*, 20, 91.

Lieber, W. and Gebauer, J. (1969). 'Einbau von Zink in Calciumsilicathydrate', *Zem.-Kalk-Gips*, 22, 161.

Mackenzie, K. J. D. and Alasti, H. (1979). 'Formation kinetics of portland cement clinker phases, II. Tetracalcium Aluminoferrite', *Trans. Br. Ceram. Soc.*, 162–67.

Majumdar, A. J. (1965). 'The ferrite phase in cements', *Trans. Br. Ceram. Soc.*, 64, 105–19.

Majumdar, A. J. (1965). 'The ferrite phase in cements', *Trans. Br. Ceram. Soc.*, 64, 105–19.

Maki, I. and Goto, K. (1982). 'Factors influencing the phase constitution of alite in Portland cement clinker', *Cem. Concr. Res.*, 12, 301–8.

Maki, I. and Chromy, S. (1978). 'Characterisation of the alite phase in Portland cement clinker by microscopy', *Il Cemento*, 75, 247–52.

Maki, I. and Chromy, S. (1978). 'Microscopic study on the polymorphism of Ca₃SiO₄', *Cem. Concr. Res.*, 8, 407–14.

Maki, I., Fukuda, K., Seki, S. and Tanioka, T. (1991). 'Impurity distribution during crystal growth of alite in Portland cement clinker', *J. Am. Ceram. Soc.*, 74, 2082–5.

Maki, I. (1974). 'Morphology of the so-called prismatic phase in Portland cement clinker', *Cem. Con. Res.*, 4, 87–97.

Maki, I. (1978). 'Characterization of the alite phase in Portland cement clinker by microscopy', *Cemento*, 3.

Maki, I. (1978). 'Microscopic study on the polymorphism of Ca₃SiO₅', *Cem. Concr. Res.*, 8, 407–14.

Maki, I. (1986). 'Relationship of processing parameters to clinker properties; influence of minor components', *Proc. 8th Int. Cong. Chem. Cem.*, VI(I-2), pp. 34–47.

Mayerhofer, W. (1996). 'Rietveld-analyse der Ferratphase: vom Laborsystem zum techn. Portlandzement', *Dipl. Arb. Erlangen*.

McMurdie, H. F. and Insley, H. (1936). 'The quaternary system CaO-MgO-2CaO · SiO₂-5CaO Al₂O₃', *J. Res. Natl. Bur. Stand.*, 16, 467–74.

Meissner, E. 'Symmetrieänderungen der Mischkristalle im binären System 8(CaO)-Na₂O-3(Al₂O₃) – 3CaO · Al₂O₃ mit besonderer Berücksichtigung der Hydratationsreaktionen. Dipl-Arbeit-Univ. Erlangen (1986).

Midgley, C. M. (1967). 'The crystal structure of beta dicalcium silicate', *Acta Crystallogr.*, 1, 1948–23; (1962) *ACCRA*, 5, 807–12.

Mondal, J. W. and Jeffrey, P. (1975). 'The crystal structure of tricalcium aluminate, Ca₃Al₂O₆', *Acta Crystallogr.*, 31, 689–97.

Moranville-Regourd, M. and Boikova, A. I. (1992). 'Chemistry, structure, properties and quality of clinker', *9th ICCC*, New Dehli, Vol. 1, pp. 3–45.

Motzet, H. F. and Pöllmann, H.: Quantitative Phase Analysis of High Alumina Cements, *Proc. 20th Int. Conf. Cem. Micr.*, Guadalajara (1998).

Mumme, W. G. (1950). 'Crystal structure of tricalcium silicate from a Portland cement clinker and its application to quantitative XRD analysis', *Neues Jb. Min.*, Monatshefte (Band–Jahr); (1995), *NJMMA*, 145–60.

Mumme, W., Cranswickl, Onakoumak, S. B. (1950). 'Rietveld crystal structure refinement from high temperature neutron powder diffraction data for the polymorphs of dicalcium silicate', *Neues Jb. Min.*, Abhandlungen (Band-Nr.); (1996) *NUMIA*, 170(2), 171–38.

Mumme, W. G., Hill, R. J., Bushnell-Wyeg, Segniter (1950). 'Rietveld crystal structure refinements, crystal chemistry and calculated powder diffraction data for the polymorphs of dicalcium silicate and related phases', *Neues Jb. Min.*, Abhandlungen (Band-Nr.); (1995) *NJMIA*, 169(1), 35–68.

Murat, M. and Sorrentino, F. (1996). 'Effect of large additions of Cd, Pb, Cr, Zn, to cement raw meal on the composition and the properties of the clinker and the cement', *Cem. Concr. Res.*, 26(3), 377–85.

Neubauer, J., Sieber, R., Kuzel, H.-J. and Ecker, M. (1996). 'Investigations on introducing Si and Mg into brownmillerite – A Rietveld refinement', *Cem. Concr. Res.*, 26, 77–82.

Neubauer, J.: Realisierand des Deponie Konzeptes der Inneren Barriere – Digs. Erlungen (1992).

Newkirk, T. F. and Thwaite, R. D. (1958). 'Pseudo-ternary system calcium oxide-monocalcium aluminate-dicalcium ferrite', *J. Res. Natl. Bur. Stand.*, **61**, 233–45.

Newkirk, T. F. and Thwaite, R. D. (1958). 'Pseudo-ternary system calcium oxide – monocalcium aluminate ($CaO \cdot Al_2O_3$) – dicalcium ferrite ($2 CaO \cdot Fe_2O_3$)', *J. Res. Natl. Bur. Stand.*, **61**, 233.

Niesel, K. and Thorman, P. (1967). 'The stability fields of dicalcium silicate modifications', *Tonindustrie Zeitung Keramische Rundschau*, **91**, 362–69.

Nishi, F. and Takeùchi, Y. (1975). 'The Al_6O_{18} rings of tetrahedra in the structure of $Ca_{8,5}NaAl_6O_{18}$', *Acta Crystallogr.*, **31**, 1169–73.

Nishi, F. and Takeùchi, Y. (1985). 'Tricalciumsilikate $Ca_3O[SiO_4]$: the monoclinic superstructure', *Z. Kristallogr.*, **172**, 297–314.

Nishi, F. and Takeùchi, Y. (1979). 'The rhombohedral structure of tricalsium silicate at 1200 degrees C', *Z. Kristallogr.*, **149**. *ZEKRD*, **168**, 197–212.

Nurse, R. W., Welch, H. H. and Majumdar, A. J. (1965). 'The CaO-Al_2O_3 system in a moisture-free atmosphere', *Trans. Br. Ceram. Soc.*, **64**, 409–18.

Nurse, R. W., Welch, J. H. and Majumdar, A. J. (1965). 'The $12CaO \cdot 7Al_2O_3$ phase in the CaO-Al_2O_3 system', *Trans. Br. Ceram. Soc.*, **64**, 323–32.

Nurse, R. W. (1954). 'The dicalcium silicate phase', *Proc. 3rd Int. Symp. Chem. Cem.*, London, 1952, 56–90.

Nurse, R. W., Welch, J. H. and Gutt, W. (1959). *J. Chem. Soc.*, 1080.

O'Daniel, H. and Hellner, B. (1950). 'Zur Structur von $3 CaO \cdot SiO_2$', *Neues Jb. Min.*, Monatshefte (Band–Jahr) (1950). *NJMMA*, 1950, 108–11.

Odler and Wonnemann, R. (1983). 'Effect of alkalis on Portland cement hydratation I. Alkali oxides incorporated into the cristalline lattice of clinker minerals', *Cem. Concr. Res.*, **13**(4), 477–82.

Odler, I. (1990) and Abdul-Maula, S. (1983). 'Polymorphism and hydration of tricalcium silicate doped with ZnO' *J. Am. Ceram. Soc.*, **66**, 1–4.

Odler, I. and Abdul-Maula, S. (1980). 'Effect of mineralizers on the burning of Portland cement clinker', *Zem.-Kalk-Gips*, **66**, 132–6, 278–82.

Odler, I. (1990) 'Improving energy efficiency in Portland clinker manufacturing', in *Progress in Cement and Concrete*, S. N. Gosh (ed.), Vol. I, Part I, 174–201, ABI Books Pvt Ltd, New Delhi.

Perez-Mendez, M., Fayos, J., Howie, R. A., Gard, J. A. and Glasser, F. P. (1985). 'Calcium fluorosilicates: the $Ca_{6-0.5x}Si_2O_{10}F_x$ – phase', *Cem. Concr. Res.*, **15**, 600–04.

Perez-Mendez, M., Howie, R. A. and Glasser, F. P. (1984). 'Ca_3SiO_5 and its fluorine stabilized aristotype: synthesis, stability and postulated structure of $Ca_{6-0.5x}Si_2O_{10-x}F_x$.' *Cem. Concr. Res.* **14**, 57–63.

Phillips, B. and Muan, A. (1958). 'Phase equilibria in the system CaO – Iron Oxide in Air and at 1 Atm.

O_2 Pressure', *J. Am. Ceram. Soc.*, **4**(11), 445–54; (1959) **42**, 414.

Pöllmann, H., Rohleder, M., Neubauer, J., Riedmiller, A. and Göske, J. (1998). 'Quantitative phase analysis of cements, Lime, gypsum, building and construction materials by using a new software for quantification', *Proc. 20th Int. Con. Cem. Micr.*, Guadalajara, 159–74.

Pöllmann, H., Neubauer, J., König, U. and Motzet, H. (1997). 'Clinker quality control using combined methods – A study using microscopy and X-ray techniques – especially Rietveld method', *Proc. 19th Int. Con. Cem. Micr.*, Cincinnati, 195–221.

Ponomarev, V. I., Kheiker, D. M. and Belov, N. V. (1971). 'Crystal structure of calcium dialuminate, CA_2. *Sov. Crystallogr.*, **15**, 995–8.

Rankin, G. A. and Wright, F. E. (1915). *Am. J. Sci.*, **39**, 31.

Regourd, M. *et al.* (1968). *Proc. 5th Int. Symp. Chem. Cem.*, Tokyo, 44.

Regourd, M. (1964). 'Determination des réseaux de cristaux microscopiques. Application aux différentes formes du silicate tricalcique', *Bull. Soc. Miner. Crist. Franc.* **LXXXVII**, 241–72.

Richartz (1986). 'Einfluβ des K_2O-Gehaltes und des Sulfatisierungsgrades auf Erstarren und Erhärten des Zementes', *Zem.-Kalk-Gips*, **12**, 678–87.

Richartz, B. (1994). 'Reaktion und Abscheidung von Spurenelementen beim Brennen des Zementklinkers', *Schriftenreihe der Zementindustrie*, Heft, **56**.

Rietveld, H. M. A. (1969). 'Profile refinement method for nuclear and magnetic structures', *J. Appl. Crystallogr.* **2**, 65–71.

Robson, T. D. (1967). High Alumina Cements and Concretes, John Wiley and Sons, New York.

Sahu, S. and Majling, J. (1993). 'Phase compatibility in the system CaO-SiO_2-Al_2O_3-Fe_2O_3-SO_3 referred to sulphoaluminate belite cement clinker', *Cem. Concr. Res.*, **23**, 1331–9.

Sieber, R., Pöllmann H. and Brunner, P. (1996). 'Investigation on linker microstructure, hydration characteristics and quality coming from varying burning conditions in the kiln using waste fuels', *Proc. 18th Int. Con. Cem. Micr.*, Houston, 284–303.

Singh, V. K. and Glasser, F. P. (1988). 'High temperature reversible moisture uptake in calcium aluminate $Ca_{12}Al_{14}O_{33-x}(OH)_{2x}$,', *Ceram. Int.*, **14**(1), 59–62.

Smirnoff, G. S., Chatterjee, A. K. and Zhmoidan, G. I. (1973). 'The phase equilibrium diagram of the ternary sub-system CaO-$CaO \cdot Al_2O_3$-$11CaO \cdot 7Al_2O_3 \cdot CaF_2$', *J. Mater. Sci.*, **8**, 1278–82.

Smith, D. K., Majumdar, A. and Ordway, F. (1965). 'The crystal structure of γ-dicalcium silicate', *Acta Crystallogr.*, **18**, 787–95.

Smith, D. K. (1962). 'Crystallographic changes with the substitution of aluminium for iron in dicalcium Ferrite', *Acta Crystallogr.*, **15**, 1146–52.

Smith, D. K., Majumdara, A. and Ordway, F. (1967). 'The structure of gamma dicalcium silicate', *Acta Crystallogr.*, 1, 1948–23, (1965). ACCRA, 18, 787–95.

Sorrentino, F. and Glasser, F. P. (1976). 'Phase relations in the system CaO-Al₂O₃-SiO₂-FeO. Part. II The system CaO-Al₂O₃-Fe₂O₃-SiO₂', *Trans. Br. Ceram. Soc.*, 75, 95–103.

Sorrentino, F. P. (1998). 'Dicalcium silicate in special cements', *Proc. 20th Int. Conf. Cem. Micr.*, Guadalajara.

Sprung, S. and Schmidt, O. (1980). 'Structure and properties of Portland cement clinker doped with zinc oxide', *J. Am. Ceram. Soc.*, 63(1/2), 13–6.

Stemmermann, P. and Pöllmann, H. (1992). 'The system CaO-SiO₂-CaCl₂-phase equilibria and polymorphs below 1000°C – An interpretation on garbage combustion ashes', *Neus. Jb. Min. H.* 9, 409–31.

St.Pierre, P. D. S. (1956). *J. Am. Ceram. Soc.*, 39, 96.

Strunge, J. (1986). Einfluß der Alkalien und des Schwefels auf die Zementeingenschaften', Dissertation, Phillips-Universität Marburg/Lahn.

Surana, M. S. and Joski, S. N. (1990). 'Use of mineralizers and fluxes or improved clinkerization and conservation of energy', *Zem.-Kalk-Gips* 43, 43–5.

Swayze, M. A. (1946). A report on studies of: 1. Ternary system CaO-C₅A₃-C₂S. 2. The quaternary system CaO-C₅A₃-C₂F-C₂S. 3. The quaternary system as modified by saturation with magnesia. Part II', *Am. J. Sci.*, 244, 65–94.

Swayze, M. A. (1946). A report on studies of: 1. The ternary system CaO-C₅A₃-C₂F. 2. The quaternary system CaO-C₅A₃-C₂F-C₂S. 3. The quaternary system as modified by 5 per cent magnesia. Part I', *Am. J. Sci.*, 244, 1–30.

Swayze, M. A. (1946). 'A report on studies of CaO-C₅A₃-C₂F', *Am. J. Sci.*, 244(1), 1–14.

Takashima, S. (1958). *Rev. 12th Gen. Meet.*, Cement Association of Japan, Tokyo, pp. 12–13.

Takeuchi, Y., Nishi, F. and Maki, I. (1980). 7th *Int. Con. Chem. Cem.*, Paris, Vol. VII, p. 426.

Taylor, H. F. W. (1990). *Cement Chemistry*, pp. 45–50, Academic Press, New York.

Taylor, H. F. W. (1990). *Cement Chemistry*, p. 89, Academic Press, New York.

Taylor, H. F. W. (1971). 'The crystal structure of kilchoanite, Ca₆(SiO₄)(Si₃O₁₀), with some comments on related phases', *Miner. Mag.*, 38, 26–31.

Trömel, G. (1943). *Stahl und Eisen*, 63, 21.

Tsuboi, T., Ito, T., Hokinoue, Y. and Matsuzaki, Y. (1992): 'Die Einflüsse von MgO-SO₃ und ZnO auf die Sinterung von Portland klinker', *Zem.-Kalk-Gips*, 9, 426–31.

Tsurumi, T., Hiranoy, Katoh, Kamiya, T., and Daimon, M. (1994). 'Crystal structure and hydration of belite', *Ceram. Trans.*, 13; *Supercond. Ceram. Supercond.*, CETRE, 40, 19–25.

Tumidajski, P. J. and Thomson, M. L. (1994). 'Influence of cadmium on the hydration of C₃A', *Cem. Concr. Res.*, 24(7) 1359–72.

Udagawa, S. and Urabek (1978). 'Crystal structure of modifications of calcium silicate Ca₂SiO₄ and their phase transformation', *Semento Hijutsu Nembo*, SGNEA, 32, 35–38.

Udagawa, S., Urabek and Yanot (1977). Crystal structure analysis of alpha-Ca₂SiO₄. *Semento Hijutsu Nempo*, SGNEA, 31, 26–9.

Udagawa, S., Urabek, Yanot, Natsumem (1979). 'Studies on the dusting of calcium silicate (Ca₂SiO₄). The crystal structure of alpha-Ca₂SiO₄', *Semento Hijursu Nempo*, SGNEA, 33, 35–8.

Udagawa, S., Urabek, Natsumem, Yanot (1980). 'Refinement of the crystal structure of gamma-Ca₂SiO₄', *Cem. Concr. Res.*, CCNRA, 10, 139–44.

Vaniman, D. T. and Bish, D. L. (1990). 'Yoshiokaite, a new Ca, Al-silicate mineral from the moon', *Am. Mineral.*, 75, 676–86.

Von Lampe, F. *et al.* (1986). *Cem. Concr. Res.* 16, 624.

Walenta, G. (1997). 'Synthesen und Rietveldverfeinerungen der Einzelphasen von Tonerdezementen', Dissertation University Erlangen.

Welch, J. H. and Gutt, W. (1960). 'Effects of the minor components on the hydraulicity of the Ca silicates', *Proc. 4th Int. Symp. Chem. Cem.* Monograph no. 43, 1, pp. 59–68.

Welch, J. H. and Gutt, W. (1961). 'High temperature studies of the system calcium oxidephosphorus pentoxide', *J. Chem. Soc.*, 874, 4442–6.

Welch, J. H. and Gutt, W. (1959). 'Tricalcium silicate and its stability within the system CaO-SiO₂', *J. Am. Ceram. Soc.* 42, 11–5.

Wenxi, H., Yurong, Y., Guangren, Q. and Yangyoun, Q. (1992). 'The research on C₂AₓF₁₋ₓ in iron-rich fluoraluminate cement', *9th ICCC*, New Dehli, Vol. 3, pp. 9–15.

Woermann, E., Hahn, T. H. and Eysel. W. (1963). 'Chemische und strukturelle Untersuchungen der Mischkristallbildung von Tricalciumsilikat', *Zem.-Kalk-Gips*, 9(1) Bericht S.370–5.

Woermann, E., Hahn, T. H. and Eysel, W. (1967) 'Chemische und strukturelle Untersuchungen der Mischkristallbildung von Tricalciumsilikat', *Zem.-Kalk-Gips*, 9(2) Bericht S.385–91.

Woermann, E., Hahn, T. H. and Eysel, W. (1968). 'Chemische und strukturelle Untersuchungen der Mischkristallbildung von Tricalciumsilikat', *Zem.-Kalk-Gips*, 6(3). Bericht S.241–51.

Woermann, E., Hahn, T. H. and Eysel, W. (1969). 'Chemische und strukturelle untersuchungen der Mischkristallbildung von Tricalciumsilikat', *Zem.-Kalk-Gips*, 5(4) Bericht S.235–41.

Woermann, E., Hahn, T. H. and Eysel, W. (1969). 'Chemische und strukturelle Untersuchungen der- Mischkristallbildung von Tricalciumsilikat', *Zem.-Kalk-Gips*, **9**(5) Bericht S.414–22.

Woermann, E., Hahn, T. H. and Eysel, W. (1979). 'The substitution of alkalies in tricalcium silicate', *Cem. Concr. Res.*, **9**, 701–11.

Woermann, E., Hahn, T. H. and Eysel, W. (1979). 'The substitution of alkalies in tricalcium silicate', *Cem. Concr. Res.*, **9**, 701–11.

Yamaguchi and Takagi (1969). 'Analysis of Portland cement clinker', *5th ISCC*, Tokyo, Vol. 1, p. 181.

Yimin, C., Liping, S., Muzhen, S. and Jun'an, D. (1992). 'Kinetics of aluminoferrite formation at different conditions', *9th ICCC*, New Delhi, Vol. 3, pp. 262–7.

Zhmoidin, G. I. and Chatterjee, A. K. (1972) 'Phase diagram of the subsystem $CaO-CaO.Al_2O_3-CaF_2$ Transl.f. Izv.Ak.Nauk.SSSR', *Neorg. Mat.*, **10**, 1846–51.

Chapter three

Hydration of Portland cement

E. M. Gartner, J. F. Young, D. A. Damidot and I. Jawed

3.1 Introduction

Since the chemical reactions of Portland cement with water are responsible for the setting and hardening of mortar and concrete, the underlying chemistry must be investigated and described in order to understand these processes. However, this is not an easy task because cement is a complex mixture of numerous compounds. In this chapter, the hydration of C_3S, C_2S, C_3A, and the ferrite phase is first discussed and then the hydration of cement. Interactions between the individual hydrating components, and the relationships between hydration and setting and hardening of concrete are considered. An attempt has been made to briefly identify the role of chemical admixtures and minor components (such as alkalis) in hydration; however, lack of space precludes any consideration of the hydration of blended cements. A short account of the microstructure of hardened cement paste and the structure and properties of the hydration products is included because of their importance in determining hardened paste properties.

Our knowledge of the hydration processes has expanded considerably since this chapter first appeared, mostly due to the application of modern instrumental methods of analysis, but a universally accepted view of hydration still awaits detailed formulation. While we are able to describe the hydration processes in considerable detail now, there are still important issues, which await universal agreement. This chapter attempts to address the debatable issues, while still providing an overarching description of the hydration processes.

3.2 Hydration of tricalcium silicate

Tricalcium silicate (or alite) is the most important constituent of Portland cement. Since the hydration of alite usually controls the setting and early strength development of Portland cement pastes, mortars and concretes, the hydration of relatively pure C_3S has been extensively investigated. It is widely recognized that there are important differences between the hydration kinetics of C_3S in water and of alite in a Portland cement – water mixture, primarily due to interactions with the other reacting constituents (see Section 3.3). Nevertheless, these differences can in the main be considered to be perturbations on the C_3S–water system, which do not greatly change the fundamental reaction mechanisms. Studies on pure C_3S–water mixtures have proved invaluable in delineating and understanding the effects of temperature, dilution, admixtures, and other compounds found in concrete.

There continues to be considerable divergence of opinion expressed in the literature concerning the mechanism of C_3S hydration, which has been discussed in many previous reviews [1–6]. Two problems make it difficult to directly compare experimental results from different laboratories: first, different preparations of C_3S can vary widely in their reactivity; and second, a complete set of experimental data does not exist for any one C_3S sample. Therefore, comparisons of results between separate groups can only be made in a general way with considerable latitude for error and misinterpretation. In addition, studies have been made using widely different water-to-solid ratios (w/s) (see Appendix) which have a large effect on the course of hydration. Nevertheless, we believe that the hydration mechanism can be reasonably well explained by a series of hypotheses, only some of which have been clearly proven. In this section, we attempt to provide a unified theory, drawing attention to specific points of disagreement which need further investigation.

3.2.1 Overview of C₃S hydration

The hydration of C_3S under ambient conditions can be described by reaction (1)

$$C_3S + (3-x+n)H \rightarrow C_xSH_n - (3-x)CH$$
$$[\Delta H = -121 \, kJ \, mol^{-1}] \tag{1}$$

However, the calcium silicate hydrate (C-S-H) is an apparently amorphous phase of variable composition, and hence is usually written as C-S-H, which does not imply a specific stoichiometry. When the contact solution is saturated with respect to crystalline CH (portlandite) $x = 1.7$ and $n = 4$. The value of ΔH is estimated for $x = 1.7$ and $n = 2.7$, [7,8] although it should not change much with increasing water content as the additional water is very weakly bound. The enthalpy of dissolution of C_3S at infinite dilution is estimated to be $-132 \pm 10 \, kJ/mol$ [9], which is consistent with the observation that the dissolution of CH is slightly exothermic.

An initial w/s of about 0.42 is necessary for complete reaction. If more water is added, then there will be an excess of liquid at the end of the reaction, but if w/s exceeds about 280, the amount of CH formed will not be enough to saturate the solution, and [CH] will depend on the

value of w/s. Since the value of x decreases with decreasing [CH] [9], the C-S-H formed under these conditions will have $x < 1.7$, and n may also vary. If w/s > 2200, the solubility limit of all of the potential solid reaction products will be exceeded, and C_3S will dissolve completely. If w/s < 0.42 then either some of the C_3S will not react, or else the C-S-H formed will have $n < 4$. The results obtained at low w/s apparently depend not only on temperature, but also on the way in which the reaction is conducted, indicating that it is very difficult to establish a true equilibrium in this system. In addition, the C-S-H formed is probably a metastable product in most cases. Crystalline afwillite ($C_3S_2H_3$) may be the thermodynamically stable product in this system at room temperature [10]; however, it only forms if deliberately nucleated [11,12].

The kinetics of reaction (1) are rather complex. Traditionally the early stages of the reaction have been followed by monitoring the rate at which heat is evolved; preferably by using a pseudo-isothermal conduction calorimeter, which can be a very sensitive technique because the hydration of C_3S is highly exothermic. The instantaneous thermal power output of such a calorimeter (properly corrected for instrumental factors) is a good measure of the rate of enthalpy change of the reaction, and thus of the rate of C_3S hydration. An idealized calorimetric curve is given in Figure 3.1a which has usually been divided into five distinct stages [13,14] but, in the light of recent studies, we have added an initial 'Stage 0', as shown in Table 3.1. The choice of the calorimetric curve to represent the reaction is purely arbitrary, other experimental observations such as the rate of release of Ca^{2+} into solution (drawn schematically in Figure 3.1b) could be used. However, the calorimetric curve has the advantage that it is usually closely related to the actual rate of reaction, and when calibrated on an absolute scale, can be integrated to give the total degree of reaction at any given time. Qualitatively it has the advantages of relative simplicity, sensitivity and accuracy, which have made it the most frequently used method, although the exact shape and position of the peaks will depend on

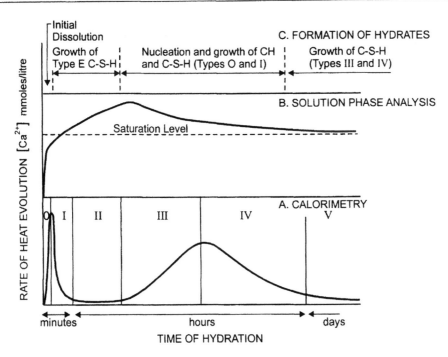

FIGURE 3.1   Schematic representation of changes taking place during the hydration of C$_3$S pastes (w/s < 1.0).

the particular instrument used. However, without the capability of mixing all of the materials inside the calorimeter after they have reached equilibrium temperature, Stages 0 and I cannot be measured. Note that the quality of the mixing obtained in typical laboratory calorimeters varies widely, and can lead to differences in the hydration kinetics. (Also the time-constants of most calorimeters do not allow Stage 0 to be resolved even with *in situ* mixing.)

The duration of each of the stages shown in Figure 3.1 varies widely depending on the reaction conditions as well as the 'inherent reactivity' and particle size distribution (psd) of the C$_3$S sample used; for example, Stage 0 perhaps lasts for only a fraction of a second for a typical C$_3$S of average 'cement' fineness at the low w/s values typical of concrete, but can represent essentially the whole hydration process (of the order of 1 h) in extremely dilute stirred suspensions.

The typical rates of hydration (dα/dt) during all of the stages except Stage 0 are shown in Figure 3.2, plotted against the degree of reaction, (α), rather than against time. The induction period (Stage II)

is represented by the only minimum in the curve. The transition from Stage IV to Stage V is somewhat arbitrary; it can be considered to be represented by the decrease in the rate of decline of dα/dt (occurring for this sample at about 25 per cent hydration or one day's hydration). The slow steady decline in rate during Stage V continues up to complete hydration, which can take years for a sample with a typical cement-like psd. In this respect, C$_3$S hydration is very different from that of plaster (calcium sulphate hemihydrate), which shows no decrease in the rate of decline, and ends abruptly with complete hydration after, typically, only about one hour [3].

3.2.2 The early hydration periods

3.2.2.1 The initial fast reaction

The very early reactions between water and C$_3$S have been the most difficult to elucidate because of the high rates of the processes involved and the difficulty of stopping the hydration rapidly

TABLE 3.1   Typical sequence of the hydration of C_3S at low w/s

Reaction stage	Corresponding stages in concrete-making	Chemical processes	Overall kinetic behaviour
0. Initial fast reaction	First wetting	Surface hydrolysis and release of ions into solution.	Very rapid, exothermic dissolution; can only be observed at very high w/s.
I. First deceleration	Wetting, mixing	Formation of hydrate coating on C_3S surface, retarding dissolution.	Chemical control (probably by rate of heterogeneous nucleation)
II. Induction period	Transporting, agitating, placing, and finishing, (especially in presence of 'retarders').	Retarded nucleation of final hydrates; slow consumption of 'retarders.'	Chemical control (by consumption of 'retarders' and/or nucleation of hydrates)
III. Acceleration period	Setting, initial curing.	Accelerating growth of principal hydration products.	Auto-catalytic growth of principal hydrates (chemical control)
IV. Second deceleration	Curing, demoulding.	Continued growth of hydration products into large empty spaces.	Onset of diffusion control; products otherwise similar to stage III.
V. Final slow reaction	Continued slow hardening as long as moist curing is continued.	Gradual densification of microstructure around residual unhydrated C_3S; CH re-crystallization.	Diffusion controlled, but not necessarily by the same species as in IV. Slow changes in C-S-H?

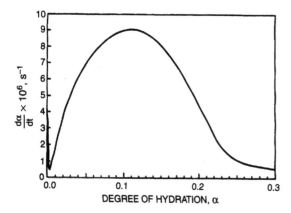

FIGURE 3.2   Rate versus degree of hydration for a typical C_3S sample (23 C, *w/s* = 0.5) [3].

enough without creating artefacts. Recent studies [15], have made effective use of the method of dilute suspensions to lengthen the early reaction steps and thus resolve them more clearly.

On first contact of C_3S with liquid water, the calcium, oxide and silicate ions in the surface layers of the anhydrous phase hydrate and pass rapidly into solution, presumably as simple hydrated ('aq.') ionic species, represented here by the symbols Ca^{2+}, OH^- and $H_2SiO_4^{2-}$.

$$C_3S\ (3Ca^{2+}:O^{2-}:SiO_4) + \text{excess water}$$
$$\rightarrow 3Ca^{2+}(aq.) + 4OH^-(aq.) + H_2SiO_4^{2-}(aq.) \quad (2)$$

The initial reaction is rapid and highly exothermic, due primarily to the protonation of the extremely basic oxide ions. All of the anhydrous ions, but especially Ca^{2+}, solvate by forming a shell of oriented H_2O molecules. Reaction (2) represents congruent dissolution (similar to the dissolution of a simple salt), so that the rate should depend on the specific surface area of the C_3S and its density of 'active sites.'

Thermodynamic calculations [16] indicate that C_3S would be extremely water-soluble if no other phases precipitated. An advantage of working with dilute stirred suspensions is that the ions created at the surface of the C_3S can be removed rapidly enough to prevent immediate precipitation of any reaction products close to the surface. It is then possible to estimate the very high initial rate of dissolution of C_3S. For example, the data shown in Figure 3.3 indicate that, at w/s = 20 $[SiO_2]^1$ is above 1 mM at the time of the first measurement (< 1 min). If this rate were maintained for the whole reaction, it would correspond to complete dissolution of C_3S in only

about 4 h; but this is certain to be an underestimate since the rate is decreasing even before the first measurement. Damidot and Nonat [9] showed that a C_3S sample of 310 m²kg⁻¹ Blaine, stirred in pure water at w/s = 5000, dissolves completely in about 30 min, from which we calculate the initial dissolution rate of this sample to be at least 10^{-3} mol s⁻¹, giving a rate of dissolution per unit surface area of at least 1.4∗ 10^{-5} mol s⁻¹ m⁻². Even at this extremely high w/s, the initial dissolution rate is not properly resolved by calorimetry, suggesting that even the best-performing calorimeters currently available probably have time-constants that are too long to resolve such a rapid reaction. At 'normal' w/s values, the initial fast reaction may be almost over even before the C_3S powder is completely wetted by mixing with water, and the observed hydration rates are always much lower. This proves that the dissolution of C_3S is strongly inhibited at all times after the first rapid dissolution, either by a protective surface hydrate coating, or by poisoning of its dissolution sites.

3.2.2.2 The first deceleration period

Under most experimental conditions it is observed that once the mean concentrations of calcium and silica in the solution reach a critical solubility product, defined by Barret *et al.* [17–20] as the 'maximum supersolubility,' nucleation and precipitation of an initial product occurs. In principle, it is this transition that must terminate the initial fast reaction, since the surface of C_3S becomes altered in such a way as to be less reactive. The formation by local precipitation of a protective coating of a type of calcium silicate hydrate is not the same as the concept of the 'superficially hydrolysed' layer, C_3S_{sh}, proposed by Barret *et al.* [20,21]. Both hypotheses seem at first sight to differ only in the detailed atomic structure of the surface layer. However, the C_3S_{sh} proposed by Barret *et al.* is not a true surface phase, but a charge-separated transition state on the surface that is expected to occur if the first step in the hydrolysis is protonation of O^{2-} and SiO_4^{4-}, to give OH^- and $H_3SiO_4^-$,

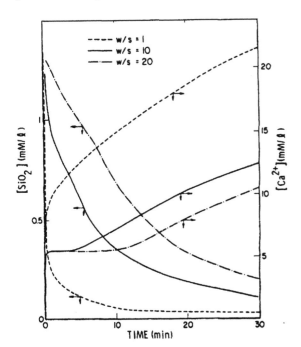

FIGURE 3.3 Changes in calcium and silica in solution during early hydration of C_3S suspensions.

respectively. Clearly, this results in a protonated C$_3$S surface, with a net positive charge, which must be balanced by the hydroxide ions in the adjacent solution. Although Barret *et al.* calculated a theoretical solubility curve for this C$_3$S$_{sh}$ which is appreciably lower than that of anhydrous C$_3$S, it seems questionable to calculate the solubility of a charge-separated phase which cannot be more than one molecular layer thick, and is only one of many possible transition states that might occur during the dissolution process. Furthermore, recent surface charge measurements on C$_3$S in dilute suspensions [22] suggests that it is negative, rather than positive as had been previously reported [23]. We conclude that the initial rapid dissolution is terminated by the formation of a protective coating of a type of C-S-H that is probably somewhat different from the C-S-H that forms at later ages [24–26] (which must be more stable).

The question of whether changes in the 'intrinsic reactivity' of the C$_3$S sample used results in changes in the rate of initial rapid dissolution has not been clearly answered. Fierens and Verhaegen [27] studied the hydration of C$_3$S preparations, which had been subjected to different thermal treatments, and found that the apparent reactivity, based on the rate at which a given level of CH supersaturation could be attained, increased with the intensity of a thermo-luminescence peak which corresponded with the probability of escape of trapped electrons. However, their results appear to show that this difference in reactivity was manifested largely during the acceleration period, and not during the initial fast reaction and deceleration, in contrast to earlier studies [28]. These differences may relate to the stability of the protective layer or to the amount of protective hydrate formed during the first two stages of the reaction. If the substrate C$_3$S contains more reactive sites, this could presumably increase the mean rate of thickening of the protective coating, which could, in turn, lead to generation of more nuclei for the final hydrates.

During the 'initial heat peak' of a typical calorimetric curve (which is a combination of Stages 0 and I), calcium concentrations in solution increase rapidly (Figure 3.3), but silica concentrations are only transitory, reaching maximum values approaching 1.7 mM and then decreasing again. The Ca/Si ratio in solution at early ages depends strongly on w/s (Figure 3.4) [29], and on time. At very high w/s, or short times, it approaches the value of three associated with congruent dissolution of C$_3$S. The [Ca^{2+}] profile in the first hour of hydration is extremely sensitive to the specific C$_3$S preparation used; frequently, but not always, a [Ca^{2+}] plateau is observed, especially at high w/s (Figure 3.5) [30].

The surfaces of C$_3$S samples after very short periods of immersion in water have been analysed by ESCA [31–34]. (This high-vacuum technique measures the composition to a depth of only 10–20 Å, but averaged over a large area, while changes in binding energy can also be detected by measuring the energy shift of ESCA peaks.) It was found that the binding energy of some oxygen atoms increases slightly after wetting, and also that even fresh 'anhydrous' C$_3$S samples show signs of surface hydration which have to be removed by sputtering under high vacuum. The apparent surface Ca/Si ratio versus hydration time curves (Figure 3.6), show an initial dip within a few seconds, followed by a brief increase

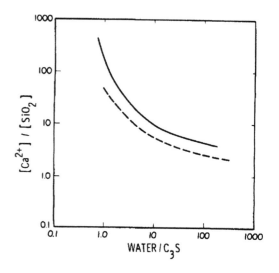

FIGURE 3.4 Variation of [Ca^{2+}]/[SiO$_2$] ratio in solution after 1 min hydration, as a function of w/s ratio [23,29].

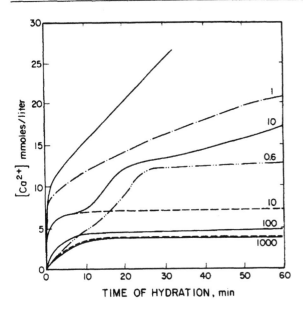

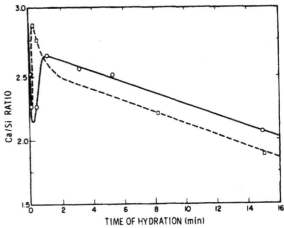

FIGURE 3.6 ESCA analysis of the hydrated C_3S surface during Stage I hydration.

FIGURE 3.5 Release of calcium into solution in the first hour of hydration for different w/s ratios (numbers on curves) and different C_3S samples [23,27,29,30].

(but always to a value < 3), followed again (after about 1 min) by a slower but steady decrease. Although the initial dip seems to indicate incongruent dissolution from the very start, it could arise from surface contamination (such as traces of silica in the water or removal of calcium during rinsing). The subsequent rise does not occur when very high w/s values (> 3000) are used [34], and may also be an artefact of sample preparation, although it could be due to some re-adsorption of calcium by the surface layer, as predicted by the 'double-layer' theory [23]. The steady decrease after the first minute can readily be explained on the assumption of a steadily increasing thickness of a low Ca/Si hydrated layer, and is less likely to be due to artefacts of sample preparation. Thus, ESCA data support the concept that a fairly uniform coating of some kind (with Ca/Si ≪ 3) develops within the first 10 min even at w/s = 10.

Although the C/S ratio of the initially precipitated material was originally assumed by Ménétrier [30] to be 1.0, while Vernet et al. [35] used a value of 1.5, the C/S ratio should vary

with the composition of the solution [18,19]. Direct experimental estimation of the C/S ratio in these early stages is almost impossible to obtain other than by the ESCA measurements, which can only be used to estimate Ca/Si if they represent a continuous C-S-H layer whose thickness exceeds the depth that the ESCA method can sample. In the case of partial surface coverage, the precipitated superficial phase would have a lower Ca/Si than is actually observed, but since it cannot have a ratio of less than zero, the data in Figure 3.6 imply that the surface coverage is at least 30 per cent after about 15 min. Parallel SEM studies [34] show the rapid precipitation of a few isolated globular particles which is completed within the first minute. Thereafter, the surface becomes progressively damaged by selective pitting on certain crystal faces [33,36,37]. The initial surface hydrate layer detected by ESCA must presumably be fairly homogeneous and uniform in thickness over the surface, so as not to be visible by scanning electron microscope (SEM). Later, thin flakes or foils of C-S-H are seen to develop and grow rapidly at the surface forming honeycomb-like structures [33,36,37] or an oriented 'fish scale' morphology [38]. Similar morphologies have been detected by TEM observations in a 'wet-cell', which at 30 min show very variable distributions of identifiable concentrations of hydrates over the surface, with a wide range of compositions ranging from almost pure amorphous silica up to C-S-H with Ca/Si = 1.2 [39].

Skalny, Young and co-workers [23,30] have proposed a mechanism for the early hydration that does not involve congruent dissolution, and thus essentially omits Stage 0 entirely. Initial dissolution is considered to be incongruent with Ca^{2+} and OH^- moving rapidly into solution leaving a silica-rich surface layer. Subsequent re-adsorption of Ca^{2+} on the now negatively charged surface creates an electrical double layer and gives a positive zeta potential. This helps explain the initial dip in the ESCA curve, and why the surface of C_3S appears to be initially free of hydration products, (assuming that the silica-rich surface layer parallels the morphology of the original anhydrous surface.) The silica-rich layer is considered to be a disorganized layer containing protonated SiO_4^{4-} ions and possibly some silicate dimers, but since it is also likely to contain some calcium for charge balance it could be considered as a C-S-H of low Ca/Si ratio. Continued incongruent dissolution should increase the thickness of the amorphous layer, which eventually reorganizes into C-S-H possibly via a 'topotactic' transformation of the type that had been proposed by Fujii and Kondo [40]. If the concept of incongruent and topotactic formation of a silica-rich layer is maintained, this need not exclude subsequent dissolution precipitation processes [41]. Dent Glasser [42] has suggested that this layer can be a quasi-solid, rich in water, which will subsequently form C-S-H.

More recent studies making use of very dilute stirred suspensions have shown that the transition from rapid congruent dissolution to a period of deceleration (PQ in Figure 3.7) [43], during which the calcium and silica concentrations decrease from the curve of maximum supersolubility to a curve for a slightly lower solubility product, referred to as the 'associated path.' This can be considered to represent the rapid precipitation of a form of C-S-H from the solution (which at high w/s acts a significant reservoir of ions). This period essentially marks the end of the free

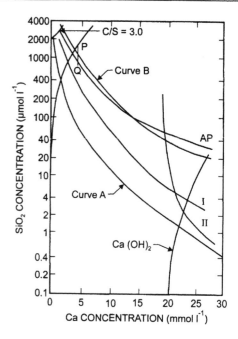

FIGURE 3.7 Idealized representation of the dissolution–precipitation processes in C_3S suspensions plotted on a C-S-H diagram, showing the associated path (AP) and curves for C-S-H(SI) and C-S-H (SII) (I and II respectively) [44], and curves for C-S-H(s) and C-S-H (m) (A and B respectively) [26].

dissolution of C_3S, but there is still some disagreement about exactly what happens at this point. The period PQ can be seen to be associated with a rapid deceleration in the rate of hydration. At the end of this brief precipitation, the solution composition follows the 'associated path' for a long time, while calcium concentrations increase and silica concentrations decrease continuously. In this respect the 'associated path' has all the hallmarks of a solubility curve associated with a C-S-H coating forming on the surface of C_3S as proposed by Jennings [26], Barret and Bertrandie [44] propose that the associated path is generated by a kinetic balance between C_3S dissolution and the precipitation of 'normal' C-S-H (and thus, it must lie between the C-S-H supersolubility curve and the normal C-S-H solubility curve.) In principle, a kinetic balance cannot give a unique curve on the lime-silica solubility diagram; however, the data show considerable scatter about the 'associated path'. Gartner and Jennings [16] developed a more refined thermodynamic analysis of the C-S-H system and introduced the terminology 'C-S-H(m)' to indicate that 'curve B' was the solubility curve of a 'metastable' form of C-S-H that was probably the protective coating on the C_3S surface. The fact that 'curve B' was obtained from solution data for preparations still containing some unreacted C_3S indicates that the system must be reacting slowly and is not, therefore, at equilibrium, but it is likely that some kind of pseudo-equilibrium can exist between whatever is on the surface of the C_3S particles (nominally, 'C-S-H(m),') and the solution during this induction period.

The relative positions of the different proposed solubility curves and the 'associated path' are summarized in Figure 3.7. The concept of protective coatings that are in pseudo-equilibrium with their environment, and which protect an underlying phase of high intrinsic reactivity with the environment, is certainly not restricted to cement systems.

3.2.2.3 The induction period

The induction period in the unretarded C_3S–water system is just the minimum in the curve between Stages I and III (see Figure 3.2), and the idealized calorimetric curve shown in Figure 3.1 is in fact only observed when chemical retarders are added to the system. A more typical calorimetric curve

for the early hydration period, given by Brown et al. [45] is shown in Figure 3.8, where it can be seen that there is no true induction period, either in water or in solutions of salts that accelerate the hydration.

During this period of low heat evolution, the trends established at the end of Stage I continue. $[Ca^{2+}]$ and $[OH^-]$ continue to increase, surpassing the theoretical level of saturation for CH (Figure 3.9). C-S-H ('Type E') continues to develop on the C_3S surface, but the total degree of hydration by the beginning of the acceleration period is typically only 0.1–1 per cent [6]. A maximum saturation factor (defined as the ionic activity product for the relevant solid, divided by its value at saturation) of well over two with respect to CH is reached before it begins to crystallize at a significant rate. The time at which this maximum is reached is strongly dependent on both the w/s (Figure 3.9) and the reactivity of the C_3S sample used. At about the same time as the crystallization of CH begins [30], a new C-S-H morphology (Type 0/Type I) can often be observed by electron microscopy [46]. For moderate values of w/s the correlation between the onset of the acceleration period, the formation of crystalline CH, and the formation of Type 0/Type I C-S-H is often quite close, when due allowance is made for the difficulties in detecting the new phases. This was one of the factors that led Young et al. [30] to conclude that it was the rapid removal of CH from solution at that point that drove the renewed

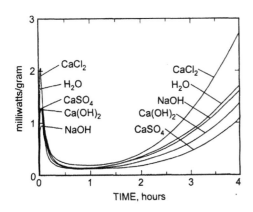
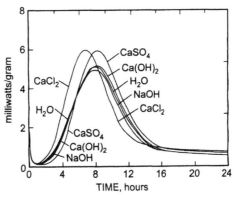

FIGURE 3.8   Rate of heat evolution from hydrating C_3S with inorganic salts [45].

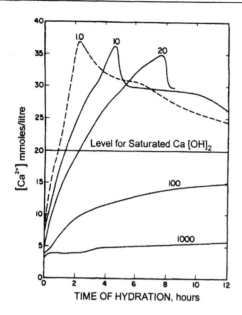

FIGURE 3.9 Release of calcium into solution as a function of w/s ratio (indicated for each curve) [29,30].

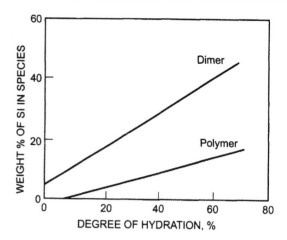

FIGURE 3.10 Silicate polymerization during hydration.

acceleration of the reaction. However, the observation that addition of crystalline CH to the mixture frequently had no accelerating effect, and often actually retarded the onset of the acceleration period, [47] is more difficult to explain. It was assumed that the high concentration of silica in the solution at early ages poisoned the growth of CH [23,30], which could not resume until either silica concentrations became much lower or CH supersaturation much higher. The presence of C_3S has been shown to inhibit crystallization of portlandite from supersaturated solutions of CH [48].

For many years, the theory of Young *et al.* [30], based on portlandite crystallization as the driver for the re-acceleration of C_3S hydration at the end of the 'induction' period, co-existed somewhat uneasily with the alternative theory first proposed by Stein and Stevels [24], that it was the nucleation and growth of a more stable form of C-S-H that caused the acceleration. Jenning's observations [26] on the lime-silica solubility diagram supports a modified version of Stein and Stevels' mechanism, and Gartner and Gaidis [3] explained in detail how the conversion of the metastable surface coating, [C-S-H(m)] to a

more stable form of C-S-H[C-S-H(s)] by a nucleation and growth process could give rise to the observed kinetics of Stages II and III in C_3S hydration. Once the more stable product has nucleated, either within the surface layer or in close proximity, growth of the new phase occurs at the expense of the coating. The coating can apparently reform very rapidly to reach an 'equilibrium thickness'. Thus, in effect, it is the rate of growth of the stable hydrates that controls the rate of dissolution of the anhydrous C_3S. A second important factor in the Gartner–Gaidis mechanism is the suggestion that the initial number of nuclei of C-S-H(s) is actually proportional to the total amount of C-S-H(m) precipitated during the first deceleration period, which explains why there is no true induction period in the absence of retarders. The importance of the presence of C-S-H(s) nuclei is highlighted by experiments in which an artificially-prepared C-S-H was added resulting in a very early transition from the first deceleration period to the acceleration period (see Figure 3.13A).

More recently, the mechanism has been probed further using stirred suspensions at high w/s (often with CH solutions of various concentrations replacing water) allowing a greater variation in the times of onset and the duration of the different stages of the reaction. In water at very high w/s, the rate of formation of C-S-H increases well before the onset of rapid CH crystallization, which is detected by its endothermic calorimetric peak [15]. Damidot and Nonat [43,49] explained these results in terms of two different types of C-S-H, termed 'SI' and 'SII', stable at CH concentrations below and above about 20 mM, respectively. (These types may be the same as those observed by Ramachandran and Grutzek [50] in studies of C_3S under controlled pH). The initially-formed C-S-H(SI) becomes metastable, and it is then the nucleation and growth of C-S-H(SII) controls the kinetics. This mechanism is fairly similar to that proposed by Gartner and Gaidis [3], except for the details of the structure and composition of the two types of C-S-H. C-S-H(SI) and C-S-H (SII) have very differently-shaped solubility curves which cross at an invariant point (at about 20 mM $[Ca^{2+}]$ and 0.01 mM $[SiO_2]$) and therefore must have significantly different compositions. NMR spectroscopy showed C-S-H(SII) to have a Ca/Si ratio > 1.7 and to contain fewer highly-polymerized silicate anions than C-S-H(SI) (Ca/Si = 1.0–1.5). This is in contrast to earlier NMR studies [51,52], while the Ca/Si ratios are also different from those of the C-S-H(s) and C-S-H(m) phases as calculated by Gartner and Jennings [16] based on the two solubility curves observed by Jennings [26].

These differences are still not resolved; the compositions calculated by Gartner and Jennings are very sensitive to both the assumed shapes of the solubility curves and the method of computation of the activities of the various ions in solution. Damidot and Nonat emphasize the need for a high lime concentration in solution, close to that of CH saturation, for the onset of a phase transition in C-S-H which destabilizes the initial protective C-S-H(SI) and allows it to convert to C-S-H(SII), at the same time as permitting an acceleration period in overall hydration. In this respect, their theory helps form a bridge between the 'CH nucleation' and 'C-S-H nucleation' theories.

3.2.2.4 The acceleration period

At the beginning of this period, typically about 1 h after mixing, well under 1 per cent of the C_3S

is hydrated, but, at its end, typically at about 5–10 h, this amount has increased to over 10 per cent (see Figure 3.2). The detailed kinetics of the acceleration period are still a subject of some debate; various kinetic models have been proposed to account for the whole process of hydration, but relatively few deal well with the acceleration period. One of the most popular models is the equation first proposed by Avrami [53] for phase changes in initially homogenous media.

$$-\ln(1-\alpha)=kt^m \qquad (3)$$

where t = time and k is a rate constant. For small values of α, this equation is equivalent to $\alpha = kt^m$. The physical significance of the exponent m depends on the following three factors, according to the relationship

$$m=[(p/s)+q] \qquad (4)$$

In this equation p represents the morphology of the growing phase: ($p = 1$ for growth as needles, 2 for sheets and 3 for isotropic growth; s represents the rate-limiting process: $s = 1$ for pure phase-boundary control and 2 for pure diffusion control; and q relates to the rate of nucleation; $q = 0$ for no nucleation and 1 for continuous nucleation at constant rate).

This equation has been successfully applied to Stages II and III, but we find a wide range of values quoted for m in the literature, from below or around 1 [54] to 3 [55,56]. The values obtained clearly depend on the source of C_3S used and the time period over which the analysis is applied. If the data are taken too late in the acceleration period, the value of m will decrease due to the onset of diffusion control, while at earlier ages it may include the first deceleration period. Gartner and Gaidis [3] analysed their calorimetric data in the form of a plot of $d\alpha/dt$ versus α (Figure 3.2), which allows for the early hydration without explicitly re-defining the time axis. It is apparent from Figure 3.2 that the initial part of the acceleration stage on this plot is almost a straight line passing through the origin, approximating an exponential form:

$$d\alpha/dt = k\alpha \qquad (5a)$$

or

$$\alpha = \alpha_0 \exp(kt) \qquad (5b)$$

Similar growth kinetics have also been observed at early ages in the hydration of calcium sulphate hemihydrate [57]. To fit Avrami's equation to the same data we must eliminate time, which at low values of α, results in the following form:

$$d\alpha/dt = mk\{\alpha/k\}^{\{(m-1)/m\}} \qquad (6)$$

This equation also comes close to fitting the early-age results shown in Figure 3.2, if $m > 3$. There are relatively few good sources of kinetic data on the very early hydration of C_3S, but the calorimetric study of Brown *et al.* [46] (Figure 3.8a) gives additional data which agree very well with the data of Gartner and Gaidis when reanalysed in this way [58], giving $m > 3$ during the first 4 h.

Since nuclei must normally be present at the beginning in order for the process to start, we represent the initial nuclei by an initial value of α, (α_0), and integrate (6) to obtain:

$$(\alpha)^{1/m} - (\alpha_0)^{1/m} = k't \qquad (7)$$

An exponential (or almost exponential) growth rate is indicative of a branching process, (which continually creates more active growth sites) consistent with an exponent of $m > 3$ in the Avrami equation. A hypothesis capable of explaining this type of growth mechanism and relating it to the structure of the C-S-H found in cement pastes has recently been proposed by Gartner [59].

Not surprisingly, this very high order growth process cannot continue for long and after about 4 h of hydration, the rate of increase of the heat production begins to decrease noticeably. Nevertheless, between 1 and 4 h the rate has increased by more than an order of magnitude, indicating a doubling time of under 1 h. The decrease in rate is universally agreed to be due to the onset of some kind of diffusion-controlled process.

An alternative non-calorimetric approach to understanding the nucleation and growth processes involved in C_3S hydration at early ages has been taken by Gauffinet [38], who used a modification of the dilute stirred suspension approach (w/s ~ 50) in which the calcium hydroxide concentration in the liquid phase is initially adjusted to a desired value and then remains fairly constant. The degree of hydration of the C_3S could be followed by conductimetric measurement of the increasing CH concentration because the system contained no other components and CH did not precipitate during the period of interest. C_3S hydration profiles were observed to be sigmoidal and to depend significantly on the initial CH concentration chosen (Figure 3.11), indicating a growth process starting from a relatively small number of nuclei formed at the very beginning. At [CH] = 16 mM (below saturation) it occurs at about 12 per cent hydration, very similar to what is observed by calorimetry at low w/s in Figure 3.2. However, at lower [CH] it occurs earlier and at lower degrees of hydration, while at higher [CH] it occurs later and at higher degrees of hydration, the explanation proposed being that the morphology of the C-S-H formed changes with [CH]. Gauffinet *et al.* [37] used a computer simulation to allow for the possibility of different aspect ratios in the resulting C-S-H product. They concluded that the apparent early termination of the acceleration period, which was prevalent at low initial CH concentrations, was due to the preferentially laminar growth of the C-S-H parallel to the C_3S surface under these conditions, which led to earlier surface coverage and subsequently, the earlier onset of diffusion control. Re-analysis of Gauffinet's data for the early acceleration stage (up to about 5 per cent hydration) gives an Avrami exponent $(m) > 3$, similar to results obtained by calorimetry at low w/s [58].

This dilute suspension study demonstrated clearly that the overall shape of the kinetic curve is similar for a wide range of CH concentrations; only the actual values of rate versus degree of hydration change (Figure 3.11). Furthermore the kinetics are not very sensitive to [CH] or the precipitation of CH.

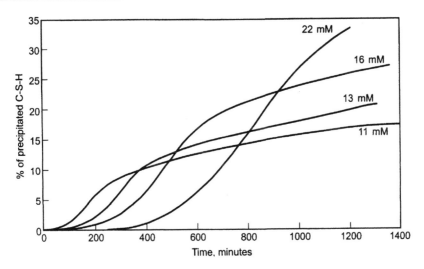

FIGURE 3.11 Precipitation of C-S-H over time from C_3S suspensions in lime solutions of different concentrations [36].

3.2.2.5 Summary of hydration mechanisms for the early hydration period

While the number of proposed schemes in the literature are legion they can now be grouped into a few major categories, as summarized in Table 3.2. The most recent work tends to support the concept that it is one form or another of C-S-H which controls the rate throughout the early period, either by its growth or its dissolution. The close correlation between the end of the induction period, the attainment of maximum $[Ca^{2+}]$, and the nucleation of CH [30] does not seem to be true in dilute suspensions, which nevertheless follow very similar hydration kinetics. Delayed growth of CH has been attributed to the adsorption of silicates on the [001] faces [61] and delayed precipitation of portlandite in the presence of soluble silicate has been demonstrated experimentally [23,48]. When C_3S is added to saturated portlandite solution immediate precipitation of small crystals occurs [47], but they do not grow larger with time because of silica 'poisoning' explaining why CH growth is retarded in the C_3S–water system even when CH nuclei are deliberately added. The accelerating influence of carbon dioxide [62] can be explained well by Gartner–Gaidis' mechanism, since it consumes CH (by the precipitation

of calcium carbonate) at early ages and thus creates more early C-S-H nuclei, just as does dilution in a large access of pure water. Calcium carbonate is itself quite effective as a nucleus for C-S-H [63], which should augment the effect.

Thus, the latest evidence tends mainly to support theories based on C-S-H control, although the influence of retarders can still be explained fairly convincingly by their influence on CH nucleation (see below). The concept of one poorly-ordered phase (C-S-H) nucleating from another requires an abrupt change in short-range order from the intermediate C-S-H structure. It is clear that the high lime concentration in solution is required for this phase transition in C-S-H to occur. If the fundamental structure of C-S-H is considered to be derived from 1.4 nm tobermorite, then the basic structural unit for both C-S-H and CH is a layer of edge-connected Ca–OH octahedra. (In the case of C-S-H the layer is distorted by the condensation of silicate dimers.) Thus, both hydrates develop from a common nucleus of $Ca(OH)_2$ with adsorbed silicate ions (probably, but not necessarily, dimeric). The higher degree of supersaturation with respect to C-S-H, and the greater variation in local composition that is allowed, means that most nuclei would grow into

TABLE 3.2   Main theories for the onset and termination of the induction period

Hypothesis	Brief description	Key references
Cause of Stage I		
"C-S-H" barrier	Formation of a 'first' or 'primary' hydrate (probably a specific early form of C-S-H) which forms a physical diffusion barrier around the C_3S grains. This hydrate is "metastable."	6, 13, 16–15 17, 26
Lattice defects	The rate of C3S hydration and the length of the induction period depend on the number and type of its lattice defects.	18, 19
Boundary layer reaction	C_3S dissolves congruently by via a protonated layer "C_3S_{sh}", which is less soluble than anhydrous C_3S, allowing the rate to be retarded by accumulation of Ca^{2+}, SiO_2 and OH^- in solution.	13, 20–25
Electrical double layer	Incongruent dissolution of C_3S forms a silica-rich layer which readsorbs Ca^{2+} to create an electrical double layer at the surface that impedes further dissolution.	1, 8, 10
Cause of Stage III		
Nucleation of Portlandite	Nucleation and growth of portlandite become rate-controlling, (and thus indirectly control rate of growth of C-S-H.)	1, 8, 10
Nucleation of C-S-H	Nucleation and growth of C-S-H become rate-controlling, (and thus indirectly control the rate of growth of portlandite.)	15–19, 25
Rupture of initial barrier	C-S-H barrier layer is semi-permeable. Solution inside is close to saturation with respect to C_3S).Osmotic pressure leads to its rupture.	60
Growth of C-S-H	Nuclei of "stable" C-S-H, already formed during stage I, grow at a nearly exponential rate. There is no stage II in the pure system	6, 38

C-S-H, and relatively few into crystalline $Ca(OH)_2$. Such a conceptual view removes the confrontation between 'CH nucleation' and 'C-S-H nucleation' theories by making both part of a single process.

3.2.3 The later hydration periods

It is apparent from the calorimetric curves (Figure 3.8) that the change from the first acceleration period to the second deceleration period is part of a continuous downward trend in $d^2\alpha/dt^2$ that actually begins in the middle of the acceleration period. This trend must be due to either a change in the chemistry, which might provoke a reduction in growth rates, or the onset of some kind of 'diffusion control' due to precipitation of hydrates around the hydrating C_3S particles. All of the available evidence points to the latter hypothesis as being the more likely. Simple observation by SEM is enough to show that a dense layer of C-S-H forms around the hydrating C_3S grains (Figure 3.12a). 'Wet cell' TEM studies show that this C-S-H has no definite form in the wet state [46], and this has recently been confirmed by AFM studies under wet conditions (Figure 3.12b) [64]. It is not yet clear exactly which diffusing specie is responsible for limiting the rate of hydration in Stage IV, or what exactly is the form of the diffusion barrier.

(a)

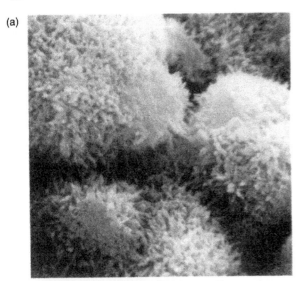

(b)

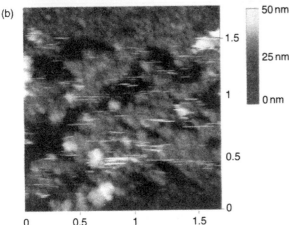

FIGURE 3.12 Growth of C-S-H around C₃S particles: (a) SEM micrograph, × 500; (b) atomic force microscopy image (photograph courtesy of Dr. E Gauffinet and Dr. A. Nonat.).

3.2.3.1 The second deceleration period

Several reviews of kinetic models have been published [65,66]. A wide variety of mathematical expressions have been used to fit the rate versus time curves in this period, but equation (8) is the most commonly used: [66,67]

$$[1-(1-\alpha)^{1/3}]^2 = k_D t \qquad (8)$$

where α is the degree of hydration at time t, and k_D is the rate constant for the diffusion process.

This equation was developed by Jander for diffusion control in solid-state sintering. Several investigators have also satisfactorily fitted kinetic data for the deceleration period to a more general form of this equation:

$$[1-(1-\alpha)^{1/3}]^x = k_D t \qquad (9)$$

Frequently, $x > 2$, which suggests that the reaction is complicated by the changing thickness of the diffusion barrier or by a changing value of the diffusion coefficient (D), which is contained in k_D. Ginstling and Brounshtein [68] modified equation (9) to allow for the changing thickness of the diffusion barrier (e.g. in this case, the C-S-H coating) around the C₃S grain. Their derivation results in the following equation:

$$[1-(1-\alpha)^{1/3}]^2 - 2/3[1-(1-\alpha)^{1/3}]^3 = k_D t \qquad (10)$$

Furthermore, C-S-H appears to form with different morphologies at different times and locations, and it is likely that Types I and III C-S-H (see Table 3.3) have different diffusion characteristics from Type IV C-S-H. Taplin [69] was the first to make a kinetic distinction between 'inner' and 'outer' hydration products, and so derived the following equation on the assumption that the two give different diffusion co-efficients, (D_i and D_o, respectively):

$$(1/D_i)[1-(1-2\alpha/3)-(1-\alpha)^{2/3}]$$
$$+ (1/D_o v)[1+2v\alpha/3 -(1+v\alpha)^{2/3}] = kt \qquad (11)$$

where v is the volume ratio of outer to inner product. The differences between D_o and D_i do not imply differences in chemical composition, since they could be due to changes in physical properties, for example, density or porosity.

The rate constants contain terms related to particle size, so that in polydisperse samples, such as are obtained by normal milling procedures, different sized particles may reach a particular rate-determining situation at different times. The apparent overall kinetics will be a complex function and lead to erroneous conclusions about rate constants and rate mechanisms if corrections are not made. The problem was first pointed out by

TABLE 3.3 Types of C-S-H morphology observed during the various stages

Microscopy tool	Stage				
	0, I, II (Early products)	III, IV (Middle products)		V (Late products)	
TEM					
Designation	E	0	1(1')	3	4
Morphology		Amorphous	Needles radiating from grain	Crumpled foils	Dense gel
SEM					
Designation	II	I		III	IV
Morphology	Reticulated (lace-like)	Needles radiating from grain (acicular)		Indefinite	Spherical agglomerates

Taplin [69], but it required extensive computations [66]. The problem can be approached by treating the particle radius (r) as another variable in the rate equation, solving for each particle size obtained from an experimentally determined psd curve, and summing over all sizes:

$$\alpha = \sum (m_i \alpha_i) \qquad (12)$$

where m_i is the mass fraction of particles of initial radius r_i, and α_i is their corresponding degree of hydration. At any given time the thickness of the C-S-H coating will be greater around grains of large r_i, and diffusion control will thus be reached at lower values of α_i. Knudsen [70] used a psd equation combined with either parabolic or linear kinetics. The exact kinetic equation used is not critical.

The question of whether the effects of psd are equally important at all ages is less easily answered. Many studies have tried to combine the whole hydration process into a single set of equations, but this is intrinsically difficult because it is the early-age reactions that are less 'localized' than the later reactions. Brown [71] found that it required as many as 50 different size classes to simulate the influence of psd of a typical ball-milled C_3S on the hydration kinetics. His results indicated that the effects of psd were much more important at early ages than at later ages, and that no single expression could account for the kinetics over the whole period of hydration, although the hydration kinetics beyond the end of the acceleration period could be represented by a

TABLE 3.4 Diffusion rate constants for two samples of alites [67]

Rate constants (units)	Alite A[a]	Alite B[a]
K_N (h^{-1})	0.134	0.110
K_I (μm h^{-1})	0.082	0.067
K_D (μm^{2}h^{-1})		
Initial	0.040	0.026
Subsequent	0.025	0.014

[a] A and B are different preparations of monoclinic alite.

relatively simple and (apparently) psd-independent form of the Ginstling–Brounshtein equation:

$$[1 - 2\alpha/3 - (1-\alpha)^{2/3}] = kt/r^2 \qquad (13)$$

It is not clear how an average value of r is estimated for use in equation (13) or how the model allowed for the transition from the acceleration to the deceleration period. Since acceleration-stage kinetics are implicitly surface-area dependent, since phase boundary reactions are involved, the transition time from Stage III to Stage IV is psd-dependent, even though the subsequent diffusion-controlled reaction is not very sensitive to the original psd. A typical example of the effects of psd on rate constants for two different alites is shown in Table 3.4.

3.2.3.2. The final slow reaction

It is not likely that there is any very fundamental change in mechanism between the second deceleration period and the final slow reaction. During

Stage V, C-S-H that has already formed may densify because less empty space is available. Taylor [72] has suggested that the hydration is probably at least partly 'pseudomorphic' during this stage, and, indeed, apparent relics of original alite particles (which can presumably be associated with Taplin's 'inner product') are frequently observed in microscopic examinations of hydrated pastes. [71,73] The degree of silicate anion polymerization in the C-S-H increases steadily with time during Stage V (Figure 3.10) which implies that the average C-S-H in the pastes is probably tending toward a lower Ca/Si ratio. It is not known if there is a compositional difference between the two principal C-S-H morphologies (Types III and IV; see Table 3.3) which are observed at this time.

Portlandite crystals continue to grow larger, filling large water-filled voids, and can apparently totally engulf zones of C-S-H and even some of the hydrating C₃S grains, thereby limiting their potential for complete hydration. These CH crystals always contain small quantities of silica and show quite distinct cleavage on the [001] plane during fracture which may be indicative of the concentration of silica on these planes.

Although the kinetics of the diffusion-controlled period have been modelled in detail, because of its importance in understanding how best to estimate the degree of hydration of real cement systems as a function of time, the mechanism itself is still poorly understood. Taylor [8,72] has investigated this issue in detail, and has made an interesting comparison between the atomic concentrations of the major elements in alite, belite and C-S-H. In order for C_3S to convert pseudomorphically to C-S-H, it must lose about 65 per cent of its original calcium atoms, 40 per cent of its silicon atoms, but only about 6 per cent of its oxygen atoms, while gaining about one proton for every oxygen atom. Clearly, large migrations of Ca and Si must be involved in Stage V hydration and it is reasonable to assume that the rate-limiting process in Stage V must involve diffusion of one of these ions through the hydrated coating. Calcium may be able to diffuse as Ca^{2+}, but silicon is most likely to diffuse as the larger partially protonated orthosilicate monomer. There is evidence that Ca^{2+} is the more mobile of the two species. The main difference in Stage V is that there is no longer much empty pore space available around each hydrating grain, due to the overlap of the hydrated coatings of neighbouring grains, thereby resulting in a rapid increase in the mean diffusion distance that an atom has to travel between a water-filled pore in which fresh C-S-H and CH can still deposit, and the anhydrous surface of a C_3S particle from where the Ca and Si must originate.

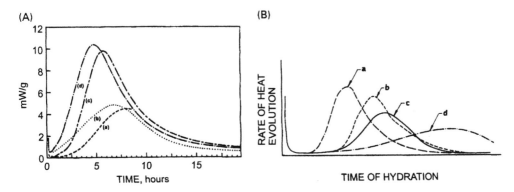

FIGURE 3.13 (A) Effect of nucleating phase on the heat evolution of hydrating C₃S: (a) unmodified curve; (b) + C-S-H; (c) + CaCl₂; (d) both added [3]. (B) Effect of set-modifying admixtures on the heat evolution of hydrating C₃S: (a) set accelerator (e.g. CaCl₂ or high temperature); (b) hardening accelerator (e.g. NaCl); (c) unmodified curve; (d) set retarder (or low temperature).

3.2.4 Types of C-S-H formed during the different hydration periods

3.2.4.1 Morphologies

In the course of hydration, several different morphological types of C-S-H have been identified. Various descriptive adjectives have been used to describe the morphologies observed, which depend on the technique used. Two major classifications have been proposed based on SEM [74] and TEM [46] studies; they are compared in Table 3.4, and have already occasionally been referred to in the preceding discussions. There are no demonstrated compositional or structural differences to support either classification, and both schemes must be considered arbitrary. There is clearly a need for a better understanding of the differences between the types of C-S-H formed at different times and locations within the microstructure, based less on morphology than on true atomic structure and nanostructure.

The problem with conventional SEM and TEM techniques is that they are very sensitive to the methods used to prepare the sample, which involve drying to very low relative humidities, and consequent deformation of the microstructure. With TEM, at high magnification, there is also a serious risk of significant damage due to the electron beam while the specimen is under observation. On drying or fracturing, the dense C-S-H layer around hydrating C_3S particles usually splits and consolidates into splines or needles (Types 1/1') due to crumpling and rolling up of sheet-like material. The saturated material extends into and fills the space between hydrating grains.

3.2.4.2 Silicate polymerization

Since the first edition of this book, new techniques such as solid state NMR spectroscopy, have provided a wealth of information about the atomic structures and nanostructures developed during hydration. It is not possible to discuss this area in detail, but the highlights are reviewed and reference 8 can be consulted for more information. The use of ^{29}Si CP-MAS NMR spectroscopy

has revealed that a hydrated orthosilicate (monomer) forms in cement pastes during the induction period [51,52] and persists in small quantities (~ 2 per cent of Si) for at least one month. During Stage III C-S-H contains primarily dimeric silicates. Later (1–3 days or 20–30 per cent hydration) dimers polymerize to give pentamers, octomers, etc. [51,52,75]. This process can be modelled as a progressive insertion of bridging tetrahedra into the gaps between dimers to form the '*dreierketten*' structure of tobermorite. Similar insertion has been observed with alumina [76]. This model suggests that in the initial formation of C-S-H, a dimeric silicate species is being laid down on a $Ca(OH)_2$ layer template. The process of gradual polymerization (Figure 3.12) is an indication of slow chemical change in C-S-H, which is concomitant with, but independent, of its formation by hydration. Rather different results were obtained on examining C-S-H formed by hydration of C_3S in dilute suspensions; C-S-H(SI) which formed first was more highly polymerized [43,49]. Since this forms when [CH] < 20 mM it is possible that this form has a radically different structure.

Regardless of the model adopted, a picture has emerged which supports the concept of the initial C-S-H which forms the protective hydrate layer having a different local structure from the C-S-H that forms after the induction period. Such a viewpoint helps support the concept of the nucleation of stable C-S-H discussed earlier. It could also explain, why, under the right conditions afwillite might nucleate as the dominant hydration product in the system.

3.2.5 Effects of temperature

Temperature has a marked effect on Stages II and III of the hydration periods of C_3S. The induction period is shortened as temperature increases, and the rate of heat evolution increases during Stage III. Hydration can occur at temperatures as low as $-20°C$ (if liquid water is present, e.g. in salt solutions), albeit slowly.

Measurements have been made of the apparent activation energy based on degree of hydration-time curves. Values obtained during Stages III and IV are typically close to 40 kJ/mol, whereas during Stage V the apparent activation energy approaches 20 kJ/mol, which is close to the value expected for a diffusion-controlled reaction. It should be remembered that the true activation energy of a reaction can only be measured by means of an 'Arrhenius' plot if a single, clearly-defined reaction step is the only process influencing the change of rate with temperature; a situation seldom encountered in cement chemistry.

The effects of temperatures significantly removed from ambient condition on C-S-H properties have not been as extensively studied. Morphologies do not appear to differ extensively, but the degree of silicate polymerization increases with temperature. Up to about 80°C, C-S-H appears to have essentially the same properties as that formed at room temperature. However, there are some indications that the Ca/Si ratios of C-S-H change with temperature [77–80].

3.2.6 Influence of admixtures

Admixtures that have a very marked influence on the hydration kinetics can be used to probe the mechanism of hydration. Such admixtures are generally classified as either 'accelerators' or 'retarders,' depending on how they affect the time of set. The effects of certain 'typical' admixtures, (and of temperature,) are shown schematically in Figure 3.13B. However, it is important, (especially in the case of Portland cement,) to note that admixtures that modify the hydration of pure C_3S will not necessarily show the same effect when used with portland cement, and vice versa. Furthermore, many admixtures may change their behaviour completely as a function of dosage.

There are at least two major classes of accelerator: soluble salts and fine powders, which appear to work by completely different mechanisms. Fine particles that accelerate C_3S hydration do so by nucleation of C-S-H. Silica fume and laboratory preparations of C-S-H can effectively shorten the induction period without changing the maximum rate of hydration. On the other hand, accelerators based on soluble salts generally increase the

maximum rate without accelerating the onset of Stage III to any great extent. Because the two mechanisms operate on different steps in the process, their effects are almost completely separable, as can clearly be seen from the results in Figure 3.13B.

Although many soluble salts act to accelerate the hydration of C_3S, especially during Stages I, II and/or III, there are many exceptions, for example, soluble phosphates, fluorides, and salts of Zn, Sn and Pb, which are retarders. On the other hand, almost all organic compounds act as retarders. The most thorough survey to date of the effects was made by Kantro [81] who studied the heat liberation rates of C_3S, hydrated in different salt solutions of equal normality. His results confirmed that the anion and the cation influence the hydration rate differently. For the chlorides of alkali and alkaline earth metals at a dosage equivalent to 2 per cent $CaCl_2 \cdot 2H_2O$ by mass of C_3S, the order of effectiveness as an accelerator of Stage III at 27°C was influenced by the cation as follows:

$$Ca^{2+} > Mg^{2+} > Sr^{2+} > Ba^{2+} \sim Li^+$$
$$> K^+ \sim Rb^+ \sim Cs^+ > Na^+ > NR_4^+ > H_2O \quad (14)$$

where NR_4^+ denotes a quaternary ammonium ion and H_2O means that no additive was used.

Clearly, these results indicate that calcium occupies a special position in the ranking, which is, perhaps, not surprising. Apart from that, the data suggest that both the charge and radius of the ions are important, those with high charge and small size being the most effective. A roughly similar trend is observed for the anions, which also make a very important contribution. For highly soluble salts of calcium at the same equivalent concentration, the order of effectiveness was influenced by the anion as follows:

$$Br^- \sim Cl^- > SCN^- > I^- > NO_3^- >$$
$$ClO_4^- > H_2O \quad (15)$$

Kantro's data also indicate that alkali sulphates are very effective accelerators, and others have reported on the efficacy of calcium thiosulphate (which is much more soluble than calcium sulphate). These observations imply that divalent anions which do not precipitate calcium can be very effective catalysts, and support the idea that an electrostatic shielding effect may be important, although the mechanism is by no means clearly understood. Figure 3.8 suggests that it is the rate of increase in the rate of heat liberation during the early part of Stage III, (roughly equivalent to $d^2\alpha/dt^2$), that is influenced by soluble calcium salts. Fierens *et al.* [56] found that the rate constant determined from Avrami kinetics with $m = 3$ (equation 8) is linearly dependent on $CaCl_2$ concentration up to about 0.6 M. Hence $CaCl_2$ must be influencing the growth of C-S-H, which is assumed to be the rate-determining step for Stages II and III. This is in contrast to earlier work of Kondo *et al.* [77] who correlated the overall influence of different salts with ionic mobilities and diffusion characteristics, and concluded that acceleration is determined by the rate of ionic diffusion, the non-participating electrolytes simply providing ions for counter-diffusion. However, a chemically-controlled surface reaction seems inherently more likely to explain the rather complicated influence of different salts during Stage III [6]; and it seems likely that the incorporation of Ca^{2+} into the growing hydrate is part of the rate-determining step, which would account for the strong sensitivity to $[Ca^{2+}]$. The presence of a mobile anion, however, clearly plays an important role, which is difficult to understand unless related to charge shielding. If, for example, the attachment of a single calcium ion to a growth site on the C-S-H is the rate-determining step, it is apparent that the surface charge will increase by $+2$ during this step. In order to minimize the free energy of a hypothetical positively-charged transition state, it is presumably necessary to provide sufficiently mobile anions in the close vicinity (mobile in the sense that they can arrive and depart easily). Chloride and bromide are the smallest and most mobile of the monovalent anions in the list above.

Both soluble calcium salts and alkali hydroxides can influence the initial dissolution of C_3S because of the common ion effect. It should be remembered that crystallization of CH is controlled by the solubility product K_{CH} which is affected by the activities of the two ions involved,

as represented by (Ca^{2+}) and (OH^-):

$$K_{CH} = (Ca^{2+})\cdot(OH^-)^2 \qquad (16)$$

If we apply the dilute solution approximation (activity $\sim$ concentration) and we start with a solution of either a soluble calcium salt or an alkali hydroxide, it can be shown that less CH needs to dissolve (either from free lime or from C_3S) to saturate the solution relative to CH than would be the case in if we had started with pure water. Thus, although the apparent rate of solubility of the CH from the C_3S often appears to decrease, [30,82] high concentrations of a soluble calcium salt or alkali hydroxide in the mix water can allow the system to reach saturation with respect to CH earlier, after a smaller amount of C_3S has dissolved. It seems that a high alkali hydroxide concentration can give significant acceleration [83], although 0.04 M NaOH has no accelerating effect [45].

Alkali salts that precipitate insoluble calcium salts will generally increase the early rate of dissolution of C_3S by keeping the calcium concentration low, as has been demonstrated for sodium carbonate and sodium fluoride [30]. As a consequence of the hydroxide and silicate ions which dissolve along with the calcium, the pH increases and a higher concentration of dissolved silica may also be observed, which should precipitate as C-S-H once the calcium concentrations begin to rise again. The length of the induction period appears to depend on how rapidly the calcium concentration rises to reach the maximum CH supersaturation, supporting the idea that a certain minimum CH concentration, probably fairly close to CH saturation, is required for the onset of the true Stage III. The more the calcium consumed by early precipitation, the more 'early' C-S-H with a low Ca/Si ratio is likely to form, and this may act as a thickened protective barrier around the C_3S. Similar considerations apply when C_3S dissolves in solutions of metallic salts from which insoluble hydroxides are precipitated. Additionally, if the metal hydroxide can react further with hydroxyl ions to form a complex oxyanion, which may precipitate as an insoluble calcium salt, then rapid hydration will be retarded until these reactions are complete because both $[Ca^{2+}]$ and $[OH^-]$

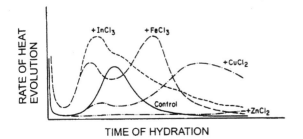

FIGURE 3.14   Effect of metal salts which form reactive hydroxides on the heat evolution of hydrating C_3S. The first peak is associated with the precipitation and reaction of metal hydroxides, while the second is the main peak of C_3S hydration. [After Ref. 81, with permission.]

remain low for a long time. This was first recognized in the case of zinc salts (equation 17), but is a universal phenomenon (Figure 3.14):

$$ZnCl_2 \Rightarrow Zn(OH)_2 \Rightarrow Zn(OH)_4^- $$
$$\Rightarrow CaZnO_2 \qquad (17)$$

Retardation of this kind is evidence that a certain level of $K_p(CH)$ is required in the solution in order for the stable form of C-S-H to nucleate and grow at a reasonable rate. There is, additionally, the possibility of forming insoluble coatings around the C_3S grains that can continue to retard the rate even during the acceleration period [84].

The effects of adding acids are somewhat similar to those of adding metal salts that precipitate insoluble hydroxides. In both cases, hydroxide ions are consumed and the level of K_{CH} is depressed. However, in the case of acids that give soluble calcium salts, no precipitates are formed, and the salt itself is generally an accelerator, so little, if any, retardation is observed. Additionally, the excess silica precipitated during Stage I can provide a large number of nuclei for the growth of C-S-H in Stage III. In the case of acids that give insoluble calcium salts the same rules apply [85], with the complication that the precipitating calcium salt may also be a retarder for portlandite or C-S-H growth. This double effect can often lead to very strong retardation.

The ability to retard seems to be strongly linked to the formation of calcium complexes. Thus, soluble calcium salts of simple sulphonic acids, which do not complex calcium to any great extent, are generally non-retarding, although they are not particularly good accelerators. The (soluble) calcium salts of simple carboxylic acids (acetic, propionic, etc.) which form slightly stronger calcium complexes than sulphonic acids, are either weak accelerators or weak retarders, the exception being calcium formate, which is a moderately good accelerator. However, as soon as the complexity of the anion is increased by the addition of a second group (either carboxylic or hydroxyl) that can interact with calcium, the tendency lies strongly in favour of retardation.

It has been observed that many organic acids are capable of inhibiting the growth of CH, apparently due to adsorption onto the [001] face, thus preventing normal crystal growth in the same way as postulated for silica poisoning [61]. Many compounds that strongly chelate calcium also have strong retarding capabilities. The strong retardation caused by many organic acids containing the β-hydroxy-carboxylate grouping, such as citric, tartaric and gluconic acids, have been attributed to the formation of a strong six-membered chelate ring with calcium:

$$M^{n+} + R\text{-}C(OH)\text{-}CH_2\text{-}CO_2H \rightleftharpoons \underset{\underset{M^{n+}}{\overset{HO\qquad O\text{-}H^+}{\underset{|}{R\text{-}C}\quad \underset{|}{C=O}}}}{\overset{CH_2}{\diagup \quad \diagdown}} \qquad (18)$$

The size and number of chelate rings that can be formed by an organic molecule apparently affect the retarding power but electronic effects are also important for chelating ability. Most inorganic ions do not have this ability for strong chelation. The formation of a chelate complex, such as shown in reaction (18) above, typically occurs in solution, reducing the activity of the calcium ion; however, this alone is not enough to cause severe retardation; rather, it has a similar effect to the addition of an acid, that is, it requires that more calcium dissolves in order to saturate the solution with respect to CH, as shown for the case of a neutral chelating ligand (L) below:

$$C_3S \Rightarrow CH \Rightarrow Ca^{2+} + 2OH^- (+L)$$
$$\Rightarrow (1-x)L.Ca^{2+} + (x)Ca^{2+} + 2OH^- \qquad (19)$$

where $0 < x < 1$, and $x \ll 1$ for a strong ligand, that is, almost all of the calcium is bound to the ligand if there is an excess of ligand available.

Since the complex ion L.Ca^{2+} does not behave as Ca^{2+}, more CH must be released by the source, (in this case, C$_3$S) to combine with the free ligand in order to reach any given value of the solubility product. This can give rise to retardation of the type discussed above for acids or compounds that precipitate calcium. However, the retarding effect is frequently multiplied by the ability of certain chelated calcium ions to adsorb onto the growth sites normally occupied by free calcium ions, thereby 'poisoning' these growth sites. Evidence for such strong selective retardation comes from the effectiveness of many other organic retarders, (such as many sugars), which do not contain the β-hydroxy-carboxylate grouping, and do not necessarily form very strong complexes with calcium in solution.

There have been relatively few studies of the solution chemistry in retarded C$_3$S systems. Jennings *et al.* [84] observed that the concentrations of free Ca^{2+} (corrected for complexation) and total silica in solution during C$_3$S hydration at early ages, in the presence of three different retarders (at 2 per cent concentration in the water, and w/s = 2,) always fell on or close to the C-S-H(m) solubility curve that was believed to represent the surface coating on the C$_3$S. The data obtained with the strong tridentate ligand EDTA showed lower free-calcium concentrations as expected for strong complexation. The data suggests that calcium chelation may not be necessary *per se* for retardation, but that the retarder may stabilize the initial C-S-H coating against conversion to the more stable final C-S-H. The fact that free-calcium concentrations can remain low for a long time in such retarded systems implies that the protective layer reaches a fairly stable thickness, beyond which it allows very little extra calcium into solution. Moreover, an increase in the amount of free calcium in solution at early ages may not lead to reduced retardation. Sucrose retards much longer than EDTA, despite maintaining free [Ca^{2+}] close to CH saturation. This may also imply that the C-S-H coating is itself

stabilized directly by adsorption of the retarder. These data support the concept that retardation is caused mainly by inhibition of the growth of the more stable type of C-S-H that forms during Stage III. The concept of C-S-H and CH sharing common nuclei, discussed earlier, may also be the key to reconciling the behaviour of retarding admixtures.

3.3 Hydration of dicalcium silicate

Dicalcium silicate (or belite) is less extensively studied than C$_3$S because its hydration contributes mostly to long term properties of Portland cement pastes, mortars and concrete. The β polymorph has been the most extensively studied since it is most commonly found in Portland cements. The much slower reactions of β-C$_2$S lead to very slow strength development and for this reason cements formulated to be rich in belites usually contain small quantities of alite or calcium aluminates, such as C$_4$A$_3\bar{\text{S}}$, to provide rapid setting and early strength development. The slower rate of hydration of C$_2$S will make its early hydration more amendable to study than that of C$_3$S. The large differences in reactivity between C$_3$S and C$_2$S can be ascribed to the presence of O^{2-} ions in the structure of the former, in addition to SiO$_4^{4-}$. If the initial attack is by water molecules acting as Bronsted acids [2], then, the higher basicity of O^{2-} is clearly driving the process. However, other factors must be operating to explain the distinct differences in reactivity between different polymorphs of C$_2$S, as discussed more fully in Section 3.2. For example, Jost and Ziemer [87] argue that the sharing faces of CaO$_x$ polyhedra is an important feature controlling reactivity.

3.3.1 Overview of hydration

The reaction of dicalcium silicate is described by equation (20), which is analagous to equation (1):

$$2C_2S + 4.3H \rightarrow C_{1.7}SH_4 + 0.3CH$$
$$\Delta H = -43 \, \text{kJ mol}^{-1} \tag{20}$$

The composition of C-S-H is similar to that formed from C$_3$S. Hydration and the development

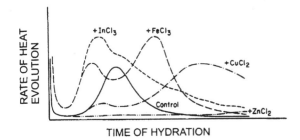

FIGURE 3.14 Effect of metal salts which form reactive hydroxides on the heat evolution of hydrating C_3S. The first peak is associated with the precipitation and reaction of metal hydroxides, while the second is the main peak of C_3S hydration. [After Ref. 81, with permission.]

remain low for a long time. This was first recognized in the case of zinc salts (equation 17), but is a universal phenomenon (Figure 3.14):

$$ZnCl_2 \Rightarrow Zn(OH)_2 \Rightarrow Zn(OH)_4^-$$
$$\Rightarrow CaZnO_2 \qquad (17)$$

Retardation of this kind is evidence that a certain level of $K_p(CH)$ is required in the solution in order for the stable form of C-S-H to nucleate and grow at a reasonable rate. There is, additionally, the possibility of forming insoluble coatings around the C_3S grains that can continue to retard the rate even during the acceleration period [84].

The effects of adding acids are somewhat similar to those of adding metal salts that precipitate insoluble hydroxides. In both cases, hydroxide ions are consumed and the level of K_{CH} is depressed. However, in the case of acids that give soluble calcium salts, no precipitates are formed, and the salt itself is generally an accelerator, so little, if any, retardation is observed. Additionally, the excess silica precipitated during Stage I can provide a large number of nuclei for the growth of C-S-H in Stage III. In the case of acids that give insoluble calcium salts the same rules apply [85], with the complication that the precipitating calcium salt may also be a retarder for portlandite or C-S-H growth. This double effect can often lead to very strong retardation.

The ability to retard seems to be strongly linked to the formation of calcium complexes. Thus, soluble calcium salts of simple sulphonic acids, which do not complex calcium to any great extent, are generally non-retarding, although they are not particularly good accelerators. The (soluble) calcium salts of simple carboxylic acids (acetic, propionic, etc.) which form slightly stronger calcium complexes than sulphonic acids, are either weak accelerators or weak retarders, the exception being calcium formate, which is a moderately good accelerator. However, as soon as the complexity of the anion is increased by the addition of a second group (either carboxylic or hydroxyl) that can interact with calcium, the tendency lies strongly in favour of retardation.

It has been observed that many organic acids are capable of inhibiting the growth of CH, apparently due to adsorption onto the [001] face, thus preventing normal crystal growth in the same way as postulated for silica poisoning [61]. Many compounds that strongly chelate calcium also have strong retarding capabilities. The strong retardation caused by many organic acids containing the β-hydroxy-carboxylate grouping, such as citric, tartaric and gluconic acids, have been attributed to the formation of a strong six-membered chelate ring with calcium:

$$M^{n+} + R\text{-}C(OH)\text{-}CH_2\text{-}CO_2H \rightleftharpoons \begin{array}{c} CH_2 \\ R\text{-}C \quad\quad C{=}O \\ | \quad\quad\quad | \\ HO \quad\quad O\text{-}H^+ \\ \diagdown\quad\diagup \\ M^{n+} \end{array} \qquad (18)$$

The size and number of chelate rings that can be formed by an organic molecule apparently affect the retarding power but electronic effects are also important for chelating ability. Most inorganic ions do not have this ability for strong chelation. The formation of a chelate complex, such as shown in reaction (18) above, typically occurs in solution, reducing the activity of the calcium ion; however, this alone is not enough to cause severe retardation; rather, it has a similar effect to the addition of an acid, that is, it requires that more calcium dissolves in order to saturate the solution with respect to CH, as shown for the case of a neutral chelating ligand (L) below:

$$C_3S \Rightarrow CH \Rightarrow Ca^{2+} + 2OH^- (+L)$$
$$\Rightarrow (1-x)L.Ca^{2+} + (x)Ca^{2+} + 2OH^- \qquad (19)$$

where $0 < x < 1$, and $x \ll 1$ for a strong ligand, that is, almost all of the calcium is bound to the ligand if there is an excess of ligand available.

Since the complex ion $L.Ca^{2+}$ does not behave as Ca^{2+}, more CH must be released by the source, (in this case, C_3S) to combine with the free ligand in order to reach any given value of the solubility product. This can give rise to retardation of the type discussed above for acids or compounds that precipitate calcium. However, the retarding effect is frequently multiplied by the ability of certain chelated calcium ions to adsorb onto the growth sites normally occupied by free calcium ions, thereby 'poisoning' these growth sites. Evidence for such strong selective retardation comes from the effectiveness of many other organic retarders, (such as many sugars), which do not contain the β-hydroxy-carboxylate grouping, and do not necessarily form very strong complexes with calcium in solution.

There have been relatively few studies of the solution chemistry in retarded C_3S systems. Jennings *et al.* [84] observed that the concentrations of free Ca^{2+} (corrected for complexation) and total silica in solution during C_3S hydration at early ages, in the presence of three different retarders (at 2 per cent concentration in the water, and w/s = 2,) always fell on or close to the C-S-H(m) solubility curve that was believed to represent the surface coating on the C_3S. The data obtained with the strong tridentate ligand EDTA showed lower free-calcium concentrations as expected for strong complexation. The data suggests that calcium chelation may not be necessary *per se* for retardation, but that the retarder may stabilize the initial C-S-H coating against conversion to the more stable final C-S-H. The fact that free-calcium concentrations can remain low for a long time in such retarded systems implies that the protective layer reaches a fairly stable thickness, beyond which it allows very little extra calcium into solution. Moreover, an increase in the amount of free calcium in solution at early ages may not lead to reduced retardation. Sucrose retards much longer than EDTA, despite maintaining free $[Ca^{2+}]$ close to CH saturation. This may also imply that the C-S-H coating is itself stabilized directly by adsorption of the retarder. These data support the concept that retardation is caused mainly by inhibition of the growth of the more stable type of C-S-H that forms during Stage III. The concept of C-S-H and CH sharing common nuclei, discussed earlier, may also be the key to reconciling the behaviour of retarding admixtures.

3.3 Hydration of dicalcium silicate

Dicalcium silicate (or belite) is less extensively studied than C_3S because its hydration contributes mostly to long term properties of Portland cement pastes, mortars and concrete. The β polymorph has been the most extensively studied since it is most commonly found in Portland cements. The much slower reactions of β-C_2S lead to very slow strength development and for this reason cements formulated to be rich in belites usually contain small quantities of alite or calcium aluminates, such as $C_4A_3\bar{S}$, to provide rapid setting and early strength development. The slower rate of hydration of C_2S will make its early hydration more amendable to study than that of C_3S. The large differences in reactivity between C_3S and C_2S can be ascribed to the presence of O^{2-} ions in the structure of the former, in addition to SiO_4^{4-}. If the initial attack is by water molecules acting as Bronsted acids [2], then, the higher basicity of O^{2-} is clearly driving the process. However, other factors must be operating to explain the distinct differences in reactivity between different polymorphs of C_2S, as discussed more fully in Section 3.2. For example, Jost and Ziemer [87] argue that the sharing faces of CaO_x polyhedra is an important feature controlling reactivity.

3.3.1 Overview of hydration

The reaction of dicalcium silicate is described by equation (20), which is analagous to equation (1):

$$2C_2S + 4.3\,H \rightarrow C_{1.7}SH_4 + 0.3CH$$
$$\Delta H = -43\,kJ\,mol^{-1} \qquad (20)$$

The composition of C-S-H is similar to that formed from C_3S. Hydration and the development of microstructure proceed considerably more slowly than for C_3S, when both are prepared by solid-state sintering at 1400–1450°C (i.e. normal clinkering conditions).

The hydration processes of β-C_2S show considerable similarities with that of C_3S, but monitoring hydration by pseudo-isothermal conduction calorimetry is more difficult because of the lower enthalpy and slower kinetics of hydration.

Nevertheless, the same profile of heat evolution observed for C_3S occurs also for β-C_2S [29], and it may be appropriate to divide the reaction into the same six stages. The rate of release of heat in Stage 0 has been reported [88] to be similar to C_3S, suggesting that this process involved only heat of wetting. But a higher rate of exothermic dissolution of C_3S might be expected due to the greater basicity of O^{2-}, as discussed above. Immediately upon contact with water non-uniform etching is observed and grain boundaries are revealed [89]. Maycock *et al.* [28] studied the early heat evolution of both C_2S and C_3S, but unfortunately do not provide direct quantitative comparisons that might provide the answer to this question. Surprisingly there appear to be few studies of the early release of ions into solution, in stark contrast to the extensive studies of C_3S. Barret and Bertrandie [44] have shown that rapid release of calcium and silicate ions into solution also occurs with β-C_2S, but $[Ca^{2+}]$ is lower than in the case of C_3S and increases more slowly.

However, the same problems that bedevil the study of C_3S also occur here, namely, that different investigators work with C_2S preparations that may have different reactivities, using different experimental techniques. Therefore direct comparisons are not always easy to make and it is difficult to assess the validity of apparently contradictory conclusions. A particular problem concerns the stabilization of β-C_2S with respect to γ-C_2S. The most common means to prepare stable β-C_2S in the laboratory is the use of B_2O_3 but in cements, stabilization is much more complex (see Section 3.3.2). The way the β-polymorph is stabilized can have a significant influence on its reactivity.

The first deceleration (Stage I) can also be attributed to the formation of a protective layer

of a metastable form of C-S-H. This conclusion is based on ESCA studies [31,88,90], SEM studies [89] and solution analysis [44]. While initial precipitation of C-S-H may occur as early as 15 s, subsequent growth is slower than in the case of C_3S. The C-S-H at this early stage appears to contain hydrated monomeric silica ($Q°$) [91,92], so it seems reasonable to assume that the nature of the protective coating is similar to that formed by C_3S. ESCA studies support this conclusion.

The induction period (Stage III) appears to be extremely variable in length with extreme values reported from 9 h [90] to over 20 days [92]. Since the second heat peak is very weak and difficult to measure, factors influencing the length of the induction period have not been probed. During Stage III there is a steady increase in $[Ca^{2+}]$ and $[OH^-]$ in solution [93], but the rate of increase is much slower than is observed for C_3S. The maximum CH concentration coincides with the onset of Stage III. A large degree of supersaturation with respect to $Ca(OH)_2$ is not observed and this is consistent with the slow hydration rate and the observation that relatively few portlandite crystals form; but those that are seen are generally very large [94].

To a first approximation, the growth, morphology and composition of C-S-H during subsequent hydration (Stages III–V) are not very different from what is observed in C_3S pastes, but there are some interesting differences [94]. Type II morphology is seen well beyond Stage II, whereas this form is not seen in C_3S pastes after the induction period unless admixtures are used [95]. The implications of this observation are not clear, but do perhaps point to the importance of $[Ca^{2+}]$ in solution when C-S-H is precipitated as proposed by Damidot and Nonat [43,49]. Second, hydration shells have been observed to pull away from the unhydrated core during fracture [94], which is a phenomenon seldom observed in C_3S pastes, although common in Portland cement pastes. However, the formation of 'hollow shells' in the latter case is believed to be due to the formation of AFt networks, as discussed in Section 3.7.1. The C-S-H formed in the later stages of hydration has a much more variable morphology than seen in comparable C_3S pastes [94].

3.3.2 Effects of polymorphs

Whereas the rates of hydration of the different modifications of C_3S stabilized by various impurities are not radically different, the same is not true of C_2S. At room temperature, γ-C_2S is the thermodynamically stable form, but it is only weakly hydraulic. The other polymorphs are metastable at room temperature and must be stabilized with impurity oxides. While the mechanism of stabilization has been implicitly attributed to the effects of accommodating the impurity ions within the crystal lattice, the situation is more complicated. Recently it has been shown that other phases rich in impurity ions, and which are partially amorphous, can be formed at grain boundaries [96,97]. It has been suggested [98] that these phases can provide a physical component to stabilization of the martensitic $\beta \rightarrow \gamma$ transformation, as well as limiting the size of β crystals to below a critical size. The rate of cooling is therefore important [99] since it can control the effects of intermediate transformation, the amount and distribution of second phases and the size of the crystals. For a given stabilizer, α'-C_2S (which has a higher level of impurities) has a greater hydraulic activity than does β-C_2S, but the actual degree of reactivity depends on the kind of stabilizer used, so that some β-forms are more reactive than some α'-forms [100]. Some kinetic data illustrate these points (Table 3.5). The variable reactivities of these polymorphs have been attributed to differences in ionicity of M–O bonds, (where M is a cation) or to different levels of crystal imperfection including, perhaps, twinning [87] and incoherent precipitates [96]. Highly reactive β-C_2S has been obtained using special low-temperature syntheses (see Section 3.3.4). The reactivity of α-C_2S in relation to the other polymorphs is not well established; its reactivity may lie between α' and β. The stabilization of this polymorph, which occurs only at the highest temperatures, is not as easy as it is for β and α'.

While γ-C_2S is generally described as being non-hydraulic it does show very weak hydraulicity [102,103], about 25 per cent hydration being reported after 5 years [102], or 40 months [103]. Trettin *et al.* [88] reported that γ-C_2S is hydraulically active when first mixed with water, but that the protective coating of CSH formed early on inhibits subsequent hydration. Thermodynamic calculations suggest that the free energy of hydration of γ-C_2S to give C-S-H(m) plus CH is very small [16,44]. However, γ-C_2S is also highly reactive towards carbonation [104].

The reason for the differences in reactivity between the different polymorphs is still a matter for debate. Jost and Ziemer [87] emphasize the importance of face-sharing CaO_x polyhedra and the mean Ca–Ca distances in the crystal. Both α' and β have distorted CaO_7 polyhedra, while γ has a relatively symmetrical closed-packed CaO_6 octahedra which do not share faces. Thus the symmetry and strength of the primary coordination shell of Ca is clearly important. However, as pointed out earlier, the complex microstructures that develop during phase transformation must play a significant role.

3.3.3 Reactive dicalcium silicate

Over the years there has been a continuing interest in the development of belite cements, which have

TABLE 3.5 Comparison of rate constants of some C_2S preparations [101]

Modification	Amount of Stabilizer	Nucleation and crystal growth $K_N \times 10^3$ (h^{-1})	Phase boundary $K_I \times 10^3$ $(\mu m^2 \cdot h^{-1})$	Diffusion $K_D \times 10^3$ $(\mu m^2 \cdot h^{-1})$	
				Initial	Subsequent
α'	11% BaO	—	31	5.4	1.4
α'	3% BaO	9.1	5.5	5.0	1.1
β	1% BaO	7.3	1.4	3.5	1.8
Alite		11.0	67	26	14

the advantage of low heat of hydration and improved durability. However the sluggish reactivity of β-C$_2$S, as formed in Portland cement clinkers, means that the rate of strength development is too low for commercial applications. Much research has centred on the stabilization of reactive α'-C$_2$S, mostly through the addition of stabilizing elements [100]. Rapid cooling has also been a useful strategy [99,105], which has been developed for belite-rich clinkers at a pilot plant scale [106]. The effect of particle size on the $\beta \rightarrow \gamma$ transition is known to be important and can be described in terms of a critical particle size effect controlling nucleation of the transformation [107].

In recent years there has been renewed interest in reactive forms of β-C$_2$S synthesized by chemical processes [108], spurred by the study of β-C$_2$S formed by topotactic dehydration of hillebrandite [109]. However, the earliest chemical synthesis was reported by Roy and co-workers in the 1970s [110] and details of the early work are summarized by Ishida and Mitsuda [108] and Young [100]. Preparations from hillebrandite have surface areas of 2–8 m^2/g (depending on calcination temperature) and react completely within 28 days. Subsequently β-C$_2$S was prepared by the Pechini process [111], a widely used chemical synthesis for many ceramic phases which utilizes a mixture of citric acid and ethylene glycol as the organic carrier phase. (The organic mixture can be replaced by poly(vinyl alcohol) [112] for better yield, but slightly lower phase purity.) In all these preparations β is the stable polymorph and γ is difficult to obtain except when calcination takes place at high temperatures. This confirms the concept of a critical particle size for preventing the $\beta \rightarrow \gamma$ transition.

The Pecchini process gives β-C$_2$S with surface areas as high as 40 m^2/g, depending on the calcination temperature [113], and hence with proportionally faster reaction (Figure 3.15). The hydration kinetics are similar irrespective of the method of synthesis, and diffusion-controlled kinetics are only observed in the latter stages with samples that have the lowest critical surface area. The activation energy is comparable to that obtained for conventionally sintered C$_2$S [108]. The C-S-H produced has a C/S molar ratio approaching 2.0

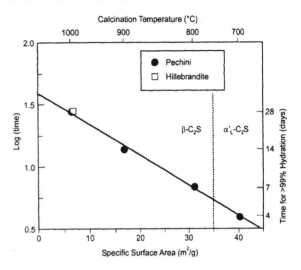

FIGURE 3.15 Effect of surface area on the time required for complete hydration. [Ref. 105, with permission.]

(i.e. little crystalline Ca(OH)$_2$ is formed) but the silicate structure of C-S-H does not appear to be changed significantly [114].

3.4 Hydration of tricalcium aluminate

Among the minerals present in Portland cement, C$_3$A is the most reactive and is known to have a significant influence on the early hydration and rheology of Portland cement and concrete. The hydration behaviour of C$_3$A, like that of other clinker minerals, depends on several factors such as temperature, w/s ratio, specific surface area of C$_3$A, mixing procedure, size of the hydrating sample, and presence of admixtures. A review of the hydration of C$_3$A and C$_4$AF has been published by a RILEM Committee [115].

3.4.1 Hydration in the absence of sulphate

Breval [116] made an extensive study of the sequence of product formation in C$_3$A hydration both in water vapour and in liquid, and confirmed the overall findings of numerous earlier studies. A gel-type material is apparently the precursor of the hexagonal hydrates which grow first as poorly crystalline, thin, irregular flakes. With time, these flakes grow to better crystallized hexagonal plates with compositions C_2AH_8 and C_4AH_{19}. These crystals have a layered structure consisting of positively-charged sheets with $Al(OH)_4^-$ or OH^- in the interlayer region to balance the net positive charge of the layer, together with additional H_2O. Some of the typical hydrate morphologies are shown in Figure 3.16 (see Section 3.7.4.2). The hexagonal hydrates are metastable and are converted to stable forms of cubic C_3AH_6.

$$2C_3A + 27H \rightarrow C_4AH_{19} \text{ (or } C_4AH_{13}) \quad (21)$$
$$+ C_2AH_8 \text{ (hexagonal hydrates)}$$

$$C_4AH_{19} + C_2AH_8 \rightarrow 2C_3AH_6 \quad (22)$$
$$\text{(cubic hydrates)}$$

C_4AH_{19} loses part of its interlayer water to form C_4AH_{13} at relative humidities below 88 per cent, so that the latter phase is detected when dried pastes are examined by X-ray diffraction. Also, in the presence of atmospheric CO_2 part of the interlayer OH^- is replaced by CO_3^{2-} to form

FIGURE 3.16 SEM micrographs of the hydration products of C_3A: (a) gel, consisting of irregular foils; (b) hexagonal flakes of AFm; (c) AFt in Portland cement paste.

stable hemicarboaluminate [$C_4A(1/2OH^-$, $1/2\ CO_3^{2-})H_{12}$].

Figure 3.17 shows the heat-evolution profile of C_3A hydrating in water. The time at which the conversion of hexagonal hydrates to the cubic phase takes place is strongly dependent on the conditions of the reaction, and is accelerated by the addition of C_3AH_6 nuclei. The high heat evolution, accompanying the formation of hexagonal hydrates, raises the temperature of the hydrating system appreciably, and can accelerate the conversion. Above 30°C the conversion of C_4AH_{19} (or C_4AH_{13}) and C_2AH_8 is very rapid, but once C_3AH_6 has nucleated, crystal growth can readily occur even below 30°C. Direct formation of C_3AH_6 from C_3A occurs at temperatures higher than 80°C, and can result in a strong matrix. However, conversion of hexagonal hydrates to C_3AH_6 leads to an increase in porosity and disruption of microstructure, which results in loss of paste strength.

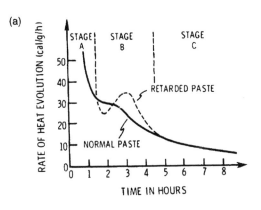

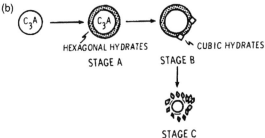

FIGURE 3.17 Hydration of C_3A: (a) heat evolution curve; (b) sequence of hydration.

3.4.2 Hydration in the presence of sulphate

More relevant to the cement system is the hydration of C_3A in the presence of gypsum, which is added to control setting. The ideal reactions are given in equation 23–25:

$$C_3A + 3C\bar{S}H_2 + 26H \rightarrow C_6A\bar{S}_3H_{32} \quad (23)$$
$$\text{(ettringite)}$$

$$C_6A\bar{S}_3H_{32} + 2C_3A + 4H \rightarrow 3C_4A\bar{S}H_{12} \quad (24)$$
$$\text{(monosulphoaluminate)}$$

$$C_4A\bar{S}H_{12} + xC_3A + xCH + 12H$$
$$\rightarrow C_4A\bar{S}_xH_{12+x} \quad (25)$$
$$\text{(solid solution)}$$

Ettringite is the first stable hydration product (equation 23) and forms as long as sulphate ions are available for reaction. Once the sulphate in solution drops below a critical value when all the gypsum is consumed, ettringite becomes unstable and further hydration yields the tetracalcium monosulphoaluminate hydrate (monosulphoaluminate) as the stable hydration product (equation 24). However, should the monosulphoaluminate become exhausted before the C_3A is fully consumed then equation (25) will commence which describes a solid solution between $C_4A\bar{S}H_{12}$ and C_4AH_{13} where x lies between 0.4 and 1.0.

The establishment of a solid solution is due to the isostructural relationship between $C_4A\bar{S}H_{12}$ and C_4AH_{13} in which the ions SO_4^{2-} and $2OH^-$ occupy interlayer positions in a random way. Other anions, for example, CO_3^{2-} can also occupy interlayer positions, and since Fe^{3+} can also substitute for Al^{3+} in the layer, there is the possibility of wide compositional variations. Therefore, the compounds are collectively known as AFm phases. Similarly, the designation AFt is used to denote ettringite-type phases with similar (but more limited) variations in composition, that occur when portland cement hydrates.

The reaction can be followed calorimetrically as shown in Figure 3.18. The initial rapid hydrolysis of C_3A and precipitation of ettringite is followed

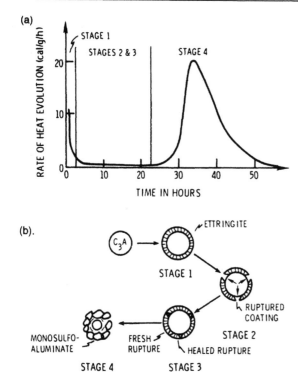

FIGURE 3.18   Hydration of C_3A in the presence of gypsum: (a) heat evolution curve; (b) classical mechanism of hydration proposed by Schwiete *et al.* [117].

by a period of slower reaction, as ettringite forms more slowly by a diffusion-controlled process. The renewed activity accompanies the transformation to monosulphoaluminate according to equation 24. The onset of the second peak depends on the amount of gypsum added. Figure 3.16 shows typical morphologies of ettringite and $C_4A\bar{S}H_{12}$.

3.4.3 Mechanisms of C_3A hydration

3.4.3.1 In the absence of gypsum

When C_3A is brought into contact with water, its high initial reactivity is followed by a period of slow reactivity. This is generally believed to be due to the initial hydration products, the hexagonal hydrates (C_4AH_{13} and C_2AH_8), forming a protective barrier on the C_3A grain surface (Figure 3.17b). The conversion of these hydrates to cubic hydrates (C_3AH_6) disrupts the barrier

and the hydration proceeds rapidly again, accompanied by a large heat evolution. Supporting this mechanism are the findings that compounds, both organic and inorganic, that stabilize the hexagonal hydrates with respect to the cubic hydrates retard the C_3A hydration, whereas those that increase the rate of conversion, accelerate C_3A hydration [118–121].

Corstanje *et al.* [122] further developed this concept of retardation by proposing the formation of amorphous $Al(OH)_3$ on the reacting surface, due to localized concentration of Al^{3+} in solutions close to the surface of the C_3A grains. This is visualized as occurring when the hexagonal hydrates form an isolating layer to prevent rapid ion exchange with the bulk solution. As the hexagonal hydrates form, more Ca^{2+} is removed from the solution than is Al^{3+} and hence $Al(OH)_3$, eventually precipitates. Periodic breakdown of the isolating layer, for example, by conversion to cubic hydrates, would allow the retarding $Al(OH)_3$ layer to be attacked by Ca^{2+} and OH^- ions in the bulk solution, and C_3A would continue to hydrate. Skalny and co-workers [123,124] also attributed the retardation of initial C_3A hydration to the formation of an Al-rich layer on the hydrating grain surface based on their observation of incongruent dissolution of C_3A in water and dilute HCl. This layer, according to these authors, eventually re-arranges either to $Al(OH)_3$ or to polynuclear aluminum complexes such as $[Al_8(OH)_{20}]^{4+}$.

Both these hypotheses, while ultimately predicting the formation of some kind of amorphous layer, approach the problem from quite different perspectives and were derived using data obtained under widely differing experimental conditions. The initial formation of an amorphous gel, first observed by Breval [116], is perhaps pertinent here. It can be considered to be either $Al(OH)_3$ or a co-precipitate of $Al(OH)_3$ and $Ca(OH)_2$ which, through further reaction with Ca^{2+} and OH^- ions from solution, forms the hexagonal hydrates. The exact role of an Al-rich layer in the overall hydration of C_3A is not yet clear. Does it persist through the hydration process, or is it no longer capable of forming once C_3AH_6 nucleates and grows? To further understand the consequences of initial hydrolysis and early formation of precipitates, we need to investigate this part of the reaction in considerably greater detail.

3.4.3.2 In the presence of gypsum

Studies on the mechanism of C_3A hydration in the presence of gypsum are of great practical value because of the influence of this reaction on the early reactions of Portland cement and concrete. Since ettringite is usually detected at the surface of the C_3A, rather than on the gypsum particles, Schwiete and co-workers [117] proposed that it forms a continuous coating on the reacting C_3A surface which impedes diffusion of SO_4^{2-}, OH^- and Ca^{2+} ions, slowing down the reaction and resulting in an induction period. The concept is shown schematically in Figure 3.18b. The more gypsum present in the system, the longer the induction period and the kinetics of ettringite formation indicate diffusion control [125,126]. When the liquid phase of the hydrating system becomes deficient in Ca^{2+} and SO_4^{2-} ions, the protective coating is disrupted, which results in the renewed hydration of C_3A to form AFm phases.

Subsequently, the role of ettringite in the retardation of C_3A hydration in the presence of gypsum was the subject of much discussion [115,122,127–131]. From their dissolution and electrokinetic data on C_3A suspensions in acid media, Skalny and co-workers [123,124] concluded that the retarding effect of sulphate is due to its adsorption on reactive sites on the surface of C_3A, a conclusion reached earlier by Feldman and Ramachandran [132]. Zeta-potential measurements show that the positive charge on the hydrating C_3A particles (apparently due to adsorption of Ca^{2+} on the Al-rich surface) is decreased by the adsorption of sulphate ions; chloride ions do not have any significant effect. The effect of sulphate ions in decreasing the number of active dissolution sites is probably achieved via a blocking effect of sulphate ions on coordination sites which would otherwise be occupied by hydrogen or hydroxyl ions, and which are known to catalyse the dissolution of alumina and other oxides. Collepardi et al. [131] disagreed with the above conclusions because Na_2SO_4 does not have the same retarding effect as gypsum.

Clearly, the ideal behavior as outlined in the equations 23–25 is more simple than the observed behaviour, which depends on the forms of sulphate available, the reaction of C_3A and other factors. Evidence has been presented [130] for the formation of amorphous precipitates at the surface of C_3A prior to the nucleation and growth of ettringite. Furthermore, monosulphoaluminate can co-exist with ettringite in the very early stages of hydration. Obviously equilibrium conditions are not established quickly and depend on the relative rates of solubility of C_3A and the sulphate phase to establish the solution composition from which the initial products form. The composition of initial amorphous product will reflect the composition of the solution from which it precipitates and both will determine whether monosulphoaluminate will occur as a transient phase. Similarly, the presence of other sulphate phases such as hemihydrate ($C\bar{S}H_{1/2}$), arcanite ($K\bar{S}$), or calcium langbeinite ($C_2K\bar{S}_3$)) can lead to the formation of gypsum or syngenite ($CK\bar{S}_2H$)) and other intermediate products in hydrating cements. However, the sulphate ions eventually end up in AFt or AFm phases. Corstanje et al. [122] noted that at $\bar{S}/A$ molar ratios where ettringite can be detected retardation was observed, while at $\bar{S}/A$ molar ratios where ettringite could not form, gypsum tended to promote C_3A hydration. This they attributed to the 'isolating ability' of AFm, which does not promote $Al(OH)_3$ precipitation to the same extent as do sulphate-free AFm phases, because Ca^{2+} ions are supplied by the gypsum in the system. Chatterji [133] disagrees and claims that because ettringite hinders the transport of sulphate ions from the bulk solution to the C_3A surface, a retarding layer of sulphate-free AFm can form adjacent to the C_3A surface.

The above apparently contradictory observations and conclusions present a confusing situation. It must, however, be remembered that conditions of hydration were not the same in all studies; some were looking at dilute suspensions, while others were studying pastes of C_3A or Portland cement. The same processes may not occur in pastes and in suspension, and if they do, the times at which different processes occur are likely to be affected by the time at which critical ion concentrations are reached in solution. Ionic concentrations will also be affected by differing reactivities of different C_3A preparations and, also by the presence of other cement minerals. Since AFm may form even when gypsum is still present in the system, [130,134] the reactivity of C_3A compared to the rate of release of sulphate ions from gypsum crystals is a critical factor.

Despite its importance in controlling the behavior of Portland cement pastes immediately after mixing, the hydration of the $C_3A–C\bar{S}H_2$ system has not been studied in the same depth as C_3S. Nevertheless we anticipate many common elements between the behaviour of these two reactive phases. On that basis we propose the following scenario.

1. Initial rapid dissolution of C_3A.
2. Almost immediate precipitation of an amorphous calcium aluminate hydrate gel, which acts as a protective coating. This gel probably incorporates a significant amount of anions, such as from the surrounding solution, as well as organic species if present.
3. Nucleation of crystalline products from the amorphous layer: AFm or AFt depending on the particular circumstances. Minor components, such as alkalis, chlorides and admixtures, may also enter the gel and modify the crystal growth by random substitution in the lattice of the hydrates.

3.4.3.3 C_3A hydration under alkaline conditions

In cements, C_3A and sulphate ions react in the presence of saturated $Ca(OH)_2$ solutions derived from C_3S hydration, and in many cases in the presence of alkali sulphates will raise the pH still further. When both $Ca(OH)_2$ and gypsum are present the hydration of C_3A is more retarded than by gypsum alone [131,132,135]. The result is slower formation of ettringite, and the conversion to AFm can be delayed for several days. The slower reaction has

been attributed to the development of a more effective diffusion barrier: colloidal-sized ettringite [127] or C_4AH_x [138]. In such systems a gel-like phase precedes the growth of crystalline ettringite [130].

Alkali compounds in general accelerate the hydration of C_3A. Spierings and Stein [136] observed that in a sulphate-free environment $<0.4\,M$ NaOH decreased early hydration (ascribed to decreased solubility of Ca^{2+}), while $>0.4\,M$ NaOH accelerated hydration. This latter effect was attributed to the OH^- catalysed hydrolysis of the Al–O network and a reduced stability of AFm with respect to C_3AH_6. However, in cements, alkalis are usually present as sulphates. Tang and Gartner [137] blended a low alkali clinker with various combinations of gypsum and alkali sulphates which are found in most high alkali clinkers. The reaction of C_3A was enhanced by aphthitalite ($K_3N\bar{S}_4$), but reduced by syngenite or calcium langbeinite.

Alkalis are also incorporated into the C_3A structure during the clinkering process. The presence of varying amounts of Na_2O in C_3A can result in as many as four different crystalline modifications: cubic, orthorhombic, tetragonal and monoclinic. There is general agreement that the hydraulic reactivity of C_3A markedly decreases as the Na_2O content in the crystal increases; [136,138–142] for example, cubic C_3A (pure phase) is more reactive than the monoclinic polymorph (5.7 per cent Na_2O), as shown in, Figure 3.19) [139]. For Na_2O-doped C_3A, the time of conversion of hexagonal hydrates to cubic hydrates during hydration, as observed by calorimetry, increases considerably and the size of the heat-evolution peak decreases markedly [139]. Other 'impurities', such as Cr_2O_3, TiO_2, MgO, Fe_2O_3 and SiO_2, have a similar effect [140]. The reactivity of a C_3A polymorph has been found to depend on the extent to which the lattice holes are occupied by the 'impurities' and it decreases with increasing occupancy [140,141].

3.4.4 Hydration in the presence of other species

C_3A hydration in Portland cement should not be expected to be independent of the presence of

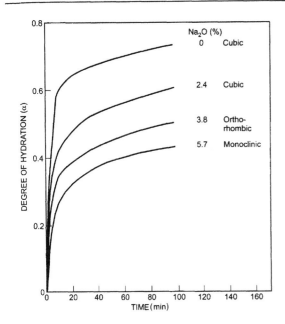

FIGURE 3.19 Hydration of Na_2O-doped C_3A polymorphs [138].

either C_3S or C_2S, and indeed an interaction exists. Regourd and co-workers [142,143] found that when C_3A was hydrated in the presence of C_3S, mono- and trisilicoaluminates ($C_3A \cdot C\bar{S} \cdot H_{12}$ and $C_3A \cdot 3C\bar{S} \cdot H_{31}$ respectively) were formed around C_3A by diffusion of silicate ions from C_3S through the solution. The monosilicoaluminate first formed outside the original C_3A grain boundary; after 7 days, poorly, crystalline trisilicoaluminate was detected. These silicoaluminates were unstable in the presence of sulphate ions and did not appear in samples with 20 per cent gypsum. The interaction of C_3S was greater with cubic C_3A than with tetragonal C_3A. The addition of other forms of silica, for example, ground quartz accelerates C_3A hydration [144,145], although natural pozzolans retard [146]. Effects at later ages may differ from those observed initially. The addition of chloride accelerates the formation of ettringite in C_3A-gypsum mixtures [147]. Only when gypsum is exhausted does the chloride AFm phase (calcium chloroaluminate hydrate) form. (This phase also will form as chloride-based deicing salts penetrate concrete.) Other species present in chemical admixtures (e.g. nitrate and organic carboxylates) may

form modified AFm phases. Calcium carboaluminate hydrate forms from C_3A–$CaCO_3$ mixtures [148] and almost certainly forms during the hydration of Portland cements containing ground limestone, although it is not easily detected.

3.5 Hydration of the ferrite phase

C_4AF is not a stoichiometric compound, but part of a solid-solution series called the ferrite phase. It has the general formula $Ca_4Fe_{(2-x)}Al_xO_{10}$, where x varies from 0 to 1.4, C_4AF is just one point in the series where $x = 1.0$. [It can be considered as a limited solid solution between C_2F and hypothetical C_2A.] Many cements have a ferrite phase with a composition close to C_4AF and so this is the most common composition studied. However, the reactivity of the ferrite phase decreases as the iron content increases [115,149].

The course of hydration of the ferrite phase is very similar to that of C_3A [149–157]. In the absence of gypsum, the metastable iron-substituted AFm phases $C_4(A,F)H_{19}$ and $C_2(A,F)H_8$ form, which convert to $C_3(A,F)H_6$. These crystalline phases have a higher A/F ratio than does the unhydrated material. Thus, to maintain stoichiometry an iron-rich hydroxide must form. This phase is amorphous to X-rays, but has been detected by SEM [152] and Mossbauer spectroscopy [154]. Thus the reactions can be written as:

$$3C_4AF + 60H \rightarrow 2C_4(A,F)H_{19} + 2C_2(A,F)H_8 \\ + 4(F,A)H_3 \qquad (26a)$$

$$\rightarrow 4C_3(A,F)H_6 + 2(F,A)H_3 + 24H \qquad (26b)$$

Although the exact composition of the crystalline phases appears to be highly variable, a complete disproportionation of iron is unlikely. However, $C_3(A,F)H_6$ is unstable at high temperatures, decomposing to C_3AH_6, and α-Fe_2O_3 (haematite) [151].

When gypsum is present the hydration sequence for the ferrite phase is the same as for C_3A. Iron-substituted AFt first forms and later converts to AFm when the gypsum is exhausted. An amorphous phase, presumably aluminum-substituted $Fe(OH)_3$ is also formed. Disproportionation of iron between the crystalline and amorphous phases is observed here too, and is attributed to the low mobility of iron. The formation of iron-rich pseudomorphs of the original C_4AF grain has been observed both in pure systems [158], and in cement [159]. The acceleration of C_4AF hydration with tertiary alkanolaminines is attributed [160] to an increase in the mobility of iron, through the formation of Fe^{3+}-amino complexes. Brown [161] has concluded that C_2AH_8 or alumina gel precedes the formation of AFt in the C_4AF-gypsum system and that early AFt is essentially iron-free. Ramachandran and Beaudoin [157] found that the sequence of reactions and interconversions of hydration products at low w/s ratios and high temperatures, were different from those occurring at the normal w/s ratio and temperature. The conversion of ettringite to AFm phases could occur even in the presence of gypsum. Also, at very low w/s ratios and higher temperatures, AFm phases did not necessarily result from reactions involving ettringite.

It is generally assumed that C_4AF is less reactive than C_3A, but this may not necessarily be true in actual cements. It must be remembered that impure phases are present in Portland cement. The reactivity of C_3A depends on alkali content and it may contain up to 10 per cent Fe_2O_3 (corresponding to $C_9A_{8/3}F_{1/3}$). The ferrite phase in Portland cement contains SiO_2, MgO and alkalis. The effect of these substitutions might be to lower the reactivity of C_3A and increase that of the ferrite phase [154], thereby lessening the difference between their reaction rates. C_4AF is much more strongly retarded by the addition of gypsum or a combination of gypsum and lime than is C_3A (Figure 3.20) [150,154], although the reasons for it are not clear. The persistence of iron-rich amorphous coatings that do not transform to crystalline products may be the key to retardation, although the formation due to stable coatings of AFt have been proposed [154].

3.5.1 Hydration in the presence of C_3A

Since their hydration is so similar, C_3A and the ferrite phase must compete for gypsum during the

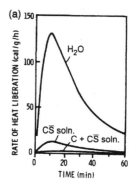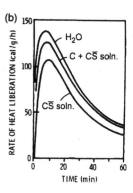

FIGURE 3.20 Heat evolution of (a) C_3A and (b) C_4AF during hydration in water, in saturated solution of gypsum, and in saturated lime–gypsum solutions [149].

hydration of Portland cement. This interaction was studied by Jawed *et al.* [149], who added small quantities of C_3A to mixtures of C_4AF and gypsum. The conversion of ettringite to AFm occurred sooner, depending on the amount of C_3A added (Figure 3.21). Ettringite is formed more rapidly indicating that C_3A competes more efficiently for sulphate ions. Once depletion of gypsum occurs, which happens earlier in the presence of C_3A, C_4AF reacts to form monosulphoaluminate.

3.5.2 Hydration of calcium ferrite

Calcium ferrite ($Ca_2Fe_2O_5$ or C_2F) can occasionally occur in Type V cements. The hydration of C_2F has been shown [156] to produce C_4FH_{13} (which initially forms as C_4FH_{19}) and amorphous $Fe(OH)_3$. The C_4FH_{13} converts to C_3FH_6 within a day, but the latter rapidly breaks down to CH and haematite. Teoreanu *et al.* [154] did not observe either C_4FH_{19} or C_3FH_6 in their study of C_2F hydration. However, it has been pointed out that these hydrates give the same Mossbauer spectra as $Fe(OH)_3$, and that X-ray diffraction (XRD) is not sensitive unless a good monochromator is used to filter out iron fluorescence [161]. Rogers and Aldridge [101] also studied the hydration of 'C_3F' and 'C_4F' which are actually intimate mixtures of C_2F and C. As expected, they behave similarly. CF also hydrates in the same way, but much more slowly.

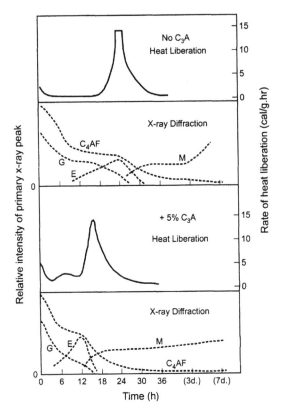

FIGURE 3.21 Hydration of C_4AF in the presence of C_3A with either gypsum or hemihydrate [149] G = gypsum, H = hemihydrate, E = AFt, M = AFm.

3.6 Hydration of Portland cement

3.6.1 The hydration process

The hydration of Portland cement is a sequence of overlapping chemical reactions between clinker components, calcium sulphate and water, leading to setting and hardening. In the major application, that is in concretes (or mortars, etc.) the setting process is the consequence of a change from a concentrated suspension of flocculated particles to a viscoelastic skeletal solid capable of supporting an applied stress, and can be monitored by rheological measurements [162]. The continued development of the solid skeleton is called hardening – a physico-chemical process, which leads to the development of ultimate mechanical properties. The principal stages of hydration are summarized

in Table 3.6. The hydration reactions proceed until either a lack of reactants (cement components, water), or a lack of space to deposit the hydration products, causes the reactions to cease. The hydration of a Portland cement is much more complicated than the hydration of individual 'pure' clinker phases, because the reactions of the different phases proceed simultaneously at differing rates and thus, can influence each other in complex ways. However, the general principles are the same: the dissolution of anhydrous phases leads to the precipitation of much less soluble products, typically colloidal and micro-crystalline hydrates, that form the hardened paste.

An oversimplified but convenient way of following the progress of cement hydration reactions is by the use of heat evolution curves (Figure 3.22), such as are used in studying the pure components, they are useful in correlating the known experimental data to a single parameter, heat evolution. There are at least two limitations to this approach: (1) the rate of heat evolution is neither simply proportional to the degree of hydration of any one phase (or even to the average of all of the phases, nor to the development of physical properties; (2) isothermal hydration studies do not properly represent hydration processes in most engineering applications where concrete develops mechanical properties in semi-adiabatic rather than psuedo-isothermal conditions.

3.6.1.1 The first minutes of hydration

As soon as the cement contacts water, a series of rapid reactions begins which involve mainly the clinker interstitial phases (i.e. the aluminates and aluminoferrites, alkali sulphates, and free lime), plus the calcium sulphates (gypsum, hemihydrate and/or anhydrite) which have been interground with the cement. The aluminate phases (including the aluminoferrites in most cases) react very rapidly and exothermically, giving a flush of calcium and aluminate ions into solution. The only component which is not usually observed going through solution is the iron, which probably remains mostly undissolved in the form of a hydrated oxide coating on the surface of the aluminoferrite phase [161].

TABLE 3.6 The four principal hydration stages of Portland cement concretes

Processing stage (in concrete)	Chemical processes	Physical processes	Relevance to physical properties of concrete
First minutes (wetting and mixing)	Rapid dissolution of free lime, sulphate and aluminate phases; Immediate formation of "AFt"; Superficial hydration of C_3S. Hemihydrate dissolves but gypsum or syngenite can form.	Large initial burst of heat, mainly from dissolution of aluminate phases, plus some from alite and CaO. (Dissolution of alkali sulphates is endothermic.)	Rapid formation of aluminate hydrates, plus gypsum and syngenite influences rheology and may also affect the subsequent microstructure.
Induction period (agitation, transport, placing and finishing)	Nucleation of "C-S-H(m)"; Rapid decrease in $[SiO_2]$ and $[Al_2O_3]$ to very low levels; [CH] becomes supersaturated and portlandite nucleates; $[R^+]$, $[SO_4^{2-}]$ stay fairly stable.	Low heat evolution rate; slow formation of early C-S-H and more AFt leads to continuous increase of viscosity (in absence of admixtures)	Continued formation of AFt and AFm phases. can influence workability, but it is formation of C-S-H that usually leads to the onset of normal set.
Acceleration period (setting and early hardening.)	Hydration of C_3S (to C-S-H and portlandite) accelerates and reaches maximum; CH supersaturation decreases. $[R^+]$, $[SO_4^{2-}]$ stay fairly stable.	Rapid formation of hydrates leads to solidification and decrease in porosity; high rate of heat evolution.	Change from plastic to rigid consistency (initial and final set); early strength development.
Post-acceleration period: (demoulding; continued hardening)	Decelerating rate of formation of C-S-H and portlandite from both C_3S and C_2S; $[R^+]$ and $[OH^-]$ increase but $[SO_4^{2-}]$ falls to very low levels; renewed hydration of aluminates to give (mainly) AFm phases. AFt may redissolve and/or recrystallize.	Decrease in rate of heat evolution. Continuous decreasing in porosity Particle-to-particle and paste-to-aggregate bond formation.	Continuous strength increase due to decreasing porosity, but at an ever-diminishing rate. Decrease in creep capacity. Hydration continues for years if water is available. paste will shrink due to drying.

However, the high aluminate concentrations do not last long. In a matter of seconds there is precipitation of an aluminate hydrate layer over the surfaces of the cement particles, this layer being generally referred to as being 'ettringite' or 'AFt.', although it is often poorly crystalline and difficult to detect by XRD at very early ages. As expected for an AFt phase, it contains a lot of sulphate, which comes from the rapidly soluble sources such as alkali sulphates, hemihydrate and gypsum. Free lime in the clinker usually dissolves rapidly and exothermically, and, if present in sufficient quantity, can lead to portlandite supersaturation almost immediately; however, some may

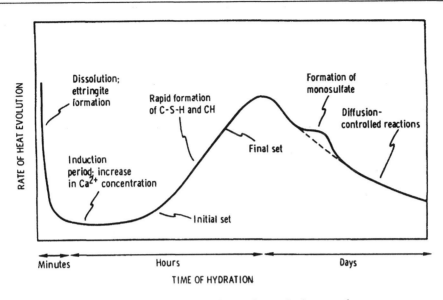

FIGURE 3.22 Schematic representation of heat evolution during hydration of a cement.

also be consumed by reaction with the aluminate phases. Reactive alite particles also release CH rapidly into solution, so it is not always easy to separate the influence of alite and free lime at early ages. Alite usually contributes much less to the initial heat release than do the aluminates, but the amount coming from free lime can vary widely. Dissolution of calcium sulphates is only weakly exothermic and relatively little dissolves in the first few minutes; on the other hand, dissolution of alkali sulphates is very rapid but endothermic. The solution usually reaches a fairly constant alkali concentration within less than a minute, when all of the exposed crystals of alkali sulphate have dissolved completely. Alkali metal ions are not precipitated in any of the early reactions except for the formation of syngenite ($CK\overline{S}_2H$), which only occurs at high potassium concentrations. Rapid precipitation of sulphate always occurs, principally as aluminate hydrates, but often also as gypsum, and/or syngenite. To balance the dissolved cations (mostly alkalis) this missing sulphate is replaced by hydroxide ions: thus, the pH increases very rapidly in the first minute, and soon becomes dependent almost entirely on the total alkali metal concentration in the liquid phase, (denoted by $[R^+]$) the anion

concentrations in most typical cases being determined by a state of near equilibrium with gypsum and 'supersaturated' CH [163].

The initial dissolution reactions can be represented chemically by the following equations, in which all of the ions shown are intended to represent fully solvated aqueous species:

$$C_3A \Rightarrow 3Ca^{2+} + 2Al(OH)_4^- + 4OH^- \quad (27)$$

$$C_4AF \Rightarrow \text{``}FH_3\text{''} + 4Ca^{2+} + 2Al(OH)_4^- + 6OH^- \quad (28)$$

$$CaO \Rightarrow Ca^{2+} + 2OH^- \quad (29)$$

$$(C, N, K)S \rightarrow (Ca^{2+}, R^+) + SO_4^{2-} \quad (30)$$

$$C_3S \Rightarrow \text{C-S-H(m)} + xCa^{2+} + 2xOH^- \quad (31)$$

We have chosen to represent the initial coating that forms on alite surfaces (and probably also on belite surfaces) as 'C-S-H(m)' by analogy with the pure C_3S system, and ignore the fact that it probably forms via a rapid solution-precipitation reaction and so may incorporate other ions; similarly, we represent the iron-rich coating that forms on the aluminoferrite phase as 'FH_3'. However, the actual compositions and structures of these superficial phases are unknown, and their spatial extent may be influenced by the adjacent phases.

91

These dissolution reactions are followed immediately by rapid precipitation of aluminate hydrates:

$$6Ca^{2+} + 2Al(OH)_4^- + 3SO_4^{2-} + 4OH^-$$
$$\Rightarrow C_6A\bar{S}_3H_{32} \qquad (32)$$
$$4Ca^{2+} + 2Al(OH)_4^- + 2X^- + 4OH^-$$
$$\Rightarrow C_4AX_2H_n \qquad (33)$$

It is by no means certain that all of the aluminate precipitates as ettringite as shown in equation (32); AFt phases in which some sulphate is replaced by silicate or hydroxide have been reported [164,165]. In cases of calcium sulphate deficiency AFm phases such as the hydroxyaluminate ($X = OH^-$) or monosulphoaluminate ($2X = SO_4^{2-}$) (or a solid solution of both) can form instead, as shown in equation (33). There may also be a little iron substitution in these phases. As stated earlier, the AFt phase that forms early on often shows very weak XRD peaks and is rather difficult to characterize morphologically; it is sometimes claimed to be amorphous.

In addition to the hydrated aluminate phases, which precipitate in significant amounts within the first minutes (of the order of 1–5 per cent by mass of cement), there is often substantial precipitation of hydrated sulphate phases (gypsum or syngenite, or both,) from cements that contain high levels of 'metastable' sulphate sources (principally calcium sulphate hemihydrate and calcium langbeinite) [136]. Precipitation of these phases occurs at about the same time as that of the aluminate hydrates, and all of these precipitation reactions can lead to observable changes in workability if they continue after the concrete has left the mixer.

3.6.1.2 The induction and acceleration periods

The hydration of C_3S has often been used as a model for the hydration of Portland cement. This practice is based on the fact that Portland cement clinkers typically comprise about 75–80 per cent alite and belite (impure C_3S and C_2S respectively), and that C-S-H produced in the early stages of hydration is predominantly due to alite hydration.

Using this analogy, it is common to refer to the period between mixing and set as the 'induction' period, although it is not a period of complete inactivity. In fact, most Portland cements show a significant level of heat evolution throughout this period, but it is difficult to determine how much of this heat comes from the silicate phases and how much from the aluminates.

The mechanism necessary to explain this period of slow activity in portland cements need not be very different from that of the pure systems. In most cases, the order of events appears to be similar to a simple combination of those observed in the two separate systems: C_3S–water and C_3A–C$\bar{S}$–CH–water. Both systems show an induction period after an initial fast reaction, although, in the aluminate–sulphate system, the length of this induction period depends very strongly on the ratio of sulphate to C_3A [146], and is also sensitive to the presence of CH [128,135,166]. It is quite reasonable to suppose that the two systems behave fairly independently from a chemical standpoint. Microanalyses suggest that initially C-S-H is admixed with AFt, AFm and CH at a nanometer level, but that C-S-H itself has relatively low degrees of aluminate or sulphate contamination [8]. Thus, the behaviour of the two systems probably become inter-related physically [167] and this could have a big influence on the actual rates observed in combined systems, even if the chemical mechanisms are not fundamentally changed.

Bearing this in mind, there is fairly good evidence that the acceleration period observed in pure C_3S hydration occurs very similarly in most normal Portland cements, and it is usually the formation of C-S-H during this stage that is responsible for the set and early hardening of concrete. According to Bezjak and Jelenic [67], the presence of C_3A and soluble sulphates significantly extends the acceleration period, but, once the diffusion stage begins, the rates become slower than in the pure C_3S system. However, the hydration of the aluminate phases is probably even more strongly modified from that observed in the C_3A–C$\bar{S}$–CH–water system. In most Portland cements, there is sufficient sulphate available, in principle, to prolong the C_3A induction period well beyond the first day; thus, very little C_3A hydration is occurring at the time at which the set occurs, except in specially modified Portland cements. Kanare and Gartner [168] suggested that it was necessary to assume that calcium sulphate was significantly adsorbed (or, possibly, physically entrapped) by the C-S-H produced during rapid C_3S hydration, in order to explain why the re-acceleration of C_3A hydration usually occurs much earlier than expected in portland cements.

Electron microscopy studies have not always aided in the comprehension of the hydration mechanism, due to the complexity of the microstructure and the difficulty in analysing small volumes of hydrate phases. Double *et al.* [60], using an analogy with 'silicate gardens,' proposed an 'osmotic pump' caused by the formation a semi-permeable silicate hydrate envelope, separating the bulk liquid from the anhydrous surface, after mixing cement with water. This envelope eventually ruptures under an increasing osmotic pressure forming hollow tubules. Later studies [169] showed these tubules to be AFt. Osmotic mechanisms are not observed for pure C_3S and the presence of alumina (and perhaps sulphate or other species) may be necessary for the formation of the proposed semi-permeable membrane. This mechanism has not stood the test of time.

However, frequent observations of 'hollow shell' hydration [170] (see Figure 3.23) in most portland cement pastes seems to be consistent with a modification of the hydration processes, at least for some of the particles in the cement. The excellent early work of Pratt and co-workers [171,172] was followed by a more detailed analysis of the sequence of events leading to the formation of hollow shells [158,167]. Since hollow shells are not observed in pastes of pure C_3S they probably originate from multi-component grains due to concomitant precipitation of AFt and C-S-H. Dalgleish *et al.* [172] attempted to correlate the main heat evolution peak of cement hydration to a new point in the process of hardening, which they termed the 'cohesion point', at which time the bonds between particles supposedly become strong enough to fracture, under stress, the thickening shells and thus expose unhydrated particles. In our

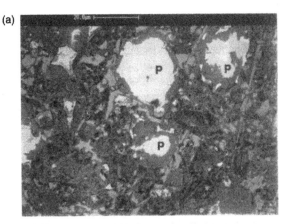

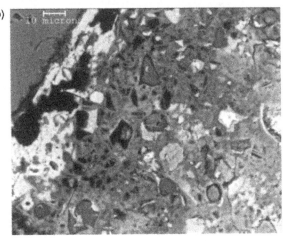

FIGURE 3.23 Back-scattered electron images of cement paste, × 1000 (a) showing phenograins (p). (Micrograph courtesy of Dr. P. Stutzman.) (b) showing hollow hydration shells and calcium hydroxide (white) adjacent to an aggregate particle. (Micrograph courtesy of Dr. A. Crumbie and Dr. C. Famy.)

view, the above correlation is questionable because the top of the heat evolution peak represents the onset of diffusion control. Cements with coarser particles have a tendency to form a different shell structure from that formed by finer cements, and this may have an effect on the early strength development and Young's modulus.

3.6.1.3 Overall kinetics of hydration

Although most of the cement reacts with water within the first few days, the rates at which the individual clinker compounds react are different. Ranking them in terms of their average rate of disappearance over the first few days or weeks of hydration, their overall reactivities usually decrease in the order: $C_3A > C_3S > C_4AF > C_2S$. However, the effective reactivities of two cements of similar bulk chemical and mineralogical compositions may differ significantly, because the rates at which the individual components react with water depend on the processing history, primarily on the clinker-burning rate and temperature, and the cooling rate [173]. These factors affect the distribution of mineral phases in clinker; their crystal size, concentration of crystal imperfections particularly at phase boundaries crystallinity, and polymorphic form of some of the clinker components. Clinker composition and reactivity also depend on the mineralogy of the cement raw mix components.

Additional differences in effective cement reactivity result from comminution – principally differences in psd. Because of the different hardness and fracture toughness of the various clinker components, differential grinding leads to uneven distribution of individual components amongst the various size fractions. This effect is particularly marked with calcium and alkali sulphates, which are much softer than the clinker phases, and tend to be found preferentially in the finest fractions of the cement. However, such particles can have a tendency to self-agglomerate, which means that they may behave as much coarser particles unless they are well dispersed in the cement. The distribution of the calcium sulphate phases in the cement, as well as their solubility rates, can have a very significant influence on the course of the rapid early reactions of the aluminate phases [136,174]. Tang [174] found that a cement with poorly distributed gypsum particles (produced by inter-blending) is less well controlled than one of identical mineralogical composition and apparently similar psd, but with well-distributed gypsum particles (produced by inter-grinding). The aluminate phases in the latter cement only reacted to about the half extent of those in the former during the first two minutes of hydration. This is consistent with the concept that formation of a

thin protective hydrate layer on the aluminate phases requires a very high initial flux of calcium and sulphate ions close to the surfaces of the aluminate particles [136].

These observations are just one facet of the general observation that the rate of hydration of clinker minerals depends on the composition of the liquid phase surrounding them, especially if dissolved species (either those originating from the other cement phases, or admixtures deliberately added to modify the hydration reactions) are involved in the reaction in question. There are also cases in which the presence of fine, but essentially insoluble, particles can have strong influence, if they are well distributed. One such is the observation that fine calcite particles generally accelerate early cement hydration [175–177], probably by acting as potential nucleation sites for the reaction products. A similar example is the increased rate of reaction of C_2S in the presence of C_3S [178,179], suggesting that the C-S-H produced by the hydration of the C_3S can serve as a local growth site for the products of C_2S hydration. C_3S controls the solution composition even when only small amounts are blended with C_2S [93]. Differences in typical rates of hydration of pure clinker minerals and those of the same compounds in cement are illustrated in Figure 3.24. The differences are not very great, the two principal effects in the cement paste being, as explained above, the shorter than expected induction period for the C_3A, relative to the pure C_3A–gypsum system, and the faster than expected hydration of C_2S.

The temperature of hydration is another important kinetic factor. Generally, increased temperature accelerates the early rate of cement hydration (Figure 3.25), although the degrees of hydration and strength development at later ages are often decreased. In mass concrete, where near-adiabatic conditions are maintained, the evolution of heat of hydration will raise the temperature to well above ambient. The generation of high internal temperatures occurs primarily over the first few days when the concrete is still relatively plastic, and will not cause immediate damage, but potential effects on strength development are significant [180,181], illustrated in Figure 3.26. On

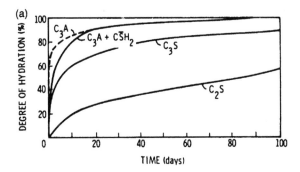

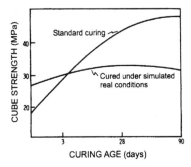

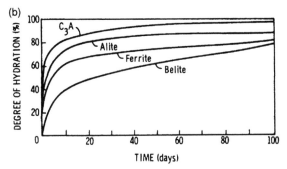

FIGURE 3.26 Compressive strength development of concrete cured under standardized laboratory conditions, and under a temperature regime mimicking internal concrete temperatures in the field [182].

commonly, heat evolution is controlled by the addition of mineral admixtures such as fly ash or blast furnace slag [183].

3.6.1.4 Energetics of hydration

Hydration of cement involves mainly exothermic reactions, and most of them also have positive 'activation energies.' Thus, if cement hydration is conducted under adiabatic or semi-adiabatic conditions, it will self-accelerate due to the increase in internal temperature, regardless of the specific chemistry. The initial burst of heat produced mainly by aluminate and free-lime hydration is usually enough to raise concrete temperatures in the mixer by about 2°C. Once alite hydration enters its acceleration period, the temperature of the concrete rises rapidly, typically by 20–40°C.

FIGURE 3.24 Rate of hydration of the cement compounds: (a) in pastes made with pure compounds; (b) in a Portland cement paste [180].

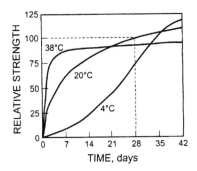

FIGURE 3.25 Effect of curing temperature on compressive strength development.

subsequent cooling when the concrete has gained rigidity, thermal stresses resulting from temperature gradients may well lead to tensile cracking. Temperature rise can be controlled by lowering the total heat of hydration of a cement through a reduction in C_3S and C_3A contents, because these compounds have the highest heats of hydration and the highest rates of heat evolution. More

Selected heats of hydration for the main clinker components are given in Table 3.7 in joules per gram (rather than joules per mole) as a more practical measure for the cement technologist. The enthalpy of hydration, ΔH, represents the differences in the residual energy in the system after breaking the old and forming the new chemical bonds. Using these ΔH values, together with the mineralogical and chemical compositions of the cement and the degree of hydration of individual clinker minerals one could, in principle, estimate the heat of hydration of a cement at any given time. However, ΔH for many hydration reactions is still not known with certainty, and it

TABLE 3.7 Heats of hydration of the cement compounds

Reaction	$-\Delta H$ (Jg^{-1}) for complete hydration			
	Pure compounds		Clinker measured[a]	Cement measured[a]
	Calculated	Measured		
$C_3S \rightarrow C\text{-}S\text{-}H + CH$	~380	500	570	490
$C_2S \rightarrow C\text{-}S\text{-}H + CH$	~170	250	260	225
$C_3A \rightarrow C_4AH_{13} + C_2AH_8$	1260	—	—	—
$C_3A \rightarrow C_3AH_6$	900	880	840	—
$C_3A \rightarrow AFm$	—	—	—	~1160
$C_4AF \rightarrow C_3(A,F)H_6$	520	420	335	—
$C_4AF \rightarrow AFm$	—	—	—	~375

[a] Determined from regression analyses of heats of hydration of ground clinker or cement.

is experimentally difficult to determine the exact clinker composition and the actual degrees of hydration. The ΔH values given in Table 3.7 for cement and clinker have been determined by multiple regression analyses of heat of hydration conducted on a series of cements and clinkers hydrated for 1 year, and thus do not necessarily represent complete hydration.

3.6.2 Setting of cement paste

One of the first physical consequences of the chemical processes between cement components and water (described earlier) is the transformation of a workable plastic cement paste (fresh paste) into a rigid, brittle material (hardened paste). This solidification process is the result of the formation of interlocking hydration products and is commonly described as setting. In general terms, however, setting depends on the relative rate at which the ionic species released from the cement components into the liquid phase react to form initial hydration products, which can be either colloidal or crystalline in nature.

3.6.2.1 Forms of setting

Setting encompasses 'initial' and 'final' set, as specified by various national standards (e.g. BS12, ASTM C150, DIN 1164) using penetration methods. These terms do not have a well-defined physical meaning. However, recent studies have shown that initial set can be determined rheologically

from the rate of change of yield stress [162] or ultrasonically from the rate of propagation of shear waves [184]. These measurements emphasize that setting is not dependent on hydration *per se*, but is a microstructural parameter representing the onset of percolation of a network of agglomerated solid particles. Thus, setting time depends on w/c and this is a major reason for poor correlation between the setting of cement paste and the setting of concrete. Most investigators have concluded that normal setting is controlled by the hydration of alite. Initial set under normalized conditions is reported [182] to correlate with the end of the induction period, although this time is not always well defined. Final set is said to occur about mid-way through the acceleration period.

3.6.2.2 Abnormal setting

Abnormal setting behaviour can, in most instances, be traced to chemical reactions involving the aluminates (C_3A and C_4AF) and sulphate phases (calcium and alkali sulphates). In some cases, however, highly reactive or unreactive alites (a consequence of burning/cooling kiln conditions or other factors) may be the cause. The most common types of abnormal setting are discussed briefly below. Chemical admixtures can result in abnormal setting behaviour in some cases, because they may change the delicate balance between the aluminate and sulphate precipitation reactions in an otherwise normally setting cement.

False set (also referred to in the literature as early set or premature stiffening), is defined as a rapid increase in paste rigidity without excessive heat evolution, which can be dispelled by immediate further mixing to regain plasticity. False set generally results from rapid crystallization of gypsum from the paste liquid phase supersaturated in sulphates by excessive dissolution of calcium sulphate hemihydrate (plaster of Paris) or soluble anhydrite present in cement. Hemihydrate usually forms as a result of the dehydration of gypsum during grinding or during storage, especially if the cement is hot. Occasionally, the formation of ettringite or syngenite will cause false set.

Flash set (also called quick or grab set) is defined by a strong heat evolution and a very severe stiffening, which cannot be dispelled by remixing. It is the result of an excessive reaction of C_3A with water in the earliest stage of hydration to form AFm phases, due to an excessive rate of dissolution of the aluminate phases relative to the rate at which sulphate and calcium ions can be transported to the reacting aluminate surfaces. It has been attributed to an inadequate content of rapidly soluble sulphate; however, calcium hydroxide is also required, and cements with very unreactive alites or low free-lime levels can thus be susceptible.

A special case of abnormal setting is *pack set* (also called warehouse or silo set) is characterized by a decrease in the ease of flowability of the unhydrated cement powder. Although this type of abnormal set happens during cement storage rather than after mixing cement with water, pack set can influence the subsequent hydration reactions and thus, is of interest to this discussion. Pack set is related to the degree of gypsum dehydration during cement grinding and to silo temperature and humidity, which, in turn, are functions of mill temperature and the amount of water added to the cement during grinding. It is most commonly due to strongly attractive surface forces (probably multipolar electrostatic interactions) between cement particles, and these can be reduced greatly by certain small organic molecules, such as glycols, acetic acid, etc., that can adsorb on the particle surfaces and make them

more mutually repulsive. Such additives are referred to, in this application, as 'pack-set inhibitors;' but they have a similar dry-dispersant function during grinding and so are normally added as grinding aids to improve powder flow and dispersion in the mill and separator.

Gypsum in the warmer parts of the silo may dehydrate, (to give hemihydrate, or to react with potassium sulphate to form syngenite), releasing water vapour that can superficially hydrate other clinker phases such as the aluminates and free lime. Condensation of water near the colder silo walls may also cause formation of wall coatings via cement hydration. These silo reactions generally lead to a cement with a reduced rate of aluminate phase dissolution and, at the same time, an increased the rate of solubility of calcium sulphate, (because of the higher solubility rate of hemihydrate compared to gypsum.) This can lead to false set in concrete, and also to poor early-strength development due to slow hydration of the clinker phases.

3.6.2.3 Mechanisms of setting

In the classical literature [185], the setting or stiffening of cement paste is recognized to be caused by the increasing volume of hydration products, leading to a decrease in the distance between individual particles until plastic flow is restricted by cohesive forces. Although most researchers accept the importance of alite hydration, some believe that setting is controlled by recrystallization of ettringite. Rhebinder and co-workers [186,187] proposed setting to be the result of the formation of a coagulated thixotropic structure from the dispersion of unhydrated cement components by the initial colloidal hydration products (setting), with products from the aluminate phases being the dominant factor.

Locher [130,188] concluded that set control is not the result of a chemical reaction, but rather of the form and size of crystalline hydrates produced, which depends on the relative availability of aluminate, sulphate, and calcium ions in the solution. When sulphate is present, minute ettringite particles deposit on the cement grains.

Because of their fine structure, plasticity of the system is maintained and normal set is the consequence of re-crystallization, during the induction period, of fine-grained ettringite into large needle-like particles, which produce a rigid structure by bridging the adjacent cement particles. Only a small portion of the C_3A, dissolved in the first minutes of hydration, participates in these reactions. Normal setting is achieved when the amount and rate of availability of aluminates and sulphates enables complete coverage of the available surface by fine-grained ettringite. Inadequate sulphate causes the formation of large tabular crystals of AFm in addition to ettringite, leading to quick set. An excess of sulphate in the solution may lead to false or quick set, depending on the amount of secondary gypsum formed.

Other authors [123,189,190] have shown, however, that retardation of C_3A hydration by gypsum does not necessarily involve the formation of ettringite. Also, as noted earlier, it has been demonstrated that the presence of soluble calcium sulphates can often have a strong influence on the amounts of C_3A and C_4AF that hydrate in the first few minutes [136]. The amounts of these two phases consumed in the first two minutes can be reduced by at least a factor of three on going from a pure clinker with no added sulphate, to a cement with a suitable combination of soluble sulphates. The re-crystallization of ettringite is probably not the usual cause of abnormal set. However, it is often observed in cement pastes about the time that the free sulphate disappears, which is after set [167]. This transformation can be attributed to the increased solubility of aluminate ions when sulphate concentrations are reduced, allowing some of the ettringite to re-dissolve and either re-crystallize locally (probably mainly outside the hydration shells which surround cement particles at that time), or contribute sulphate ions for continued formation of AFm phases (probably mainly inside the shells) [167].

We concur with the view that normal set is the consequence of the formation of C-S-H resulting from C_3S hydration. The formation of fine-grained ettringite from a solution of proper aluminate–sulphate–calcium proportions is desirable to maintain a fluid paste through the induction period until the formation of C-S-H causes the system's mobility to decrease. All forms of abnormal set result from imbalance of ionic species in the liquid phase, and are the result of either improper proportions of C_3A and sulphate in the cement, or differing reactivities (solubility) of sources supplying the aluminate and sulphate ions.

3.6.3 Liquid phase composition

It is clear from the above discussion that the dissolution of ionic species and composition of the liquid phase play an important role in cement setting. The absolute amount of C_3A in the clinker is of less importance than the reactivity of C_3A and the solubility of calcium sulphate phase present in cement (gypsum hemihydrate soluble anhydrite). The concentrations of sulphate ions in the pore solution are very much influenced by the presence of soluble alkalis in the cement clinker [191] (see Figure 3.27). Gartner *et al.* [163] showed that the composition of the aqueous phases extracted from different cement pastes could be explained roughly on the basis of established solubility. The solutions were initially supersaturated with respect to gypsum and undersaturated with respect to CH, but fell close to gypsum saturation levels after the first few minutes, while slowly becoming supersaturated by a factor of 2–3 with respect to CH after a slightly longer period. High-potassium cements gave solutions close to syngenite saturation

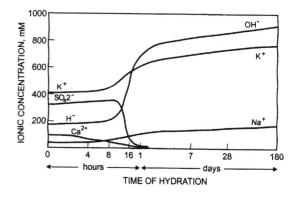

FIGURE 3.27 Composition of pore solution in a high-alkali Portland cement paste (w/s = 0.5).

most of the time, so that the liquid phase at early ages can be regarded as a perturbed portlandite–gypsum (–syngenite) system. Their data for early ages were consistent with previous results obtained over a wider range of hydration times, as summarized in Figure 3.27 [129,192]. The disappearance of sulphate ions from solution, due to consumption of all of the readily-soluble sulphate sources to form AFt phases, is accompanied by a decrease in calcium concentrations and often an increase in alkali concentrations as the aluminate phases react rapidly, thus releasing the alkalis bound up in them. The alkali metal ions must be balanced by hydroxide ions, causing a significant rise in pH, which accounts for the precipitation of most of the calcium as portlandite. Silicate concentrations remain very low but fairly constant throughout the whole period.

3.6.4 Optimum sulphate contents

Calcium sulphates are not added to Portland cement for set regulation alone. They also influence volume stability and play an important role in the acceleration of alite hydration, thus influencing the rate of strength development and volume stability. At the 'optimum gypsum' content,

the strength is maximized and the drying shrinkage is minimized. However, the 'optimum' is actually a variable and is not the same for strength and shrinkage, so that a compromise value must be adopted. Furthermore the optimum sulphate level varies with specimen age, the form in which the sulphate is added, the temperature of curing, the amounts and reactivities of the aluminate phases, the water/cement ratio, the presence of admixtures, and the specific test method used. It is also observed that increased alkali levels usually increase the optimum sulphate requirement [168] (see Figure 3.28). The presence of mixed alkali-calcium sulphates such as syngenite or calcium langbeinite in the cement can affect the results, even when the total alkali and sulphate contents are kept constant [136]. Thus, the rate of availability of the sulphate ions at very early ages dictates the path of later hydration and may have a significant influence on microstructure (and hence on strength and other physical properties).

The scientific basis of the optimum sulphate content is unclear. One explanation is that high gypsum content leads to the excessive formation of ettringite after the paste sets, causing disruption of the paste microstructure. A low gypsum content, on the other hand, leads to the formation

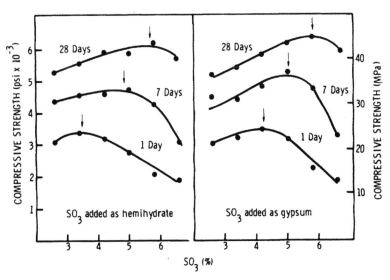

FIGURE 3.28 Optimization of sulphate additions for compressive strength at different ages. (With this cement quite high SO_3 levels were needed to obtain maximum strength.)

of AFm phases prior to the renewed accelerated hydration of alite, so that the resulting consumption of Ca^{2+} ions from the paste–liquid phase prevents nucleation of CH and C-S-H, thus leading to retardation of strength development. Another viewpoint explains the optimum gypsum content by the fact that gypsum accelerates alite hydration but, at the same time, lowers the intrinsic strength of C-S-H due to the incorporation of sulphate ions into its structure. Both mechanisms may contribute to the phenomenon.

3.6.5 Influence of admixtures

Many different types of admixtures are available to modify cement hydration and concrete properties, but space does not permit a detailed discussion here. However, we will briefly address the question of the influence of simple accelerating and retarding admixtures on C_3S and C_3A hydration, which has already been discussed, and their influence on cement can be interpreted on this basis. The accelerating or retarding action of an admixture is, in most cases, due to its effect on C_3S hydration. However, the strong sorption by hydrating C_3A effectively reduces the amount of admixture in solution that can influence C_3S [193]. We can consider C_3A to have a 'buffering' action, mitigating the effects of overdosing the admixture. Most admixtures have similar effects on both C_3S and C_3A, that is, they retard or accelerate both. But this is not always the case, for example the accelerating action of triethanolamine is due to its tendency to accelerate the hydration of C_3A even though it may actually retard C_3S [194]. A recent development in hydration-modifying admixtures is the discovery that certain related alkanolamines can have a catalytic effect on cement hydration at very low dosages, increasing strengths of some cements at ages up to at least several months [160,195]. This has been explained in terms of the increased hydration of the ferrite (due to iron solubilization by the alkanolamine) which, in turn, frees the surfaces available for the hydration of otherwise slow-reacting silicate phase particles.

Flash-setting admixtures (e.g. certain fluorides, carbonates, aluminates, etc.,) used in sprayed concrete owe their rapid-setting abilities to renewing the flash setting tendencies of C_3A [196]. By altering the rate of the early C_3A reactions, admixtures may cause premature stiffening or early loss of workability. The exact influence of an admixture depends on the temperature and the composition of the cement used.

3.7 Microstructure of hardened paste

Since the properties of concrete frequently depend on paste properties, knowledge of the architecture of the hardened paste is critical for the establishment of structure–property relationships. The advent of the scanning electron microscope (SEM) and energy-dispersive spectra for microanalysis has resulted in a comprehensive qualitative description of paste microstructure [167,199–201]. However the subsequent development of back-scattered electron (BSE) imaging [167,202] and computer-simulation modeling [203,204] has enabled reliable quantitative microstructural analysis to be performed relatively easily. Much greater detail can be found in these references, since space only permits a broad overview of this topic.

3.7.1 Basic microstructure

Hardened cement paste (HCP) is a composite material whose properties ultimately depend on the properties of the components and their spatial relationship to each other (the microstructure). The composition of a typical cement paste is given in Table 3.8, from which it can be seen that C-S-H gel is the predominant constituent, comprising about two-thirds of the solid phase. Characteristics of the hydration products are compared in Table 3.9.

A typical back-scattered SEM image is shown in Figure 3.23a. Large particles consisting of unreacted cement grains surrounded by a dense coating of C-S-H can be clearly seen, which have been designated as 'phenograins' by Bonen and Diamond [203]. These are embedded in a porous 'ground mass' of rather undifferentiated material containing calcium hydroxide, C-S-H gel and capillary pores.

TABLE 3.8 Composition of hardened cement paste made from Type I Portland cement (w/c = 0.5)

Component	Approximate volume %	Remarks
C-S-H	50	Amorphous, includes microporosity
CH	12	Crystalline
AFm[a]	13	Crystalline
Unreacted cement	5	Depends on hydration
Capillary pores	20	Depends on w/c ratio

[a] Considered to be the final hydraton product of C_3A and C_4AF.

A higher magnification micrograph of the 'ground mass' showing a characteristic cellular structure containing porosity which is thought to derive from 'hollow-shell' hydration [166,169] is shown in Figure 3.23b. Examples of hollow-shell hydration were first published by Barnes and co-workers [169] and are now considered to be ubiquitous in cement hydration, although they do not occur during the hydration of C_3S. This suggests that the early formation of C-S-H is influenced by the concomitant hydration of the aluminates. A scenario for the development of the C-S-H coating around cement grains, which was developed by Scrivener, [166,205] is shown schematically in Figure 3.29, and briefly described below.

TABLE 3.9   Properties of hydration products

Compound	Specific gravity	Typical morphology	Typical dimensions	Resolved by[a]
C-S-H	1.9–2.1	Variable	~0.1 μm	SEM, TEM
CH	2.24	Equant prisms or thick plates	0.01–0.1 mm	OM, SEM
AFt	~1.75	Prismatic needles	10 × 1 μm	OM, SEM
AFm	1.95	Thin hexagonal platelets or irregular 'rosettes'	1 × 0.1 μm	SEM

[a] SEM: scanning electron microscopy; TEM: transmission electron microscopy; OM: optical microscopy.

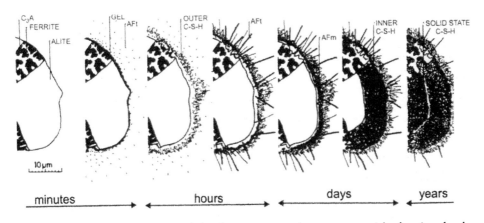

FIGURE 3.29   Summary of microstructural development around a cement particle showing development and filling in of a hollow shell [167].

Most cement grains are multi mineralic and early reactions are dominated by C_3A and C_3S reactions. Within a few minutes an aluminate-rich amorphous coating, (probably also containing silica) forms, from which small AFt needles crystallize. After the induction period is passed, C-S-H from C_3S hydration is deposited on a network of AFt needles and grows outwards into the water-filled space. At the end of the acceleration period a narrow ($<1\,\mu m$) space has opened up. Later AFt re-crystallizes as much larger needles and thereafter, the thickening of the shell by continued C-S-H formation occurs due to the interior surface advancing inwards rather than the outside surface advancing outwards. Grains smaller than $5\,\mu m$ probably completely hydrate before the gap can be bridged leaving permanent hollow shells as large capillary pores which are accessible only through smaller ones. Grains larger than $15\,\mu m$ will hydrate to close the gap and form phenograins. (Hollow shells between 5 and $15\,\mu m$ are probably due to unreacted cores falling out during specimen preparation.) Taylor [8] has discussed this scenario in more detail.

3.7.2 C-S-H gel

3.7.2.1 Classification

The literature has references to C-S-H and C-S-H gel. It is recommended [206] that C-S-H nomenclature be confined to quasi-crystalline or amorphous calcium-silicate hydrates phases (including foreign ions in the structure), whereas C-S-H gel designates an assemblage of C-S-H on a colloidal scale which may be intermixed with other cryptocrystalline phases. Thus C-S-H gel is identified by BSE imaging while local microanalyses give information about the distribution of phases within the gel. The literature has also classified C-S-H gel according to its location in the microstructure or its morphology and this can lead to considerable confusion. Table 3.10 provides a correlation between these different classifications, but it should be noted that different morphologies do not necessarily imply different compositions. Indeed, a specimen can show different C-S-H morphologies depending on the EM method used for imaging [207]. The changes in C-S-H morphology during hydration are discussed in Section 2.4.

TABLE 3.10   Classification of C-S-H gel

Microstructural Designation	Morphological designations		Time of formation
	SEM[a]	TEM[b]	
Outer (product) Groundmass Early product	Type II[c] (reticulated)	Type E (globular)	During the induction period
Undesignated product First few days up	Type I (acicular)	Type O (foil-like)	to the diffusion-controlled stage
		Type 1 (acicular)	Dried form of Type O
Inner (product) Phenograins Late product	Type III (indeterminate)	Type 3 (fibrous)	Mid-stages, during first 2–3 weeks
	Type IV (spherical)	Type 4 (indeterminate)	Late stages, weeks to months

[a] Ref. 209; [b] Ref. 46; [c] Type II can persist beyond the induction period when certain admixtures are used. Thus it may be more closely related to Type O than Type E.

3.7.2.2 Chemical composition

C-S-H is essentially amorphous when formed in paste hydration, but longer range order may exist when C-S-H is formed in a relatively pure system. This is unlikely in cement pastes where C-S-H can adsorb considerable quantities of impurity oxides to form solid solutions. It has been known for decades that sulphates, alumina, iron, alkalis, and possibly other species are found in C-S-H [208, 209]. It has been estimated [209] that as much as 50 per cent of the total sulphate, alumina and iron present in the cement is found in C-S-H gel rather than in the AFt and AFm phases estimated from the hydration of the pure cement compounds. The composition of C-S-H gel is therefore quite variable depending on the extent of substitution. Microanalyses using either SEM on polished samples [73,74,210] and TEM [214–217] on dispersed pastes or ion-thinned sections show an extreme variability of composition from point to point (or particle to particle) in the microstructure. Compositional maps (Figure 3.30) show that the composition of the C-S-H gel coatings

around the large phenograins (inner product) has a relatively constant composition, and may represent a single C-S-H phase [215,216]. However, the outer product in the groundmass is highly variable and the analyses tend to lie along the lines between inner product C-S-H and other hydrate phases such as AFm, AFt and Ca(OH)$_2$, suggesting that C-S-H gel is an intimate mixture of C-S-H with other cryptocrystalline hydrates. According to these studies C-S-H has compositional ratios of: Ca:Si = 1.7–2.1, Ca:Al = 20–30, Ca:S~25, giving an approximate composition of $C_{1.7-2.1}S(A_{0.03-0.05}\bar{S}_{0.04})H_{xd}$. The microanalyses indicate that most chemical species are distributed widely throughout the matrix. The exceptions are iron and magnesium which tend to be located in the space that was formerly occupied by the ferrite phase [8,167]. The former is probably present as an iron-substituted hydrogarnet (or hydrogarnet precursor) and the latter as hydrotalcite, which is a magnesium aluminate layer structure similar to, but distinct from, AFm.

3.7.2.3 Molecular structure

It is not the purpose of this review to discuss details of the molecular structure of C-S-H, which is still under active debate. Since C-S-H is amorphous, and therefore poorly organized on the molecular scale, drawing structural analogies with crystalline compounds of known structure can be useful. Based on chemical analysis [218] and ^{29}Si NMR spectroscopy [219,220] it is known that the organization of the silica is in the form of linear chains of silicate tetrahedra of variable length, that most probably represent fragments of *dreierketten* (which are a structural feature of many calcium silicates) due to random omission of bridging tetrahedra. Therefore the hypothesis that C-S-H gel contains structural elements of the layer structure of tobermorite [8] appears reasonably valid. An early hypothesis was that amorphous C-S-H formed in calcium silicate hydration was most likely a solid solution between tobermorite and Ca(OH)$_2$, but Taylor [222] later proposed that C-S-H gel contains elements of both 1.4 nm tobermorite and jennite

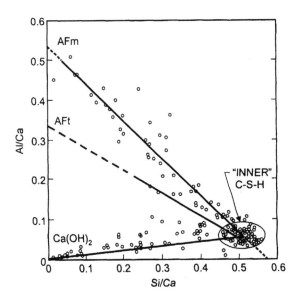

FIGURE 3.30 Composition plots (Al/Ca versus Si/Ca) for individual microanalyses of C-S-H gel in a typical cement paste. The cluster of points labelled C-S-H represents the composition range of 'inner C-S-H' in a phenograin coating [8].

structures, in which the infinite chains of *dreier-ketten* are degraded by omission of some bridging tetrahedra. Richardson and Groves [220] proposed a generalized formulation:

$$[Ca_xSi_{(3n-1)}O_{(9n-2)}H_{(6n-2x)}] \cdot zCa(OH)_2 \cdot mH_2O$$

where the formula in square brackets represents silicate chain fragments attached to the central Ca-O sheet. The $Ca(OH)_2$ component in this formula can either be interpreted as accounting for the presence of degenerate jennite elements, or solid solution with amorphous (or crypstocrystalline) $Ca(OH)_2$, or indeed perhaps both. 'Amorphous $Ca(OH)_2$' can be regarded as regions in the C-S-H structure that are highly deficient in silica.

It would appear from such diverse experiments, such as the leaching of lime from C-S-H [221] and the synthesis of quasi-crystalline C-S-H [222], that the limiting composition is C/S ~ 1.25. This value corresponds to the composition of a degenerate 1.4 nm tobermorite structure when $n = 1$ (i.e. bridging tetrahedra are missing from all *dreierketten* giving dimeric silicate species). However, the degree of silicate polymerization of C-S-H increases with age towards a mean value of $n = 4-5$, making it necessary to postulate the presence of jennite-like structures to account for the high C/S ratios. Since there are random chain lengths, and perhaps omission of entire sections of chains, the distinction between degenerate jennite and amorphous $Ca(OH)_2$ may be more a matter of semantics. Therefore in Portland cement pastes the inner C-S-H is best regarded as a solid solution (or mixture at the nanolevel) between at least two phases. The presence of aluminum can be accounted for by isomorphous substitution for Si in bridging tetrahedra of the *dreierketten*. The presence of sulphur (as sulphate ion) is less easily accounted for, but may be adsorbed on surfaces or present in the pore solution.

3.7.2.4 Water content and physical properties

Values for the water content (expressed as the H/S ratio) can be even more variable, because there is no sharp distinction between water present as part of C-S-H structure and water held in micropores, isolated mesopores, and as multilayer adsorbed films. Thus, the H/S ratio depends not only on the equilibrium drying conditions used, but also on the time at which the material is held under these conditions. (Equilibrium is approached asymptotically under any drying conditions.) C-S-H that has been D-dried (vacuum drying over the water vapour pressure of dry ice at $-79°C$) has an H/S ratio of about 1 (approximately 0.5 less than the corresponding C/S ratio) while drying at 11 per cent r.h. gives an H/S ratio close to 2.0. In saturated paste H/S is estimated to be close to 4 [8].

The continuum of interactions between water and C-S-H also causes problems with determinations of densities, surface areas, and pore size distributions. It is now recognized that the microstructure of C-S-H changes as water is removed during drying. These changes are largely reversible, but some irreversibility is observed, its magnitude depending on the method used to remove the water. The least disruptive method is to replace the water by a miscible, non-aqueous solvent such as methanol or isopropanol [223]. The pore size distribution also depends on the extent and mode of drying. In addition, both surface area and porosity depend on the methods used to measure them. Similarly, the density of C-S-H is dependent upon the moisture content (Table 3.11) and the method of drying.

TABLE 3.11 Variation of properties of C-S-H with moisture content[a]

Property	Saturated	Dried at 11% rh	D-dried[b]	
			VAC	SR
Surface area ($m^2 \cdot g^{-1}$)	~700[c]	—	<100[d] ~400[e]	~400[d]
Specific gravity	<2.0	2.2	2.7	—
C/S ratio	~1.7	~1.7	~1.7	~1.7
H/S ratio	~4.0	~2.0	~1.0	~1.0

[a] Present in cement or C_3S pastes.
[b] VAC: water removed directly be vacuum; SR: water first replaced by methanol, then vacuum-dried.
[c] Measured by small-angle X-ray scattering (SAXS).
[d] Measured by nitrogen adsorption.
[e] Measured by water vapour adsorption.

Jennings and co-workers [224,225] have discussed the measurement and reporting of surface areas of pastes by different methods and summarized the effect of different sample preparation methods prior to measurement, the influence of various paste parameters, and the problems of reconciling the disparate values. They conclude that the data can best be interpreted by assuming that C-S-H gel exists in both 'low-density' and 'high-density' forms. Only the former has surface area accessible to nitrogen. The available evidence suggests that low-density C-S-H gel may be the outer product (ground mass), and high-density C-S-H gel is the denser coatings of inner product in the phenograins. This classification of C-S-H has been incorporated into the calculation of volume fractions of the individual components of hardened cement paste from first principles [226].

3.7.3 Calcium hydroxide

Calcium hydroxide forms relatively massive crystals: two or three orders of magnitude larger than C-S-H particles. The crystals grow within the water-filled capillary pores, surrounding and sometimes completely engulfing partially hydrated grains (Figure 3.31) and eventually may form a percolating network [227]. The crystals may grow up to 0.1 mm in size in the paste. Initially, growth

occurs preferentially on the [001] face causing prismatic habits along the c-axis, but later, at lower Ca^{2+} concentrations, growth occurs on the other faces. In the presence of admixtures, prismatic or platy crystals can be formed [228]. The size and shape of $Ca(OH)_2$ crystals are also affected by the hydration temperature. In SEM micrographs of fracture surface $Ca(OH)_2$ crystals often have a characteristic striated appearance due to some slip along the weakly-bonded [001] planes.

3.7.4 Calcium sulphoaluminate hydrates

3.7.4.1 AFt phase

Ettringite commonly forms thin prismatic crystals of hexagonal cross-section with aspect ratios > 10 in many instances. The crystal structure is based on close-packed columns of composition $[Ca_3Al(OH)_6 \cdot 24H_2O]^{3+}$ running parallel to the long axis of the crystals (c-axis) (Figure 3.32). In the channels between the columns are charge-balancing SO_4^{2-} and additional water molecules giving the formula $[Ca_3Al(OH)_6 \cdot 24H_2O]^{6+} \cdot 3SO_4^{2+} \cdot 2H_2O$ or ($C_6A\bar{S}_3 H_{32}$)). On strong drying (e.g. D-drying or heating to 105°C) the water molecules within the column are removed and the structure becomes X-ray amorphous. However, rehydration above 60 per cent rh will restore crystallinity. Other ions can partially

FIGURE 3.31 Calcium hydroxide crystal in a cement paste, × 6000. [Micrograph courtesy of Dr. P. Stutzman.]

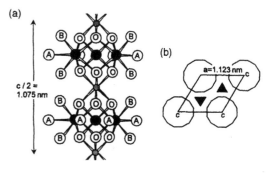

FIGURE 3.32 Crystal structure of ettringite. (a) Composition of the columns. filled circles = Ca^{2+}; hatched circles = Al^{3+}; open circles: O = bridging OH, and A, B = terminal H_2O. (b) Arrangement of columns. Filed triangles represent the location of SO_4^{2-} in the intercolumnar space.

(or in certain cases completely) replace SO_4^{2-} (e.g. OH^-, CO_3, Cl^-, $H_2SiO_4^{2-}$), while Fe^{3+} and perhaps Si^{4+} may partially substitute for Al^{3+} in the columns. Thus, ettringite in cement paste may not have the exact stoichiometry indicated by $C_3A \cdot C\bar{S} \cdot H_{32}$ and the more general designation the AFt phase, should be used.

AFt formation is often accompanied by a significant increase in volume and this is the reason for one common form of sulphate attack. Expansion due to controlled formation of ettringite during hydration is the basis of shrinkage-compensating cements. However, there is not an increase in *total* volume when ettringite is produced (whether from C_3A or AFm) and it appears that crystal-growth pressures arising from crystallization from highly supersaturated solutions in confined spaces must occur. Ettringite is often observed in old concretes in air voids or pre-existing cracks where it has apparently deposited by harmless re-crystallization processes. The presence of free lime enhances expansion; where there is a deficiency of CaO ettringite forms more blocky crystals with much less expansion and can rapidly develop a strong matrix. This is the basis of some rapid-hardening Portland cements, and belite-rich cements.

3.7.4.2 AFm phase

Calcium monosulphoaluminate hydrate, $C_4A\bar{S}H_{12}$, is one member of an isostructural group of compounds of formula C_4AXH_n which have a layer structure (Figure 3.33) where anions balance the positive charge of the layers. Some of the more common members of the group are given in Table 3.12, but compounds with other anions have been prepared. The end product in cement hydration is a solid solution containing both OH^- and SO_4^{2-} in the structure. Recent evidence suggests that silicate ions (and perhaps some carbonate from atmospheric carbonation) may also be present. Partial substitution of Fe^{3+} and Si^{4+} for Al^{3+} will also occur, and the designation AFm is used.

3.7.5 Pore structure

HCP contains a wide range of pore sizes, with effective diameters ranging over several orders of magnitude. Thus, no one method of analysis can give information about the whole range of pores. Characteristics of the pore system are summarized in Table 3.13. Porosity affects every major property of concrete and, although only total porosity is often determined (by loss of water on heating), size distribution is a much more important parameter. Large pores are extrinsic to the HCP microstructure, but limit the strength of a specimen because they act as flaws to initiate failure. However, since they do not form a continuous network their influence on permeability is not large.

Intrinsic HCP porosity can be divided into the IUPAC classification of macro-, meso- and microporosity, but a different size differentiation (0.1 μm v 0.05 μm) between macropores and mesopores better represent their influence on paste properties. Capillary pores are considered remnants of the

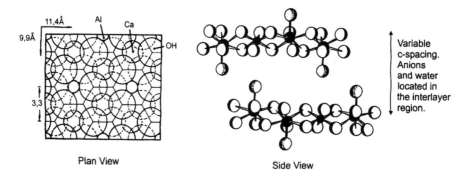

FIGURE 3.33 Crystal structure of AFm. In side view filled circles are Ca^{2+}, hatched circles Al^{3+}, and open circles OH^-.

TABLE 3.12 Structure of hexagonal hydrates $[Ca_2Al(OH)_6]_2^{2+} X_n^{m-}{}_{n \cdot y}H_2O$

Compound	Anion	n	Moles of water (y)	Basal spacing (Å)	Occurrence
C_4AH_{19}	OH^-	2	12	10.6	Forms from C_3A
C_4AH_{13}	OH^-	2	6	7.9	Dehydration of C_4AH_{19}
Hemicarbo-aluminate	OH^- CO_3^{2-}	1 1/2	6	8.2	Carbonation of C_4AH_{19}
Carboaluminate ($C_4A\overline{C}H_{11}$)	CO_3^{2-}	1	5	7.6	In the presence of carbonates
Monosulpho-aluminate ($C_4A\overline{S}H_{12}$)	SO_4^{2-}	1	6	9.0	In the presence of sulfates
Solid solution	OH^- SO_4^{2-}	Variable	6	8.2–8.9	Final hydration product of C_3A in Portland cement
C_2AH_8	$Al(OH)_4^-$	2	6	10.6	Forms from C_3A
Haloaluminate (C_4AXH_{10})	X^-	2	6	7.8 (Cl^-) 8.2 (Br^-) 8.8 (I^-)	Can form at high concentrations of halide

TABLE 3.13 Pores in hardened Portland cement paste

Designation	Diameter	Description	Origins
Macropores	10,000–50 nm (10–0.05 m)	large capillaries interfacial pores	Remnants of water-filled space
Mesopores	50–10 nm	medium capillaries	Remnants of water-filled space
	10–2.5 nm	small, isolated capillaries	Part of outer product C-S-H
Micropores	2.5–0.5 nm		Intrinsic part of C-S-H
	<0.5 nm	interlayer spaces	Intrinsic part of C-S-H

water-filled space between hydrated cement grains. Hence, their size and volume depend on the initial w/c ratio and the degree of hydration, and occur in the outer, 'low density' C-S-H gel. When macropores form an inter-connected network they will provide pathways for easy mass transport, but frequently they are isolated by mesopores (e.g. cavities formed by hollow-shell hydration). While mercury-intrusion porosimetry does not give an absolute measure of pore size distributions it is an effective way to measure the degree of connectivity of the pore system since it measures size access to a given intrudable pore volume.

Mesopores are of the size where capillarity becomes dominant leading to the development of internal stresses as water is removed, or high suction pressures when water is added. Micropores are considered to be an intrinsic part of C-S-H gel representing spaces in which water is confined between two closely-spaced surfaces. They are located primarily within the inner product C-S-H. Interlayer regions could be considered as part of the microporosity. However, Groves [229] has reported the presence of small mesopores (<10 nm) in ion-thinned sections of inner product C-S-H. While this may be an artefact of drying it does lend credence to the classical view that C-S-H gel always forms with pores <10 nm (gel porosity).

3.8 Appendix – Glossary of terms

AFm	The phase formed in the hydration of Portland cement which is derived from pure mono-sulphoaluminate with the partial substitution of A by F, and SO_4^{2-} by other anions
AFt	The phase formed in the hydration of Portland cement which is derived from pure ettringite with the partial substitution of A by F, and SO_4^{2-} by other anions
Alite	Impure C_3S as found in Portland cement containing other oxides in solid-state substitution
α	Degree of hydration
Belite	Impure C_2S as found in Portland cement containing other oxides in solid-state substitution
C_4AF	Ferrite solid solution (Fss). This represents the iron-containing phase which varies in most cements between C_6A_2F and C_6AF_2 with a mean close to C_4AF
C-S-H	Amorphous calcium silicate hydrate formed in hydration of calcium silicates, or in an impure form in hydration of Portland cement. In the text different types of C-S-H formed during hydration are identified by different nomenclatures; e.g. type I C-S-H, C-S-H(m), etc.
ESCA	Electron spectroscopy for chemical analysis (or X-ray photo-electron spectroscopy)
Ettringite	$C_6A\bar{S}_3H_{32}$ (or $C_3A \cdot 3C\bar{S} \cdot H_{32}$)
Monosulphoaluminate	$C_4A\bar{S}H_{12}$ (or $C_3A \cdot C\bar{S} \cdot H_{12}$)
Portlandite	Calcium hydroxide [$Ca(OH)_2$]
r.h.	Relative humidity
Syngenite	$CaK_2(SO_4)_2 \cdot H_2O$ (or $CK_2\bar{S}_2H$)
w/s	Water-to-solid ratio

3.9 References

1. Jennings, H. M., in *Advances in Cement Technology*, S. N. Ghosh (ed.), p. 349, Pergamon, Oxford (1983).
2. Taylor, H. F. W., *Proc. 8th Int. Cong. Chem. Cem.*, Rio de Janiero, Vol. I, p. 82 (1986).
3. Gartner, E. M. and Gaidis, J. M., in *Materials Science of Concrete I*, J. Skalny (ed.), p. 95, American Ceramic Society, Westerville, Ohio, USA (1989).
4. Massazza, F. and Daimon, M., *Proc. 9th Int. Cong. Chem. Cem.*, New Delhi, Vol. I, p. 383 (1992).
5. Scrivener, K. L. and Wieker, W., *Proc. 9th Int. Cong. Chem. Cem.*, New Delhi, Vol. I, p. 449 (1992).
6. Taylor, H. F. W. *et al.*, *Mater. Struct.*, 17, 457 (1984).
7. Fuiji, K. and Kondo, W., *J. Am. Ceram. Soc.*, 66, C220 (1983).
8. Taylor, H. F. W., *Cement Chemistry*, 2nd edn, Thomas Telford, London (1997).
9. Damidot, D. and Nonat, A., *Proc. 9th Cong. Chem. Cem.*, New Delhi, Vol. III, p. 227 (1992).
10. Long, J. V. P. and Mcconnell, J. D. C., *Miner. Mag.*, 32, 117 (1959).
11. Kantro, D. L., Brunauer, S. and Weise, C. H., *J. Colloid Sci.*, 14, 363 (1959).
12. Davies, R. W. and Young, J. F., *J. Am. Ceram. Soc.*, 58, 67 (1975).
13. Kondo, R. and Ueda, S., *Proc. 5th Int. Symp. Chem. Cem.*, Tokyo, 1968, 11, p. 203 (1969).
14. Kondo, R. and Daimon, M., *J. Am. Ceram. Soc.*, 52, 503 (1969).
15. Damidot, D., Nonat, A. and Barrett, P., *J. Am. Ceram. Soc.*, 73, 3319 (1990).
16. Gartner, E. M. and Jennings, H. M., *J. Am. Ceram. Soc.*, 70, 743 (1987).
17. Barret, P., Menetrier, D. and Bertrandie, D., *Rev. Int. Hautes Temp. Refract.*, 14, 127 (1977).

18. Barret, P., Menetrier, D., Betrandie, D. and Regourd, M., *Proc. 7th Int. Cong. Chem. Cem.*, Paris, Vol. II, II-279 (1980).
19. Barret, P., Betrandie, D. and Menetrier, D., *Proc. 7th Int. Cong. Chem. Cem.*, Paris, Vol. II, II-261 (1980).
20. Menetrier, D. and Barret, P., *Cem. Res. Prog.*, **1981**, 223 (1982).
21. Barret, P. and Menetrier, D., *Cem. Concr. Res.*, 10, 521 (1980).
22. Nachbaur, L., Nkinamubanzi, P.-C., Nonat, A. and Mutin, J. C., *J. Colloid Interface Sci.*, 202, 261–8 (1998).
23. Tadros, M. E., Skalny, J. and Kalyoncu, R. S., *J. Am. Ceram. Soc.*, 59, 344 (1976).
24. Stein, H. N. and Stevels, J. M., *J. Appl. Chem. (London)*, 14, 338 (1964).
25. Stein, H. N., *Il Cemento*, 74, 3 (1977).
26. Jennings, H. M., *J. Am. Ceram. Soc.*, 69, 614 (1986).
27. Fierens, P. and Verhaegen, J. P., *Cem. Concr. Res.*, 6, 287, 337 (1976).
28. Maycock, J. N., Skalny, J. and Kalyoncu, R., *Cem. Concr. Res.*, 4, 835 (1974).
29. Menetrier, D., DSc. Thesis, University Dijon (1977).
30. Young, J. F., Tong, H. S. and Berger, R. L., *J. Am. Ceram. Soc.*, 60, 344 (1977).
31. Thomassin, J. H., Regourd, M., Baillif, P. and Touray, J. C. C. R., *Hebd. Seances Acad. Sci., Ser. C.*, 93, 288 (1978).
32. Regourd, M., Thomassin, J. H., Baillif, P. and Touray, J. C., *Cem. Concr. Res.*, 10, 223 (1980).
33. Menetrier, D., Jawed, I., Sun, T. S. and Skalny, J. P., *Cem. Concr. Res.*, 9, 473 (1979).
34. Stadelman, C., Trettin, R., Wieker, W. and Ramm, M., *Cem. Concr. Res.*, 15, 145 (1985).
35. Vernet, C., Demoulian, E., Gourdin, P. and Hawthrone, F., *Proc. 7th, Int. Cong. Chem. Cem.*, Paris, Vol. II, II-267 (1980).
36. Gauffinet, S., Finot, E. and Nonat, A., *Proc. 2nd Int. RILEM Wkshp. Hydr. Settg.*, Dijon (1997).
37. Nonat, A., Lecoq, X. and Gauffinet, S., *Proc. 10th Int. Cong. Chem. Cem.*, Gotenborg, Vol. 2, Paper 2ii055 (1997).
38. Gauffinet, S., DSc. Thesis, University Dijon (1998).
39. Henderson, D. and Bailey, J. E., *J. Mater. Sci.*, 28, 3681 (1993).
40. Fujii, K. and Kondo, W., *J. Am. Ceram. Soc.*, 57, 492 (1974).
41. Dent Glasser, L. S., Lachowski, E. E., Mohan, K. and Taylor, H. F. W., *Cem. Concr. Res.*, 8, 733 (1978).
42. Dent Glasser, L. S., *Cem. Concr. Res.*, 9, 515 (1979).
43. Damidot, D. and Nonat, A., *Adv. Cem. Res.*, 6, 27 (1994).

44. Barret, P. and Bertrandie, D., *J. Am. Ceram. Soc.*, 71, C113 (1988).
45. Brown, P. W., Harner, C. L. and Prosen, E. J., *Cem. Concr. Res.*, 16, 17 (1986).
46. Jennings, H. M., Dalgleish, B. J. and Pratt, P. L., *J. Am. Ceram. Soc.*, 64, 567 (1981).
47. Jawed, I. and Skalny, J., *J. Colloid Interface Sci.*, 85, 235 (1982).
48. Young, J. F. and Wu, Z.-Q., *J. Am. Ceram. Soc.*, 67, 48 (1984).
49. Damidot, D. and Nonat, A., *Adv. Cem. Res.*, 6, 83 (1994).
50. Ramachandran and Grutzeck, M., *J. Am. Ceram. Soc.*, 71, 91 (1988).
51. Rodger, S. A., Groves, G. W., Clayden, N. J. and Dobson, C. M., *J. Am. Ceram. Soc.*, 71, 91 (1988).
52. Brough, A. R. et al., *J. Mater. Sci.*, 29, 3926 (1994).
53. Avrami, M., *J. Phys., Chem.*, 7, 1103 (1939); 8, 212 (1940).
54. Brown, P. W., Pommersheim, J. and Frohnsdroff, G., *Cem. Concr. Res.*, 15, 35 (1985).
55. Fierens, P., Kabuema, Y., Orszagh, J., Tenoutasse, N. and Tirlocq, J., *Proc. 8th Int. Cong. Chem.*, Rio de Janiero, Vol. III, 17 (1986).
56. Klyusov, A. A., *Cem. Concr. Res.*, 24, 127 (1994).
57. Ridge, M. J. and Beretka, J., *Rev. Pure Appl. Chem. (London)*, 19, 17 (1969).
58. Gartner, E. M., unpublished work.
59. Gartner, E. M., *Cem. Concr. Res.*, 27, 665 (1997).
60. Double, D. D., Hellawell, A. and Perry, S. J., *Proc. R. Soc. London*, 359, 435 (1978).
61. Young, J. F., *Cem. Concr. Res.*, 2, 415 (1972).
62. Goodbrake, C. J., Young, J. F. and Berger, R. L., *J. Am. Ceram. Soc.*, 62, 488 (1979).
63. Ramachandran, V. S., *Il Cements*, 82, 129 (1985).
64. Gauffinet, S., Finot, E., Lesniewska, E. and Nonat, A., *C. R. Acad. Sci. Paris, Earth Planet. Sci.*, 327, 231 (1998).
65. Pommersheim, J. and Clifton, J. F., *Cem. Res. Prog.*, **1979**, 281 (1980).
66. Jelenic, I., in *Advances in Cement Technology*, S. N. Gosh (ed.), p. 397, Pergamon, Oxford (1983).
67. Bezjak, A. and Jelenic, I., *Cem. Concr. Res.*, 10, 553 (1980); *Proc. 7th Int. Cong. Chem. Cem.*, Paris, Vol. II, II-111 (1980).
68. Ginstling, A. M. and Brounshtein, B. I., *J. Appl. Chem. VSSR*, 23, 1327 (1950).
69. Taplin, J. H., *Proc. 4th Int. Symp. Chem. Cem.*, Washington, 1960, Vol. I, p. 263 (1962).
70. Knudsen, T., *Proc. 7th Int. Cong. Chem. Cem.*, Paris, Vol. II, I-170 (1980).
71. Brown, P. W., *J. Am. Ceram. Soc.*, 72, 1829 (1989).
72. Taylor, H. F. W., *Mater. Sci. Monogr.*, 28A, 39 (1985).
73. Bonen, D. and Diamond, D. *J. Am. Ceram. Soc.*, 77, 1875 (1994).

74. Diamond, S., *Hydraulic Cement Pastes: Their Structure and Properties*, p. 55, Cement and Concrete Association, Slough, UK (1976).
75. Hirljac, J., Wu, Z.-Q. and Young, J. F., *Cem. Concr. Res.*, 13, 877 (1983).
76. Richardson, I. G., Brough, A. R., Brydson, R., Groves, G. W. and Dobson, C. J., *J. Am. Ceram. Soc.*, 76, 2285 (1993).
77. Kondo, R., Daimon, M., Sakai, E. and Oshiyama, H. *J. Appl. Chem. Biotechnol.*, 27, 191 (1977).
78. Kantro, D. L., Brunauer, S. and Weise, C. H., *J. Phys. Chem.*, 66, 1804 (1962).
79. Locher, F. W., *Symp. Struct. Portland Cem. Paste Concr.*, Highway Res. Bd. Spec. Rpt. 90, 300 (1966).
80. Odler, I. and Skalny, J., *J. Appl. Chem. Biotechnol.*, 23, 661 (1973).
81. Kantro, D. L., *J. Test Evaln.*, 3, 312 (1975).
82. Menetrier, D., Jawed, I. and Skalny, J., *Silicates Ind.*, 12, 243 (1980).
83. Ramachandran, V. S., *Il Cemento*, 90, 73 (1993).
84. Jennings, H. M., Taleb, H., Frohnsdorff, G. and Clifton, J. F., *Proc. 8th Int. Cong. Chem. Cem.*, Rio de Janiero, Vol. III, p. 239 (1986).
85. Odler, I. and Dorr, H., *Cem. Concr. Res.*, 9, 277 (1979).
86. Thomas, N. L., Jamieson, D. A. and Double, D. D., *Cem. Concr. Res.*, 11, 143 (1981).
87. Jost, K. H. and Ziemer, B., *Cem. Concr. Res.*, 14, 177 (1984).
88. Trettin *et al.*, *Cem. Concr. Res.*, 21, 757 (1991).
89. Menetrier, D., McNamara, D. K., Jawed, I. and Skalny, *J. Cem. Concr. Res.*, 10, 107 (1980).
90. Menetrier, D., Jawed, I. Sun, T. S. and Skalny, *J. Cem. Concr. Res.*, 10, 425 (1980).
91. Tong, Y., Du, H. and Fei, L., *Cem. Concr. Res.*, 20, 986 (1990).
92. Cong X. and Kirkpatrick, R. J. *Cem. Concr. Res.*, 23, 1065 (1993).
93. Tong, H.-S. and Young, J. F., *J. Am. Ceram. Soc.*, 60, 321 (1977).
94. Young, J. F. and Tong, H.-S., *Cem. Concr. Res.*, 7, 627 (1977).
95. Berger, R. L., Young, J. F. and Lawrence, F. V., *Cem. Concr. Res.*, 3, 689 (1973).
96. Chopra, S., Ghose, A. and Young, J. F., *J. Mater. Sci.*, 18, 2905–14 (1983).
97. Chan, C. J., Kriven, W. M. and Young, J. F., *J. Am. Ceram. Soc.*, 71, 713–9 (1988).
98. Chan, C. J., Kriven, W. M. and Young, J. F., *J. Am. Ceram. Soc.*, 75, 1621–7 (1992).
99. Gawlicki, M. and Nocum-Wczelik, W., *Proc. 7th Int. Cong. Chem. Cem.*, Paris, Vol. I, p. 161–5 (1980).
100. Young, J. F., in *Concrete: From Material to Structure*, J.-P. Bournazel and Y. Malier (eds), p. 1–15, RILEM, Paris (1998).
101. Jelenic, I. and Bezjak, A., *Cem. Concr. Res.*, 7, 627 (1977).
102. Bensted, J., *Cem. Concr. Res.*, 8, 73–6 (1978).
103. Sun, G. K., Young, J. F., Matkovic, B., Paljevic, M. and Mikoc, M., *Adv. Cem. Res.*, 6 (24) 161–4 (1994).
104. Bukowski, J. and Berger, R. L., *Cem. Concr. Res*, 11, 467 (1981).
105. Chromy, C., *Zem.-Kalk-Gips*, 8, 382–8 (1970).
106. Muller, A., Stark, J. and Rumpler, K. R., *Zem.-Kalk-Gips*, 38, 303–4 (1985).
107. Kriven, W. M., Chan, C. J. and Barinek, E. A., *Adv. Ceram.*, 24A, 145–55 (1988).
108. Ishida, H. and Mitsuda, T., in *Materials Science of Concrete*, V, J. Skalny and S. Mindess (eds), pp. 1–44 and references therein, American Ceramic Society, Westerville, OH (1998).
109. Ishida, H., Sasaki, K. and Mitsuda, T., *J. Am. Ceram. Soc.*, 75, 353–8 (1992).
110. Roy, D. M. and Oyefesobi, S. O., *J. Am. Ceram. Soc.*, 60, 178–80 (1977).
111. Nettleship, I., Shull, J. L. Jr., and Kriven, W. M., *J. Eur. Ceram. Soc.*, 11, 291–8 (1993).
112. Lee, S. -J., Benson, E. A. and Kriven, W. M., *J. Am. Ceram. Soc.*, 82(8), 2049–55 (1999).
113. Hong, S. H. and Young, J. F., *J. Am. Ceram. Soc.*, 82, 1681–6 (1999).
114. Okada Y., Ishida, H., Sasaki, K., Young, J. F. and Mitsuda, T., *J. Am. Ceram. Soc.*, 76, 1313–8 (1994).
115. Brown, P. E. *et al.*, *Mater. Struct.*, 19, 137 (1986).
116. Breval, E., *Cem. Concr. Res.*, 6, 129 (1976); 7, 297 (1977).
117. Schwiete, H. E., Ludwig, U. and Jager, P., *Symp. Struct. Portland Cem. Paste Concr.*, Washington, 1965, Highway Res. Bd. Spec. Rpt. 90, 353 (1966).
118. Young, J. F., *Transp. Res. Rec.*, 564, 1 (1976).
119. Tashiro, C. and Oba, J., *Cem. Concr. Res.*, 9, 253 (1979).
120. Milestone, N. B., *Cem. Concr. Res.*, 6, 89 (1976).
121. Ramachandran, V. S., *Cem. Concr. Res.*, 3, 41 (1973).
122. Corstanje, W. A., Stein, H. N. and Stevels, J. M., *Cem. Concr. Res*, 3, 791 (1973); 4, 193, 417 (1974).
123. Skalny, J. and Tadros, M. E., *J. Am. Ceram. Soc.*, 60, 174 (1977).
124. Tadros, M. E., Jackson, W. Y. and Skalny, J., in *Proc. Int. Conf. Colloids and Surfaces*, Puerto Rico, M. Kerker (ed.), Vol. IV, p. 211, Academic Press, New York (1976).
125. Brown, P. W. and Lacroix, P., *Cem. Concr. Res.*, 19, 879–84 (1989).
126. Pommersheim, J. and Chang, J., *Cem. Concr. Res.*, 18, 911–22 (1988).

127. Mehta, P. K., *J. Am. Ceram. Soc.*, **56**, 315 (1973); *Cem. Concr. Res.*, **3**, 1 (1973).
128. Fierens, P., Verhaegen, A. and Verhaegen, J. P. *Cem. Concr. Res.*, **4**, 381 (1974).
129. Locher, F. W., Richartz, W. and Sprung, S., *Zem.-Kalk-Gips*, **29**, 435 (1976).
130. Tinnea, J. and Young, J. F., *J. Am. Ceram. Soc.*, **60**, 387 (1977).
131. Collepardi, M., Baldini, G., Pauri, M. and Corradi, M., *Cem. Concr. Res.*, **8**, 571 (1978); *Il Cemento*, **75**, 169 (1978).
132. Feldmand, R. F. and Ramachandran, V. S., *Mag. Concr. Res.*, **18**, 185–96 (1966).
133. Chatterji, S., *Proc. 7th Int. Cong. Chem. Cem.*, Paris, 1980, Vol. IV, p. 465 (1981).
134. Traetteberg, A. and Grattan-Bellew, P. E., *J. Am. Ceram. Soc.*, **58**, 221 (1975).
135. Brown, P. W., Lieberman, L., Frohnsdforff, G., *J. Am. Ceram. Soc.*, **67**, 293–5 (1984).
136. Spierings, G. A. C. M. and Stein, H. N., *Cem. Concr. Res.*, **6**, 265, 487 (1976).
137. Tang, F. and Gartner, E. M., *Adv. Cem. Res.*, **1**, 67–74 (1988).
138. Boikova, A. I., Domansky, A. I., Paramonova, V. A., Stavitskaja, G. P. and Nikushchenko, V. M., *Cem. Concr. Res.*, **7**, 483 (1977).
139. Boikova, A. I., Grischenko, L. V. and Domansky, A. I., *Proc. 7th Int. Cong. Chem. Cem.*, Paris, 1980, Vol. IV, p. 459 (1981).
140. Regourd, M., *Il Cemento*, **75**, 323 (1978); *Proc. Eng. Found. Conf.*, Rindge, 1979, J. Skalny (ed.), p. 41, (1980).
141. Bilanda, N., Fierens, P., Tenoutasse, N. and tir-locq. J., *Proc. 7th Int. Concr. Chem. Cem.*, Paris, 1980, Vol. IV, p. 607 (1981).
142. Regourd, M., Hornain, H. and Mortureaux, B. *Cem. Concr. Res.*, **6**, 733 (1976).
143. Mortureaux, B., Hornain, H. and Regourd, M. *Proc. 7th Int. Concr. Chem. Cem.*, Paris, 1980, Vol. IV, p. 575 (1981).
144. Klavins, Z., Alksnis, F. and Kauke, A., *Vopr. Stroit.*, **6**, 93 (1978); *Neorg. Stekla Pokrytiya Mater.*, **4**, 117 (1979).
145. Holton, C. L. M. and Stein, H. N., *Cem. Concr. Res.*, **7**, 291 (1977).
146. Collepardi, M., Baldini, G., Pauri, M. and Corradi, M., *Cem. Concr. Res.*, **8**, 741 (1978).
147. Tenoutasse, N., *Proc. 5th Int. Symp. Chem. Cem.*, Tokyo, Vol. II, P. 372 (1968).
148. Ramachandran, V. S. and Zhang, C. M., *Mater. Struct.*, **19**, 437 (1986).
149. Negro, A. and Staffieri, L. *Zem.-Kalk-Gips*, **32**, 83 (1979).
150. Jawed, I., Goto, S. and Kondo, R. *Cem. Concr. Res.*, **6**, 441 (1976).
151. Bensted, J., *Il Cemento*, **74**, 45 (1976).
152. Rogers, D. E. and Aldridge, L. P., *Cem. Concr. Res.*, **7**, 399 (1977).
153. Tamas, F. D. and Vertesm A., *Cem. Concr. Res.*, **3**, 575 (1973).
154. Teoreanu, I., *et al.*, *Il Cemento*, **76**, 19 (1979).
155. Sanzhaasuren, R. and Andreeva, E. P., *Colloid J. USSR*, **39**, 197, 494, 738, 802 (1977).
156. Collepardi, M., Monosi, S., Moricani, A. and Corradi, M., *Cem. Concr. Res.*, **9**, 431 (1979).
157. Ramachandran, V. S. and Beaudoin, J. J., *Proc. 7th Int. Cong. Chem. Cem.*, Paris, Vol. II, II-25 (1980).
158. Fukuhara, M., Goto, S., Asaga, K., Daimon, M. and Kondo, R., *Cem. Concr. Res.*, **11**, 407 (1981).
159. Scrivener, K. L., *Proc. 8th Int. Cong. Chem. Cem.*, Rio de Janiero, Vol. III, 389 (1986).
160. Gartner, E. M. and Meyers, D. F., *J. Am. Ceram. Soc.*, **76**, 1521–30 (1993).
161. Brown, P. W., *J. Am. Ceram. Soc.*, **70**, 493 (1987).
162. Struble, L. J. and Lei, W-G., *Advd. Cem.-bsd. Mater.*, **2**, 244 (1995).
163. Gartner, E. M., Tang, F. J. and Weiss, S. J., *J. Am. Ceram. Soc.*, **68**, 667 (1985).
164. Midgley, H. G. and Rosaman, D., *Proc. 4th Int. Symp. Chem. Cem.*, Washington, 1960, Vol. I, 259 (1962).
165. Poellman, H. and Kuzel, H. J., *Cem. Concr. Res.*, **20**, 941 (1990).
166. Scrivener, K. L., in *Materials Science of Concrete* J. Skalny (ed.), Vol. I, p. 127, American Ceramic Society, Westerville, Ohio, USA (1989).
167. Kanare, H. and Gartner, E. M., *Cem. Res. Prog.*, 213 (1984).
168. Barnes, P., Ghose, A. and Mackay, A. L., *Cem. Concr. Res.*, **10**, 630 (1980).
169. Barnes, B. D., Diamond, S. and Dolch, W. L. *Cem. Concr. Res.*, **8**, 263 (1978).
170. Pratt, P. L. and Jennings, H. M., *Annu. Rev. Mater. Sci.*, **11**, 123 (1981).
171. Dalgleish, B. J., Ghose, A., Jennings, H. M. and Pratt, P. L., *Proc. 7th Int. Cong. Chem. Cem.*, Paris, 1980, Vol. IV, 137 (1981).
172. Ono, Y. *Microscopy of Clinker and Cement* (short course), Portland Cement Association, Chicago (1978).
173. Tang, F. J., *J. PCA R & D Bull. RD105J* Portland Cement Association, Skokie, IL (1992).
174. Soroka, I. and Stern, N., *Cem. Concr. Res.*, **6**, 367 (1976).
175. Ramachandran, V. S. and Zhang, C.-M., *Proc. 8th Int. Cong. Chem. Cem.*, Rio de Janiero, Vol. VI, 178 (1986).
176. Tezuka, Y, Gomes, D., Martins, J. M. and Djanikan, J. G., *Proc. 9th Int. Cong. Chem. Cem.*, New Delhi, Vol. V, 53 (1992).
177. Kawada, N. and Nemoto, A., *Zem.-Kalk-Gips*, **21**, 65 (1967).

178. Odler, I. and Schuppstuhl, *J. Cem. Concr. Res.*, **12**, 13 (1982).
179. Mindess, S. and Young, J. F., *Concrete*, Prentice Hall, Englewood Cliffs, NJ (1981).
180. Blackie, A. D., 'The use of PFA in concrete', in *Proc. Int. Symp.*, Leeds, J. G. Cabrera and A. R. Cusen (eds), 289 (1982).
181. Verbeck, G. and Helmuth, R. *Proc. 5th Int. Symp. Chem. Cem.*, Tokyo, 1968, Vol. III, 1 (1969).
182. Price, W. H., *J. Am. Concr. Inst.*, **47**, 417 (1951).
183. Boomiz, A., Vernet, C. and Tenjoudji, F. C., *Advd. Cem.-bsd. Mater.*, **3**, 94 (1996).
184. Adams, L. D., *Cem. Concr. Res.*, **6**, 293 (1976).
185. Lea, F. M., *The Chemistry of Cement and Concrete*, 3rd edn, Edward Arnold, Glasgow (1970).
186. Rhebinder, P. A., *Symp. Chem. Cem.*, State Publications of Literature on Structural Materials, 125, Moscow (1956).
187. Segalova, E. E., Rhebinder, P. A. and Lukyanova, O. I., *Vestink Moscow Univ.*, **9**, *Ser. Phys. Math. Nat. Sci.*, **1**, 17 (1954).
188. Locher, F. W., *Proc. 7th Int. Cong. Chem. Cem.*, Paris, 1980, Vol. IV, 49 (1981).
189. Parker, T. W., *J. Soc. Chem. Ind. London*, **58**, 203 (1939).
190. Greene, K. T., *Proc. 4th Int. Symp. Chem. Cem.*, Washington, 1960, Vol. I, 359 (1962).
191. Hansen, W. C. and Pressler, E. E., *Ind. Eng. Chem.*, **39**, 1280 (1947).
192. Lawrence, C. D., *Proc. Symp. Struct. Port. Cem. Paste Concr.*, Washington DC, 1965, Spec. Rpt. 90, Highway Res. Bd., 378 (1966).
193. Collepardi, M., in *Concrete Admixtures Handbook*, V. S. Ramachandran (ed.), p. 116, Noyes Publication. (1984).
194. Ramachandran, V. S., *Cem. Concr. Res.*, **6**, 623–31 (1976).
195. Chiesi, C. W., Gartner, E. M. and Myers, D. F., *Proc. 14th Int. Conf. Cem. Microsc.*, Costa Mesa, CA, pp. 388–401, Int. Cem. Microsc. Assoc., Duncanville, TX (1992).
196. Schultz, R. J., *Shotcrete for Ground Support*, SP-54, p. 45, American Concrete Institute, Farmington Hills, Michigan, USA (1977).
197. Young, J. F., in *Creep and Shrinkage in Concrete Structures*, Z. P. Bazant and F. Wittman (eds), p. 3, John Wiley & Sons, Chichester (1981).
198. Taylor, H. F. W., *Proc. 8th Int. Cong. Chem. Cem.*, Rio de Janeiro, Vol. I, 82 (1986).
199. Diamond, S., *Proc. 8th Int. Cong. Chem. Cem.*, Rio de Janeiro, Vol. I, 122 (1986).
200. Scrivener, K. L. *World Cem*, **28**(9), 92 (1997).
201. Garboczi, E. J. and Bentz, D. P., in *Materials Science of Concrete II*, J. Skalny and S. Mindess (eds), p. 249, American Ceramic Society, Westerville, OH (1991).
202. Garboczi, E. J. and Bentz, D. P., in *Materials Science of Concrete II*, J. P. Skalny and S. Mindess (eds), pp. 249–77, American Ceramic Society, Westerville, OH (1989).
203. Diamond, S. and Bonen, D., *J. Am. Ceram. Soc.*, **76**, 2993–9 (1993).
204. Scrivener, K. L. and Pratt, P. L., *Br. Ceram. Soc. Proc.*, **35**, 207 (1984).
205. Scrivener, K. L., Ph.D. Thesis, University of London (1984).
206. Rilem Committee 1C170-CSH, Unpublished Draft Recommendations.
207. Dalgleish, B. J., Pratt, P. L. and Moss, R. I., *Cem. Concr. Res.*, **10**, 665 (1980).
208. Kalousek, G. L., *Mater. Res. Stand.*, **5**, 292 (1965).
209. Copeland, L. E. and Kantro, D. L., *Proc. 5th Int. Symp. Chem. Cem.*, Vol. II, 387–421, Cement Association of Japan, Tokyo (1968).
210. Rayment, D. L. and Majumdar, A., *J. Cem. Concr. Res.*, **12**, 753 (1982).
211. Harrison, A. M., Winter, N. B. and Taylor, H. F. W., *Proc. 8th Int. Cong. Chem. Cem.*, Rio de Janeiro, Vol. 4, 170 (1986).
212. Richardson, I. G. and Groves, G. W., *J. Mater. Sci.*, **28**, 265 (1993).
213. Lachowski, E. E., Mohan, K., Taylor, H. F. W. and Moore, A. E., *J. Am. Ceram. Soc.*, **63**, 447 (1980); **64**, 319 (1981).
214. Taylor, H. F. W., Mohan, K. and Moir, G. K., *J. Am. Ceram. Soc.*, **68**, 680 (1985).
215. Taylor, H. F. W., *J. Am. Ceram. Soc.*, **69**, 494 (1986).
216. Taylor, H. F. W., *in Advances in Cement Manufacture and Use*, E. M. Gartner (ed.), p. 295, Engrg. Found., NY (1989).
217. Richardson, I. G., *Cem. Concr. Res.*, **29**, 1131 (1999).
218. Sun, G. K. and Young, J. F., in *Handbook of Analytical Techniques in Concrete Science and Technology*, V. S. Ramachandran and J. J. Beaudoin (eds), pp. 629–57, Noyes Publication, Park Ridge New Jersey, USA. (2000).
219. Grimmer, A. R., in *Applications of NMR to Cement Science*, P. Colombet and A. R. Grimmer (eds), p. 77, Gordon and Breach, London (1994).
220. Richardson, I. G. and Groves, G. W., *Cem. Concr. Res.*, **22**, 1001 (1992).
221. Stade, H. and Wieker, W., *Z. Anorg. Allg. Chem.*, **465**, 55 (1980).
222. Cong, X. and Kirkpatrick, R. J. *Advd. Cem.-bsd. Mater.*, **3**, 144 (1996).
223. Litvan, G. G., *Cem. Concr. Res.*, **6**, 139 (1976).
224. Rarick, R. L., Bhatty, J. I. and Jennings, H. M., in *Materials Science of Concrete IV*, J. Skalny and S. Mindess (eds), pp. 1–40 (1995).

225. Thomas, J. J., Jennings, H. M. and Allen, A., *J. Concr. Sci. Engr.*, **1**, 45–64. (1999).

226. Jennings, H. M. and Tennis, P. D., *J. Am. Ceram. Soc.*, **77**, 3161–72 (1994).

227. Bentz, D. P. and Garboczi, E. J., *Cem. Concr. Res.*, **21**, 325 (1991).

228. Berger, R. L. and Macgregor, J. D., *Cem. Concr. Res.*, **2**, 43 (1972).

229. Groves, G. W., '*Microstructural Development During Hydration of Cement*', in *Symp. Proc. L. J. Struble and P. W. Brown* (eds), Vol. 85 3–12, Materials Research Society, Pittsburg, PA (1987).

Chapter four

Calcium aluminate cements

John Bensted

4.1 Introduction

The name 'high alumina cement' (HAC) first started to become superseded by the alternative 'calcium aluminate cement' (CAC) in the late 1970s. It is sometimes erroneously stated that HAC is simply the dark grey/black variety (Al_2O_3~35–40 per cent) whilst the white refractory grades (with Al_2O_3 contents ranging from ~50 per cent to ~80 per cent) are other members of a family of CACs. In fact, the white CACs were – and still are, like the dark grey/black product – often referred to as HACs and more specifically as white HACs. This was exemplified in Robson's classic text [1] and at the 1970 Belfast Meeting on Structures and Properties of Cements [2]. The name 'calcium aluminate cement' is employed in this chapter because of its increasingly widespread usage, including within the forthcoming European Standard. However, the name 'high alumina cement' is still widely utilized in the construction industry and is given in British Standard 915: 1972 (amended 1995) [3]. Thus the names 'calcium aluminate cement' and 'high alumina cement' are in reality interchangeable and in common use.

The origins of CAC, otherwise date back as far as the work of Frémy who reported in 1865 that good hydraulic properties were obtained from various melts of lime and alumina [4]. Commercialization of CACs really began in 1910 with the work of Bied [5,6] at the laboratories of the *Société J. and A. Pavin de Lafarge* in Le Teil, France. These investigations originally centred upon the production of cements resistant to sulphate attack and followed on from work undertaken on mortar and concrete deterioration in sulphate-containing ground waters and sea water. High early-strength properties and good sulphate resistance were found for CACs. Refractory applications followed in due course.

CACs are based upon the presence of hydratable aluminates, principally monocalcium aluminate CA (see Table 4.1 for the Cement nomenclature system) and to some extent mayenite $C_{12}A_7$.

There are two major types of CACs:

1. the 'normal product', which is dark grey or black in colour, and can be employed over a wide temperature range; and
2. the white varieties that are utilized primarily for refractory purposes at high temperatures and sometimes also for appropriate decorative usage.

TABLE 4.1 Cement chemical nomenclature[a]

A = Al$_2$O$_3$	H = H$_2$O	$\overline{C}$ = CO$_2$
C = CaO	M = MgO	$\overline{S}$ = SO$_3$
F = Fe$_2$O$_3$	S = SiO$_2$	
f = FeO		

Examples

Anhydrous phases

CA	= CaOAl$_2$O$_3$	= CaAl$_2$O$_4$
C$_{12}$A$_7$	= 12CaO · 7Al$_2$O$_3$	= Ca$_{12}$Al$_{14}$O$_{33}$
CA$_2$	= CaO · 2Al$_2$O$_3$	= CaAl$_4$O$_7$
C$_4$AF	= 4CaO · Al$_2$O$_3$ · Fe$_2$O$_3$	= Ca$_2$AlFeO$_5$
β-C$_2$S	= β-2CaO · SiO$_2$	= β-Ca$_2$SiO$_4$
C$_2$MS$_2$	= 2CaO · MgO · 2SiO$_2$	= Ca$_2$MgSi$_2$O$_7$
C$_2$AS	= 2CaO · Al$_2$O$_3$ · SiO$_2$	= Ca$_2$Al$_2$SiO$_7$
CT	= CaO · TiO$_2$	= CaTiO$_3$
f	= FeO	= FeO
C$_6$A$_4$(M,f)S	= 6CaO · 4Al$_2$O$_3$(Mg · Fe)O · SiO$_2$	= Ca$_6$Al$_8$(Mg · Fe)SiO$_{21}$

Hydrated phases

CA	= CaO · Al$_2$O$_3$ · 10H$_2$O	= Ca[Al(OH)$_4$]$_2$6H$_2$O
C$_2$AH$_8$	= 2CaO · Al$_2$O$_3$ · 8H$_2$O	= Ca$_2$[Al(OH)$_5$]$_2$3H$_2$O
C$_3$AH$_6$	= 3CaO · Al$_2$O$_3$ · 6H$_2$O	= Ca$_3$[Al(OH)$_6$]$_2$H$_2$O
C$_4$AH$_{13}$	= 4CaO · Al$_2$O$_3$ · 13H$_2$O	= Ca$_2$[Al(OH)$_7$]3H$_2$O
C$_2$ASH$_8$	= 2CaO · Al$_2$O$_3$ · SiO$_2$·8H$_2$O	= Ca$_2$Al$_2$SiO$_7$8H$_2$O
AH$_3$	= Al$_2$O$_3$ · 3H$_2$O	= Al(OH)$_3$

[a] In the cement chemical nomenclature system, capital letters are used to represent the oxides of the elements. Where overlap can arise a bar is placed over the capital letters to avoid ambiguity. C-S-H is used to represent amorphous calcium silicate hydrate, which is non-stoichiometric. The hyphens indicate that no precise chemical composition is employed. It is often represented approximately as 'C$_3$S$_2$H$_3$' or 'Ca$_3$Si$_2$O$_7$3H$_2$O'.

In chemical terms, these two types are broadly similar in their main cementing behaviour, the white CAC having more alumina (also present as calcium dialuminate CA$_2$) – and the dark grey/black CAC more iron-containing phases.

White CACs normally contain more alkalis than dark grey/black CACs, unlike the relative position for white and ordinary Portland cements (OPC). Total K$_2$O + Na$_2$O quantities for dark grey/black CACs should be lower than 0.4 per cent for optimum performance and are normally well below this figure. For white CACs, alkali limits can be tolerated to higher levels such as 0.5–0.7 per cent maximum. White CACs normally have higher alkali levels than dark grey/black CACs, because the purer white grades of bauxite utilized in white CAC manufacture commonly have higher alkali contents than the less

pure grades employed in dark grey/black CAC production [7].

However, SO$_3$ levels are closer to each other in these two main classes of CAC. The dark grey/black CACs usually contain 0.02–0.07 per cent SO$_3$, whereas the white CACs normally have around 0.03 per cent SO$_3$. Such low intrinsic sulphate levels are beneficial from the viewpoint of resistance to sulphate attack (see under Section 4.11: Durability) [7].

4.2 Manufacture

Dark grey/black CAC is usually produced by fusing a mixture of calcium carbonate (normally limestone) and ferruginous bauxite to completion in reverberatory furnaces at 1500–1600°C. These furnaces are fired by pulverized coal, oil, natural gas or a combination of some of these fuels. The totally molten product is allowed to pour out of a tap hole at the base of the furnace onto moving trays or pans and permitted to cool down relatively slowly. The clinker thus produced is then crushed in ball mills to the required surface areas – for instance not less than $225\,m^2/kg$ in accordance with BS 915 [3]. No material other than water should be added subsequent to clinker production. Appropriate grinding aids are allowed to facilitate pulverization during grinding. Calcium sulphate addition is not permitted since, if present in any of its forms (gypsum, hemihydrate/soluble anhydrite, insoluble anhydrite) in appreciable quantities within the CAC, it would cause an undesirable quick set [8]. Some typical physical and chemical data for dark grey/black CAC are given in Table 4.2. Dark grey/black CAC is sometimes alternatively manufactured in rotary kilns by sintering or in electric arc furnaces.

White CAC is produced from selected purer raw materials (both limestone and bauxite) than for grey CAC. Also, the ratio of limestone to bauxite is varied for producing different grades of white CAC. These white CACs are usually

TABLE 4.2 Typical data for dark grey/black calcium aluminate cement

Compressive strength		Setting time (min)	
Time	Value (MPa)	Initial	Final
6 h	4.8	200–300	< 360
24 h	70.4		
3 days	77.8		
7 days	80.2		

Chemical analysis

	(%)		(%)
SiO_2	3.11	Na_2O	0.09
Al_2O_3	39.72	K_2O	0.15
Fe_2O_3	13.12	CO_2	0.18
FeO	3.17	SO_3	0.03
CaO	37.10	LOI @ 550°C	0.53
MgO	0.58	Free CaO	0.08

Specific Surface Area (Lea-Nurse)
$338\,m^2/kg$

LOI: Loss on ignition.

produced in small kilns with fuel oil firing in order to avoid contamination of the white product. They are normally utilized as refractory products, but increasingly now there is a demand for decorative products such as garden ornaments. Coloured CAC products are made by blending pigments with the white CACs. Occasionally dark grey/black CACs are blended with pigments to produce dark shades of coloured products.

A CAC manufacturing plant is depicted in Figure 4.1.

4.3 Phase composition

Dark grey/black CAC consists primarily of the following phases: monocalcium aluminate CA (the principal hydraulic constituent), mayenite $C_{12}A_7$, melilite (gehlenite C_2AS – åkermanite C_2MS_2 solid solution), belite (larnite) β-C_2S, pleochroite glass (sometimes represented approximately as C_6A_4 $(M,f)S$,) ferrite C_4AF, wüstite f, perovskite CT, and traces of sulphides, phosphides and metallic iron. Pleochroite has a very variable composition, but imparts very little hydraulicity. Perovskite has no hydraulic properties.

Small quantities of β-C_2S are usually formed in CAC melts, which solidify upon cooling. Despite reducing conditions being normal during the manufacturing process, there are no reports of the feebly hydraulic γ-C_2S forming instead of, or in addition to, β-C_2S. Although reducing conditions might be considered to encourage γ-C_2S (shannonite) formation, good raw material homogenization favours production of the β-variety, which has a superior hydraulicity to shannonite.

White CAC, due to the different processing conditions and use of purer raw materials, also commonly contains calcium dialuminate CA_2 and α-alumina (α-A). CA is the main constituent, although it is often present in smaller relative amounts than in dark grey/black CAC. Dark phases such as ferrite C_4AF exist in negligible quantities in white CACs.

None of the phases is pure, since other ions present can substitute for the major oxides. For example, CA in a grey CAC showed the following

FIGURE 4.1 View of CAC manufacturing plant at Dunkerque, France. Photograph supplied by Lafarge Aluminates.

analysis: CaO – 34.5%, Al_2O_3 – 59.5%, Fe_2O_3 – 4.9%, SiO_2 – 0.5%, TiO_2 – 0.1%.

A high CA content is desirable for high early strength development, with 60–70 per cent being quite common in dark grey/black CAC. $C_{12}A_7$ is normally present in quantities ~2–5 per cent and causes a greater rate and degree of hydration of CA (Chakraborty et al. [9]). Ferrite may be present ~10–20 per cent or more, but melilite, belite, wüstite and perovskite are normally found in less than 10 per cent amounts. In white CAC, CA may actually be lower than 60–70 per cent, with α-alumina and CA_2 being significant. $C_{12}A_7$ is also present in white CAC, but the other aforementioned phases are mostly much reduced [10].

Phase determinations are more difficult than for Portland cement, since – despite numerous attempts – there is as yet no equivalent to the Bogue calculation for the approximate estimation of the main phases. X-ray diffraction and optical microscopy are the main techniques utilized for determining the principal phases present in CAC. Care needs to be exercised with these determinations if just one technique is employed [10].

4.4 Physical properties

The specific gravity of the various types of CAC may vary from just under 3.0 for the white CAC to *circa* 3.25 for ordinary dark grey/black HAC. The iron compounds present in the cement have specific gravities which can range from 3.7 to 4.9, whilst the average for the aluminates and silicates present is under 3.9. Hence the iron content primarily influences the specific gravity of the CAC. Table 4.3 gives published data for specific gravity

TABLE 4.3 Densities of cement materials [8,16]

Unhydrated		Hydrated	
CA	2.98	CAH_{10}	1.72
CA_2	2.91	α-C_2AH_8	1.95
C_4AF	3.77	β-C_2AH_8	1.97
β-C_2S	3.28	C_3AH_6	2.52
C_2AS	3.04	C_4AH_{13}	2.02
Glass	3.0	AH_3	2.44
$C\bar{C}$	2.71	'$C_3S_2H_3$'	2.44
		CAC@18°C	2.11
		CAC@45°C	2.64

TABLE 4.4 Particle size distribution – fractions (wt%) remaining after sieving dark grey/black CACs

Fraction passing (µm)	Cement			
	A	B	C	D
100	98	97	98	99
64	86	87	86	90
32	69	69	70	70
16	49	51	52	51
8	33	36	36	33
4	21	23	24	20
2	11	12	12	10

BS 915 only requires that not more than 8 per cent is retained on a 90 µm sieve, or that the specific surface is not less than 225 m²/kg.

(relative density) of the main cement compounds [8,11]. For comparison, the specific gravity of Portland cements lies in the range ~2.95–3.20.

Fineness of CAC is usually measured by the Blaine surface area method and is typically *circa* 315 m²/kg for dark grey/black CAC. Particle size distribution is different from that of Portland cements with *circa* 50 per cent of CAC consisting of particles below 30 microns in size. For the refractory grades of CAC, the surface areas tend to be higher, namely 400 m²/kg for 50 per cent Al_2O_3 content, 420 m²/kg for 70 per cent Al_2O_3 and above 800 m²/kg for 80 per cent Al_2O_3. Some typical particle size distribution figures for dark grey/black CAC are given in Table 4.4.

Soundness tests are easily satisfied by CAC. In BS 915: 1972 (amended 1995) [3], for example, the maximum expansion in the Le Chatelier test is limited to 10 mm, but actual values tend to be in the range 0–1 mm. The low values arise from the lack of free calcium and magnesium oxides in these fused cements and these are advantageous for most practical purposes.

CACs are importantly relatively slow setting but rapid-hardening. The compressive strength achieved at 1 day often exceeds that of Portland cement at 28 days. The slow setting characteristics allow suitable time for placement in most instances without any additional need for set retarders. Slow setting also permits plenty of time for the initiation of regular hydrate growth, which facilitates the high initial strength development observed. Setting can be attributed to the loss in mobility of water in the aqueous phase as

hydrate formation begins to take place appreciably, which then leads on to the development of compressive strength.

4.5 Hydration

The reactions of the anhydrous phases with water usually appear to arise by congruent dissolution followed by precipitation of hydration products. CA is the principal cementing component. In addition $C_{12}A_7$ (mayenite), C_4AF (ferrite) and β-C_2S (belite) exhibit good hydraulicity. Melilite (C_2AS-C_2MS_2 solid solution), CA_2 and pleochroite glass are weakly hydraulic. CT is non-hydraulic [10].

Consider firstly the hydration behaviour of CA, the principal hydraulic phase.

- Below *circa* 15°C the initial hydration product is CAH_{10}, which has excellent binding properties:

$$CA + 10H \rightarrow CAH_{10}$$

or in chemical terminology:

$$CaAl_2O_4 + 10H_2O \rightarrow Ca[Al(OH)_4]_2\,6H_2O$$

- Above 15°C, C_2AH_8 is formed along with CAH_{10}, the proportion of C_2AH_8 increasing with temperature rise, until above *circa* 25°C, C_2AH_8 is the dominant calcium aluminate hydrate formed:

$$2CA + 11H \xrightarrow{25°C+} C_2AH_8 + AH_3$$

or

$$2CaAl_2O_4 + 11H_2O \xrightarrow{25°C+} Ca_2[Al(OH)_4]$$
$$[Al(OH)_6]3H_2O + 2Al(OH)_3$$

Alumina gel $Al_2O_3xH_2O$ is formed along with C_2AH_8 and changes with time from the amorphous state to form hexagonal crystals of gibbsite $Al(OH)_3$.

$$AH_x + (3-x) \rightarrow AH_3$$

$$Al_2O_3xH_2O + (3-x)H_2O \rightarrow 2Al(OH)_3$$

- Above *circa* 60°C, CA hydrates to form the hydrogarnet C_3AH_6 during early hydration:

$$3CA + 12H \rightarrow C_3AH_6 + 2AH_3$$

or

$$3CaAl_2O_4 + 12H_2O$$
$$\rightarrow Ca_3[Al(OH)_6]_2 + 4Al(OH)_3$$

Again, alumina gel is initially formed before changing to crystalline gibbsite. Very occasionally, small quantities of the aluminium hydroxide polymorphs bayerite and nordstrandite are produced during manufacture along with much larger amounts of gibbsite.

- A fourth calcium aluminate hydrate C_4AH_{13} is sometimes observed at ordinary temperatures, particularly with grey CAC in high C/A areas of the hydrating matrix, as characteristic hexagonal plates, which can be represented symbolically as follows:

$$CA + 3[C] + 13H \rightarrow C_4AH_{13}$$

or

$$CaAl_2O_4 + 3[CaO] + 13H_2O$$
$$\rightarrow 2\{Ca_2[Al(OH)_6]OH\ 3H_2O\}$$

More C_4AH_{13} is observed in the presence of admixtures like lithium salt accelerators because of high lime areas in the aluminous cementing matrix being more readily available to promote its formation.

The use of brackets in the above equation indicates that it is not necessarily actual free lime that is reacting, but the lime-rich matrix. More work is required to characterize the precise formation of $CaAH_{10}$. The hydrates CAH_{10}, C_2AH_8 and C_4AH_{13} are all metastable hexagonal plate-type phases that transform, or convert, with time to the more stable and denser cubic hydrogarnet C_3AH_6. These are the well-known conversion reactions:

$$3CAH_{10} \rightarrow C_3AH_6 + 2AH_3 + 18H$$
$$3C_2AH_8 \rightarrow 2C_3AH_6 + AH_3 + 9H$$
$$C_4AH_{13} \rightarrow C_3AH_6 + [CH] + 6H$$

The conversion reactions produce more porosity and permeability and also less strength [8,10].

The lack of free lime produced during CAC hydration means that there is strong resistance to attack by sulphates, vegetable oils and sugar solutions, and is also beneficial from the workability viewpoint. However, resistance to alkalis, particularly strong alkalis, is weak.

By way of contrast, C_4AH_{13} is actually stable in hydrated Portland cements, because the high lime media effectively prevent any significant conversion to C_3AH_6 at ordinary temperatures, with only a little such conversion (involving substitution of some SiO_4^{4-} for OH^-) arising at higher temperatures. Large increases in relative density take place upon conversion of the metastable hexagonal hydrates in CAC to the cubic form C_3AH_6 as demonstrated in Table 4.3 [10–15].

CAH_{10} and C_2AH_8 are the main strength-giving phases in hardened CAC at ordinary temperatures. C_4AH_{13} is usually insignificant in this respect because of the very small quantities by comparison normally formed during hydration. Formation of the dense C_3AH_6 by conversion results in increased porosity of the structure and the release of water, which can cause a marked deterioration in strength. The strength deterioration can be minimized by keeping the original water/cement ratio to 0.40 or less, so as to give a denser structure and to stop the porosity from substantially increasing upon hydration. C_2AH_8 has been shown to be a ubiquitous transitory phase that initiates the formation of C_3AH_6 in both the hydration and conversion processes over the temperature range 40–90°C, when examined by synchrotron energy-dispersive diffraction [11]. At 40°C the transformation progression $CAH_{10} \rightarrow \alpha\text{-}C_2AH_8 \rightarrow \beta\text{-}C_2AH_8 \rightarrow C_3AH_6$ with increasing density at each stage is indicative of a solid-state reaction with the penultimate $\beta\text{-}C_2AH_8$ phase acting as a nucleating agent for C_3AH_6 growth [12]. This is quite different from the concept of direct $CAH_{10} \rightarrow C_3AH_6$ transformation which is invariably described as a through-solution mechanism [13]. Differences between the α-(main form) C_2AH_8 and the β-variety are small and essentially of the displacive type.

When C_3AH_6 is produced from CAH_{10} and C_2AH_8 there are volume decreases to 47 per cent and 75 per cent respectively of the original value [9]. This normally leads to strength decrease. At low water/cement ratios (below 0.40) there is usually insufficient space to permit all the CA to hydrate, so that even after conversion, hydration occurs in the pore spaces and fills them up more effectively. As a consequence, although the converted strength is lower than the initial transitory strength, in relative terms the converted strength is greater than at higher water/cement ratios. C_3AH_6 is not *per se* a weak binder – it gives strong bonding when formed directly, as at temperatures above *circa* 60°C [2].

$C_{12}A_7$, which is present in much smaller quantities than CA, reacts rapidly with water to produce similar hydration products to CA, predominantly C_2AH_8 (prior to conversion) and alumina gel that gradually crystallises to gibbsite AH_3:

$$C_{12}A_7 + 51H \rightarrow 6C_2AH_8 + AH_3$$

Like CA, $C_{12}A_7$ hydration rapidly gives rise to C_3AH_6 at elevated temperatures, especially above 60°C:

$$C_{12}A_7 + 33H \rightarrow 4C_3AH_6 + 3AH_3$$

Some CAH_{10} can also be formed at ordinary temperatures:

$$C_{12}A_7 + 69H \rightarrow 6CAH_{10} + AH_3 + 6CH$$

The CH formed transiently is then available to react with CAH_{10} and C_2AH_8 to form more C_2AH_8 (from CAH_{10}) and also C_3AH_6 from both of these hydrates.

CA_2, which is more commonly present in white CAC, is only weakly hydraulic, but also does react with water at a much slower rate than CA or $C_{12}A_7$ to produce CAH_{10}, C_2AH_8 and/or C_3AH_6 according to the temperature, as for CA or $C_{12}A_7$.

Ferrite C_4AF is an enigmatic phase in CACs. Although present in quantities *circa* 10–20 per cent, its hydraulicity has commonly been ignored in studies of CAC hydration [14]. One study showed that in grey CAC, ferrite played no significant part in early hydration at 20°C, but at 30–38°C over 80 per cent was found to have reacted in 2 months [15]. Ferrite in CAC normally has a wider compositional range than in Portland cement [16], but like the latter, can be represented for simplicity as C_4AF. Ferrite can be considered as an iron (III) oxide substituted C_2A which mainly hydrates thus:

$$C_2A_{0.5}F_{0.5} + 8H \rightarrow C_2A_{0.5}F_{0.5}H_8$$

There can also be some formation of $CA_{0.5}F_{0.5}H_{10}$ as well:

$$C_2A_{0.5}F_{0.5} + 11H \rightarrow CA_{0.5}F_{0.5}H_{10} + CH$$

Any CH formed again only has a brief transitory existence, as it reacts in the high aluminous environment to help form more of the higher lime calcium aluminate hydrates.

The iron (III) hydrate $C_2A_{0.5}F_{0.5}H_8$ in reality forms a solid solution and hence a continuous mass with C_2AH_8 produced from CA and $C_{12}A_7$, and for practical purposes is not independently identifiable. The same is true of $CA_{0.5}F_{0.5}H_{10}$ with CAH_{10}. Both hydrates can also convert to $C_3A_{0.5}F_{0.5}H_6$, which for similar reasons does not independently exist in the hydrating cement matrix. Therefore, ferrite plays an integral part in the setting, strength determining and conversion processes of CAC. In white CAC, where very little is present, its influence is minimal.

The melilite phase in CAC is normally closer in composition to the gehlenite (C_2AS) end of the gehlenite–åkermanite solid-solution series and, to a first approximation, can be regarded as gehlenite. The gehlenite hydrates slowly to form strätlingite C_2ASH_8, which is also known as gehlenite hydrate, although it may only slowly achieve equilibrium [17]:

$$C_2AS + 8H \rightarrow C_2ASH_8$$

or

$$Ca_2Al_2SiO_7 + 8H_2O \rightarrow Ca_2Al_2SiO_7 8H_2O$$

This hydration reaction to form strätlingite is becoming increasingly recognized as important for long-term strength development, where it is often found in structures many years old. Strätlingite forms stable assemblages with hydrogarnet and AH_3 or C–S–H [17].

Belite β-C_2S slowly hydrates to form the amorphous non-stoichiometric calcium silicate hydrate C-S-H and transiently calcium hydroxide (portlandite) CH:

$$2C_2S + 4H \rightarrow \text{‘}C_3S_2H_3\text{’} + CH$$

$C_3S_2H_3$ is an approximate composition for C-S-H. The C-S-H arises here in a similar way to its formation in Portland cement from β-C_2S. It has strong binding power and can make a significant contribution to the compressive strength of CAC at 28 days and beyond in particular. This can partially offset some of the strength lost through conversion.

However, a major difference from the Portland cement environment is the more aluminous medium during long-term hydration, which can react with some of the C-S-H to form strätlingite C_2ASH_8. Strätlingite formation, as previously indicated, also favourably assists long-term strength. There is no clear-cut proof at present as to whether reaction of at least some of the C-S-H to form C_2ASH_8 is *per se* beneficial or otherwise to long-term strength performance, although current indications suggest that it is not detrimental. However, since the original quantities of the belite phase present have been very small anyway, such effects are not likely in practice to be of profound importance. The portlandite CH formed from belite hydration only has, at best, a transitory existence, as reported for other hydrating phases earlier in this section. It reacts with the aluminous medium to produce more calcium aluminate hydrates, and usually also some more strätlingite, which can augment the long-term strength of the cement, which is lowered anyway by the conversion reactions.

Wüstite (FeO) – a minority constituent – can hydrate to iron (II) hydroxide, some of which, at least, is likely to become oxidised to iron (III) hydroxide.

$$FeO + H_2O \rightarrow Fe(OH)_2 \xrightarrow{+OH^- -e} Fe(OH)_3$$

Both of these hydroxides can react and become incorporated in the other hydrated phases. These effects are unlikely to influence significantly the ongoing hydration and strength development of the CAC.

Overall, hydration appears to be easier to predict than for Portland cement in many instances, even though a Bogue style equation cannot, at present, be reliably formulated for the mineral content of the anhydrous cement.

CA hydration initially takes place through the formation of a layer whose composition lies close to C/A = 1 below 27°C, and closer to alumina gel $Al_2O_3 xH_2O$ above 27°C. $C_{12}A_7$ hydrates in a straightforward manner to C_2AH_8 and CAH_{10}. A gel of possible zeolitic structure is produced in the presence of silica, which can be identified through scanning electron microscopy and nuclear magnetic resonance (NMR) studies. ^{27}Al magic angle spinning nuclear magnetic resonance spectroscopy (MAS NMR), in combination with proton decoupling, has shown that the CAC hydration process can be followed quantitatively in terms of both the degree of hydration and the formation of the individual hydrate phases especially that of C_3AH_6 (hydrogarnet), the product of conversion [18].

Minerals formed during the early hydration of dark grey/black CAC are depicted for clarity in Table 4.5. The setting process is relatively slow in CAC, although the subsequent strength development is rapid [16,19] at 25–30°C, which is dependent primarily upon the presence of CA. Shortening of the setting time is usually observed between 0°C and *circa* 20°C above which it lengthens until at 28–30°C it can be up to 8 times as long as 20°C. The setting times shorten very rapidly above 30°C. Changes in the setting time with temperature have been monitored with conduction calorimetry [20]. The maximum observed in the setting time can be attributed to increasingly difficult nucleation of CAH_{10} above 20°C until it becomes nearly impossible above 29°C, and subsequent formation of a gel phase which acts as a slowing barrier to further dissolution and/or precipitation. The combination of C_2AH_8 and AH_3 presents a difficult nucleation and slows the growth of nuclei whilst temperature *per se* acts as an accelerator. When this latter reaction becomes predominant, the setting time accelerates very quickly (Capmas *et al.* [9]). At higher temperatures, the formation of the cubic hydrate C_3AH_6 by nucleation is the prime cause of setting. Only a minor role in setting is actually played by the structureless alumina gel that is formed concomitantly with hydrates such as C_2AH_8 and C_3AH_6. The alumina gel tends to crystallize with time to form gibbsite, as mentioned already.

C_2AH_8 can exist in two closely related polymorphic forms α- and β-. The α-C_2AH_8 is the most commonly encountered variety, but β-C_2AH_8 can exist transitorily during subsequent conversion to C_3AH_6 [12]. It is not known if these two forms differ significantly in their hydraulicity.

Setting results in a loss in mobility of the water in the mix and compressive strength development involves a continuation of the hydration reactions. CAH_{10} forms distinctive networks of interlocking prisms, C_2AH_8 produced well-formed hexagonal plates that are normally clustered into aggregates and C_3AH_6 gives aggregates of equidimensional nodules. Figures 4.2 and 4.3 both show scanning electron micrographs of C_3AH_6,

TABLE 4.5 Minerals formed during early hydration of dark grey/black ratio CAC at water/cement 0.27

Curing temperature (°C)	Time (h)	Minerals formed (wt%)				
		Gel	CAH_{10}	C_2AH_8	AH_3	C_3AH_6
5	24	—	5	tr	2	4
10	24	—	21	3	1	1
15	24	5	8	—	tr	tr
20	8	—	10	6	2	1
25	24	—	—	12	3	—
30	16	—	—	2	tr	tr
30	24	—	—	15	4	2
33	22	—	—	11	3	—
35	4	—	—	2	0.5	—
35	8	—	—	10	4	5

tr: trace.

FIGURE 4.2 SEM of cubic crystals of C_3AH_6 produced at water/cement ratio of 0.5 at magnifications of (a) 2000 and (b) 10,000, respectively. (Photographs supplied by Lafarge Aluminates.)

FIGURE 4.3 SEM indicating nodular spheroidal prisms of C_3AH_6 (magnification 5800) in a dark grey/black calcium aluminate cement paste of water/cement ratio 0.40 at 24 h and 60°C.

which is the most stable calcium aluminate hydrate produced in CAC hydration, either as an early hydration product at elevated temperatures or the product of conversion of the other metastable hydrates initially formed at later ages. C_4AH_{13} is not very important in setting and strength development because of the relatively small quantities involved in neat cement pastes; it forms characteristic hexagonal plates.

Initial set takes place after *circa* 5–6 h for dark grey/black CAC at ambient temperature, after which compressive strength development is so rapid that strengths of *circa* 60 MPa or more are commonly found after 24 h (see Table 4.2). There is some later strength retrogression caused by

conversion. However, this is not deleterious if the water/cement ratio is kept suitably low, as discussed under Section 4.11: Durability. There is some partial offsetting of the strength retrogression by the formation of the amorphous binder C-S-H from the belite β-C_2S hydration and also the aforementioned hydration of melilite directly to strätlingite.

Knowledge of the hydration reactions is important from the viewpoint of safe usage and good durability. In this context it is necessary to consider long-term converted strengths rather than the more conveniently measured high transitory early strengths. These points are further discussed in the Sections 4.11 and 4.12 on Durability and Safe usage.

4.6 Admixtures

Admixtures can modify the properties of CAC, but, very importantly, may sometimes behave differently from Portland cements. Setting is accelerated by lithium salts (like the carbonate Li_2CO_3 and the chloride LiCl), and the hydroxides of sodium, potassium and calcium, whilst many common inorganic salts like calcium chloride retard the set. Hydroxylic organic compounds that retard Portland cement setting also retard CAC setting, namely lignosulphonates, sugars, citric, gluconic and tartaric acids [10,14]. Indeed, depending upon the particular admixture and its concentration, for example, setting can be adjusted from about one minute to indefinite. Mechanisms of acceleration and retardation have been described in the technical literature [10,16,21,22]. Retardation of setting has been ascribed to chelation of Ca^{2+} ions in solution with the retarder and acceleration to heterogeneous nucleation. Superplasticizers for Portland cement tend to act as plasticizers for CAC (S. M. Gill *et al.* [9]), namely SMFC (sulphonated melamine formaldehyde condensate) and SNFC (sulphonated naphthalene formaldehyde condensate).

True superplasticizing for CAC is difficult, but sulphonated phenol formaldehyde condensate (SPFC) has shown some promise in this direction [23,24]. It is possible, with new generations of superplasticizers that proper superplasticizing properties for CACs will arise in the future. This would then allow better workability and even lower water/cement ratios to be employed in mortar and concrete mixes, which should aid long-term durability significantly.

4.7 Blends with other materials

4.7.1 CAC–OPC mixes

CAC–Ordinary Portland Cement (OPC) mixes are widely utilized to produce cements that set and harden rapidly. Such mixes have reduced setting times and compressive strengths compared with CAC and OPC alone. For certain ranges of mixes, setting is very short and flash set can arise; these ranges vary with particular CAC–OPC types and cannot be predicted without trial experiments. CAC setting is accelerated by gypsum and calcium hydroxide (high pH) from the OPC. OPC setting is accelerated by preferential reaction of gypsum with CAH_{10}, C_2AH_8 and C_3AH_6 from the CAC to give ettringite $C_3A \cdot 3C\bar{S} \cdot H_{32}$. Hydration products include C–S–H, CH, $C_2A\bar{S}H_8$, C_4AH_{13} and ettringite. Later strength of CAC–OPC mixes decreases progressively as CAC is replaced by OPC until a minimum is reached that is lower than for OPC alone. The strength developed over the first few hours is sometimes greater for certain mixes than with CAC alone, even though later strength is lower. The mix proportions giving the higher early strength and minimum later strength can differ with different CAC and OPC batches. Published compressive strength data from Lea [8] for CAC–OPC mixes, but expressed in MPa, are given in Table 4.6 for illustrative purposes. (See also Section 4.8 and Chapter 8:Gypsum in cements.)

The actual sulphate resistance of CAC–OPC mixes will depend upon the smoothness of the hydration and setting, since these influence the microstructure developed. Irregular fast setting gives a more heterogeneous microstructure, which will produce more large pores and thus greater ingress of extraneous salts like sulphates. Where sulphate resistance was specifically studied, CAC–OPC mixes ranging from 5:95 to 20:80 weight ratio had lower sulphate resistance than

TABLE 4.6 Mixes of CAC and OPC: compressive strength of 1:2:3 concrete a water/cement 0.55

CAC (%)	OPC (%)	Compressive strength (MPa)				
		3h	6h	24h	7 days	28 days
100	0	0	8.3	46.9	62.1	70.3
90	10	0	10.3	37.9	51.7	56.6
75	25	1.7	11.0	22.6	40.0	42.8
60	40	2.1	6.9	17.2	21.4	22.0
40	60	5.5	6.2	8.3	10.3	11.0
25	75	3.4	4.1	5.5	9.7	12.4
10	90	0.3	0.7	2.1	7.6	11.7
0	100	0	0	4.8	20.7	34.5

OPC alone, whilst CAC–OPC mixes from 80:20 to 100:0 had sulphate resistance like CAC [8].

CAC–OPC mixes, often containing additional calcium sulphate, are frequently used in minepacking, where ready formation of ettringite assists in the development of acceptable high early strength (see Brooks and Sharp [9]).

Some CAC–OPC mixes show expansive behaviour in the presence of calcium sulphate. A simple rapidly setting mix might contain 15–20 per cent OPC, 10–30 per cent dark grey/black CAC, 10–25 per cent CH and 40–50 per cent sand. CH can be replaced by gypsum to give an expansive mix. Self-levelling mortars can be produced from more complex formulations containing proprietary admixtures. As indicated above, the setting times of these depend greatly on the reactive proportions, the characteristics of the individual cements, the mixing time and the temperature – only laboratory tests can predict the precise behaviour of a specific mix [16].

Conversion is not normally important in the context of CAC–OPC mixes, because of the combined siliceous-aluminous hydration environment, coupled with the specialist application of these mixes.

4.7.2 Calcite

Use of calcareous aggregates or additions of finely-ground calcite to CAC would prevent strength loss by formation of the calcium monocarboaluminate hydrate $C_3A \cdot C\bar{C} \cdot H_{11}$ ('monocarbonate') which appeared to limit or stop conversion. Later work showed that at 25°C and below, CAH_{10} was the main hydration product; conversion was not prevented but merely delayed. $C_3A \cdot C\overline{C} \cdot H_{11}$. was formed at the expense of C_2AH_8 at 25–60°C but was unstable above 60°C. Long term and high temperature durability of this carboaluminate variant of CAC is unknown [25,26]. Some 'monocarbonate' can also form quickly on CAC, when surface dusting takes place.

4.7.3 Ground-granulated blastfurnace slag (ggbs)

Blends of 50:50 ggbs with CAC slow down the conversion rate and give no noticeable strength decrease up to at least one year unlike control pastes of CAC. Strätlingite C_2ASH_8 forms in abundance in the slag pastes and appears to be a stable hydration product. Improvements in sulphate resistance of the slag blends (BRECEM^{TM}) have also been noticed by Majumdar *et al.* [9], and Osborne [27]. Temperature durability is not yet known. Ten-year studies of BRECEM^{TM} hydration have been reported and are noted in the Section 4.17: Further comments.

4.7.4 Macro defect free (mdf) cements

With 'macro defect-free cements', which are in reality cement – polymer composites, CAC can be used instead of Portland cement to give high flexural and compressive strengths. These high strengths are obtained by incorporating a water-soluble polymer such as polyvinylacetate or polyvinylalcohol in the cement mix to permit water/cement ratios as low as 0.10 to be achieved and then treating the dough produced by pressure moulding, extrusion or calendaring (as in plastics technology) [28,29]. Flexural strengths of 40–150 MPa, compressive strengths of 100–300 MPa, Young's moduli of 35–50 GPa and fracture energies of 40–200 J/m^2 can be achieved. These 'mdf cements' are not normally civil engineering materials in the conventional sense, but are more akin to ceramics and plastics. Durability is an area that needs more investigation here.

The flexural strength of CAC is also greatly improved by addition of various sodium-based inorganic polymers and calcium phosphate at low water/cement ratios. Values of over 40 MPa in the short term compared with 14 MPa for the pure cement were achieved. After 10 months curing, the strengths of the modified samples continued to be more than twice as high as that of the pure CAC. These phosphate-based compounds do not form crystalline hydration products; an aluminate-phosphate gel appears to be produced, which provides structural integrity to the CAC by reducing the pore size and porosity [30].

4.7.5 Metakaolin

Metakaolin can stabilize the strength of CAC-metakaolin blends when stored under conditions that ordinarily accelerate the conversion reactions to C_3AH_6, namely being stored under water at 40°C [31,32].

4.7.6 Gasifier slag

Similar benefits to those of CAC-metakaolin blends were found for CAC-gasifier slag blends [31,32].

4.7.7 Pulverized fuel ash (pfa)

CAC–pfa blends have to-date mostly been used in oilwell cementing formulations in Arctic environments for cementing through permafrost without damage (see Chapter 7: Developments with Oilwell Cements).

4.7.8 Condensed silica fume

Although some research has been done on the hydration CAC–csf blends, [33] clear-cut benefits have not yet been established. No lower rates of conversion have been found during curing than for neat CAC pastes (Bentsen *et al.* [9]). Little CH and alkalis are present in CAC and pozzolanic properties are either non-existent or negligible.

125

4.7.9 Calcium sulphate

CAC and calcium sulphate proprietary blends are employed in mining and tunnelling. They give rapid setting and hardening properties, assisted by the early formation of ettringite.

4.8 Low-temperature applications

Calcium aluminate cements can be employed at temperatures below ambient down to −10°C, because of the speed of hydration. The actual heat of hydration is ~325–400 kJ/kg, which is somewhat less than that of the majority of OPC. However, with the CAC most of this heat is dissipated within 24 h, which assists the furtherance of hydration at low temperatures. Typical low-temperature applications include winter construction work and repairs to cold stores [14,16]. Even under very cold conditions, there is a normal type of setting followed by rapid strength development, as CAH_{10} in particular is being formed.

CAH_{10} is the main hydration product at low temperatures. For instance, at 5°C solutions obtained from CA rapidly become very supersaturated in CAH_{10}, but much less so in C_2AH_8 or AH_3. Whereas CAH_{10} is the major hydration product at 5°C, it is not normally detectable above 30°C, whilst at 50°C supersaturation is high only in C_3AH_6 which is rapidly formed here [13]. At 5°C, conversion to C_3AH_6 may take many years, whereas at 50°C formation of C_3AH_6 and AH_3 is almost immediate.

CAC diluted with 50 wt% pfa or a natural pozzolan is used to cement surface conductor casings in wells that have been drilled through permafrost in Arctic regions, because it can set and harden suitably around 0°C. The aforementioned 50 per cent dilution lowers the heat of hydration liberated and hence minimizes the risk of damage to the permafrost [34,35].

4.9 High-temperature applications

CAC is extensively utilized at high temperature as a refractory cement, particularly the white varieties. Dark grey/black CAC is sometimes utilized in particular high temperature applications up to *circa* 1100°C. Aggregates such as crushed firebrick are employed for producing refractory concrete. White CAC is normally used for the highest temperature applications – with pure alumina aggregate a concrete is made that can be utilized up to 1800°C. CAC is also used to line fireflood and thermal recovery wells up to 1000°C or above; CAC has the advantage here in being able to withstand the considerable fluctuations in temperature that are normally experienced in such wells without deleterious high shrinkage (see Chapter 7: Developments with Oilwell Cements). The absence of calcium hydroxide (portlandite) CH is a prime reason why CAC maintains its integrity under such conditions.

When CAC is heated, the hydration products CAH_{10} and C_2AH_8 convert to C_3AH_6 and dehydrate. The main hydrate produced above 100°C is C_3AH_6, the dense cubic hydrogarnet phase. This gradually loses its water content above 225°C; between 550°C and 950°C recrystallization to phases such as CaO and $C_{12}A_7$ takes place. Compressive strength decreases with temperature until a minimum value is reached at ~900–1100°C. It is essential that the cement content of the concrete be sufficient to maintain adequate strength at this minimum. Above that temperature, the strength starts to increase, as the dehydration products begin to react with the aggregate to produce ceramic bonding, largely as a consequence of tight intergrowth of $C_{12}A_7$ crystals. As a result, there is a rise in both compressive strength and abrasion resistance. The phases formed under these conditions depend very much upon the precise compositions of the CAC and aggregate, and commonly encompass anhydrous calcium aluminates like CA and CA_2 as well as magnesium aluminate (spinel) MA [16,18].

Aggregates used in refractory castables containing CAC include fused alumina, chromite, sintered bauxite, chamotte and exfoliated vermiculite. It should be remembered that for refractory usage, CAC concrete is mixed and placed in the usual way. In refractory concrete the tendency is to have a minimum cement content (e.g. low cement castables) and allow the aggregates to give the strength through dense packing.

4.10 Hydrophobic applications

Additions of hydrophobic agents to CACs, which enable the cements to perform in adverse climatic conditions, have been made in Russia and other CIS States, but have not been commonly undertaken elsewhere.

If CAC be ground with suitable hydrophobic agents such as lauric, stearic and oleic acids in quantities ~0.05 per cent by weight of cement, a hydrophobic seal forms around the cement particles. This gives a hydrophobic cement which can be stored for weeks in damp circumstances without combining with water and giving air set lumps. It has been reported that hydrophobic agents such as the aforementioned affect early strengths (~8 h) but have little influence upon later compressive strength or indeed upon setting [1]. By increasing the hydrophobic agent concentration to ~0.10–0.25 per cent by weight of cement, the water-repellent properties rise considerably but both setting times and compressive strength values up to *circa* 7 days are retarded. Intensive mixing of the concrete is required to break the hydrophobic seal [1]. Waterproofers have not been recommended for addition to CACs since they might seriously affect the strength developed [8].

However, it must be emphasized that there are no readily available results on the effects of hydrophobic and waterproofing agents upon the long-term converted strengths developed by CACs. Nor are there any standards available for the utilization of hydrophobic CACs, whereas some do exist for hydrophobic Portland cements, which can be employed for construction work in adverse climatic conditions [36,37].

4.11 Durability

Following three well-publicized failures of buildings containing CAC beams in 1973–1974 in the UK, there has been much concern and interest in CAC and its usage, particularly concerning strength loss after conversion. The details of these failures have been given and in each instance poor workmanship was reported to be the prime cause of blame (Midgley [9]). Recommendations were given on how CAC should be used as a construction material and these will be discussed.

Deleterious strength losses have taken place when the water/cement ratio has been above 0.40. CAC is not detrimental in its long-term compressive strength when the water/cement ratio is 0.40 or less. Indeed, CAC is very predictable in its behaviour. Its long-term strength after conversion depends primarily upon porosity, that in turn depends upon the original water/cement ratio employed. The occurrence of large pores is minimized by keeping the water/cement ratio to 0.40 or less, which ensures that the residual compressive strength is more than adequate to maintain the structure in question (see Midgley and George [9]). The originally quoted value of 0.4 for water/cement ratio, has subsequently been refined to 0.40 and is given here to avoid any confusion. Also, where CAC had been utilized previously in structural applications, investigations did not detect any adverse effects of corrosion upon steel tendons removed from prestressed concrete beams in normal buildings [38].

The new European standard for CAC being developed will contain guidance for users in its Annex A, where the maximum water/cement ratio is being tightened up not to exceed 0.40 [39].

Long-term durability of CAC has occurred in numerous structures. Examples of such structures are illustrated in Figures 4.4 and 4.5.

A highly porous CAC has a low resistance to alkaline hydrolysis, probably because the hydrous alumina is dissolved in the process [16]. A water/cement ratio of above 0.8 is normally needed for deterioration, even in the most accelerated conditions. There is often confusion regarding the role of alkalis in the hydration of CACs. Actually, there are two different effects due to

1. the alkali metal ions on the hardening process;
2. an alkaline environment (high pH) on hardened concrete.

The first type of *alkali effect* is due to 'releasable alkalis'. The alkalis, if present in the fine fractions of the aggregates, may release sodium (Na^+) or potassium (K^+) ions into the water and these may increase the rate at which the conversion reaction

FIGURE 4.4 Montrose Bridge in Angus, Scotland, built in 1931, showing the piers made of CAC concrete supporting the main Portland concrete structure of the bridge. Interestingly, whilst the piers are still structurally sound after 70 years of service, the Portland concrete structure that it supports is suffering from alkali-silica reaction. As a result of the latter phenomenon, the bridge is likely to be demolished and rebuilt before long. (Photograph supplied by Lafarge Aluminates.)

may proceed. No data have been published to quantify this effect, although it has been considered to add to the vulnerability of CACs to attack by deleterious chemicals or weaken the CAC concretes by too rapid conversion.

Alkaline hydrolysis is a manifestation of the second type of alkali effect. It arises where alkali carbonates decompose the calcium aluminate hydrates formed, namely with CAH_{10}:

$$K_2CO_3 + Ca[Al(OH)_4]_2 + 6H_2O$$
$$\rightarrow CaCO_3 + 2K[Al(OH)_4]_2 + 6H_2O$$

Such attack is accentuated if carbon dioxide is present because the OH^- ions act as carriers of atmospheric CO_2, whilst the CAH_{10} is decomposed to give hydrous alumina $Al_2O_3xH_2O$ that can recrystallize not only as the stable end product gibbsite AH_3 but also in significant quantities

sometimes as the metastable AH_3 polymorphs nordstrandite and/or bayerite:

$$CO_2 + 2K[Al(OH)_4]_2$$
$$\rightarrow K_2CO_3 + 2Al(OH)_3 + H_2O$$

The alkali carbonates like K_2CO_3 are regenerated by the action of atmospheric CO_2. The alkaline hydrolysis reactions can be continuous and may disrupt the concrete. However, if the concrete is well made and dense, the reaction may be imperceptible due to the lack of effective penetration of the CO_2 and alkalis into the concrete.

It needs to be remembered that the qualitative presence of bayerite and/or nordstrandite per se is not a sign that alkaline hydrolysis has necessarily occurred. Small quantities of the polymorphs are periodically detected in normal hydration along with gibbsite. The precise reasons for bayerite

FIGURE 4.5 Pier B at the Ocean Terminals, Halifax Harbour, in Nova Scotia, Canada, built during 1930–31 mostly containing CAC concrete in the pace slabs shown, with the majority of CAC concrete in the caissons below the waterline, and still in active service. (Photograph supplied by Lafarge Aluminates.)

and/or nordstrandite to form during CAC hydration are not yet known.

However, carbonation in the absence of alkalis increases the strength of CAC concrete [16]. Within sufficiently porous concrete where conversion has occurred, some of the C_3AH_6 reacts to form calcite and hydrous alumina. Conversely, in dense concrete the degree of CO_2 penetration is low, presumably since the pores do not contain the water to promote the reaction and become blocked by any calcite and alumina gel formed. Such observations support the view that the high resistance of dense CAC concrete to various forms of attack is at least partly due to the effects which impede or prevent ingress of the attacking agent. In addition to blocking pores by reaction

products, these effects could include the rapid removal of water by continued hydration [16].

Alkaline hydrolysis has been described as a catalytic process which leads to the total destruction of the affected concrete. The minimum parameters required for this hypothetical alkaline hydrolysis to occur have been established, but the process has not yet been proven, especially the role played by K^+ ions. At 4°C with K^+ ions present, different hydration products are formed according to the ionic concentration, whereas at 40°C the hydrated phases are the same irrespective of the mixing liquid. When the K^+ ion concentrations are high, the hydration slows down – the process is better defined at 4°C than at 40°C. K^+ ions enable the formation of hydration products in the gel phase, which have a high K^+ and iron oxide content. When CAC is hydrated with water, isomorphous substitution arises in the calcium aluminate hydrates, such as Al^{3+} by Fe^{3+}. When K^+ is present, not only is there substitution of Al^{3+} by Fe^{3+}, but also for Ca^{2+} by K^+ ions. With admixtures in water, and especially when KOH is present, different morphologies are found for C_3AH_6, which have a subsequent conclusive influence on the mechanical behaviour of the material [40].

From the viewpoint of durability, carbonation is inevitable as with Portland cement concrete. If good concrete is made, the lower porosity leads to improved compressive strength. However, if bad concrete is made, carbonation causes that concrete to deteriorate faster.

CAC shows good sulphate resistance in magnesium and sodium sulphate environments, even where the water/cement ratio exceeds the recommended dosage [41]. In long-term studies of CAC in sulphate-bearing environments [42] corrosion was found to start in the centre of the cube or beam where conversion had taken place (Crammond [9]). This often leads to a 'skin' effect, where hard, durable outer layers encompass a soft, usually highly converted interior. As long as the outer layers remain unpunctured, the structure may last for years. The high sulphate resistance of CAC appears to be due to the formation of a protective coating of alumina gel coupled with the absence of calcium hydroxide in the cement system [8]. The absence of the latter also produces strong resistance to the potentially damaging effects of vegetable oils and sugar solutions.

CAC also demonstrates good acid resistance in comparison with Portland cements and can usually be safely employed down to pH levels as low as 4. (This normally augments good sulphate resistance as well). Acidic corrosion can set in at lower pH values and is governed by various factors:

(a) Thermodynamic stability of the phases – conversion improves the relative resistance to corrosion.
(b) Solubility of the calcium salts formed.
(c) Neutralizing effect of the initial phases.
(d) Porosity is important where the salts are very soluble.

Limestone filler addition increases sulphate resistance by the formation of the hexagonal plate phase calcium monocarboaluminate hydrate $C_3(A,F) \cdot C\overline{C} \cdot H_{11}$ and related phases. The very low solubility of this hydrate severely limits reaction with sulphate ions SO_4^{2-} within the pore solutions of CAC pastes in comparison with ettringite $C_3(A,F) \cdot 3C\overline{S} \cdot H_{32}$ in Portland cement systems. This effect is observed to disappear when conversion takes place, even though the conversion is slightly retarded under these conditions (Piasta [9]).

$C_3(A,F) \cdot C\overline{C} \cdot H_{11}$ is also found on CAC surfaces where surface dusting occurs. Fibres of calcium tricarboaluminate hydrate $C_3(A,F) \cdot 3C\overline{C} \cdot H_{30}$ have occasionally been found to form initially at low to medium temperatures in CAC hydration, but are not intrinsically stable.

Conversion is not normally a problem with refractory CACs, since C_3AH_6 is formed very early in hydration before the structure has appreciably hardened. Refractory CACs are satisfactorily durable when the structures are designed to accommodate the minimum strength experienced at ~900–1100°C.

With CACs at lower temperatures, durability is improved if concrete designs are based upon the ultimate strength likely to be achieved after conversion, rather than in trying to prevent conversion. After all, the pre conversion hydrates CAH_{10}

and C_2AH_8 are not coherently stable and, sooner or later, conversion with its concomitant strength reduction will take place. This subject is dealt with more fully in Section 4.12: Safe usage for CACs. Examples of long-term compressive strengths obtained with dark grey/black CAC at different water/cement ratios have been illustrated [10].

The permeability, thermal expansion and modulus of elasticity of CAC mixes are essentially the same as those of equivalent Portland cement mixes. However, owing to the use of much lower water/cement ratios for CACs, the permeability will also be much lower in practice at those low water/cement ratios where the effects of conversion are not deleterious. Indeed, the permeability of CAC depends to a large extent on the porosity. Usually the lower the permeability the lower the porosity is and vice versa. This is not always the case, as with Portland cements [43] (for comparable reasons). Creep of CAC concretes is essentially of the same magnitude as for similar Portland cement concretes. Since creep is proportional to the stress/strength ratio, if there is a reduction in strength resulting from conversion then there may well be an increase in creep. Good resistance is normally shown towards corrosion, abrasion and freeze–thaw cycles. These properties, along with others, have been reviewed [44].

4.12 Safe usage of CACs

The main concern of civil engineers is that the 'conversion' reaction is usually accompanied by strength reduction and increased porosity and permeability. Such alterations in properties arise from the morphological changes to the hydrates. The converse is not always true, namely, the conversion reaction may not necessarily be accompanied by a reduction in strength, such as when high temperature CAC concrete curing is undertaken, and is better described as a chemical transformation.

The rate of conversion depends upon a wide range of factors (see Midgley [9]). These include:

- temperature (temperature rise produces faster conversion),

- humidity (fall in relative humidity below saturation gives slower conversion),
- water/cement ratio (higher w/c results in faster conversion),
- choice of aggregate (effects depend upon permeability),
- mineralogical composition of cement (higher $CA/C_{12}A_7$ and lower alkalis favour slower conversion),
- a series of results illustrating some of the above factors are given in Tables 4.7–4.11, namely rates of conversion, effect of relative humidity on degree of conversion, effect of water/cement ratio on rate of conversion, compressive strengths at different water/cement

ratios 1 day–5 years cured at 18°C and 38°C, and compressive strengths at different water/cement ratios 52 h–10 years cured outdoors, respectively.

To summarize, CAC can be a useful construction material, if utilized for appropriate applications under the right conditions, and shows rapid-hardening and strong resistance to corrosion. Despite the few well-publicized failures of CAC concrete beams in the 1970s, if recommended guidelines are followed, then there should be no problems with its normal usage (Midgley [9]).

These recommendations [9,45] are essentially as given: For the manufacture of good

TABLE 4.7 Rate of conversion with a dark grey/black CAC paste at water/cement 0.27 for various temperatures

Curing temperature (°C)	Degree of conversion @ 1 month (Dc)[a]	Rate of conversion during 1 month (R)[b]
18	20	0.238
30	22	0.253
35	24	0.268
40	32	0.291
45	35	0.325
50	60	0.326
55	80	0.335
60	100	0.73
70	100	0.78
80	100	0.93
90	100	1.15

[a] Degree of conversion (Dc) $= \dfrac{\text{Amount AH}_3}{\text{Amount CAH}_{10} + \text{amount AH}_3} \times 100\%$.

[b] Rate of conversion (R) $= \dfrac{Dc}{\text{age}}$ (with Dc as a percentage and age as years $= 10^4$).

TABLE 4.8 Effect of relative humidity on the degree of conversion of dark grey/black CAC pastes cured at 50°C

Relative humidity (%)	Degree of conversion (Dc) (%)
30	3
50	19
65	54
75	41
92	34

TABLE 4.9 Effect of water/cement ratio for dark grey/black CAC pastes on the rate of conversion

Water/cement ratio	Rate of conversion (R)
0.27	0.238
0.30	0.231
0.35	0.240
0.40	0.230
0.45	0.250
0.60	0.268

TABLE 4.10   Compressive strengths of CAC concretes 1:2:3 at 18°C and 38°C

Total water/cement ratio	Compressive strengths (MPa)					
	1 day	7 days	28 days	3 months	1 year	5 years
18°C curing						
0.30	68	84	89	88	92	82
0.32	62	75	84	88	90	83
0.33	63	75	83	87	87	80
0.34	57	75	81	87	78	81
0.35	58	72	73	87	85	74
0.40	53	69	74	85	85	78
38°C curing						
0.30	70	82	64	46	50	53
0.32	66	80	57	41	44	47
0.33	65	77	52	41	44	46
0.34	61	78	53	34	35	40
0.35	61	74	50	35	37	42
0.40	56	71	41	27	30	33

TABLE 4.11   Compressive strengths of CAC concretes 1:2:3 with different water/cement ratios cured outdoors

Total water/cement ratio	Compressive strengths (MPa)					
	52 h	3 months	1 year	3 years	5 years	10 years
0.33	90	90	91	54	52	74
0.40	77	90	82	46	50	71
0.50	72	65	58	41	36	54
0.67	55	66	46	30	28	36

quality durable CAC concrete, the fundamental requirement is that the total water/cement ratio shall not exceed 0.4.[1]

Although at low water/cement ratios CAC concrete exhibits better workability than an equivalent mix based on Portland cement, the adoption of a minimum cement content of $400\,kg/m^3$ is a normal practical safeguard against errors in the control of water/cement ratio in practice. As an added precaution against the possibility of alkaline hydrolysis, aggregate and sands containing releasable alkalis are not permitted.

- Spraying or soaking in water is necessary during the initial curing period to prevent drying out of the concrete, as a result of heat evolution, whenever conditions of placing lead to substantial increase in temperatures.

- When these requirements are met, experience shows that CAC concrete, even when rapidly and fully converted, is a reliable structural material. While exceptional high strengths can be obtained under special circumstances, prudence dictates that structural design should be based on minimum converted strength.

- With the possible exception of prestressed beams, there is every reason to use CAC as a construction material, especially where its special properties, such as resistance to chemical attack or high early strength can be used to advantage.

[1]Current practice has redefined the total water/cement ratio not to exceed 0.40.

A fresh look at the problems of durability and conversion has indicated that from the durability angle, it is better to instigate conversion early rather than to delay, so that the self-healing capacity of CAC concrete becomes a valuable practical asset; examples of highly durable CAC concrete are given [46]. Research into the long-term properties of CAC concrete has demonstrated that where structures are designed to cope with strength loss through conversion on a long-term basis, CAC concrete can be safely employed. However, wet CAC concrete was found to continue reducing in strength even when highly converted and to reach lower minimum strength than concrete under dry conditions [47]. What needs to be emphasized is that CAC gives a high early transient strength, but that the long-term strengths achieved are the relevant data. Consequently, predictions of attainable long-term strengths need to be as reliable as possible.

Appropriately low water/cement ratios are essential for good long-term durability [41]. Any later development of strength and durability that might arise following loss in strength by conversion can be associated mainly with the formation of strätlingite (or gehlenite hydrate) C_2ASH_8. There is also a contribution from calcium silicate hydrate C-S-H, which arises through the hydration of the minor cement constituent belite β-C_2S present, that has not undergone subsequent change in the aluminous cement medium to C_2ASH_8, which is itself similarly beneficial. After all, the reaction between alumina gel and aggregates with siliceous concretes results in the formation of C_2ASH_8, which appears to contribute positively towards compressive strength and longer-term durability.

Also, long-term studies of sulphate resistance of buried concrete have shown good durability for CAC concretes [42].

4.13 Concrete society report in the UK on CACs

In 1997, the Concrete Society in the UK published a technical report on a reassessment of CACs in construction [48]. This reviewed the current status for employing CACs in construction and proposed a basis for their safe usage, in the light of Codes of Practice having banned the use of CAC in structural units. The nature of CACs, including the manufacture, science and technology, engineering properties, use of admixtures and blends was considered together with applications and field experiences. The background to the 1970s ban in structures together with proposals for a new way forward were set out.

Designs were recommended to be made on the basis of a converted strength, and that a prediction of a long-term minimum hardened strength can be made by curing concretes at 38°C for 5 days, following previous investigations (Midgley [9]) and test work by the Building Research Establishment (as in BRE CP 34/75 and BRE Digest 392 [49,50]) and also by Lafarge Aluminates (George [9]). In addition, concretes were to have a minimum cement content of 400 kg/m^3 and maximum total water/cement ratio of 0.4 *(now 0.40)*.

Designs for durability need to be undertaken and evaluations made of previously encountered poor performances, such as those described by Neville both before the 1970s ban on structural usage for CAC in the U.K., when he drew attention to the dangers posed by conversion [49], and then after problems had arisen [50,51].

However, the Concrete Society Report did not recommend the use of CAC in prestressed concrete or in concrete pipes for the conveyance of drinking water. Manufacturers do not recommend the latter use because under certain circumstances the leach rate of aluminium from CAC may be 100-fold that of OPC.

Recommendations made were as follows [48]:

1. Specifiers, users and clients should be encouraged to consider applications where CACs would have technical and commercial benefits either in conventional concrete form or as specialist proprietary products.
2. A change of emphasis should be considered for the *Approved Documents to the Building Regulations* to reflect more fully selection on the basis of the demonstration of suitability contained in the regulation itself.

3. To underpin recommendations 1 and 2, further coherent, detailed and independent guidance should be developed as a safe basis for determining the minimum *in situ* strength of CAC concrete for particular structures.
4. Further to 3, research should be undertaken and guidance developed, which is devoted to understanding more fully the nature and behaviour of CAC concrete in aggressive service conditions. In particular, the role of the various cement hydrates and microstructure in influencing performance should be examined further. The studies should include examples of both good and bad performance.

4.14 CAC and structures

New Civil Engineer, in their review of a CAC Seminar in 1992 [52], summarized the views of Neville, who had previously drawn attention to the potential dangers that could be associated with the conversion reaction in the 1960s [49]. This was prior to the structural failures of the 1970s. Conversion occurs particularly in damp and warm environments. The end result is a porous cement paste which is vulnerable to further attack and up to 40 per cent strength loss. Neville does not, nevertheless, advocate that the material should be banned form all structural uses. However, the risk of using CAC concrete could not still be ignored. It is the engineers who are responsible for their structures and not the scientists who should be allowed to make the decision on whether to use CAC again [52].

Neville advocated that much more work needed to be carried out on long-term properties like ductility, the capacity to accommodate deformation before failure [49–51]. After all, the Building Research Establishment had shown that the strain capacity of precast concrete CAC beams at failure was two-thirds less than that of beams made with OPC. The residual strength is much more sensitive to variations in the water/cement ratio than the initial strength; under wet conditions for water/cement ratios above 0.5, there is no levelling off in the downward trend of strength development curves. CAC can be a useful

cement to employ in flue blocks and refractories and for many specialist purposes, but should not be employed in precast concrete.

Attention was also drawn to the original recommendations that the water/cement ratio for safe usage should not exceed 0.4. The value should be 0.40. Although most people would be likely to interpret 0.4 as being 0.40, mathematically speaking 0.4 would refer to 0.44 or may be even 0.45 which could be unacceptably high [50,51]. In order to avoid any ambiguities, the most recent recommendations regard 0.40 as the desirable maximum total water/cement ratio for practical purposes [39]. In reality the views expressed by Neville [49–52] and those of the Concrete Society Working Group on CAC and its uses [48] are not on balance very far apart and hence can be readily accommodated for the benefit of users of CAC, so that CAC can continue to be safely employed in suitable construction work.

4.15 Further comments on sulphate resistance

CAC shows good sulphate resistance even in $MgSO_4$ and Na_2SO_4 environments where the water/cement ratio exceeds the recommended dosage (Crammond [9]) [53]. With longer-term studies of CAC in sulphate-bearing environments, corrosion was observed to begin at the centre of the cube or beam where conversion had arisen. Good quality CAC concrete produced in accord with approved guidelines gives a greater resistance to sulphate attack than ordinary or rapid-hardening PC concrete. However, should the porosity or permeability be high as a consequence of rapid conversion in concrete made with deleteriously high water/cement ratios and/or poor workmanship during placement, CAC might well become vulnerable to attack by sulphates, alkalis and other aggressive substances, due to easier ingress into the hardened structures under such unacceptable conditions [54].

In practice, situations where there is sulphate attack often involve chloride attack as well [55]. CAC has good durability in chloride solutions

(Kurdowski *et al.* [9]). Sulphate and chloride attack are both reduced relatively, or effectively overcome, by the use of low water/cement ratios, which should not exceed 0.40. Under these conditions, when conversion does take place, CAC should still give good sulphate and chloride resistance.

Delayed or secondary ettringite formation, which is well known in Portland cement-based systems, can also arise in CACs. Normally, as already mentioned, CACs give good sulphate resistance, but at high water/cement ratios sulphate attack can occur. CAC mortars with $3:1$ sand/cement ratio were cured initially at 70°C at the high water/cement ratio of 0.6 and stored in 5 per cent wt sodium sulphate solutions. An examination of the mortar specimens at 1 year showed only slight cracking, with clusters of well formed pseudohexagonal prisms of ettringite in the cracks and voids. There was no evidence that these crystal clusters had caused any cracking of the mortar during their formation and growth (see Figure 4.6). The initially high water/cement ratio utilized here, assisted by extensive conversion of CAH_{10} and C_2AH_8 into C_3AH_6, would have ensured that internal transport of ions like Ca^{2+}, Al^{3+} and SO_4^{2-} would have occurred within the aqueous medium as hydration and conversion had both taken place [14].

This ettringite formation was delayed and therefore could be described as delayed ettringite formation (DEF). Even the clusters of ettringite crystals formed resembled those clusters given by DEF in Portland cement concretes or mortars. However, some purists prefer to consider DEF as being self-contained (encompassing no external entry of sulphate or other ions) and situations where external ions like sulphate participate as secondary ettringite formation. The ettringite crystals formed – from internal sulphate, external sulphate, or both – are usually very similar, so that employment of different definitions for delayed and secondary ettringite formation can be an arbitrary method of conjecture. What is not arguable is that the phenomenon of ettringite

FIGURE 4.6 SEM (magnification 1800) depicting pseudohexagonal prisms of ettringite formed in clusters in a void after one year, during a sulphate attack study on a CAC mortar at the unacceptably high water/cement ratio of 0.6 in 5 per cent wt sodium sulphate solution. These clusters represented delayed/secondary ettringite formation under the conditions experienced.

being formed in CAC systems is secondary ettringite formation, because CACs do not produce primary ettringite, unlike Portland cements [14].

Interestingly, the ettringite crystals formed from the CAC in sulphate solution showed very little iron (III) content, despite the significant ferrite phase (C_4AF) content of the CAC used *circa* 15 per cent wt). In consequence, ferrite in CACs is more reluctant to form ettringite than the main calcium aluminate phases CA and $C_{12}A_7$, presumably because of a dearth of available calcium hydroxide to facilitate its formation. This suggests that the C_4AF content of CACs is beneficial against sulphate attack and is thus a useful component of these cements [14].

4.16 Thermal analysis methods for assessing CAC conversion

Measurements of the degree and rate of conversion by thermal analysis methods have been made (see Tables 4.7–4.9). Such methods encompass DTA (differential thermal analysis), DTG (differential thermogravimetry), DSC (differential scanning calorimetry), QDTA (quantitative differential thermal analysis) and EGA (evolved gas analysis). DTA has been extensively employed [56–58]. Quantitative determination of the degree of conversion of CAC by the different thermal analysis techniques has also been discussed with particular reference to DTA [59]. Numerous shortcomings in the interpretation of thermal analysis data were identified, such as when sulphate attack or alkaline hydrolysis of CAC had arisen, or if C_2AH_8 formation was extensive at the expense of CAH_{10} [59]. In practice, the rate of conversion is more important than its extent, because a faster rate of conversion normally leads to a greater relative diminution in compressive strength with time in mortars and concretes containing CAC.

Attention has been drawn to sampling, since drilling out CAC samples from hardened mortar or concrete could increase the temperature considerably and therefore alter the extent of conversion by facilitating further conversion of CAH_{10} and C_2AH_8 to C_3AH_6 [50,60,61]. A view was

expressed that only selected samples of chipped material, ground without heating, could reasonably be expected to give meaningful results [61]. BRE replied to these doubts in the form of a press release [62], stating that experimental results from samples taken from carefully drilling small holes according to their published procedure [60] compared with results from crushed specimens from the interior of beams showed negligible increases in conversion by drilling and denying that it was better to utilize chipped samples. Neville [50] reported that tests undertaken at Birmingham showed good agreement between the results of DTA on drilled and chipped samples and that there were problems with the reproducibility of the DTA results. Bensted [59] concluded that the true value of thermal analysis tests was qualitative rather than quantitative in being able to distinguish CAC samples as being little, moderately or highly converted.

Infrared spectroscopy (IR) can be employed as an alternative technique for identifying the state of conversion of CAC concretes. Like the thermoanalytical techniques discussed above, the IR results are best interpreted qualitatively, owing to extensive waveband overlap in the IR spectra obtained [63]. Nevertheless, IR is a useful technique to utilize for studying the hydration and conversion of CAC [64].

4.17 Further comments

CACs are useful in being able to prevent or at least minimize efflorescence, the unsightly white blemishes caused by transportation of salts in solution, like calcium hydroxide in the form of ions Ca^{2+} and OH^- through the structure to the external surfaces where crystallization takes place, such as to calcium hydroxide and thence to calcium carbonate by reaction with atmospheric carbon dioxide in the presence of moisture:

$$Ca^{2+} + 2OH^- \rightarrow Ca(OH)_2$$
$$Ca(OH)_2 + CO_2 \rightarrow CaCO_3 + H_2O$$

At low water/cement ratios (not exceeding 0.40), the effects of conversion upon lowering compressive strength and increasing porosity and permeability are minimized and efflorescence is normally prevented or at least very much minimized, as compared with the situation in Portland cement concretes and mortars [65].

In situ X-ray powder diffraction studies on the hydration of monocalcium aluminate CA and on commercial dark grey/black CAC at different water/solid ratios of 0.3, 0.4 and 0.5 at ambient temperature yielded some interesting results [66]. The specimens were studied both uncovered and covered with a 4 micron *Mylar* film covering C_2AH_8 formed prior to CAH_{10} irrespective of the water/solid ratio employed.

Increasing the water/solid ratio delayed hydrate formation, this delaying effect being greater for the covered samples. Results from the CA hydration for the covered samples showed:

- water/solid ratio 0.3: C_2AH_8 forms within 10 h and CAH_{10} within 12 h.
- water/solid ratio 0.4: C_2AH_8 forms within 12 h and CAH_{10} after 15 h.
- water/solid ratio 0.5: C_2AH_8 appears after 16 h and CAH_{10} after 18 h.
- At all three water/solid ratios the transition α- to β-C_2AH_8 [12] was detected during the course of the hydration.
- Retention of water within the sample, aided by the *Mylar* film, increases the CA dissolution, which modifies the CaO/Al_2O_3 ratio, thus inducing C_2AH_8 formation.

However, with uncovered CA, water is lost from the hydrating system by evaporation and CAH_{10} is produced as the initial hydration product.

Similar trends were observed for both uncovered and covered dark grey/black CAC. It was demonstrated that by employing the *Mylar* film, the hydration behaviour of cured CAC is simulated and that the hydration reactions under these conditions favour the formation of C_2AH_8 [66] rather than CAH_{10} as previously reported [1].

Both amorphous and cryptocrystalline portions of CAH_{10} have been clearly identified during investigations of the hydration of CA [67]. The relative effects of formation of these two different morphologies of CAH_{10}, together with changes from the amorphous to the cryptocrystalline state,

upon the setting process and the development of high transient strength remain to be elucidated in detail.

Wideline ^{27}Al nuclear magnetic resonance studies have shown that the hydration of CA, CA_2 and C_3A can readily be followed. (C_3A is, of course, not found in CACs.) This technique has potential for characterizing structural changes during hydration of CAC and its components in an analogous way to studies on Portland cement-based systems [68].

Results of 10-year trials for BRECEM™ at water/cement ratios of 0.45 and 0.56 at 20°C and 38°C have shown that at 20°C there has been a small increase in compressive strength from the 5-year old specimens previously examined at 20°C, but a slight strength reduction at 38°C. Strätlingite C_2ASH_8 appears to be a stable phase [69]. These results suggest cautious optimism for the use of such CAC-ggbs blends in longer-term situations.

High performance concretes from CACs have been highlighted. They show resistance to acid attack and especially to biogenic corrosion, and also resistance to abrasion in hydraulic structures. The high abrasion resistance makes these products useful for repairs to hydraulic dams, industrial floors and pipe linings [70].

4.18 Conclusion

CACs have been discussed from a broad perspective, including the chemistry of hydration and conversion. The manufacture, nature and durability of these cements have been considered. CACs are indeed very versatile products. Understanding the details of the hydration process enables the cements to be employed for a wide range of construction applications. If used correctly, CACs can give both good technical performance and durability, and can be safely utilized in both non-structural and certain structural applications.

The Concrete Society Report [48], which has also been briefly summarized [71], should offer a good basis for the future, hopefully with the technical guidance notes within the European Standard for CAC which is currently being developed.

Observing the recommended guidance for use of CACs is essential. Most of the basic science and technology is now well established [10,16,48, 50,72], but needs to be better understood and disseminated. In the event of any doubt, the supplier or manufacturer should be consulted for further advice. With this proviso, there is no reason, given the right conditions, why CACs should not have a continuing bright future for cementing in those areas where they have been successfully utilized and given good durability in the past [1,48–51,73].

4.19 Acknowledgements

The author wishes to thank the following: Lafarge Aluminates for the provision of the photographs in Figures 4.1, 4.2, 4.4 and 4.5, and Ron Montgomery and Tony Newton in particular for useful comments on the draft manuscript and illustrations; Tom McGhee of Castle Cement, UK (Heidelberger Zement Group) for helpful advice; Juliet Munn, formerly of the School of Crystallography, Birkbeck College, University of London for the photographs in Figures 4.3 and 4.6 and for useful comment; Professor Adam Neville, A&M Neville Engineering, for valuable discussion and assistance; and Josephine R Smith of The Hannington Group for useful discussion.

The author pays special tribute to: the late Dr C.M. (Mike) George, formerly of Lafarge Aluminates – for useful discussion, and the late Dr H.G. (Midge) Midgley, formerly of Ilminster, Somerset – for provision of data given in Tables 4.4–4.11.

4.20 References

1. Robson, T. D., *High Alumina Cement and Concrete*, Contractors Record Ltd, London (1962).
2. British Ceramic Society, *Basic Science Section Meeting on Structure and Properties of Cements held at Queen's University*, Belfast, September 10–11 (1970).
3. British Standards Institution, Specification for High Alumina Cement. Part 2. Metric Units. BS 915: 1972 (amended 1995).
4. Frémy, E., *Compt. Rend. Acad. Sci. Paris*, 60, 993 (1865).
5. Bied, J., *Br. Pat.*, 8193 (1909).
6. Bied, J., *Recherches Industrielles sur les Chaux, Ciments et Mortier*, Dunod, Paris (1926).
7. George, C. M., personal communication (1997).
8. Lea, F. M., *The Chemistry of Cement and Concrete*, 3rd edn, Edward Arnold (Publishers) Ltd, London (1970).
9. Mangabhai, R. (ed.), *Calcium Aluminate Cements – Proc. Int. Symp.*, Queen Mary and Westfield College, University of London, July 9–11, E.&F.N. Spon, London (1990).
10. Bensted, J., *Zem.-Kalk-Gips*, 46(9), 560 (1993).
11. Rashid, S., Barnes, P. and Turrillas, X., *Adv. Cem. Res.*, 4(14), 61 (1991/1992).
12. Rashid, S., Barnes, P., Bensted, J. and Turrillas, X., *J. Mater. Sci. Lett.*, 13, 1232 (1994).
13. Capmas, A. and Ménétrier-Sorrentino, D., *UNITECR Proc. Conf. Global Advances in Refractories*, L. J. Trostei (ed.), Vol. 2, p. 1157, American Ceramic Society, Westerville, OH (1989).
14. Bensted, J. and Munn, J., *L'Ind. Ital. Cemento*, 715, 806 (1996).
15. Cottin, B. and George, C. M., *Int. Sem. Calcium Aluminates*, M. Murat *et al.* (eds), Politecnico di Torino, Turin (1982).
16. Taylor, H. F. W., *Cement Chemistry*, 2nd edn, Thomas Telford, London and San Diego (1997).
17. Quillin, K. C. and Majumdar, A. J., *Cem. Concr. Sci. Abs. (Oxford)*, 48 (1994).
18. Müller, D., Rettel, A., Gessner, W., Bayoux, J. P. and Capmas, A., *Proc. 9th Int. Cong. Chem. Cem.*, New Delhi, Vol. 6, p. 148 (1992).
19. Rutle, J., *Cementer-Fremstilling og Egenskaper-Andre Hjelpestoffer for Betongarbeider*, Fuglseth & Lorentzen A/S, Oslo (1981).
20. Bensted, J., *Proc. 9th Int. Cong. Chem. Cem.*, New Delhi, Vol. 4, p. 220 (1992).
21. Rodger, S. A. and Double, D. D., *Cem. Concr. Res.*, 14, 73 (1984).
22. Bensted, J., *Cem. Concr. Res.*, 24(2), 385 (1994).
23. Banfill, P. F. G. and Gill, S. M., *Adv. Cem. Res.*, 5(20), 131 (1993).
24. Bensted, J., *Adv. Cem. Res.*, 7(25), 39 (1995).
25. Cussino, I. and Negro, A., *Proc. 7th Int. Cong. Chem. Cem.*, Paris, Vol. 3, p. V/62 (1980).
26. Fentiman, C. H., *Cem. Concr. Res.*, 15, 622 (1985).
27. Osborne, G. J., *Proc. Inst. Civil Eng.*, Structures and Buildings, 104(2), 93 (1994).
28. Birchall, J. D., *Philos. Trans. R. Soc. London A*, 310, 31 (1983).
29. Kendall, K., Howard, A. J. and Birchall, J. D., *Philos. Trans. R. Soc. London A*, 310, 139 (1983).
30. Ma, W. and Brown, P. W., *Cem. Concr. Res.*, 22(6), 1192 (1992).

31. Majumdar, A. J. and Singh, B., *Cem. Concr. Res.*, **22**(6), 1101 (1992).
32. Osborne, G. J., 'Ash – Valuable Resource,' in *2nd Int. Symp. S. African Coal Ash Assoc.*, J. E. Krüger and R. A. Krüger (eds), Vol. **2**, p. 535, SACAA, Johannesburg (1994).
33. Marcdargent, S., Testu, M., Bayoux, J. P. and Mathieu, A., *Proc. 9th Int. Cong. Chem. Cem.*, New Delhi, **4**, 651 (1992).
34. Nelson, E. B., *Well Cementing*, Schlumberger Educational Services, Houston, TX (1990).
35. Bensted, J., *World Cem.*, **23**(11), 40 (1992).
36. Bensted, J., *World Cem.*, **23**(7), 30 (1992).
37. Bensted, J., *World Cem.*, **24**(5), 34 (1993).
38. Lewis, A., *Concrete*, **26**(5), 27 (1991).
39. Comité Européen de Normalisation (CEN), *Calcium Aluminate Cement: Composition, Specifications and Conformity Criteria*. (European Standard in preparation.)
40. Fernández-Carrasco, L., Blanco-Varela, M. T., Puertas, F., Vázquez, T., Glasser, F. P. and Lachowski, E., *Adv. Cem. Res.*, **12**(4), 143 (2000).
41. George, C. M., *Proc. 7th Int. Cong. Chem. Cem.*, Paris, **1**, Vol. p. V-1/3 (1980).
42. Harrison, W. H., 'Sulphate resistance of buried concrete. The third report on a long-term investigation at Northwick Park and on similar concretes in sulphate solutions at BRE', BRE Report BR 164. Building Research Establishment, Garston, Watford (1992).
43. Bensted, J. *World Cem.*, **22**(8), 27 (1991).
44. Capmas, A. and Scrivener, K., *Lea's Chemistry of Cement and Concrete*, 4th edn, Arnold Publishers, London (1998).
45. George, C. M., *The Structural Use of High Alumina Cement Concrete*, Lafarge Fondu International, Neuilly-sur-Seine (1975).
46. George, C. M. and Montgomery, R. G. J., *Mater. Const. (Madrid)*, **44**(228), 33 (1992).
47. Collins, R. J. and Gutt, W., *Mag. Concr. Res.*, **40**(145), 195 (1988).
48. Cather, R., Bensted, J., Croft, A., George, C. M., Hewlett, P. C., Majumdar, A. J., Nixon, P. J., Osborne, G. J. and Walker, M. J., *Concrete Society Technical Report No. 46: Calcium aluminate cements in construction – a re-assessment*, The Concrete Society, Slough (1997).
49. Neville, A. M., *Proc. Inst. Civil Eng.*, **25**(7), 287 (1963); and discussion thereof; **28**(5), 57 (1964).
50. Neville, A. M., *High Alumina Cement Concrete*, The Construction Press, Lancaster (1975).
51. Neville, A. M., *Conc. Int.*, **20**(8), 51 (1998), and discussion thereof; **21**(3), 7 (1999).
52. Anon., *New Civil Eng.*, pp. 5, 10–11, October 5 (1992).
53. Building Research Establishment, 'Sulphate and acid resistance of concrete in the ground,' *BRE Digest* 363, Garston, Watford (1991).
54. Bensted, J., *World Cem.*, **26**(7), 47 (1995).
55. Bensted, J., *World Cem.*, **26**(8), 12 (1995).
56. Midgley, H. G. and Midgley, A., *Mag. Concr. Res.*, **27**, 59 (1975).
57. Wilburn, F. W., Keattch, C. J., Midgley, H. G. and Charsley, E. L., 'Recommendations for the testing of high alumina cement concrete samples by thermoanalytical techniques', pp. 11, Thermal Methods Group, Analytical Division of the Chemical Society, Lyme Regis, Dorset (1975).
58. Midgley, H. G., *Proc. 1st Eur. Symp. Thermal Anal.*, Salford, September 20–24, D. Dollimore (ed.), p. 378, Heyden & Son Ltd, London (1976).
59. Bensted, J., *Proc. 7th Int. Conf. Chem. Cem.*, Paris, 1980, **4**, Vol. p. 377 (1981).
60. Department of the Environment: Collapse of Roof Beams – The Sir John Cass's Foundation and Red Coat Church of England Secondary School, Stepney, BRA/1068/2, May 30 (1974).
61. Double, D. D. and Hellawell, A., *Building*, September 13, p. 82 (1974).
62. Building Research Establishment, BRE statement on high alumina cement concrete testing methods, p. 2 October (1974).
63. Bensted, J., *World Cem. Technol.*, **13**(3), 117 (1982).
64. Bensted, J., *World Cem. Technol.*, **13**(2), 85 (1982).
65. Bensted, J., *Constr. Rep.*, **8**(1), 47 (1994).
66. Rashid, S., Turrillas, X., Bensted, J. and Barnes, P., *Cem. Concr. Sci. Abs. (Oxford)*, 29 (1994).
67. Guirado, F., Galt, S. and Chinchón, J. S., *Cem. Concr. Res.*, **28**(3), 381 (1998).
68. Kudryavtsev, A. B., Kouznetsova, T. V., Linert, W. and Hunter, G., *Cem. Concr. Res.*, **27**(4), 501 (1997).
69. Singh, B., Majumdar, A. J., and Quillin, K., *Cem. Concr. Res.*, **29**(3), 429 (1999).
70. Scrivener, K. L., Cabiron, J. L. and Letourneux, R., *Cem. Concr. Res.*, **29**(8), 1215 (1999).
71. Cather, R., Bensted, J., Croft, A., George, C. M., Hewlett, P. C., Majumdar, A. J., Nixon, P. J., Osborne, G. J. and Walker, M. J., *Concrete*, **31**(4), 17 (1997).
72. Chistyakov, V. V., Serbin, V. P., Smirnov, M. A. and Zavarina, N. T., *Tsement (St. Petersburg)*, 3/4, 34 (1991).
73. Grice, L. and Grice, M., *Three Score Years and Ten. A Personal View of Lafarge Aluminous Cement Co. Ltd 1923–1993*, L. & M. Grice, West Thurrock (1993).

Chapter five

Properties of concrete with mineral and chemical admixtures

S. Chandra

5.1 Introduction

Concrete is like a human body which works well depending upon the functioning of different parts which do their job and have synergistic effects, but how each part performs its work is far from being understood. The human body is one of the best automatic machinery systems on earth, with good synchronization effects of the roles played by different organs. In case of some defect the organ does not perform its role, the balance is disturbed and one feels pain, uneasy or uncomfortable. Analogous to this is concrete which is composed of various ingredients. These have simultaneous or subsequent effect. But how exactly they work is far from being understood.

In spite of the intensive research of the last fifty years, the hydration process of pure Portland cement-based concrete is not clear. Then, one cannot expect to understand the mechanism or mechanisms involved in the hydration process of the binder which often consists of mineral admixtures; fly ash, blast furnace slag, silica fume, rice husk ash etc., and chemical admixtures; water reducers, air entraining agents etc. These interact both physically and chemically during the hydration process. The microstructure depends very much upon the behaviour of these constituents, which control the properties of concrete.

In recent years, high strength concrete is developed in many countries. It is accomplished by the use of mineral and chemical admixtures which help in making the concrete with low water-to-binder ratio. It thus necessitates the need to understand the recently used cementitious materials, and their way of working in concrete production. This ultimately will help in producing concrete of high strength with high performance. In this chapter, cementitious materials, mineral admixtures, and chemical admixtures are discussed in relation to their effect on the microstructure formed during the hydration process and on the properties of the concrete. Concrete is exposed to very hostile atmospheric conditions and the demand for durability properties has increased; but the test methods used to evaluate this higher quality of concrete are the same. The need for developing test methods for long-term durability evaluation is emphasized.

Concrete is a composite material made of different ingredients which have to play a specified role failing which concrete of good quality will not be produced. It will be like a story of the four people who were assigned to perform a job. The job was not done as nobody was responsible for it. This is exactly what is happening in today's concrete:

This is a story of the four people,
Everybody, Somebody, Anybody and Nobody.
There was an important job to be done
and Everybody was asked to do it.
Everybody was sure that Somebody would do it.
Anybody could have done it.
Somebody got angry about that
because it was Everybodies job.
Everybody thought Anybody could do it,
but Nobody realized that Everybody would not
  do it.
It ended up that Everybody blamed Somebody
when actually Nobody asked Anybody.

5.2 Binders

Concrete consists of aggregates and a matrix which holds the aggregates together and produces a monolithic structure. This matrix is composed of binder and water. In order to produce a good quality of concrete, besides good quality materials, a good technique is needed. It is actually another very important and hidden component necessary to produce satisfactory concrete. Binders have been developed with time. In ancient times, concrete was made using lime-based binders, crushed red bricks as aggregates and water. Lime-based binder being not so durable, depending upon the value of the construction, organic admixtures have been used. These were locally available and were in the natural form. These natural admixtures significantly enhance the durability properties of concrete (Chandra and Ohama 1994). With the development of

Portland cement by Aspdin, lime was replaced by Portland cement. In the beginning of the nineteenth century other pozzolanic materials such as fly ash, silica fume, blast furnace slag, rice husk ash etc. have been used. These are also known as cement-replacement materials, supplementary cementing materials or mineral admixtures. In the American Concrete Institute terminology these are termed as 'mineral admixtures'. Such terminology is misleading and gives an impression that the Portland cement is the cement, the pure product and is the sole binder. The other products which have direct or latent pozzolanic properties are named as the supplementary cementing materials or cement replacement materials. Being the industrial by-products, the composition of these supplementary cementing materials is very wide and differs greatly. Nevertheless, Roy (1989) has presented the compositions in a simplified three-phase diagram (Figure 5.1). This is composed of the products generally used as the pozzolanic materials and belong to the system $CaO–Al_2O_3–SiO_2$. The composition of natural pozzolanic materials like santorin earth, pumice or volcanic tuff are not included here.

The term mineral admixture runs in parallel to the chemical admixtures. Such classification is

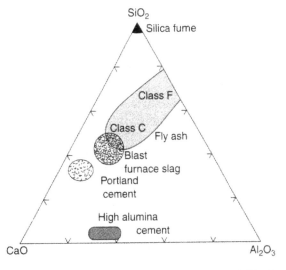

FIGURE 5.1 Compositions of cements and blending materials; simplified representation of components as $CaO–Al_2O_3–SiO_2$ compositions (Roy 1989).

hardly sound, considering the fact that the mass of chemical admixture represents a fraction of one per cent of the mass of binder while the mass of mineral admixture represents anything from a few percent to even 100 per cent. Most of them thus cannot be described as mineral admixtures. All these materials form part of what ASTM C-1157-92 calls 'blended hydraulic cements'. The various materials whether hydraulic or latent hydraulic like blast furnace slag or pozzolanic, will be referred to as cementitious material.

5.3 Interrelation of cement properties to concrete

Concrete is made of various components. It has not only two phases; binder and aggregates, but has more than that. In this situation, the question arises: whether cement properties highlight the properties of concrete? It seems logical to think that the properties of concrete will be determined by the properties of the binder and the aggregates used. This statement has considerable validity as the properties of the cementing materials control and determine the properties of the concrete to a great extent. It is previously understood that there is a clear interrelation between the cementing material and concrete. The other view point is that the interrelation is validated by important features of concrete which are absent in the cement paste, the binder phase. Nevertheless, the interrelation may be present in the ideal conditions, but in actual practice the validity of relationship decreases because of the non-existence of the ideal conditions in the field. The extensive study of a variety of cementitious materials has improved our understanding of the properties of these materials and of their behaviour. Still there is a problem in extrapolating these observations to concrete.

There are many variables in concrete which make it difficult to understand the mechanisms of interaction. In order to obtain a clear picture, the influence of a single variable or occasionally of two or three variables is to be considered. This instantly leads to the elimination of aggregates,

especially the coarse aggregates as the presence of aggregate makes for large variability and requires the use of relatively large specimens leading to thermal strains and inhomogeneity. So it is convenient to conduct tests on very small specimens of neat cement paste or at most on sand cement mortar. There is, of course, the temptation to generalize the results by putting the word 'concrete' in the title of the resulting paper.

Some researchers are aware of the difference between the behaviour of cement and the behaviour of concrete, but have an underlying belief that a better study of cement will some how reveal all that happens in concrete. One of the examples (Struble 1992) says 'Unfortunately there are other aspects of concrete performance whose links to cement behaviour are not understood. Still it is difficult to imagine that any aspect of concrete performance is not in some way related to cement behaviour'. It is further stated that 'For a given aggregate and aggregate content, one would expect a correlation between the consistency of cement paste at the same water-to-cement ratio and including mineral or chemical admixtures with the workability of concrete using that paste. If there were such a relationship between cement paste and concrete workability, then cement workability could be tested, controlled, and perhaps specified in order to provide predictable and uniform concrete workability'. However, such a correlation has eluded researchers mainly because it is too weak to be of practical value.

Specifically, as far as the workability is concerned, a correlation 'for a given aggregate and aggregate content' would be limited in application. This is because the various properties of the aggregate involved govern the workability. It should be remembered that concrete is not just a diluted cement paste, but a composite material of which cement paste is only one component. Furthermore, the Portland cement used is not pure. It is blended cement which is made with the addition of mineral and chemical admixtures. The properties of concrete made with this cement depend greatly upon the technique of mixing and curing and need precision. Due to this, the concrete made in practice differs from the one made in the laboratory and consequently has different properties.

There are at least three principal reasons why the observations in the cement paste cannot in some cases be automatically considered to be valid in concrete. First, in the hydrated cement paste, the inhomogeneity arising from the differential properties of the cement paste and of the aggregate with respect to the coefficient of thermal expansion or contraction, modulus of elasticity and Poissons ratio is absent. This inhomogeneity affects the stress field not only under load, but also in the drying conditions. The inhomogeneity can be harmful, for example when shrinkage cracking develops at the surface of large aggregate particles, but it can also be beneficial, for example when aggregate particles act as crack arresters. It is to be noted that the chemical shrinkage cannot be avoided whereas drying shrinkage can be controlled. Thus, the interaction of aggregates and cement paste affects the behaviour of concrete with respect to the microcracking or even in the development of larger cracks, and also with respect to permeability.

The second reason is the amount and distribution of air in concrete. In cement paste, the air content is very little. But in the case of concrete when aggregates are used, the air content increases. It is more influenced when chemical admixtures, especially the air-entraining agents (AEA), are used. It is not only the total amount of air which is important, but the size and uniform distribution of the air bubbles is of immense importance as it controls the microstructure of concrete and thus plays a decisive role in the engineering and durability properties.

The third reason originates from the fact that the concrete consists of not only hydrated cement paste and aggregates, but also contains the interface between these two components, Interfacial Transition Zone (ITZ). It is sometimes ignored or forgotten, yet it creates a 'wall effect' with a consequent differential local packing of dry cement and a differential microstructure of the hydrated cement paste in the ITZ. This zone has considerable consequences on the durability and permeability. It is of paramount importance when considering the action of mineral admixtures. This gives sufficient ground to confirm that the behaviour of cement paste cannot be necessarily related to the behaviour of concrete.

5.4 Hydration of cementitious materials

In the ancient concrete, quick lime (CaO) was used as a binder which was produced by burning limestone. It hydrated on contact with water, and subsequently carbonated on contact with the CO_2 from the atmosphere producing calcium carbonate, the starting raw material for producing lime. This, being a natural mineral found in nature, is stable. Thus, it makes a complete cycle.

$CaCO_3$ (Limestone) Burnt $\rightarrow$ CaO (quick lime)
CaO (quick lime) $+ H_2O \rightarrow Ca(OH)_2$
$Ca(OH)_2 + CO_2$ (atmosphere) $\rightarrow CaCO_3$

In this process, what is done is from a stone of random shape (limestone used for making lime) a stone of desired shape after hydration and carbonization of burnt lime is produced which being a natural mineral is stable in a wide range of environmental conditions. This is the clue as to why the ancient lime-based masonry is durable compared to that made with cement as a binder.

The carbonization process can be considered to take place in stages. Initially CO_2 diffusion into the pores takes place, followed by dissolution in the pore solution. It forms carbonic acid, which interacts with $Ca(OH)_2$ and forms very soluble calcium bi-carbonate $Ca(HCO_3)_2$ and finally $CaCO_3$ with further interaction. This precipitates out. Carbonic acid also dissolves part of the $CaCO_3$ formed. Besides this, reaction with the very soluble alkali metal hydroxide probably takes place first. This reduces the pH and more $Ca(OH)_2$ goes into the solution, and the process repeats.

$H_2O + CO_2 \rightarrow H_2CO_3$
$H_2CO_3 + Ca(OH)_2 \rightarrow Ca(HCO_3)_2$
$Ca(OH)_2 + Ca(HCO_3)_2 \rightarrow CaCO_3$

A solution of CO_2 can dissolve $CaCO_3$, with the formation of additional HCO_3

$CaCO_3 + CO_2 + H_2O \rightarrow Ca^{2+} + 2HCO_3$

and can similarly dissolve CH or Ca^{2+} and OH^- ions from C-S-H or hydrated calcium aluminate phases. The term 'aggressive CO_2' is properly defined as the quantity of CO_2 in unit volume of solution that can react with $CaCO_3$ according to the above mentioned equation.

The dissolved calcium carbonate recrystallizes with further interaction and with lowering in the moisture or rise in the temperature. In this way, larger crystals of preferred orientation are produced. Sometimes it inherits the inclusions of some impurities. This decreases the permeability of the masonry. Consequently, the strength and durability increases with time.

In other cases, when pozzolans were used together with the lime to enhance the strength and durability properties, the hydration products produced are: calcite, analcimes and zeolites. Still the minerals which are found in nature and are thus stable. X-ray diffraction analyses of ancient lime–pozzolanic concretes from Italy, Greece and Cyprus indicate that the calcite is the predominant crystalline phase together with some weak crystalline phases of analcime, $Na_2O \cdot Al_2O_3 \cdot 4SiO_2 \cdot 2H_2O$ (Langton and Roy 1984).

In the case of Portland cement, the hydration products are calcium hydroxide, calcium silicate hydrates, ettringite, monosulphate, calcium carbonate etc. It does not make a complete cycle. Besides, these hydration products except calcium carbonate not being natural minerals found in nature, are not stable. Due to this, there is always chemical interaction causing differentials leading to the formation of cracks, and development of osmotic and hydrostatic pressures. These consequently make the concrete structure weak. Beside the physical destruction, there is also chemical degradation of concrete due to the chemical transformations of the hydration products. Lime, lime–pozzolan and cement cycles are shown by schematic diagrams in the Figures 5.2–5.4 respectively.

Investigations have revealed that, while ancient concrete in Roman structures has remained unaffected by severely corrosive conditions such as flowing water and salt laden air, in a period of two thousand years, modern Portland cement concrete has suffered extensive damage in the

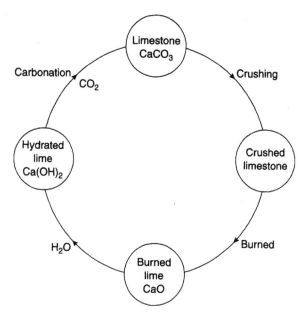

FIGURE 5.2 Lime cycle.

same localities and under the same conditions in a period of ten years (Malinowski 1979; Malinowski et al. 1961). This confirms the theory of recrystallization, stable mineral formation as described before, lowering the permeability with time and strengthening the structure and enhancing the durability properties, in contrast to the Portland cement-based concrete where the permeability increases and strength decreases with time.

As in Portland cement, this research assumed that the binding properties of the ancient cements are due to the production of calcium silicate hydrates (C-S-H gel), which are colloidal precipitates and have varying composition. Indeed, the chemical composition of these cements significantly differs from that of the Portland cement due to the presence of a very high amount of amphoteric oxides (Al_2O_3, Fe_2O_3), acidic oxides (SiO_2) and the presence of the oxides of the alkaline earth metals, for example, sodium and potassium. It is erroneous to think that the C-S-H gel accounts for the durability of ancient concrete. It has been observed that the hydrated products of modern Portland cements composed of C-S-H gel deteriorate under conditions in which ancient cements remain intact. This is explained by the

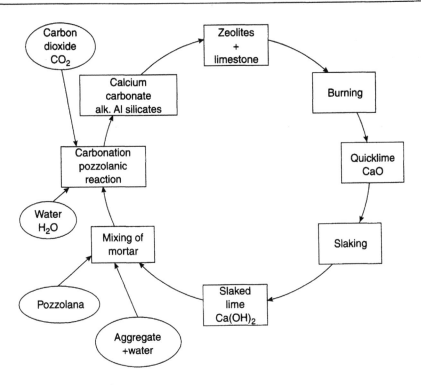

FIGURE 5.3   Lime–pozzolan cycle.

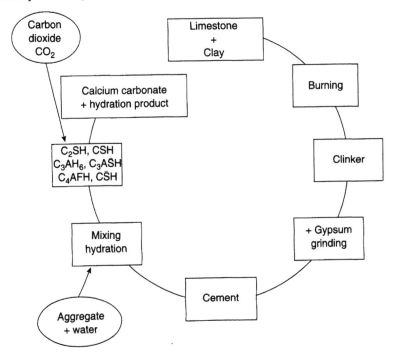

FIGURE 5.4   Cement cycle.

fact that the durability and strength of the latter is due to the presence of alkaline silico–alumino hydrates, and that one of the structural formation components in these concretes is the amorphous-crypto-crystalline zeolite, sodium–potassium analcime, and not the carbonation of lime (Davidovits 1986; Davidovits and Courtois 1981).

Recarbonation of lime in such concretes takes place very slowly, over several days and months, while the synthesis of analcime can occur in a few hours. Therefore, the recarbonated lime appears to act as an inert filler when combined with an analcime matrix.

Modern concrete is the concrete based on Portland cement saturated with the basic minerals C_3S, C_2S, C_3A and C_4AF. Though Portland cement concrete has a lot of good properties, there are disadvantages also associated with it, such as high energy required for producing cement, poor or lack of reaction with clay and dust particles, chemical interactions and formation of undesirable deleterious products, the necessity of using both coarse and fine aggregates for producing rich concrete etc. All these disadvantages are due to the fact that Portland cement contains a great amount of calcium oxide (CaO 63–67 per cent). As a result, the products of hydration together with CSH act as physically and chemically active compounds/$Ca(OH)_2$; (3–4) $CaO \cdot Al_2O_3$ (10–13) H_2O, (3–4) $CaO \cdot Fe_2O_3$ (10–13) H_2O, $3CaO \cdot Al_2O_3 \cdot 3CaSO_4 \cdot 31H_2O$; (1.5–2) $CaO \cdot SiO_2 \cdot nH_2O$ etc., that is, the minerals that practically do not exist in nature. These interact actively with the surrounding environment and are thus subjected to various physical, chemical changes during the process of exploitation and alteration.

The hydration process of Portland cement is well known and will not be discussed here in detail. However, based upon isothermal calorimetry, a classification of hydration has been done. It is shown schematically in Figure 5.5 (Roy 1989).

5.4.1 Hydration with mineral admixtures

The most widely used mineral admixtures are fly ash, blast furnace slag, and silica fume. Some

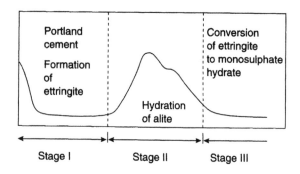

FIGURE 5.5 Classification of hydration stages by isothermal calorimetry (Roy 1989).

times rice husk ash is also used. Colloidal silica, a new product with the trade name 'Cembinder' has recently been developed. It is man made and contains tiny pure silica particles in dispersion form. It is highly pozzolanic and has shown very promising results. The fly ash is generally used at replacement levels of up to 30 per cent. For slag replacements, levels at around 40 per cent and 70 per cent are used for different applications. With these two minerals, the development of microstructure during hydration is slightly modified from the pattern obtained without their addition. The replacement levels for silica fume are much lower, typically around 10 per cent, but the effect on the microstructure development are very pronounced (Scrivener 1989). In the case of Cembinders, the amount used is very low 2–4 per cent to the weight of the binders, and it is thus used as a modifier. It spontaneously interacts with calcium hydroxide and thus also has significant influence on the microstructure development (Chandra and Bergqvist 1997).

In the case of mineral admixtures, the hydration process differs from Portland cement depending upon the material used. Blast furnace slag, for example, has latent hydraulic property. It has to be activized. Fly ash F, rich in iron, has pozolanic character, whereas fly ash C is rich in calcium hydrates itself. On the other hand, silica fume, rice husk ash has huge amount of amorphous silica, and Cembinder is pure amorphous silica and thus behaves like reactive pozzolan. The hydration processes involved are given below.

5.4.1.1 Hydration of fly ash cement

The effect of fly ash on the hydration of cement paste and concrete is still not very well understood (Roy 1989). This is because fly ash is very inhomogeneous and its chemical composition may differ between two or even within one single particle. Fly ash particles which are essentially inert in the early age, are reported to enhance hydration of cement at a later stage. These are slow-reacting pozzolanic materials. For example, the amount of calcium hydroxide and non-evaporable water produced are higher for pastes containing fly ash (Larbi and Bijen 1990). This phenomenon which also occurs in the other finely-divided non-reactive mineral powders is attributed to the existence of these grains between the cement (Yamazaki 1989), which results in increased space available for the growth of the hydration products.

In addition to enhanced cement hydration, fly ash by itself has the ability to react typically with Ca^{2+} and OH^- ions released from the cement to form secondary products. In this sense, as has been mentioned before, class F fly ash is a pozzolanic material. Class C fly ash, on the other hand, may contain a higher amount of CaO and some C_3A, C_2S etc. is able to hydrate by itself, and is more of a cementitious nature.

McCarthy and Tishmack (1994) have investigated fly ash from Shawnee Pilot Plant, Paducah, Kentucky (KY), USA; bubbling bed atmospheric fluidized bed combustion technology (AFBC); and high-sulphur bituminous coal: mixed fly ash and char. They have studied a paste of KY AFBC and char at a water-to-solid ratio of 0.45 cured at 100 per cent RH for three months. It is shown that all the lime has been hydrated to portlandite, and much of the anhydride has hydrated to gypsum. The peaks in the XRD pattern were very sharp and distinct, which indicates good crystal formation. Ettringite is a prominent hydration reaction product. There is also some calcite which formed by partial carbonation of portlandite.

Besides the chemical interactions fly ash also densifies the microstructure which is measured by the Mercury intrusion porosimetry. One of the example of such densification is shown in the Figure 5.6 (Pieterson 1993) for three types of fly ashes.

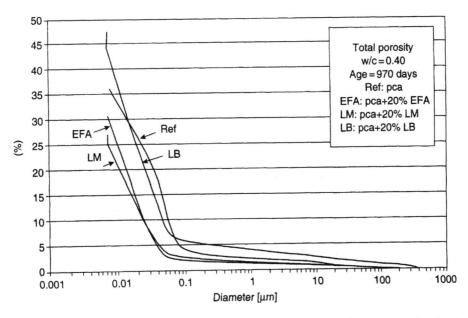

FIGURE 5.6   Pore size distribution measured by the mercury porosimetry of an ordinary Portland cement and three fly ash composite cements respectively using fly ashes EFA, LM and LB in the order of increasing particle size (Pietersen 1993).

5.4.1.2 Hydration of blast furnace slag cement

The principal hydration products formed in blast furnace slag cement are similar to those formed in Portland cement, CSH gel, C_3AH_6 and Aft. Increasing slag contents gradually results in the formation of C_2ASH_8 (incompatible with CH) and silicious hydrogarnets (Atkins *et al.* 1991). Pieterson (1993) studied the interaction of slag with cement. His observations are shown in Figures 5.7 and 5.8. Figure 5.7 shows a back-scattered image of a reacted zone enveloping a slag grain in blast furnace slag cement hydrated for 90 days. Here the differences in composition along the slag edge are closely visible. The two smaller maps represent XRD analyses of Mg and Al respectively, and indicate the relative concentration of these elements at the slag edge due to the diffusion of calcium and silica. In Figure 5.8, it is seen that the Inner Slag Hydrate (ISH) indeed seems to have crystallized to well-formed platelets, presumably a hydrotalcite (HT)-like phase which is known to be capable of being formed in water-deficient

regions. The apparent high internal porosity is noteworthy.

It clearly shows that a net transport of Ca, Si, and Al from the slag centre has taken place; magnesium seems to act as a chemical 'goalpost' as suggested by Feng *et al.* (1989). It is also evident that chemically different zones are formed approximately at the position of the original slag boundary; apparently these zones remain, once they are formed at their place, and may act as a chemical barrier or membrane. Feng and Glasser (1990) have introduced the term ISH which reflects the inside of the reacting slag grain, and the term Outer Slag Grain (OSG) which stands for the additional hydrates formed outside the original slag grain boundary.

The element transport mentioned above results in relatively porous but isolated areas formerly occupied by the interior of the slag grain (ISG). The densification of the slag cement paste outside the slag grains results in an overall reduced permeability. The total porosity of the blast-furnace slag cement pastes changes only little in this process relative to pastes without blast-furnace slag. The densification of the microstructure of the paste is, of course, well known from the pore size distribution, as, for instance, measured by mercury

FIGURE 5.7 Back-scattered image of a reacted zone enveloping a slag grain in blast furnace cement hydrated for 90 days (Pietersen 1993).

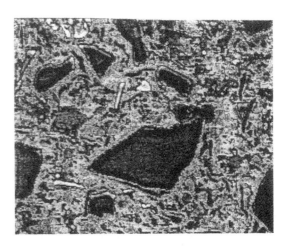

FIGURE 5.8 Scanning Transmission Electron Microscopy micrograph of a well-crystallized hydrotalcite-like phase which has been developed within the original contours of a fully hydrated slag particle (ISH) (Pietersen 1993).

porosimetry as is shown in the case of fly ash (Figure 5.6).

5.4.1.3 Hydration of cement with silica fumes

Silica fume contains particles as fine as 0.1 μm or less and has high surface energy; it partially dissolves in saturated CH solution within 5–15 min, and a SiO_2-rich hydrate is deposited in layers or films on the silica fume particles. C-S-H having a C/S ratio of about one has been reported to form at 20°C in 24 h after mixing with water and at 38°C in only 6 h (Grutzeck *et al.* 1982). The reaction of silica fume with CH is affected by the specific surface area, and the surface energy. Typically the C-S-H produced has a lower C/S ratio than that of fly ash. Since C-S-H with a C/S ratio of less than 0.8 is considered to be thermodynamically unstable, calcium silicate hydrate once produced is converted to a C-S-H variety having the lower limit of C/S which coexists with excess silica containing very little calcium. This lower C/S ratio tends to be confirmed in reports of Taylor (1986). It is noticed that there is a change in the composition of the pore solution (Diamond 1982; Glasser *et al.* 1987; Roy 1986) with time. This shows that the hydration process is not only in the early age, but continues with time. After the early stage reaction Li *et al.* (1985) have shown that much silica fume remains unreacted after seven days. The silica fume particles surround each cement grain, densifying the matrix and filling the voids with strong hydration products. Grutzeck *et al.* (1983) has reported that the Ca ion is actively dissolved from alite, its adsorption on the silica surface lowers the calcium ion concentration in the liquid phase and the hydration of alite in the early age is accelerated. At the same time, the $CaSO_4$ content is also lowered and the hydration of the interstitial phases C_3A and C_4AF is therefore accelerated. C-S-H formation by pozzolanic reaction is initiated at around ten hours and it progresses in the period from 1–7 days. C-S-H with low C/S ratio is produced, and the amount of $Ca(OH)_2$ remaining in the hardened paste is decreased.

5.4.1.4 Hydration of cement with colloidal silica: cembinders

Silica colloid: cembinders, are aqueous silicic acid suspensions. These contain silicic acid particles expressed as silica, in an amount in the range from about 8 to 60 per cent silica by weight of solution. The specific surface area of the particles is in the range of 50 to about $200 \, m^2/g$. Generally, the aqueous silicic suspension employed includes silicic acid particles which have a wide particle size distribution with the particles ranging from 5 to 200 nm. Cembinder reacts chemically with the calcium hydroxide produced during hydration of Portland cement and produces C-S-H. The reaction is spontaneous. Thus, its addition increases the early strength of concrete (Chandra and Bergqvist 1997). This reaction is marked by the consumption of calcium hydroxide determined after a particular interval by XRD and thermogravimetric analyses in a mixture of cement and Portland cement at 0.35 water-to-cement (w/c) ratio. There is a significant decrease in the peak of calcium hydroxide after one day of hydration. This decreases with the increase in the amount of cembinder and time (Figure 5.9). Similar observations were made by thermogravimetric analyses (Figure 5.10) where weight loss between 105–400°C is related to the non-evaporable water representing C-S-H formation and the weight loss between 400–500°C is related to calcium

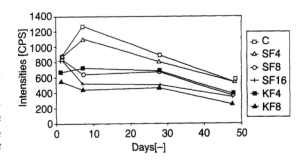

FIGURE 5.9 Quantity of calcium hydroxide calculated as counts per second (at $2\theta = 18.10°$) (Chandra and Bergquist 1997). C=Reference; SF4, SF8, and SF16 represents mixtures with 4, 8, and 16 per cent condensed silica fume 4; KF and KF 8 are 4 and 8 per cent colloidal silica respectively, w/c 0.35.

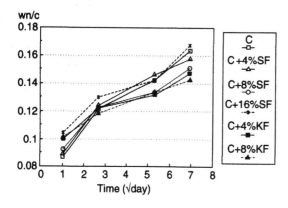

FIGURE 5.10 Weight loss between 105°C and 400°C to the weight of cement with respect to time (Chandra and Bergqvist 1997).

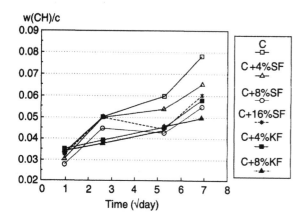

FIGURE 5.11 Weight loss between 400°C and 500°C to the weight of cement with respect to time (Chandra and Bergqvist 1997).

hydroxide. It is shown that the amount of non-evaporable water increased up to seven days with colloidal silica addition. Later the increase is not substantial (Figure 5.11). It shows that the reaction becomes slow as most of the calcium hydroxide is already consumed.

Reactivity of condensed silica fume marked SF in this case and colloidal silica marked KF with CH was done by mixing them in particular percentages to the standard Portland cement at 0.35 w/c ratio. Free CH was measured at particular

intervals. It is seen from the Figure 5.8 that the amount of CH substantially decreased in all the cases with respect to the reference sample except in the case of 4 per cent SF where the difference is not so very big at seven days. This tendency continued also even at the later stage whereas with the addition of 8 and 16 per cent SF, there is not much decrease at one day but at later stage substantial decrease in the free CH is observed. In the case of colloidal silica KF, the amount of free CH is significantly low even after one day. Four per cent KF addition consumed 60 per cent more CH after seven days than 4 per cent SF. Colloidal silica has shown more reactivity in all the cases. This is attributed to its fine particle size and purity.

5.5 Superplasticizers

Superplasticizers (SP) are surface active agents and have water-reducing character. These are also known as high-range water reducers, fluidifiers, plasticizers etc. This implies that it is possible to produce a concrete with lower water-to-cement ratio with their addition. High-performance concrete (HPC) and high-strength concrete (HSC) are generally made with high-cementitious materials. It causes difficulties in obtaining good workability and in homogeneous dispersion of the cementitious material. This problem can be overcome by adding more water, but this decreases the strength. This creates the necessity of using water-reducing admixtures. High-cementitious materials and low water-to-binder ratio may create other problems such as early age cracking. It occurs due to the loss of workability and the very quick drying at the open surface caused by the dual effects of lack of bleeding, and the lack of bleed water to move up to the surface. High heat of hydration is another major factor causing thermal cracking with high cement concrete. It is reported that the slag reduces the temperature rise during hydration. But it does not give early high strength to the concrete. However, a combination of high-range water reducer and fine ground slag which significantly improves the strength development can be a possible solution (Swamy 1999).

5.5.1 Influence of cement type on superplasticizing admixture

It is now recognized that the superplasticizers behave differently with the same cement and different cements behave differently with the same superplasticizer. The various superplasticizers available are neither equivalent in their chemical characteristics nor in their functional properties. Even if the superplasticizers meet standard specifications, it does not mean that they are the same. The standard specifications only define minimum or maximum values of certain parameters so that a particular superplasticizer may meet the minimum standard, while an other may exceed this minimum specification by a large margin. This is the reason why superplasticizers sometimes work and sometimes do not.

Superplasticizers' behaviour with different cements mainly depends upon the sulphate minerals in the cement. The type of sulphate in the cement has a major effect on the viscosity and yield. Claisse *et al.* (1999) have shown that the anhydride gives substantially greater yield values than the gypsum. The type of superplasticizer has

influence on the workability of cement paste. For example, sulphonated naphthalene formaldehyde condensates (SNF) make the paste more workable than sulphonated melamine formaldehyde condensates (SMF) followed by Lignosulphonates (LS). Influence of variation in the composition of clinker mineral is elaborated by Kinoshita and Okada (1999). They have tested cements which varied in the clinker composition; C_3S, C_2S, C_3A, C_4AF and in fineness, with the superplasticizers synthesized in the laboratory (Methacrylic Water-Soluble Polymers; MSPs). These varied in molecular weight. Comparison is done with γ-naphthalene-based polymer. Cements used were low heat Portland cement L1, belite rich Portland cement L2, normal Portland cement N and early high strength Portland cement H. Composition of the polymers and cements are shown in the Tables 5.1 and 5.2 respectively.

It has been shown that the apparent saturated adsorption of MSPs was significantly lower than that of NSF, and MSPs with longer polyethylene graft chains led to the lower adsorption. This may be because MSPs with graft chains form a bulky, stereo-structural layer when adsorbed by cement

TABLE 5.1   Properties of the water soluble polymers, POE is polyoxyethylene (Kinoshita and Okada 1999)

Mark	Kind of superplasticizer	Mol. wt. of polymer	Mol wt. of POE, graft chain	Graft proportion of POE (%)
MSP 1	Methacrylic water-soluble polymer	37000	1000	240
MSP2	Methacrylic water-soluble polymer	43000	2000	340
MSP3	Methacrylic water-soluble polymer	43200	3000	340
NSF	γ-Naphthalene based polymer	2500	0	0

TABLE 5.2   Composition of cements (Kinoshita and Okada 1999)

Mark	Type of cement	C_3S	C_2S	C_3A	C_4AF	Specific surface area (cm²/g)	Specific gravity
L1	Low-heat Portland	27	58	2	8	3350	3.22
L2	Belite-rich Portland	35	46	3	9	4080	3.20
N	OPC	52	23	9	9	3250	3.16
H	High early strength	64	11	8	9	4340	3.13

S. Chandra

particles. This is in contrast with the NSF; a copolymer in the shape of stiff chains, forming a flatter layer on cement particles. In other words, MSPs are considered to produce strong stearic repulsion (entropy effect), rather than electrostatic repulsion, by being adsorbed by cement particles and forming a barrier, which provides an excellent cement-dispersing capability. It is also shown from the exothermic rate curves of cement pastes made with Portland cement and a constant dosage of MSPs that the longer the polyoxyethylene (POE) chain is, the weaker is the set-retarding effect on hydration.

Flow values of cement paste made with normal Portland cement and MSPs at w/c ratio 0.30 were similar. But when the w/c decreased to 0.25 and 0.20, the differences between the cement dispersing capabilities of MSPs became significant. Polyoxyethylene with the longest chain showed the highest dispersing capability. Further, it is reported that it needs lower dosage of MSPs compared to NSF for good cement dispersion. With the use of MSP1 and MSP3, the fluidity of cement L1 was the highest followed by N and H. It was observed that at w/c = 0.50, the fluidity-retaining capability of polymers with shorter graft chains is higher than that of the polymers with longer chains. However, the differences are narrowed in the range of low w/c ratio. MSPs impart the highest fluidity to low heat cement followed by normal cement and high early strength cement. In the low w/c range, low heat cement requires lower dose of MSP than normal cement. This curbs the adiabatic temperature rise of concrete without retardation in setting. It is concluded that MSP3 and low heat cement used in combination make possible the production of high performance concrete having high fluidity, low heat of hydration, and high strength. Further, it is reported that the incompatibilities of cement/superplasticizers can be overcome by the use of high molecular weight SP for normal alkali cement (Page *et al.* 1999). For low alkali cements, the use of an SP having a high residual sulphate content improves the rheology with time and decreases the SP dosage.

The time at which the SPs are added also plays important role. It has been shown that the difference of flow of cement paste with SNF and amino sulphonic acid base (AS) between simultaneous addition and later addition was larger than those with polycarboxylic acid base (PC) and lignosulphonate (LS) (Uchikawa 1999; Uchikawa *et al.* 1995). Particularly, the flow of cement paste with SNF added by later addition was larger than that added by simultaneous addition. This increases slump and causes bleeding and segregation problems. This can be overcome by using some thickening agents. For coarsely ground cements having a low C_3A content, a small amount of a viscosity-enhancing agent such as Welan gum is recommended to prevent bleeding and segregation in high slump concretes (Ambroise *et al.* 1999). In such cases, higher dosage of SP can be used to have better rheological properties.

5.5.2 Fluidity and temperature

The fluidity and the setting time are influenced by the ambient temperature. An increase in the ambient temperature leads to an increase in the fluidity of cement paste over the range 5–45°C (Roncero *et al.* 1999). However, the saturation dosage of the SP is not affected by the change in the temperature. This confirms that the adsorption of SP molecules on the cement particles depends on the surface area. Once this surface area has been covered, any further increase in the dosage does not lead to an improvement of the fluidity. Furthermore, it is shown that the copolymer of oxyethylene–oxypropylene is more effective than the SNF. Since the former acts principally by generating a repulsive force between the cement particles through steric hindrance. This is built up due to the adsorption of the large molecules that prevent the proximity between the cement particles. While the latter acts due to the electrostatic repulsion due to the electrical charge of the adsorption layer.

The water demand of the cement at normal paste consistency increases with temperature and decreases at a given SP dosage (Roncero *et al.* 1999). This decrease occurs until a certain dosage beyond which the water demand is practically constant. It is the same for all temperatures. SP leads to the retardation of the setting process which increases significantly with the increase in SP dosage up to the saturation point.

The slump of concrete is also greatly influenced by the ambient temperature. It is reported that the slump loss of mixes prepared at 30°C/40 per cent RH was faster than those prepared at 20°C/65 per cent RH (Ben-Bassat and Baum 1999). The setting time was also influenced by increase in the temperature. However, the presence of SP resulted in extended setting time, compared to the reference mixes. Bleeding has been reported less in dryer exposure conditions than at 20°C/65 per cent RH. Extended setting time has influence upon the hydration process, so much so that the concrete containing SP gained higher compressive strength than the reference mixes. Furthermore, it improves the microstructure, the capillarity become lower. Consequently, mixes containing SP show lower total water absorption than the control mixes. This difference was more pronounced at higher temperature.

5.5.3 Mechanism of interaction of SP with cement

Commercially known superplasticizers are: SNF, SMF, and alkali salts of LS. Newly developed SPs are based on PC, and AS, which are known as new age superplasticizers. A brief description of these new age superplasticizers is given here (Uchikawa 1999; Uchikawa *et al.* 1995).

5.5.3.1 PC based admixtures

There are two types of PC: (i) Those containing mainly copolymers of unsaturated chain hydrocarbon olefins with carboxylates including maleic acid (CHOOC–HC–CH–COOH): Admixtures composed of copolymer of olefin with maleic acid, however, have limited water-reducing ability. Therefore, it is not widely used at present but the admixtures composed of acrylic acid with acrylic ester occupy the position of the popular high range water reducer; (ii) This includes a copolymer of an acrylic acid (CH_2–CH·COOH)

with acrylic acid ester with many ester bonds (–C–O–C–) in the side chain forming so called graft polymer. This structure increases the steric repulsive force and contributes to the dispersion of solid particles. Recently, a new type of admixture composed of water-soluble methacrylic polyoxyethelyne graft polymers containing both carboxyl groups and sulphone groups is developed. However, the slump loss using this admixture is extremely small though the range of dosage without separation of material is narrow.

5.5.3.2 AS based admixtures

AS base admixtures are three-dimensional polycondensation products of an aromatic aminosulphonic acid with trimethyl phenol.

5.5.4 Superplasticizer adsorption on solid particles

Mechanism of interaction of superplasticizers with cement is a complex process. However, adsorption of SP on the solid particles is thought to be mainly responsible for dispersion of cement. These form an adsorption layer on the solid surface. The absorption in the water medium depends upon the kind of solid particles including cement compounds, blending components, cement hydrates and the type and molecular weight of the polymer.

The typical state of adsorption is illustrated by train-and-loop system in Figure 5.12. The train portions are large in a high adsorption energy system and /or at low temperature. Hence the stability of this dispersion system is high, while the loop and tail portions become large in a low

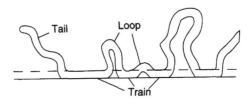

FIGURE 5.12 Train–loop–tail adsorption of organic admixture to solid (Uchikawa 1999).

adsorption energy system and/or at high temperature, hence the stability of the system is low. The ratio of the train portion to the whole-admixture molecules adsorbed is called the 'bound function' which is used as an indication of the stability of the dispersion system.

The motive force for the adsorption of the admixture to the surface of the cement particle includes van der Waals force, the electrostatic forces by the ion exchange, formation of ion pairs and polarization of electron, the hydrogen bonds of atoms in the admixture with the atoms on the surface of cement particles, formation of complexes with Ca-atoms on the surface of cement particles and the hydrophobic bonds with cement particles by the affinity of the hydrophobic group in the admixture (Uchikawa *et al.* 1992). It has been reported that the mechanism of the adsorption of the admixture by the cement hydrates, in addition, includes the substitution of the OH groups in the cement hydrates for sulphonic and carboxylic groups in the molecule of the admixture, and the entering of the admixture between the layers of the hydrates with laminated structure.

5.5.4.1 Adsorbed amounts of admixture to cement compounds, blending components and cement hydrates

The adsorption is different on different clinker minerals. It is more on C_3A, C_4AF and CaO compared to those on C_3S, C_2S etc. This subsequently depends upon the type of cement and the type of SP used. It is generally in the following order: $NS = AS > PC > LS$.

Since calcium sulphates and alkali sulphates contained in cement are rapidly dissolved in the liquid phase in the form of SO_4^{-2}, the ion adsorption of an admixture containing a sulphonic acid group to the interstitial material is decreased and the calcium concentration increases. The dispersion of the solid particles depends upon the interactive force between cement particles coexisting with organic admixture which depends upon the distance between them, and is related to the interactive energy.

5.5.5 Influence of the type of admixture on the morphology of hydrates

Besides influencing the dispersion process of the solid particles, SP also influences the hydration process. Uchikawa (1999) has reported that at 0.27 w/c, the Ettringite (Aft) crystals are of the same dimension as in the case without SP. After 2 h, the crystal size became a little larger but had the same morphology. At the age of three days, the type III C-S-H crystals produced at the surface of unhydrated cement particles are about 0.2–0.3 mm while the large crystals of CH are not seen. On the other hand the equidimensional irregular Aft crystals of the size as large as 0.2–0.3 mm are observed in the case of NS after 5 min. After 2 h the structure of the hydrates is similar to the one produced after 5 min. After three days, the surface of the unhydrated cement particles is covered with type III C-S-H crystals; their agglomerates and large crystals of CH are not seen. With the increase in the w/c ratio the crystals of hydration products including AFt, C-S-H and CH are enlarged, and the surface of unhydrated cement particles becomes almost covered with type I C-S-H crystals. The crystallinity is improved.

C-S-H's are formed when Portland cement hydrates. These are intermediate in crystallinity and have different morphologies. It is therefore essential to mention the difference between them. Four types of C-S-H gels are distinguished (Diamond 1976); Type I is prominent at the early ages, it is a fibrous material, the fibres being up to about 2 μm long. Type II, forms honeycombs or reticular networks. It is considered to be very rare in the pure calcium silicate paste, but later work showed that the material resembling it is a normal early product. The most clearly defined of these semi-crystalline phases are C-S-H I and C-S-H II, which are structurally imperfect forms of 1.4 nm tobermorite and jennite respectively. Type III, is prominent in the older pastes, is more massive and consists of tightly-packed equant grains up to 300 nm across. Type IV, is still more featureless and massive.

5.5.6 Mechanism of change in performance of fresh concrete by cements

The rapid decrease of fluidity and increase of slump loss with time often observed in the fresh concrete using admixtures are not clearly understood. It is reported that the depositing time and amounts of ettringite (AFt) and monosulphate (AFm) in the early stages of hydration of cement and the thickness and density of the adsorption layer of admixture on the surface of the cement particles and cement hydrates are responsible for change in the performance of concrete in the early stage. AFt has a stronger influence on stiffening than AFm because of its morphology and the amount produced in the early stage of hydration.

Production of AFt is more accelerated by calcium sulphates than alkali sulphates. Na_2SO_4 gives higher acceleration effect than K_2SO_4, although the morphology of AFt varies according to the kind of admixture used and the w/c ratio. The small crystals produced during the hydration with the use of SP increase the specific surface area of hydrated cement.

SO_4 concentration in the mixing water of fresh concrete in the initial stage is higher when the cement contains large amounts of hemihydrate, particularly γ-form and alkali sulphates. In this case, SO_4 is preferentially adsorbed and combined with the interstitial material of cement-forming AFt and thereby reducing the adsorption of the admixture. As a result, repulsion by the adsorption of admixture to the cement particles and cement hydrates is reduced and a large amount of AFt is produced, causing rapid decrease of fluidity and the increase of the slump loss with time.

In the PC admixture COOH is located on the outer side of the polymer and thus has a good possibility of coming into contact with the cement particles and cement hydrates. However, in general cases, the calcium sulphates and alkali sulphate coexist; the COOH group is embedded in the inner side of polymer because alkali sulphate decreases the polymer rotation radius, that is, the spread of the polymer. Therefore, the opportunity for the COOH group to make contact with the cement particles and cement hydrates is decreased. Consequently, the adsorbed amount and thickness of the adsorption layer of polymer is remarkably reduced and the steric hindrance of the adsorbed admixture is weakened. This brings depression of cement particles and cement hydrates and decreases the fluidity.

The influence of the molecular weight of polynaphthalene sulphonate (PNS) SP on the rheological behaviour of cement pastes depends on the alkali content in cement. It is reported that the high molecular weight of PNS is more effective in fluidizing pastes made with high alkali cements than low molecular weight PNS (Kim *et al.* 1999). However, PNS molecular weight has little effect in fluidizing pastes made with low alkali cement. The adsorption is much more with low alkali cements than with high alkali cements. In high alkali cement, the addition of PNS SP initially retards cement hydration in first few hours and then accelerates it. The difference in the molecular weight of PNS SP do not appear to significantly affect cement hydration in high alkali cement. In low alkali cement, the induction period is increased by the addition of low molecular weight PNS, but not by the addition of relatively high molecular weight PNS.

5.5.7 Functionality of superplasticizer

Functionality of concrete can be enhanced by retarding the setting time of fresh concrete and thus retain the slump for a longer time. It will make possible long transportation of fresh concrete without change in the flowability. Takeuchi and Nagataki (1999) have reported that by the use of gluconate SPB (polycarboxylate type) as an SP, slump can be maintained for 3–4 days. These do not adversely affect the compressive strength of concrete and the final setting time. It is further reported that compared with the four superplasticizers based upon amino sulphonate type (SPA), naphthalene sulphonate (SPC) type and melamine sulphonate type (SPD), the percentage of SPB remaining in the liquid phase is highest followed by SPA, SPD and SPC in that order. This agrees with the order of high slump retention of mortar. The higher the percentage of SP in the liquid phase, the higher will be the slump retaining power.

5.6 Interfacial transition zones (ITZ)

Generally, the engineering properties of the concrete are governed by the characteristics of the bulk cement paste matrix, as has been pointed out by Farran (1956) some forty years ago. Extensive research in the last few decades has shown that the paste structure in concrete is not identical with the structure of the bulk paste due to the formation of interfacial transition zone (ITZ) around inclusions in the cement paste, for example, aggregates, fibers etc. The structure of the ITZ depends upon many parameters amongst others on the composition of the binder. The microstructure of the paste varies with the distance from the inclusions. Different methods have been developed for studying the composition and structure of the ITZ and its changes as a function of the distance.

ITZ are more influenced in the high strength concrete where the inclusions in the cementitious matrix are not only the inert fillers, but are active components. The modern cementitious materials are essentially composites, therefore the ITZ plays a much more important role than in traditional concrete.

Generally, it is understood that the concrete is a two phase system: the matrix and the aggregates. The composition and the microstructure of the matrix varies between the aggregate surface and the bulk paste. The heterogeneity is very pronounced in the vicinity of the aggregate particles and dilutes going away from the aggregate (Ollivier et al. 1995). This phase is known as the ITZ and can be taken as the third phase. A schematical illustration of the ITZ is given in Figure 5.13 (De Rooij et al. 1998).

When water is mixed in mortar and concrete, a thin film is formed directly at the surface of the aggregates. This film is characterized as a 'duplex film' (Barnes et al. 1978). It consists of calcium hydroxide, CH, and a thin single layer of short fibers of C-S-H gel in an open parallel array on its paste-facing surface. Some researchers have observed only the network of fine ettringite crystals which have been deposited on the aggregate surface in the initial hydration (upto the age of 10 h) (Zimbelman 1978). Next to the duplex film there is a contact layer mainly of CH crystals with the c-axis of their hexagonal unit cells perpendicular to the aggregates surface (Strubble et al. 1980;

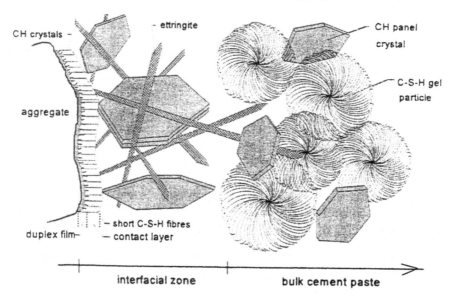

FIGURE 5.13 Schematic illustration of the ITZ (De Rooij et al. 1998).

Zimbelman 1985). This contact layer serves as an intermediate layer between the duplex film and the bulk paste (Diamond 1987). This contact layer develops in a day or two after mixing. On the side of the contact layer opposite to the aggregate, much solution-filled space remains in which special forms of hydration product develop. One often finds relatively large, hexagonal CH crystals, sometimes tens of microns across but only one or two microns thick, and clusters of ettringite needles (Scrivener and Pratt 1986). The panel-shaped crystals and ettringite needles are anchored in a unit of hydrates of the neighbouring cement grains. The needle and rod-shaped ettringite rub against and adhere to the contact layer (Hoshino 1989).

It is generally considered that the ITZ cement paste has a significantly higher porosity and is weaker than the paste further away from the aggregates. An exception to this statement is in the case of HPC where the thickness of ITZ is reduced due to the interactions of the supplementary cementing materials and superplasticizers used.

The ITZ is considered different from the bulk cement paste in the same concrete in the following way (Diamond and Huang 1998):

1. a significantly greater content of capillary pore space, much of it interconnected, and thus has higher permeability;
2. a reduced content of unhydrated cement grains;
3. a larger content of calcium hydroxide with substantial orientation parallel to the local surface of the aggregate;
4. a larger content of ettringite; and
5. significantly lower strength and stiffness.

There are several theories on the mechanisms of ITZ formation, but these have not led to any internationally accepted view. The simplest is that immediately after mixing water, all the solid particles (aggregates, cement etc) are covered at once by a water film. The thickness of this water film is constant at all particles of about $10\,\mu m$. Another explanation is based upon localized bleeding (Scrivener and Pratt 1986). They reported that the relative movement of sand and cement grains during mixing, and possibly settling of the sand grains before the cement paste sets, may lead to

the regions of low paste density at the interface. Bleed water accumulates beneath the larger aggregate particles, creating additional planes of weakness. This has been confirmed by the quantitative studies of the microstructure (Hoshino 1989). However, the ITZ is also present at the 'upper' side of the aggregates. Thus, the localized bleeding may help but is, in fact, not the only mechanism for the formation of ITZ.

The most often used explanation of ITZ formation is due to the wall effect (Ollivier et al. 1994). Mehta and Monteiro (1988) used this term for the problem of particle packing against the wall which is the aggregate surface here. In fresh concrete, the concentration of the anhydrous grains decreases from the surface of aggregate particles. This wall effect is also confirmed by Escadeilas and Maso (1991) (Figure 5.14). Cement grains have a wide range of sizes ranging from $1\,\mu m$ up to even $100\,\mu m$ sometimes. The smaller particles fill up the space in between the bigger particles. This leads to a gradient in the particle density at the interface, but only to an extent of $10\,\mu m$. Thus, the wall effect also cannot explain the thickness of $50\,\mu m$ at the interfacial zone.

Thus, none of the theories can explain the formation of ITZ. In recent years, during the work done at the Delft University of Technology, it was concluded that the ITZ formation can be better

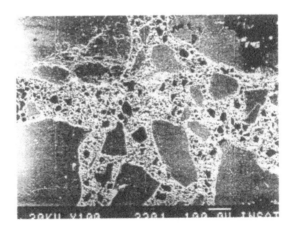

FIGURE 5.14 Packing of cement grains against the sand grains into fresh mortar: the excess of porosity in the vicinity of the sand surface is explained by wall effect (Escadeilas and Maso 1991).

explained based upon colloid science. A DLVO theory has already been introduced to explain this (Yang *et al.* 1997). It consists of van der Waals attraction and the electrostatic double-layer repulsion, on cement systems. They have shown that neat cement paste of OPC has an ionic concentration which is well above the critical coagulation concentration and therefore, is subject to rapid coagulation.

Another way to explain ITZ is by the syneresis process. When a colloid sol becomes coagulated rapidly, a very open structure results, in which most of the particles are linked to only two or three other particles (Figure 5.15a). It contains a great deal of entrapped solvent water (Hunter 1993). After the first rapid setting of a gel, the particle still retain some freedom of motion. A slow flocculation and coagulation continues, against the strength of the gel, to increase the number of contacts per particle in order to decrease the free energy of the system. This entails a contraction of the disperse phase (Figure 5.15b), the volume decreases and solvent is spontaneously pressed out. This phenomenon was first termed 'syneresis' by Thomas Graham (1949). The syneresis process is explained below (De Rooij *et al.* 1998);

During syneresis, the cement gel shrinks and the water comes out of the gel structure. This leads to a redistribution in the originally homogeneous mass of cast cement in water rich and solid rich mass. Besides the expulsion of water, contraction of cement gel itself has also been observed during their experiment. The total shrinkage should be due to two effects: the chemical shrinkage which is known to occur during the hydration of cement (Geiker 1983) and the colloidal syneresis. The shrinkage observed was more than the chemical shrinkage.

These conditions, local variations of water-to-cement ratio, translate into variations in the cement-hydration process which follow at filling the form work and which eventually determine the structure of cement and concrete. The formation of ITZ can thus be explained.

Moreover, during the vibration of concrete, water flows out under the aggregates. The first effect creates a gradient of porosity and of the w/c ratio between the bulk paste and in the vicinity of the aggregates. Due to the second effect another heterogeneity can appear around the aggregate particles. Ollivier and Massat (1994) have described the microstructure of the ITZ between the OPC and the inert filler and with the addition of mineral admixture. The characteristics of the ITZ may be modified by a chemical reaction between the aggregate and the cement paste.

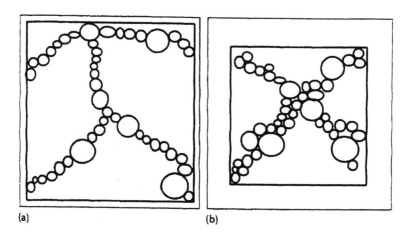

(a)   (b)

FIGURE 5.15 Schematic presentation of the effect of syneresis (De Rooij *et al.* 1998). (a) Open structure particles, (b) increased number of contacts only linked to two or three other, contracted structure.

In the neighbourhood of the aggregates, the increase of water-to-cement ratio and the lower concentration of anhydrous grains modify locally:

- the porosity,
- the hydration process.

Evaluation of the porous microstructure, hydration of the ITZ and the morphology of the hydrates are discussed here (Ollivier and Massat 1994).

5.6.1 The porosity of the ITZ

The porosity of the ITZ is an important parameter to be considered while assessing the engineering properties of concrete. There is no general rule or understanding but it seems logical to think that the porosity is highest at the contact surface of the paste and the aggregate and decreases with the distance from the contact surface. However, it depends upon the porosity of the aggregate itself. If the aggregates are porous, the binder will partially penetrate inside the aggregate. Then, the adhesion will not be just on the contact surface but will move to the inside of the aggregate. This phenomenon is very clearly observed in the case of lightweight aggregate concrete. The crack passes through the aggregate while testing the compressive strength, whereas the concrete cracks at the interface of the aggregate and the matrix in the case of normal concrete. Of course, this will depend upon the strength of the matrix.

Several studies have been made to highlight the difference between the top and the bottom porosity around the aggregate. However, the results reported are contradictory. One group of researchers have reported the existence of a gradient around the aggregate (Goldman and Bentur 1992; Hoshino 1988). The other group noticed very little difference in the porosity around the aggregate. This difference depends upon many factors but essentially on the casting conditions. The results can be compared only if the casting conditions are the same or if the casting conditions are to be described precisely.

The porosity may be quantified microscopically by means of image analyses of flat polished surfaces by SEM using back-scattered imaging and macroscopically using mercury intrusion porosimetry (MIP).

The first technique developed by Scrivener and Pratt (1986) allows the porosity as a function of the distance to the aggregate surface to be determined in a 2D image. The results plotted in Figure 5.16 (Scrivener *et al.* 1988) confirm the existence of porosity gradient in young OPC paste. The porosity decreases when the distance from the aggregate increases. It is shown that the variation of the porosity diminishes with age due to the filling of the empty space with the hydration products.

The second technique, MIP, allows pore size distribution to be quantified. Some measurements have been carried out by Bourdette *et al.* (1995). Figure 5.17 presents the incremental porosity as a function of pore diameter for pure cement paste and a mortar paste (w/c = 0.4, hydrated for three months in lime water). These results showed that in the mortar paste a coarser pore system is developed than in the pure cement paste. Bourdette *et al.* have proposed an approach to calculate the porosity of ITZ in mortars based upon a geometrical model of percolation through ITZ and upon the results obtained by the MIP. Aggregate particles are modelled by spheres and ITZ by a layer around the aggregates. The calculations highlight the greater porosity of ITZ 19 ± 1 per cent for the bulk paste and 48 ± 2 per cent for the ITZ. In

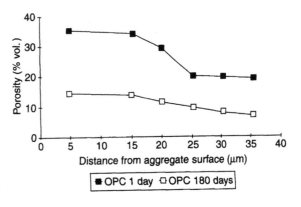

FIGURE 5.16 Effect of age on the porosity in the ITZ (Scrivener 1989).

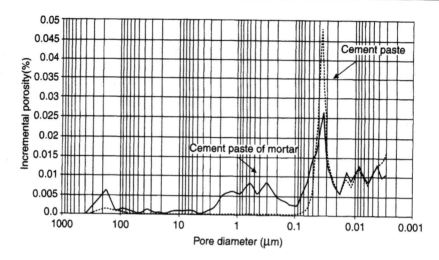

FIGURE 5.17 Incremental porosity distribution of pure cement paste w/c = 0.4 and cement paste of mortar (3-months-old). A new porous volume in excess for the pores bounded by 0.045 and 5 μm may be observed (Bourdette *et al.* 1995).

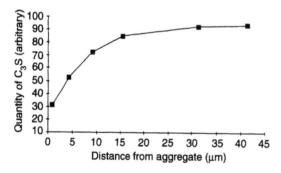

FIGURE 5.18 Variations in C_3S concentration in OPC paste cast against a flat aggregate (w/c = 0.5 after 10 days) (Ollivier 1981).

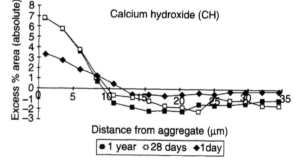

FIGURE 5.19 Redistribution of CH into the ITZ (Uchikawa *et al.* 1995).

conclusion, the porosity is higher in the ITZ than in the bulk paste and the pores are coarser in the ITZ.

5.6.2 The hydration into ITZ

Farran (1956) was the first to reveal the difference in the mineralogical composition of the interface of aggregates. Ollivier carried out the measurements by XRD analyses on specimens composed by cement paste cast against a flat face of a cylindrical aggregate. He confirmed the existence of a gradient of anhydrous cement grains (Figure 5.18),

which is further confirmed by other researchers (Ollivier 1981; Scrivener and Gartner 1987).

After the dissolution of anhydrous grains, the existence of a concentration gradient induces a transport of ions by diffusion from the bulk to the ITZ. In OPC, Na^+, K^+, SO_4^-, $Al(OH)_4^-$ and Ca^{2+} ions are the most mobile ions. Numerous studies have been carried out comparing the amount of hydrates as a function of the distance to the aggregates. The results highlight the ion diffusion process (Breton *et al.* 1993; Monteiro *et al.* 1985; Scrivener *et al.* 1988). Figures 5.19 and 5.20 present the evolution of the amounts of CH and

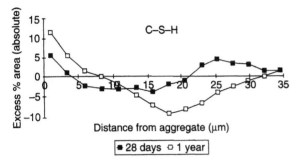

FIGURE 5.20 Redistribution of C-S-H into the ITZ (Uchikawa *et al.* 1995).

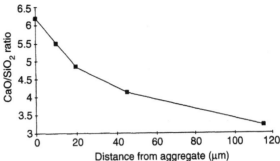

FIGURE 5.21 Average CaO/SiO_2 molar ratio of the hydrated C_3S paste as a function of the distance from the marble–paste interface (Chen and Odler 1987).

C-S-H calculated as being in excess of the amounts expected if all the hydration products were precipitated in the vicinity of the anhydrous cement grains from which they are formed. These results confirm the redistribution of CH and C-S-H. The evolution of profiles between 28 days and one year are not similar for the two hydrates: the diffusion of calcium seems to be more influenced by the packing in the ITZ, than the silicate ion does.

Chen Zhi Yuan and Odler (1987) have studied the C/S ratio with respect to the distance from the marble-paste interface. It is seen from his results presented in the Figure 5.21 that the C/S ratio is much higher than three in the vicinity of aggregates, the C/S ratio of the anhydrous silicate (C_3S) was used in this study.

5.6.3 Morphology of hydrates in the ITZ

Several studies have been carried out on the morphology of hydrates in the ITZ. A few of them are reviewed here. Barnes *et al.* (1978) have detected a duplex film, a film around the aggregates. This film is typically 1–1.5 μm thick. The inner part of this film which is formed in direct contact with the aggregate is CH (portlandite) and the outer part is C-S-H gel. The CH film is preferentially oriented with the c-axis normal to the aggregate. SEM investigations made by these and several other authors have shown that the hydrate crystals are well formed and are bigger than those formed in the bulk paste. The

preferential orientation of portlandite in the ITZ has been studied by DRX, a method proposed by Hadley (1972) and developed by Grandet and Ollivier (1980). In this method, the CH orientation is characterized by means of an orientation index defined from the ratio of two DRX reflections. According to the method, the ITZ thickness is the distance to the aggregate in which the orientation index is higher than in the bulk. For OPC paste the typical thickness is about 50 μm.

5.6.3.1 Influence of mineral additions on the ITZ

The characteristics of the ITZ may be affected by mineral additions in two ways:

- the packing of the particles may be densified if the size of the additives is much finer than the size of the cement grains;
- the hydration process may be modified if the addition grains are reactive or if they act as crystallization nuclei.

The first way may be only encountered with ultra fine additions like silica fume, whereas the second one depends upon the chemical composition.

5.6.4 Influence of ultra fine additions

Silica fume is considered as a microfiller. It densifies the poor packing of the cement grains in the

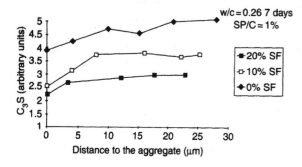

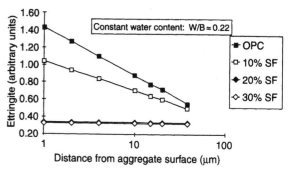

FIGURE 5.22 Concentration profile of C_3S into the ITZ as a function of silica fume addition (Ollivier 1986).

FIGURE 5.23 Concentration profile of ettringite into the ITZ as a function of silica fume substitution (Hanna 1987).

ITZ (Hanna 1987). Nevertheless, there is a gradient in anhydrous concentration (Ollivier 1986). In cement paste cast against a flat aggregate, the C_3S profile is similar with and without silica fume (Figure 5.22). When silica fume is used, w/c ratio is reduced due to the use of superplasticizers. Consequently, the w/c gradient is decreased in spite of the wall effect and thereby the possibility of diffusion are also reduced. Such an analysis may be confirmed by the modification in the ettringite profile with silica fume content (Figure 5.23, Hanna 1987).

These facts may be considered as a proof of the existence of a wall effect. When using very fine alumina grains as another microfiller, Hanna (1987) has shown that the filling effect into the ITZ as well as in the bulk paste is due to the size of the finer particles. This effect occurs if the particles are deflocculated, for example, well dispersed with the help of dispersing agents or superplasticizers. A similar conclusion has been drawn by Goldman and Bentur (1992) using very fine carbon black grains.

Moreover, a pozzolanic reaction takes place with silica fume, and calcium hydroxide is consumed in the bulk paste as well as in the ITZ (Bentur and Cohen 1987; Ollivier 1986). It produces C-S-H; a component which provides extra binding. This modification in the ITZ has been proposed to explain the enhancement of the compressive strength of concrete. However, it seems that the main factor could be the filling effect because the strength increase is found to be similar with silica fume and inert alumina or carbon black grains (Carles-Gibergue et al. 1989).

5.6.5 Influence of fine additives

The influence of fly ash and slag additions on the ITZ microstructure has been studied by Carles-Gibergues (1981). During the first days, the influence of these additions are poor, because the wall effect is not noticed. Some miner effect can however be underlined, for example micro-bleeding can be more important with slag additions increasing the ITZ thickness. These additions may act as the nucleation sites and therefore, the crystallization of portlandite is more disoriented. The pozzolanic reaction in the presence of fly ash has no effect in the ITZ. The influence of calcarious fillers on the ITZ has not been extensively investigated. It could be assumed that the packing around aggregate particles is not affected, the size of these additions being the same as cement grains. With the higher w/c ratio in the ITZ, calcium aluminate hydrates are formed more easily instead of ettringite, and the reactivity of calcarious fillers is enhanced (Carles-Gibergue et al. 1989). A positive consequence of this reaction is the formation of calcium carbo-aluminate which fills in the porous space of the ITZ.

5.7 Durability properties

Durability properties of concrete mainly depend upon its microstructure, which in turn is related to its pore structure. Pore structure controls the moisture movement which is one of the major cause in the concrete deterioration. Water is essential for making concrete. At the same time water has a negative effect on the durability properties. Basically, it can be said that the durability properties of concrete are mostly controlled by the pore structure. Denser pore structure enhances the durability properties whereas open pore structure has a deleterious effect.

5.7.1 Influence of concrete porosity on its durability

The porosity of the concrete is of prime importance and is a subject of its own. This is due to the fact that the pores are mainly responsible for the permeability of concrete which allows the penetration of water and gases into the concrete. Thus, a better understanding of the mechanism through which the pore structure characteristic influence on the durability of concrete structure exposed to the aggressive environment is important from both theoretical and practical point of view.

The term 'pore structure' which is generally encountered in the concrete technology refers to the total pore volume and pore size distribution, their specific surface, geometrical and functional morphology. Considering the volume location of pores in a concrete body, the pores are regarded as pores in cement paste, pores in aggregates and the pores in the ITZ. Additional concrete porosity is formed by an incomplete compacting of the mixture, entrapped air or by entrained air.

The mechanism of liquid and gas transfer through the pores is assumed to be as follows: in pores with radii (r) less than $100\,nm$; molecular diffusion takes place; when $r = 100–10000\,nm$ a molecular flux is observed, and if $r > 10000\,nm$ the viscous flux is seen (Schmalfeld 1977).

Being a heterogeneous body, concrete is capable of incorporating three mechanisms simultaneously. It may be concluded that the porosity with

$r > 100\,nm$ is mainly responsible for the decreasing of concrete resistance against corrosive agents. Some pores with $r > 100\,nm$ are connected by the smaller pores, the latter being not able to filtrate liquids or solutions. Disruption of capillary continuity, however, does not necessarily destroy the continuity of the pore system as a whole because water flow can still occur through the gel pores (Popovics 1980). The pore in the aggregates do not change with time, although their size and quantity can influence the corrosion resistance of concrete. The pores in the ITZ and those caused by incomplete compacting or due to the large air bubbles formed when AEA are used, can allow easy access of aggressive agents to the inner structure.

The permeability of concrete is usually taken as the indicator of the abilities of the material to resist corrosion. Many researchers have made efforts to correlate porosity and permeability, but it is not so simple as it is necessary to take into account not only the complexity of the pore structure itself, but also the combined capacity in the zones with higher and lower permeability (Wiggs 1958). In this context, three zones of pore structure can be identified:

- low permeability zone surrounded by a zone with higher permeability,
- high permeability zone surrounded by a zone of lower permeability,
- the presence of low and high permeability zones together.

It is believed that continuous porosity is mainly responsible for the moving of liquids or gases through the structure of concrete. Both the continuous porosity and permeability decrease when concrete hardens under water (Barovsky 1983). During hardening, the total pore volume decreases and a certain redistribution of pore sizes takes place, the latter being characterized by a relative increase of the quantity of fine pores and decrease of larger pores.

The main factor controlling the rate of the corrosive process in concrete is the internal diffusion of ions into the pores with subsequent chemical reactions. The process of diffusion transfer under various conditions is described usually by equations that are analogous to those used for heat transmission (Fick's law). Theoretical investigations on the kinetics of corrosion processes when concrete and liquid aggressive media interacts have been carried out by Polak (1971). The Cl^- ion penetration into concrete is of prime importance for corrosion in concrete itself and on the steel reinforcement. The value of the effective coefficient of diffusion of these ions varies from 0.52×10^{-7} to $1.95 \times 10^{-7} cm^2/sec$ (Ivanov 1977). More detailed data about the effective values for diffusion coefficients and the coefficients of chemical reaction are published by Ushiyama and Goto (1974). Barovsky has tested the sulphate ion diffusion (Barovsky 1993). It is shown that the coefficient of sulphate ion diffusion in hydrated cement pastes varies from 0.5×10^{-8} to $17 \times 10^{-8} cm^2/sec$ depending upon the w/c ratio and the hardening time.

5.7.2 Freeze–thaw resistance

Freeze–thaw resistance of concrete is basically related to the freezing of water in the concrete structure producing 9 per cent volume expansion. Thus, it is controlled by the pore structure of concrete. The pore structure, subjected to combined attack of freezing and thawing and salt solutions have been investigated by Hochstetter (1973). He established that there is a good relationship between the proportion of the pore volume and the degree of damage as well as between the degree of damage and pore size distribution within the group of cement bounded materials. The temperature of freezing of the water in a porous material depends upon two factors:

- the specific surface of the solid phase (porosity), and along with this, the quantity of adsorbed water,
- the quantity of isolated (relatively closed pores and connected pores.

The larger the specific surface (larger number of fine pores) and quantity of relatively closed pores, the lower is the freezing temperature. Migration of water under temperature and moisture gradient may also affect the process of freezing.

Different concretes have different pore structures depending upon the cementitious materials used and the water-to-binder ratio. Especially there is significant difference in the pore structure of normal and high strength concrete. These will be dealt with separately.

5.7.2.1 Freeze–thaw resistance of normal concrete

There is a lot of discussion about the freeze–thaw resistance of normal concrete. Fagerlund (1986) states that 6 per cent air entrainment in the concrete is prerequisite to have good freeze–thaw resistance. Some other researchers insist on pore-spacing factor of 0.2 mm. When one talks about the air entrainment; the amount of total air entrained can be 6 per cent, but its homogeneous dispersion is difficult to guarantee. The dispersion can vary from batch to batch. When it is not possible to reproduce the concrete of the same pore characteristics, it is not realistic to guarantee the freeze–thaw resistance. Regarding the bubble spacing factor of 0.2 mm, it is not easy to introduce and disperse such tiny pores homogeneously in the concrete. Even in the case of foam, the small bubbles dissolve in water, and it does not help to improve the freeze–thaw resistance. Nevertheless, there is evidence that the concrete having 4 per cent air with a bubble spacing factor less than 0.2 mm was damaged due to frost action. Mather (1986) has the opinion that 'in order to be immune to frost action; the concrete is to be properly cured and have a strength of 3500 psi (25 MPa) and it is protected from freezing and thawing in a critical saturated state until it matures enough so that the fraction of volume of the remaining mixing water is small enough so that the air void system can handle the increase in the volume upon freezing'. In fact in normal concrete, the maximum amount of freezable water is such that the excess volume needed to accommodate ice is less than 1 per cent of the concrete volume; then the question arises why we need more than 1 per cent extra air in the concrete for frost resistance. There are many researchers who believe that there is something more than the bubble-spacing factor and air entrainment which is responsible for the frost damage of concrete, like, for example, adhesion between the matrix and the aggregate. Apart from this, the thermal behaviour and compatibility of aggregates, rate of cooling etc. are also to be considered (Chandra 1998b). Apart from the material properties and the pore size distribution, the mechanical properties of the concrete are also important. These depend very much upon the curing conditions and the type of cement used to make the concrete, and the test method used for evaluation (Aavik and Chandra 1995; Chandra and Xu 1992). It is reported that the concrete to be tested for freezing should have a minimum strength of 28 MPa. Freezing and thawing depends greatly upon the sample preparation and the method used for the test. Thus, for comparing the results, it is essential that the sample preparation and the freeze–thaw test method implied is the same, failing which it is not possible to compare the result.

Nevertheless, it is emphasized here that the mechanism of freeze–thaw resistance of normal concrete, in spite of so many decades of intensive research, is still a matter of discussion.

5.7.2.2 Freeze–thaw resistance of High strength concrete

High strength concretes (HSCs) are made with a very low water-to-binder ratio (0.20–0.40) which is accomplished with the addition of SP. Very often, these contain mineral admixture like condensed silica fume, fly ash and slag. These concretes have denser pore structure than that of ordinary concrete, and possess a very small amount of freezable water. Thus, HSC has lower total porosity and has refined pore structure. This refinement is not only due to the reduction in total porosity, but is related to the densification of the paste aggregate interface (Ollivier et al. 1994; Scrivener et al. 1988). This subsequently decreases the continuity of its pore structure which has direct influence on the freezing and thawing.

The air entrainment in such a low water-to-binder ratio concrete is not easy. But first of all, the question is if it is necessary to have air entrainment in HSC? The pore structure in HSC is influenced by the cement type and by the type and length of curing. These are the major factors which control the capillary porosity of the cement paste. However, it can also be influenced by the aggregate characteristics because of the higher porosity of the paste at the aggregate-paste interface (Maso 1980).

The chemical admixture like AEA, SP and corrosion inhibitors are often not compatible and hinder each other's influence. This hindrance is more pronounced with the use of supplementary cementing materials. Obla et al. (1999) have reported that the NSF based high range water reducer (HRWRA) has a tendency to decrease the air content in concrete. None of these specialty admixtures cause a synergistic effect with a low dosage of AEA. At an intermediate dosage of AEA, in the absence of HRWRA, some of these admixtures tend to entrain more air while others detrain air. Very similar results are obtained when the combination of specialty admixtures were used.

It is emphasized that the dosage of AEA needs to be increased by 150–300 per cent to obtain 4–6 per cent air entrainment in the presence of HRWRA. Phase mineralogy of cement appears to have no distinctive effect on air entrainment, and the above conclusions are valid even for the cement containing high C_3A, high alkali cements and plain cement concrete without fly ash.

Microstructure studies did not show any distortion in the shape of air voids, but the size varied from one concrete mixture to the other, with the air voids tending to grow coarser in some mixtures. Concentration of air voids at the paste-aggregate interface can occur in some concrete with high air content.

Pigeon et al. (1991) have tested non-air-entrained HSC having compressive strength of 74.8–89 MPa, and air-entrained concrete having a strength of 60–69 MPa. It is shown that with the use of ASTM type III cement, which is very finely ground (specific surface; 545 kg/m^2), air entrainment is not required for good resistance to freezing and thawing at a water-to-binder ratio of 0.30 even after one day of curing. Thus 0.30 is to be taken as the limiting value of water-to-binder ratio below which air entrainment is not needed for adequate freeze–thaw resistance. Concrete having a spacing factor of 948 μm has cleared 500 freeze–thaw cycles (ASTM 666) without deterioration. Comparing these results with the other published data (Gagne *et al.* 1990) which has shown that the non-air-entrained concrete made with silica fume had excellent freezing and thawing resistance, it can be inferred that at water-to-binder ratio of 0.30 the use of silica fume does not significantly improve freeze–thaw resistance.

It is also shown that the use of dolomitic gravel instead of granitic gravel had no significant influence on the freeze–thaw resistance.

Pigeon *et al.* (1991) further report that the type I cement behaves completely different from ASTM type III cement. Concrete samples made with 6 per cent silica fume and ASTM type I cement need air entrainment for good freeze–thaw resistance even at a water-to-binder ratio of 0.30. Results about the spacing factor, on the other hand, are very controversial. For some of the samples the critical value of the spacing factor is reported to be 203 μm and for others 503 μm (air-entrained concrete) for good freeze–thaw resistance. In the case of non-air-entrained HSC the spacing factor is usually in 600 μm to 1.0 mm range, and still the concrete is freeze–thaw resistance. It does not agree with the hypothesis and the results reported above about the bubble spacing.

In another test (Pigeon *et al.* 1991), concrete was made with ASTM type I cement, 6 per cent silica fume with 0.26 water-to-binder ratio and a pore-spacing factor of 690 μm. It has shown excellent freeze–thaw resistance. For the ASTM type I cement, the limiting value of water-to-binder ratio is thus about 0.25 below which no air entrainment is needed for good freeze–thaw resistance. Similar results have been reported by Foy *et al.* (1988) for the concrete made without the addition of silica fume and cured for seven days. Hammer and Sellevold (1990), on the other hand, have shown that a non-air-entrained concrete (with a water-to-binder ratio of 0.26) containing almost no free freezable water was severely damaged by freezing and thawing. They suggested that it was due to the difference between the coefficient of thermal expansion of the paste and of the aggregate and not due to the freezing effect.

Freeze–thaw resistance of concrete is related to the ice formation. Hammar and Sellevold (1990) have shown using a low temperature calorimetry technique that in HSC there is a very small amount of freezable water which is less than 10 per cent of the evaporable water content. The authors further postulate that the ice formed did not propagate as a front through a continuous pore system, but was formed in isolated areas that might have been filled up with water during the vacuum resaturation of the sample before test.

Low water-to-binder ratio leads to self desiccation which helps in emptying the water-filled pores. Subsequently, the number of unsaturated voids are increased. This may enhance the resistance to microcracking.

According to Fagerlund's (1993) calculations a small amount of water 0.7 per cent is sufficient to exceed the tensile strength of the paste at $-20°C$. Hence a very little amount of freezable water is sufficient to damage a saturated concrete.

In the case of HSC, the freeze–thaw damage does not appear as surface scaling-off leading to mass loss as in the case of normal concrete. The concrete may appear to be in an undamaged state, but it can be full of microcracks. Jacobsen (1994) has shown in his work the microcrack formation in non-air-entrained HSC which has a very low amount of freezable water. It confirms Fagerlund's theoretical calculations while the other researchers (Jacobsen 1994; Mitsui *et al.* 1989; Okada *et al.* 1981; Pigeon *et al.* 1991) show that the non-air-entrained concrete can be frost resistant.

Freeze–thaw resistance does not only depend upon the water-to-binder ratio, it also depends upon the type of cement used. For non-air-entrained HSC, very controversial results have been reported. Air entrainment reduces the strength; besides it is not easy to entrain air in concrete having such a low water-to-binder ratio, as a very high dose of air-entraining agent will be necessary. Its influence on the long time durability is to be tested.

5.7.3 Freeze–salt resistance

Freeze–salt resistance of concrete is different from the freeze–thaw resistance. Deicing salt used in the test produces a concentration gradient. It influences on the freezing temperature of water at different levels. It subsequently produces extra stresses and the concrete destroys at a much faster rate in comparison with those tested without using deicing salt. A comparative study of salt-frost resistance of normal and HSC with condensed silica fume has been done by Petersson (1988). It has been shown that in the case of ordinary Portland cement concrete the damage due to the freeze–thaw attack was gradual while in the case of silica fume concrete the destruction was abrupt after 100 cycles. The concrete is destroyed without any prior signs or indications (Figure 5.24). The concrete had 0.11 and 0.19 silica-to-cement ratio and 0.54 and 0.35 water-to-binder ratio. The reference concrete used had 0.35 w/c ratio. The concrete was not air entrained. The specimens (40 × 40 × 160 mm prisms) were stored in water (+20°C) during the first 14 days after casting and 14 days in the climate room at 20°C and 50 per cent RH. Thereafter these were immersed in 3 per cent NaCl solution and were subjected to repeated freezing and thawing cycles (16 h freezing at −20°C and 8 h thawing at +20°C). The question is, was it due to the freezing and thawing, or was it something else. One possible reason can be that the

concrete, because of its low water-to-binder ratio was not critically saturated at the start of the test. Gradually, with time during the test, it became critically saturated and gave way after particular number of cycles (100 in this case). The other possibility is the alkali–silica reaction developing microcracks, weakening the structure. Now if the concrete is to be evaluated for freeze–salt, concrete with silica fume will be declared as excellent because it did not show damage up to 56 cycles (according to Swedish standards, concrete is frost resistant if the mass loss upto 56 cycles is less than 1 kg/m²), but later, it failed abruptly without any warning. It is a big risk. The freeze–thaw resistance test with the deicing salt component is more severe than the test without it.

Jacobsen and Sellevold (1992) have tested freeze–salt resistance of concrete with and without condensed silica fume according to the Swedish standard SS 13 72 44 (Borås method). It is shown that the concrete made with 8 per cent CSF led to reduced scaling at all the curing temperatures upto 60°C. The concrete was tested upto 280 cycles. There was no abrupt collapse. Based upon this work, the concrete will be evaluated as freeze-salt resistant. This is in contradiction to the results reported by Petersson. Freeze–thaw resistance of concrete using deicing salt is to be further investigated.

It can be concluded that the freeze–thaw resistance of HSC made with the addition of SP at low water-to-binder ratio but without air-entrainment are controversial. It seems that the air-entrainment is necessary for adequate freeze–thaw resistance of HSC with and without deicing salt. It is not easy to entrain air in concrete with low water-to-binder ratio. Besides all AEA do not work with every type of SP. A system has to be worked out where both SP and AEA will work together. Laboratory tests are to be performed and reproducibility of pore structure is to be ascertained prior to the full-scale concrete mixing.

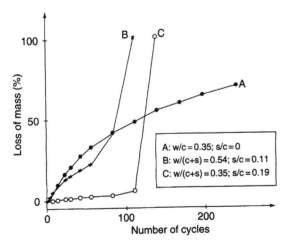

FIGURE 5.24 Scaling resistance curves (loss of mass) for concretes without (A) and (B and C) with silica fume as functions of the number of freeze–thaw cycles (w-water, c-cement, s-silica fume) (Petersson 1988).

5.7.4 Efflorescence

Sometimes, concretes made with the addition of sulphonated naphthalene or melamine condensate superplasticizers show white or yellow spots. This is known as efflorescence and is thought to appear due to the impact of superplasticizers. It creates some uneasiness among the producers of superplasticizers. The efflorescence is caused by the formation of calcium salts, sodium sulphate (thenardite) and potassium sulphate (aphitalite, glaserite). Calcium efflorescence appears when the cement used is rich in alite C_3S and concrete is fluid. With the evaporation of water, the widespread lime-based efflorescence appears on the surface and may cause problems. In cases where the alkali content is high and the NaOH/KOH ratio is high, thenardite efflorescence can appear. Thenardite crystals expand by absorbing water and may cause peeling off. The efflorescence observed with the use of superplasticizers is different from that appearing due to the crystallization of calcium and sodium salts. Crystals formed on the surface are localized and spotty. It is reported that these crystals do not reduce the compressive strength of the concrete and can be washed away (Aakman and Cavdar 1999). Nevertheless, addition of 1 per cent cumene sulphonate to melamine sulphonate reduces the formation of glasserite. Reducing of bleeding by the addition of mineral additives is another alternative to reduce efflorescence. The efflorescence caused by the SP is not deleterious, but it is advisable to avoid use of high alkali cement.

5.8 Engineering properties

Superplasticizers or high-range water-reducing admixtures have led to the development of very low w/c ratio; HSC and HPC both for new construction and for repair and for the rehabilitation of deteriorated infrastructure. Innovative concretes have been developed; superplasticized HPC containing high volume fly ash and slag, ultra HSC incorporating silica fume, fiber reinforced concrete, heavyweight and lightweight concretes, rice husk ash and meta kaolin concretes, low-heat HPC and reactive powder mortars. Apart from the new construction, it is reported that the use of superplasticized silica fume concrete performs better as a repair material as it provides higher

strength and lower shrinkage. Properties of concrete depend very much upon the mixing technique as it influences upon the water release process which is entangled between the binder particles, a typical behaviour of SP.

5.8.1 Mixing technique

Mixing technique also plays an important role on the properties of concrete made with SP. The concretes with SP and fly ash and micro silica made with two-stage mixing technique (SEC) is reported to have better strength and durability properties (Al-Nageim and Lewis 1999). It is attributed to the decrease in the transition zone, the weakest point existing between the aggregates and the cement paste. It is further reported that the settlement of the coarse aggregates becomes negligible. Besides the non-segregating characteristics noted, SEC concretes have higher workability than conventionally made concrete with the same overall w/c ratio. Microstructure analyses has shown smaller pore size distribution of the paste at the age of 28 days in addition to the continuous and dense structure. Calcium silicate hydrates are enriched with small homogeneous calcium hydroxide crystals well attached to the glassy side. Besides the mixing technique, mixing time also plays an important role upon the workability, and dispersion of the binder, the factors which influence the strength of the concrete. Chandra (1994) has shown that the compressive strength increases with the increase in mixing time.

5.8.2 Bond strength

The overall strength of concrete is a function of the strength of matrix, aggregate, and the bond between the aggregate and matrix. Thus, it will be influenced by the type of the cementitious material, water-to-binder ratio, curing conditions, and the type and physical characteristics of the aggregates.

ITZ is usually regarded as the weakest region in the concrete influencing both the mechanical properties and the durability. The importance of the interfacial bond between the cementitious matrix and the aggregates has been emphasized by many researchers (Alexander *et al.* 1965; Hsu and Slate 1963). It has been suggested that the bond between the aggregate and the cementitious matrix depends upon three different mechanisms (Alexander *et al.* 1965; Farran 1956; Monterio and Mehta 1986; Yuan and Guo 1987; Zimbleman 1985). These are:

1. The mechanical keying of the hydration products of cement with the rough surface of aggregate (often covered with fine cracks);
2. Epitaxial growth of hydration products at some aggregate surfaces;
3. Physical–chemical bond between the hydrating cement paste and the aggregate, due to the chemical reaction.

The physical and chemical interfacial processes affect the overall strength beyond that of the aggregate strength. The physical process identified was densification of the interfacial transition zone, specially with the use of pozzolanic materials, and the chemical process involved in the adhesion of cementitious material to the aggregate. The latter depends upon the characteristic of the aggregate surface and its chemical and mineralogical character. The decrease of the rock surface relief makes for decrease of the bond strength. It is also influenced by the rock porosity. Generally, with the increasing of porosity, bond strength increases as the binder impregnates the aggregate making the adhesion not only on the contact surface, but in the pore structure of the aggregate. It is naturally influenced by the character of the cementing material (Tasong 1997).

Microstructure of the cementitious system related to ITZ has been investigated by many researchers. It is indicated that the use of pozzolanic material such as silica fume may improve mainly the microstructure and the strength of the bulk paste. It is also shown that the transition-zone densification could not serve as the major process leading to the concrete strengthening. The transition zone is referred to as a result of interaction between paste and the aggregate in a fresh matrix, that is, a zone of paste in a close aggregate vicinity where the paste is not as dense as in the bulk. The total volume of paste in mortar or concrete is simply a sum of bulk paste and transition zone.

Goldman and Cohen (1998) have shown that the silica fume mortar when used with dolomite limestone exhibited very low sensitivity to the variation of surface area in a wide range. The same results were obtained for quartzite aggregates. On the other hand, the reference mortars demonstrated high sensitivity in the same range of aggregate surface area. Silica fume paste that represented bulk paste, as not denser than the reference pastes. It is postulated that the drastic reduction of sensitivity was therefore a result of paste densification in the transition zone due to the presence of silica fume. The densification took place already at 1 day, which should be related to the microfiller effect of silica fume (Goldman and Bentur 1992; 1993). Their results showed substantial decrease of the modulus of elasticity, E, along with the increase in aggregate size in the reference mortar (without silica fume). This relationship indicated the weakening effect of the transition zone on mortars without silica fume. On the contrary, mortars with silica fume showed very low, if any, sensitivity to changes of aggregate size which could be related to the modification of the ITZ.

Effect of different cements and aggregates on the bond strength was studied by Bazantova and Modry (1998). The bond strength at 28 days is shown in the Table 5.3. It is seen from the table that the bond strength of the rock and gypsum free cement is high in compressive strength with normal Portland cement. Value of bond strength of gypsum free cement, ASTM type III attains higher values than bond strength of gypsum free cement ASTM type 0. The specific surface area of ASTM type III cement is double (708 m^2/kg) that of the ASTM type 0 (309 m^2/kg). However, the higher value of bond strength in ASTM type III cement is not attributed only to its high specific area but to the other properties as well, like higher microhardness and lower porosity, achieved due to the use of chemical admixture. Larger differences between bond strength values for different

TABLE 5.3 Bond strength at the age of 28 days. Quarts-Svedlar; Slovenia, Granite; Hudcice, Limestone; Vraca; Bulgaria, Marble; Silvenec (Bazantova and Modry 1998)

Cements	Specific surface area (m^2/kg)	Bond strength (MPa)			
		Quarts	Granite	Limestone	Marble
OPC, PC400	365.4	3.46	3.01	4.63	3.70
Sul. Res. PZ 2/30	371.3	3.32	3.13	3.42	3.30
Gypsfree ASTM type 0	309.0	5.49	4.83	8.85	3.30
Gypsfree ASTM type III	708.6	7.82	6.21	10.05	4.26
OPC PC400 + Plasticizer		3.69	3.86	4.12	5.94
PZ 2/30 + Plasticizer		3.43	4.11	3.28	3.70

cements appeared only for Portland cement PC400 and PZ 2/30 (which has different C_3A content) in the case of limestone. The bond strength was higher with limestone aggregates. This may be due to the interaction of the pastes with the aggregate. Simultaneously, the difference in bond strength for low aluminate cement PZ 2/30 and the rocks is negligible. These results are in agreement with Chatterji and Jeffery (1971) who showed the influence of C_3A content on the mechanical properties of mortars produced with siliceous and calcarious aggregates. Looking into the composition of rock, it is seen that the bond strength increases with increase in SiO_2 content, whereas in the case of calcarious rocks, this relationship between the CaO content is the reverse. Nevertheless, it is necessary to consider the porosity and its influence on character of contact between rock and cement binder (Bazantova 1985).

The bond strength of quartz, granite and marble with Portland cements depends only a little on relief changes. In the case of limestone with surface properties close to the surface properties of granite, bond strength of limestone is higher in comparison with granite. This fact is valid for PC400 and for both types of gypsum-free cements but not for sulphate-resistant Portland cement PZ 2/30. Difference in bond strength between limestone and granite is negligible. This is related to the C_3A content of cement (Bazantova 1985). Dependence of bond strength of gypsum-free cements on relief character of rock surface is pronouncedly higher and almost linearly increases in sequence as follows; marble, quartz, granite.

There are some indirect evidences of the bonding mechanisms due to mechanical interlocking aided by the aggregate surface texture (Alexanderson et al. 1968; Barnes et al. 1979). It was investigated by comparing the bond strength of the fractured rock surfaces with that of the polished rock surface. However, in these investigations the true surface area that each aggregate particle presents for bonding was either not adequately quantified, or not estimated at all. Two-thirds of the volume of concrete consists of aggregates. The quality of concrete will also be governed by the quality of the aggregates used, especially in the case of HSC where strength is the main criteria. Therefore, the quality and type of aggregate play important role. Hence it is very important to understand the properties of the aggregates. Tasong et al. (1998) have investigated bond strength and the mode of failure of concrete. The aggregates used for making concrete were basalt, limestone, and quartzite, and the cement used was ordinary Portland cement complying with the BS12.

5.8.2.1 Effect of aggregate surface roughness on the nature of the bond at ITZ

Surface properties of the aggregates are very important for good adhesion between the cement matrix and the aggregates. Differences in the rock crystalline texture, hardness and intergranular bonding between the mineral grains are possibly the main reasons for the differences in the surface texture characteristics of the rocks. Four different modes of bond failure in direct tension tests were observed due to the difference in mineralogy, structure, and surface features of the rocks used in the investigation and the resulting variation in their interaction with the cement paste:

1. Mode A: Failure cracks passed cleanly at the ITZ;
2. Mode B: Failure cracks ran through the ITZ at some points and continued through the paste;
3. Mode C: Failure cracks ran through the ITZ at some points and continued through the aggregate;
4. Mode D: Failure cracks passed through the ITZ as in mode A, but with loose fragments of aggregate particles peeling off to the cement paste.

Basalt aggregates. Unlike limestone and quartzite, the basalt-cement paste interfacial bond strength was observed to increase consistently with increasing surface roughness of the aggregate. The ground aggregate surface achieved the lowest bond strength, followed by the sawn surface, and both types have shown the type A mode of failure. On the other hand, the fractured surface achieved higher 'apparent' bond strength and shown the type B mode of failure. A similar mode of failure was reported by Alexander *et al.* (1994) during an investigation into cement paste–andesite rock composite fracture properties. The basalt rock has a firmer or stronger crystalline texture with no zones of weakness and hence the high 'apparent' interfacial strength achieved by its fractured surface.

Limestone aggregates. Unlike the basalt aggregates, the limestone-aggregate cement interfacial bond strength did not increase consistently with increasing surface roughness of the aggregate. The fractured surfaced aggregate achieved the least bond strength followed by the ground surface aggregate, which showed the type A failure. On the other hand, limestone and basalt aggregate cement-paste composites achieved similar bond strength with the sawn surfaces. The slightly rougher surface of the basalt aggregate coupled with the high porosity of the limestone

(Tasong 1997) may have compensated for the chemical influence of the limestone, and this could possibly explain why both the aggregates achieved similar bond strength with sawn surface. In contrast to basalt aggregate, the fractured surface of limestone achieved an 'apparent' bond-strength value lower than that for its sawn surface and showed a type C mode of failure. A similar behaviour was reported for fractured limestone when cement paste–limestone was tested for bond strength (Monterio and Andrade 1987). Their observations indicated that although the fractured limestone surface provided larger surface area for bonding with the cement paste compared with the sawn and the ground surfaces, a lower bond strength was achieved. This was because of the presence of large crystals of calcite, which have low bond and tensile strength on their cleavage surfaces in this rock.

Quartzite aggregates. The nature of the quartzite aggregate–cement paste interfacial bond was significantly different from those of the basalt and limestone aggregates. The ground surface achieved the least bond strength and showed a type A mode of failure. This is probably due to the fact that its ground surface was smoother than the ground surfaces of basalt and limestone due to its strong internal texture. Unlike basalt and limestone, the saw-cut surface of quartzite achieved the highest 'apparent' interfacial bond strength and showed a type B mode of failure. A pozzolanic reaction between silica, leached out from quartzite (Tasong 1997), and the CH may possibly is the reason for this high bond strength. On the other hand, the fractured surface of quartzite achieved a lower 'apparent' bond strength and showed a type D mode of failure. A combination of the smooth fractured surface and the presence of an already cracked surface or partially loose rock fragments at the bonded surface of this aggregate could be the reasons for low 'apparent' bond strength achieved by the fractured surface. Fragments of this rock were observed on the cement paste after the test. In contrast, basalt aggregate has a firmer or stronger crystalline texture with no zones of weakness and hence higher bond strength was achieved with its fractured surface.

It is generally believed that the interfacial bond increases with increasing surface area available for bonding (i.e increasing roughness); however the aggregates should be strong enough at its bonded surface to withstand this increase in bond strength. It is further shown that the surface characteristics alone could not be used successfully to predict bond strength where a variety of aggregate types and surfaces are used. This suggests that the differences in mechanical strength of aggregates, internal structure and chemical interactions are also important parameters controlling bond strength and mode of failure at the interface (Alexander *et al.* 1965; Tasong 1997).

It is suggested that the ITZ in concrete should not only concern the transition zone between the bulk cement paste and the aggregate, as referenced here, but should also include the outermost part of the aggregate including undulations and crevices. Hence, the interface should be modelled as a composite in its own. This has significant influence on the engineering properties of concrete. It further implies that though the weakest region is thought to be the ITZ, surface properties and the type of aggregate may be need to be considered as another weak area while concentrating on the high strength, high performance concrete.

5.8.3 Shrinkage and creep

Drying shrinkage and creep of concrete have been studied for a long time, because they greatly influence the development of cracks, deformation properties, or loss of prestress of prestressed concrete members. Various factors influencing drying shrinkage and creep were studied (Lyse 1960; Picket 1956) and some theories were proposed regarding the mechanisms. With regard to the drying shrinkage, there are the capillary tension theory, surface adsorption theory and interlayer water theory (Powers 1965). While regarding creep, there are the visco-elastic theory, seepage theory, and viscous flow theory (Ali and Kesler 1964). In most of these theories, the behaviour of

moisture contained in concrete is considered to be the principal factor causing drying shrinkage and creep; moisture existing in the voids in cement gel especially is dominant. However, all these theories are for concretes of relatively high w/c ratio with the compressive strengths in the range 50–60 MPa.

With the development of the SP, it is possible to significantly cut down the w/c ratio, still maintaining the workability. As a result, it is possible to produce concretes with w/c = 0.20–0.30 having compressive strength of 80–100 MPa. Being low w/c ratio it is conceivable that the sizes and quantity of fine pores in cement paste differs from those in conventional concretes.

The pore structure is governed by the curing conditions, for example, the size and quantities of pores of concrete produced by accelerated curing such as steam curing and autoclave curing are different from those obtained by standard curing (Czernin 1960).

Drying shrinkage occurs chiefly in the cement paste portion, while the aggregate portion restrains shrinkage. Accordingly, it may be said that the drying shrinkage is greater for higher unit water content or unit cement paste volume. However, Nagataki and Yonekura (1982) have reported that the drying shrinkage of HSC of unit cement content $700 \, kg/m^3$ is smaller than that of normal-strength concrete of unit cement content $300 \, kg/m^3$ in spite of the approximately 1.5 times larger cement paste. It implies that the drying shrinkage of concrete is closely related not only to unit cement paste volume, but also to the thickness of cement paste, and, in effect, to the water-to-cement ratio. It is shown that the drying shrinkage per unit cement paste volume for HSC (100 MPa) is half that for 40 MPa concrete. However, in the case of autoclave curing it is half that of the standard cured specimens of the identical compressive strength. In case of steam curing $\mu s/\mu$ is slightly lower than that of the normal cured concrete, and practically no difference is seen in HSC. It is concluded that the drying shrinkage of concrete is influenced by the compressive strength and curing conditions of the concrete. Strength of concrete is related to the w/c ratio and the degree of hydration, which in turn are related to the porosity or the pore volume. Pore volume becomes smaller for lower w/c ratio and higher degree of hydration, while strength becomes higher. Since it is a linear relationship between strength and pore volume, it can be said that the pore volume has close relationship with the shrinkage. This differs from the curing conditions especially because of changes in the pore size distribution and the differences in the reaction products, gel structure and gel volume.

5.8.3.1 Relationship between the drying shrinkage and diffused water

It is reported that even if the quantity of diffused water at early stage of drying is extremely large in case of normal strength concrete compared with the HSC, drying shrinkage strains are roughly the same for the two units for 10 days of drying. Subsequently, the increase in drying shrinkage strain of normal concrete becomes greater than that of the HSC. This increase is independent of the strength and curing conditions. However, the increase after 100 days of drying differs and is dependent upon the curing conditions.

It is considered that in case of normal strength concrete the proportion of large pore diameter is greater than in case of HSC. But since the water is lost from relatively large voids at the early stage of drying, the capillary tensions produced are low, and because of this, the proportion of drying shrinkage is small. In the case of HSC, relatively large voids are not considered to exist because the proportion of drying shrinkage in relation to the quantity of water diffusion is large. In case of autoclave curing, the crystals become coarser and large voids are produced. Due to this, the quantity of diffused water becomes large and shrinkage is small at the early stage of drying. Tazawa and Miyazawa (1997) have shown that the shrinkage of cement paste is dependent upon the composition of the cement. It increases with the increase in the C_3A and C_4AF content.

It is reported that the creep strain of HSC (unit cement content $C = 700\,kg/m^3$) is smaller than that of normal concrete ($C = 300\,kg/m^3$) at the same age of drying in spite of double stress intensity of 30.6 MPa and larger unit paste cement

volume about 1.5 times as large as the former. The relationship between the creep strains μ_c of concretes of various mix proportions (Figure 5.25) and loading logarithm of period (t) (Figure 5.26), is approximately linear up to 200 days, but the subsequent proportion of increase in creep strain becomes extremely small in case of air-cured

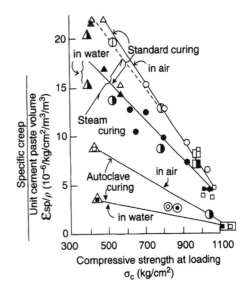

FIGURE 5.25 Relation between $\mu s/\mu$ and μ_c prism samples were used. The period of drying was 1250 days (Tazawa and Miyazawa 1997).

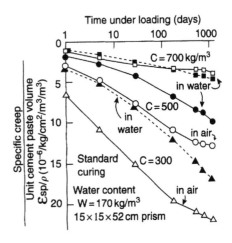

FIGURE 5.26 Relation between creep strain and time under loading (Tazawa and Miyazawa 1997).

concrete. On the other hand, in case of water-cured concrete, it shows a linear relationship up to 1250 days of loading. Consequently, the difference in the creep strains at the same age under loading becomes larger up to about 200 days of loading, but later becomes gradually smaller both for the cases of air and water cure when $C = 300$ and $500 \, kg/m^3$. In contrast for $C = 700$, the difference between the two is very small.

Pentala and Rautanen (1990) have studied the creep and shrinkage of HSC. Four types of concrete at the age of one year were tested. Silica concrete: strength 115–121 MPa at one year, blast furnace slag concrete, 105 MPa, and HSC made by using rapid-hardening Portland cement had compressive strength of 95–101 MPa. The compressive strength of normal concrete used for comparison was 66 MPa. All creep specimens were loaded to 30 per cent of their compressive strength at seven days. The shrinkage and creep deformations of the HSC take place, with the greatest part occurring during the first forty days. The decrease thereafter is rather small (Pentala 1987). The shrinkage and creep of normal strength concrete occur more gradually and the curves level off after 100 days.

The creep deformations of HSC are considerably smaller than the values obtained using CEB (CEB 1973) creep formulae. The creep of silica fume concrete and the HSC produced by rapid-hardening Portland cement was about 70 per cent of the CEB values in the average. The creep of blast furnace slag concrete was only 41–45 per cent of the CEB values, while the creep of the normal-strength comparison concrete was 85–99 per cent of the CEB values. The shrinkage deformations of the blast furnace slag concrete were 85–98 per cent of the CEB values. The shrinkage of silica fume concrete was 112 and 113 per cent of the CEB value, and the shrinkage of the rapid-hardening Portland cement concrete was 135 and 151 per cent of the CEB values. Even the normal concrete exceeded the CEB values the corresponding shrinkages were 155 at 45 per cent RH and 173 at 60 per cent RH.

O'Moore and Dux (1999) have studied four code prediction models generally used amongst design engineers for predicting time-dependent deformations in HSC. The chosen models were: ACI 209, CEB-FIP 1990, Bazant-Panula BP, and AS3600. As 3600, CEB-FIP 1990 and the ACI 209 creep models all predict the creep behaviour of concrete on the basis of the creep coefficient. This was used as the basis for all comparisons. Because the BP method calculates the creep-plus-instantaneous strain using an effective modulus of elasticity rather than the true Young's modulus, it is impossible to divide the predicted strain into its two components. Thus the BP method is reviewed in terms of the creep-plus-instantaneous strain per applied stress. Strains were used to review the performance of the shrinkage component of the different models. The test conditions are given in the Table 5.4

They have drawn some general conclusions from the contents of the Table 5.4. AS3600, with

TABLE 5.4 Test conditions used in time-dependent deformation predicting models (O'Moore and Dux 1999)

	Series 1A	Series 2A	Series 3A
Fc at 28 days	88 MPa	69 MPa	69 MPa
Age at loading	31 days	29 days	29 days
Sustain stress	0.4 fcm (35 MPa)	0.4 fcm (28.2 MPa)	0.4 fcm (28.2 MPa)
Specimen size	150 mm diameter 1500 mm tall	150 mm diameter 15000 mm tall	380 mm diameter 1500 mm tall
Applied force (kN)	625 ± 5%	500 ± 5%	3200 ± 3%
Exposure conditions	Drying	Drying	Drying
Temperature	23 ± 2°C	23 ± 2°C	23 ± 2°C
Humidity	60 ± 5%	60 ± 5%	50 ± 6%

its current limitations could not be recommended for either the large or the small test columns. Despite a significant additional degree of numerical complexity, the BP model offered no greater accuracy than the simpler models. It would therefore be difficult to support its use, as the increased number of calculation steps has the undesirable effect of implying greater accuracy which has been shown to be unfounded.

The authors have drawn the following conclusions: based upon the results obtained Tables 5.4 and 5.5, none of the prediction models can be recommended for use within HSC. The inconsistencies between the prediction performance for series 2 (69 MPa and 380 mm diameter) highlights the need for further work to establish appropriate methods for incorporating size effects in HSC specimens. If the design engineer desires to predict the time-dependent behaviour of HSCs within acceptable accuracy range, a test program is the only viable option. The use of current prediction models without verification or modification, is undesirable.

It is further shown that the drying creep and shrinkage are caused by the evaporation of water from the larger pores of the cement paste. The correlation coefficient for the silica concrete is 0.90 for shrinkage and 0.95 for the creep. The corresponding coefficients for blast furnace slag concrete are 0.63 and 0.60 respectively. The coefficients for the HSC are 0.92 and 0.91, while for the normal strength concrete are 0.96 and 0.99 in shrinkage and creep.

The test results appear to show that the creep deformation of HSC does not squeeze out pore water from the specimen, but narrows the pore entrances and thus hinders the evaporation relative to the shrinkage specimens.

In comparing the creep deformations of the test concretes, it is convenient to use specific creep values where the deformations are divided by the applied stress. The shrinkage and specific creep values indicate that blast furnace slag concrete has clearly the smallest shrinkage and creep deformations. Silica fume concrete has about 15 per cent smaller deformations than the HSC produced by rapid-hardening Portland cement. Shrinkage and specific creep deformations of the normal strength concrete are about twice as large as the deformations of the slag concrete. Blast furnace slag concrete had very small pore volumes from the beginning of the test and the weight-loss values were similarly very small so it is very natural that this concrete had the smallest deformations. Silica fume concrete had smaller meso and capillary pore volumes than the HSC made by rapid-hardening Portland cement. The smaller pore volume seems to dominate the evaporated water amount because silica concrete had smaller creep and shrinkage deformations despite the larger weight-loss values.

5.8.3.2 Influence of unit cement paste volume, thickness of cement paste and curing conditions

Specific creep: It is defined as the deformation divided by the applied stress. Similar to the drying shrinkage, specific creep increases with the increase of unit cement paste volume, but the specific creep at equal unit cement paste volume is greater in case of normal concrete than in case of HSC. This is extremely small for HSC of $C = 700 \, kg/m^3$. Therefore it is inferred that similar to the drying shrinkage, the creep is affected not

TABLE 5.5   Overall performance of the prediction models at the completion of the testing

Model	Creep			Shrinkage		
	Series 1A	Series 2A	Series 2	Series 1A	Series 2A	Series 2
ACI209	x	y	y?	x?	Y?	X?
AS3600	x	x	y?	x	x	x
BP	x	y	x	x	x	x
CEB-FIP	x	y	x	x	x	y

y – Acceptable predictions – within $\pm 30\%$; x – Unacceptable predictions; ? – Potential adverse predictions/ divergence at later ages.

only by the unit cement paste volume, but also by the thickness of cement paste.

At equal strength, the difference in the values of μsp/μ between the air and water-cured concrete; drying creep is larger for lower strength. It is larger in the case of air-cured concrete, but in the case of HSC (100 MPa), the values are the same or slightly less than the water-cured concrete. The values of autoclaved concrete on the other hand are extremely small for the same loading compared with the case of standard curing, and for normal-strength concrete it is 1/3–1/5, and for HSC it is less than 1/10 because almost no creep strain is produced.

Cement replacement materials, and SP addition modifies the pore structures and produces concrete of high strength. It is reported (Ramazaniapour 1999) that the highest shear strength was obtained with the addition of 7 per cent silica fume and 1 per cent SP. It is followed by the concrete having only SP. It is also reported that the silica fume addition has significantly decreased the drying shrinkage. Similar results have been reported by Alsayed (1998). It is further shown that replacing SP with the plasticizer increases the drying shrinkage of the specimens containing silica fume and subjected to controlled laboratory-curing conditions as well as to the field conditions by 21 and 27 per cent respectively. This may be attributed to the fact that the plasticizer is less efficient in dispersing the cement paste particles in water than the SP, and therefore concrete containing plasticizer has lower rate of hydration and slow building up of strength. Consequently, it is less resistant to the drying shrinkage.

Pavlenko and Bazhenov (1999) have also shown that at 365 days the drying shrinkage and creep strains of concrete were much lower than those of the ordinary heavy concrete based on coarse aggregate, fine-grained concrete, and ash-sand concrete.

There is a misconception that HSC is highly durable. There is evidence that the HSCs produced with high-cementitious materials and low water-to-binder ratio, high dosages of superplasticizers and high proportions of very fine pozzolanic materials are highly susceptible to early

age cracking. Further, internal microcracking interlinked with external cracking is known to be a major cause of deterioration, even when the concrete is properly cured (Bloom and Bentur 1995). The exact mechanism of early age cracking of plastic concrete is not clear, but this seems to be a combined effect of plastic shrinkage, thermal gradients and autogenous shrinkage or the early volume contraction of the binder phase.

5.8.4 Long-term strength of HSC

HSC is made by the addition of reactive pozzolan, cement and gap graded aggregates. The technique involved is especially the use of low water-to-binder ratio which is achieved by use of SP and the mixing procedure. The high strength of the concrete with silica fume is achieved due to the combined effect of certain factors; (i) pozzolanic reaction of silica fume, (ii) filler effect and (iii) improvement in the bond between the aggregate and the binder phase. These are controversial. Some researchers are of the opinion that the enhancement of the strength is due to the improvement in the paste phase as a whole (Cong et al. 1992; Darwin et al. 1988), whereas other researchers hypothesized that the improvement is due to the improvement in the bond strength between the paste and the aggregates (Bentur and Goldman 1989; Rosenburg and Gajdis 1989; Scrivener et al. 1988; Toutanji and El-Korchi 1995). The principle of making HSC is well described by Chandra et al. (1990). It is clearly shown that the criterion for making the HSC is its mixture design where stress is given to the grading of the aggregate (gap-grading) and low water-to-binder ratio.

5.8.4.1 Strength enhancement

The concrete strength is influenced by the curing condition. High performance concrete (HPC) is more susceptible to this as it has a low water-to-binder ratio. The strength is enhanced by moist-curing compared to concretes which are not moist-cured. Decrease in strength is due to the microcracking induced by drying in the concrete and it does not occur when the samples are wet-cured (De Larrad and Aitcin 1993). The strength increase in the continuously wet-cured concrete is not necessarily due to the more hydration of concrete, at least not due to the hydration alone. Immaterial of the length of wet-curing of the specimen, microcracking will occur when these are subsequently dried for some time, and the long-term strength will decrease. Besides, long-term wet-curing used in the laboratories has no meaning because in practice it never occurs except in the case of submerged structures. For practical purposes, the concrete is seldom cured for more than two weeks.

5.8.4.2 Strength retrogression

Another controversial subject of concrete made by using condensed silica fume is the long-term strength retrogression (Caratte and Malhotra 1993; De Larrad and Bostvironois 1991; Detwiler and Burg 1996).

In a recent paper (Kjellsen *et al.* 1999) it is reported that the strengths of the concretes and pastes with 10 per cent CSF appeared to decrease a few MPa between three and nine months. The strength retrogression appears to be temporary, as at two years the strength has been regained and is higher than after three months. Between two and four years, the strength of the 0.25 water-to-binder ratio (W/B) concrete appears to decrease somewhat. The small decrease in the strength between two and four years may be due to the testing variability. However, the strength retrogression of the system with SF between three and nine months was observed independently in both pastes and concretes. Strength retrogressions of this type have also been reported previously. Detwiler and Burg (1996) reported results on strength development of HSC that were moist-cured for five years. While the reference mix without SF gained strength continuously over time, the mixes containing SF appeared to lose strength over a period of time before it increased again. Five year strength was higher than the strength at any other age. These differences were explained due to the variability in the testing

method. De Larrard and Bostvironious (1991) suggested that the moisture gradient due to drying would lead to the surface stresses and thereby reduce the strength. However, the results reported by Caretta and Malhotra (1993) were not influenced by this as the samples were seal-cured. There is no adequate explanation of the strength retrogression. However, the results obtained by the Environmental Scanning Electron Microscopy (ESEM) have shown that the density of microcracks caused by self-desiccation is considerably higher when the paste contain SF. Most probably the strength retrogression of the SF concrete is linked with microcracking or stresses induced by self-desiccation.

Other researchers have also reported loss in strength in the concrete containing SF. Some of the references are given below:

Atlassi (1990): Atlassi presented results on continuous cured concretes which displayed a drastic decline in compressive strength at 250 days, compared with that measured at 90 days. She explains this to be a result of low workability and inhomogeneity of the concrete.

Mak *et al.* (1990): Mak *et al.* showed a significant strength decrease in the 28–60 day concrete strengths of cylinders stored in lime-saturated water baths. The main reason given for this decrease is inhomogeneity.

Sriviatnannon *et al.* (1990): They have reported decrease in strength between 90 and 365 days though the curing was continued over that period.

Hinczak (1996): Hinczak has studied the influence of different proportions of microsilica fume addition on the compressive strength of concrete. It is shown that 5 per cent microsilica binder produced the highest compressive strength at all ages up to 180 days. However, a decrease in strength of about 10 per cent was noticed between 3 and 180 days. The concrete compressive strength with 5 per cent microsilica was more than twice that for 20 per cent microsilica at seven days. At 180 days both microsilica mixtures indicated a decline in compressive strength of 6 per cent. Twenty per cent microsilica binder produced low strengths at all ages. At three days the strength was only 54 per cent of the precursor Portland cement and only 74 per cent at 180 days. This is probably due to two factors: the first is higher water-to-binder ratio of 0.85 instead of 0.68 and 0.65 for the 5 per cent microsilica and precursor Portland cement respectively and secondly the dispersion problems associated with microsilica reported by Lagerblad and Utkin (1993).

Chandra S (1998c): He has reported decrease in strength over 90 days–2 years. The samples were cured in the laboratory at room temperature.

With these few citations, it is observed that researchers working independently of each other have reported the real strength loss from the highest strength achieved. It should not be lightly dismissed. Generally it is reported that the cause of this strength loss is associated with the moisture loss, which may be one of the factors.

The other cause, which seems more plausible, is the volume change which occurred by late hydration reaction. It is because of the inhomogeneous dispersion of condensed silica fume especially at the low water-to-binder ratio (Perraton *et al.* 1994). It is clear from the tests performed by Chandra (1998c) that there is less decrease in the strength of concrete with the same amount of silica fume in the case of concrete made with higher water-to-binder ratio (30 MPa concrete) compared to that made with low water-to-binder ratio (50 MPa concrete).

The concrete made with the addition of a small amount of silica colloid (2 per cent) in dispersion form, cembinder, did not show such deleterious effect on the strength upto two years. It is attributed to the homogeneous dispersion of the silica colloid, and its spontaneous reaction with the calcium hydroxide produced resulting in high early strength concrete. This decreases the possible volume changes which occurred otherwise due to the delayed hydration (Chandra 1998a, c).

The contractors apparently think that a higher than designed strength would help to speed up the construction work. It is reported that 100,000 concrete bridge decks in the USA showed transverse cracking even before the structures were less than one-month-old (Rogalls *et al.* 1995). A combination of thermal shrinkage and drying shrinkage caused most of the cracks, not the traffic loads or the vibrations during the hardening of the concrete.

HSC typically contain more cement, therefore they shrink more and produce higher temperature during early hydration. Modern cements are apt to cause cracking because they are finer and contain higher sulphate and alkali content (Mehta 1987).

5.9 Need for development of long-term test methods

Much work has been done for the development of HSC, numerous international congresses have been organized, shelves are full of the proceedings, but the test methods used for evaluating the HSC are the same as used for the normal traditional concrete. There is no distinction in the test methods for traditional concrete and the HSC. HSC passes the specifications of the test method made for traditional concrete, but that does not mean in many cases that it is a durable concrete in the long term. The mechanism of deterioration is different. In some cases, there is abrupt deterioration of concrete due to the internal micro cracking which is not visible. It can be caused by many reasons:

1. micro cracking in virgin concrete due to the fast cooling,
2. micro cracking due to drying,
3. micro cracking due to the combined effect of cooling and drying,
4. micro cracking due to the freeze–thaw cycling of concrete,
5. micro cracking due to the mechanical loading,
6. micro cracking due to the thermal incompatibility between the aggregates and the matrix.

Micro cracks are the veins in the durability as they take part in the transportation of liquid and gaseous material into the concrete structure. It will depend upon the size of the micro cracks. For example, if the material being transported is in a gaseous state, for example, CO_2, SO_2, Cl_2 etc., water and acid vapour and the source and sink is at atmospheric pressure, then cracks with widths down to a few micrometer may take part in the material transport process. If, on the other hand, the material being transported is in liquid state, for example, liquid water, dilute solution of acids from acid rain, and the source of the penetrating liquid is in an atmospheric pressure, then micro cracks of widths lower than about 1 μm may be of less importance. At higher pressure, a lower crack size may become effective. Besides the width, there are other micro crack parameters which are of importance. These are length, depth of penetration inside the concrete structure, density of micro cracks per unit area and finally their interconnectivity. However, there seems to have been very little work done on these aspects especially in relation to the durability of concrete structures. Details of the micro crack formations are discussed in a state of the art report of the TC-122-MIC: 'Micro-cracking and life time performance of concrete' (Jensen and Chatterji 1996).

5.10 Concluding remarks

Strength and durability are two different things. It is logical to think that the increase in strength increases the durability properties also. But it is not always true. Increase in strength increases the brittleness of concrete. Concrete becomes more susceptical to the micro crack formation due to the decrease in elasticity. This is why it is not correct to say that the HSC is high performance concrete. It may or may not be a high performance concrete.

Test methods are to be developed which will differentiate the concrete type like concrete up to the strength of 50 MPa, upto 80 MPa and exceeding 80 MPa. Durability properties are to be tested for longer period or accelerated test methods are to be developed which correspond to the longer period of prediction of the durability. Of course, the problem arises, what is a longer period? How it is to be stipulated?

Another problem is that in some countries, the test methods followed are taken from some other countries without taking into consideration the difference in the environmental conditions of their own and the country where the standard is

made and used. One of the examples is from United Arab Emirates. During the early days of the boom in the Gulf states, European consultants and contractors were employed to build up infrastructural facilities in the region. They used European standards and construction methods not fully realizing their significance *vis-à-vis* the harsh Gulf environment. The most common problem faced is corrosion of reinforcement concrete which results in cracking and spalling and consequential loss of bond and structural integrity. The factors which make concrete structures prone to deterioration are high humidity, high temperature, presence of chloride and sulphate in the atmosphere and ground water (Agarwal 1991). Similar examples are seen in many other countries. In Asian countries, water temperature for making concrete can go up to 40°C or even more in the summer period. The concrete is even then made using British standards or American standards. Concrete made in such a way does not display with the properties specified when the concrete is made with the water temperature of 25–30°C. These are very important aspects which are to be taken into consideration when choosing a method for the making and testing of the concrete. These are to be clearly specified in the code of practice.

Another factor which is also very relevant while evaluating the concrete is the synergistic effect of many factors present in the atmosphere. This is very aggressive especially in the marine environment. The concrete is tested in the laboratory where very few factors are considered. Chandra (1999) has designed a test method where the concrete is exposed to the combination of gases in a gas chamber for 6 h, 2 h under water spray, 16 h drying. Samples after particular exposure cycles are subjected to freezing and thawing. The concrete is evaluated by measuring the weight loss and by the compressive strength test.

This test is closer to the environmental conditions as there is variation in the humidity and temperature, due to the gas exposure there are changes in the environmental conditions. This will give a more realistic picture of the durability of concrete.

5.11 Acknowledgements

The author is thankful to Leif Berntsson for reading the text and giving valuable comments. The author is thankful to the authors and publishers of different journals and proceedings and books who have given permission for the reproduction of the tables and figures specified in the text.

5.12 References

Aavik, A. and Chandra, S. (1995). 'Influence of organic admixtures and testing method on freeze-thaw resistance of concrete', *ACI Mater. J.*, January–February, 10–3.

Aakman, M. S. and Cavdar, Z. (1999). 'Efflorescence problem in superplasticized concrete', in RILEM Int. Conf.; The Role of Admixtures in High performance Concrete, PRO 5, Monterrey, Mexico, Cabrera and Villarreal (eds), March, pp. 316–33.

Agarwal, S. N. (1991). 'Corrosion in reinforced concrete', Inst. Eng. (India), Bulletin, 41, August 38.

Alexander, K. M., Wardlaw, J. and Gilbert, D. J. (1965). In *The Structure of Concrete and its Behavior Under Load*, A. E. Brooks and K. Newman (eds), pp. 59–81, Cement and Concrete Association, London.

Alexanderson, M., Wardlaw, J. and Gilbert, D. J. (1968). In *The Structure of Concrete and its Behavior Under Load*, A. E. Brooks and K. Neuman (eds), pp. 59–81, Cement and Concrete Association, London.

Alexander, M. G. and Stamatiou, A. (1994). 'Further studies on fracture properties of paste/rock interfaces', in *Proc. Mater. Res. Soc. Symp.*, Boston, Massachusetes, USA, Diamond *et al.* (eds), Vol. 370, pp. 367–376.

Ali, I. and Kesler, C. E. (1964). 'Mechanism of creep in concrete', *Symp. Creep in Concr.*, ACI Publication, SP-9, pp. 15–63.

Al-Nageim, H. K. and Lewis, R. (1999). 'The improvement of the properties of concrete containing PFA, Micro silica enhanced by using two stage mixing technique', in RILEM Int. Conf.; The Role of Admixtures in High performance Concrete, PRO 5, Monterrey, Mexico, Cabrera and Villarreal (eds), March, pp. 471–82.

Alsayed, S. H. (1998). 'Influence of superplasticizer, plasticizer, and silica fume on the drying shrinkage of high strength concrete subjected to hot dry field conditions', *Cem. Concr. Res.*, 28(10), 1405–16.

Ambroise, J., Chabanet, M., Pals, S. and Pera, J. (1999). 'Basic properties and effects of starch on self levelling of concrete', in RILEM Int. Conf.; The Role of Admixtures in High Performance Concrete, PRO 5,

Monterrey, Mexico, Cabrera and Villarreal (eds), March, pp. 377–86.

Atkins, M., Bennet, D., Dawes, A., Glasser, F. P., Kindness, A. and Read, D. (1991). 'A thermodynamic model for blended cements', DoE Report No DoE/HMIP/RR/92/005, December.

Atlassi, E. (1990). 'High strength concrete – the influence of mix proportions on some mechanical and physical properties', Concrete for the Nineties, in Int. Conf. on the Use of Fly Ash, Slag, Silica fume and other Silicious Materials in Concrete, W. B. Butler and I. Hinczak (eds), pp. Atlassi1–20.

Barnes, B. D., Diamond, S. and Dolch, W. J. (1978). 'The contact zone between Portland cement paste and glass "aggregate" surfaces', *Cem. Concr. Res.*, 8, 233–44.

Barnes, B. D., Diamond, S. and Dolch, W. L. (1979). 'Micromorphology of the interfacial zone around aggregate in Portland cement mortar', *J. Am. Ceram. Soc.*, 62(1–2), 21–4.

Barovsky, N. (1983). 'Pore structure of cement stone and its influence on the mechanical properties of concrete', Ph.D. Thesis, Bulgarian Academy of Sciences, Sofia.

Barovsky, N. (1993). 'Modelling of sulphate ion diffusion in cement stone', *Mater. Eng.*, 3(3).

Bazantova, A. Z. (1985). 'Bond between cement binder and rock', Ph.D. Thesis, Faculty of Civil Engineering, Czechoslovakia University, Prague, Czechoslovakia.

Bazantova, Z. and Modry, S. (1998). 'Properties of gypsum free cement rock composite', *RILEM 2nd Int. Conf. on Interfacial Transition Zone*, Haifa, pp. 264–70.

Bentur, A. and Goldman, A. (1989). 'Bond effects in high strength silica fume concrete', *ACI Mater. J.*, 86(5), 440–7.

Bentur, A. and Cohen, M. D. (1987). 'Effect of condensed silica fume on the microstructure of the interfacial zone in Portland cement mortars', *J. Am. Ceram. Soc.*, 70(10), 738–43.

Ben-Bassat, M. and Baum, H. (1999). 'Effect of superplasticizers on some properties of fresh and hardened concrete', in RILEM Int. Conf.; The Role of Admixtures in High performance Concrete, PRO 5, Monterrey, Mexico, Cabrera and Villarreal (eds), March, pp. 482–92.

Bloom, R. and Bentur, A. (1995). 'Free and restrained shrinkage cracking in normal and high strength concretes', *ACI Mater. J.*, 92(2), 211–17.

Bourdette, B., Ringot, R. and Ollivier, J. P. (1995). 'Modelling of the transition zone porosity', *Cem. Concr. Res.*, 25(4), 741–51.

Breton, D., Carles-Gibergues, A., Ballivey, G. and Grandet, J. (1993). 'Contribution to the formation mechanism of the transition zone between rock-cement paste', *Cem. Concr. Res.*, 23, 335–46.

Carles-Gibereus, A. (1981). 'Les ajouts dans les micro-betons, influence sur l'aureole de transition et sur les proprietes mecaniques', Ph.D. Thesis, INSA-UPS, Toulouse, France.

Carles-Gibergue, A., Ollivier, J. P. and Hanna, B. (1989). 'Ultra fine admixtures in high strength paste and mortars', in 3rd Int. Conf. CANMET/ACI, Fly ash, silica fume, slag and natural pozzolana, Trondheim, Norway, V. M. Malhotra (ed.), ACI SP 114, pp. 1101–17.

Caretta, G. G. and Malhotra, V. M. (1993). 'Long term strength development of silica fume concrete', in Proc. 4th Int. Conf. on Fly ash, Slag, silica fume and natural pozzolan in concrete, Istanbul, V. M. Malhotra (ed.), ACI SP 132, Vol. 2, pp. 1017–44.

CEB Manuel de Calcul CEB (1973). 'Effects structuraux du fluage et des déformations différées du béton', Com. Eur. Béton Bull. Inf., 94, 62.

Chandra, S., Kutti, T. and Berntsson, L. (1990). 'Principle of making high strength concrete', Concr. Int., pp. 59–62.

Chandra, S. and Xu, A. (1992). 'Influence of presaturation and freeze-thaw conditions on length changes of Portland cement mortars', Cem. Concr. Res., 22, 515–24.

Chandra, S. (1994). 'Influence of superplasticizers mixing technic on the strength of mortars', Associated Cement Company, Thane, India, Tokten Report, (UNDP Project), p. 35.

Chandra, S. and Ohama, Y. (1994). Polymers in Concrete, Chap. 1, CRC Press, Boca Raton, USA.

Chandra, S. and Bergqvist, H. (1997). 'Interaction of colloidal silica and Portland cement' in Proc. 10th Int. Conf. Chem. Cem. Göteborg, Sweden, June 2–6, H. Justnes (ed.), pp. 3 ii 106–12.

Chandra, S. (1998a). 'Mortars and concretes with silica colloids having improved properties', Indian Concr. J., February, (2), 73–5.

Chandra, S. (1998b). 'Freeze thaw resistance of polymer mortars', 9th Int. Conf. on Polymer in Concrete, Bologna, Italy, September, F. Sandrolini (ed.), pp. 733–9.

Chandra, S. (1998c). 'Long term strength testing of concrete having silica fume and colloidal silica', Unpublished Work, Chalmers University of Technology, Göteborg, Sweden.

Chandra, S. (1999). 'Laboratory and field test of concrete damaged by pollution', in Proc. 24th Cong. on the World of Concrete and Structures, Singapore, August 24–26, J. Tam (ed.).

Chatterjee, S. and Jeffery, J. W. (1971). 'The value of bond between different types of aggregates and Portland cement', Indian Concr. J., 45, 346–9.

Chen Zhi Yuan and Odler, J. (1987). 'The interfacial zone between marble and tricalcium silicate paste', Cem. Concr. Res., 5, 784–92.

Claisse, A. P., Lorimer, P. and Al-Omari, M. H. (1999). 'The effect of changes in cement on the properties of cement grouts with superplasticizing admixtures', in RILEM Int. Conf.; The Role of Admixtures in High performance Concrete, PRO 5, Monterrey, Mexico, March, Cabrera and Villarreal (eds), pp. 19–33.

Czernin, W. (1960). Chemistry of Cement Concrete for Construction Engineers, Bauverlag, Germany.

Cong, X., Gong, S., Darwin, D. and McCabe, S. I. (1992). 'Role of silica fume in compressive strength of mortar and concrete', ACI Mater. J., 89(4), 375–87.

Darwin, D., Zhenjia, S. and Harsh, S. (1988). 'Silica fume, bond strength, and compressive strength of mortar', in Proc. Bonding in cementitious composites, Mater. Res. Soc. Symp. Boston, December 1987, Material Research Society, Pittsburgh, Vol. 114.

Davidovits, J. and Courtois, L. (1981). 'D.T.A. detection in intra-ceramic geopolymeric setting in archaeological ceramics and mortars', Abstracts, 21st Symp. on Archaeometery, Brookhaven, p. 22.

Davidovits, J. (1986). 'X-ray analyses and X-ray diffraction of casing stones from the pyramids of Egypt and the limestones of the associated queries', Science in Egyptology, Manchester University Press, UK, pp. 511–20.

De Rooij, M. R., Bijen, J. M. and Frens, G. (1998). 'Introduction of syneresis in cement paste', in RILEM 2nd Int. Conf. on the Interfacial Transition zone in cementitious composites, Haifa, Israel, A. Katz (ed.), pp. 4949–56.

De Larrad, F. and Bostvironois, J. L. (1991). 'On the long term strength losses of silica fume concrete', Mag. Concr. Res., 43(155), 109–19.

De Larrad, F. and Aitcin, P. C. (1993). 'The strength retrogression of silica fume concrete', ACI Mater. J. 90, November–December, pp. 511–20.

Detwiler, R. J. and Burg, R. G. (1996). 'Long term strength tests of high strength concrete, concrete technology today', Portland Cem. Assoc., 17(2), 5–7.

Diamond, S. (1976). 'Hydraulic cement pastes; their structure and properties', p. 2, Cement and Concrete Association, Slough, UK.

Diamond, S. (1982). 'Effect of microsilica (silica fume) on the pore solution chemistry of cement pastes', J. Am. Ceram. Soc., 66, C82–4.

Diamond, S. (1987). 'Cement paste microstructure in concrete', Mater. Res. Soc. Symp., 85, 21–31.

Diamond, S. and Huang, J. (1998). 'The interfacial transition zone, reality or myth' in RILEM 2nd Symp. on the Interfacial Transition Zone in Cementitious Composites, Haifa, Israel, A. Katz et al. (eds), March, pp. 1–40.

Escadeillas, G. (1988). 'Les ciments aux fillers calcaires, contribution à leur optimisation parlende des propriétés mécaniques et physiques des bétons fillerises', Ph.D. Thesis, INSA-UPS, Toulouse, France.

Escadeilas, G. and Maso, J. C. (1991). 'Approach of the initial state in cement paste, mortar & concrete', in Ceramics Transaction, American Ceramic Society., S. Mindess (ed): Proc. Advances in Cementitious Materials, Vol. 16, pp. 169–84.

Fagerlund, G. (1986). 'Effect of air entraining and other admixtures on the salt scaling resistance of concrete', International Seminar on Durability of concrete: Aspects of Admixtures and Industrial By-products, D1:1988, Swedish Council of Building Research, Stockholm, Sweden, pp. 233–66.

Fagerlund, G. (1993). 'Frost resistance of high strength concrete – Some theoretical considerations', Report TVBM-3056, Division of Building Materials, Lund Institute of Technology, Sweden, p. 38.

Farran, J. (1956). 'Construction minéralogique a l' Etude de l'Adhérence entre' les constituents hydratés des ciment et les matériaux', Revue des Materiaux de Construction, 490–I, July–August, 155–72, and 492 September, 191–209.

Feng, Q. L., Lachowski, E. E. and Glasser, F. P. (1989). 'Densification and migration of ions in blast furnace slag-Portland cement pastes', Mater. Res. Soc. Symp., 136, 263–72.

Feng, Q. L. and Glasser, F. P. (1990). 'Mass transport and densification of slag cement pastes', Mater. Res. Symp. Proc., 178, 57.

Foy, C., Pigeon, M. and Banthia, N. (1988). 'Freeze thaw durability and deicer salt scaling resistance of a 0.25 water to cement ratio concrete', Cem. Concr. Res., 18(4), 604–14.

Gagne, R., Pigeon, M. and Aitcin, P. C. (1990). 'Durabilite au gel betons de hautes performances mecanques', Mater. Struct., 23, 103–9.

Geiker, M. (1983). 'Studies of Portland cement hydration by measurement of chemical shrinkage and a systematic evaluation of hydration curves by means of the dispersion model', Ph.D. Thesis, Technical University of Denmark, Lyngby, Denmark.

Glasser, F. P., Diamond, S. and Roy, D. M. (1987). 'Hydration reactions in cement paste incorporating fly ash and other pozzolanic materials', Mater. Res. Soc. Proc., 86, 139–58.

Goldman, A. and Bentur, A. (1992). 'Effect of pozzolanic and non reactive microfillers on the transition zone in high strength concrete', RILEM International Conference: Interfaces in Cementitious Composites, RILEM Proceedings 8, Tolouse, France, pp. 53–61.

Goldman, A. and Bentur, A. (1993). 'The influence of microfillers on enhancement of concrete strength', Cem. Concr. Res., 23, 962–72.

Goldman, A. and Cohen, M. D. (1998). 'The role of silica fume in mortars with two types of aggregates', in 2nd RILEM Int. Conf. Interfacial Transition Zone, A. Katz et al. (eds), pp. 315–22.

Grandet, J. and Ollivier, J. P. (1980). 'New method of study of cement aggregate interface', in 7th Int. Cong. Chem. Cem., Paris, III(VII), pp. 85–9.

Grutzeck, M. W., Roy, D. M. and Wolfe-Confer, D. (1982). 'Mechanism of hydration of Portland Cement composites containing ferro-silicon dust', in Proc. 4th Int. Conf. Cem. Micros., Assn. Mtg., Las Vegas.

Grutzeck, M. W., Atkinson, S. D. and Roy, D. M. (1983). 'Mechanism of hydration of condensed silica fume in calcium hydroxide solutions', in Proc. CANMET/ACI First Int. Conf. On the use of fly ash, silica fume, slag and other mineral by products in concrete, Detroit, V. M. Malhotra (ed.), ACI-SP-79, Vol. ii, pp 643–64.

Hadley, D. W. (1972). 'The nature of the paste aggregate interface', Purdue University, W. Lafayette, USA.

Hammer, T. A. and Sellevold, E. J. (1990). 'Frost resistance of high strength concrete', 2nd Int. Symp. on Utilization of High Strength Concrete, Berkeley, p. 17.

Hammer, T. A. and Sellevold, E. J. (1990). 'Frost resistance of high strength concrete', ACI Special Publication, SP-121, 457–89.

Hanna, B. (1987). 'Contribution a' l'etude de la structuration des mortiers de ciment portland contenant des particules ultra fines', INSA-UPS, Ph.D. Thesis, Toulouse, France.

Hinczak, I. (1996). Ph.D. Thesis, University of Sydney, Australia.

Hochstetter, R. (1973). 'The pore structure of some building materials and their resistance against the combined attack of frost and salt solutions', in Proc. Int. Symp. on Pore Structure and Properties of Materials, Part IV, Prague, Czechoslovakia.

Hoshino, M. (1988). 'Difference of the W/C ratio, porosity and microscopical aspect between the upper boundary paste and the lower boundary paste of the aggregate in concrete', Mater. Struct., 21, 336–40.

Hoshino, M. (1989). 'Relationship between bleeding, coarse aggregate and specimen height of concrete', ACI Mater. J., 86, 125–90.

Hsu, T. T. C. and Slate, O. F. (1963). 'Tensile bond strength between aggregate and cement paste and mortars', Am. Concr. Inst. J., 60, 465–85.

Hunter, R. J. (1993). Introduction to Modern Colloid. Science, Oxford University Press.

Ivanov, F. M. (1977). 'Investigation on the salt diffusion in cement mortars', J. Appl. Chem. (in Russian) XVI, 12.

Jacobsen, S. and Sellevold, E. J. (1992). 'Frost-salt scaling of concrete, effect of curing temperature and condensed silica fume on normal and high strength concrete', in 4th CANMET/ACI Int. Conf. on fly ash, silica fume and natural pozzolans in concrete, V. M. Malhotra (ed.), pp. 369–84.

Jacobsen, S. (1994). 'The frost durability of high performance concrete', Ph.D. Thesis, Norwegian Institute of Technology, Trondheim, Norway.

Jensen, A. D. and Chatterji, S. (1996). 'State of the art report on micro-cracking and lifetime of concrete', *Mater. Struct.*, **29**, 3–8.

Kinoshita, M. and Okada, K. (1999). 'Performance of methacrylic water soluble graft polymers with different types of cement', in RILEM Int. Conf.; The Role of Admixtures in High performance Concrete, PRO 5, Monterrey, Mexico, Cabrera and Villarreal (eds), March, pp. 34–47.

Kim, B.-G., Jiang, S. P. and Aitcin, P. C. (1999). 'Influence of molecular weight of polynaphthalene sulfonate, superplasticizers on the properties of cement pastes containing different alkali content', in RILEM Int. Conf.; The Role of Admixtures in High performance Concrete, PRO 5, Monterrey, Mexico, Cabrera and Villarreal (eds), March, pp. 97–111.

Kjellsen, Wallevik and Hallgren (1999). 'On the compressive strength development of high strength concrete and paste- effect of silica fume', *Mater. Struct.*, **32**, 63–9.

Kobayashi, M., Nakakuro, K., Kodama, S. and Negami, S. (1981). 'Frost resistance of superplasticized concrete', ACI Special Publication, SP-68, 269–82.

Kruyt, H. R. (1949). *Colloid Science*, *II*, Elseveir Publishing Company, Inc., New York, p. 573.

Lagerblad, B. and Utkin, P. (1993). 'Silica granules in concrete-dispersion and durability aspects', CBI report, 3:93. Swedish Cement and Concrete Institute, Stockholm, Sweden, p. 44.

Langton, C. A. and Roy, D. M. (1984). 'Longevity of bore hole and shaft sealing materials; characterization of ancient cement based building materials', Symp. Proc. No. 26, Material Research Society, Pittsburgh, USA, pp. 543–9.

Larbi, J. A. and Bijen J. M. (1990). 'Evolution of lime and microstructural development in fly ash Portland cement systems', *Mater. Res. Soc.*, **178**, 127–38.

Li, S., Roy, D. M. and Kumar, A. (1985). 'Quantitative determination of pozzolans in hydrated systems of cement or $Ca(OH)_2$ with fly ash or silica fume', *Cem. Concr. Res.*, **15**, 1079–86.

Lyse, I. (1960). 'Shrinkage and Creep of Concrete', *J. Am. Concr. Inst.*, February, 775–82.

Mak, S. L., Ho, D. W. S., Darvall, P. and Le, P. (1990). 'The effect of moist curing on the compressive strength of some very high strength concrete' in Proc. Concrete for Nineties, Int. Conf. On the Use of Fly-ash, Silica Fume, and Other Silicious Materials in Concrete Australia, W. B. Butler and I. Hinczak (eds), pp. Mak1–12.

Malinowski, R. (1979). 'Concretes and mortars in ancient aqueducts', *Concr. Int., Design Constr.*, **1**(1), January 66–76.

Malinowski, R., Slatkine, A. and Ben Yair, M. (1961). 'Durability of Roman mortars and concretes for hydraulic structures at Caesarea and Tiberias', in RILEM Int. Symp. Durab. Concr., Prague.

Maso, J. C. (1980). 'La liason entre les granulats et la pate de ciment hydrate', in 7th Int. Symp. Chem. Cem., Paris, Vol. i, VII-I pp.1–14.

Mather, B. (1986). 'Discussions'; International Seminar on Durability of concrete: Aspects of Admixtures and Industrial By-products, 1986, D1:1988, Swedish Council of Building Research, Stockholm, Sweden, pp. 307–13.

McCarthy, G. J. and Solem Tishmack (1994). 'Hydration mineralogy of cementitious coal combustion by products', Advances in Cement and concrete, In Proceedings, Engineering Foundation Conference, Grutzeck and Sarkar (eds), May, pp. 103–21.

Mehta, P. K. (1987). 'Durability-critical issue for the future', *Concr. Int.*, July, 27–33.

Mehta, P. K. and Monterio, P. J. M. (1988). 'Effect of aggregate, cement and mineral admixtures on the transitional zone', *Mater. Res. Soc. Symp.*, Vol. 114, 65–75.

Mitsui, K., Kasami, H., Yoshika, Y. and Kinoshita, M. (1989). 'Properties of high strength concrete with silica fume using high range water reducer of slump retarding type' (ed. V. M. Malhotra), ACI Special Publication SP-119, pp. 215–31.

Monterio, P. J. M. and Andrade, W. P. (1987). 'Analyses of the rock cement paste bond using probabilistic treatment of brittle strength', *Cem. Concr. Res.*, **17**, 919–26.

Monterio P. J. M. and Mehta, P. K. (1986). 'Interaction between carbonate rock and cement paste, *Cem. Concr. Res.*, **16**, 127–34.

Monterio P. J. M., Maso, J. C. and Ollivier, J. P. (1985). 'The aggregate-mortar interface', *Cem. Concr. Res.*, **15**(6), 953–8.

Nagataki, S. and Yonekura, A. (1982). 'Properties of drying shrinkage and creep of high strength concrete', *Trans. Jpn. Concr. Inst.*, **4**.

Obla, K., Hill, R. and Sarkar, S. L. (1999). 'Impact of chemical admixtures on air entrainment of high performance concrete', in RILEM Int. Conf.; The Role of Admixtures in High performance Concrete, PRO 5, Monterrey, Mexico, Cabrera and Villarreal (eds) March, pp. 255–70.

Okada, E., Hisaka, M., Kazama, Y. and Hattori, K. (1981). 'Frost resistance of superplasticized concrete', ACI Special Publication, SP-68, pp. 215–31.

Ollivier, J. P. (1981). 'Contribution a l'Etude de l'Hydratation de la Pâte de Ciment Portland au Voisinage des Granulats', Ph.D. Thesis, INSA-UPS, Toulôuse, France.

Ollivier, J. P. (1986). 'Rôle des additions minérales sur la formation de l'aureole de transition entre gros granulat et mortier', *8th Int. Cong. Chem. Cem.*, Rio de Janeiro, Vol. VI, pp. 189–97.

Ollivier, J. P. and Massat, M. (1994). 'Microstructure of the paste, aggregate interface including the influence of mineral admixture', Advances in cement and concrete, in Proc. of Engineering Foundation Conference, Durham, publ. Am. Soc. of Civ. Eng., New York, Grutzeck and Sarkar (eds), pp. 175–85.

Ollivier, J. P., Maso, J. C. and Bourdette, B. (1994). 'Interfacial transition zones in concrete', *Adv. Cem. Based Mater.*, 13.

Ollivier, J. P., Maso, J. C. and Bourdette, B. (1995). 'Contribution minéralogique a l'Etude de l'Adhérence entre les constituents hydratés des ciment et les matériaux', *Adv. Cem. Based. Mater.*, 2, 30–8.

O'Moore, L. and Dux, P. (1999). A comparison of predicted and measured time-dependent deformations in high strength concrete columns, in proceedings', 24th Int. Conf. Our World in Concrete & Structures, 25–26 August Singapore, CI-Premier PTE LTd, C. T. Tam (ed.), pp. 283–90

Pentala, V. (1987). 'Mechanical properties of high strength concrete based on different binder combination', in Proc. Symp. Utilization of High Strength Concrete, Stavanger, Norway, pp. 123–34.

Pentala, V. and Rautanen, T. (1990). 'Microporosity, creep and shrinkage of high strength concretes', 2nd Int. Symp. High Strength Concrete, SP 121–21, Weston T. Hester (ed.), pp. 409–32.

Petersson, P. E. (1988). 'Influence of silica fume on salt frost resistance of concrete', Int. Sem. Durab. Concr.: Aspects of admixtures and industrial by-products, 1986; D1:1988, Swedish Council of Building Research, Stockholm, Sweden. pp. 179–89.

Page, M., Nkinamubanzi, P.-C. and Aitcin, P.-C. (1999). 'The cement/superplasticizer compatibility, a headache for superplasticizer manufacturers', in Proc. RILEM Int. Conf. Role of Chemical Admixture in High Performance Concrete, Monterrey, Mexico, Cabrera and Villarreal (eds), pp. 48–56.

Pavlenko, S. and Bazhenov, Yu. (1999). 'Complex chemical admixture improving properties of fine grained slag ash concrete', in Proc. RILEM Int. Conf. Role of Chemical Admixture in High Performance Concrete, Monterrey, Mexico, Cabrera and Villarreal (eds), pp. 112–20.

Perraton, D., De Larrard, F. and Aitcin, P. C. (1994). 'Additional data on the strength retrogression of some air cured silica fume concretes', CANMET/ACI Conference, Nice, France, supplementary papers.

Picket, G. (1956). 'Effect of aggregate on shrinkage of concrete and hypothesis concerning shrinkage', *J. Am. Concr. Inst.*, 52(5), 581–90.

Pieterson, H. S. (1993). ' Reactivity of fly ash and slag in cement', Delft University of Technology, September 1993.

Pigeon, M., Gagne, R., Aitcin, P. C. and Banthis, N. (1991). 'Freezing and thawing tests of high strength concrete', *Cem. Concr. Res.*, 21, 844–52.

Popovics, S. (1980). Concrete Making Materials, HPS, Washington, USA.

Polak, A. F. (1971). 'Corrosion of reinforced concrete structures', Moscow (in Russian).

Powers, T. C. (1965). 'Mechanism of shrinkage and reversible creep of hardened cement paste', Proc. Int. Conf. Struct. of Concr., London, pp. 319–44.

Ramazaniapour, A. A. (1999). 'Superplasticzers to improve mechanical properties of concrete for repair', in RILEM Int. Conf.; The Role of Admixtures in High performance Concrete, PRO 5, Monterrey, Mexico, Cabrera and Villarreal (eds), March, pp. 461–70.

Rogalls, E. A., Kraus, P. D. and Macdonald D. B. (1995). 'Reducing transverse cracking in new concrete bridge decks', *Concr. Construct.*, 40, 735–8.

Roncero, J., Gettu, R., Vazquez, E. and Torrents, J. P. (1999). 'Effect of superplasticizer content and temperature on the fluidity and setting of cement paste', in RILEM Int. Conf.; The Role of Admixtures in High performance Concrete, PRO 5, Monterrey, Mexico, Cabrera and Villarreal (eds), pp. 343–56.

Rosenberg, A. M. M. and Gajdis J. M. (1989). 'A new mineral admixture for high strength concrete', *Concr. Int.*, 11(4), 31–6.

Roy, D. M. (1989). 'Fly ash and silica fume chemistry and hydration', in Proc. 3rd Int. Conf. on Fly Ash, Slag, Silica Fume and Natural Pozzolan (ICFSS), Trondheim, ACI SP 114, Vol. I, pp. 119–38.

Roy, D. M. (1986). 'Mechanisms of cement paste degradation due to chemical and physical factors', Proc. 8th Int. Cong. Chem. Cem., Brazil, Vol. I, 362–80.

Schmalfeld, V. (1977). 'Untersuchungen des Einflusses verschiedener Salze auf den Erhaetungsprozess von Trikalciumaluminat in Anwesenheit von Gips', *Silkattechnik*, 28, p. 3.

Scrivener, K. L. and Pratt, P. L. (1986). 'A preliminary study of the microstructure of the cement/sand bond in mortars', in Proc. 8th Int. Cong. Chem. Cem., Rio de Janeiro, Vol. III, pp. 466–71.

Scrivener, K. L. and Gartner, E. M. (1987). 'Microstructural gradients in cement paste around aggregate particles', in Proc. Bonding in Cementitious Composites, Pittsburgh, USA, S. Mindess and S. P. Shaw (eds), 114, pp. 77–85.

Scrivener, K. L., Crumble A. K. and Pratt, P. L. (1987). Bonding in Cementitious Composites, Pittsburgh, USA, Mindess S and S. P. Shaw (eds), 114, pp. 87–8.

Scrivener, K. L., Bentur, A. and Pratt, P. L. (1988). 'Quantitative characterization of transitional zone in high strength concretes', *Adv. Cem. Res.*, 1(4), 230–9.

Scrivener, K. L. (1989). 'The microstructure of concrete', Materials Science of Concrete, American Ceramic Society, J. P. Skalny (ed.), pp. 127–57.

Srivivatnanon, V., Baweja, D., Cao, H. T. and Hassel, D. (1990). 'Production Australian silica fume and its utilization in concrete' in Proc. Concrete for the

Nineties, International Conference on the Use of Fly Ash, Slag, Silica fume and other Silicious Materials in Concrete, CSIRO, Australia, W. B. Butler and I. Hinczak (eds), pp. Sirivivatnanon 1–15.

Struble, L. (1992). 'The performance of Portland cement', ASTM Standardization News, January, 38–45.

Struble, L., Skalny, J. and Mindess, S. (1980). 'A review of the cement aggregate bond', *Cem. Concr. Res.*, 10, 277–86.

Swamy, R. N. (1999). 'The magic of synergy: Chemical and mineral admixtures for high durability concrete', in Proc. RILEM Int. Conf. The Role of Admixtures in High performance Concrete, PRO 5, Monterrey, Mexico, Cabrera and Villarreal (eds), March, pp. 3–18.

Takeuchi, T. and Nagataki, S. (1999). 'Effect of set retarders and superplasticizers on the long term slump retardation' in Proc. RILEM Int. Conf. The Role of Admixtures in High performance Concrete, PRO 5, Monterry, Mexico, Cabrera and Villarreal (eds), March, pp. 403–18.

Tasong, W. A. (1997). 'Department of civil and structural Engineering', Ph.D. Thesis, University of Sheffield, UK.

Tasong, W. A., Cripps, J. C. and Lynsdale, C. J. (1998). 'Aggregate-cement paste interface: Influence of Aggregate Physical Properties', *Cem. Concr. Res.*, 28(10), 1453–66.

Taylor, H. F. W. (1986). Chemistry of cement hydration, Paper 2.1, in Proc. 8th Int. Conf. On the Chemistry of Cement, Rio de Janerio, Vol. 1, pp. 82–110.

Tazawa, E. and Miyazawa, S. (1997). 'Influence of cement composition on autogenous shrinkage of concrete', in Proc. 10th Int. Conf. on the Chemistry of Cement, Sweden, H. Justness Gothenburg (ed.) Vol. 2, pp. 2ii071.

Toutanji, H. A. and El-Korchi, T. (1995). 'The influence of silica fume on the compressive strength of cement paste and mortar', *Cem. Concr. Res.* 25(7), 1591–602.

Uchikawa, H. (1999). 'Function of organic admixture supporting high performance concrete', in Proc. RILEM Int. Conf. The Role of Admixtures in High performance Concrete, PRO 5, Monterrey, Mexico, Cabrera and Villarreal (eds), March, pp. 69–96.

Uchikawa, H., Sawaki, D. and Hanehara, S. (1995). 'Influence of kind and added timing of organic admixture on the composition, structure and property of fresh cement paste', *Cem. Concr. Res.*, 25, 353–64.

Uchikawa, H., Hanehara, S., Shirasaka, T. and Sawaki, D. (1992). ' Effect of admixture on hydration of cement, Adsorptive behavior of admixture and fluidity and setting of fresh cement paste', *Cem. Concr. Res.*, 22, 1115–29.

Ushiyama, H. and Goto, S. (1974). 'Investigation of various ions in hardened cement paste', Supplementary paper, Section II-3, Proc. VI the Int. Cong. on 'Chemistry of Cement', Moscow, Russia.

Wiggs, P. K. C. (1958). 'The relation between gas permeability and pore structure of solids', in Proc. Int. Conf. Structure and Properties of Porous Materials, London.

Yamazaki, K. (1989). 'Fundamental studies of the effects of mineral fines on the strength of concrete', *Trans. Jpn. Soc. Civil Engg.*, 85, 15–44.

Yang, M., Neubaur, C. M. and Jennings, H. M. (1997). 'Interparticle potential and sedimentation behavior of cement suspensions', *Adv. Cem. Based Mater.*, 5, 1–7.

Yuan, C. Z. and Guo, W. J. (1987). 'Bond between marble and cement paste', *Cem. Concr. Res.*, 17, 544–52.

Zimbelman, R. (1978). 'The problem of increasing the strength of concrete', *Betonwerk und Fertigteil-Teknik*, 2, 89–96.

Zimbelman, R. (1985). 'A contribution to the problem of cement aggregate bond', *Cem. Concr. Res.*, 15, 801–8.

Chapter six

Special cements

A. K. Chatterjee

6.1 Introduction

Construction as a human activity is believed to have started from the Neolithic age. Since then, till to-day, there is a continuous search for newer and more effective construction materials. Clay, lime, gypsum and lime–pozzolana mixes were known as effective binders prior to the advent of Portland cement and served mankind for perhaps five millennia. Compared to this, Portland cement has so far had a short history of less than two centuries, but within this time frame, it has not only grown to a production volume of about 1.5 billion tonnes, perhaps the largest amongst man-made materials, but also diversified into a host of derivatives with specific properties and application areas. Despite such a phenomenal growth, the generic Portland cement is known to suffer from several shortcomings, restricting its usage in many ways. Hence, scores of new cementitious systems have been and are being investigated all over the world. In this melée of Portland cement, its derivatives and new cementitious systems, it is difficult,

if not impossible, to define what is a 'special cement'. Notwithstanding this fundamental problem, an attempt has been made in this chapter to dwell on cements that are basically hydraulic in nature and are perceptibly different from the traditional varieties of Portland cement in composition and use. The basic attempt is to deal with cements that hold high potential for such specific applications that cannot be easily addressed by Portland cement and its usual varieties.

6.2 Prior reviews and past definitions

Amongst the reviews of special cements, the one authored by Kurdowski and Sorrentino in the eighties [1] is a comprehensive one. In an attempt to define and classify special cements in relation to normal ones through transitions of composition, manufacturing technology and application characteristics, the authors primarily classified cements into the following four groups:

1. Cements with special properties like high early strength, fast setting, low heat of hydration, colour, dimensional change, water-proofing characteristics, etc., which included, *inter alia*, products like calcium sulpho-aluminate, calcium fluoroaluminate, calcium aluminates, calcium sulphate and free CaO bearing complex formulations, white and decorative cements, etc.
2. Cements produced with low energy consumption to cover such products as blended cements, masonry cement, belite cement, alinitic cement, iron-rich cements, etc.

3. Chemically hardening cements covering oxy-chloride, oxysulphate, phosphatic and dental cements.
4. Cements for non-constructional applications like machine grouting, nuclear entrapment, refractory furnace linings, etc.

Similar comprehensive reviews of new and special cements were attempted by the present author in the nineties [2,3]. In these reviews, an attempt was made to highlight the following four broad categories:

(i) Current commercial production range that covered high-strength Portland cement, blended Portland cements, oil-well and geo-thermal cements, expansive cements and regulated-set cements.
(ii) Cements on the threshold of commercialization which included reactive belite cement, geopolymeric cements like alkali-activated slag and metakaolinitic varieties, etc.
(iii) New cement composites for non-constructional engineering applications so as to cover such products as DSP and MDF families.
(iv) New cement systems to encompass products like phosphatic cement, calcium aluminate glass cement, sol–gel cement, blended high-alumina cement and organo-mineral cement.

In an attempt to deal with the management strategy in cement technology for the new century, Uchikawa [4] emphasized the need to create high-performance multifunctional cements to meet the challenges of a more discerning and sophisticated market as well as modifying the existing cements for better performance including compliance with environmental demands. The author emphasized that for creating special cements it would be necessary to work on reaction-control functions, structure-control functions and intelligent functions like auto-controls by pH, temperature, atmosphere and light. While elaborating on the approaches for such functional controls, the author touched upon certain special cements like belite cement, calcium sulphoaluminate cement, alinite cement, etc.

In recent years, an equally comprehensive review has been presented by Bensted [5], in which the oil-well cement and its additives have been dealt with in great detail, keeping in view the paucity of information on this particular cement and its applications. In addition, the author categorized other special cements under the following headings:

1. Decorative Portland cements including white and coloured cements as well as cement paints.
2. Chemical cements covering fifteen different varieties such as oxychloride, oxysulphate, phosphate, water glass, borate types.
3. Special Portland-type and other cements which included experimental products like non-calcareous and non-siliceous cements, replacement of gypsum as set-retarder in Portland cement, alinite and belinite phase synthesis, microfine cements, Portland-polymer cement, high early strength cement, expansive cement, hydrogarnet-type cement, hydrophobic Portland cement, ferrite cement and thermoplastic cement.

In the year 2000, a comprehensive volume has been published on special inorganic cements under the authorship of Odler [6]. The book is organized in 30 chapters devoted to chemico-mineralogical composition and physico-mechanical performance of various cements.

One of the salient features of this publication is the wide coverage of special cements and their best applications as fast setting cements, low heat evolution cements, oil-well cements, cement for high-temperature applications, etc.

In addition to such comprehensive reviews as illustrated above, many authors had attempted sectoral coverage of special cements. One such early compilation on energy-saving cements was presented by Mehta [7]. Here, the effort was to focus on the energy-saving potential of calcium sulphoaluminate–belite type of cements termed as 'modified Portland cements' and blended Portland pozzolana cement containing 'plaster of Paris'. Similarly, an effective summary of recent developments in the field of ultra high-strength cementitious materials as well as synthesis and characterisation of low-temperature materials can be found in the articles of Roy [8,9]. An informative compilation of the varieties of chemically resistant cements has been given by Fenner [10]. A fairly comprehensive coverage on high-belite cements has been provided by the present author in the recent past [11]. While enlisting such selected sectoral reviews, it may be pertinent to mention the special issue of Advances in Cement Research [12] that was devoted to the production and use of calcium sulphoaluminate cements.

On the whole, from all the previous reviews and write-ups, it can only be reiterated that there is no single acceptable definition of special cements. It is also evident that during the past few decades, although there has been great progress in our understanding of the constitution of Portland cement, the development of high-potential or successfully commercialized special cements is less impressive. It should also be acknowledged that a cement may be of 'commodity' status in one country and yet be considered as 'special' in another, where climatic or economic conditions are different. It may as well be recognized that the use and production of special cements would seem to be more important in industrialized countries than in the developing ones. Finally, one may have to concede that when it comes to 'special' construction materials, the dividing line between cement and concrete tends to disappear and, hence, the scope and coverage appear unwieldy. The text that follows in this article has to be seen in this perspective.

6.3 Drivers for special cement development

As already mentioned, the primary driver for the development of special cements is to overcome the following shortcomings of Portland cement:

(i) not being a finished product by itself, Portland cement serves as an intermediate material for concrete, which is a more complex material system;
(ii) being weak in tension, plain concrete is used only in compression;
(iii) use of steel-reinforced concrete with relatively improved flexural properties is confined to fairly large elements with adequate cover;

(iv) Portland cement-based concrete suffers from poor toughness or crack resistance on impact;
(v) it has a tendency to shrink and crack on drying or cooling;
(vi) generally, it has high permeability to fluids, particularly in aggressive environment, affecting its durability;
(vii) it needs long time to mature;
(viii) it has problems of usage at low temperatures.

Despite the above deficiencies, structural reinforced concrete made with Portland cement has become the most universal and indispensable building material. In parallel, two distinct directions are being adopted to overcome the above deficiencies:

1. How to make cementitious materials more durable in normal and aggressive environments;
2. How to improve the tensile, bending and impact-resistance properties of the resultant finished concrete.

Further, in recent years, the pressures of energy cost and environmental degradation have added new dimensions to the development of special cements. Taking into consideration these pulls and pushes in the cement field, the promising special cement formulations can conveniently be clubbed into the following four categories:

- construction cements with enhanced durability potential;
- cement formulations with improved engineering properties;
- cementitious products for environmental benefits;
- cements with high energy conservation potential.

It is likely that a given cementitious product may qualify for being categorized under more than one of the above groups. Any special cement of that type has been dealt with in this article under the category in which its role appears to dominate.

6.4 Construction cements with enhanced durability potential

6.4.1 Genesis of blended cements

It should be understood at the very outset that the attribute of durability really belongs to concrete and not to cement, although the cement characteristics contribute to the achievement of durability.

This can be appreciated if one looks carefully at the concrete microstructure which presents a composite view of aggregates, cement paste and void (Figure 6.1). The part surrounding the aggregate is almost always rich in CaO and porous in its physical form. This part is called the 'transition zone'. The void includes the space below the aggregate formed by the bleeding water, the capillary pore space remaining unfilled with solid substance in the hardened cement paste and the gel pore space corresponding to the interlayer space of C-S-H. This kind of microstructure permits the ingress of deleterious agents such as chlorides and sulphates. Such permeation leads to the deterioration of the paste matrix and consequent damage of the embedded steel reinforcement.

The most effective, technically sound and economically attractive solution to the durability problems of reinforced Portland cement concrete is the use of finely divided siliceous or latently hydraulic cementitious materials in concrete directly or through pre-mixed cement blends. It is now well established that the incorporation of industrial byproducts such as fly ash, ground granulated blast furnace slag and silica fume in concrete directly or through cement can significantly enhance its basic properties in both the fresh and hardened state [13,14]. In particular, these materials greatly improve the durability of concrete through control of high thermal gradients, pore refinement, depletion of cement alkalies and capability for continued long-term hydration or pozzolanic reaction.

This breakthrough has resulted in production, use and standardization of blended cements as there is a school of thought that the factory-blended cements have reasons to perform better in concrete than when blended materials are used directly in site-mixed ready-mixed concrete [15]. The blended cements are being increasingly manufactured by various countries in accordance with the national and international standards (Table 6.1).

It may be relevant to mention here that after having reached the concluding phase of understanding to accommodate all cements manufactured and used in Europe, the European Standard EN-197-1 today includes twenty-five different cements that are classified as 'common cements' (Table 6.2). Although all these varieties are being treated as common cements, one may well like to raise an issue if such cements like Portland silica fume type or Portland limestone variety or composite cement should, as of now, be universally treated as 'common' or 'special' cements. In the standardization exercise of the European Union, cements with special properties such as high sulphate-resisting properties, or low heat of hydration, alumina cement for constructional purposes, masonry cement, hydraulic road binders and grouting mortars are still under various stages of testing and examination in various countries prior to harmonization and firm adoption.

In a similar way, if one were to look at the ASTM and AASHTO standards for basic and blended cements (ASTM C 150 and C 595, AASHTO M 95 and M 240), one would realize that the types and varieties covered are quite

FIGURE 6.1 A typical scanning electron micrograph of a normal Portland cement concrete. 1: aggregate; 2: calcium hydroxide; 3: transition zone; 4: hydrated cement paste; 5: porosity.

TABLE 6.1   Blended cements standardized in various countries

Country	National symbol	Designation	Composition Slag (%)	Pozzolana (%)	Fly ash (%)
Argentina	PS	IRAM-1503	⩽10	—	—
	BLF	IRAM-1636	10–35	—	—
	PSL	IRAM-1630	35–75	—	—
	POZ	IRAM-1651	—	15–50	—
Canada	10 SM	CAN/CSA	<25	—	—
	10 FM	A 362–M 88	—	—	⩽15
	10 S	CAN/CSA	25–70	—	—
	10 F	A 362–M 88	—	—	15–40
	10 SF	A 362 – M 88	—	⩽10	—
China	PC	GB 1344-85	—	—	20–40
	BLF	GB 1344-85	20–70	—	—
	BLF-LH	GB 200-89	20–60	—	—
	PB	GB 175-85	⩽15	⩽15	⩽15
France*	CPJ (C-L) PM	NF P 15-317	—	⩽20	⩽20
	CPJ (L) ES	P 15-319	⩽35	—	—
	CPJ (Z) PM	NFP 15-317	—	⩽ 20	—
	CHF	NFP 15-301	40–75	—	—
	CHF ES	P 15-319	60–75	—	—
	CLK	NFP 15-301	⩾80	—	—
Germany*	EPZ	DIN 1164-1	6–35	—	—
	TrZ	DIN 1164-100	—	20–40	—
	HOZ	DIN 1164-1	36–80	—	—
	HS (HOZ)	DIN 1164-1	70–80	—	—
	NW (HOZ)	DIN 1164-1	36–80	—	—
	ZPZ 35 A-MS	TGL 28101/01	—	—	20–25
India	PPC	IS 1481 Part I	—	—	15–35
		IS 1481 Part II	—	15–35	—
	PSC	IS 455	25–65	—	—
Japan	BSC-A	JIS R 5211	5–30	—	—
	FAC-A	JIS R 5213	—	—	5–10
	FAC-B	JIS R 5213	—	—	10–20
	FAC-C	JIS R 5213	—	—	20–30
	PPC-A	JIS R 5213	—	5–10	—
	PPC-B	JIS R 5212	—	10–20	—
	PPC-C	JIS R 5212	—	20–30	—
	BSC-B	JIS R 5211	30–60	—	—
	BSC-C	JIS R 5211	60–70	—	—
South Korea	PZA	KS L 5401	—	5–10	—
	PZB	KS L 5401	—	10–20	—
	PZC	KS L 5401	—	20–30	—
	PFA	KS L 5211	—	—	5–10
	PFB	KS L 5211	—	—	10–20
	PFC	KS L 5211	—	—	20–30
	BLF-S	KS L 5210	25–65	—	—
	BLF-1	KS L 5210	25–65	—	—
Mexico†	PUZ-1	DGNC-2	—	15–40	—
	PUZ-2	DGNC-2	—	15–40	—
	SL Li	NOMC-184	65–90	—	—

TABLE 6.1 *Continued*

Country	National symbol	Designation	Composition		
			Slag (%)	Pozzolana (%)	Fly ash (%)
South Africa	OPC-15 SL	SABS 831	5–15	—	—
	RHPC-15 SL	SABS 831	5–15	—	—
	OPC-15 FA	SABS 831	—	—	5–15
	RHPC-15 FA	SABS 831	—	—	5–15
	OPC-15 CS	SABS 831	—	5–15	—
	RHPC-15 CS	SABS 831	—	5–15	—
	PBFC	SABS 626	15–70	—	—
	PFA	SABS 1466	—	—	25–36
United Kingdom*	PBF	BS-146	$\leqslant 65$	—	—
	PPFA	BS 6588	—	—	15–35
	LHPBF	BS 4246	50–90	—	—
	POZ	BS 6610	—	—	35–50
	SS	BS 4248	$\geqslant 75$	—	—
USA	I (SM)	ASTM C-595	$\leqslant 25$	—	—
	I (PM)	ASTM C-595	—	$\leqslant 15$	—
	IS	ASTM C-595	25–70	—	—
	IP	ASTM C-595	—	15–40	—
	S	ASTM C-91	$\geqslant 70$	—	—

*Also governed by CEN standards; †Two different strength clauses indicated by Puz 1 and Puz 2.

extensive and many of these cements have special properties for special applications (Table 6.3). However, a specific survey revealed [16] that the Portland pozzolana cements (ASTM C 595 IP or P) are more popular than type IS or ISM or IPM type with the producers in the USA.

On a global plane, it appears that blended Portland cements have been accepted as more durable types and so far the blended cements with fly ash, ground granulated blast furnace slag and pozzolana (natural or calcined) as binary ingredients have been widely accepted as common general purpose cements.

6.4.2 Special binary blended cements

Amongst the special essentially two-component blended cements, one may mention the following in terms of growing acceptance:

(i) Portland silica fume cement,
(ii) Portland limestone cement,
(iii) Portland cement mixed with kiln/raw mill dust.

Some characteristics of these specially blended cements are furnished below.

6.4.2.1 Portland silica fume cement

This cement probably received a product status on its own merit in 1979, thanks to the geo-environmental conditions of Iceland [17]. Due to raw material constraints, the cement clinker composition in this country is low in silica, high in iron oxide, and very high in alkalies. In parallel, Icelandic aggregates are mostly alkali-reactive. Up to 1979, attempts were made to contain alkali reactivity in concrete by blending pulverized rhyolite glass, locally available, as a pozzolanic material. Although this blend did bring down the AAR expansion, it could not eliminate the problem. In 1979, the Icelandic building code introduced an amendment that mandated the dredged aggregates to be washed to a minimum chloride content and incorporation of a minimum quantity of five per cent of silica fume in cement through intergrinding.

The submicron high-surface area ($\geqslant 15\,\mathrm{m^2/g}$) silica fume is a spherical form of amorphous silica ($\geqslant 85$ per cent) obtained as a by product in the

TABLE 6.2 Cement types and composition specified in EN(V) 197-1 for common cement

Cement type	Designation	Notation	Main constituents						Minor additional constituents[1,2]
			Clinker (K)	Blastf. slug (S)	Silica fume (D)[1,3]	Pozzolana Natural (P)	Pozzolana Industrial (Q)[4]	Fly ash silceous (V)	
CEM 1	Portland cement	CEM I	95–100	—	—	—	—	—	0–5
	Portland-slag cement	CEM II/A-S	80–94	6–20	—	—	—	—	0–5
		CEM II/B-S	65–79	21–35	—	—	—	—	0–5
	Portland-sillica fume cement	CEM II/A-D	90–94	—	6–10	—	—	—	0–5
	Portland-pozzolana cement	CEM II/A-P	80–94	—	—	6–20	—	—	0–5
		CEM II/B-P	65–79	—	—	21–35	—	—	0–5
		CEM II/A-Q	80–94	—	—	—	6–20	—	0–5
		CEM II/B-Q	65–79	—	—	—	21–35	—	0–5
CEM II	Portland-fly ash cement	CEM II/A-V	80–94	—	—	—	—	6–20	0–5
		CEM II/B-V	65–79	—	—	—	—	21–35	0–5
		CEM II/A-W	80–94	—	—	—	—	—	0–5
		CEM II/B-W	65–79	—	—	—	—	—	0–5
	Portland-burnt shale cement	CEM II/A-T	80–94	—	—	—	—	—	0–5
		CEM II/B-T	65–79	—	—	—	—	—	0–5
	Portland-limestone cement	CEM II/A-L	80–94	—	—	—	—	—	0–5
		CEM II/B-L	65–79	—	—	—	—	—	0–5
	Portland-composite cement	CEM II/A-M	80–94			6–20[5]			0–5
		CEM II/B-M	65–79			21–35[5]			
CEM III	Blast furnace cement	CEM III/A	35–65	36–65	—	—	—	—	0–5
		CEM III/B	20–34	66–80	—	—	—	—	0–5
		CEM III/C	5–19	81–95	—	—	—	—	0–5
CEM IV	Pozzolanic cement	CEM IV/A	65–89	—		11–35			0–5
		CEM IV/B	45–64	—		36–55			0–5
CEM V	Composite cement	CEM V/A	40–64	18–30	—		18–30		0–5
		CEM V/B	20–39	31–50	—		31–50		0–5

[1]The value in the table refer to the sum of the main and minor additional constituents.
[2]Minor additional constituents may be filler or may be one or more of the main constituents unless these are included as main constituents in the cement.
[3]The proportion of silica fume is limited to 10 per cent.
[4]The proportion of non-ferrous slug is limited to 15 per cent.
[5]The proportion of filler is limited to 5 per cent.

manufacture of silicon and ferrosilicon. The amorphous character helps in the reactivity, while the particle shape improves the workability. The loss on ignition for acceptance of silica fume is kept below four per cent. It also reportedly absorbs the alkalies from the basic cement paste and ultimately, helps to form the tobermorite gel. Silica fume in the Icelandic Portland cement has been limited to 5–7.5 per cent as the particle packing effect is seen at this level of addition while higher percentages increase the bulkiness of the cement and its water demand. Every third year, the field effect of the use of such cement is reviewed through core study, which has so far brought out

TABLE 6.3 Classification of Portland cements by ASTM standards [16]

Designation	Distinctive property	Salient compositional characteristics
Type I	'Ordinary' Portland cement	Any Portland cement not meeting the criteria for one of the other types or subtypes
Type 1 MA	Moderate alkali	As Type I and equivalent alkalies greater than 0.60% but do not exceed 0.80%
Type 1 L/A	Low alkali	As Type I and equivalent alkalies do not exceed 0.60%
Type IA	Air entraining Type I	As Type I and air content of mortar by ASTM C 185 is 16% or greater
Type IA LA	Low alkali	As Type IA and equivalent alkalies do not exceed 0.60%
Type II	Moderate sulphate resistance	C_3A is 8% or less, SiO_2 is 20% or more, Al_2O_3 is 6% or less, and Fe_2O_3 is 6% or less
Type II MA	Moderate alkali	As Type II above plus equivalent alkalies do exceed 0.60% but do not exceed 0.80%
Type II L/A	Low alkali	As Type II but equivalent alkalies do not exceed 0.60%
Type II MH	Moderate heat of hydration	As Type II above and C_3S plus C_3A does not exceed 58%
Type II MHLA	Moderate heat of hydration, low alkali.	As Type MH above and equivalent alkalies do not exceed 0.60%
Type II FG	Fine grind	As Type II above, but designated, 'fine grind' by the producer. Air permeability fineness is generally greater than for the other Type II cements ($420\,m^2/kg$ or less), but less than for more Type III cements ($500\,m^2/kg$)
Type IIA	Air entraining Type II	As Type II above and also ASTM C 185 air content of mortar is 16% or greater
Type IIA LA	Low alkali	As Type IIA and equivalent alkali content is 0.60% or less
Type III	High early strength	The defining criterion is the stated intent of making a Type III cement. Type III cement generally has air permeability fineness greater than $450\,m^2/kg$. one-day cube strength greater than 19 Mpa and 3-day strength greater than 30 Mpa although there are exceptions
Type III MA	Moderate alkali	As Type III above but equivalent alkalies greater than 0.60% and not exceeding 0.80%
Type III LA	Low alkali	As above but equivalent alkalies not exceeding 0.60%
Type III MS	Moderate sulphate resistance	As for regular Type III plus C_3A not exceeding 8%
Type III MSLA	Both low alkali and moderate sulphate resistant	As Type III MS but equivalent alkalies do not exceed 0.60%
Type III HSLA	High sulphate resistance and low alkali	As Type III above but C_3A 5% or less
Type IIIA	Air entraining	As Type III, and also ASTM C 185 air content of mortar is 16% or greater or manufacturer designated the cement as air entraining
Type IV	Low heat of hydration	Cements designated low heat of hydration by their manufacture and showing both low C_3A (max. 7%) and relatively low C_3S (max. 35%) C_2S min. 40%
Type V	High sulphate resistance	C_3A less than 5% and either SO_3 less than 2.3% or data given in accordance with NOTE B to Table 1 of C 150
Type V LA	Low alkali	Criteria as for Type V above and equivalent alkalies 0.60% or less

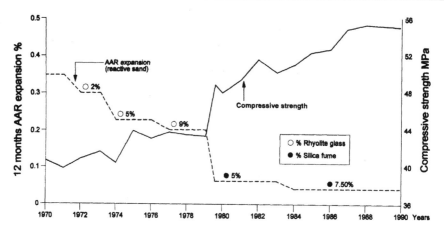

FIGURE 6.2 Development of Icelandic cement: AAR expansion and compressive strength versus pozzolans [17].

the beneficial effects of containing the AAR expansion and increased strength of cement (Figure 6.2).

Backed by this experience, it is reported [16] that two US companies began marketing cement containing silica fume upto 10 per cent. The CEN standard has also adopted the Portland silica fume cement with silica fume ranging from 6 to 10 per cent.

6.4.2.2 Portland limestone cement

Cements containing limestone were first developed in France. Historically speaking, Guyot and Ranc [18] described cement containing up to 25 per cent limestone as a technical and economic solution to obtaining medium-class cements which provide all the guarantees of performance, workability and durability required of other cements. As already mentioned, this variety of cement has already been accepted as a standard product under EN 197-1 with two classes – one with addition of 6–20 (II/A-L) and another with 21–35 per cent (II/B-L) limestone. However, in Portland limestone cements produced on an industrial scale conforming to CEM II/A-L, the average content of limestone is 15–17 per cent by weight. A large number of limestone powders used and cements produced have been analysed at the Research Institute of the German Cement Industry and the results are given in Table 6.4 [19]. In the US, limestone filler is not permitted under current ASTM provisions. Under the cur-

rent standards in Canada, however, addition of limestone at the clinker grinding stage is permitted only up to five per cent.

The specification of limestones that can be used in the production of Portland limestone cement has been given in EN 197-1. According to this standard, limestones should meet the following requirements:

$CaCO_3 \geqslant 75$ per cent by mass
Clay content: methylene blue adsorption
  $\leqslant 1.20/100\,g$
Total Organic Carbon $(TOC) \leqslant 0.20$ per cent
  by mass

The TOC content is an empirically determined indicator of the behaviour of concrete, when exposed to freeze–thaw conditions. It may, however, be understood that the effect of organically combined carbon on the durability of mortar and concrete is not fully deciphered and is still under investigation in various laboratories. According to CEN, limestones with TOC between 0.20 per cent and 0.50 per cent by mass may also be suitable for producing cement with acceptable performance, however, more performance tests are necessary.

One physical variable that could describe the behaviour of limestone with varying TOC contents in Portland limestone cements is their zeta potential, which determines the tendency to form agglomerates when particles are in suspension. Initial measurements of zeta potential on limestone powders and cement–limestone mixes using an

TABLE 6.4 Chemical composition of limestone powders used and Portland limestone cement produced in Europe

Component	Limestone			CEM II/A-L		
	N	M	H	N	M	H
SiO_2	0.02	9.45	22.82	18.11	19.81	22.52
Al_2O_3	0.02	2.13	6.52	3.54	4.81	5.83
TiO_2	0.05	0.14	0.27	0.20	0.25	0.45
Fe_2O_3	0.04	0.87	2.92	1.24	2.45	3.64
Mn_2O_3	0.01	0.04	0.17	0.03	0.06	0.10
P_2O_5	0.02	0.08	0.21	0.03	0.10	0.20
CaO	69.46	84.47	98.94	63.69	66.82	69.25
MgO	0.24	1.56	4.46	0.68	1.60	5.11
SO_3	0.01	0.31	1.01	2.16	3.01	3.60
K_2O	0.04	0.54	1.81	0.35	0.90	1.68
Na_2O	0.01	0.06	0.27	0.03	0.15	0.35
Na_2O-Eq.	0.04	0.41	1.32	0.31	0.75	1.33
LOI	36.78	40.90	44.63	2.91	7.35	9.41
C	0.03	0.11	0.26	n.d.	n.d.	n.d.
CO_2	35.48	40.01	44.91	2.10	6.62	8.86
H_2O	0.15	0.75	2.28	0.41	0.73	1.53

H = highest value; M = average value; N = lowest value [19]; n.d. = not defined.

electro-acoustic vibration test method have shown that frost-sensitive mortars are characterized by higher zeta potentials of the limestones used, which were indicative of a high dispersion tendency [19].

From the studies carried out on concretes prepared from 32.5 R strength class Portland limestone cement, it was found that the strength characteristics, static modulus of elasticity and freeze–thaw resistance of such concretes did not show any appreciable difference from control concretes made of normal Portland cement. The Portland limestone cement concrete, in the laboratory, however, showed marginally greater depths of carbonation than the reference concrete, which did not have any practical significance. The results of extensive approval investigations and the practical experience gained over ten years have paved the way for wide acceptance of Portland limestone cement as a satisfactory product.

Needless to emphasize the limestone additive is no more considered as an inert filler but is now widely recognized as contributing to the hydrate phase assemblage through the formation of calcium mono-carboaluminate ($C_3A \cdot CaCO_3 \cdot 11H_2O$) [20]. Other beneficial effects reported include

improved sulphate resistance [21] although the deleterious formation of thaumasite can occur under certain circumstances [22]. A study involving partial replacement of Canadian Type 10 cement by up to 50 per cent pulverized limestone showed increased resistance of cement pastes to attack by sulphuric acid at pH 4 [23].

6.4.2.3 Portland cement mixed with kiln/raw mill dust

In broad principle, this particular blend may be treated as a variant of Portland limestone cement, since in most cases, the kiln-raw mill ESP or baghouse dust is a fine (less than 5 μm) lime-rich powder with some proportions of clinker dust and free lime particles. The survey on the production of blended cements in the US mentioned earlier [16] brought out that a Portland cement mixed with baghouse dust is produced there, although it does not strictly conform to the ASTM Standard.

A cement of this type, produced from a mineralized clinker with addition of 15–20 per cent limestone-rich ESP dust has been described by Borgholm *et al.* [24]. The combined effect

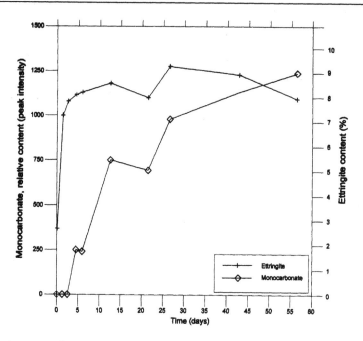

FIGURE 6.3 Development of ettringite and carboaluminate hydrate (monocarbonate) in paste containing mineralized cement with 19% limestone-bearing ESP dust (w/c = 0.40) [24].

resulted in compressive strengths of cement mortars significantly greater than what is found with other combinations of clinker and inert or less reactive hydraulic fillers.

The formation of ettringite and carboaluminate in the hydrated cement paste with time for 19 per cent addition of ESP dust is shown in Figure 6.3. The rapid formation of carboaluminate hydrates and the simultaneous presence of ettringite contribute to better strength development.

It is claimed that this type of cement is marketed in Denmark under the name of 'Basis-Cement' as an alternative to rapid hardening Portland cement.

6.4.3 Multicomponent blended cements

One of the avenues of improving the quality of cement has been the optimization of its particle size distribution. But, this approach has not yielded the results of reducing the water content in concrete close to or below the theoretical requirement.

Further, it is well established today that the properties of concrete are to a large extent governed by the fines in the cement. The granulometry of the fines determines the packing density. The denser the packing of cement, the less is the demand of mixing water of the cement paste and the denser is the resultant paste, which ultimately contributes to the durability of concrete. The basic principles of mixing solid powders indicate that denser packing can be achieved when the average particle size of two components to be mixed differs appreciably from each other. With mixing materials of approximately equal fineness, the packing density is increased if one of the two materials has a significantly wider particle size distribution than the other. It should also be borne in mind that with increasing fineness of particles ($< 1 \mu m$), interparticulate adhesive forces occur which can lead to agglomeration.

Relying on these basic principles, Nagataki and Wu [25] have shown that it is possible to produce densely-packed binder by blending Portland cement, silica fume and blast furnace slag. The particle size distribution of these three materials and the opportunity available to achieve denser packing in triple-blend cements are shown in Figure 6.4. The resultant reduction in pore volume and improvement in compressive strength of such cements are demonstrated in Figures 6.5 and 6.6a. The results of another study, showing the influence of microsilica and hydrated lime on a blast furnace slag cement are furnished in Figure 6.6b [26].

A comparable approach was also suggested by Uchikawa [4] in arriving at the particle size distribution of a multi-blend cement formulation designed to produce high-strength concrete. The particle size distribution patterns of the ingredients for high-strength concrete cement are shown in Figure 6.7. The relationship between mixing proportions of cement materials and slump, yield value, plastic viscosity and strength of concrete is shown in Figure 6.8.

The feasibility of wider acceptance of such multi-blend cements by the producers and consumers has, in recent times, improved considerably with the perceptible movement towards the adoption of performance-oriented standards in place of prescriptive ones. For example, ASTM C 1157 entitled, 'Standard Performance Specification for Blended Cements' offers greater flexibility than the prescriptive standard C 595. With full compliance to such standards, it would now be possible to produce cements with both flyash

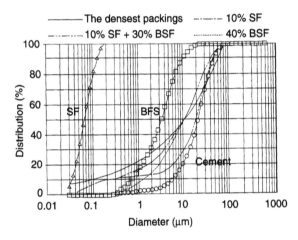

FIGURE 6.4 Particle size distribution of Portland cement, silica fume (SF), blast furnace slag (BSF) and blended cement [25].

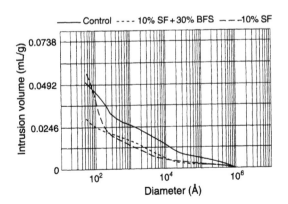

FIGURE 6.5 Pore size distribution of hardened cement pastes [25].

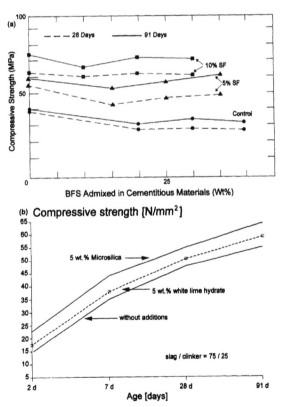

FIGURE 6.6 (a) Compressive strength of hardened mortars [25]. (b) Influence of microsilica and white lime hydrate on the strength development of blast furnace slag cement [26].

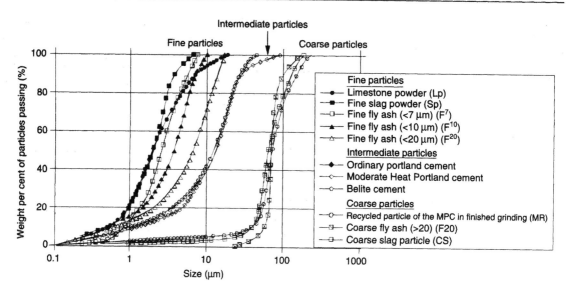

FIGURE 6.7 Particle size distribution of materials used to formulate cements for high-strength concrete [4].

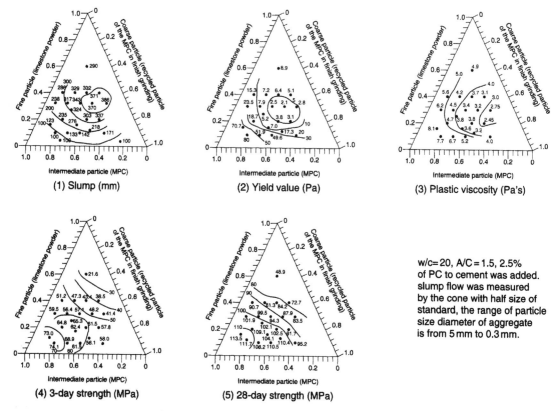

FIGURE 6.8 Relationship between cement blends and slump, yield value, plastic viscosity and compressive strength of concrete [4].

and silica fume or slag and kiln dust or slag and silica fume or more complex formulations. Similarly the standardization of cements like 'composite cement' under EN-197-1 has also paved the way for popularizing multi blend cements.

6.4.4 High-belite portland cement

A Portland type of cement with alite phase restricted to 20 per cent or so and belite phase close to or in excess of 60 per cent with a partially glassy matrix containing low proportions of tricalcium aluminate and fine crystallites of aluminates and ferrite can be categorized as the high-belite or reactive-belite cement as summarized in [11].

From this review it can be seen that these kinds of binders can be produced by three alternative routes: chemical stabilization through conventional firing or fast firing but always with rapid quenching, sol–gel type of processing, and hydrothermal treatment. In the last two routes the operating temperatures are low and no chemical stabilization is called for, while in the first route oxides like K_2O, Na_2O, SO_3, B_2O_3, Cr_2O_3 and BaO have received preference as dopants in multifarious studies.

Interest in high-belite cements was initiated by the energy crunch of the early seventies as its production was expected to be less energy-intensive. Subsequently, the interest was reinforced by the potential of raw material conservation, but ultimately, the growing interest in the product deflected towards its capability of producing a more durable hydrated matrix with a very steady rate of strength gain with time. But the realization of this goal continues to elude the scientific community for two reasons: first, the search for a practical means of improving the hydration rate of the belite phase, and, second, establishing the technology to overcome the strong resistance of belite to grinding and surface generation.

In recent times, a promising approach has been reported by Fukuda and Ido [27]. The authors have proposed a possible new type of belite cement suitable for mass production, named 'remelted belite cement'.

When a belite crystal with impurities is cooled from the stable temperature region of the α-phase, it decomposes to a liquid and the α'_H-phase that is lower in impurity concentration than the parent α-phase crystal. This process, according to the authors, is called the remelting reaction of belite. The rate of the re melting reaction depends on the Al/Fe ratio of the parent α-phase and the temperature.

In a recently conducted experiment by the authors [27] with industrially produced belite-rich clinker (Table 6.5), two reheating and quenching regimes were adopted for two fractions of the same clinker so as to stabilize α-C_2S in one and α'-C_2S in the other. When these two stabilized fractions were subjected to grindability and hydration studies under comparable conditions, the results indicated that the grindability (Figure 6.9) and hydration activity (Figure 6.10) of belite-rich cement were markedly improved by the remelting technique.

So far as the durability issues are concerned, it has been seen that activated belite cement mortars and concretes follow the behaviour of alitic cements. However, the belitic cements show lower carbonation and capillary porosity of 7–14 per cent as compared to 15–24 per cent in alitic cements. Further, the belitic cements can be used quite effectively to produce blended cements. Some comparative results generated in the author's laboratory, are furnished in Table 6.6, which clearly demonstrate that for the cement-substitute materials

TABLE 6.5 Chemical composition of industrial belite-rich clinker [27]

Substance	Composition (wt%)
LOI	0.22
SiO_2	29.32
Al_2O_3	3.02
Fe_2O_3	2.81
CaO	62.71
MgO	0.83
SO_3	0.35
Na_2O	0.13
K_2O	0.24
TiO_2	0.14
MnO	0.07
P_2O_5	0.05
Total	99.89

$Al_2O_3 : Fe_2O_3 = 1.07$

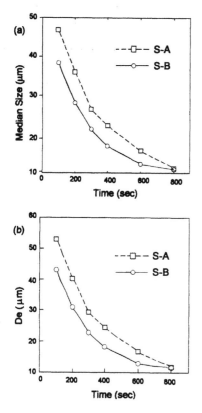

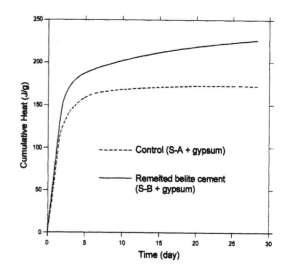

studied, the 28-day strength gains in blended high-belite cements, with respect to the control sample, are more than in the case of high-alite cements. From the table one may also observe that the relative advantage of simultaneous blending of slag and silica fume is retained in belitic cements as in the alitic variety, but to a different degree.

FIGURE 6.9 Change in (a) median size and (b) De of Rosin–Rammlar distribution with grinding time (S-A and S-B contain α- and α'-belite phases respectively) [27].

FIGURE 6.10 Cumulative heat evolution for 28 days at w/c $=0.4$ and 20°C (S-A and S-B contain α- and α'-belite phase respectively) [27].

TABLE 6.6 Comparison of response of high-alite and high-belite clinkers to cement substitute materials

Cements	NC (%)	Setting time (min)			Compressive strength (Mpa)			% increase over control in 28 days
		Initial	Final	1 day	3 days	7 days	28 days	
High-alite clinker (belite content 30%)								
Control OPC	36.3	40	65	36.0	40.0	46.0	56.0	100
PPC+15% FA	30.0	40	95	25.5	36.5	42.0	40.0	89
PSC+45% Slag	29.0	65	120	19.0	33.5	44.0	55.0	98
PSC+40% Slag +5% CSF	25.5	50	90	24.5	40.5	47.0	61.0	109
High-belite clinker (belite content 50%)								
Control OPC	26.8	40	110	19.5	28.5	34.5	39.5	100
PPC+15% FA	29.0	45	90	15.0	23.0	27.0	40.0	101
PSC+45% Slag	29.5	90	150	13.0	23.0	30.5	41.5	105
PSC+40% Slag +5% CSF	25.3	40	65	20.0	28.5	35.5	48.5	123

In another study [28], an attempt has been made to understand the concrete properties of belite cements with or without chemical and mineral admixtures (Table 6.7). It can be seen from the data that the low early strength, the largest drawback of belite cement, may be markedly improved by combining the adjustment of particle size distribution through the use of mineral admixtures and the reduction of water–cement (w/c) ratio by the use of an organic admixture.

The comparative behaviour of mortars and concretes prepared from ordinary Portland cement, moderate heat cement and belite cement is presented in Table 6.8 [29]. The improvement in later age strength with lower heat of hydration is noticeable for the belite cement.

6.4.5 Portland cement with spherical particles

The concept of producing a variant of Portland cement with spherical particles owes its origin to the expectation that concrete with such cement particles would demand lower quantity of water. In a very novel study conducted by Yamamato et al. [30], the particle morphology of cement was altered with the help of a high-speed impact-blender. The powder characteristics of spherical cement as compared to the normal cement are given in Table 6.9 and the morphological features can be seen in Figure 6.11.

The above two types of cements were converted into concretes under identical conditions and by using the same admixtures, coarse and fine aggregates. It was found that SC concrete was more fluid than NPC concretes. The slump value of concrete using SC was higher than that of NPC with the same w/c ratio. As a result, the w/c ratio for SC concrete could be reduced by 6–8 per cent with the resultant benefits of durability.

The technology for producing spherical cement is still experimental in nature. Apart from the first experimental approach mentioned earlier, the other attempt includes a process of rapidly heating, melting and cooling the raw material using a plasma in a high-velocity air stream. It is claimed that this process has been tried at the pilot scale;

the development of an economically viable commercial process is still in the realm of future [31].

6.4.6 Blended high-alumina cement

Calcium aluminate or high-alumina cement (HAC), known since 1908 when the first patent was taken, has been in commercial production since 1913; but it is not in use in structural applications in many countries including UK and India at the present time, as concrete made from such cement shows reduction in strength with time under hot and humid conditions. This reduction in strength is associated with conversion of the hydrate phase CAH_{10}, belonging to the pseudo-hexagonal system, into cubic C_3AH_6. The cement is also known to be prone to high alkali attack.

The Building Research Establishment in UK has developed a blended cement, tradenamed BRECEM, based on HAC and ground granulated blast furnace slag. The addition of slag (typically 40–60 per cent) alters the course of hydration reactions in HAC with the formation of gehlenite hydrate C_2ASH_8 (strätlingite), not seen in plain HAC concrete in significant amounts. This change in hydration chemistry of blended HAC has been studied under various conditions of temperature, humidity and environmental conditions, some details of which are summarized in [3,32,33], which provide enough basis to conclude that the blended HAC concrete would demonstrate:

- uninterrupted gain in compressive strength over five years,
- stability and long-term durability in a variety of environments including a temperature of 38°C,
- superior sulphate resistance even under relatively high w/c ratio (>0.40).

Some illustrative properties of concretes made from HAC and 50 per cent HAC+50 per cent GGBFS are given in Table 6.10.

In order to establish the behaviour of HAC with a high titanium oxide content of 6.64 per cent and alumina content of 48.3 per cent in the formulation of a blended cement with ground granulated blast furnace slag having relatively high alumina content of 19.6 per cent, an investigation was

TABLE 6.7 Properties of belite cement concrete with low water–cement ratio [28]

	Mixing proportion					w/c	Super plasticizer (%)	Air (%)	Slump (m)	Slump flow (cm)	Compressive strength (N/mm²)			
	Water (kg/m³)	Cement (kg/m³)			Sand (kg/m)	Coarse aggregate (kg/m³)						3 days	7 days	28 days	91 days
		Cement	Coarse particle[a]	Fine particle[b]											
1. Ordinary Portland cement	200	400			790	869	0.50	—	4.7	18.5	—	15.0 (147)	23.5 (230)	35.6 (349)	40.5 (397)
2. Belite cement C	140	400			871	951	0.35	0.7	4.0	18	—	19.9 (195)	35.9 (352)	59.2 (584)	88.8 (868)
3. Belite cement C	174	870 609	174	87	540	856	0.20	1.62	2.8	22.0	45 × 46	58.1 (570)	72.9 (714)	110.9 (1087)	138.9 (1361)

[a] Coarse particle of ordinary Portland cement clinker ranging from 50 to 200 μm.
[b] Fine particle of calcite powder of which particle size is less than 3 μm.
( ): Data expressed in kgf/cm².

TABLE 6.8 Composition of ordinary, moderate heat and belite cements and properties of paste, mortar and concrete prepared using them [29]

Cement	Character of cement	Chemical composition (%)								F.CaO (%)	Mineralogical composition (%)[a]				B[b] (cm²/g)
		SiO_2	Al_2O_3	Fe_2O_3	CaO	MgO	SO_3	Na_2O	K_2O		C_3S	C_2S	C_3A	C_4AF	
OPC		21.6	5.3	3.4	64.5	1.4	2.1	0.46	0.23	1.3	51	23	8	10	3450
MPC		23.3	3.9	4.1	63.4	1.3	1.9	0.05	0.52	0.5	44	34	3	13	3250
BPC		27.9	2.8	3.2	62.6	1.1	1.6	0.29	0.02	0.3	21	67	2	10	3400

Properties	Setting time of cement		Heat of hydration of cement (J/g)			Compressive strength of mortar[c] (MPa)				Unit amount of concrete[d] (kg/m³)				Slump (cm)	Compressive strength of concrete (Mpa)			Adiabatic temp. rise of concrete (°C)
	Initial	Final	7d	28d	91d	3d	7d	28d	91d	W	C	S	G		7d	28d	91d	
OPC	2–40	3–40	313	356	400	15.3	25.3	41.7	48.2	158	300	803	1045	10.5	20.3	39.1	45.6	46.0
MPC	3–30	4–35	265	308	349	11.3	16.3	33.2	53.4	158	300	803	1045	11.2	13.8	35.5	46.8	38.5
BPC	6–05	6–50	187	251	322	6.8	9.7	32.0	63.2	158	300	803	1045	12.7	8.5	32.6	58.9	29.5

OPC: Ordinary Portland cement; MPC: Moderate heat cement; BPC: Belite cement.
[a] Calculated from Bogue's equation.
[b] Blaine specific surface area.
[c] Setting time, heat of hydration and compressive strength of mortar was measured according to JIS R5201, JIS R5203 and JIS R5201, respectively.
[d] Concrete was prepared at w/c = 0.527, s/a = 0.434, Admixture (LS 0.75% cement) < Maximum size of aggregate was 25 mm.

TABLE 6.9 Comparative powder characteristics of spherical cement [30]

Characteristics	Spherical cement (SC)	Normal OPC (NPC)
Degree of roundness (sphere: 1)	0.85	0.67
Mean diameter (μm)	11.3	13.5
Specific surface area (cm²/g)	2700	3360
n-value from RR plot	1.18	0.97

(a)  (b)

FIGURE 6.11 Visible difference in particle morphology of ordinary Portland cement (a) and spherical cement (b) [30].

TABLE 6.10 Properties of HAC and blended HAC 1 m concrete cubes [33]

Cement type	HAC	1:1 HAC+GGBS
Mix proportion	1:1.75:2.91	1:2.50:3.50
Cement content (kg/m³)	390	320
w/c ratio	0.45	0.55
Maximum temperature reached (°C)	91.2	58.0
Compressive strength (MPa) (100 mm core)		
9 days	41.0	19.5
29–32 days	36.0	21.0
230 days	—	24.5
Hydrated phases (in cores)		
9 days	C_3AH_6, AH_3	C_2ASH_8, $C_3AS_{0.25}H_{5.5}$, AH_3
29–32 days	C_3AH_6, AH_3	C_2ASH_8, $C_3AS_{0.25}H_{5.5}$, AH_3

sponsored by the author's laboratory with BRE. The programme of work included hydration studies, measurement of heat output of cement pastes, chemical resistance of mortar bar prisms and a limited study on concrete in water at ambient and higher temperatures for strength development up to one year. The results of this investigation re established the basic virtues of blended HAC explained earlier. In specific terms the conclusions of the investigation were as follows:

(i) Concrete mixes were of low workability perhaps due to high levels of $C_{12}A_7$ in the HAC causing it to act like a fast setting cement.

(ii) The trends of lower early strength and sustained strength development for 1:1 mixture of HAC+GGBFS compared with that of HAC mortar and concrete cubes (Figures 6.12 and 6.13) were consistent with the performance of UK BRECEM mixtures but the strength values were lower than those expected from BRECEM mixtures. This is presumably due to the high w/c ratios needed to produce adequate workability for the experimental mixes.

(iii) The calorimetric data confirmed that less heat was produced with the mixture of

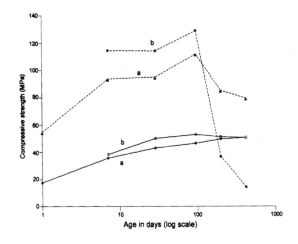

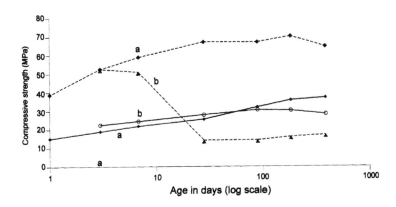

FIGURE 6.12 Compressive strength of 10 mm mortar cubes made with HAC (dotted line) and 1:1 mixture of HAC + GGBFS (solid lines) using w/c = 0.40 and stored in water at (a) 20°C and (b) 32–38°C.

HAC + GGBFS than with HAC at both the temperatures 20°C and 40°C (Table 6.11).

(iv) The sulphate resistance of mortar prisms was very good for both plain and blended HAC (Figure 6.14).

From the above results, as well as from the basic investigations conducted at BRE, it is evident that the blended HAC is quite likely to offer a cementitious system that would give moderate to high strengths, rapid rate of strength gain and extremely high chemical stability. Additionally, there would be environmental advantages on account of the use of industrial wastes.

While on the subject of HAC, it may be relevant to mention about the development of hydraulically reactive glasses in the system $CaO-Al_2O_3-SiO_2$ [34,35]. The highest strengths were obtained in the region close to pure gehlenite glass composition which resulted in the formation of strätlingite

FIGURE 6.13 Compressive strength of 100 mm mortar cubes made with HAC (dotted line) and 1:1 mixture of HAC + GGBFS (solid line) and stored in water at (a) 20°C and (b) 38°C.

TABLE 6.11 Total heat evolution for HAC and 1:1 HAC + GGBFS mixture

Mixture		w/c ratio	Maximum output time (h)		Total heat evolved (kJ/kg)	
HAC	GBBS		20°C	40°C	Up to 48 h at 20°C	Up to 24 h at 40°C
100	0	0.40	5.6	12.4	274	309
50	50	0.40	4.9	7.6	200	250

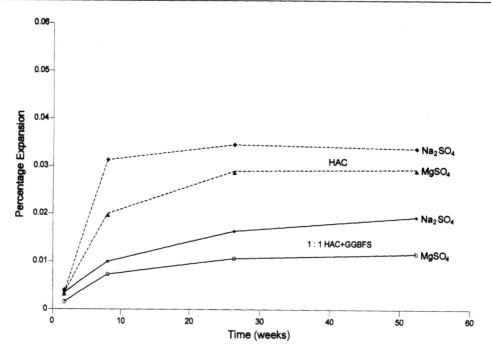

FIGURE 6.14 The percentage expansion of 1:3 mortar prisms made with HAC and 1:1 mixture of HAC + GGBFS and stored in sulphate solutions.

(gehlenite hydrate) on hydration. This cement also showed transformation of gehlenite hydrate to hydrogarnet in some compositions with decrease in strength. This finding as well as the questionable cost effectiveness of the process due to high melting temperature have not allowed this development to be commercially successful so far.

Thus, since no other solutions are now in sight to overcome the problem of strength retrogression of HAC in hot environments there is high probability of technical success of the blended HAC for constructional purposes.

6.5 Cement formulations with improved engineering properties

6.5.1 Interground fibre cement

The concept of steel reinforcement for cement concrete came within three decades of the invention of cement by Joseph Aspdin, when Joseph Lambot used wiremesh to take care of weak tensile strength and resultant cracking problem of

boats and flower pots in 1855. Later, Joseph Monier patented steel bar reinforcement in 1867. Patents on glass and synthetic fibres appeared much later in the 1960s and 1970s. Today, concrete reinforced in two or three directions has come to stay as a basic material of construction. Despite this development, cracking still continues to be the main complaint of ready-mixed concrete and precast concrete production.

With a view to reducing the cracking over rebar by reinforcing the cover with fibre-reinforced concrete, the Von Tech™ intergrinding cement has been patented [36] to produce an intermilled fibre cement. It is a dry process of finish grinding at the cement plant itself (Figure 6.15). According to the inventors, several bulk fibre feeding and metering systems are adaptable to a typical cement mill. In addition, some modifications of mill diaphragms may be required to avoid matting of fibres as they are blown through the mill. Chute and bin openings need proper engineering. Feed to the finish mill can be from the gypsum feed belt, the top of air separator or any other convenient point.

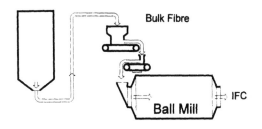

FIGURE 6.15 Schematic diagram of bulk fibre flow using the Von Tech IFC™ process [36].

The system, when properly designed, is expected to give uniformity to the extent of ± 2 per cent of fibre content. Fibrillated polypropylene or nylon fibre dosage of 0.25 per cent by weight is indicated as a general measure of intergrinding with cement clinker. It is claimed that while the smooth intermixed nylon and polypropylene fibres reduce longitudinal shrinkage by 17 per cent and 18 per cent respectively, the interground mixtures achieve reduction by 35 per cent and 25 per cent respectively compared to fibre-free control specimens, due to downsizing of fibres (29–21 to 5 mm length) and surface roughening. The technology apparently holds high promise for the future.

6.5.2 High-performance fibre-reinforced cementitious composites

The concept of fibre-reinforced cement composite is not new. The conventional fibre-reinforced concretes exhibit an increase in ductility and impact resistance while, according to some authors, improvement in their first cracking strength, tensile strain capacity and peak load are not significantly improved.

On the other hand, composites that develop a series of very fine and disconnected cracks through a special multicracking mechanism are categorized under high-performance fibre-reinforced type. The multiple cracking mechanism is also beneficial from the durability standpoint as it increases impermeability and may even lead to autogenous healing of very fine cracks.

There are two broad categories of high-performance fibre-reinforced composites – first, made with high fibre volume fractions of discontinuous short fibres and, second, made with continuous fibre-mats. The short discontinuous fibre composites have been in use for a number of years. One of the established techniques for production has been the Slurry Infiltrated Concrete (SIFCON) [37,38]. These types of composites are ideally suited for three-dimensional applications where there is a need for increased shear and moment strengths, increased ductility, better concrete confinement, etc.

On the other hand, the continuous fibre-mat composites are ideally suited for two-dimensional applications such as jacketing of existing structural elements or construction of multi layer sections. The continuous steel fibre-mat composite, called Slurry Infiltrated Mat Concrete (SIMCON) has been recently developed by infiltrating the mats with a specially designed cement-based slurry [39]. It has been possible to show that in such composites with only 5.25 per cent volume fraction of continuous fibre-mat, strain hardening can be achieved with increase in strength and toughness. The typical stress–strain behaviour of 28-day SIMCON specimens is shown in Figure 6.16, which demonstrates the achievement of tensile strength of 15.9 MPa at 1.1 per cent of strain for a fibre volume fraction of 5.25 per cent as mentioned earlier [40].

It is anticipated that with high-performance fibre-reinforced cement composites the engineering community would be in a better position to develop innovative repair, retrofit and new construction approaches ensuring seismic resistance, ductility and durability.

6.5.3 DSP formulations

The densified system of homogeneously compacted ultrafine particles (DSP), as is well known, was developed in the seventies in Denmark based on the use of very dense and strong binders with 70–80 per cent cement and 20–30 per cent ultrafine materials having an average size of only 0.1–0.2 microns along with simultaneous incorporation of effective particle dispersants [41]. It has been estimated that with addition of about 25 per cent fine material with an effective diameter ratio of

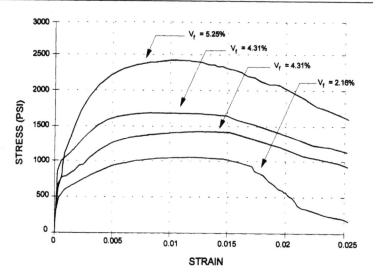

FIGURE 6.16 Typical stress–strain response of 28-day old SIMCON specimens in tension (Note that deformations beyond peak stress are associated primarily with widening of critical crack, and thus, cannot be directly translated into strain. This should be kept in mind when interpreting strain values that were obtained by measuring displacements over a 10.8 cm [4.25 inches] base length.) (1000 psi = 6.9 MPa) [39].

over 50, one may achieve an initial porosity of 17 per cent [42]. Over the years, silica fume has turned out to be the widely used ultrafine material in this formulation. Based on rheological considerations, cement with low C_3A and alkalies appear to be more effective. The ratio of silica fume to cement can be varied widely. Results obtained in various laboratories for concretes containing 5–15 per cent silica fume indicate that the compressive strength, flexural strength, Young's modulus and water permeability can reach levels like 100 MPa, 12 MPa, 34 GPa and $10^{-9}\,\mu m^2$ respectively.

The overall pozzolanic reaction occurring in DSP cement containing silica fume is:

$$C_3S + 1.7S + 10.3H \rightarrow 2.7\,(C_{1.1}\,SH_{3.8})$$

This requires about 45 per cent of silica fume for complete reaction [42], which is too high for the workability requirement. While using silica fume the general experience has been that the particle packing effect becomes visible with as little as 6 per cent and reaches a maximum at about 20 per cent [42]. Hence the optimum requirement is limited to 20 per cent or so.

The basic problem of dispersion of silica fume can be tackled by employing the high-shear mixing technique as the de-agglomeration of densified fume cannot be achieved by low-shear mixing in a pan or planetary mixer. Similarly, the undensified silica fume cannot be fully dispersed by chemical agents, although ultrasonic treatment may often help to achieve de-agglomeration.

The limiting w/(c+sf) ratio for castability of DSP cement is 0.16, when air removal is effected through vacuum treatment and vigorous vibration. For still lower ratios and further densification pressure de-watering is necessary [43]. For really low porosity, hydration can be allowed to take place if the paste is kept in a state of constant compaction [44]. A typical composition of a DSP cement matrix is given in Table 6.12 [45].

When cured above 200°C, depending on the silica fume content, either 1.1 nm tobermorite or xonotlite is formed. A combination of pressure compaction and autoclaving can produce a porosity of less than 5 vol percentage and strengths of 700 MPa. This kind of product design has led to a new class of DSP product known as reactive powder concrete (RPC) [46,47].

TABLE 6.12   Typical composition of a DSP cement [45]

	Mass (g)	Wt%	
Portland cement	750	66.1	Mean size 10–15 μm
Silica fume	190	16.7	Mean size 0.2–0.3 μm 98% SiO_2
Dispersing agent	28	2.5	Sulphonated naphthalene polymer
Water	167	14.7	

Silica\cement = 0.24; water/solids = 0.18

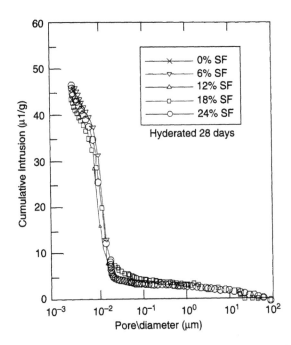

FIGURE 6.17   Capillary pore size distribution of DSP pastes [42].

6.5.3.1 Properties of DSP cement pastes

Some quantitative data on a few important properties of DSP cement pastes have been furnished in [42]. The capillary pore size distribution as measured by mercury porosimetry is shown in Figure 6.17. It is evident that the pore size distribution is essentially independent of silica fume addition. It may also be noted that the DSP cements are expected to display high chemical shrinkage, low drying shrinkage and high coefficient of thermal expansion. Chemical shrinkage is associated with both hydration and the pozzolanic reaction and the value increases with lower w/c ratios and higher

contents of silica fume. The results of drying shrinkage measurements indicate that water loss and shrinkage is a slow process (Figure 6.18). It is generally observed that equilibrium is not obtained even within 50 days, while cement pastes with reference material like high-porosity Vycor glass attain equilibrium within 10 days. So far as the thermal expansion is concerned, cement pastes showed an anomalous delayed contraction on holding at the elevated temperature (Figure 6.19), which resulted in net contraction on cooling with pronounced hysteresis. The measured coefficient of thermal expansion depends on what is taken as the basis: immediate response on heating or the net response after delayed contraction. Preliminary results indicated that adding silica fume increases the initial thermal expansion as well as the delayed time-dependent contraction, so that the net expansion is independent of fume additions.

6.5.3.2 DSP cement composites

Since the DSP cement matrix can be combined with aggregates to form mortars and concretes and since fibres could also be used for reinforcement, a variety of composites can be prepared from DSP cement. The general criteria that may be used to distinguish these products from conventional concrete are w/s ratio (<0.30) and compressive strength (>150 MPa). The illustrative properties of some of the DSP products are given Table 6.13, while the processing parameters are summarized in Table 6.14 [42].

6.5.3.3 RPC

For the recently formulated products in the family of reactive powder concrete mentioned earlier,

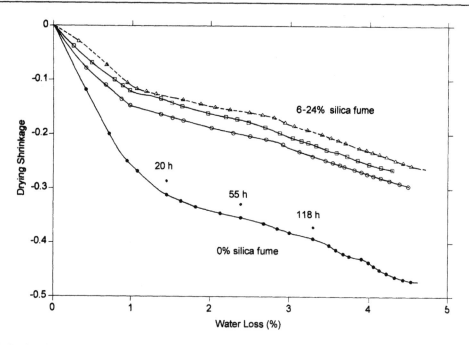

FIGURE 6.18 Drying shrinkage of thin DSP paste slab (about 2 mm thick) cured for 8 weeks, completely saturated in water and then dried out at 50% RH or 11% RH [42].

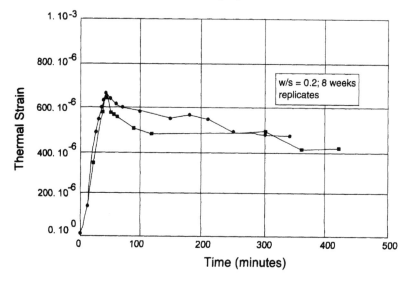

FIGURE 6.19 Anomalous delayed contraction of DSP paste on holding at elevated temperature on repeated cycles [42].

the typical composition and properties of RPC 200 and RPC 800 are given in Tables 6.15 and 6.16 [9]. It is evident from Table 6.15 that the RPC 200 variety has the options of addition of fibres and higher curing temperature, while the RPC 800 variety with either silica or steel aggregates need treatment at temperatures in the range of 250–400°C as well as high prior compacting

TABLE 6.13   Properties of some DSP products [42]

	Ultra H-S concrete 1992	Densit A/S (Denmark) 1982	Cemcom Inc. (USA)		Boygues* (France) 1993
			1984	1987*	
Compressive strength (MPa)	150	250	350	500	~600
Flexural strength (MPa)	15	20	25	75	~75
Modulus# of elasticity (GPa)	35	50	—	~70	~65
Energy# of fracture (J/m²)	~1	~2	—	~100	~1500

#Typical values taken from the literature. These figures do not necessarily represent the properties of actual commercial products.
*Fibre-reinforced composites.

TABLE 6.14   Processing of DSP cement composites [42]

System	Dispersion of fume	Casting	Curing temp. (°C)	Post-cure treatment	Aggregate	Strength (MPa)
Densit	Yes	Vibration +Vacuum	25	None	Quartz	250
DASH 47	No	Vibration	60	200°C/dry	Steel	350
	No	Vibration +Fibres	25	400°C/dry	Steel	500
RPC 200	Yes	Vibration	90	None	Quartz	250
RPC 800	Yes	Compaction +Fibres	25	400°C	Quartz	500

pressure. A typical sequence of hot curing is given in Figure 6.20 [46]. The ratio of residual water to initial water shows that it takes a long time to reach constant weight at a given temperature. In practice, higher curing temperature may accelerate the process. The relationship between compressive strength and density for both RPC 200 and 800 products is shown in Figures 6.21 and 6.22 [9]. The high compressive strength and very low porosity add promise to the future success of this concrete. From the angle of practical use, attempts are being made for cost reduction in making RPC 200 variety by eliminating the use of steel microfibre without any sacrifice of properties. It has recently been reported [48] that the requisite ductility can be obtained in an inexpensive way by confining a non-fibre RPC in thin tubes. In this case, a non-fibre-containing confined RPC can achieve 350 MPa compressive strength.

Many of the DSP composites find application in flooring, safes and storage, nuclear entrapment high-impact, resistant panels tooling and moulding, etc. [49]. An important emerging application of this class of composite is as an overlay or coating to existing concrete or a thin long-span precast element. It is reported that the French Atomic Energy Commission has attempted to characterize RPC to render it suitable for making high-intensity containers for long-term interim storage of nuclear wastes. On the whole, DSP cements seem to hold high promise in different civil engineering fields. It also has high applicability in non-constructional engineering uses in replacement of metals and plastics.

TABLE 6.15 Typical RPC compositions (by weight) [9]

| | RPC 200 | | | | RPC 800 | |
	Non-fibred		Fibred		Silica aggregate	Steel aggregate
Portland cement	1	1	1	1	1	1
Silica fume	0.25	0.23	0.25	0.23	0.23	0.23
Sand 150–600 μm	1.1	1.1	1.1	1.1	0.5	—
Crushed quartz (d50=10 μm)	—	0.39	—	0.39	0.39	0.39
Superplasticizer (Polyacrylate)	0.016	0.019	0.016	0.019	0.019	0.019
Steel fibre						
L=12 mm	—	—	0.175	0.175	—	—
L=3 mm	—	—	—	—	0.63	0.63
Steel aggregates (<800 μm)	—	—	—	—	—	1.49
Water	0.15	0.17	0.17	0.19	0.19	0.19
Compacting pressure (MPa)	—	—	—	—	50	50
Heat treatment (°C)	20	90	20	90	250–400	250–400

TABLE 6.16 Typical properties of RPC 200 and RPC 800 [9]

	RPC 200	RPC 800
Pre-setting pressurization	None	50 MPa
Heat-treating (°C)	20–90	250–400
Compressive strength (MPa)		
Using quartz sand	170–230	490–680
Using steel aggregate	—	650–810
Flexural strength (Mpa)	30–60	45–141
Fracture energy (J m^2)	20,000–40,000	1200–20,000
Ultimate elongation m m^{-1}	$50,000 \times 10^{-6} - 7000 \times 10^{-6}$	
Young's modulus (GPA)	50–60	65–75

6.5.4 Cement–polymer systems including macro-defect-free cement

The addition of polymers to cement has a long history and there are positive claims of improvements in strength, toughness, bondability and other relevant properties. Polymers used for this purpose include those of methyl methacrylate, styrene, vinyl acetate, styrene–acrylonitrite, chlorostyrene, vinyl chloride and polyester–styrene. These polymers supplement the effects of Portland or other varieties of cement. Most are attacked by strong alkalies but some such as polyester styrene give high resistance to strong acids. The mechanical strengthening of cement–polymer composites is explained through pore-filling by the polymers, which act as continuous reinforcing networks. In addition, the polymers form some stable bonds by interaction with the hydrated cement phases.

In the context of cement polymer systems, it may be pertinent to look at some recent interesting findings of Robertson [50] on the improvements that polyvinyl alcohol (PVA) in small amounts (about 1.4 per cent by weight of cement) can make in cement. The pull-out behaviour of steel fibres from 28-day hydrated cement blocks with, and without PVA is shown in Figure 6.23, which shows that the bond strength was approximately

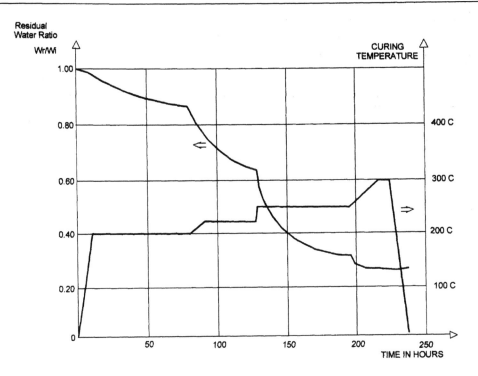

FIGURE 6.20   Typical curing sequence for RPC-800 products [46].

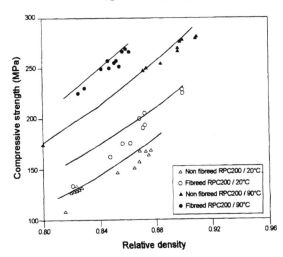

FIGURE 6.21   Envelope of optimum strength values obtained for different relative densities of RPC 200 concretes.

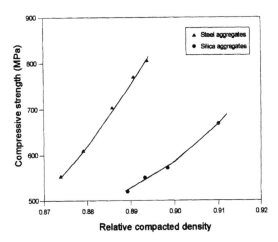

FIGURE 6.22   Variation in strength with relative density of RPC 800 concretes [9].

doubled. PVA has been found to inhibit the nucleation of calcium hydroxide crystals during the hydration of cement and evidence is there to conclude that the nucleation of C-S-H is

increased. Further, the addition of PVA seems to have some positive effect on the transition zone in concrete (Figure 6.24). Specimens with PVA have narrower gaps next to the aggregate, though some parts of the aggregate seem to be in direct

213

contact with bulk cement paste, which is more densely packed.

However, the major opportunity that emerges from the cement–polymer systems is the development of materials with high flexural strength, which could behave more like plastics. Keeping this goal in view, the macro-defect-free (MDF) cements have been developed on the basis of hydraulic cements, water-soluble polymers and high-shear mixing. The primary credit for this development goes to Birchall and coworkers of ICI, who patented the product in the early eighties

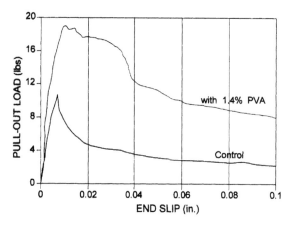

FIGURE 6.23 Effect of PVA on steel fibre pullout from a 28-day hydrated cement paste block (fibre diameter 0.016″, embedded length 1″, w/c = 0.38) [50].

[51,52]. The cement systems have been worked out in great detail – HAC with partially hydrolysed polyvinyl acetate and OPC with polyacrylamide. Other water-soluble polymers such as hydroxypropyl-methyl cellulose have also been tried. Generally the use of polymers is limited to 4–7 per cent. The material systems can be reinforced with fibrous and particulate materials. Processing is critical in developing the microstructure which is responsible for the material properties of the composite. However, the practical application of this class of material has been adversely affected by the lack of its durability due to the deterioration of mechanical properties at high relative humidities [53]. To overcome this basic infirmity of this otherwise highly potential composite material, the following approaches have been considered:

(i) Use of lower proportions of polymer: Although this approach is likely to reduce the effective cement–polymer interface, thereby reducing the impact of humidity-related durability problem, at the same time polymer concentrations below the optimum range may not help in achieving the desired mechanical properties [54].

(ii) Insolubilizing the PVA matrix with crosslinking and/or coupling agents. Some studies reveal that the epoxy-functional silane coupling agent appears to be effective, although it is necessary to expose MDF to high humidity to complete the reactions [55]. In a similar manner, the

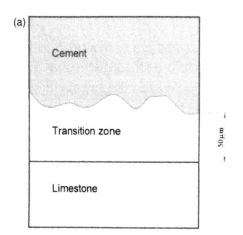
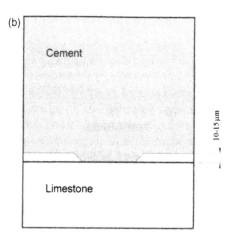

FIGURE 6.24 Drawings of transition zone of flat aggregate (a) without and (b) with PVA [50].

improvement of water-resistance of MDF sheet prepared with HAC, 3 per cent PVA, 0.3 per cent glycerol and 12.3 per cent water immersing in isocyanate has also been reported [56]. The addition of an organotitanate coupling agent during processing has also been reported as a successful approach (Figure 6.25) [57].

(iii) *In situ* reticulation of polymer. In a recent study [58], using PVA as the water soluble polymer, sodium silicate solutions as the reticulation agent and ASTM Type I Portland cement, it has been demonstrated that the *in situ* reticulation of PVA by sodium silicate in cement materials is achievable (Table 6.17). It was also established experimentally that sodium silicate has no significant effect on the hydration rate (Figure 6.26) and flexural strength (Table 6.18).

(iv) Cement–inorganic polymer system. As an alternative to organic polymers attempts are being made to develop cement composites with inorganic polymers such as $Na(PO_3)_4$ and $Na(PO_3) \cdot nH_2O$ in combination with HAC [58]. It is reported that the flexural strength in such composites is greatly improved, while the moisture-induced deterioration of strength is reduced.

(v) One of the early attempts made by the present author to produce and evaluate an MDF test piece with calcium sulphoaluminate-belite cement in comparison with other cements revealed that the flexural strength of CSA-belite MDF achieved was more that that of PC-MDF but less than the value obtained for HAC-MDF (Table 6.19). In

recent times this low-energy cement has been receiving more and more attention from the point of view of its processability into MDF composite [60,61]. A specific study by Drabik *et al.* [62] on calcium sulphoaluminate-calcium alumino ferrite-belite clinker with hydroxypropylmethyl cellulose (5 per cent by weight) and polyphosphates (5 per cent by weight) showed that these mixes have good MDF processibility across the entire composition range. This happens due to the preferential formation of clusters of Al(Fe)–O–C(P) cross-links under the

TABLE 6.17 Effect of sodium silicate on the properties of PVA-mixed cement films [58]

PVA (%, w/v)	Sodium silicate (%, v/v)	Weight Increase[a] (%)	Weight loss after drying[b] (%)
1.0	0	310 ± 21	27.1 ± 11
	0.5	270 ± 14	26.0 ± 7.2
	1.0	114 ± 13	10.2 ± 3.1
	2.0	29 ± 3.1	6.7 ± 1.9
2.0	0	294 ± 19	23 ± 10
	1.0	267 ± 19	9.2 ± 2.7
	2.0	81.0 ± 3.1	8.2 ± 4.1
4.0	0	315 ± 18	27.2 ± 13
	1.0	295 ± 27	27.2 ± 13
	4.0	50.7 ± 8.2	5.1 ± 3.9
5.0	0	320 ± 11	27.2 ± 9.9
	5.0	20.1 ± 3.7	2.7 ± 2.5

[a]Weight increase of each film as obtained after 24 h immersion in water.
[b]Weight loss after drying the immersed films.

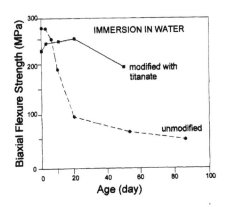
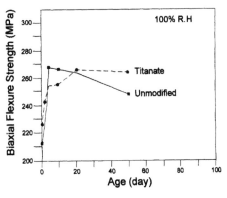

FIGURE 6.25 Effect of adding an organotitanate to the MDF formulation before processing on the moisture sensitivity of the final product [57].

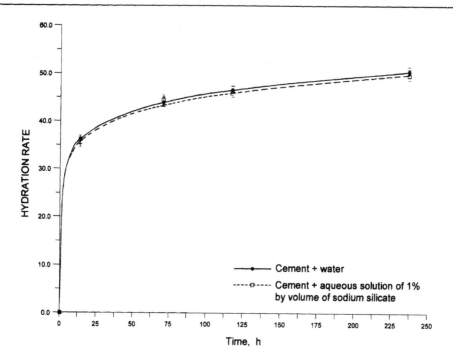

FIGURE 6.26   Comparative cement hydration rate with or without sodium silicate (w/c = 3) [58].

TABLE 6.18   Flexural strength of cement test specimens[a] as a function of age [58]

Test specimen	Flexural strength (Mpa)	
	7 days age	28 days age
Control	4.4 ± 0.2	5.7 ± 0.1
Admixtured[b]	4.4 ± 0.1	5.6 ± 0.2

[a] Control: Cement and water; [b]admixtured: cement and aqueous solution of sodium silicate 2% v/v. In both cases w/c = 0.45 and duplicates.

TABLE 6.19   Comparative MOR of MDF prepared with different cements

Composition	
Cement	78.5%
PVA	8.0%
Water	13.0%
Glycerine	0.5%
MOR (MPa)	
HAC	100.0
CSA	77.6
OPC	60.0

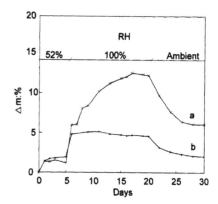

FIGURE 6.27   Mass change Δm as a function of time for MDF test pieces fabricated from SABF clinker and (a) HPMC and (b) poly-P, and treated at RH = 52% (until the 5th day), at RH = 100% (from the 6th to 20th day) and under ambient conditions (from 21st to the 30th day) [62].

conditions of synthesis. There is sufficient moisture resistance at RH = 52 per cent, while there are substantial mass changes at RH = 100 per cent. The moisture resistance at RH = 100 per cent is substantially enhanced in materials containing polyphosphates (Figure 6.27). Calcium carbonate is present in samples attacked at RH=100 per cent from the reaction with atmospheric CO_2. The cross-links themselves do not appear to be the subjects of attack by moisture in this MDF system.

6.5.5 Ultra rapid hardening (modified) Portland cement

This category of cement has been covered quite extensively in the previous reviews as regulated-set or jet cement. The clinker is composed of 55–60 per cent C_3S, 5–10 per cent C_2S, 20–25 per cent $C_{11}A_7CaF_2$ and 4–8 per cent C_4AF, and this clinker is converted into cement by adding 8–12 per cent anhydrite and small but varied amounts of hemihydrate, sodium sulphate and calcium carbonate. The SO_2/Al_2O_3 ratio is carefully adjusted to about unity [63]. The primary use of this kind of cement is for repair of highways, bridges, railways, etc.

The pattern of strength development of jet cement concrete as compared to that of ordinary and high early strength Portland cements is shown in Figure 6.28.

The flexural strength of jet cement concrete is one-fifth to one-tenth of its compressive strength – quite the same as observed in the OPC concrete. Hence, to increase the toughness, steel fibre or polymer is used. Attempts have also being made to modify the phase assemblage of regulated-set cements with more of belite or ferrite so as to achieve longer handling and setting time, the details of which have been summarized in [3,4].

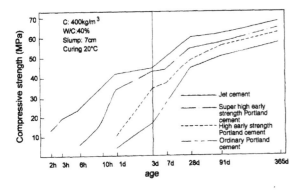

FIGURE 6.28 *Strength development of jet cement concrete and others.*

6.5.6 Expansive cements

Expansive cements are generally multi-blend formulations with Portland cement as the basic ingredient. Apart from the well-known types designated as 'K' (interground mix of Portland cement clinker+expansive clinker containing $C_3A_3CaSO_4$ +calcium sulphate as gypsum or anhydrite and gypsum mixture), 'M' (blend of Portland cement and aluminous cement or aluminous slag) and 'S' (produced from high C_3A Portland cement clinker), special types produced with the incorporation of calcium or magnesium oxide have been extensively reported in the previous literature and hence, are not reproduced here. Endeavours made to produce blended expansive cement with silica fume and superplasticizer with improved properties have also been reported earlier [3]. Even at the cost of repetition, it may be pertinent to briefly mention here that the expansive cements are used for producing shrinkage-compensating concrete, chemically prestressed concrete, expansive grouting mortars and also for non-explosive concrete or rock-disintegrating agents. The shrinkage-compensating cements are used whenever the prevention of shrinkage cracking is critical, such as in the construction or airfields, runways, pavements, water tanks, underground garages, railroads etc.

The chemically-stressing cements are better suited for making structural elements requiring low prestressing such as pipes, wall panels, shells, folded plates, etc. Expansive grouts are extensively used for heavy machinery foundations. Controlled expansion for cracking of rocks and concrete structures is also a useful application of expansive cements (Figure 6.29).

6.5.7 High fineness cement

It is believed that the use of super high early strength Portland cement, the precursor of high fineness or microfine cement, dates back to early forties, although it found a place in the Japanese Industrial Standard R5210 in 1973. In Japan, in the past, two microfine cements – one trade named as 'colloid cement' with specific surface of $6000\,cm^2/g$ and maximum particle size of $40\,\mu m$,

FIGURE 6.29 Disintegration of unreinforced cement concrete at site by the progressively hydrating and expanding cementitious product (expansive force ~50 MPa) (courtesy: ACC).

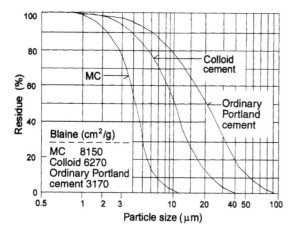

FIGURE 6.30 Particle size distribution of MC and other cements.

and another 'MC' with maximum particle size of 10 μm have been used essentially for rock grouting in tunnels. The particle size distributions of these cements are shown in Figure 6.30 [64]. Today, microfine cements are produced in classifiers or by ultrafine grinding to a high surface area, usually above 6000 cm²/g. These cements can either be of pure Portland cement composition or with blends of ground granulated blast furnace slag. This variety of cement is used in grout by batch-mixing with water and a dispersant or in the form of two-component mix with sodium silicate. The setting times are controlled from a few minutes to a few hours, depending on the application. It appears that the petroleum industry is the major user of this variety of cement in oil-well repair jobs [65].

It is reported [66] that for injection work the ultrafine cement is processed into a suspension in high-speed colloidal mixers. Because of the level of fineness and the closely-graded particle size distribution, this suspension is capable of exceptional penetration. Its stability to sedimentation is also very high. Recently microcement has been used increasingly as an alternative to the use of plastic gels for the production of grouted bottoms.

6.6 Cementitious products for environmental benefits

The cement industry is known to emit high quantities of carbon dioxide and nitrogen oxides to the environment. The manufacture of blended cements, on one hand, and of low-lime high-belite cement, on the other, has been widely accepted as effective solutions to the above problems, because of decrease in the specific consumptions of limestone and fuel respectively. Concurrently, these cements bring in the benefits of energy conservation as well. These cements have already been dealt with in the previous sections of this chapter.

Looking at the environmental issues from another perspective, it is evident that the quantities of industrial waste and municipal refuse have rapidly increased in recent years with the progress of industry as well as of the living standards of human beings. This problem is also being tackled by the cement industry through development of special cementitious formulations based on industrial and other wastes. The more promising ones out of such eco-binder systems are discussed under this category.

6.6.1 Alkali-activated cements

Alkali-activated alumino-silicate cements constitute a major family of eco-friendly materials and have also been named as geopolymeric cements based on the concept that most silicate or alumino silicate minerals, being covalantly bonded chains, layers or three-dimensional networks are known as geopolymers [67]. The alkali-activated slag cement and concrete, alkali-activated metakaolinitic cement and alkali-activated flyash cement are some examples in this respect and have been reviewed [2,3]. The salient aspects may be summarized as follows:

(i) Use of alkali-activated slag cements have been known for about four decades in the former USSR and some East European countries. Usually for activation 2–7 per cent Na_2O or 3–10 per cent K_2O of the slag content is necessary.

(ii) There has been commercial exploitation of 'F' cement in Finland and 'Pyrament' in the USA. The former is based on fine slag, fly ash and other pozzolana, activated by an admixture of various components exhibiting alkaline reactions while the latter is an alkali-activated blended Portland pozzolana cement containing organic retarders and water reducers.

(iii) The alkali-activated slag concrete of Indian origin in the author's laboratory has shown progressive gain of strength from 21.0 MPa in 3 days to 36.0 MPa in 1 year and to 40.0 MPa in 6 years for a concrete mix having binder content of 350 kg/m^3 of concrete, binder: aggregate ratio of 1 : 5.96, water-to-binder (w/b) ratio of 0.46 and MSA of 25 mm.

(iv) There is a strong revival of interest in this type of binder formulation for recycling of slag, flyash, kiln dust, etc.

(v) The potential of using such products for entrapping toxic metals and stabilizing low level radioactive wastes is being explored.

Recognizing the significance of alkali activation as a process and low-calcium flyash as a major source material, certain interesting methodological investigations are being carried out in different laboratories to assess the reactivity of flyash and mechanism of alkali activation.

Since the lime reactivity test does not always bring out the intrinsic reactivity of a fly ash in rela-

tion to its fineness, a technique has been developed in the author's laboratory to determine the alkali reactivity of a flyash [68]. This technique offers a better assessment of the quality of Indian flyash and also helps in predicting the relation between lime reactivity and the specific surface area of a given fly ash for its use in the production of pozzolanic cement or other cementitious products.

The other interesting development has been a technique for assessing the mechanism of alkali activation of fly ash [69]. This method utilizes an ethylenediamine tetra-acetic acid/triethanolamine/NaOH mixture to determine the proportion of unreacted fly ash particles present in an OPC – low-calcium fly ash mixture, with or without prior alkali activation. This study also utilized another chemical technique, an orthophosphoric acid dissolution method, to assess the amount of unreacted crystalline phases present in hydrated OPC–ash mixtures at different ages of hydration. The information obtained from these two chemical techniques was used to identify the quantity and nature of unreacted fly ash particles present in the hydrated samples. The amount of unreacted fly ash and the compressive strength data were correlated to assess the influence of alkali activation on the reactivity of the amorphous and crystalline phases in the low calcium fly ash.

6.6.2 Reactive fly ash belite cement

In the early part of the last decade, new possibilities emerged to extend the reuse of fly ash by virtue of the development of a technology of its activation and hydrothermal treatment [70]. This technology has now been tried out for some Spanish fly ash of ASTM Class F type [71].

The hydrated precursor (obtained after the hydrothermal treatment) was calcined at 900°C, 800°C and 700°C. While the product obtained at 900°C had a phase assemblage of C_2AS, α'-C_2S and β-C_2S, lower temperatures for sintering were tried out to eliminate the gehlenite phase, although the belite phase in the process turned out to be less and less crystalline (Figure 6.31). The trends of compressive and flexural strengths for the above samples are shown in Figure 6.32. The binder calcined at 800°C

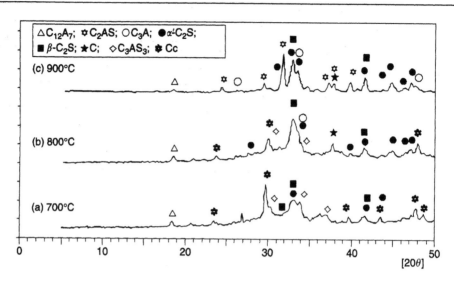

FIGURE 6.31   XRD patterns of the FABC calcined at 700, 800 and 900°C [71].

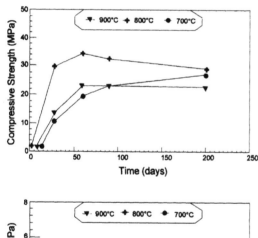

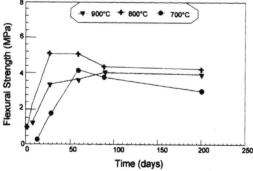

FIGURE 6.32 Development of compressive and flexural strength in FABC with time [71].

showed the best mechanical properties. The changes in the pore size distribution (Figure 6.33) mainly occurred in the range of pores having diameters more than 0.05 μm, which decreased at the same rate as the pores with less than 0.05 μm diameter increased. Thus, the total porosity was not affected to any appreciable extent. The alkalinity of the extracted pore solution of these binders is comparable to that of OPCs and this is attributed to the alkalis present in the starting fly ash.

It may be relevant to mention here that the above technology of lime activation and hydrothermal treatment of fly ash could not be made effective for Indian fly ash and hence, a somewhat different process route had to be developed [72]. The process involves activation of fly ash with alkali first, and then reaction of the activated fly ash with hydrated lime to produce a gel which is composed of calcium silicate-aluminate hydrate (C-S-A-H) and C-S-H precursors. The precipitated gel on de-watering followed by low temperature sintering produces a clinker. The critical steps in the process are illustrated in Figure 6.34. The process utilizes about 22–25 per cent fly ash. The gel characteristics including the proportioning of ingredients are given in Table 6.20, while the clinker composition and cement properties are furnished in Table 6.21.

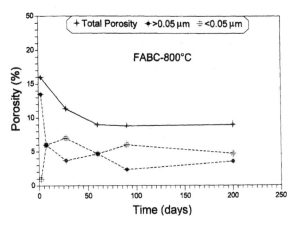

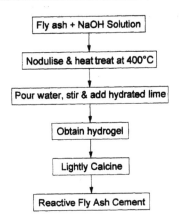

FIGURE 6.33   Changes in pore size and volume in FABC with time [71].

FIGURE 6.34   Salient steps of 'hydrogen' process for obtaining a cementitious product from fly ash [72].

TABLE 6.20   Raw material proportions and gel compositions [72]

	Hydrogel clinker I mix	Hydrogel clinker II mix	Hydrogel clinker III mix
Raw material proportions			
Activated fly ash	23.5	22.0	—
Hydrated lime	76.5	78.0	79.3
Activated fly ash +silica fume	—	—	19.0
Ferric hydroxide	—	—	Equivalent to 1.6% Fe_2O_3
Chemical composition			
% Oxides			
SiO_2	17.2	16.33	16.58
Al_2O_3	6.68	6.34	4.35
Fe_2O_3	2.02	1.97	2.96
CaO	46.14	47.03	48.01
MgO	0.17	0.17	0.13
LOI	26.19	26.67	27.11
Na_2O	0.61	0.57	0.34
Modulii conditions			
LSF	0.80	0.86	0.90
SM	1.98	1.96	2.27
AM	3.3	3.2	1.47

6.6.3 High-volume fly ash concrete (HVFC)

The generation of very large quantities of fly ash in many countries due to the use of high-ash coal in power generation has led to very concentrated efforts on utilizing significantly large quantity of fly ash in cement and concrete. There are two technical concepts for increasing the addition of fly ash in concrete – one as a replacement of cement, and another as a replacement of fine aggregate.

The mix design differences in the above two approaches are shown in Table 6.22 [4]. It is evident from the table that the 90-day strength of HVFC in case of cement replacement can reach

221

TABLE 6.21 Clinker and cement from activated fly ash by the hydrogel process [72].

% Oxides	Clinker-I	Clinker-II	Clinker-III
Chemical composition of clinkers			
SiO_2	23.36	22.27	22.75
Al_2O_3	9.05	8.64	5.97
Fe_2O_3	2.74	2.69	4.07
CaO	62.51	64.14	65.87
MgO	0.23	0.23	0.20
Na_2O	0.83	0.80	0.47
Free CaO	1.0	0.80	1.0
Modulii conditions			
LSF	0.80	0.86	0.90
SM	1.98	1.96	2.27
AM	3.3	3.2	1.47
Potential phase composition			
C_3S	8.1	25.8	45.2
C_2S	60.9	44.4	31.2
C_3A	19.4	18.4	8.9
C_4AF	8.3	8.2	12.4
C/S ratio	2.68	2.88	2.9
Physical properties of cements (Indian standards)			
Physical tests			
Blaine's specific surface (m^2/kg)	320	320	320
SO_3(%)	2.7	2.7	2.6
Consistency (%)	28.3	27.5	25.6
Setting Time (min)			
Initial	75	65	110
Final	115	105	165
Compressive strengths (MPa)			
1 day	11.0	16.5	24.5
3 days	16.0	27.5	42.0
7 days	22.0	36.0	58.5
28 days	33.0	41.0	79.5

the level of 50 MPa. Further, it is also seen that in the case of sand replacement the consumption of admixture is much higher.

In the case of cement replacement, cement containing 50 per cent or more of fly ash is used. Although the strength of such a cement on hydration would show significant reduction, the strength of concrete is prevented from any such decreases by using an appropriate water-reducing agent. The research on high-volume fly ash concrete in Canada is based on this concept [73,74].

A fairly comprehensive study [75], based on concretes containing mineral admixtures with a constant replacement level of 50 per cent by mass of cement, brought out that concrete containing lignite ash generally developed higher ultimate strengths than the control mix containing no ash. Replacement of 50 per cent cement by subbituminous ash affected the strength development only slightly whereas the use of bituminous ash reduced the ultimate strength of concrete by about 20–30 per cent. It was also established that properly designed concrete containing 50 per cent fly ash as cementitious material may develop compressive strengths at 28 days and beyond, which are equivalent to or better than the control. Furthermore, in the majority of cases the strength of fly ash concrete at 7 and 14 days is adequate for

TABLE 6.22 Mixing proportion and properties of HVFC [4]

Mixing proportion (kg/m³)						Mixing ratio				Compressive strength (Mpa)					Setting time (hour:minute)	
Water (w)	Cement (c)	Fly ash (F)	Sand (s)	Gravel (g)	Admixture (%)	F/c+F	w/c	w/c+F	Slump (cm)	1 day	3 days	7 days	28 days	90 days	Initial	Final
First case																
115	428	0	682	1211	1.14	0	0.27	0.27	18.0	37.2	51.2	61.3	71.8	83.7	5:15	6:50
116	180	248	659	1171	0.86	0.58	0.64	0.27	24.0	8.6	20.5	33.7	44.1	51.9	7:45	10:30
	428															
115	156	215	685	1218	0.75	0.58	0.74	0.31	19.0	6.8	16.0	28.0	37.1	47.1	8:25	11:25
	371															
Second case																
160	300	0	789	1044	0.25	0	0.53	0.53	12.5	7.4	18.4	27.5	40.8	44.7	5:16	7:16
180	300	180	528	1044	1.0	0.38	0.6	0.38	6.5	10.3	19.9	27.1	42.9	58.1	7:55	10:35
	480		708													
180	300	300	339	1044	2.0	0.5	0.6	0.3	24.0	16.0	26.1	34.4	56.9	75.5	14:20	18:05
	600		639													

Plastic viscosity of concrete is extremely high.

223

many practical applications. Use of high-range water reducing and air-entraining agents, either separately or in combination, reduce long-term strength of high fly ash concrete as compared to the control containing no fly ash. With proper air-entrainment, even the high fly ash concrete may perform well under freezing and thawing conditions. Resistance to sulphate attack was enhanced by the use of fly ash in all cases except for the ash containing a high calcium oxide content.

6.6.4 Alinite-based eco-cement

The historical outline of alinite synthesis as a new silicate phase with mixed O-Cl anions by Nudelman in the former Soviet Union to the industrial production of alinite cement and its large scale application in construction has been presented in [3]. This new cement has also been duly covered in other reviews on the subject mentioned earlier and more particularly in [5]. For a quick recapitulation one may take note of the following points:

(i) Although the originally accepted formula of alinite was $Ca_{21}Mg [(Si_{0.75}Al_{0.25})O_4]_8O_4Cl_2$, later work established that it can be best represented by $Ca_{10}Mg_{(1-x/2)}!_{x/2}[(SiO_4)_{3/x} (AlO_4)_{1-x} O_2Cl]$, where $0.35 < x < 0.45$ and ! refers to a lattice vacancy.

(ii) It is possible to effect isostructural substitution of Cl by S, B and F in alinite. Belinite $[Ca_8Mg(SiO_4)_4Cl_2]$ and calcium silicate sulphate chloride $[Ca_4(SiO_4)(SO_4)Cl_2]$ have also been synthesized. The latter has shown better hydraulic properties than the former.

(iii) The alinite phase is stable in the temperature range of 900–1150°C and the actual clinkerization is carried out at temperatures ranging from 1150°C to 1200°C with heat consumption of about 535 kcal/kg clinker.

(iv) The range of variation of the composition of the industrially-produced alinite clinker has been as follows: 55–65 per cent alinite, 25–35 per cent belite, 3–5 per cent calcium chloroaluminate, 3–5 per cent dicalcium ferrite, often with 5–7 per cent alite.

(v) The concentration of calcium chloride in the reaction medium is 6–7 per cent as measured inside the reactor, resulting in 2–2.5 per cent fixed chloride in clinker corresponding to 50–60 per cent alinite formation, and leaving less than 0.5 per cent as free chloride.

(vi) The alinite cement has high early strength properties. The main hydration products of alinite cements are calcium silicate hydrates that incorporate the chloride, in addition to calcium hydroxide and an AFm phase $C_3A \cdot CaY_2 \cdot 10H_2O$ ($Y = Cl^-$, OH^-, $\frac{1}{2}CO_3^{2-}$).

(vii) The wider adoption of this technology has been affected by the controversy of the durability of alinite cement due to likely chloride corrosion of steel reinforcement. However, it has been used for plain concrete in actual construction.

Since the burning temperature of alinite cement is low and chlorine-containing wastes including incinerated ash of municipal refuse might serve as a suitable raw material for the cement, renewed attention has recently been paid to this cement. A special mention may be made of the development of 'eco-cement' made in Japan [76] from fly ashes of incinerated urban wastes, containing chlorine ranging up to about ten per cent (Table 6.23). The composition and property details of eco-cements are furnished in Table 6.24. The strength properties of these cements have been compared with other cements in Figure 6.35. There are, however, still many problems awaiting solution for commercial adoption of this technology. The problems are primarily on account of the presence of chlorine and resultant corrosion of manufacturing equipment, recovery of chlorine from exhaust gases, accumulation of heavy metals in clinker, and diffusion of harmful elements into the environment by dissolution from hydrated cement mortar and concrete.

6.7 Simultaneous production of energy and cement

In the mid-sixties the Krzhizhanovskii Energy Institute and Giprotsement of the former Soviet Union conceived a technological scheme of producing

TABLE 6.23 Chemical composition and proportioning of raw materials (in %) for eco-cement production [76]

	Raw materials				
	Fly ash (1)	Fly ash (2)	Limestone	Clay	Copper slag
Chemical compositions					
SiO_2	21.4	26.7	0.32	72.9	29.6
Al_2O_3	28.1	16.7	0.15	8.8	0.0
Fe_2O_3	5.1	5.0	0.2	4.4	54.8
CaO	25.1	28.4	55.3	0.0	2.9
MgO	10.4	2.4	0.4	0.7	0.0
Cl	4.7	8.0	0.0	0.0	0.0
Proportion of raw material in different clinkers					
Ordinary type	—	3.8	75.4	19.6	1.2
Rapid hardening A	25.1	—	64.2	10.7	0.0
Rapid hardening B	—	45.4	53.6	0.0	0.2

TABLE 6.24 Typical chemical and mineralogical compositions of eco-cement [76]

	Ordinary type	Rapid hardening type A	Rapid hardening type B
Chemical compositions (%)			
SiO_2	22.6	18.4	18.6
Al_2O_3	4.1	12.1	13.1
Fe_2O_3	2.7	1.5	2.5
CaO	59.9	63.1	62.4
MgO	2.2	1.5	1.0
Cl	4.0	2.0	0.5
Mineralogical compositions (%)			
Alinite	70.0	58.0	—
Belite	23.0	18.0	9.0
Calcium chloro-aluminate	2.0	20.0	25.0
Alite	—	—	60.0
Ferrite	2.0	—	4.0

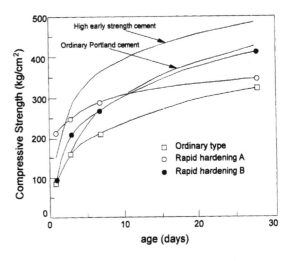

FIGURE 6.35 Compressive strength of alinitic eco-cements (Japanese standard mortar test) [76].

fused clinker-based cement at thermal power stations. Partly this process was tried at a special semi-industrial experimental set-up [77]. According to this process, prior to feeding a boiler, the fuel was mixed with other components so as to approximate the clinker composition. In the high-temperature zone the modified fuel ash melted and it was taken out as 'slag'. The clinkerization reactions were completed in the molten state. This 'slag' was named as 'electro-clinker' which, on grinding, gave rise to a cementitious product.

Another variant of this scheme was also proposed in the then Soviet Union. If the conversion of fuel ash to clinker was not feasible due to technical constraints, it could be converted by suitable adjustment of composition into an 'electroslag' which could be a more reactive additive for cement grinding or as a better raw material for clinker production in a regular cement plant.

Depending on the type of solid fuel, the yield of electro-clinker was estimated at 0.05–0.25 kg per

1000 kcal of fuel burnt [78]. However, this process scheme did not meet commercial success perhaps due to unfavourable techno-economic factors.

In recent years, a comparable technology has been patented under the title 'One Furnace, Two Functions' process for producing both steam power and cement clinker simultaneously in one apparatus. The patented technology is also known as 'Admixture of Ash-Modified Component' (AMC) referring to the addition of an ash-modified component to the pulverized coal burning furnace of a power-generating plant for the purpose of producing cement clinker, desulphurization and NO_x reduction. It is claimed that AMC contains caking modifiers and loosening ingredients that effectively inhibit caking inside the furnace. The mineral formation of AMC and coal ash is an exothermic reaction, which improves burning efficiency and reduces incomplete burning. De-sulphurization rates can reach up to 90 per cent.

The main clinker phase in this process is belite, which is ground to a specific surface of about $6000\,cm^2/g$ to produce the cement. The strength properties are given below [79].

	1-day	3-day	7-day	28-day
Compressive strength (Mpa)	15.8	35.1	45.2	66.2
Flexural strength (Mpa)	4.2	5.6	6.9	7.8

The first North American installation, using AMC to produce cement clinker inside the furnace of a utility boiler, is planned to be commissioned in the spring of this year. A pilot plant in Henan, China is reported to have established that with AMC, all coal ash and more than 90 per cent of coal sulphur are converted into cement clinker.

Needless to say, the simultaneous process of generating steam power and producing cement clinker is a welcome development both from the point of view of cement technology and environmental management.

6.7.1 Pore-reduced cements

Pore-reduced cements (PRC) are high-density, high-strength materials that are made by pressing immature OPC pastes, as a result of which there is partial removal of excess mix water [80]. The resulting reduction in the effective water/cement ratio and the significant reduction in porosity can increase paste densities to greater than $2600\,kg/m^3$ compared with $2000\,kg/m^3$ in unpressed pastes. The reduction in volume porosity causes a change in the pattern of porosity, which becomes isolated from the earlier interconnected nature. Although hydration in PRC does not stop completely, the bulk of the unhydrated clinker is sealed from any external water supply and therefore does not exhibit any tendency to dimensional instability.

The low permeability to water and other aqueous agents, coupled with its significantly improved mechanical strength, indicates that the material would be useful in waste management. Both the fabrication of containment vessels for toxic slurries and the utilization of PRC as the immobilization matrix itself, according to the authors, appear to be viable.

6.8 Cements with high energy conservation potential

The energy price increases resulting from the oil crisis in the early 1970s triggered a drive in the cement industry to reduce fuel cost by becoming more energy efficient and by switching away from petroleum products and natural gas. The gains in energy efficiency were achieved by the following approaches:

(i) adoption of energy-efficient plants and machinery;
(ii) use of selected waste materials such as spent solvents, paint residues, used oil and scrap tyres as fuel;
(iii) modification of the product composition.

Elaborating on the last approach, one may mention the following accepted practices:

(i) extending the basic cement with supplementary cementitious materials such as fly ash, granulated blast furnace slag, natural and manufactured pozzolanas, etc. to produce blended cements;

(ii) modifying the clinker composition with lower lime content, thus moving from high-alite to high-belite type of cement;

(iii) altering the clinker composition by substituting the high-temperature tricalcium silicate (or alite) phase by the relatively low-temperature calcium sulpho-aluminate phase;

(iv) moving towards the oxy chloride system for clinkerization and resultant alinite type of cements.

Of the above options, the manufacture of blended cements has been widely adopted both for the advantages of durability, and the benefits of energy saving. The high-belite cement, which was conceived as an energy-saving product, turned out to be a product requiring lower energy in pyro-processing but higher energy in grinding. Hence, its wider adoption as an energy-saving product is under further research and investigation. However, as mentioned earlier, interest in this cement is more for its observed long-term durability characteristics. The option of alinite cement certainly carries with it a high potential of energy saving but today it is of limited practical value because of its controversial status for durability in reinforced concrete structure.

This leaves only the calcium sulpho-aluminate containing cements as the potential low-energy binder group.

6.8.1 Calcium sulpho-aluminate cements

Calcium sulpho-aluminate cements (CSA) do not contain C_3S (alite) but have as a major constituent a sodalite-structured phase $Ca_4(Al,Fe)_6O_{13} \cdot SO_3$. These cements have been covered in almost all the previous reviews [1–5,7]. The broad features can be summarized as follows:

1. the cements belong to the fundamental system $CaO–Al_2O_3–SiO_2–Fe_2O_3–SO_3$ and the strength properties are sensitive to the composition and phase assemblage (Figure 6.36) [7].
2. the essential phases are $C_4A_3\bar{S}+\beta\text{-}C_2S+C\bar{S}$, while the associate phases include $C_{12}A_7$, C_4AF

and free CaO. Phases like C_2AS and $2(C_2S)$, CS are considered deleterious. It is possible to have analogues of $C_4A_3\bar{S}$ by substituting CaO by SrO and $CaSO_4$ by $BaSO_4$ or $SrSO_4$.

3. the clinkering temperature is around 1300°C and the critical compositional controls are exercised by $A/\bar{S}$, $A/\bar{S}$, free CaO and $C\bar{S} \cdot nH_2O$.

4. the paste-hydration products primarily consist of ettringite, monocalcium sulpho-aluminate hydrate (AFm), alumina gel and ferrite gel. Appearance of C_3AH_6, C_4AH_{13}, CSH gel and $C_4A\bar{C}H_{11}$ has been noted in some cases.

Considering that there is an advent of a new era in concrete technology, marked by more widespread production and use of CSA, a special issue of 'Advances in Cement Research' was recently published as indicated earlier. From the articles appearing in this special issue, one may obtain an idea about the present status of this special cement in China and East European countries [81–83]. Certain basic aspects of manufacturing and use have also been dealt with [84,85]. Some recent findings and observations are discussed in the following sections.

6.8.1.1 Nomenclature

Amongst various countries taking an interest in CSA, China to day should perhaps be regarded as having the longest experience of production and usage. In the middle of the 1970s, the sulpho-aluminate cement (SAC) series was put into industrial production and the relevant Chinese standards were published in 1981 (Table 6.25). The ferroaluminate cement (FAC) series was developed at the beginning of the 1980s and the standards were drawn up in 1987 (see Table 6.25) [82]. Despite this standardization, there is still no global consensus on whether the Chinese cements with 35–37 per cent $C_4A_3\bar{S}$ phase, less than 30 per cent belite and 10–35 per cent ferrite phase should be named as SAC or CSA cement. Similarly, it is still open for discussion if for the CSA cement with high-iron content, the Chinese nomenclature of FAC is justified [81]. An international consensus is called for.

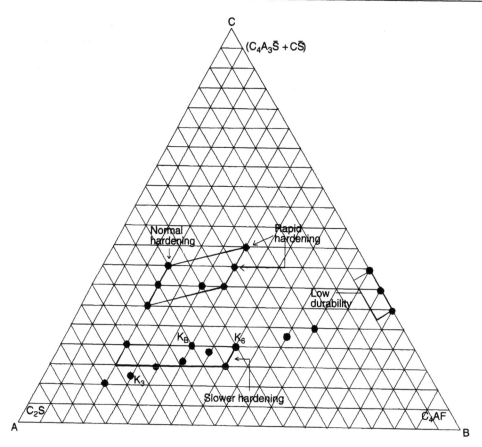

FIGURE 6.36   Ternary diagram showing different compositional areas reflecting distinctive properties [7].

6.8.1.2 Manufacture

The energy requirement and CO_2 release in the synthesis of $C_4A_3\bar{S}$ as compared to other phases are given in Table 6.26 [81], which clearly shows that cements based on sulpho-aluminate phase scores a discernible advantage in energy conservation. Generalized Bogue computations to forecast the mineralogical composition of SAC based on fly ashes have been presented in [85]. In the phase composition ranges of C_2S 60.7–83.4 per cent, $C_4A_3\bar{S}$ 4.2–27.2 per cent and C_4AF 2.6–30.5 per cent fly ash, particularly of low S/A or S/(A+F) ratios, is seen as convenient raw materials along with limestone and gypsum as other components.

It may be relevant to mention here that with the progressive maturity of the production technology, the number and size of production plants in China are on the increase. Plants of 90,000 tpa are being commissioned. At present, there are fifteen special cement plants with a total output of about one million tonnes [82]. Different types of SAC prepared with additions of gypsum, anhydrite and other admixtures are available in the Chinese market (Table 6.27). The major difference between the Chinese SAC and FAC varieties is shown up in the lower alkalinity of the former. In order to tackle the problem of short handling time of cements, special admixtures have been developed in China. These admixtures have been classified as retarding, plasticizing and superplasticizing, anti-freezing, pumping and rapid-setting varieties.

In another study [83], it is claimed that the blending of sulpho-aluminate-belite cement with

TABLE 6.25 Important property specifications for various sulpho and ferro-aluminate cements in the new Chinese standards [82]

Variety	Class	Specific surface (m^2/kg)	Setting time (min)		Compressive/flexural strengths (MPa)			Free expansion (%)	
			Initial	Final	1 day	3 days	28 days	1 day	28 days
R.SAC/FAC	425	≥350	≥2.5	≤180	34.5/6.5	42.5/7.0	48.0/7.5		
	525				44.0/7.0	52.5/7.5	58.0/8.0		
	625				52.5/7.5	62.5/8.0	68.0/8.5		
	725				59.0/8.0	72.5/8.5	78.0/9.0		
E.SAC/FAC	S.E.525	≥400	≥30	≤180	31.4/4.9	41.2/5.9	51.5/6.9	≥0.05	≤0.5
	E.525				27.5/4.4	39.2/5.4	51.5/6.4	≥0.10	≤1.0

Self-stressing value (MPa)

Variety	Class	Specific surface (m^2/kg)	Initial	Final	7 days	28 days	Free expansion % 7 days	28 days
S.SAC/FAC	30	≥370	≥40	≤240	≥2.3	≥3.0 ≤4.0	≤1.30	≤1.75
	40					≥3.1	≥4.0 ≤5.0		
	50					≥3.7	≥5.0 ≤6.0		

Compressive/flexural strengths (MPa)

Variety	Class	Specific surface (m^2/kg)	Initial	Final	1 day	7 days	Alkalinity	Compressive strength: MPa 7 days	28 days	Free expansion %
L.SAC	425	≥430	≥25	≤180	32.0/4.5	42.5/6.0	c/w+1:10 ph≤10.5	≥32.5	≥42.5	c/s=1:0.5 28 days 0–0.15%
	525				39.0/5.0	52.5/6.5				

R: rapid hardening; E: expansive; S: self-stressing; L: low alkalinity.

229

15 per cent Portland cement may extend the setting time, increase the strength as well as the dynamic modulus of elasticity, reduce expansion under 20°C/100 per cent RH moist air curing, and ensure passivation of embedded steel to the same level as of pure PC.

6.8.1.3 Mortar and concrete Properties

The degree of hydration of $C_4A_3\bar{S}$ in the CSA cements with different proportions of gypsum is shown in Figure 6.37. The hydration of $C_4A_3\bar{S}$ after one day is significantly accelerated with increased addition of gypsum. As already mentioned, ettringite is the only crystalline product in most of the hydrated CSA cements. An investigation on the pore solution composition of CSA cements has been reported in [84]. Values of pH and alkali, calcium, aluminium and anion content of pore

TABLE 6.26 Energy requirement and CO_2 released during formation of important phases [81]

Cement compound	Enthalpy of formation kJ/kg clinker	Carbon dioxide released (kg/kg clinker)
C_3S	1848.1	0.578
β-C_2S	1336.8	0.511
CA	1030.2	0.278
C_4A_3S	~800	0.216

solution at various ages are given in Table 6.28 from a paste with w/c ratio of 0.80 cured at 20°C.

In India, the CSA cement is produced on a commercial scale but the application is essentially in grouting formulations for use in heavy machinery foundation, mine bolting and emergency repairs [86]. For durability studies, an experimental structure consisting of columns (23×35 cm), roof slab (12 cm thick) and T-beams (23×45 cm) was built with $1:2:4$ concrete mix about 15 years ago. The properties of this structure are still being monitored visually and through core samples. The trend of strength development for the

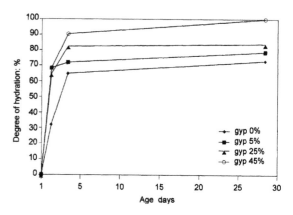

FIGURE 6.37 Degree of hydration of $C_4A_3\bar{S}$ against curing time for CSA cements with different levels of gypsum addition [81].

TABLE 6.27 The Chinese SAC/FAC types [82]

Cement series	Clinker	Cement type	Interground calcium sulphate type	Sulphate proportion	Other mineral additives
SAC series	SAC	Rapid hardening	Gypsum or anhydrite	Low	Limestone or others
		High strength	Gypsum or anhydrite	Low	Limestone
		Expansive	Gypsum	Medium	None
		Self-stressing	Gypsum	High	None
		Low alkalinity	Anhydrite	Low	Limestone
FAC series	FAC	Rapid hardening	Gypsum or anhydrite	Low	Limestone or others
		High strength	Gypsum or anhydrite	Low	Limestone
		Expansive	Gypsum	Medium	None
		Self-stressing	Gypsum	High	None

TABLE 6.28   pH and elemental compositions of pore fluids of CSA cements at various ages [84]

Age (days)	pH	Na (mM)	K (mM)	Ca (mM × 10⁻⁴)	Al (mM × 10⁻⁴)	SO₄²⁻ (mM × 10⁻⁴)	Cl⁻ (mM × 10⁻⁴)	Br⁻ (mM × 10⁻⁴)
1	13.10	18.3	106	3.8	555	6.2	10.3	ND
3	13.14	18.9	115	3.7	677	14.4	8.6	ND
7	13.04	18.6	86	3.9	381	14.9	6.3	ND
14	12.94	14.0	69	NM	166	4.2	7.9	ND
30	12.82	14.5	52	4.1	147	7.1	16.4	0.9
60	12.90	17.2	60	4.6	179	8.7	17.7	2.9

ND: not detected; NM: not measured.

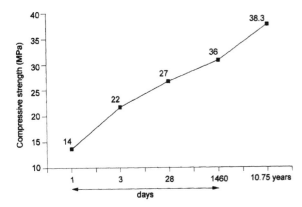

FIGURE 6.38   Long-term trend of strength development of CSA cement concrete in a load-bearing structure.

first 11 years is shown in Figure 6.38 and no deterioration has been observed so far on account of the exposure to tropical climatic conditions including heavy monsoons.

From the foregoing details, it is evident that the development of CSA-based cements seems to offer a new opportunity to the cement producers for environmental protection and energy conservation. A speedier introduction of these cements across the globe is possible through systematic generation of field data.

6.9 Miscellaneous hydraulic and chemical cements

It may be relevant to mention here that the above discourse does not claim to have covered all the special cements that are basically hydraulic or chemical in nature. The exceptions in the hydraulic variety include such special cements as oil-well cement, white and coloured cements, hydrophobic cement etc., that are produced in various countries as standard products. The details of these cements are extensively available in the previous reviews as well as in general literature and no significant or technological developments have been reported in respect of these products in recent times. Hence, no special attempts have been made in this chapter to deal with these cements. On the other hand, sketchy experimental results have been reported on such hydraulic cements as strontium and barium silicates and aluminates as well as tri- and di-calcium germanates [87–89]. Calcium stannates and plumbates have also been seen to possess hydraulic properties [5]. Possible alternatives to the use of gypsum as a set regulator in Portland cements have been found in calcium selenate [5] and calcium-lignosulphonate–sodium hydrogencarbonate mixtures [90,91]. The industrial implications of all the above developments are not evident at the present time due to their cost disadvantage, low availability, paucity of application data, etc.

The other group of chemical cements is a world by itself. Organo-mineral cements, phosphate cements, oxychloride and oxysulphate cements, sulphur cements, soluble silicate cements, phenol-formaldehyde resin cements, furan resin cements, epoxy resin cements, unsaturated polyester resin cements, vinylester resin cements, bituminous cements, etc., to name only a few out of the wide variety of products and cements, have been made use of in various applications.

Some of the above cements are obtained from the reaction between an acid and a base. The reaction product, usually a salt or a hydrogel forms the cement matrix in which the fillers are embedded. The acids can be mineral (phosphatic acid), Lewis (namely magnesium chloride or sulphate) or even an organic chelating agent (namely polyacrylic acid). The bases used for these cements are generally metal oxides such as magnesium oxide, gel-forming silicate minerals like wollastonite, etc.

A few of the above cements, using an organic (resins) or inorganic binder (silicates) and fillers like quartz and carbon, behave like mortars at normal temperature and they harden when the binders undergo a chemical reaction.

A third variety of cement like the sulphur type, demands higher temperatures for melting and subsequent cooling for hardening.

All the above chemical cements are extensively used for jointing, lining, flooring, emergency repairs, containment, storing, etc. Taking cognizance of the vastness of this subject, the present chapter was planned not to dwell on these products.

6.10 Conclusion

From the above discourse, it is perhaps quite apparent that the divide between the 'ordinary' and 'special' cements is thin and diffused. It is even more elusive when the products are seen in the context of time, place and user perception. This does not necessarily mean that OPC and its derivatives do not suffer from technical deficiencies; on the contrary, these commonly used cements, despite their stupendous growth in production and use, display serious shortcomings in terms of durability, engineering properties and application comforts. Hence, the search for special cements remains unabated. The increasingly growing concerns for energy and environment have further fuelled the ongoing efforts to develop new and special cements.

Seen in the perspective of structural durability, energy conservation and environmental amelioration, the expanding world of 'blended cements' appears to be a promising trend for tailor-made cements with special properties. Binary mixes

with microsilica, cement process dusts, limestone, etc. on one side, and ternary/multi component blended cements with desired packing density and more intense hydration behaviour are seen to be opening up newer avenues to achieve special application properties in resultant concretes.

Looking at the 'basic' cements, one finds that Portland cement with its high or moderately-high alite composition reigned in the construction industry either on its own or in binary blended forms with ground granulated blast furnace slag, fly ash, and natural pozzolans. It is interesting to note that there is an emerging perceptible change towards developing 'high-belite' cement with or without blending materials for obvious benefits of long-term durability and energy saving. The residual problems, in such development, like further improvement of hydraulicity and grindability of the belite phase, are being addressed.

It is only now, as never before, that such basic cements as 'HAC' and 'CSA' are being acknowledged for their utility in special construction purposes. The blended high-alumina cement of 'BRECEM' type and 'CSA-belite' variety stand out as promising special cements, particularly for marine structures and precast industry.

Although not possible to classify them strictly as special cements, such cement-based composites as high-performance fibre-reinforced products, DSP materials including RPCs, MDF formulations, etc., seem to be gaining progressive acceptance as special construction materials with unique engineering properties. Many of these products will find use as novel repair and renovation material and as replacements of metal and plastic.

From the perspective of environmental protection, the marriage of 'alinite cement' technology with conversion of high-chlorine incineration wastes appears as an interesting development. Similar is the nature of development of 'energy-clinker' in coal-fired steam-generating power-plant boilers. The alkali activation of industrial wastes to convert them into useful binders is also a new direction. These approaches for 'eco-cements' are of high practical significance and need to be tracked and encouraged for further development.

Finally, the search for new chemical systems with hydraulic potential such as barium and strontium silicates and aluminates, calcium germanates, calcium stannates and plumbates, etc., will continue to attract the attention of scientists and technologists. These developments will have to be taken through the logical steps from the present level of exploration.

It may be borne in mind that construction in the new millennium will require a wide range of materials and new application techniques. There will be many challenges in developing these products but, perhaps, the most significant demand would be from the speed of construction, energy conservation, environmental protection and maintenance perspectives for longer service life.

6.11 References

1. Kurdowski, W. and Sorrentino, F., 'Special Cements', in *Structure and Performance of Cements*, P. Barnes (ed.), pp. 471–554, Applied Science Publishers, London (1983).
2. Chatterjee, A. K., 'Special and new cements', *Proc. 9th Int. Cong. Chem. Cem.*, Vol. 1, NCB, New Delhi, pp. 177–212 (1992).
3. Chatterjee, A. K., 'Special and new cements in a historical perspective', *Proc. 3rd Congresso Brasilieiro de Cemento*, Vol. 2, Sao Paulo, ABCP, pp. 693–739 (1993).
4. Uchikawa, H., 'Management strategy in cement technology for the next century', *World Cem. Technol.*, Pts. 1–4, September (1994), pp. 66–76, October, (1994), pp. 49–56, November (1994), pp. 47–52, December (1994), pp. 58–63.
5. Bensted, J., 'Special cements', in *Lea's Chemistry of Cement and Concrete*, Peter C. Hewlett (ed.), 4th edn, pp. 779–835, Arnold Publisher, London (1998).
6. Odler, I., 'Special inorganic cements', E & F N Spon, London and New York, pp. 1–395 (2000).
7. Mehta, P. K., 'Investigations on energy-saving cements', *World Cem. Technol.*, May, 166–77 (1980).
8. Roy, D. M., 'New strong cement materials: chemically bonded ceramics', *Science*, **235**, 651–8 (1987).
9. Roy, D. M., *Ultra-High Strength Cementitious Materials*, MRL, The Pennsylvania State University, PA 16802, USA, September, pp. 1–16 (1998).
10. Fenner, J., 'Chemically resistant cements', in *Industrial Inorganic Chemicals and Products*, Vol. 2, pp. 931–942, Wiley-VCH, Weinheim (1999).
11. Chatterjee, A. K., 'High belite cements – present status and future technological options', *Cem. Concr. Res.*, **26**(8), Pts. I & II, 1213–37 (1996).
12. *Adv. Cem. Res.*, **11**(1) (1999).
13. Frohnsdorff, G. (ed.), *Blended Cements*, ASTM Special Technical Publication (STP) 897, Philadelphia, PA (1986).
14. Mehta, P. K., 'Role of pozzolanic and cementitious material in sustainable development of the concrete industry', *6th CANMET/ACI Int. Conf. on Fly Ash, Silica Fume, Slag and Natural Pozzolans in Concrete*, Bangkok, Vol. I, pp. 1–20 (1998).
15. Detwiler, R. J., 'Blended Cement now and for the future', *Rock Products Cement Edition*, July, pp. 27–33 (1996).
16. Gebhardt, R. F., 'Survey of North American Portland cements: 1994', *ASTM Cem., Concr. Agg.*, **17**(2), 145–189 (1995).
17. Asgeirsson, H., 'Intermilled silica fume in Icelandic cement', *Concr. Int.*, **14**(7), 56 (1992).
18. Guyot, R. and Ranc, R., 'Controlling the properties of concrete through the choice and quality of cements with limestone additions', *Int. Conf. on the Utilisation of Fly Ash, Silica Fume, Slags and other By-products in Concrete*, Montebello, Canada (1983).
19. Research Institute for the Cement Industry, German Cement Works Association, Activity Report: 1996–99, Verlag Bau+Technik GmbH, Dusseldorf, Germany, pp. 68–9 (1999).
20. Ingram, K. *et al.*, 'Carbonate additions to cement, *ASTM STP*, **1064**, 14–23. (1990).
21. Soroko, I. and Stern, N., 'Effect of calcareous fillers on sulphate resistance of Portland cement', *Ceram. Bull. Am. Ceram. Soc.*, **55**(6), 594 (1976).
22. Crammond, J. J., BRE Internal Note no. 148/91.
23. Marsh, B.K. and Joshi, R. C., 'Sulphate and acid resistance of cement paste containing pulverized limestone and fly ash', *Durab. Build. Mater.*, **4**, 67–80 (1986).
24. Borgholm, H. E., Herfort, D. and Rasmussen, S., 'A new blended cement based on mineralised clinker', *Emerging Technol. Symp. Cem. for the 21st century*, PCA SP 206, pp. 95–115 (1995).
25. Nagataki, S. and Wu, C., 'A study of the properties of Portland cement incorporating silica fume and blast furnace slag', *Proc. 5th Int. Conf. on Fly Ash, Silica Fume, Slag and Natural Pozzolans in Concrete*, Milwaukee, USA, V.M. Malhotra (ed.), Vol. 2, ACI Special Publication SP-153, pp. 1051–1068 (1995).
26. Swamy, R. N. and Bouikni, A., 'Some engineering properties of slag concrete as influenced by mix proportioning and curing', *ACI Mater. J.*, **87**, 210–220 (1990).

27. Fukuda, K. and Ito, S., 'Improvement in reactivity and grindability of belite-rich cement by remelting reaction', *J. Am. Ceram. Soc.*, 82(8), 2177–80 (1999).

28. Uchikawa, H., 'Present problems in cement making', *Proc. 9th Int. Cong. Chem. Cem.*, Vol. VII, NCB, New Delhi, pp. 23–57 (1992).

29. Uchikawa, H., 'Characterisation and material design of high-strength concrete with superior workability', *PAC RIM Meeting, SII Cem. Technol. Symp.*, Honolulu, Hawai (1993).

30. Yamamato, H. *et al.*, 'A study on spherical cement', *J. Res. Onoda Cem. Co.*, 43(125), 35–44 (1991).

31. Uchikawa, H., 'The future for cement and concrete in Japan', *J. Res. Onoda Cem. Co.*, 42(122), 1–62 (1990).

32. Majumdar, A. J. and Singh, B., 'Properties of some blended high-alumina cements', *Cem. Concr. Res.*, 22(6), 1101–14 (1992).

33. Osborne, G. J., 'BRECEM: a rapid hardening cement based on high alumina cement', *Proc. Instn. Civ. Engrs. Structs & Bldgs.*, 104, 93–100 (1994).

34. MacDowell, J. F., 'CaO-Al$_2$O$_3$-SiO$_2$ glass hydraulic cements', US Patent 4,650,443, August 12 (1986).

35. MacDowell, J. F. and Sorrentino, F., 'Hydration mechanism of gehlenite glass cement', *Adv. Cem. Res.*, 3(12), 143–52 (1990).

36. Vondran, G. L., 'Interground fibre cement in the year 2000', *Emerging Technol. Symp. Cem. for the 21st Century*, PCA SP 206, 116–34 (1995).

37. Lankard, D. R., 'Slurry infiltrated fibre concrete (SIFCON): properties and applications: potential for very high strength cement based materials', *Proc. Mater. Res. Soc.*, 42, Fall Meeting (1984).

38. Naaman, A. E. and Homrich, J. R., 'Tensile stress-strain properties of SIFCON, *ACI Mater. J.*, May–June, 255–251 (1989).

39. Krstulovic-Opara, N., 'SIMCON – a novel high performance fibre-mat reinforced cement composite for repair, retrofit and new construction', Materials for the New Millennium, in *Proc. 4th Mater. Eng. Conf.*, Washington, DC. K. P. Chong (ed.), Vol. II, pp. 288–97 (1996).

40. Krstulovic-Opara, N. and Malak, S., 'Tensile behaviour of slurry infiltrated mat concrete (SIMCON)', *ACI Mater. J.*, January–February, pp. 39–46 (1997).

41. Bache, H. H., 'Densified cement/ultrafine powder, etc. based materials', CBL Report no. 40, Aalborg Portland, Denmark (1981).

42. Young, J. F., 'Densified cement pastes and mortars', *Cementing the Future*, The Newsletter of ACBM Centre, Northwestern University, Fall (1997).

43. Borglum, B. P., Buchanan, R. C. and Young, J. F., 'Electrical properties of chemically bonded ceramics based on calcium aluminates', *Adv. Cem.-bsd. Mater.*, 1, 47–50 (1993).

44. Richard, P. and Cheyrezy, M., 'Composition of reactive powder concrete', *Cem. Concr. Res.*, 25, 1501–11 (1995).

45. Young, J. F., 'Recent advances in the development of high performance cement-based materials', Materials for the New Millennium, *Proc. 4th Mater. Eng. Conf.*, Washington, DC, K. P. Chong Vol. II, (ed.), pp. 1101–11 (1996).

46. Richard, P. and Cheyrezy, M., 'Reactive powder concretes with high ductility and 200–800 MPa compressive strength', *ACI SP*, 144, 507–18 (1994).

47. Cheyrezy, M., Maret, V. and Frouin, L., 'Microstructural analysis of RPC (Reaction Powder Concrete)', *Cem. Concr. Res.*, 25, 1491–500 (1995).

48. Dallaire, E., Bonneau, O., Lachemi, M. and Aitcin, P. C, 'Mechanical behaviour of confined reactive powder concretes', Materials for the New Millennium, *Proc. 4th Mater. Eng. Conf.*, Washington, DC, K. P. Chong (ed.), Vol. I, pp. 555–63 (1996).

49. Young, J. F. (ed.), *Very High Strength Cement-Based Materials*, Vol. 42, Materials Research Society, Pittsburg (1985).

50. Robertson, R. E., 'Poly (Vinyl Alcohol) as an admixture in cement', *Cementing the Future*, The Newsletter of the ACBM Centre, Northwestern University, Spring (1995).

51. Birchall, J. D., Howard, A. J. and Kendall, K., European Patent no. 0021682 (1981); European Patent no. 0055035 (1982).

52. Roy, D. M., 'Advanced cement systems including CBC, DSP, MDF', *Proc. 9th Int. Cong. Chem. Cem.*, Vol. I, NCB, New Delhi, pp. 357–82 (1992).

53. Russell, P. P., Shunkwiler, J., Berg, M., Young, J. E. and Young, J. F., 'Moisture resistance of macro-defect-free cement', *Ceram. Trans.*, 16, 501–18 (1991).

54. Kataoka, N. and Igarashi, H., 'Expansion properties of macro-defect-free cement in water', *Proc. MRS Int. Meeting on Advanced Materials*, 13, pp. 195–205 (1989).

55. Shah, S. P. and Young, J. F., 'Current research at the NSF Science and Technology Centre for advanced cement-based materials', *Ceram. Bull. Am. Ceram. Soc.*, 69(8), 1319–31 (1990).

56. Popoola, O. O., Kriven, W. M. and Young, J. F., 'Microstructural and microchemical characterisation of calcium aluminate – polymer composite (MDF cement)', *J. Am. Ceram. Soc.*, 74(8), 1928–33 (1991).

57. Lewis, J. A. and Boyer, M. A., 'Effects of an organotitanate cross-linking additive on the processing and properties of macro-defect-free cement', *Adv. Cem.-bsd Mater.*, **2**, 2–7 (1995).

58. Rodrigues, F. A. and Joekes, I., 'Macro-defect-free cements: a new approach', *Cem. Concr. Res.*, **28**(6), 877–85 (1998).

59. Ma, W. P. and Brown, P. W., 'Cement-inorganic polymer composites, microstructure and strength development', *Proc. 9th Cong. Chem. Cem.*, Vol. IV, NCB, New Delhi, pp. 424–9 (1992).

60. Drabik, M., Galikova, L., Kubranova, M. and Slade, R. C. T., 'Studies of model macroscopic defect-free materials: I. Investigations of the system C_4AF-C_4A_3S – HPMC-H by X-ray thermoanalytical and NMR techniques', *J. Mater. Chem.*, **4**, 265–70 (1994).

61. Drabik, M., Frtalova, M., Galikova, L. and Kristofik, M., 'Studies of model macroscopic defect free materials: II. Microstructure and open porosity of the system C_4AF-C_4A_3S – HPMC-H', *J. Mater. Chem.*, **4**, 271–4 (1994).

62. Drabik, M., Zimermann, P. and Slade, R. C. T., 'Chemistry of MDF materials based on sulphoaluminate-ferrite belitic clinkers: syntheses and tests of moisture resistance', *Adv. Cem. Res.*, **10**(3), 129–33 (1983).

63. Uchikawa, H. and Kohno, K., 'Ultra-rapid hardening cement (Jet Cement)', *New Concrete Materials*, pp. 70–136, Survey University Press (1983).

64. Sumitomo Cement Co's brochure on 'Colloid Cement' and Onoda Cement Co's brochure on 'Alofix-MC'.

65. Bensted, J., 'Oil well cement standards – an update', *World Cem.*, **23**(3), 38–44 (1992).

66. Schmidt, M., 'Innovative cements – quick-setting cement, spray cement, ultrafine cement and cements with high resistance to sulphates or acids', *ZKG Int.*, **8**, 444 (1998).

67. Davidorvits, J., 'Geopolymers of the first generation: siliface process', *Proc. 1st Eur. Cong. on Soft Mineralogy* (Geopolymer '88), France, Vol. I, pp. 49–67 (1988).

68. Chatterjee, A. K., 'Availability and use of pozzolanic and cementitious solid wastes in India', in *Concrete Technology for Sustainable Development in the 21st Century*, P. Kumar Mehta (ed.), CMA, India, pp. 556–89 (1999).

69. Materials Research Laboratory, The Pennsylvania State University, Annual Report, p. 31 (1999).

70. Jiang, W. and Roy, D. M., 'Hydrothermal processing of new fly ash cement', *Ceram. Bull. Am. Ceram. Soc.*, **71**(4), 642–7 (1992).

71. Guerrero, A., Goni, S., Macias, A. and Luxan, M. P., 'Mechanical properties, pore size distribution and pore solution of fly ash – belite cement mortars', *Cem. Concr. Res.*, **29**(11), 1753–8 (1999).

72. Khadilkar, S. A., Karandikar, M. V., Ghosh, D. and Chatterjee, A.K., 'Alkali activated fly ash cement under nonhydrothermal conditions – some results on mechanism', *Proc. 2nd Int. Conf. on Alkaline Cements and Concrete*, Kiev, Ukraine (1998).

73. Malhotra, V. M., 'CANMET investigations dealing with high-volume fly ash concrete', *Adv. Concr. Technol.*, Natural Resources, Ottawa, Canada, pp. 445–82 (1994).

74. Malhotra, V. M. and Bilodeau, A., 'High-volume fly ash system – the concrete solution for sustainable development', in *Concrete Technology for Sustainable Development in the 21st* Century, P. Kumar Mehta (ed.), CMA, India, pp. 43–64 (1999).

75. Joshi, R. C., Day, R. L., Langan, B. W. and Ward, M. A., 'Strength and durability of concrete with high proportions of fly ash and other mineral admixture', *Durab. Build. Mater.*, **4**, 253–70 (1987).

76. Tamashige, T. *et al.*, 'A new method of incorporating fly ash from municipal incineration plants into cement production', *Ecocement Programme*, Chichibu Onoda Cement Corporation, Tokyo, Japan (1996).

77. Tager, S. A., 'Use of mineral constituents of fuel at power stations', *Electricheskie Stantsii Journal (in Russian)* 12 (1965).

78. Gudkov, L. V., Kuznetsvov, B. B. and Mikhailov, V. V., Reserves for reducing energy consumption in the cement industry (in Russian), Stroiizdat, Moscow, pp. 69–70 (1971).

79. Global New Energy Inc., Canada, Private Communication (2000).

80. Israel, D. *et al.*, 'Dimensional stability of pore reduced cement (PRC)', *ZKG Int.*, **8**, 460 (1998).

81. Sharp, J.H., Lawrence, C. D. and Yang, R., 'Calcium sulphoaluminate cements, – low energy cement special cement or what ?', *Adv. Cem. Res.*, **11**(1), 3–14 (1999).

82. Zhang, L., Su, M. and Wang, Y., 'Development of the use of sulpho-and ferroaluminate cements in China', *Adv. Cem. Res.*, **11**(1), 15–22 (1999).

83. Janotka, I and Krajci, L., 'An experimental study on the upgrade of sulphoaluminate-belite cement systems by blending with Portland cement', *Adv. Cem. Res.*, **11**(1), 35–42 (1999).

84. Andac, M. and Glasser, F.P., 'Pore solution composition of calcium sulphoaluminate cement', *Adv. Cem. Res.*, **11**(1), 23–6 (1999).

85. Majling, J., Strigac, J. and Roy, D. M., 'Generalized Bogue computations to forecast the mineralogical composition of sulphoaluminate cements based on fly ashes', *Adv. Cem. Res.*, **11**(1), 27–34 (1999).

86. Pai, B. V. B, Bapat, A. D., Dhuri, S. S., Mudbhatkal, G. A. and Chatterjee, A. K., 'Role of a nonalitic cement in practical applications', *7th ASEAN Federation of Cement Manufacturers Technical Symposium on Innovations in all Aspects of the Cement Industry*, Singapore (1986).

87. Braniski, A., 'The properties of siliceous barium cements in dependence upon their clinker constituents', *Zem.-Kalk-Gips*, 31, 91 (1968).

88. Trettin, R., Odler, I. and Gerdes, A., 'The hydration of hydraulic active calcium germanates: Part I – the early hydration of tricalcium germanate', *Adv. Cem. Res.*, 7(27), 117 (1995).

89. Trettin, R., Odler, I. and Gerdes, A., 'The hydration of hydraulic active calcium germanates: Part 2 – the early hydration of dicalcium germanate', *Adv. Cem. Res.*, 7(28), 139 (1995).

90. Diamond, S., 'An alternative to gypsum set regulation for Portland cements', *World Cem. Technol.*, 11, 116 (1980).

91. Skvara, F. *et al.*, 'The cement for use at low temperature', *Proc. 7th Int. Cong. Chem. Cem.*, Vol. 3, p. V57 (1980).

Chapter seven

Developments with oilwell cements

John Bensted

7.1 Introduction

Oilwell cements are utilized primarily in the exploration for and production of oil and gas for securing the metal casings and liners, out of which the oil and/or gas eventually flows. Oilwell cements are also used sometimes to line water wells, waste disposal wells and thermal recovery wells, also to line steel tubular goods like metal piping in some parts of the world and for grouting the bases of production rigs and platforms.

Cements were first used in drilling when in 1903 a cement slurry was used to shut off downhole water just above an oil sand in the Lompoc field in California. Frank Hill of the Union Oil Company is credited with mixing and dumping a slurry consisting of fifty sacks of neat Portland cement. After 28 days, the cement was drilled from the hole and the oil sand was drilled through

to complete the well. The water zone had been effectively isolated. A. A. Perkins of the Perkins Cementing Company developed the origins of the modern well cementing process by employing a two-plug cementing method in 1910. The first plugs acted as wipers for mud on the casing. The second plugs came into play when cement was displaced from the pipe by steam by becoming stopped and thus causing a pressure increase which shut off the steam pump [1].

Construction Portland cements were used for well cementing at this time. Gradually, however, problems arose as wells were drilled deeper with these cements. The American Petroleum Institute (API) set up a Cement Committee in 1937 to study cements and testing procedures for use in well cementing. This led to the development of oilwell cements, based upon Portland cements, but manufactured to a higher level of consistency from one production batch to another, and utilizing different types of tests that were more appropriate for a cement being employed downhole. In the former Soviet Union, drilling or plugging cements, as they have been known, were also introduced, but on a very different basis from that developed under the auspices of the API [2,3].

The API classification scheme for oilwell cements has now been superseded by the so far practically identical ISO classification for international usage (see below).

Oilwell cements have been covered earlier in some detail [4]. The present chapter is designed to complement this information by focusing on important developments that have had considerable impact upon oilwell cements and well cementing in recent times. These have involved cementing in more complex wells than the traditional vertical ones and also the development of new types of cements for use in well-cementing operations.

7.2 ISO Classes and Grades

There are eight ISO Classes and three Grades – ordinary (O), moderate sulphate-resistant (MSR) and high sulphate-resistant (HSR) – of oilwell cements (referred to as well cements) in the ISO classification system [5]:

- Class A is an ordinary Portland cement like ASTM Type I.
- Class B is a sulphate-resistant Portland cement like ASTM Type II.
- Class C is a rapid-hardening Portland cement like ASTM Type III.
- Classes D, E and F are sulphate-resistant Portland cements containing set-modifying additives like retarders. The differences between cements of Classes D, E and F are performance related with Class E cements able to be employed in hotter wells than Class D cements and Class F cements in generally hotter wells than Class E cements. Class E cement has the least retardation and Class F the most retardation. Class E cements (sometimes) and Class F cements (normally always) are protected against high-temperature strength retrogression by intergrinding or blending in 35–40 per cent silica during manufacture.
- Classes G and H cements are the most commonly used oilwell cements around the world. They are sulphate-resistant Portland cements which have more stringent thickening time (setting time under simulated downhole conditions) requirements, so as to minimize batch to batch variability within each manufacturing plant. They are intended for use over a wide range of downhole conditions with a variety of additives to optimize thickening and hardening.

Their differences are performance-related in that Class G cement is tested at 44 per cent water and Class H cement at 38 per cent water in the specification requirements. Class H cement would be relatively more coarsely ground than a Class G cement from the same manufacturing plant.

The three Grades of oilwell cement – O, MSR and HSR – do not apply to all the Classes. MSR and HSR cements have maximum C_3A contents of 8 per cent and 3 per cent respectively. HSR cements, in addition have a maximum $C_4AF + 2 \times C_3A$ content of 24 per cent.

- Class A – only O type.
- Class B – only MSR and HSR types.
- Class C – O, MSR and HSR types.
- Classes D, E and F – only MSR and HSR types.
- Classes G and H – only MSR and HSR types.

Minor additional constituents of the type now permitted in the European standard for common cements used in construction (EN 197-1) are not allowed in the oilwell cements of classes A–H. Also, processing additions like grinding aids are not officially permitted in Class G and H cements, but can be utilized in Class A–F cements.

A complementary standard to this specification, which is a recommended practice for the testing of oilwell cements, is to be available as ISO 10426-2 [6]. Further standards for oilwell cements are in the process of development.

Class J cement (a dicalcium silicate–silica based composition) was withdrawn from the oilwell cement classification scheme some years ago. This cement was developed for use in deep hot well sections [7]. It is currently manufactured in the Far East for use primarily in geothermal well cementing.

Class L cement was proposed from within the API as a compromise replacement for classes G and H cements, so that there would only be one basic oilwell cement. The physical and chemical requirements would have been the same as for G and H, except that all slurry testing would be done at 42 per cent water instead of the 44 per cent for G and the 38 per cent for H. The concept was rejected, because users were broadly satisfied with the existing classifications. They did not want to convert to a new standard water requirement that would have made much information on their databases on G and H cements of less value for the future in making comparisons with Class L cementing data [8].

For successful well cementing, the basic oilwell cements of Classes G and H need to have a high level of reasonable batch to batch consistency in downhole performance. They should demonstrate a good additive response, which is encouraged by their containing a low free-lime content and have acceptable rheological characteristics, as demonstrated by rotational viscometry [9].

Cement free lime contents need to be kept low for ensuring good responses to additives in the well cementing slurries, particularly towards retarders. In practice this means that for cements with MgO contents above 1.5 per cent wt the free-lime content should not exceed 0.5 per cent wt, whilst for cements of MgO content up to 1.5 per cent wt the free-lime should not exceed 1.0 per cent wt. Sometimes higher free-lime contents up to 1.5–2.0 per cent wt can be accommodated without detriment [10]. Simulated performance tests need to be carried out on cement slurries in the laboratory in order to check these limits. MgO is normally accommodated up to *circa* 1.5 per cent wt in solid solution in the Portland clinker phases such as alite C_3S. Above this figure, it is increasingly present as

free magnesia (periclase) MgO. The effect of MgO and its hydration product brucite $Mg(OH)_2$ upon free-lime CaO and the hydroxide $Ca(OH)_2$ is to lower additive response, but less so than for these calcium salts. Hence, there is a need for two recommended limits for free-lime for oilwell cements. It would be difficult to specify a fixed maximum free-lime content for global application.

The standard test procedures used in the specification ISO 10426-1 are identical with those described earlier [4] for API Specification 10A, 22nd Edition, 1995, except for a different free fluid test method, which utilizes an Erlenmeyer flask instead of a measuring cylinder and now has a maximum free fluid of 5.90 per cent.

7.3 Relevant ISO standards

ISO standards relevant to well cementing and cementing equipment that have been or are being developed are listed in Table 7.1.

7.4 Brief comments on the ISO cementing standards

ISO 10409 utilizes oilwell cements, but not in the actual well-cementing environment downhole. This standard is basically concerned with the cementing of pipelines connected with oil and gas

TABLE 7.1 ISO standards for well cementing and cementing equipment

ISO standard	API standard developed from	Subject
10409	API RP 10E	Application of cement lining to steel tubular goods, handling, installation, joining (recommended practice)
10426-1	API Spec 10A	Cements and materials for well cementing – Part 1: specification
10426-2	API RP 10B	Cements and materials for well cementing – Part 2: testing of well cements (recommended practice)
10427-1	API Spec 10D	Bow-spring casing centralizers – Part 1: specification
10427-2	API Spec 10D	Bow-spring casing centralizers – Part 2: Recommended practice
18165	API RP 10F	Recommended performance testing of cementing float equipment

A recommended practice for cements for deepwater cementing is also being developed and will probably be designated ISO 10426-3.

exploration and production, such as those from rigs to collection points. However, ISO 10409 is in need of revision, not least because the recommended cement requirements are out of date. HSR Class C cement of zero C_3A content is described as the most preferable cement to employ for such usage, with MSR and HSR Class B and C cements also being recommended according to the level of soluble sulphates in the corrosive waters to which the cement is exposed [11]. No mention is made of MSR or HSR Class G or H cements in this context, when it is clear that the recommendations effectively predate the time when these latter two basic cements were introduced. Class G or H cements should ordinarily be employed for such usage on a worldwide basis, where the pipelines actually need to be cemented.

ISO 10426-1 is now the internationally recognized standard for oilwell cements to be applied globally and used locally. Nevertheless, in Russia and other CIS states there are still some oilwell cements being produced according to GOST 1581-85 and 1581-96, which, apart from Class G and H clinker mentioned in GOST 1581-96, were developed on a very different basis from the ISO oilwell cements [2,3]. Similarly, in China, some oilwell cements are still made according to the Chinese standard GB 202 [2]. Increasingly, in these countries, the older categories of oilwell cement are gradually being superseded by the ISO Class G (and to some extent the ISO Class H) cement for the mainstream well cementing.

ISO 10426-2 is a companion document to ISO 10426-1 and contains numerous test procedures and simulated well-cementing schedules for laboratory usage.

ISO 10427-1 is the actual specification (mandatory requirements) and 10427-2 the recommended practice (advisory) for bow-spring casing centralizers. Centralizers are mechanical devices employed in running pipe downhole, so that the metal casing is positioned more centrally in the borehole for maximizing cementing efficiency in the annulus [12]. Rigid centralizers, which are constructed with a fixed bow height and are designed to fit a specific casing or borehole size, are not covered by these two standards.

ISO 18165 describes test procedures for evaluating the performance of cementing float equipment. The latter refers to the check valves, which are incorporated into well casings to stop any flow of fluid up the casing, but to permit fluid flow down the casing and into the wellbore annulus. Placement of oilwell cement slurries in the wellbore annulus is thus facilitated by the cementing float equipment.

7.5 Effects of different conditions

Cement hydration tends to change with temperature and pressure increase. As the temperature rises, the belite C_2S and ferrite C_4AF phases tend to become more reactive with respect to the alite C_3S and aluminate C_3A phases. At *circa* 70–90°C a surge in ferrite hydration often gives rise to a threshold of a longer than expected thickening time, due to alite hydration being hindered to some extent under these conditions. However, the reactivities of all the cement phases tend to increase with temperature rise, and the normally expected behaviour begins once more after the end of the threshold period [13].

Above 100°C, the hydration chemistry changes as crystalline hydration products tend to supersede the poorly crystalline/amorphous ones and the pattern of hydration behaviour is radically changed [4] Additions of silica *circa* 35–40 per cent wt. normally need to be made to well-cementing slurries in order to avoid strength retrogression.

Initially, above 100°C, a poorly crystalline C-S-H called C-S-H (II) is produced. If there is no protection against strength retrogression, the C-S-H (II), which has high compressive strength and low permeability, changes into the crystalline material α-dicalcium silicate hydrate (α-C_2SH) that has low compressive strength and high permeability. This, in turn, above *circa* 200°C converts into tricalcium silicate hydrate ($C_6S_2H_3$), also known as jaffeite, which also has low compressive strength and high permeability. Strength retrogression caused by α-C_2SH and $C_6S_2H_3$ is not normally too severe to maintain the casings and/or liners in the annular spaces between the borehole walls and the casings/liners. A compressive strength of 5 MPa should be adequate for such maintenance and even with strength retrogression, the compressive strengths attained should normally be well above this value. However, the concomitant increases in permeability (measured in millidarcies) of the hardened cementitious products from around 0.1 md for C-S-H (II) to *circa* 10–100 md for α-C_2SH and $C_6S_2H_3$ are sufficient to have a deleterious effect on the durability of the cement placed in the annular space. Thus, the presence of significant quantities of these two hardened products is best avoided.

Additions of *circa* 35–40 per cent wt of silica flour or silica sand can prevent strength retrogression by allowing the formation of tobermorite $C_5S_6H_5$, a phase which produces both high compressive strength and low permeability. Above *circa* 150°C, tobermorite gradually changes into xonotlite C_6S_6H and some gyrolite $C_2S_3H_2$. These have lower compressive strength and higher permeability than tobermorite, but these changes are much lower than when α-C_2SH and $C_6S_2H_3$ are formed and, unlike the latter do not cause strength retrogression and durability problems *per se*. Tobermorite can remain up to *circa* 250°C, since equilibrium conditions do not normally arise under real hydrothermal situations. Above *circa* 250°C, truscottite forms from gyrolite and from any residual tobermorite, although some gyrolite may persist above this temperature. Truscottite also has lower compressive strength and higher permeability than tobermorite, but is sufficient to avoid strength retrogression and durability problems in normal circumstances [4].

Class G and Class H cements have an effective upper limit for usage under normal downhole conditions of 400°C. This arises because above 400°C, they become less stable as binders owing to the presence of high shrinkage (5–10 per cent or more) under these conditions. Truscottite and xonotlite, which are the most stable high-temperature binders formed by Class G and Class H cements, also decompose above 400°C. As a consequence of this decomposition coupled with the high shrinkage experienced, a disintegration of the hardened cement takes place [4].

Under hydrothermal conditions, there is very little ferrite or aluminate phase present and no residual calcium hydroxide. The latter is effectively absorbed into the reacting media to form the various crystalline calcium silicate hydrates. The former largely end up in solid solution within the hydrothermal calcium silicate hydrates, but some hydrogarnet and strätlingite are observed in the temperature range 100–200°C [4].

Increases in pressure also accelerate the rate of hydration [14]. These have not been as extensively studied as increases in temperature. Both pressure rise and temperature rise have a strong effect upon the hydration behaviour of oilwell cements downhole, particularly under high temperature – high pressure (HTHP) conditions [4]. Here a cocktail of chemical additives, including a heavyweight additive, a retarder, a strength retrogression inhibitor and a fluid loss controller are usually necessary to ensure that a good well-cementing job is carried out.

7.6 Some other interesting phenomena with Class G and H cements

There are a number of other interesting phenomena exhibited by oilwell cements, including the basic oilwell cements of Classes G and H [15]. Some of these are summarized hereunder.

Enhanced ettringite formation at low temperatures. Relatively larger quantities of ettringite have been produced at 5°C than at 20°C at up to 12 h hydration with an HSR Class G cement. This occurs despite the overall lower intrinsic reactivity of the cement at 5°C. The very slow rate of C-S-H formation appears to provide more space for the aluminate and ferrite phases to react with the gypsum to give rise to ettringite. This produces an apparent enhancement of the rate of formation of ettringite at 5°C than at 20°C [16].

Threshold of longer than expected thickening times at around 70–90°C. When the temperature increases, the thickening (setting) time normally decreases, as would ordinarily be expected. Nevertheless, at around 70–90°C, the thickening time unexpectedly lengthens before changing back to shortening once more as the temperature further increases. This phenomenon can usually be ascribed to a sudden outbreak of ferrite phase hydration, which interferes with the hydration of alite, the prime cause of thickening and the subsequent hardening [13].

'Anomalous thickening' activity with silica flour and silica sand under hydrothermal conditions. Under hydrothermal conditions, both silica flour and the coarser silica sand are employed as strength retrogression inhibitors. Silica flour would ordinarily be expected to react faster because it is more finely divided, and therefore produces more rapid thickening than if silica sand were present under comparable conditions. In reality, both situations are found to occur, with the somewhat surprising observation that the coarser silica sand may thicken quicker than the same cement with flour. The cause of this apparent anomaly seems to be a balancing act between the finer silica flour being intrinsically more reactive than the silica sand and the effects of retardation under the downhole conditions prevailing. Here, the finer flour can be more effectively enveloped by the retarder present in the cement slurries and hence react slower than the same cement with sand under analogous conditions [17,18].

S-curve effects during retardation with certain retarders. Raising the level of retarder lengthens the thickening time in the expected manner at first, then shortens it as if acceleration were taking place, and later lengthens the time in the usual way. This effect is due to higher than expected quantities of free CaO, MgO and/or SrO in the cement, which interferes with the normal retarding function of the retarder. With increased amounts of retarder, this unwanted phenomenon is overcome. S-curve effects have been observed with retarders like lignosulphonates, sugars and phosphonates [19,20].

S-curve effects (not related to the above) *during compressive strength development under hydrothermal conditions.* Sometimes, above 100°C, compressive strength is observed to increase at first with time and then to reduce for a while, before later rising again in the expected way. These changes are caused by differences in the types of calcium silicate hydrate being formed, due to the presence of silica in very different reactive forms like silica fume and silica sand. At first, compressive strength rises as hydrates like C-S-H (II) are formed, then falls when α-dicalcium silicate hydrate α-C_2SH (of low strength) is produced. However, the compressive strength rises again when there is sufficient available silica from the slower reacting sand to form tobermorite $C_5S_6H_5$ and other hydrates (of greater strength), as hydration progresses [21].

7.7 Cementing in hostile environments

7.7.1 Deepwater situations

Increasingly, offshore drilling activity now involves deepwater drilling and cementing. Deepwater is usually regarded as water of depths exceeding 400–500 m (1304–1630 ft), with ultra-deep water being water more than 1500 m (4891 ft) deep. The ISO 10426-2 pressure/temperature schedules are inappropriate for deepwater cementing, because as the cement slurry is pumped during passage to the seabed the temperature falls to around 5°C to −5°C, before rising as it is pumped through the formations under the seabed into the annulus. Specific pressure/temperature schedules for deepwater cementing are inappropriate, because of the considerable differences that exist from one deepwater situation to another. Temperature simulators with temperature models built into the specific cementing programmes are required.

Cement hydration is slowed dramatically at low temperatures. As a result, there are large changes in cement properties like thickening time, rate of hardening, rheology etc. The testing of oil-well cements for simulating seafloor conditions requires refrigerated equipment, such as for the HTHP consistometers, atmospheric consistometers and also ultrasonic cement analysers for continuous monitoring of compressive strength development. A new international standard (possibly ISO 10426-3: Testing of Deepwater Well Cements, Recommended Practice) is being developed. It is important to improve zonal isolation and mitigate annular flows after cementing.

The main problem areas in deepwater cementing in different parts of the world can include low seafloor temperatures, shallow water flow potential, unconsolidated formations, seafloor washout and suppressed geothermal gradients [22–26].

The desired cement properties for encountering shallow water flow in particular normally involve using Class A, C, G or H cements, or special foamed cement compositions. Foamed cements (cements mixed with nitrogen gas and a foamer) set faster than normal lightweight cement compositions and demonstrate good compressive strength and hardening properties. The nitrogen content resists the pressure of the flows by its tendency to expand. Suitable Class G or H cements should be utilized for HTHP well sections. Modified cement compositions, which contain microspheres or other precisely graded particles to give a low density system with low porosity and reduced water content, have been used to give adequate compressive strength at low temperatures – see below.

Cementing is often carried out in two stages, with the thickening time of the lead slurry being *circa* 4–6 h at the bottom hole circulating temperature (BHCT) [4] and that of the tail slurry being 3–5 h at the BHCT. Compressive strength development needs to be rapid, so as to avoid or minimize problems like gas migration. Fluid loss needs to be below 50 ml/30 min ISO at the BHCT and free fluid (free water) should be zero, with the transition time between thickening time and significant hardening preferably not being greater than 30 min. When cementing is carried out below the seabed, zonal isolation is required. When formulating the cement slurries for deepwater wells, both the mixwater temperature and the cement temperature must be accurately known.

7.7.2 Shallow-water situations

Shallow-water situations most commonly involve marshy environments and are often less than 30 m deep. Such environments are found in the United States, West Africa and in the former Soviet Union. There are logistical problems involved in supplying rigs. Wells drilled in such environments normally utilize the common oilwell cements like those of Class G or Class H for the actual well-cementing operations.

The main problem for cementing wells drilled in marshy terrain is that logistics or local environmental considerations may mean that fresh water for cement slurries either cannot be piped in, or a water well drilled into a nearby aquifer cannot be contemplated. Under such circumstances, the water supply must be obtained from the surface water of the marsh or nearby. Here, it is essential that surface water be only taken into use at a certain depth from the bottom. This depth will vary from one particular environment to another and should be such that the water taken has a minimal (ideally zero) content of environmental impurities like mud and plant roots. Such impurities normally tend to retard thickening and early compressive strength development. Cementing formulations using marsh water must be tested in the laboratory before utilization in the field to assess the optimum level of water to achieve satisfactory thickening and early compressive strength development downhole.

7.7.3 Permafrost (Arctic) environments

Normal Class G or Class H cements are unsuitable *per se* for Arctic cementing because of the slowness of their reactions at the low temperatures commonly encountered (down to *circa* −5°C at times). Also, they must not evolve too much heat of hydration, or the permafrost would be damaged. Blends of Class G or Class H cement with gypsum, ferrite phase and sodium chloride (accelerator) have been successfully employed. More commonly utilized is calcium aluminate (high alumina) cement suitably diluted with fly ash or natural pozzolan to lower the heat of hydration (see below). Portland blastfurnace cement compositions have also been employed because of the lower heat of hydration and the ability of the slag to react at low temperatures. Antifreeze agents like alcohols commonly damage the properties of the set cements and should not be employed.

Fuller details about Arctic cementing are available in the technical literature [27–29].

7.8 Cements for some given well types

7.8.1 General points

Straightforward vertical wells are increasingly becoming superseded by other types of well, like slimhole, extended reach, horizontal and multilateral. These wells are drilled for lowering overall drilling costs at given sites, for facilitating logistics and also for environmental reasons. It is often preferable to have a smaller exploration drilling set up in remote locations for dealing with logistics (slimhole), or for environmental reasons when it may be preferable to drill some distance away from the actual oil- and/or gas-fields (extended reach, horizontal and multilateral).

7.8.2 Slimhole wells

Slimhole wells have smaller diameters than conventional wells with correspondingly smaller annular widths. These wells have been variously defined, such as wells in which 90 per cent is drilled with diameters of less than 7 in (17.8 cm), or wells in which 70 per cent is drilled with diameters of less than 5 in (12.7 cm) [30,31].

Good quality Class G or H cements are normally utilized, which have all the particles finer than 1 mm and 99.99 per cent commonly finer than 500 μm, have high batch to batch consistency and low free-lime. The cement slurries must be thin and highly dispersed (low rheology), so as to survive the pressure drops across the narrow annuli encountered. Typical plastic viscosities may be of the order 10–15 cP (0.010–0.015 Pa s) with a yield point close to 5 lb/100 ft^2 (2.4 Pa).

The stability of the cement slurry must not be compromised. Fluid loss should be very low, so as to overcome any lost circulation to the formations and should not normally exceed 50 ml/ 30 min ISO. Details about suitable cement additives are given elsewhere [31].

Microfine cements have sometimes been utilized here. Apart from cost and logistical considerations, their generally greater hydraulic reactivity means that greater retardation must be employed to produce satisfactory thickening. It is vital that there are no coarse particles whatsoever in the cements. These cements have more often been employed in squeeze (repair jobs) to slimhole wells.

7.8.3 Extended reach and horizontal wells

Extended reach wells are those which end up at a highly deviated angle from the vertical axis, culminating sometimes as horizontal wells. 'Horizontal' wells in the oil and gas industries are not always precisely horizontal in the geometrical context. Normally, the term refers to those wells where some of the wellbore is inclined at 90° to the vertical. It can also refer to highly-deviated wells that are almost (but not exactly) horizontally inclined. Highly-deviated extended reach wells and, in particular, horizontal wells permit big increases in production of oil and/or gas, especially where reservoir zones are narrow.

High quality Class G or H oilwell cements of good batch-to-batch consistency and low free-lime are to be preferred for cementing these wells. Requirements for the cements are often similar to those for slimhole wells mentioned above. The cement slurries must show low rheology, zero free fluid and low fluid loss, the latter being below 50 ml/30 min ISO and preferably below 30 ml/ 30 min ISO [12].

7.8.4 Multilateral wells

Multilateral wells are literally wells within wells and are defined as two or more drainholes (branches) drilled from a primary wellbore (trunk), wherein both the trunk and the branches can be horizontal, deviated or vertical [32] (Figure 7.1). There are two classification schemes for multilateral well completions, the TAML (*Technology Advancement Multi-Lateral Group*) and the Longbottom, both of which comprise six main levels at present [33–36] (Table 7.2).

For the cementing of multilateral wells, each well must be considered individually and cemented appropriately. The junctions between the wells, where they need to have a cement seal, should be cemented carefully, so that shrinkage of the hardened cement might be minimized and its impact resistance (shock absorption) maximized. Cementing the junctions of multilateral wells needs separate consideration from cementing the rest of the well sections in the multilateral well complex due

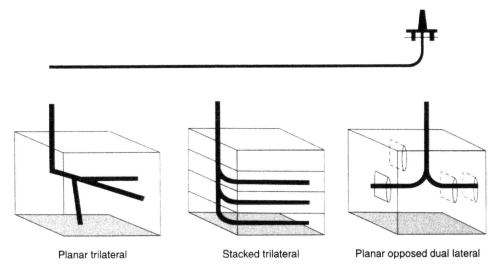

Planar trilateral   Stacked trilateral   Planar opposed dual lateral

FIGURE 7.1   Common types of multilateral wells currently in use (Diagram courtesy of BJ Services).

TABLE 7.2   Classification schemes for multilateral completions

TAML Scheme [33,34]
Level 1	Open unsupported junction
Level 2	Main bore is cased and cemented, but lateral bore is open
Level 3	Main bore is cased and cemented; lateral bore is cased but not cemented
Level 4	Both main bore and lateral bore are cased and cemented
Level 5	Pressure integrity is achieved at the junction, which is not cemented
Level 6S	Downhole splitter with pressure integrity; large main wellbore with two smaller lateral bores; the junction is not cemented
Level 6	Pressure integrity at the junction is achieved with the casing; the junction is not cemented.

Longbottom Scheme [35,36]
Levels 1, 2, 3 and 5 are the same as in the TAML scheme
Level 4 is lined lateral from cased main bore; lateral liner is supported and isolated at junction using cement *but not connected* to main bore casing
Level 4.5 is lined lateral from cased main bore; lateral liner is supported and isolated at junction using cement *and connected* to main bore casing
Level 6 is cased main bore and lateral with direct connection and mechanical seal between casing and liner at junction; *note that* there is no separate Level 6S category, since it comes under Level 6 in this scheme

to their particular requirements [32]. Where well junctions cannot or should not be cemented, then attention should be given to providing a suitable mechanical seal, so as to deal with extraneous phenomena like gas migration.

For the successful cementing of multilateral junctions where cementing is appropriate, it is necessary to design a special cementing composition using a good quality Class G or Class H cement. A latex cement, or even an expansive cement, both of which might also need fibres for increasing the tensile strength, need to be considered for such usage. Fibres (like polypropylene) can reinforce the latex where the need arises and minimizes the shrinkage. As the cement slurry hardens in position, the latex films occupy the pore spaces and hence reduce the propensity for shrinkage and other forms of cracking. This 'scaffolding effect' of the latex, which gives periodic, but intermittent, bonding into the cement matrix, increases the tensile strength attainable. Since such bonding is not regular throughout the hardened cement paste, the latex cannot truly be said to have undergone compound formation with the hardened cement in the normally accepted sense. Therefore, the cementing of the lateral-to-parent wellbore junction permits the cement seal to isolate the laterals from each other and from the parent bore [32].

Each individual well should be dealt with separately for cementing in accordance with the conditions downhole and the well type. Low fluid loss, zero free fluid (in many instances), retardation and adequate defoaming are essential. Suitable laboratory simulation of the particular downhole conditions for such cementing is essential for successful cementing of each multilateral well complex.

7.8.5 High temperature high pressure (HTHP) wells

Cements for cementing HTHP wells usually need to be high quality Class G or H cements of relatively low intrinsic reactivity containing around 35–40 per cent wt of silica flour or silica sand to prevent strength retrogression with time. The

chemistry of well cementing under hydrothermal conditions is very different from that below 100°C and will require the use of appropriate additives [4,37] to suit the downhole conditions pertaining thereto.

7.8.6 Low temperature high pressure (LTHP) wells

Very little quantitative data of the effect of pressure upon thickening and hardening has been reported in the technical literature [14]. The effects of pressure are also substantial like those of temperature. LTHP wells are becoming more commonly encountered now. Note that the term *low temperature* here is merely relative. Since the downhole temperatures are below *circa* 100°C, there is often a reluctance to use silica flour or silica sand here to prevent any strength retrogression. After all, there is very little information of phase changes of the hydrates under LTHP conditions and how they are influenced by pressure as well as by temperature. It is clear that adequate laboratory simulations of the likely downhole conditions must be carried out in the laboratory prior to formulating suitable well-cementing slurries. Compressive strength data under appropriate downhole conditions at 1 and 7 d should be undertaken. If there should be any strength retrogression observed, then it would be desirable to include *circa* 35–40 per cent wt of silica flour or silica sand so as to maximize the formation of a durable hardened cement matrix in the downhole annulus.

7.9 Alternatives to standard Class G and Class H cements

7.9.1 General points

Increasingly, cements other than the standard oil-well cements like those of Class G and Class H are being utilized for particular well-cementing jobs in order to facilitate completions more easily or more appropriately for the specific well conditions. A number of alternative cements have been briefly reviewed before [1,28,38]. Various alternative cementing systems currently in use are surveyed in this section. Here the superscript™ indicates that the name of the particular cement being discussed is actually a proprietary trade mark.

7.9.2 Calcium aluminate (high-alumina) cements

Calcium aluminate cements have a variety of applications in well cementing [29]. These include cementing wells in permafrost (Arctic) environments and the grouting of oil platform bases where the temperatures are low. They can also be utilized for lining waste-disposal wells, because of their high resistance to general, chemical and acid attack, as well as to any considerable temperature fluctuations that might arise. These cements cannot be employed for alkaline wastes that have not been neutralized prior to disposal, because of the inherently low resistance to alkaline attack that calcium aluminate cements display.

Calcium aluminate cements can be used as thermal cements and in fireflood and thermal recovery wells, because of the high resistance that these cements exhibit to the high temperature fluctuations shown in such wells. This resistance is primarily demonstrated by the absence of calcium hydroxide and the ranges of different products formed by Portland-based oilwell cements. Calcium aluminate cements give a minimum strength at 900–1000°C, which is more than adequate for maintaining the integrity of the well. They can also be employed in hot dry rock (HDR) geothermal energy recovery and for more general utilization in geothermal wells. A noteworthy point is that these cements can be utilized as a base for specially prepared calcium phosphate cements [29]. An interesting example is the proprietary Halliburton *Thermalock Cement*™, which resists carbon dioxide induced corrosion, protects pipe and casing, develops high strength and weight retention and is acid resistant. It has been employed in practical situations within the temperature range 60–370°C. Clearly calcium aluminate cements have quite a wide range of practical application in well-cementing activities.

The basic chemistry of hydration has been summarized for these different practical situations [29,39,40]. Both dark grey and white calcium aluminate cements have been used in well cementing, where they have their particular niche markets. They are not suitable, primarily for economic reasons, for applications in mainstream well cementing activities.

7.9.3 Modified Sorel cements

The principal modified Sorel cements utilized in well cementing are *Magne Plus*™ of BJ Services and *Magne Set*™ of Baker Hughes. These products are basically mixtures of magnesium and calcium oxides, carbonates and sulphates, which react with water to produce complex polyhydrates that are unaffected by normal cement contaminants [41]. They react with seawater and chloride brines to function as a modified *in situ* magnesium oxychloride (Sorel) cement to form complex carbonate chloride hydrates [4], which are very stable to downhole water. The thickening behaviour can be modified as appropriate by suitable additives.

These additives can include retarders (borax), accelerators (calcium chloride), dispersants (SMFC, SNFC), typical weighting agents (barite, haematite), lightweight additives (microspheres), fluid-loss controllers (polymeric systems) etc. However, the quantity of additive may vary appreciably from what is used with Class G or H cement slurries. For example, up to 100°C, 4–12 per cent wt borax may be required to give a suitable thickening time of 3–4 h, as compared with about 0.2–0.3 per cent by weight of cement for a conventional class G or H oilwell cement slurry [37].

Magne Plus and *Magne Set* cements can be employed up to at least 120°C. They can be used for correcting lost circulation problems, as a diverter, to provide temporary zonal isolation, or as a kick-off plug in weak formations, and can be pumped similarly to the conventional oilwell cements.

7.9.4 Microfine cements

Microfine cements are of low particle sizes and high surface areas within the range *circa* 500–1000 m²/kg or more. The most commonly used microfine cements for use in well cementing cements are very fine Portland or sulphate-resisting Portland cements [42,43]. They may need to include additional gypsum and/or a suitable retarder in order to obtain smooth thickening (setting) downhole. Any retardation should not undesirably inhibit the growth in compressive strength. Such cements must not contain any coarse particles for effective utilization.

Microfine cements are employed to repair 'tight' casing leaks by squeeze cementing. The fine particles have a greater penetration than conventional oilwell cements and can thus enter the leak unrestricted [44]. These cements have been described as needing an average particle size of 6 microns and a maximum particle size of 15 microns [45].

7.9.5 Foamed cement

Foamed cementing can produce very lightweight slurries for cementing through unconsolidated formations or other weak zones that will not tolerate conventional water-based cement slurries. Class C, G or H cement is normally preferred for the base cement, containing a foaming agent and other additives to stabilize the foam that forms, and is contained in a dispersion with nitrogen gas to lower the slurry density to the desired level, usually below s.g. 1.32 (11.0 lb/US gallon). The basic details of foamed cementing are described elsewhere [25,46]. A new international standard (possibly ISO 10426-4: Foam Cementing, Recommended Practice) is to be produced.

Foamed cements tend to thicken and harden quicker than other lightweight slurries, whilst the nitrogen gas has a tendency to expand, which can resist pressure from any flows within the formation. Sometimes with high-temperature wells there is a need to cement through shallow weak formations in the production zones and foamed cement is usually the best option available. The hydration chemistry produces the normally expected hydrates for the given well conditions.

7.9.6 Novel rubber cements

For the possible application of alternative plugging materials to conventional Portland cement-based downhole slurry formulations, two component systems of room temperature vulcanizing (RTV) silicone rubbers and fluorosilicone rubbers appear to offer potential. RTV silicone rubbers (and gels) are two component inorganic polymers. Silicone polymers merge the beneficial aspects of inorganic silicones with the qualities of organic materials. The structure of silicone rubber is based upon a long chain of alternating silicon and oxygen atoms.

Most silicone sealants are solventless and show no shrinkage upon setting. Their thermal expansion is greater than that of steel and they are temperature resistant over a wide range, some up to 300°C or more. They are also resistant to a wide range of chemicals. Both their lack of shrinkage from expansion and their strong seal militates against the likelihood of gas migration during thickening. This is unlike the situation with conventional oilwell cements like class G and H, where shrinkage upon thickening encourages gas migration and the concomitant cement channelling.

A two-component product based upon the reaction of two low viscous, environmentally-friendly base chemicals, a hydride functional cross-linking agent and a vinylsilicone polymer, offers a suitable formulation. Upon reaction, a silicone rubber or gel is formed, which does not require an external agent like water or moist air to react, nor does it generate by-products [47]. RTV silicone rubbers and gels are technically superior to alternative available thermosetting resins in applications such as well repairs or abandonments. Their thickening can be retarded or accelerated as appropriate and their rheological characteristics are suitable for coiled tubing applications. For very demanding situations, the related RTV fluorosilicones can be considered for use [47].

Well cement compositions containing 30–100 per cent BWOC (by weight of cement) rubber particles with 40–60 mesh grain sizes can give a low-density slurry with low cement permeability. These cementing compositions can be reinforced by adding cast amorphous metal fibres (1–25 per cent BWOC) of *circa* 5–15 mm length. The cements can be utilized for cementing zones subject to extreme dynamic stresses, such as perforation zones and the junctions of branches in a multi-sidetrack well [48].

7.9.7 BJ *Liquid Stone*

Liquid Stone™ is a proprietary storable oilwell cement slurry that can facilitate density control at the wellsite, making it suitable for coiled tubing squeezes, abandonment plugs and slimhole applications. Retarders and conditioning agents allow *Liquid Stone* to be stored indefinitely. Just before pumping the *Liquid Stone* cementing slurry, a liquid activator is injected to reactivate chemical hydration, thus permitting the cement to set after a predetermined amount of time [49,50]. Oilwell cements of Class C, G and H, Portland blastfurnace cement and blastfurnace slag alone have been utilized with *Liquid Stone*.

7.9.8 Schlumberger *CemCRETE* technology

CemCRETE™ slurries have more solids and less liquid than conventional slurries, so as to increase compressive strength and to reduce porosity and permeability. This allows optimization of zonal isolation and reduction in waiting-on-cement time (to complete the well-cementing job), so that further drilling can resume sooner. Such technology has been adapted from that employed in concrete building, by attending to the particle size distributions of the solids in their slurries. Particles of a number of specially selected size ranges are chosen, so as to permit them to pack together more closely. In this way, slurry gradients as high as 28 kPa/m can be achieved [51].

CemCRETE cements all have engineered particle size distributions to increase the solids content per unit volume above that of standard Portland-type oilwell cement slurries like those of Class G and H cements. This gives a higher packing volume fraction (PVF) between the cement particles independently of slurry density, which reduces porosity and gives better set-cement properties. Since the remaining fluid content is utilized more efficiently, lower concentrations of most chemical additives can normally be used [51].

Examples of the *CemCRETE* range are *Lite-CRETE*™ slurries at 1.20–1.56 s.g. (10–13 lb/US gallon), *SqueezeCRETE*™ slurries for penetrating narrow gaps, *DensCRETE*™ slurries of low fluid loss and zero free water (free fluid) at above 2.52 s.g. (21 lb/US gallon) and *DeepCRETE*™ – an alternative to foamed cements for deepwater wells.

7.9.9 Mud-to-cement conversion

Ground granulated blastfurnace slag (ggbs) with a suitable activator like sodium hydroxide NaOH or calcium hydroxide $Ca(OH)_2$ can be added to water-based drilling fluids (drilling muds) to convert the latter into a hardened cement product [52]. The main hydration product at ordinary temperatures is an amorphous/poorly crystalline calcium silicate hydrate C-S-H (containing alumina in solid solution with silica) of lower CaO/SiO_2 ratio than the corresponding product from Portland cement hydration, which can be more brittle than the latter. (The use of Portland blastfurnace cement can produce a much less brittle product than that derived from ggbs alone.) Laboratory studies need to predetermine the type and dosage of activator. These slag-containing slurries can be suitably designed with drilling fluid type rheologies, which can show high compatibility with those of water-based drilling fluids. Comparative bond results of slag cementing versus conventional cementing have given encouraging results [53].

Provided that there are no problems with hydrogen sulphide emissions from the slags at more elevated temperatures [54] or logistical constraints with the particular cementing jobs, such slag-containing slurries, appropriately designed for the specific downhole conditions, can be utilized at elevated temperatures. Ggbs for cementing should conform to appropriate national standards, like BS 6699 in the UK for example. Here, for use with cement, the slag shall contain at least two-thirds by mass of glassy slag, and shall consist of at least two-thirds by mass of the sum of $CaO + MgO + SiO_2$, with the ratio $(CaO + MgO)/ SiO_2$ exceeding 1.0.

Pulverized fuel ash (fly ash; pfa) can also be activated like ggbs by alkaline materials such as sodium silicate, potassium silicate, waterglass, sodium hydroxide and potassium hydroxide [55]. Compressive strengths of 60 MPa were given when the fly ash was cured at 85°C for only 5 h. A cementitious reaction takes place giving a similar type of amorphous aluminosilicate gel to that given by alkaline-activated ggbs or alkaline-activated metakaolin [56,57].

Alkaline-activated ggbs has its own niche area in well cementing, as also should alkaline-activated pfa and metakaolin. However, logistical constraints, in particular, are unlikely to see these types of cementitious systems replacing the use of standard oilwell cements like Class G and

FIGURE 7.2 Offshore oil production platform Brent Charlie in the North Sea. (Photo by courtesy of Shell Expro, Aberdeen, U.K.)

Class H for the majority of well-cementing jobs worldwide.

7.9.10 Expansive cements

One of the major problems in well cementing, particularly for critical wells, is gas migration, which can arise when setting and hardening oil-well cements undergo shrinkage over a prolonged time period. The document API10TR2 [58] is a useful summation of background information to develop standard test procedures for measuring expansion/shrinkage, to investigate possible impacts upon invasion of well-bore fluids like gas and water, and to standardize definitions of terms for cement hydration. This report is regarded as a good basis to develop an ISO recommended practice standard (possibly ISO 10426-5) of value to the oil industry. Such a standard would be of particular value to cementing service companies and operators, on shrinkage/expansion for combatting gas migration in the downhole annuli for a wide range of cementitious compositions.

7.10 Conclusion

This chapter has focused upon developments connected with oilwell cements in recent years. Figure 7.2 shows a North Sea production platform, which services oil extraction under difficult conditions. Included in the chapter are international standardization for oilwell cements and well-cementing equipment, different phenomena that can arise with oilwell cements, cementing in hostile environments like deepwater, shallow water and Arctic conditions, different types of wells that can make well cementing more exacting and, finally, some alternative cements to the conventional Class G and H types. As the globalization of exploration and production for oil and gas has increased, this has presented major challenges, both mechanical and scientific, for the employment of both machinery and materials (like oilwell cements). These challenges are being addressed and are enabling the industry to produce more effectively and efficiently than ever before.

7.11 Acknowledgements

The author wishes to thank Martin G.R. Bosma of Shell International Exploration and Production B.V., Rijswijk, Netherlands, Dr Ashley P. Hibbert of BP Exploration Operating Co. Ltd., Sunbury-on-Thames, UK, and Josephine R. Smith of The Hannington Group, Slough, UK, for helpful discussion.

7.12 References

1. Smith, D. K., *Cementing*, Revised edn, Society of Petroleum Engineers of AIME, Richardson, TX (1987).
2. Bensted, J., *World Cem.*, 23(3), 38 (1992).
3. Bensted, J., *World Cem.*, 24(7), 39 (1993).
4. Bensted, J., in *Lea's Chemistry of Cement and Concrete*, P. C. Hewlett (ed.), 4th edn, p. 779, Arnold Publishers, London (1998).
5. International Organisation for Standardisation: Cements and Materials for Well Cementing, Part 1, Specification, ISO 10426-1, ISO, Geneva (2000).
6. International Organisation for Standardisation: Cements and Materials for Well Cementing, Part 2, Testing of Well Cements, Recommended Practice, ISO 10426-2, ISO Geneva (due 2002).
7. Maravilla, S., *J. Petrol. Tech.*, 26, 1087 (1974).
8. Bensted, J., *World Cem.*, 27(11), 70 (1996).
9. Bensted, J., *World Cem.*, 23(4), 44 (1992).
10. Bensted, J., API Task Group on Eastern Hemisphere Cementing, Wiesbaden, pp. 6 (1995).
11. International Organisation for Standardisation: Application of Cement Lining to Steel Tubular Goods, Handling, Application, Joining – Recommended Practice, ISO 10409, ISO, Geneva (1993).
12. Bensted, J., *World Cem.*, 27(5), 76 (1996).
13. Bensted, J., *Offshore Eur. 91 Proc.*, 1, 379 (1991).
14. Farris, S. R., Ph.D. Thesis, Keele University, UK (2000).
15. Bensted, J., in *Advances in Cement Chemistry*, W. Kurdowski (ed.), p. 105, Gorniczo-Hutnicza, Krakow (1997).
16. Lota, J. S., Bensted, J., Munn, J. and Pratt, P. L., *L'Ind. Ital. Cem.*, No. 725, 776 (1997).
17. Bensted, J., *Chem. Ind. (London)*, No. 18, 702 (1992).
18. Bensted, J., API Task Group on Eastern Hemisphere Cementing, Amsterdam, pp. 6 (1995).
19. Bensted, J., *Proc. 15th Int. Conf. Cem. Micr.*, Dallas, Texas, March 29–April 1, p. 51, International Cement Microscopy Association, Duncanville, TX (1993).
20. Arens, K. H. and Akstinat, M., *Bundesministerium für Forschung und Technologie – Forschungsbericht T82-112. Verbesserung der*

Ringraumzementation in tiefen und übertiefen Bohrungen – Teil 2: Tiefbohrzemente und Additive, Fachinformationszentrum Energie-Physik-Mathematik GmbH, Karlsruhe (1982).

21. Bensted, J., *Cem. Concr. Res.*, 25(2), 240 (1995).
22. Mueller, D. Lecture to ISO Standardisation Meeting on Drilling and Completion Fluids and Well Cements, Stavanger (1998).
23. Calvert, D. G. and Griffin, T. J., Jr., *IADC/SPE Drilling Conf.*, Dallas, TX, March 3–6, IADC/SPE 39315 (1998).
24. Hart's Deepwater Technology, *A Supplement to Petroleum Engineer International*, pp. 56, Hart Publications Inc., Houston, TX (1999).
25. Wetuski, J., *Oil Gas World*, No. 1, 25 (1999).
26. Ravi, K., Biezen, F. N., Lightford, S. C., Hibbert, A. P. and Greaves, C., *SPE Ann. Tech. Conf. Exhibition*, Houston, TX, October 3–6, SPE 56534 (1999).
27. Goodman, M. A., *World Oil's Handbook of Arctic Well Completions*, Gulf Publishing Company Inc., Houston, TX (1978).
28. Nelson, E. B., *Well Cementing*, Schlumberger Educational Services, Houston, TX (1990).
29. Bensted, J., *L'Ind. Ital. Cem.*, No. 740, 150 (1999).
30. Anon., *Oil Gas J.*, 90(47), 77 (1992).
31. Bensted, J., *World Cem.*, 25(8), 45 (1994).
32. Bensted, J., *World Cem.*, 28(11), 70 (1997).
33. Hogg, W. C. and MacKenzie, A., *SPE European Petroleum Conf.*, The Hague, November 20–22 SPE 50664 (1998).
34. Hogg C., *World Oil*, No. 5, 49 (1999).
35. Russell, R. and Longbottom, J., *SPE 6th Int. Oil and Gas Conf. Exhibition*, Beijing, November 2–6, SPE 48849 (1998).
36. Longbottom, J., *Offshore Technol. Conf.*, Houston, Texas, 4–7 May 1997, OTC 8800 (1998).
37. Bensted, J., in *Concrete Admixtures Handbook – Properties, Science and Technology*, V.S. Ramachandran (ed.), 2nd edn, p. 1077, Noyes Publications Inc., Park Ridge, NJ (1995).
38. Bensted, J., *R. Soc. Chem. Spec. Publ.*, 67, 36 (1988).
39. Bensted, J., *Zem.-Kalk-Gips*, 46(9), 560 (1993).
40. Bensted, J., *World Cem.*, 21(10), 452 (1990).
41. Sweatman, R. E. and Scoggins, W. C., *Ann. Tech. Conf. Exhibition*, Houston, Texas, October 2–5, SPE 18031 (1988).
42. Bensted, J., *World Cem.* 23(11), 40 (1992).
43. Bensted, J., *World Cem.* 23(12), 45 (1992).
44. Meek, J. W. and Harris, K., *III Seminário de Cementación de Pozos*, Tomo III, 3. Técnicas de Cementación, Papel 3.2, pp. 16, Intevep S. A., Caracas (1992).
45. MacEachern, D. and Young, S. C., *Oil Gas J.*, September 7, 49, (1992).
46. Miller, I. S. and Frank, W. E., *SPE Western Regional Meeting*, Bakersfield, California, May 10–13, SPE 46215 (1998).
47. Bosma, M. G. R., Cornelissen, E. K., Reijrink, P. M. T., Mulder, G. S. and de Wit, A., *IADC/SPE Drilling Conf.*, Dallas, Texas, March 3–6, IADC/SPE 39346 (1998).
48. LeRoy-Delage, S., Dargaud, B., Baret, J. -F. and Thiercelin, M., International Patent WO/20350, April 13 (2000).
49. Rae, P. and Johnston, N., US Patent 5,447,197 (1995).
50. Rae, P. and Johnston, N., US Patent 5,547,506 (1996).
51. Boisnault, J. M., Guillot, D., Bourahia, A. *et al.*, *Oilfield Rev.*, 11(1), 16 (1999).
52. Cowan, K. M., Hale, A. H. and Nahm, J. J., *Ann. Tech. Conf. Exhibition Soc. of Pet. Engi.*, Washington DC, October 4–7, SPE 24575 (1992).
53. Silva, M. G. P., Miranda, C. R., D'Almeida, A. R., Campos, G. and Bezerra, M. T. A., *5th Latin Am. Caribbean Pet. Eng. Conf. Exhibition*, Rio de Janeiro, August 30–September 3, SPE 39005 (1997).
54. Bensted, J., *World Cem.*, 27(1), 57 (1996).
55. Palomo, A., Grutzeck, M. W. and Blanco, M. T., *Cem. Concr. Res.*, 29, 1323 (1999).
56. Davidovits, J., US Patent 4,349,386 (1982).
57. Davidovits, J., US Patent 4,509,985 (1985).
58. American Petroleum Institute, Shrinkage and Expansion in Oilwell Cements, API Technical Report 10TR2, 1st edn, API, Washington DC (1997).

Chapter eight

Gypsum in cements

John Bensted

8.1 Introduction

Candlot in Paris during experiments emanating in 1890 discovered that gypsum effectively retarded the hydration of Portland cement and that tricalcium aluminate reacted with gypsum to produce ettringite [1,2]. After 1890, gypsum increasingly started to become ground in with Portland cement to improve the set regulation and the development of compressive strength. However, clinker–gypsum grinding was not universally adopted in cement works until around 1930, when the presence of gypsum effectively became part of the basic definition of Portland cement in national standards around the world.

When Portland clinker is ground on its own and then mixed with water, the aluminate phase C_3A tends to react rapidly, the alite C_3S progressively, with the other phases ferrite C_4AF and belite C_2S contributing too [3]. It is commonly said that when this happens there is a marked increase in temperature and irreversible stiffening, which is quickly followed by setting, which corresponds to what is commonly referred to as 'quick set' or 'flash set'.

It has also been said that in order to prevent this and control the reactions leading to setting, gypsum is ground into the Portland clinker to make what we know as Portland cement today [3,4]. If this were correct in all instances, then why did gypsum take so long to become universally accepted as a component of Portland cement? Again, how was it that many good quality concrete structures were made in the late nineteenth and early twentieth centuries without gypsum in the cement, when workable mixes would have been needed? This question will be answered later, when the grinding of Portland clinker is discussed.

8.2 The calcium sulphate–water system

Gypsum is the dihydrate of calcium sulphate $CaSO_4 \cdot 2H_2O$, which has a monoclinic prismatic structure [5], and occurs widely as a naturally-occurring mineral. When subjected to heat, gypsum dehydrates to form the calcium sulphate hemihydrate $CaSO_4 \cdot 0.5H_2O$ (also known as plaster of Paris), which has a basic hexagonal structure and can occur as the rare mineral bassanite in areas mostly of volcanic activity. Complete dehydration produces 'insoluble anhydrite' $CaSO_4$, which can also exist in a mineral form (anhydrite) with a rhombic dipyramidal structure, often in association with gypsum deposits.

The calcium sulphate–water system was characterized in some extensive studies [6–12], in order to throw more light on the behaviour of gypsum and its derivatives in cementitious activity. At one time, various different calcium sulphate hydrates were predicted and the overall situation was fairly chaotic. It was therefore necessary to take a fresh look at the system in order to elucidate exactly what derivatives of gypsum actually do exist, and how they influence cement hydration and performance in the field.

Gypsum partially dehydrates upon heating above *circa* 97°C (sometimes at lower temperatures when the relative humidity is much less) to give at first the hemihydrate:

$$CaSO_4 \cdot 2H_2O \rightarrow CaSO_4 \cdot 0.5H_2O + 1.5H_2O$$

This form of the hemihydrate, when produced under dry calcined conditions is often referred to as β-hemihydrate. α-hemihydrate is made in a more moist atmosphere under saturated steam conditions. Structurally they are similar [6–9], with the α-form being better crystallized with larger grains. Fine grinding of the coarser α-hemihydrate converts it into the β-hemihydrate [7–9]. Although the formula for hemihydrate is depicted for convenience as containing $0.5H_2O$, some water can be adsorbed (more so for the β-form) giving a maximum of *circa* $0.67H_2O$. The α- and β-hemihydrate forms can readily be distinguished by differential thermal analysis (DTA) [9]. Hemihydrate can only be obtained pseudomorphically within the calcium sulphate–water system by dehydration of gypsum. It does not form as an intermediate when insoluble anhydrite (see below) reacts with water to produce gypsum.

Heating calcium sulphate hemihydrate gives the ubiquitous phase known as 'soluble anhydrite', dehydrated hemihydrate or γ-CaSO₄. Despite its name, soluble anhydrite is neither very soluble (about 2–3 times that of gypsum) nor truly anhydrous [10,11]. It is basically a quasi-zeolitic variant of the hemihydrate, which contains less water, being of the form $CaSO_4 0.001–0.5H_2O$. It forms gradually above *circa* 120°C and finally converts to the 'insoluble anhydrite' $CaSO_4$ above 200–300°C. The α-soluble anhydrite barely exists,

and then only transiently upon further dehydration of α-hemihydrate, unlike the β-soluble anhydrite formed from β-hemihydrate. These differences are clearly shown by DTA [9]. Soluble anhydrite quickly reverts to hemihydrate in moist air.

'Insoluble anhydrite', also called β-CaSO₄ is anhydrous but actually more soluble than gypsum, although having a slower rate of solution than the former. Infrared spectroscopic studies [8] of calcium sulphate heated to high temperatures revealed some slight spectral differences after heating to 400–600°C compared with the situation that arises upon heating to 700–1000°C. The product formed at 600°C and above reacts more slowly with water than that formed at 400°C. There are very slight differences in spectral features between these two varieties of insoluble anhydrite, which suggests that there may be slight structural differences between them of a second order (displacive transition) nature [8]. The more reactive form formed at 400°C appears to be metastable and gradually changes with time into the less reactive form. Anhydrite converts directly to gypsum when reacting with water:

$$CaSO_4 + 2H_2O \rightarrow CaSO_4 \cdot 2H_2O$$

This reaction does not pass through the hemihydrate/soluble anhydrite stage.

The mineral anhydrite also appears to be of the less reactive form, which is also known as Keene's cement or Keene's plaster. Keene's cement normally also contains 0.5–1.0 per cent of potash alum $K_2Al(SO_4)_2 12H_2O$ or potassium sulphate K_2SO_4 to increase its reactivity, but mixed accelerators like ferrous sulphate $FeSO_4$ (or zinc sulphate $ZnSO_4$) with potassium sulphate K_2SO_4 are also occasionally used; lime also acts as a promoter, but is generally less reactive [4].

Above *circa* 900°C calcium sulphate undergoes very slow dissociation, which increases as the temperature is raised.

High-temperature anhydrite or α-CaSO₄ only exists above 1200°C [12,13]. It is normally obtained as a monoclinic phase (observed with a high temperature microscope) at 1215–1220°C, which dissociates fully at *circa* 1450°C into calcium oxide and sulphur trioxide:

$$CaSO_4 \rightarrow CaO + SO_3$$

High-temperature anhydrite is never totally pure over its range of stability as slight dissociation is always found. It reverts to insoluble anhydrite below 1205°C and is not known to be stabilized below this temperature [12]. α-$CaSO_4$ has not to date had any practical applications over its range of stability.

A traditional slow-setting plaster called *Estrich Gips* was developed in Germany by calcining gypsum or anhydrite at 1100–1200°C and cooling. The β-$CaSO_4$ obtained contains some lime obtained by the high-temperature dissociation during calcining, which acts as an accelerative activator.

More information about the different forms of calcium sulphate obtained upon heating and the various plastering materials that can arise are available in various texts [4,14,15].

Some noteworthy points about the different forms of gypsum and its derivatives in cement are as follows:

- Both natural and chemical by-product gypsums are ground in with Portland clinker to produce Portland cements.
- During clinker-gypsum grinding much (sometimes all) the gypsum is dehydrated to the hemihydrate or even soluble anhydrite. This can at times lead to false set (see below). If the grinding mill temperature does not exceed 115°C, then some residual dihydrate gypsum is usually present. Hemihydrate so formed is in the finer β- form.
- α-hemihydrate is not ordinarily encountered in Portland cements on account of its coarseness. It is on rare occasions blended into the retarded oilwell cements of Classes D, E and F to optimize the early hydration and thickening time (see Chapter 7 on Developments with oilwell cements).
- Natural anhydrite is found in some gypsums and added to others at the clinker-gypsum grinding stage in order to prevent/minimize the occurrence of false set in the resulting cement.

- Natural anhydrite is not normally ground in with Portland clinker alone without some gypsum being present on account of its higher grindability and its slower rate of solution, which might sometimes promote some initial flash setting.

8.3 The role of gypsum in set regulation of Portland cement-based systems

Gypsum reacts with the aluminate phase C_3A primarily and also with the ferrite phase C_4AF ($C_2A_{0.5}F_{0.5}$) to form the calcium trisulphoaluminate hydrate known as ettringite $C_3(A,F)$. $3CaSO_4.32H_2O$ or $Ca_6[Al,Fe(OH)_6]_2(SO_4)_326H_2O$, which is a hexagonal prism type of phase. These reactions can be represented simply as:

From aluminate phase:

$$Ca_3Al_2O_6 + 3\{CaSO_4 \cdot 2H_2O\} + 26H_2O$$
$$\rightarrow Ca_6[Al(OH)_6]_2(SO_4)_326H_2O$$

From ferrite phase:

$$Ca_2AlFeO_5 + Ca(OH)_2 + 3\{CaSO_4.2H_2O\} + 25H_2O$$
$$\rightarrow Ca_6[Al_{0.5}Fe_{0.5}(OH)_6]_2(SO_4)_3 \cdot 26H_2O$$

Aluminates react faster with gypsum than ferrites in ordinary Portland cements. Details of their hydration chemistry have been given in greater detail elsewhere. Owing to the effects of solid solution and the presence of impurity ions in the clinker phases, and as a consequence of the hydrated phases as well, ettringite formed from the aluminate phase and ettringite formed from the ferrite phase are in reality indistinguishable. They form a continuous hydrated phase [16].

After some 8–16h or more, as the gypsum becomes increasingly used up and hydrocalumite forms in significant amounts, much of the ettringite normally converts in the presence of hydrocalumite, that can most appropriately be represented as $C_4(A,F)H_{13}$, to the hexagonal plate type phase known for short as 'monosulphate' $C_3(A,F) \cdot CaSO_4 \cdot 12H_2O$, or constitutionally as $Ca_4[Al,Fe(OH)_6]_2SO_4 \cdot 6H_2O$, which is a calcium monosulphoaluminate hydrate. Hydrocalumite $C_4(A,F)$

H_{13}, which can also be represented as $C_3(A,F)$. $Ca(OH)_2 \cdot 6H_2O$, is also a hexagonal plate-type phase like monosulphate, with which it enters into solid solution. Hydrocalumite is stabilized by the high lime media of hydrating Portland cement. This phase can take in water on a quasi-zeolitic basis, that is, water can enter and leave the structure without significantly affecting the basic crystal lattice. Compositions up to $C_4(A,F)H_{19}$ of this quasi-zeolitic variant are known and perform similarly in practice to structures with less water like $C_4(A,F)H_{13}$.

Formation of monosulphate from ettringite can be effectively represented by the following equations:

From aluminate:

$$Ca_6[Al(OH)_6]_2(SO_4)_3 26H_2O$$
$$+ 4\{Ca_2Al(OH)_7 \cdot 3H_2O\}$$
$$\rightarrow 3\{Ca_4[Al(OH)_6]_2SO_4 6H_2O\}$$
$$+ 2Ca(OH)_2 + 20H_2O$$

From ferrite:

$$Ca_6[Al_{0.5}Fe_{0.5}(OH)_6]_2(SO_4)_3 26H_2O$$
$$+ 4\{Ca_2Al_{0.5}Fe_{0.5}(OH)_7 \cdot 3H_2O\}$$
$$\rightarrow 3\{Ca_4[Al_{0.5}Fe_{0.5}(OH)_6]_2SO_4 6H_2O\}$$
$$+ 2Ca(OH)_2 + 20H_2O$$

This reaction is normally incomplete and significant quantities of ettringite formed during the earlier part of the hydration usually remain in the hardened cement product. Also, only minority quantities of C_3A actually react to form ettringite. For eight OPCs hydrated to the times of initial and final set, no more than *circa* 30 per cent at most of the sulphate had combined to form ettringite [17]. Most of the sulphate (~70–80 per cent) actually ends up being incorporated into the C-S-H phase [3,15] and helps to raise the compressive strength.

Although addition of gypsum to Portland cement clinker is undertaken to ensure smooth set regulation, the formation of ettringite is not the *actual cause of setting in Portland cement systems*. It was formerly thought that Portland cement setting was due to the recrystallization of microcrystalline ettringite [18,19]. Setting is in reality caused by the onset of C-S-H formation [17,20,21]. At best, ettringite has only a minor

role to play in Portland cement setting by supplementing to a small extent the physical effects of the C-S-H as the latter is being produced. The evidence for this situation is given below [17]:

- Not even a rough correlation exists between the initial and/or final setting times and the quantity of ettringite formed. Were setting to be due to recrystallization of microcrystalline ettringite into long needle-like forms, then the greater the quantities of ettringite produced, the greater the potential should be for such recrystallization to take place as hydration proceeds.
- Tricalcium silicate (alite) preparations set and harden in their reaction with water in a manner akin to that observed with Portland cement pastes. Were hydration of C_3S not to be involved in setting, then such results would be difficult to explain.
- Many Portland cement clinkers without added gypsum, such as those with lowish tricalcium aluminate contents, set within the time limits specified in national standards, as demonstrated by the former BS 12 requirements of initial setting time not less than 45 min and final setting time not greater than 10 h [22].
- Portland cements with non-gypseous set-regulators set in the normal way, even though ettringite is not formed upon hydration [23,24].
- Cements containing retarders frequently produce more ettringite and less C-S-H during early hydration up to set [25–27].

The common message that clinkers always flash set if no gypsum is present, is not true. Some do set in the normal way, and others do not. This depends on the aluminate (C_3A) level in the clinker [22], and on the surface area. Experiments on six clinkers ground to $325 \, m^2/kg$ showed that where the C_3A level was below *circa* 9 per cent the clinkers tended to set in the way a cement would be expected to. Above that level flash set was found to occur in high C_3A Portland clinkers. The exact level at which flash set occurs would be dependent upon the precise clinker composition and its inherent hydraulicity and is not expected to be the same for all Portland clinkers. Such clinkers are more difficult to grind than clinker-gypsum combinations (gypsum is a grinding aid for Portland clinker), give inferior workability and lower compressive strength development (due to lower sulphate content) than Portland cements containing ground-in gypsum. At higher surface areas, Portland clinkers will show flash set at lower C_3A levels, but then set regulation with gypsum requires greater additions to avoid flash set, as with ultra rapid-hardening Portland cement, where surface areas are often in excess of $700\,m^2/kg$.

An illustration of the influence of gypsum in providing set regulation of Portland cements is given in Figure 8.1, which shows a scanning electron micrograph of an ordinary Portland cement (CEM I type, in the European standard EN197-1) after 30 min hydration at a water/cement ratio of 0.5. At this stage of the hydration, there is very little clear structure forming. There are a few gypsum crystals and hexagonal platelets of calcium hydroxide. The latter would primarily have originated from free lime hydration at this early stage of the hydration process. There are thin layers of amorphous/poorly crystalline material on the clinker surfaces. Much of this material on the clinker surfaces can (from EDAX analysis), be attributed to ettringite having formed significantly

with some other due to amorphous silicate, the precursor of calcium silicate hydrate C-S-H.

However, much longer time periods are required before the characteristic pseudohexagonal prisms of ettringite would have been formed from this largely amorphous mass. This 30 min hydration time occurs within the 'dormant period' of low chemical reactivity. The amorphous silicate and ettringite mass on the surfaces would play a part *in toto* in helping to reduce chemical reactivity until sufficient C-S-H can form to allow normal setting to take place.

8.4 Gypsum quality

Natural gypsums sources used for Portland cement manufacture can contain pure deposits of gypsum or impure deposits where gypsum occurs together with other minerals like anhydrite, quartz, calcite, dolomite, clays, etc.

In the pure deposits, where the $CaSO_4 2H_2O$ content may be 98 per cent or more, quality control at the cement plant is facilitated by the absence of sizeable impurities in the gypsum delivered. However, there should be low clinker-gypsum grinding temperatures (facilitated by good modern cooler systems) to stop the gypsum from fully dehydrating to the hemihydrate/soluble anhydrite stage, otherwise false set (see below) is likely to be severe.

In a number of countries (e.g. Britain, Germany, Ireland) impure deposits are common and have been utilized in Portland cement manufacture for many years. Common impurities can affect cement performance in areas like setting and workability [28].

For instance, natural anhydrite can have a beneficial effect in alleviating false set by effectively acting as a diluent when the hemihydrate/soluble anhydrite produced by the heat generated during grinding rehydrates to gypsum. At numerous cement plants, natural anhydrite is ground in with gypsum to improve the workability characteristics of the resulting cement by alleviating false set (see later in this chapter).

Calcite, dolomite and magnesite in small quantities tend to have a more neutral effect, but may

FIGURE 8.1 A view of the hydration of an ordinary Portland cement at 30 min hydration during the set regulation period.

accelerate setting somewhat if present in large amounts within the impure gypsum. Quartz also tends to have a more neutral effect, and normally has a negligible effect upon setting and workability.

Clay minerals can be detrimental in increasing water demand and the propensity for false setting because of their competition with the clinker phases for water. Montmorillonitic clays can sorb water strongly, more so in relative terms than other clays like illite and kaolinite, and this is reflected in their relative effects upon water demand in cements containing gypsum with a high montmorillonite content. Any detrimental effects can normally be minimized or alleviated by appropriate dilution with natural anhydrite being ground in with gypsum and clinker to produce the finished Portland cement.

Chemical by-product gypsums (discussed below) may also contain impurities that affect cement performance. For instance, phosphates and fluorides in both phosphogypsums and fluorogypsums, and orthoboric acid H_3BO_3 in borogypsum, can cause set retardation. Such retardation can be substantial ($\sim$7–10 h or more) with some unbeneficiated borogypsums, which can make them unacceptable for use in normal Portland clinker-gypsum grinding, unless substantially beneficiated. Even if borogypsum gives a very long setting time, the actual setting is rapid and compressive strength development once begun tends to undergo an apparent 'delayed acceleration', so that at 3 days and beyond, acceptable strength values can be given [29].

8.5 By-product gypsums

Many by-product gypsums are produced by the chemical industry and are commonly disposed of by landfill. Some of these are now being employed for Portland cement production for environmental reasons. Large quantities of gypsum are produced by flue gas desulphurization at coal-fired power stations and are increasingly being utilized in Portland cement manufacture.

By-product gypsums are generally named after the chemical process from which they have been obtained, for example, phosphogypsum from phosphoric acid manufacture, fluorogypsum from hydrofluoric acid manufacture, formogypsum from formic acid manufacture, desulphogypsum (or FGD-gypsum) from flue gas desulphurization etc. Other by-product gypsums include borogypsum, citrogypsum, montmorillonogypsum (from the acid-activated montmorillonite process), oxalogypsum, tartarogypsum, desalinogypsum (from water-desalination plants), aluminogypsum (from aluminium chloride $AlCl_3$ plants) and pharmacogypsum (from effluent neutralization within the pharmaceutical industry, such as from vitamin C production).

Some of these by-product gypsums in the raw form available are unsuitable for Portland cement manufacture at present and would need to be cleaned up before such usage. For instance, desalino-, alumino- and pharmacogypsum can be prone to having appreciable chloride contents, which might promote corrosion in reinforcement unless reduced to negligible proportions. Desalinogypsum may also contain small amounts of organics like soluble starch (used as a nucleating agent for effective deposition of calcium sulphate at 100°C upon concentrating [30]). Pharmacogypsum can include small quantities of organics that could have deleterious or toxic properties. Similarly, oxalogypsum can contain small quantities of calcium oxalate, which is poisonous, and hence renders such gypsum unsuitable as such for use in cement manufacture for health and safety reasons.

Although not strictly speaking a by-product gypsum, *in situ-gypsum* (sometimes also called synthetic gypsum) has been produced in countries or regions where there is a dearth of available gypsum, like South Africa and Brazil. *In situ*-gypsum is made by mixing concentrated sulphuric acid or even oleum (superconcentrated sulphuric acid) with ground limestone, usually in a cement mill to produce gypsum *in situ*. Clearly, careful health and safety procedures are needed for the production of *in situ*-gypsum. The purity of this gypsum is dependent upon the purity of the limestone used in its production. Good quality gypsum can be made by this means, and it has been successfully (technically and economically) employed in Portland cement manufacture.

In situ-gypsum has, for example, been produced and used by the Pretoria Portland Cement Company in South Africa [31]. Their optimum conditions were grinding limestone (*circa* 88 per cent wt $CaCO_3$ and 12 per cent wt SiO_2) to a fineness of 80 per cent passing through a 150 μm screen and diluting the concentrated sulphuric acid used from 98 per cent to 65 per cent mass. The mixer discharges into a chamber called *the Den*, which has moving sides and floor, all of which consist of teflon-lined wooden slats. A variable speed controller allows the *denning* time to be varied between 10 and 40 min. A 5 per cent wt excess of limestone is maintained to ensure full neutralization of the 1–2% free acid left in the mix. The free moisture content of the *Den* product is maintained at about 12 per cent. The product (consisting of anhydrite and hemihydrate) is transferred to a store for 2 days' curing before use. This allowed for rehydration to gypsum to take place. The *in situ*-gypsum thus manufactured has a purity of 88–90 per cent $CaSO_4 2H_2O$.

Japan has for many years utilized by-product gypsums in Portland cement manufacture, because only a relatively small number of impure gypsum deposits were available in the country, and previously, much of the natural gypsum previously used had to be imported. At first, both phosphogypsum and fluorogypsum were extensively used. Following doubts about the radioactive content of many supplies of phosphogypsum, from traces of radioactivity in the original phosphate deposits, a switch was made away from phosphogypsum to desulphogypsum, which can be a very pure white product. There are some phosphate deposits in the world that are essentially non-radioactive and can thus give rise to a safe phosphogypsum, for example, Kola (Russia), South Africa and the Pacific island of Nauru. Phosphogypsum is utilised in South Africa for Portland cement manufacture. White phosphogypsum derived from Kola phosphate has been employed in white Portland cement manufacture as an alternative to pure natural gypsum.

The UK National Radiological Protection Board [32] reported that phosphogypsum could be safely utilized in cement manufacture when derived from sources that did not give radium concentrations significantly greater than 25 picocuries per gram. This particular dosage was chosen, because it is less than the difference between the background in regions of low and high natural background radiation within the UK. This limit permits the use of phosphogypsum in cement manufacture when derived from two-stage phosphoric acid processes, such as the Fisons, Nissan and NKK that have much lower radioactivity and, of course, from the Kola, South African and Nauru sources already mentioned.

The hydration chemistry can be considerably affected by having by-product gypsum in place of pure natural gypsum in Portland cements. For instance, some by-product gypsums (e.g. fluoro-, phospho- and formogypsum) can form up to *circa* 2–3 times the quantity of ettringite formed in the first 2 h of hydration than if a pure natural gypsum were present, whereas for desulphogypsum there is commonly no significant enhanced ettringite formation [3,33]. The actual level of enhanced ettringite formation will, of course, vary from one consignment of a particular by-product gypsum to another, depending upon its purity, processing and history before delivery. Most by-product gypsums give enhanced ettringite formation, normally at the expense of early C-S-H formation. Some, like desulphogypsum, may often give no appreciable enhancement of ettringite formation during early hydration. Each particular by-product gypsum consignment for use in Portland cement manufacture needs to be checked out on its individual merits for overall suitability and optimization of cement quality.

By-product gypsums have not commonly been employed in oilwell cement manufacture, although they are not excluded by the international standard [34]. They have, in fact, been utilized periodically in the Far East. This situation has arisen because with the basic grades of oilwell cements (Classes G and H) in particular, there is a need to produce as consistent a product as possible in a given cement plant. Wide variations in quality and composition, can give rise to significant differences in setting times through differing retardability and in strength growth patterns, unless great care is taken to beneficiate the material sourced for this specialist application. With construction Portland cements, such variations are less likely to be important in their manufacture to given national and international standards in respect of setting and compressive strength requirements. The engineering properties need to be checked out first, before utilizing a given source of by-product gypsum in given types of Portland cements, so as to ensure that satisfactory cement performance is likely to be given.

8.6 Flash set

When Portland clinker is ground alone and mixed with water, the aluminate (C_3A) phase initially reacts rapidly, the alite phase progressively and the other main phases (ferrite and belite) contribute as well. If the C_3A level is appreciable (*circa* 9 per cent or more at clinker surface areas around $325 \, m^2/kg$), or the surface area is high, then a so-called flash set or quick set is likely to ensue. The C_3A reacts quickly with water to form C_4AH_{13}, a hexagonal plate phase, which is stable in the high-lime environment. This reaction is accompanied by hydration of the other phases as well. There is a considerable increase in temperature, which is followed by irreversible stiffening, which is referred to as a flash set or quick set. Since plasticity of the mix is not restored after flash setting, it is clearly deleterious to concrete production. In order to prevent flash set from taking place, and to ensure smooth set regulation prior to normal setting, gypsum is ground into cement.

The reaction that normally triggers flash set is the hydration of the aluminate phase in the high-lime environment to hydrocalumite:

$$C_3A + CH + 12H \rightarrow C_4AH_{13}$$

Hydrocalumite is a hexagonal plate type phase, which is stabilized by the presence of the high-lime environment.

Flash set can also arise at very high surface areas, as with ultrarapid-hardening Portland cements [35], unless there is adequate set regulation by gypsum and sometimes adequate retardation by a suitable retarding admixture as well.

8.7 False set

False set is sometimes also known as early stiffening, premature stiffening or gum set. It refers to cement which when gauged with water and mixed for a short while, stiffens up and appears to set. Remixing breaks up this stiffening and the cement proceeds to a normal set. Unlike flash set, where considerable heat is evolved, comparatively little heat is released in false set. There are two established procedures for measuring false set, the ASTM C359 false set mortar test, and the ASTM C451 false set paste test. Test results from given cements using both procedures do not necessarily correlate with each other, since each test measures a somewhat different property of the hydrating cement [28].

False set is primarily caused by the rehydration of hemihydrate/soluble anhydrite (formed by dehydration of gypsum during clinker-gypsum grinding at *circa* 115°C or above) back to gypsum. Since the former have solubilities *circa* 2–3 times that of gypsum, they are supersaturated with respect to gypsum in solution. As a result, when they rehydrate back to gypsum: $CaSO_4xH_2O + (2-x)H_2O \rightarrow CaSO_4 \cdot 2H_2O$ ($x = \sim 0.001$–0.67 in practice), the gypsum being formed is precipitated to some extent from solution, giving the observed stiffening effect. Remixing allows the remaining hemihydrate/soluble anhydrite to rehydrate rapidly back to gypsum. Once this is achieved, the cementitious hydration proceeds in the normally expected smooth manner. Stiffening effects from soluble anhydrite, because of its lower water content than hemihydrate, tend to be more severe than those experienced with hemihydrate.

Syngenite can also give some stiffening, since soluble potassium sulphate K_2SO_4 can react with gypsum (or its derivatives hemihydrate/soluble anhydrite) to form syngenite, which has a comparable solubility to gypsum [3] and thus abstracts some K_2SO_4 from solution, thus augmenting any stiffening effects that might take place.

False set can be a problem in practice, as in the precast concrete industry, for example, where some modern manufacturing units require short mixing times and the facility for remixing may not always be present. Modern closed-circuit grinding mills and adequate clinker-cooling systems in cement works tend to ensure that most cement plants can produce Portland cements containing at least some residual gypsum. This is important, because the presence of at least some residual gypsum can act as a nucleating site for the hemihydrate/soluble anhydrite to rehydrate smoothly back to gypsum.

Natural anhydrite is often used to alleviate false set by being ground in with the clinker and gypsum, usually in simple proportions like 50/50 with gypsum. Natural anhydrite functions by its rehydrating directly to gypsum without any passage through the hemihydrate/soluble anhydrite stage itself, and thus dilutes the effect of any hemihydrate/soluble anhydrite rehydration to gypsum, ideally to the point at which there is no observable stiffening. Some natural gypsums contain appreciable quantities of natural anhydrite and their use in cement manufacture can be an added bonus for alleviating false set [28].

In practice, false set is not always so simple in its appearance and effects. Rehydration of hemihydrate/soluble anhydrite may arise quickly, progressively or occur suddenly after a longer time period (e.g. 1 h or more) after a period of quiescence. Fuller details about the manifestation of false set have been given elsewhere [36,37].

8.8 Air set

Air set is a type of setting, usually caused by exposure to moist air and sometimes also when stored at moderate humidity in silos etc. It is traditionally noticed by the appearance of lumpiness. Such aeration, caused by abstraction of moisture and carbon dioxide from the atmosphere causing partial hydration, leads to a reduction in compressive strength, which at limited levels of ingressing moisture may not produce lumps to any significant degree. A typical situation quoted is where an increase in the loss on ignition of a Portland cement of 0.5 per cent lowers the 28-day strength by $3 \, N/mm^2$ (3 MPa) [38]. Upon further aeration, lumps develop and the cement becomes difficult to disperse in the cement,

mortar or grout. These lumps normally contain significant levels of the mineral syngenite, a potassium calcium sulphate hydrate ($K_2SO_4 \cdot CaSO_4 \cdot H_2O$ or $K_2Ca(SO_4)_2 \cdot H_2O$). No sodium analogue of syngenite is known.

Syngenite is formed by reaction of small amounts of potassium sulphate and/or calcium langbeinite from the Portland clinker ($K_2SO_4 \cdot 2CaSO_4$ or $K_2Ca(SO_4)_3$) with gypsum (and/or other forms of calcium sulphate present like hemihydrate/soluble anhydrite or insoluble anhydrite) and water. The syngenite forming reactions can be represented thus:

$$K_2SO_4 + CaSO_4 \cdot 2H_2O$$
$$\rightarrow K_2Ca(SO_4)_2 \cdot H_2O + H_2O$$

and

$$K_2SO_4 + K_2Ca_2(SO_4)_3 + 2H_2O$$
$$\rightarrow 2\{K_2Ca(SO_4)_2 \cdot H_2O\}$$

Formation of syngenite provides a strong initial bonding between the particles of Portland cement and assists, along with the partial hydration of the clinker phases, with increased setting and lumpiness. Subsequent carbonation of some of the hydrating phases may also militate against lumpiness.

Bagged cement is more prone to aeration than bulk cement, presumably on account of the higher surface to volume ratio of the former, and to the effects of compaction when the cement bags are piled on top of each other. Traditionally, air set has been associated with syngenite formation. However, there is no correlation between the extent of air setting and the quantities of syngenite formed [39]. Syngenite is merely an intermediary and reacts on, with the alkali and sulphate largely entering the other hydrate phases like C-S-H and ettringite, as they are being formed upon further hydration. Also, the partial hydration of the main clinker phases, as mentioned above, must be taken into account.

Syngenite formation is thus more likely to arise where there is a higher alkali (potassium) content in the cement, and can play a part in accentuating the onset of air setting. Once the syngenite-forming reaction starts with a little water, it can utilize hydrate water from the calcium sulphate and become self-perpetuating. However, free lime can offset or sometimes inhibit this reaction by preferentially reacting with the water present to form calcium hydroxide [3]. In addition, the mere presence of syngenite is not *per se* an indication that air setting is underway. The vast majority of Portland cements, even if to a very small extent, contain syngenite, which does not ordinarily cause air setting. Only when the conditions for aeration mentioned above prevail (including storage in moderate or high humidity conditions), will syngenite become a potential problem in the onset of air setting. It has sometimes been said that if a cement consignment is found to contain syngenite, then it should be discarded, because the cement has already air set. From the foregoing comments, this is undoubtedly nonsensical as a generalization. If any doubts arise, then the cement should be checked for air setting. If no air setting is found, then the cement consignment is safe to employ in construction.

8.9 Portland cement–calcium aluminate cement compositions

OPC–CAC mixes have been around for a long time [40–42]. Such cementitious mixes are important for rapid setting repair mixes, like the sealing of leaks in mortar or concrete and in expansive situations. These mixes usually have lower setting times and compressive strengths compared with OPC and CAC alone. For certain mixes the setting is very short and flash set can take place [4]. These mix ranges vary with particular OPC–CAC types and cannot be predicted without trial experiments. Gypsum and its derivatives accelerate the CAC setting, particularly CA and $C_{12}A_7$ hydration. Calcium hydroxide from the hydrating OPC also accelerates CAC hydration. Meanwhile the OPC setting is accelerated by reaction of gypsum and its derivatives with the calcium aluminate hydrates CAH_{10}, C_2AH_8, C_3AH_6 and C_4AH_{13} to produce ettringite. The later strength of OPC–CAC mixes decreases progressively as CAC is replaced by OPC until a minimum is reached which is much lower than for Portland cement

alone. The compressive strength developed over the first few hours is sometimes greater than for CAC alone, even though the later compressive strength is lower. It is noteworthy that mix proportions giving the higher early compressive strength (1 day) and the minimum later strength (28 days) can differ significantly with different production batches of CAC and OPC [42].

The quantities of gypsum and/or its derivatives present are critical for the practical performance of these CAC–OPC mixed cements.

8.10 Calcium sulphoaluminate cements

Calcium sulphoaluminate cements (CSAs) containing gypsum or anhydrite as a key raw material are increasingly being utilized for various applications. In China, millions of tonnes of these cements are now being produced annually, because of their flexibility in application, ability to utilize successfully industrial by-product materials in their manufacture, and also because they are produced at lower firing temperatures than Portland cements. Hence, CSA cements come into the more environmentally-friendly category of being 'low-energy cements'.

CSAs are used in expansive cement compositions. An example is Type K expansive cement, which can be manufactured by intergrinding Portland cement clinker with gypsum or a gypsum-anhydrite mixture and an expansive clinker containing kleinite (otherwise known as ye'elimite) $C_3A \cdot CaSO_4$ or $C_4A_3\bar{S}$. The expansive clinker is normally made by sintering (at temperatures up to *circa* 1300°C) limestone, bauxite or other alumina-containing materials and gypsum in rotary kilns. Such an expansive clinker can contain kleinite, some alite, belite, ferrite, anhydrite and some free lime [43].

CSAs can also be manufactured from industrial by-product materials, such as bauxite fines, by-product gypsums like phosphogypsum, pulverized fuel ash (pfa) and/or ground granulated blast furnace slag (ggbs) producing the components of these cements. These, which, when mixed in the proportions $C_3A.3CaSO_4/\beta\text{-}C_2S/C\bar{S}$ of $1.5:1:1$

by weight, have given products of acceptable compressive strength after 1 and 28 days [43,44]. The expansivity of the basic hydration reaction:

$$3CaO.3Al_2O_3 \cdot CaSO_4 + 8CaSO_4 + 6CaO$$
$$+ 96 H_2O \rightarrow 3 \{Ca_6 [Al(OH)_6]_2 (SO_4)_3 26H_2O\}$$

is considered to be due to negatively charged colloidal grains of ettringite with a high specific surface area being formed by a through solution mechanism. These colloidal grains attract the polar water molecules that surround the crystals and cause interparticle repulsion, perhaps of the double-layer type, which results in overall expansion of the system. This reaction plays an important role in setting and compressive strength development, primarily due to the abstraction of water from solution [45].

Some of the most effective low-energy cements to be developed are based around the two low-lime phases kleinite and belite. Four variations of CSA cements have been classified thus [46]:

- Ordinary CSA: C_2S, C_4AF, $C_4A_3\bar{S}$, $C_{12}A_7$
- Active belite cement: α'-C_2S, $C_4A_3\bar{S}$, C_3A
- Belite sulphoaluminate cement: β-C_2S, $C_4A_3\bar{S}$, CA, $C_{12}A_7$
- Belite sulphoferrite cement: β-C_2S, C_4AF, $C_4A_3\bar{S}$, CS

These cements can achieve rapid-hardening and high early strengths comparable with OPCs, due to the rapid hydration of the calcium sulphoaluminate phase. Later strength development is due to hydration of belite and the formation of C-S-H [47].

The particular setting and strength characteristics of these CSA cements together with their relatively high heats of hydration produced during the reaction of the kleinite constituent in particular, [47] permit them to be utilized for repair work at temperatures below 0°C [48,49]. These cements demonstrate good resistance to corrosive ions like SO_4^{2-} and Cl^- and good anti-seepage properties because of their low porosity, good frost resistance and controlled expansion [50]. As a consequence they can be widely employed in construction. For example, as cements where fast set is needed, water tank repairs, framework-joints and emergency repairs such as airport runways can be readily undertaken [15,51]. In concrete, CSA cements can be used in various moulded products like electric poles, railway sleepers, floor slabs, beams, pillars and large trusses [50].

Compressive strength can be further enhanced by producing barium-substituted calcium sulphoaluminates by utilizing waste barium-containing materials in the raw meal for clinker production. A clinker containing barium-substituted kleinite $C_{4-x}B_xA_3\bar{S}$, β-C_2S and C_4AF gave much higher early and late compressive strengths than a similar clinker with no barium in the raw meal and hence not in the kleinite. Optimum Ba^{2+} substitution for Ca^{2+} was found to be 1.25 mol. With 1.25 mol barium substitution 1 and 28 day compressive strengths of the cement were 79.4 and 98.9 MPa respectively, whereas at zero substitution by barium, the corresponding strengths were respectively 44.0 and 47.6 MPa respectively. Clinker production was optimized above 1300°C, because below this temperature, free lime was formed and the strengths decreased [52].

It is clear that CSAs, which involve utilization of gypsum and anhydrite (natural or by-product) in their manufacture have considerable potential to be developed further in the future for a wide range of applications.

8.11 Conclusion

Gypsum is a very versatile material in cementitious compositions. It has been shown how important gypsum is for Portland cement set regulation and how sulphate from gypsum affects cement properties like setting and compressive strength development. The role of gypsum in the production of different CSAs is described and numerous applications for these cements have been indicated. Whether the gypsum is natural or a chemical by-product, it will continue to play an important role in cement technology both now and in the future.

8.12 References

1. Candlot, M., *Bull. Soc. Encour. Ind. Nat.*, 682 (1890).
2. Candlot, M., *Ciments et Liants Hydrauliques*, Dunod, Paris (1906).

3. Bensted, J., in *Advances in Cement Technology*, S.N. Ghosh (ed.), p. 307, Pergamon Press, Oxford, New York, Toronto (1983).

4. Lea, F.M., *The Chemistry of Cement and Concrete*, 3rd edn, Edward Arnold Ltd, London (1970).

5. Hinz, W., *Silicat-Lexicon*, VEB Akademie-Verlag, Berlin (1985).

6. Bensted, J. and Prakash, S., *Nature (London)*, 219(5149), 60 (1968).

7. Bensted, J. and Varma, S.P., *Nat. Phys. Sci.*, 232(34), 174 (1971).

8. Bensted, J. and Varma, S.P., *Z. Naturforsch.*, 26 B, 690 (1971).

9. Bensted, J. and Varma, S.P., *Cem. Tech.*, 3(2), 67 (1972).

10. Bensted, J., *Chem. Ind. (London)*, 398 (1975).

11. Bensted, J., *Il Cemento*, 72(3), 139 (1975).

12. Bensted, J., *Zem.-Kalk-Gips*, 37(9), 401 (1975).

13. Gutt, W. and Smith, M. A., *Trans. Br. Ceram. Soc.*, 66, 337 (1967).

14. Schwiete, H.E. and Knauf, A.N., *Gips – Alte und neue Erkentnisse in der Herstellung und Anwendung der Gipse*, Merziger Druckerei und Verlags-GmbH, Merzig/Saar (1969).

15. Hewlett, P.C. (ed.), *Lea's Chemistry of Cement and Concrete*, 4th edn, Arnold Publishers, London (1998).

16. Bensted, J., *Cements – Past, Present and Future*, Inaugural Lecture Series, The University of Greenwich, Greenwich University Press, Dartford (1997).

17. Bensted, J., *Silicates Ind.*, 48(9), 167 (1983).

18. Locher, F.W., Richartz, W. and Sprung, S., *Zem.-Kalk-Gips*, 29(10), 435 (1976).

19. Locher, F.W., Richartz, W. and Sprung, S., *Zem.-Kalk-Gips*, 33(6), 271 (1980).

20. Bensted, J., *Silicates Ind.*, 45(6), 115 (1980).

21. Bensted, J., *Characterisation and Performance Prediction of Cement and Concrete*, Engineering Foundation Conference, Henniker, New Hampshire, 25–30 July 1982, J.F. Young (ed.), p. 69, United Engineering Trustees Inc., Washington DC (1983).

22. Bensted, J., *Il Cemento*, 92(2), 87 (1995).

23. Diamond, S., *World Cem. Technol.* 11(3), 116 (1980).

24. Diamond, S. and Gomez-Toledo, C., *Il Cemento*, 75(3), 189 (1978).

25. Bensted, J., *Silicates Ind.*, 43(6), 117 (1978).

26. Bensted, J. Shaunak, R., *Proc. 11th Int. Conf. Cem. Micr.*, New Orleans, Louisiana, 10–13 April 1989, p. 198, International Cement Microscopy Association, Duncanville, TX (1989).

27. Bensted, J. and Aukett, P.N., *World Cem.*, 21(7), 308 (1990).

28. Bensted, J., *World Cem.* 26(9), 97 (1995).

29. Bensted, J., *Proc. 6th Int. Conf. Cem. Micr.*, Albuquerque, New Mexico, 26–29 March 1984, p. 232, International Cement Microscopy Association, Duncanville, TX (1984).

30. Estefan, S.F., *Chem. Ind. (London)*, (16), 535 (1979).

31. Mantel, D.G. and Liddell, D.G., *World Cem.*, 19, 404 (1988).

32. National Radiological Protection Board, Report no. 7, Her Majesty's Stationery Office, London and Norwich (1972).

33. Bensted, J., SCI Lecture Paper Series no. 69, (1996).

34. International Organisation for Standardisation: Petroleum and Natural Gas Industries – Cements and Materials for Well Cementing, Part 1, Specification, ISO 10426-1, ISO, Geneva (2000).

35. Bensted, J., *Il Cemento*, 78(2), 81 (1981).

36. Frigione, G., in *Advances in Cement Technology*, S.N. Ghosh (ed.), p. 485, Pergamon Press, Oxford, New York, Toronto (1983).

37. Bensted, J. and Bye, G.C., *Cem. Concr. Res.*, 16(1), 115 (1986).

38. Jackson, P.J., in *Lea's Chemistry of Cement and Concrete*, 4th edn, p. 25, Arnold Publishers, London (1998).

39. Bensted, J., *Il Cemento*, 77(3), 169 (1980).

40. Lafuma, H., *Le Ciment*, 30, 174 (1925).

41. Gu, P., Fu, Y., Xie, P. and Beaudoin, J.J., *Cem. Concr. Res.*, 24(4), 682 (1994).

42. Bensted, J., *Cem. Concr. Res.*, 25(1), 221 (1995).

43. Bensted, J., in *Lea's Chemistry of Cement and Concrete*, 4th edn, p. 779, Arnold Publishers, London (1998).

44. Beretka, J., de Vito, B., Santora, L., Sherman, N. and Valenti, G.L., *Cem. Concr. Res.*, 23, 1305 (1993); 24, 393 (1994).

45. Mehta, P.K., *Cem. Concr. Res.*, 3(1), 1 (1973).

46. Johansen, V., Kouznetsova, T.V., *Proc. 9th Int. Cong. Chem. Cem.*, New Delhi, Vol. 1, 49 (1992).

47. Bunford, J., Ph.D. Thesis, Staffordshire University, Stoke-on-Trent (2000).

48. Ivashchenko, S., *Proc. 9th Int. Cong. Chem. Cem.*, New Delhi, Vol. 1, 222 (1992).

49. Mudbhatkal, G.A., Parmeswaran, P.S., Heble, A.S., Pai, B.V.B. and Chatterjee, A.K., *8th Int. Cong. Chem. Cem.*, Rio de Janeiro, Vol. 4, 364 (1986).

50. Sahu, S., *Silikaty*, 38, 191 (1994).

51. Taylor, H.F.W., *Cement Chemistry*, 2nd edn, Thomas Telford Publishing, London (1997).

52. Cheng, X., Chang, J., Lu, L., Liu, F. and Teng, B. *Cem. Concr. Res.*, 30(1), 77 (2000).

Chapter nine

Alkali–silica reaction in concrete

D. W. Hobbs

9.1 Introduction

Alkali–silica reaction (ASR) has caused expansion and cracking to concrete structures exposed to external moisture in a number of countries. In this chapter the reaction, the diagnosis of ASR as the prime cause of the visual cracking and the factors which influence the cracking and expansion it may induce are briefly discussed.

9.2 The reaction

ASR is a reaction between the hydroxyl ions in the pore solution of a concrete and certain forms of silica occasionally present in significant quantities in the aggregate, the most reactive forms of silica being the most disordered forms, namely, opaline silica and volcanic glass. The product of the ASR is a gelatinous hydrate containing silica, sodium, potassium, calcium and water and its volume is much greater than the silica consumed. The composition of the alkali silica gel varies widely – SiO_2 from 28 to 86 per cent, CaO from 0.1 to 60 per cent, K_2O from 0.4 to 19 per cent and Na_2O from 0 to 20 per cent [1]. The presence of calcium hydroxide is essential for the gel to form and the composition of the gel has a major influence on the mobility of the gel and hence its ability to induce internal stress [2]. Low and high calcium gel are non-expansive [3]. A consequence of the gel formation and growth is that they occasionally induce internal stresses within the concrete which are sufficient to crack the 'reacting aggregate' particles, and to induce an internal network of fine cracks resulting in expansion, and visual macro-cracks. ASR can occasionally induce pop outs. Figure 9.1 shows ASR gel infilling a fine crack running through the paste and into two aggregate particles.

When the hydroxyl ions are produced solely as a consequence of the hydration of a Portland cement (PC), the rate of reaction is greatest at the time when the reactants first come into contact and thereafter, it declines. The reaction ceases when either of the reactants are depleted or when the hydroxyl ion concentration is reduced to a threshold level. In a high cement content concrete with an original alkali content of $5\,kg/m^3$ expressed as equivalent Na_2O, the reaction of about 2.5 per cent by mass of total aggregate is sufficient to reduce either the hydroxyl ion concentration or alkali

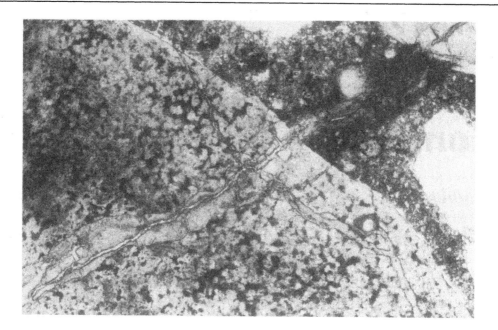

FIGURE 9.1   Gel filled cracks and gel saturated paste. (Magnification 100.) (With permission from British Cement Association.)

concentration to a threshold level or, alternatively, to deplete all the available sodium and potassium alkalis [4].

Normally, for expansion and cracking to result from ASR, an external source of moisture is required, expansion only occurring when the external and internal relative humidities are in excess of about 90 and 95 per cent respectively [5]. If the reaction is of sufficient intensity to induce expansion and the concrete is maintained moist, the rate of expansion is controlled by the rate of the chemical reaction [4]. Expansion ceases when the reaction is complete or when physical equilibrium is established. If the concrete is not maintained moist, the rate of expansion and the time taken for expansion to reach completion, are controlled by the rate at which water vapour diffuses into the concrete and/or the rate at which water is drawn into the concrete by capillary action.

9.3 Mechanism of expansion

Two main theories have been proposed to explain the mechanism of expansion caused by ASR [3].

In one, the induced stresses within the concrete are attributed to the growth of the gel caused by absorption of pore fluid and in the other, known as the osmotic cell pressure theory, to a hydraulic pressure developed across an impermeable membrane.

In the absorption theory the expansion, if induced, will depend on the volume concentration of the gel, its rate of growth and its physical properties. If the rate of gel growth is low, the internal stresses may be dissipated by migration of the gel through the concrete. If the rate of gel growth is high, the internal stresses may be able to build up to a high-enough level to cause cracking and expansion of the concrete.

In the case of the osmotic cell pressure theory, it is suggested that the cement paste acts as an impermeable membrane towards the silicate ions. Thus, the membrane allows water, hydroxyl ions and the alkali metal ions to diffuse through it but will not permit the diffusion of the silicate ions through it. Under these conditions, any reacting site would exert an increasing pressure against the restraining paste. According to Hansen, the diffusion of pore water through the membrane would also tend to accelerate the reaction.

9.4 Concretes affected

Concrete elements adversely affected by ASR are normally high alkali content concretes exposed to rain or groundwater [4]. Table 9.1 gives brief details of the alkali levels in UK concrete elements which have and have not been adversely affected by ASR according to the procedures outlined in Section 9.9. In these particular concretes, the gravels and sands contained chert and evidence of alkali–silica reactivity was found in all of the sections examined. The alkali levels of the cracked concretes range from 5.0 to 9.0 kg/m^3. A similar range of alkali levels have been observed to induce abnormal expansion in concrete blocks, made using a number of UK gravels and sands containing chert or flint, stored externally in the South East of the UK [6]. Lower alkali levels in concretes adversely affected by ASR have been

reported by other investigators, for example, concretes containing certain German greywackes [7] and concretes containing South African Malmesbury Hornfels [8].

9.5 Visual and internal cracking induced by ASR

Examples of visual cracking induced by ASR are shown in Figures 9.2 and 9.3. The cracks begin to appear at an expansive strain of about 0.05 per cent. In sections which are lightly reinforced and lightly loaded, map cracks are formed but when expansion is subject to restraint by reinforcement or loading, the ASR cracks tend to form parallel to the direction of restraint (Figure 9.3). Spalling or scaling is rarely associated with deleterious ASR. It is common for cracks formed by ASR to be bordered by broad zones of light-coloured concrete and where the cracks meet they often give the appearance of permanent dampness. The crack widths are generally less than 1 mm and

TABLE 9.1   Alkali levels in affected and non-affected concretes

Structure	Aggregate		Alkali content (kg/m^3)	Did the diagnosis indicate deleterious ASR?
	Coarse	Fine		
Beam, multi-storey carpark [3]	Mendip limestone	Sea dredged sand	Within the range 5.6–7.2	Yes
Retaining wall, multi-storey building [9]	Mendip limestone	Sea dredged sand	3.8 6.3	No Yes
Beam, multi-storey building [10]	Mendip limestone	Sea dredged sand	4.9–6.6	Yes
Prestressed columns[a][11]	Mendip limestone	Sea dredged sand	Within the range 6.0–9.1	Yes
Foundation bases [12]	Mendip limestone	Sea dredged sand	Within the range 5.0–7.0	Yes
Sections of four bridge wing walls [13]	Partially crushed natural gravel		5.15, 5.4, 5.5, 5.6, 6.0, 6.6 5.5 3.8, 3.9, 4.1, 4.1, 4.3, 4.4, 5.1	Yes Possibly No

[a] Cement content up to 650 kg/m^3, Na$_2$O$_{eq}$ of cement in the range 1.05–1.4 per cent by mass.

FIGURE 9.2   ASR cracking in a beam in a multi-storey car park. Note that the cracking ceases where the beam is not exposed to direct rain.

FIGURE 9.3   ASR cracks in unreinforced support region of a bridge and differential thermal cracks in the reinforced section of the same concrete pour which have widened as a consequence of ASR.

often between 20 and 40 mm in depth. However, much wider and deeper cracks resulting from ASR have been reported. When alkalis are contributed only by the PC, the visual cracks generally take between 1 and 10 years to appear and often the cracks cease growing in width at between 8 and 20 years [4]. When alkalis are additionally contributed from other sources, for example the aggregates, both of these ages can be substantially increased [14]. It should be noted that all processes leading to excessive shrinkage or excessive expansion can lead to visual cracks of similar character to those induced by ASR except that cracks formed at an early age (< 1 month) are regular, whilst those formed at later ages, with the exception of those induced by steel corrosion, are irregular.

ASR is a potentially expansive mechanism, and in adversely affected structures the following movements have been observed: closing of joints, relative displacements of adjacent concrete sections, hogging caused by reinforcement restraint of expansion, and the closing of flexural cracks in a partially-cracked section. Load tests on affected structures and load tests to failure on affected concrete members have led to the conclusion that deleterious ASR rarely has a major adverse effect upon structural performance [14].

Figure 9.4 shows fine cracking in a ground concrete section taken at depth from an exposed beam which exhibited cracking due to ASR. The fine cracks have been highlighted by impregnating the sections with a coloured resin. At depths in excess of about 40 mm a proportion of the chert particles in the coarser end of the sand fraction are cracked with a network of fine cracks interconnecting the reacting aggregate particles. In concretes exhibiting expansions in the range 0.15–0.25 per cent the internal fine cracks have widths of between 10 and 35 μm. At depths of less than 25–40 mm, the fine cracks are absent. It is probable that leaching of hydroxyl ions out of the surface region has prevented deleterious ASR from occurring in the surface region. As a consequence, the heart concrete expands more than the outer layer inducing tensile stresses in the outer layer and macro-cracks perpendicular to the exposed surface of the beam. Thus, the expansion of the heart concrete manifests itself in the macro-cracks; between the cracks the expansion of the surface region of the concrete is low due to its low tensile failure strain capacity. This form of internal cracking is unique to ASR.

9.6 Pessimum behaviour

The expansion and severity of cracking induced by ASR depends upon the form of reactive silica, the proportion of accessible reactive silica present in the aggregate, the porosity of the aggregate and the available alkali content. The effect of opaline silica content upon the expansion of mortar bars is shown in Figure 9.5. An explanation for the form of the relationship between expansion and reactive silica content, known as pessimum behaviour, is given in Figure 9.6. At higher available alkali contents, the pessimum reactive silica content moves to higher values and the curve broadens [4].

If a 'reactive aggregate' contains a slowly reacting form of silica or a highly reactive form of silica not readily accessible to hydroxyl ions, and the reaction is of sufficient intensity to induce expansion, then no pessimum will normally be observed and expansion increases with increasing proportion of reactive aggregate. An apparent pessimum can sometimes be observed when in a concrete containing a 'reactive sand', a low-porosity

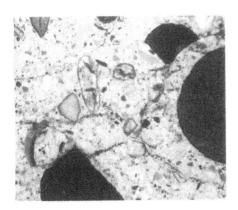

FIGURE 9.4 Characteristic internal fine crack pattern induced by ASR.

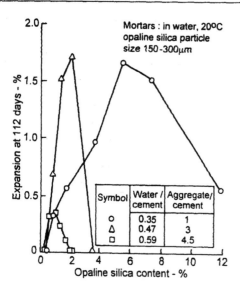

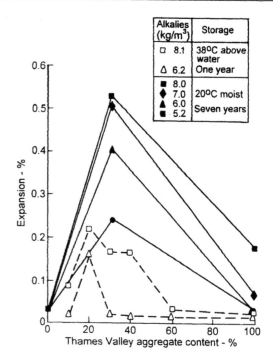

FIGURE 9.5 Influence of mix proportions on the relationship between expansion at 112 days and reactive silica content; acid-soluble alkali content of cement 1.05 per cent by mass [3]. (With permission from Thomas Telford Limited.)

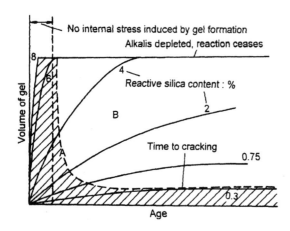

FIGURE 9.6 Possible relationship between volume of gel and reactive silica content: Region A: reaction but no cracking and no expansion; region B: cracking, expansion related to volume of gel formation which occurs after cracking.

coarse aggregate is replaced by a higher-porosity coarse aggregate (Figure 9.7), this is because expansion reduces as the porosity of the aggregate goes up [3].

FIGURE 9.7 Variation in 1- and 7-year expansion with Thames Valley aggregate content for concretes stored at 38°C [15] and 20°C respectively. TV aggregate replaced by low-porosity inert limestone.

9.7 Sources of alkali

Alkali compounds in a PC clinker are alkali sulphates, alkali aluminates and alumino ferrites and alkali silicates. The alkalis combine preferentially with the clinker SO_3 and any remaining alkalis is present in the silicates, aluminates and alumino ferrites [3]. When cement is mixed with water the alkali sulphates go rapidly into the liquid phase converting to alkali hydroxides which increase the hydroxyl ion concentration in the liquid phase. When high amounts of SO_3 are present, a higher fraction of the total alkali goes into solution within a few minutes. The total equivalent acid-soluble alkali content of a PC is conventionally calculated as

$$Na_2O_{eq} = Na_2O + 0.658(K_2O)$$

where 0.658 is the molecular ratio of Na_2O to K_2O. The ratio of potassium oxide to sodium

TABLE 9.2 Total alkali levels in concrete constituents.

Constituent	Alkali (Na_2O_{eq}) range(%)
Portland cement	0.3–1.6
Fly ash	0.7–7.8
Slag	0.3–2.6[c]
Silica fume	<5.5
Pozzolan	1.0–6.0
Artificial glass	~10%
Bronzite andesite[b]	4.4; 5.65
Witwatersrand quartzite[a]	0.1–0.7

[a] Blight [16]; [b] Kawamura et al. [17]; [c] Generally below 1.0%.

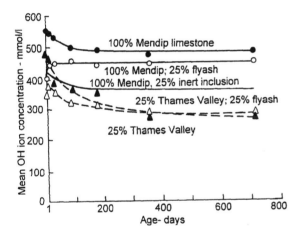

FIGURE 9.8 Variation of mean OH ion concentration with age. Binder content 400 kg/m³ [18].

oxide in most Portland cements ranges from 2:1 to 10:1 by mass, but in some US and Australian cements, the sodium oxide content is greater. It is possible that sodium oxide is more deleterious than potassium oxide [3].

Potassium and sodium alkalis released into the concrete pore solution from sources other than a PC will enhance the hydroxyl ion concentration and consequently, in the presence of sufficient quantities of calcium hydroxide, increase the risk of deleterious ASR. However, the magnitude of their physical effect is likely to be influenced by their rate of release. Possible sources of 'alkali' are fly ash, ground granulated blast furnace slag, silica fume, pozzolans (Table 9.2), external

sodium chloride (sea water or deicing salt) and certain aggregates, in particular, volcanic glass.

Extracted concrete pore solution data and deductions from expansion data indicate that fly ash (Figure 9.8), ground granulated blastfurnace slag and silica fume used as PC replacements [19] and certain aggregates [17] can contribute alkalis to the reaction, but these are probably less damaging than those released by a PC either because they are released more rapidly and depleted more rapidly (silica fume) or released less rapidly than a PC (fly ash and slag).

9.8 Reactive silica

Aggregates are termed reactive if they contain sufficient quantities of reactive silica to induce cracking in concrete. According to ASTM [20] the alkali-silica reactive constituents which may be present in aggregates include opal, cristobalite, tridymite, siliceous and intermediate volcanic glass, chert, glassy crypto-crystalline volcanic rock, artificial glasses, some argillites, phyllites, schists, gneisses, gneissic granites, vein quartz, quartzite and sandstone. Chalcedony is also considered to be reactive. Finding alkali-reactive constituents in an aggregate does not allow a prediction to be made of the likely performance of that aggregate in concrete. To do this a judgement needs to be based on field performance or a laboratory expansion test.

According to McConnell et al., and Gaskin et al., the silica minerals can be listed in decreasing order of deleterious reactivity as follows: opal, chalcedony, volcanic glass, cristobite, trydimite, and crypto-crystalline quartz [3]. Micro-crystalline quartz was found to be reactive but not deleteriously reactive.

Reactive forms of silica can sometimes be present as constituents of chert, flint, sandstone, limestone, greywacke and other rock types. Chert is a common constituent of many affected concretes. In some deleterious cherts, opaline silica has been found, whilst in others the deleterious constituents are associated with micro- and crypto-crystalline silica and chalcedony. The proportion of accessible reactive silica constituent present in an aggregate largely determines the aggregate's reactivity. If the proportion is low, the aggregate is likely to be classified as of low reactivity, and if it is high, the aggregate is likely to be classified as of high reactivity.

Proportions of reactive silica minerals as low as 0.3 per cent by mass of total aggregate can result in expansion due to ASR when present in combination with a low-porosity inert aggregate. Figure 9.9 shows expansion results obtained on concretes in which the main aggregate was a limestone and in which each size fraction of fine limestone aggregate was replaced by an equal volume of a partially-crushed reactive UK sand of similar size. The sand contained up to 20 per cent of fibrous chalcedony with crypto-crystalline quartz and more than 20 per cent chert comprising mainly micro-crystalline/crypto-crystalline quartz together with chalcedony inclusions together with ferruginous impurities. Dissolution tests on a ground fine sample of the aggregate in a I-M alkali solution containing sodium and potassium oxide in the ratio 0.44 : 0.85 for two months at 38°C indicated that only approximately 4 per cent

was reactive and X-ray analysis before dissolution and after dissolution indicated that the chalcedony was not reactive. From Figure 9.9, it follows that less than 0.3 per cent of reactive silica by mass of total aggregate is sufficient to induce deleterious expansion in a UK concrete of alkali content 7 kg/m^3.

9.9 Diagnosis of ASR as the cause of visual cracking

In the author's experience, ASR has often been incorrectly diagnosed as being the cause of cracking particularly when the prime causes of deterioration are freeze–thaw attack or delayed ettringite formation. The reverse can also apply, for example, the cracking in sections of the McPherson test road is now attributed to ASR [21] whilst in the 1950s it was attributed to a different reaction [22].

Gel and associated cracked aggregate particles can be found in uncracked concretes, in concretes which have cracked due to ASR and in concretes which have cracked due to other causes (Figure 9.1). To establish that ASR is the probable cause of the visual cracking it is necessary to establish

- that expansion of the concrete has occurred (Figure 9.4)
- that the visual cracking is characteristic of that induced by ASR (Section 9.5)
- from thin-section examination that there is considerable evidence of the ASR [23]
- that the internal cracking within the concrete is characteristic of that induced by ASR (Section 9.5).

If freeze–thaw attack (or some other deterioration process) has occurred, then it is necessary to eliminate such 'interference' by examining concrete sections taken from appropriate parts of the structure not exposed to freeze–thaw attack or if this is not possible, to carry out expansion tests on similar concrete to confirm whether or not ASR expansion can be induced at the alkali level existing in the concrete.

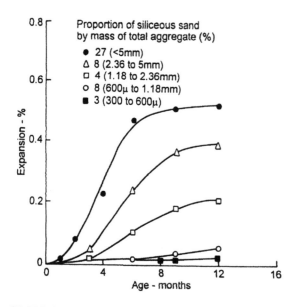

FIGURE 9.9 Variation of expansion with age for concretes containing various proportions and sizes of a partially-crushed reactive UK sand. Main aggregate Mendip limestone. 38°C, Na$_2$O$_{eq}$ 7 kg/m^3.

Additional factors or observations to consider which can assist diagnosis are:

- The age at which the cracks first appeared. ASR cracks generally take 1–7 years to appear, plastic cracks 2–10 h, thermal cracks 2–10 days, drying shrinkage cracks less than a year, cracks due to delayed ettringite formation 2–20 years.
- When a core is taken from a concrete element, the restraint to expansion is removed and consequently, there is a possibility that expansion due to ASR and the resulting internal fine cracking could result from the removal of this restraint [6].

9.10 Factors influencing expansion

9.10.1 Concrete alkali content and the aggregate combination

In earlier sections, it was shown that ASR expansion can sometimes be induced in high alkali content concretes. If the alkali content is lowered, then a level is reached when the reaction is no longer of a sufficient intensity to induce cracking and expansion. In specifications to minimize the risk of visual cracking occurring, a limit of $3\,kg/m^3$ or 0.6 per cent by mass of Na_2O_{eq} is sometimes placed on the alkali content of the concrete or cement respectively. Figure 9.10 shows expansion results plotted against concrete alkali content for concretes containing opaline silica in which the main aggregates were either Thames Valley gravel or Mountsorral granite. Also shown plotted are some results obtained on two Canadian aggregates [24] and a South African aggregate [25]. Note that the expansion is higher with the low porosity Mountsorral granite than when the main aggregate is the higher porosity Thames Valley. In Figures 9.11 and 9.12; one year expansion or maximum expansion of concretes stored moist at 38°C containing a number of chert-containing aggregates and crushed greywacke aggregates respectively, is shown plotted against alkali content. The critical alkali above which cracking is induced varies

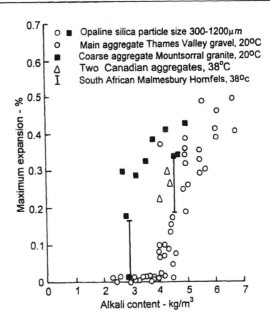

FIGURE 9.10 Dependence of expansion upon alkali content for a number of particularly deleterious aggregate combinations.

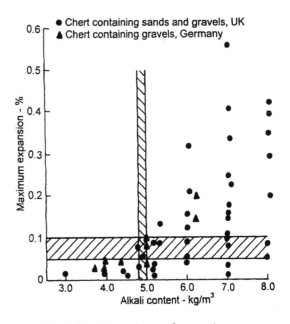

FIGURE 9.11 Dependence of expansion upon concrete alkali content. Chert containing 'gravels' (38°C or 40°C). PC concretes • German data from reference [26].

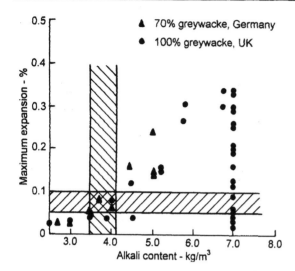

FIGURE 9.12 Dependence of expansion upon concrete alkali content. Greywackes • PC concretes (38°C or 40°C) [26, 27].

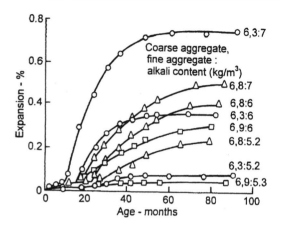

FIGURE 9.13 Relationship between expansion and age for a number of concretes made using sands containing chert and flint stored moist at 20°C. 6: Mendip limestone: 8: Thames Valley; 9: Sea dredged; 3: Somerset.

widely from 2 to 8 kg/m³ or higher. Clearly when minimizing risk in new construction, account should be taken of these differences [28].

Similar conclusions result from tests carried out at 20°C. Figure 9.13 shows the variation of expansion with age for a number of concretes made using aggregates containing chert and flint. Note that the age to deleterious expansion increases and the maximum expansion decreases as the alkali content goes down. Greywackes are generally assumed to react at a slower rate than chert and flint-containing sands and gravels; this view is not supported by the data shown in Figure 9.14 nor by data obtained on some other greywackes [29].

9.10.2 Temperature

A common assumption regarding chemical reactions is that the rate at which a chemical reaction occurs doubles for each 10°C rise in temperature. In the case of deleterious ASR, the age to abnormal expansion and the level of expansion broadly follow this rule for concretes stored in the temperature range 10–60°C. For concretes made with chert-containing aggregates stored moist at 60°C, 38°C, 20°C and externally in the south-east of the UK the ages to cracking are in the ratio 1:4:16:28

[6,30] (see also Figure 9.14). A similar temperature effect has been observed on concretes containing greywackes from the southern region of the new federal states of Germany (Figure 9.14) [26] and Malmesbury hornfels from the Cape province of South Africa [8].

In Figure 9.15 expansion measured on concretes stored at 38°C, 20°C and externally, is shown plotted against an age normalized to external exposure for one concrete which, when stored externally in the south-east of the UK, exhibited abnormal expansion [6]. In making this plot, it has been assumed that the rate at which the alkali–silica reaction occurs at 38°C is four times as fast as the rate at 20°C and seven times the rate for external exposure. From this figure it can be seen that when the expansion is unrestrained, a single expansion – age normalized relationship very broadly fits the expansion data, so the expansion behaviour of field concretes can be broadly predicted from accelerated expansion tests carried out on the same concretes.

9.10.3 Binder type

There is no universal agreement on the manner and extent to which binder type influences ASR expansion. Very few examples have been reported of

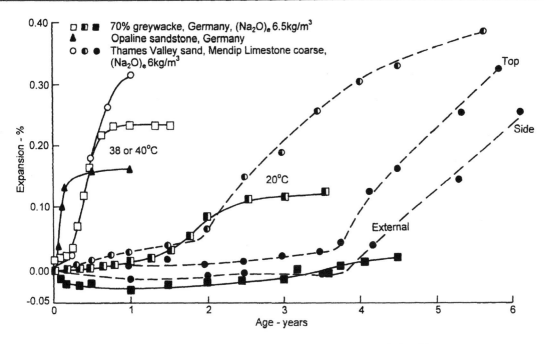

FIGURE 9.14   Influence of storage temperature upon the expansion–age relationship.

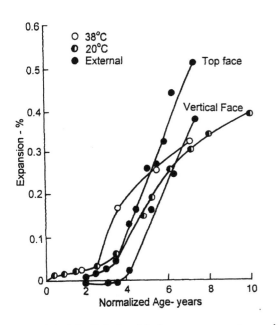

FIGURE 9.15   Relationship between expansion and normalized age (Thames Valley sand, Mendip limestone coarse) Na_2O_{eq} 6 kg/m^3.

field concretes containing fly ash, slag or pozzolans which have cracked as a consequence of ASR. In the case of a binder containing fly ash the following approaches have been adopted in specifications:

Canada	Fly ash > 15 to 30 per cent, fly ash total alkali < 3.0–4.5 per cent, PC alkali < 1.8, 2.4 or 3.0 kg Na_2O_{eq}/m^3 depending upon the level of preventative action required [31].
Germany	Effective alkali content is taken to be one-sixth of its total alkali content, with the effective alkali content not exceeding 0.6 kg/m^3 [32].
Netherlands	Fly ash > 25 per cent, alkali content of PC plus fly ash < 1.1 per cent [33].
South Africa	1.5 per cent Na_2O_{eq} equivalent limit according to ASTM C 311-91 [34]. When fly ash < 20 per cent, 40 per cent of its alkalis are taken to be reactive [35].

275

UK Fly ash >25 per cent alkali contribution ignored, 20–25% one-fifth of the alkalis are taken as reactive, <20 per cent all the alkalis are taken as reactive. Fly ash alkali limited to a maximum of 5.0 per cent [28].

On the other hand, in the case of a binder containing slag, the following approaches have been adopted in specifications:

Canada Slag $\geqslant 25$ to 50 per cent, slag alkali <1.0 per cent, PC alkali contribution <1.8, 2.4 or 3.0 kg Na_2O_{eq}/m^3 depending upon the level of preventative action required.

Germany and Netherlands Classed as a low alkali cement when (i) the slag content >50 per cent, Na_2O_{eq} of binder <1.0 per cent and (ii) the slag content >65 Na_2O_{eq} of binder <2.0 per cent.

South Africa Classed as a 'low alkali cement' when the slag content >40 per cent; when slag content <40 per cent, 42 per cent of its alkalis are taken to be reactive.

UK Slag >40 per cent alkali contribution ignored, 25–40% per cent, half of alkalis are taken as reactive, <25 per cent all the alkalis are taken to be reactive. Slag alkali limited to a maximum of 1.0 per cent.

In Canada, an alternative approach is to check whether the concrete is non-deleterious by a 2-year concrete prism expansion test. In the USA, there is no standard requirement. The most common requirement in the USA is that the binder containing fly ash or slag is checked for its effectiveness in reducing expansion due to ASR in mortar bars as compared to plain PC mortar bars.

9.10.4 Fly ash (low lime) and slag

The effectiveness of fly ash or slag in reducing expansion, when used as partial replacement for a high alkali PC [$Na_2O_{eq} > 1$ per cent], is dependent upon their total alkali content and the proportion of PC replaced. At replacement levels above 20 per cent, expansion is generally reduced but at lower replacement levels ($\leqslant 10$ per cent) expansion can be increased, indicating an effective alkali contribution from the fly ash or slag which is greater than that of a high-alkali PC. In the case of fly ash, this accords with results obtained by Dalziel et al. [36] which indicate that, in mortars stored at 20°C for 1 year, the hydration of about 20 per cent fly ash by mass of PC could be sufficient to deplete most of the calcium hydroxide resulting from the hydration of the PC.

Although at replacement levels above 20 per cent expansion is reduced, the results obtained by some investigators indicate a positive effective alkali contribution from fly ash and slag of up to 1 kg/m^3 [19,37]. This is illustrated in Figure 9.16 for concretes containing crystobalite and two UK reactive aggregates. In concretes which exhibit abnormal expansion, the use of fly ash and slag delays expansion. This is illustrated in Figures 9.17 and 9.18 for PC/fly ash concretes containing aggregates of two differing deleterious reactivities. Note that in these examples the concretes containing fly ash only begin to expand after 12 months which is sometimes the age at which expansion testing is terminated.

A positive effective alkali contribution from pulverized fly ash (pfa) and slag is noted particularly when concretes and mortars are wrapped in moist towels which are not replaced during the duration of the expansion tests, rather than when mortars or concretes are stored above water. In the case of concretes containing fly ash, it has been argued that a positive alkali contribution is only made when the PC alkali level necessary to induce abnormal expansion is less than 5 kg/m^3 [39].

9.10.5 Natural and artificial pozzolans

In expansion tests on mortars, it has been found that high silica and low-alkali content pozzolans, when used as partial replacements for PC, are most effective in minimizing expansion due to ASR [19]. It can be argued that the use of pozzolans promotes the formation of a calcium–alkali–silicate

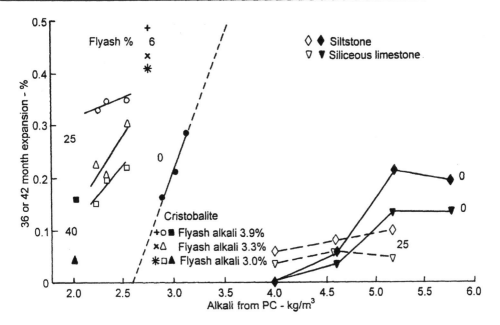

FIGURE 9.16 Relationship between expansion and PC alkali contribution. Cristobalite, 42 months, PC Na$_2$O$_{eq}$ 0.6 per cent, 20°C; natural aggregates, 36 months, PC Na$_2$O$_{eq}$ 1.15 per cent, fly ash total alkali 3.9 per cent, 38°C.

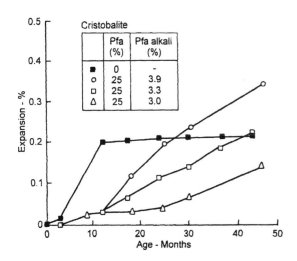

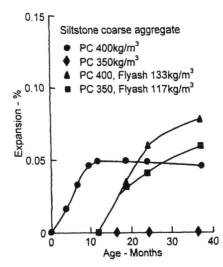

FIGURE 9.17 Influence of fly ash upon relationship between concrete expansion and age. 20°C, mixes 17 and 15A [38].

FIGURE 9.18 Influence of fly ash upon relationship between expansion and age. 38°C, fly ash total alkali 3.9 per cent, PC Na$_2$O$_{eq}$ 1.15 per cent, [39].

hydrate whilst the concrete is in a fresh state. This rapid reaction can be attributed to the presence of fine particles containing opaline or amorphous silica. Pozzolans used in sufficient quantities, can

inhibit ASR expansion in either of two ways: first, by rapidly depleting sodium and potassium alkalis from the pore solution and reducing the hydroxyl ion concentration to a threshold level prior to the concrete developing strength [19], and second, by complete depletion of the calcium hydroxide produced by the hydration of the PC [40]. It should be noted that reactive silica is still available within the aggregate and consequently external alkalis subsequently ingressing into the concrete could, if the concrete is of small section size, result in abnormal expansion due to ASR. If it is assumed that the alkali–silica ratio in the hydration product is similar to that indicated by a pessimum expansion plot (see Figure 9.5), then the partial replacement levels of PC by 'amorphous' silica, necessary to ensure that the reaction is essentially complete prior to the concrete developing strength, can be estimated from the expansion curves for mortars and concretes containing opaline silica (see Table 9.3). The replacement levels are dependent upon the mix proportions, the proportion of reactive silica in the aggregate and the alkali content of the cement. For a pozzolan containing 70 per cent 'amorphous' silica, with an equivalent sodium-oxide content of 3.0 per cent by mass all of which is present in its 'amorphous' silica fraction, the higher percentages given in Table 9.3 should be roughly doubled.

9.10.6 Silica fume

The action of silica fume is similar to that of a pozzolan with the minimum cement-replacement level by silica fume necessary to prevent expansion, depending upon the alkali content of the cement, the mix proportions, the reactive silica content of the aggregate, and the proportion of

TABLE 9.3 Minimum percentage of 'amorphous' silica required to prevent cracking due to ASR. Cement Na_2O_{eq} 1.05%

Water/ cement	Aggregate/ cement	'Amorphous' silica content[a]/ opaline silica content[b]		
		0.5	2	4
0.35	1	11.5	10	8
0.47	3	9.5	5	0
0.59	4.5	7	0	0

[a] % by mass of cement; [b] % by mass of aggregate.

amorphous silica and equivalent sodium oxide in the silica fume [19]. High silica and low alkali-silica fumes are most effective in reducing abnormal expansion due to ASR. At low-replacement levels of a PC by silica fume (3–10 per cent), and depending upon the composition of the concrete, expansion can be delayed, increased or eliminated, replacement levels of between 7 and 20 per cent being necessary to eliminate expansion [41], the replacement level being dependent upon the alkali content of the concrete (Figure 9.19).

9.10.7 Alkalis released by aggregate

Several examples exist of concretes adversely affected by ASR within which it is considered that the alkali level was enhanced by alkali release from the glassy phases in certain rhyolites and andesites and from artificial glass when they were attacked by hydroxyl ions [17,19]. For example, concrete pavements near the south-west coast of the USA where low-alkali cements have been used in combination with certain rhyolites and andesites, in bridges in Japan where an andesite has been used with a high-alkali cement and in a small industrial building in London where an artificial glass aggregate was used in combination with a low-alkali white PC. Further examples are buildings in South

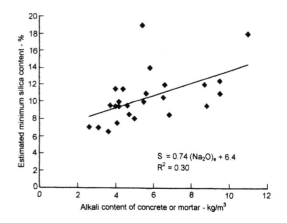

FIGURE 9.19 Dependence of estimated minimum silica fume content to prevent expansion, upon the alkali content contributed by both the PC and silica fume. Data normalized to a silica fume containing 100 per cent amorphous silica.

Africa where Witwatersrand quartzite was used as both the coarse and fine aggregate. If abnormal ASR expansion is induced, then the use of reactive aggregates which release alkalis prolong the expansive period of the reaction [14].

9.10.8 External salt

Alkalis ingressing from an external source can influence the magnitude of expansion due to ASR. Of particular concern is salt ingress from seawater, airborne salt and deicing salt, the sodium chloride concentration in sea water being approximately 0.6 M. When sodium chloride ingresses into concrete, it may react with tricalcium aluminate forming the compound $C_3 A . CaCl_2 . H_{10}$ and sodium hydroxide, the latter increasing the hydroxy ion concentration. Alternatively, the sodium chloride may simply react with calcium hydroxide forming calcium chloride and sodium hydroxide, again increasing the hydroxyl ion concentration.

In Denmark, the addition of salt to concrete swimming pools is believed to have contributed to deterioration of the concrete [3]. However, Brown [42], in a survey of a number of concrete bridges, found no evidence of deterioration which could be attributed to ASR in regions of the bridges where salt had concentrated.

Several investigators have shown that the immersion, at an early age, of mortar and small concrete specimens in a strong salt solution maintained at 20°C and 38°C, can increase the expansion induced by ASR [43] but that immersion in a 1 M NaCl solution may have little effect upon expansion [44].

The effect upon the risk of expansion due to ASR of exposing mature concrete or mortar to a sodium chloride solution is unclear. It can be argued that the effect will depend upon the following:

- Concentration of sodium chloride. At low concentrations, hydroxyl ions will diffuse out of the surface layers of the concrete at a greater rate than they are produced by ingress of sodium ions into the concrete.
- Concrete member size. Since salt ingress is normally restricted to the surface layers of the concrete, its effect upon member expansion, if any, will decrease as the member size is increased.
- Reactive silica content. If the concrete contains a small quantity of reactive silica, then the concrete contains an excess of hydroxyl ions and ingress of sodium ions from salt into the concrete will be of little consequence. However, if the quantity of reactive silica is high, any increase in the hydroxyl ion concentration resulting from ingress of sodium ions into the concrete will, in small concrete sections, be expected to result in additional expansion. Data has been reported which supports this hypothesis [43].

9.10.9 Lithium compounds

In 1951, test data was published which indicated that lithium salts could be effective in reducing expansion due to ASR [45], expansion being almost completely suppressed by lithium carbonate at a Li : Na ratio of 2 : 3 by mass. Since 1980, additional test work on mortars has been reported which support the conclusion resulting from McCoy *et al.*'s work [46], and additionally shows that at half this dosage, the expansion is not inhibited. A similar conclusion has been reached from concrete prism expansion tests carried out at 20°C using 12 per cent cristobalite by mass of aggregate [46]. The effects observed are explained by the pore solution work of Diamond and Ong [47]. This work led to the conclusion that at low levels of lithium dosage, the lithium enters into the calcium silicate hydrate and hence, is not effective in influencing expansion but at higher dosages, a sufficient quantity enters the ASR gel to render it non-expansive, indicating that the level of lithium to suppress ASR expansion is dependent on both the cement content and alkali content of the concrete.

9.10.10 Air-entrainment

Several investigators have found that air-entrainment reduces expansion due to ASR [49]. In expansion tests on concretes using Beltane opal, cristobalite, and three UK natural reactive aggregates it was found that air-entrainment reduced expansion and in the concretes tested containing the UK reactive aggregates, it was found that 5 per cent air-entrainment increased the alkali content above which abnormal expansion occurred by between 0.3 and 1.0 kg/m^3 [48].

9.11 Concluding remarks

The subject of ASR in concrete has a high profile and since 1940 more than 2,500 technical papers have been published dealing with various aspects of ASR. Often visual deterioration occurring in concrete elements has been wrongly attributed to ASR. The cracking which ASR induces is visually deceptive and the cracking and associated expansion rarely have a major adverse effect upon structural integrity. In new concrete construction, the risk of deterioration can be minimized by control of concrete alkali content or by selecting aggregates with good performance records at the concrete alkali levels to be employed.

9.12 References

1. Lombardi, J., Perruchot, A. and Massard, P. *Proc. 10th Int. Conf. Alkali-Aggregate Reaction in Concr.*, A. Shayan (ed.), pp. 934–41 (1996).
2. Diamond, S., *Proc. 8th Int. Conf. Alkali-Aggregate Reaction*, K. Okada, S. Nishibayashi and M. Kawamura (eds), pp. 83–94 (1989).
3. Hobbs, D. W., *Alkali-silica Reaction in Concrete*, Thomas Telford, London (1988).
4. Hobbs D. W., *Struct. Eng. Rev.*, 2, 65–79 (1990).
5. Blight, G. E., *Concr. Beton Third Quarter*, 49, 21–6 (1988).
6. Hobbs, D. W., *Proc. 10th Int. Conf. on Alkali-Aggregate Reaction in Concr.*, A. Shayan (ed.), pp. 316–23 (1996).
7. Siebel, E., Reschke, T. and Sylla, H. M., 'Alkali reaction with aggregates from southern region of the new federal states', in *Betontechnische Berichte 1995–1997*, G. Thielen (ed.), pp. 133–43, Verlag Bau & Technik (1998).
8. Oberholster, R. E. 'Inhibiting alkali-silica reaction; the role of Portland cement, milled granulated blast furnace slag, fly ash and silica fume', *Symposia on Practical Guidelines on the Selection and Use of Portland Cement, Ggbs, Fly Ash and Silica Fume in Concrete*. Portland Cement Institute, South Africa (1987).
9. Concrete Society, Report of a subcommittee, 'Investigation of structures affected by asr', *Concrete*, 31(3), 25–7 (1997).
10. Hobbs, D. W., *Proc. 10th Int. Conf. Alkali-Aggregate Reaction in Concr.*, A. Shayan (ed.), pp. 209–18 (1996).
11. Palmer, D. *Proc. 4th Int. Conf. Effects of Alkalis in Cem. Concr.*, School of Civil Engineering, Purdue University, pp. 285–94 (1978).
12. Cement and Concrete Association now the British Cement Association (1976–1982) Unpublished work.
13. Blackwell, B. G., Building Research Establishment, Private communication, UK (1994).
14. Hobbs, D. W., *Proc. 4th Rail Bridge Centenary Conf. Dev. Struct. Eng.* B. H. V. Topping (ed.), E & F. N., Spon, pp. 798–816 (1990).
15. Nixon, P. J. and Bollinghaus, R., *Proc. 6th* Int. Conf. Alkalis in Concr., G. M. Idorn and S. Rostam (eds), Danish Concrete Association, pp. 329–36 (1983).
16. Blight, G. E., 'The effects of alkali-aggregate reaction in reinforced concrete structures made with Witwatersrand quartzite aggregate', *Proc. 5th Int. Conf. Alkali-Aggregate Reaction in Concr.*, National Building Research Institute Pretoria, Paper 5252/15 (1981).
17. Kawamura, M., Koike, M. and Nakono, K., *Proc. 8th Ind. Conf. Alkali-Aggregate Reaction in Concr.*, K. Okada, S. Nishibayashi and M. Kawamura (eds), pp. 271–8 (1989).
18. Thomas, M. D. A., Building Research Establishment, Private Communication, UK (1994).
19. Hobbs D. W., *Proc. 8th Int. Conf. Alkali-Aggregate Reaction*, K. Okade, S. Nishibayashi and M. Kawamura (eds), pp. 173–86 (1989).
20. American Society for Testing and Materials Standard guide for petrographic examination of aggregates for concrete, ASTM C 295–90. *Annual Book of ASTM Standards. Section 4. Construction. 04. 02. 1998.*
21. Stark, D., Construction Technology Laboratories, Inc, Private Communication, USA, (1993).
22. Lerch, W., 'A cement-aggregate reaction that occurs with certain sand-gravel aggregates', *J. Portland Cem. Association Res. Dev. Lab.*, 3, 42–50 (1959).
23. Diagnosis Working Party, *The diagnosis of alkali-silica reaction*, British Cement Association, Crowthorne, (1992).
24. Duchesne, J. and Bérubé, M. A., *Cem. Concr. Res.*, 24, 73–82, (1994).
25. Oberholster, R. E. *Proc. 6th Int. Conf. Alkalis in Concr.*, G. M. Idorn and Steen Rostam (eds), Danish Concrete Association, pp. 419–33, (1983).
26. Siebel, E. and Reschke, T., 'Alkali reaction with aggregates from the southern region of the new

federal states', *Betontechnische Berichte 1995–1997*, G. Thielen (ed.), pp. 117–32, Vertag Bau & Technik (1997).

27. Blackwell, B. Q., Thomas, M. D. A., Pettifer, K. and Nixon, P. J., *Proc. 10th Int. Conf. Alkali-Aggregate Reaction in Concr.* A. Shayan (ed), pp. 492–9 (1996).

28. Building Research Establishment, *BRE Digest 330*, Part 2, 1–8 (1997).

29. Shayan, A., Quick, G. W., Lancucki, C. J. and Way, S. J., *Proc. 9th Int. Conf. Alkali-Aggregate Reactivity*, The Concrete Society, pp. 958–79 (1992).

30. Rayment, P. L. and Haynes, C., 'Suitability of proposed French test to avoid damage to alkali-silica reaction – An interim report', *Building Research Establishment, Client Report CR 23/96* (1996).

31. Canadian Standards Association, *Concrete Materials and Methods of Concrete Construction – Appendix B, CAN/CSA A23.2 - 2000* (2000).

32. German Committee for Reinforced Concrete, *Preventive Measures to Counter Harmful Alkaline Reaction in Concrete (Alkali guideline)*, Beuth Verlag GmbH, Berlin and Köln, No. 65027 (1997).

33. Conmissie voor Uituoering Van Research, *Measures to Prevent Concrete Damage Due to Alkali-Silica Reaction (ASR)*, CUR-Recommendation 38.

34. American Society for Testing Materials, 'Standard test method for sampling and testing fly ash or natural pozzolans for use as a mineral admixture in portland-cement concrete, ASTM C 311-98a'. *Annual Book of ASTM Standards*. Section 4. Construction. 04.02 (1998).

35. Oberholster, R. E 'Alkali-silica reaction', in *Fulton's Concrete Technology*, pp. 181–207, Portland Cement Institute, South Africa (1994).

36. Dalziel, J. A and Gutteridge, W. A., *The Influence of Pulverized – Fuel Ash upon the Hydration Characteristics and Certain Physical Properties of a Portland Cement Paste*, British Cement Association, Technical Report 560 (1986).

37. Moir, G. K. and Lumley, J. S., Proc. *8th Int. Conf. Alkali-Aggregate Reaction*, K. O. Kade, S. Nishibayashi and M. Kawamura (eds), pp. 199–204 (1989).

38. Hobbs, D. W., *Mag. Concr. Res.*, **46**, 167–75 (1994).

39. Thomas, M. D. A., Blackwell, B. Q. and Nixon, P. J., *Mag. Concr. Res.*, 48(177), 251–64 (1996).

40. Berube, M. A. and Duchesne, J., *Proc. 9th Int. Conf. Alkali-Aggregate Reaction in Concr.*, The Concrete Society, Crowthorne, UK, **1**, pp. 71–80 (1992).

41. Hobbs, D. W., *Silica fume as a pozzolanic constituent of a cement and a type II addition*, in Specification of Durable Concrete: Alternative methods for minimising the risk of damaging ASR in concrete, to be published by The Concrete Society (2002).

42. Brown, J. H., *The performance of concrete in practice. A field study of highway bridges*. Transport and Road Research Laboratory, Contractor Report No 43 (1987).

43. Hobbs, D. W., *Proc. of the Sec. CANMET/ACI Int. Symp. on Advances in Concr. Technol.*, V. M. Malhotra (ed.), ACI SP154, pp. 489–508 (1995).

44. Duchesne, J. and Bérubé, M. A., *Proc. 10th Int. Conf. Alkali-Aggregate Reaction in Concr.*, A. Shayan (ed.), pp. 830–7 (1996).

45. McCoy, B. J and Caldwell, A. G., *Proc. Am. Concr. Ins.*, 47(9), pp. 693–706 (1951).

46. Lumley, J. S., *Cem. Concr. Res.*, 27(2), 235–44 (1997).

47. Diamond, S. and Ong, S., *Proc. 9th* Int. Conf. Alkali-Aggregate Reaction in Concr., The Concrete Society, pp. 269–78 (1992).

48. Hobbs, D. W., 'Effect of air-entrainment upon expansion induced by ASR', *Concr. Communication Conf. 98*, British Cement Association, Crowthorne, 59–72 (1998).

Chapter ten

Delayed ettringite formation

C. Famy, K. L. Scrivener and H. F. W. Taylor

10.1 Introduction

Delayed ettringite formation (DEF) can be defined as formation of ettringite in a paste, mortar or concrete by a process beginning after hardening is substantially complete, and in which no sulfate comes from outside the cement paste. It can cause damage, which may be apparent only after a period of months or years. Except with seriously oversulfated cements, damage is known to have occurred only in materials that have been subjected to temperatures above about 70°C. Claims that DEF has caused damage in materials not thus subjected are unconvincing.

The term 'delayed ettringite formation' is widely accepted, but other terms, such as secondary or late ettringite formation, are also in use. The adjective 'secondary' is ill chosen, as it has an established meaning in petrology, denoting material that has formed by recrystallization. Some of the delayed ettringite that forms after curing at an elevated temperature is almost certainly a product of recrystallization, and thus also secondary, but secondary ettringite may originate in other ways. DEF has been defined as the formation of ettringite in a material that has been subjected to a temperature sufficient to destroy any that was previously present [1]. This definition is unsatisfactory, as ettringite is sometimes present at the end of the heat treatment.

This chapter deals primarily with the chemical and microstructural changes taking place at elevated temperatures, after cooling to ordinary temperatures, and with their relation to expansion. This topic is prefaced by a summary of relevant field observations and a discussion of the role of ettringite in concrete more generally.

10.2 DEF in field concretes

A number of cases of damage attributable wholly, or more often partly to DEF have been described [2–6]. In most of them, alkali-silica reaction (ASR) had also occurred and a case originally attributed to DEF was later shown to have been due to ASR [7]. Almost all related to precast, and often prestressed concrete products, notably including railway sleepers, produced by elevated-temperature curing below 100°C, but some [6] related to mass concretes of high cement content in which temperatures of 85–95°C were estimated to have arisen from the heat of hydration.

It has been claimed from field evidence that damage from DEF can occur in concrete that has not been subjected to an elevated temperature [8–10]. In principle, ettringite might form in normally cured concrete through carbonation [11], but no indication of damage from this cause has been reported. In an account of one case, it was stated that some sleepers made at weekends and cured at ordinary temperature had been damaged by DEF [8], but substantial evidence contradicted this [5]. It was further suggested that high contents of sulfate in the clinker could be present in phases from which they were only slowly released, such as anhydrite or the silicate phases. However, clinker anhydrite reacts relatively rapidly [12,13] and the amounts of SO_3 present in belite do not exceed a few tenths of a per cent on the mass of cement, and are balanced by approximately equimolar amounts of Al_2O_3. If this sulfate enters a sulfoaluminate phase at all, it is more likely to be monosulfate than ettringite [14]. Parallel tests of materials cured at ordinary and elevated temperatures have shown that cements meeting normal specifications that give rise to damaging expansion when cured at elevated temperatures do not do so when cured at ambient temperature [15–22].

10.3 Coarse microstructure of materials damaged by DEF

In concretes and mortars that have expanded, bands of ettringite up to ~30 μm wide are typically seen around aggregate grains and running through the paste. Pastes that have expanded show similar cracks and bands, mainly concentrated in regions near the surface [17]. These features have sometimes been regarded as diagnostic, but could also have arisen in other ways. Ettringite is present in many undamaged concretes that have not encountered elevated temperature, and that present in the paste readily recrystallizes to give larger crystals in any available spaces, provided that the water needed for transport is available. The detection of ettringite in a distressed concrete by petrography or other means therefore does not necessarily mean that it has caused the distress. Even where ettringite formation *has* caused the distress, it may not be the ettringite which is most obviously detectable, but that present as much smaller crystals within the paste. Erlin [23] considered that the presence of ettringite or other phases at aggregate interfaces shows only that something

has caused the paste to expand, and Stark *et al.* [24] concluded that the occurrence of large ettringite crystals in concrete cracks is generally a consequence and not a cause of the cracks.

10.4 Macroscopic properties associated with DEF

Many studies have been made of expansion or cracking associated with DEF, and in some cases, also of accompanying water uptake and changes in mass, resonance frequency or compressive strength [2,15–22,25–33]. Lawrence [32] reviewed work up to 1994. The following points appear well established:

- Expansion does not occur unless the temperature of heat treatment exceeds ~70°C and subsequent storage has been, intermittently or permanently, in water or air of $\geqslant 95$ per cent R.H. Within the 70–100°C temperature range, expansion increases with the temperature of heat treatment, and with its duration up to a few hours. With very long heating, expansion is limited or absent [19,20,31,32].
- If storage is in moist air, expansion is slower and less extensive than if it is in water [15,20, 21], but macrocracking can be greater [27].
- Expansion, when plotted against time, follows an S-shaped curve. There is frequently an induction period. For $40 \times 40 \times 160$ mm mortar prisms made with a quartz aggregate and stored in water, expansion is typically significant by ~3 months and tends to level off by 1–2 years [16,19,25,26]. In field concretes, it can be much slower.
- Expansion begins near external surfaces and spreads inwards. It is more rapid when determined using thin samples [22,29,33].
- Expansion is influenced by the nature and size-grading of the aggregate [19,26,27,33,34]. It is much slower if the aggregate(s) are entirely of limestone than if they are of quartz, and is also very slow with pastes [17]. In these cases, it may not reach a significant level for several years.

- Expansion is favoured by pre-existing weaknesses, including microcracks induced mechanically [35] or by freezing and thawing [26,36]. Ettringite is often associated with ASR gel. Laboratory experiments show that ASR is more rapid and extensive at elevated than at ordinary temperatures, and that, where it occurs together with DEF, it is the initial and major cause of damage, though DEF may increase the damage at later ages [30,37]. This agrees with observations on field concretes.

- Lowering of water-to-cement (w/c) ratio from 0.45 to 0.375 greatly retards expansion, but usually increases its ultimate value [19]. Increase to 0.7 accelerates it [27].

10.5 Effect of cement composition

Many attempts have been made to relate damage from DEF to cement composition. The complexity and variety of the relations proposed show that no simple relations exist. The implication of ettringite in the process has focused attention on SO_3 and Al_2O_3. Early work suggested that expansion increased with SO_3/Al_2O_3 ratio, or with the ratio of $(SO_3)^2$ to Al_2O_3 in the Bogue content of C_3A [26,27]. Later work indicated rather that there was a pessimum SO_3/Al_2O_3 mass ratio of ~0.8 [29], that high contents of both SO_3 and C_3A were needed to produce expansion [15], or that expansion increased with C_3A content and that there was a pessimum SO_3 content of ~4%, increasing with alkali content [18]. A pessimum in Bogue C_3A content has been reported [38].

The effects of alkali are complex and varying conclusions have been reported [16,18,20,21,28, 29,31,32]. Wieker and Herr [28] considered that expansion increased with Na_2O^E (Na_2O + $0.66K_2O$), but this quantity was highly correlated in their cements with SO_3/Al_2O_3 and recent data suggest that the effect of alkali increases with SO_3 level [16,20,39]. Expansion also tends to increase with MgO content, Bogue C_3S content and fineness [18,19]. Anything promoting rapid development of strength thus appears to increase the tendency to expansion. Sulfate resisting Portland cements give little or no expansion [18,29]. Substitution of sufficient proportions of cement by slag, flyash, or a natural pozzolana decreases or eliminates expansion [18,20,27,29].

Lawrence [31,32] and Hobbs [6] used multiple regression analysis to predict expansion from a number of cement variables. Taylor [1] queried the applicability of this statistical technique, and noted that Lawrence's data included many cases for which expansion was insignificant up to critical values of SO_3, Na_2O^E and MgO. A more complex expression, due to Kelham [39], gave satisfactory results when applied to a limited range of production clinkers, but results for a wider range of data were less encouraging. The results summarized in this section indicate that one cannot be certain of avoiding expansion by placing special limits on cement composition.

10.6 Cement hydration at 70–100°C

10.6.1 General aspects

Kalousek and Adams [40] found that, in cement pastes hydrated at 70–100°C, the ettringite formed initially was replaced by monosulfate or a related solid solution, which later also became undetectable. The cement was, by modern standards, very low in SO_3. They concluded that the Al^{3+} and SO_4^{2-} were incorporated into what would now be called the C-S-H gel. These changes occurred more rapidly with increasing temperature and also occurred, though more slowly, at ordinary temperatures. Further work on pastes made with cements higher in SO_3 showed that some monosulfate persisted after 19 h at 82°C, and that at sufficiently high SO_3 contents, some ettringite also remained [41]. The amounts of AFm and AFt phases found by DTA never accounted for as much as one-half of the total SO_3.

Subsequent work by many investigators has largely confirmed these results. In some cases, monosulfate has been detected by XRD or DTA directly after heat treatment, and in others it has not. The amount, if any, that remains decreases with the temperature and, as Sylla [2] noted, increases with C_3A content. Heinz and Ludwig's

[26] observations are typical for modern high-early strength cements: they found that ettringite was undetectable by XRD or DTA after several hours curing at 70°C and that monosulfate was also undetectable, or nearly so, by 100°C. Especially at the higher temperatures in this range, small proportions of a hydrogarnet were also found. With a mortar made using a cement containing 8.6 per cent of SO_3, ettringite was present immediately after heat treatment at 90°C. This confirmed that, if sufficient SO_3 is present, ettringite remains stable in a cement paste at this temperature.

These results are readily explainable by the relevant equilibria and empirical studies on pure systems. The solubility of ettringite increases markedly with temperature and alkali content [42]. Ettringite can be prepared from mixtures of C_3A, calcium sulfate and water at temperatures up to at least 90°C under atmospheric pressure [41,43], but an attempt to make it at 80°C in the presence of C_3S and 0.5M KOH yielded C-S-H and CH as the only detectable solid products [44]. Ettringite is not intrinsically unstable in aqueous systems below ~100°C, but can disappear through high pH and presence of C-S-H, which competes for Al^{3+} and SO_4^{2-}.

10.6.2 Monosulfate and hydrogarnet

The monosulfate is normally detected as the 12-hydrate, characterized by a strong XRD powder spacing of 0.89 nm. In one investigation, the 10-hydrate, with a layer spacing of 0.82 nm, was found [45]. Much of the monosulfate is very poorly crystalline, and failure to detect it by XRD does not prove its absence. ^{27}Al nuclear magnetic resonance (NMR) is more sensitive [20,46]. U-phase ($C_4A_{0.9}\bar{S}_{1.1}N_{0.5}H_{16}$?) could possibly form in pastes of high-alkali cements [47,48].

Hydrogarnet is commonly detected by XRD in materials hydrated at 80–110°C [20,22,26,45,49]. It is C_3AH_6 with substantial substitutions of Al^{3+} by Fe^{3+} and of OH^- by SiO_4^{4-} and smaller ones of Al^{3+} by Mg^{2+} and of $4OH^-$ by SO_4^{2-} [50]. The apparent content found by XRD increases with temperature; in pastes hydrated in sealed containers at 110°C for 12 h–14 days, it was the only hydrated aluminate phase detected [49]. Many investigators have found it in materials autoclaved at higher temperatures. In pastes cured at ordinary temperature, a poorly crystalline phase similar in composition to a hydrogarnet has been observed [51], and may be the major hydration product of the ferrite.

10.6.3 Al substitution in the C-S-H

SEM microanalyses of the inner product C-S-H in materials cured at 60–90°C show mean Si/Ca and Al/Ca ratios of 0.44–0.51 and 0.03–0.06, respectively [16,17,20–22,52]. These values do not differ consistently from those found in comparison materials cured at 10–20°C. Studies by electron probe microanalysis, transmission electron microscope (TEM) microanalysis and ^{29}Si NMR [20,53] indicate that most of the Al in the inner product C-S-H is present in that phase itself and not in closely admixed phases, the Al^{3+} substituting for Si^{4+} as suggested originally by Copeland *et al.* [54].

10.6.4 Sulfate in the C-S-H and pore solution

At the end of the heat treatment, the sulfate is distributed between C-S-H, pore solution, and any monosulfate, ettringite or hydrogarnet. Very small amounts may remain in clinker phases. Data from X-ray microanalyses for the C-S-H formed as inner product of clinker grains in cement pastes within 24 h of hydration for 15–24 h at 70–100°C typically show S/Ca atom ratios of 0.04–0.10 [16,20–22,45]. These values are much higher than those of 0.01–0.03 commonly found in pastes hydrated at ordinary temperature. They probably include a small contribution from the pore solution, as this evaporates during specimen preparation and subsequent examination in the microprobe or SEM.

Copeland *et al.* [54] considered that S^{6+} could substitute for Si^{4+} in C-S-H, but this is unlikely from a crystal–chemical standpoint. The sulfate is, moreover, very easily removed and has been described as adsorbed [55]. The rate at which gypsum reacts in C_3S pastes increases with temperature [56]. Sulfate is also sorbed by a more highly ordered form of C-S-H precipitated in aqueous suspensions [57]. In this case, the amount sorbed was shown to increase with the sulfate concentration in solution in accordance with the Langmuir isotherm, and also to increase with pH and temperature. The effects of pH and temperature are consistent with the observations showing that the S/Ca atom ratio in the C-S-H gel of cement pastes is much higher directly after hydration at elevated temperatures than at ordinary temperature, and that alkali promotes decomposition of ettringite.

If C_3S is hydrated at ambient temperature in the presence of gypsum in proportions up to an S/Si atomic ratio of 1:6, the gypsum becomes undetectable by XRD or DTA [54–56]. Charge balance for incorporation of SO_4^{2-} can therefore be obtained by incorporation also of Ca^{2+}, and this is perhaps the most likely mechanism in cement pastes.

Reported analyses of pore solutions squeezed out of cement pastes within a few hours or days of curing at 80–90°C show SO_3 concentrations of 15–200 mmol l^{-1} [20,28,58,59]. The wide variation in results may be associated with experimental factors, thereby obscuring significant conclusions relating them to cement composition. Calculations based on the estimated quantities and compositions of the C-S-H and of the pore solution at the end of the heat treatment show that, even assuming the highest observed value for the SO_3 concentration in the pore solution, the latter contains much less sulfate than the C-S-H.

10.7 Chemistry of changes after cooling to ambient temperatures

Ghorab *et al.* [25] were the first to describe the chemical and physical behaviour of mortars that had been cured at an elevated temperature and subsequently stored at ambient temperature. For that made with a high early strength cement, monosulfate was detected by XRD directly after the heat treatment. Over a period of months, its

content decreased and ettringite was formed; the changes were accompanied by expansion and decreases in strength and resonance frequency. Results for mixtures with a sulfate-resisting Portland cement and with flyash, slag or trass were also reported.

Many other investigators have confirmed and extended these results. With many Portland cements, including typical high early strength cements, the main change detectable by XRD or thermal methods is the replacement of monosulfate by ettringite, but with cements higher in C_3A or lower in sulfate, previously undetectable monosulfate can be formed [2,49]. Hydrogarnet, if present directly after the heat treatment, remains, but its XRD peaks do not become more intense during storage, indicating that no more is formed [20,21,26,49].

Varying observations have been reported on the changes in the pore solution [28,58,59]. All show either decreases in composition with time or values remaining constant for varying periods but, as with the values reported directly after the heat treatment, no consistent pattern has yet emerged.

The mean S/Ca ratio of the C-S-H formed during the heat treatment decreases during storage. Typically, the ultimate values for the inner product are 0.02–0.03 [16,17,20,21,36,45,60]. In contrast, the mean Al/Ca ratio of the inner product C-S-H usually decreases only slightly or not at all, typically remaining around 0.05 or falling to ~0.04 [16,20,21,36,45,60].

The data summarized above indicate that the principal reaction during storage is probably:

C-S-H provides	Ca^{2+}, SO_4^{2-}, OH^-, H_2O	
Monosulfate provides	Ca^{2+}, SO_4^{2-}, OH^-, $Al(OH)_4^-$, H_2O	which react to form ettringite
Pore solution provides	H_2O, minor SO_4^{2-}	

The relative constancy of the Al/Ca ratio of the C-S-H, and the low or zero reactivity of the hydrogarnet, if any [61,62], suggest that neither of these phases is a major source of Al^{3+}. The hydrotalcite-type phase ($\sim M_4AH_x$), which accommodates much of the MgO released from the clinker phases, does not appear to react with sulfate [63], and this is probably true also of any carbonate-containing phases that may be present. However, unreacted clinker phases may provide an additional source of aluminate and, to a very minor extent, of sulfate.

All changes are likely to proceed through the pore solution, though the distances that the ions travel could be very small. Passage of sulfate through the pore solution during and after the heat treatment might account for the varying patterns of concentration change that have been reported.

10.8 Paste microstructure in materials cured at elevated temperature

The inner product C-S-H present in pastes stored after curing at elevated temperatures shows a characteristic 'two-tone' structure in backscattered electron images, the outer zone being lighter than the inner [17,20–22,60,64,65]. The effect is due mainly to differences in gel porosity and water content, both of which are lower for the material formed at elevated temperature [20]. The differences in composition are small. Similar effects have been reported in pastes cured entirely at ordinary temperature [9,66]. We have observed this in normally cured mortars containing silica fume and in ones that have suffered carbonation, but in these cases the differences in grey level were accompanied by large differences in Ca/Si ratio.

The inner product C-S-H present directly after the heat treatment typically has a S/Al ratio of ~ 1.5 for materials that subsequently expand, and one of ~ 0.5 for ones that do not. This was originally attributed to admixture of a substantially unsubstituted C-S-H with ettringite or monosulfate, respectively [45], but the virtual absence of these phases shows that the relation to S/Al ratio is accidental. A more valid indicator may be the S/Ca ratio, values above ~ 0.06 suggesting that subsequent expansion may, but will not necessarily, occur [20].

Scrivener and Taylor [45] concluded from a study using backscattered electron imaging and XRD that the ettringite forms initially as crystals of sub-micrometre dimensions intimately mixed with C-S-H and subsequently tends to recrystallize in Hadley grains and other cavities of similar size (5–10 μm). No cracking was observed around these cavities, but expansion was unlikely to have been appreciable at the ages studied (up to 135 days). Yang et al. [17] studied pastes of the same cement cured at 100°C and subsequently stored for 4 years. The prisms had expanded significantly and were heavily cracked and sometimes distorted. BSE images showed cracks and bands of ettringite some 20 μm wide. The extent of cracking and the apparent content of ettringite (XRD) were greater near the surface of the prism than in the interior. Microanalyses of the outer product showed mixtures of ettringite or monosulfate with C-S-H in varying proportions.

Patel et al. [60] observed the presence of ettringite and monosulfate in Hadley grains in concretes cured at elevated temperatures. The proportion of ettringite was higher for material cured at 80°C than for that cured at 60°C.

Famy [20] studied the C-S-H gel present in the outer product in greater detail. For an expanding mortar, directly after heat treatment at 90°C, C-S-H with mean Al/Ca ~ 0.05 and mean S/Ca ~ 0.06 was mixed on a micrometre scale with varying proportions of monosulfate (Figure 10.1a). After a 200-day storage, the S/Ca and Al/Ca ratios of the C-S-H had dropped and the C-S-H was mixed with ettringite, monosulfate, and an aluminate phase or phases containing little or no S, shown by XRD to include monocarbonate. Almost all of the ettringite in the outer product was mixed at a sub-micrometre level with a much higher proportion of C-S-H. For a non-expanding mortar, the situation directly after the heat treatment was similar to that of the expanding mortar (Figure 10.1b); after storage the S/Ca ratio of the C-S-H had dropped, and mixtures with monosulfate and the low sulfate material, but not ettringite, were observed. However, for both the expanding and the non-expanding mortars, clusters

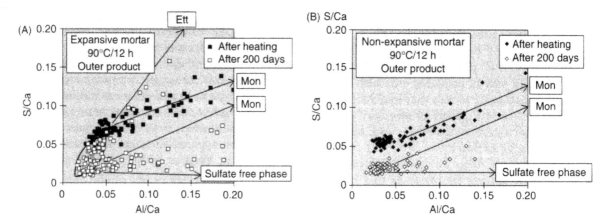

FIGURE 10.1 X-ray microanalyses of the outer product in (A) an expansive and (B) a non-expansive mortar [21]. Analyses within the ringed area are attributed to C-S-H either alone or mixed on a sub-micrometre scale with small proportions of ettringite.

of ettringite crystals were observed in hollow shells and other cavities; these are not represented in Figures 10.1a and b.

10.9 Expansion

10.9.1 The mechanism of expansion

The formation of ettringite entails a substantial increase in the solid volume, but this is comparable to that occurring in the hydration of alite, and cannot in itself account either qualitatively or quantitatively for the expansion. Although other mechanisms have been suggested, expansion most probably results from crystal growth pressure. There are differences of opinion as to whether expansion in mortars or concretes is driven by growth of ettringite crystals at aggregate interfaces or by processes occurring in the paste. If the latter expands, gaps will be formed around aggregate particles [67], and ettringite or other phases may recrystallize in them, simultaneously or subsequently.

The main arguments for the paste expansion theory are that:

(i) the widths of peripheral cracks are proportional to aggregate size;
(ii) cracks at the interfaces are initially empty;

(iii) assuming that expansion occurs through crystal growth pressure, significant growth pressures could not be obtained in relatively large cracks; and
(iv) pastes expand, albeit slowly.

Proportionality between crack width and aggregate size, which can only be explained by paste expansion, was first reported by Johansen *et al.* [67]. Hobbs [6] observed a qualitative relation. In contrast, Diamond [9] and Yang *et al.* [22,33] found no relationship and also found cracks passing through the paste. Against the paste expansion theory, Yang *et al.* [22] reported that the gaps around aggregate particles were rarely filled incompletely with ettringite; they tended to be either full or empty.

However, Famy [20] observed that gaps around aggregate particles were sometimes partly filled with ettringite or CH (Figure 10.2) and that empty gaps preceded filled gaps. The onset of significant expansion coincides approximately with the deposition of ettringite at aggregate interfaces [20,33]. This has been regarded as evidence against the paste expansion hypothesis, but this argument is only valid if recrystallization is slow relative to expansion.

The heterogeneity of any mortar or concrete leads to unavoidable ambiguities in the

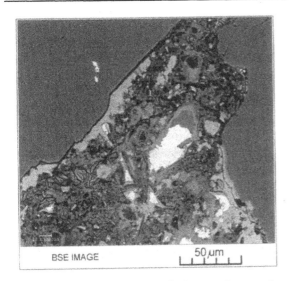

FIGURE 10.2 Backscattered electron image of a mortar cured at 90°C and subsequently stored in water for 80 days, producing a linear expansion of 0.15 per cent [20]. Gaps have started to form around aggregates; in places they contain ettringite or calcium hydroxide and in others they are empty.

interpretation of microstructure. On the paste expansion hypothesis, strict proportionality between gap width and grain size would be expected only if expansion is uniform. However, many locally anisotropic features and restraints of both internal and external origin exist in a real material, which would render expansion non-uniform except perhaps in localized regions. The microstructural arguments against paste expansion are largely ones against *uniform* paste expansion.

Thermodynamics provides strong arguments for paste expansion. The driving force for growth from solution is supersaturation, and any growth pressure produced cannot exceed a value set by the degree of supersaturation. A crystal will not grow against an external load if the material can be deposited elsewhere where free space is available [67,68]. The conditions for high growth pressure are therefore met by formation of very small, compact crystals in a region of high supersaturation in a confined space of minimal dimensions. If a crystal growing in a pore is in equilibrium with solution at the pore entrance, growth can occur at the pore walls because of the difference in curvatures, but the pressures thus arising are very small unless the pore radii are well below 100 nm [68]. Higher pressures can arise with larger crystals if these are not in equilibrium with solution at the pore entrance.

The conditions that could produce high growth pressures exist in the outer product of the paste, where particles of C-S-H and monosulfate are close together and access to larger spaces further away is restricted. They are likely to be particularly well developed close to the inner product, where the pores are especially small and the supply of sulfate from the inner product high; further out, the S/Ca ratio of the C-S-H is also high, but the C-S-H is more widely dispersed and the pores larger and more highly connected. Especially during wet storage, the conditions for high growth pressure are very unfavourable at aggregate interfaces, where supersaturation is low and there is much accessible space. The view that ettringite crystals of colloidal dimensions, but not those forming long needles, can produce large expansions was earlier expressed by Mehta [69], whose theory of the expansion mechanism was, however, different from that suggested here. A conclusion [35] that nucleation of ettringite crystals at the tips of large pre-existing cracks could generate large expansive pressures has been disputed [68]. The following mechanism, shown in Figures 10.3 and 10.4, is suggested.

Ettringite forms in pores of sub-micrometre size in the outer product. The principal reactants are monosulfate, C-S-H and the pore solution, and the ettringite crystals are formed close to those of the reacting monosulfate. When this process has proceeded to a sufficient extent, an uneven expansion of the paste begins. This produces cracks, both within the paste and at aggregate interfaces. Ostwald-type ripening causes ettringite and CH to recrystallize in accessible voids of all kinds. These include both pre-existing ones, such as Hadley grains and larger cavities, and the cracks resulting from expansion. All these processes are in varying degrees simultaneous rather than sequential. The growth of crystals by recrystallization makes, at most, a minor contribution to expansion.

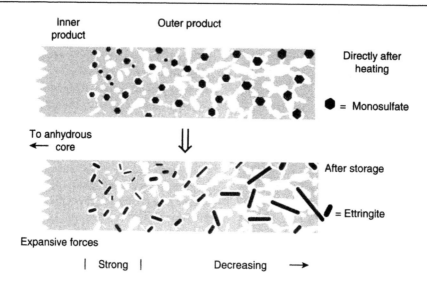

FIGURE 10.3 Suggested microstructures of the near outer product directly after the heat treatment and after subsequent wet storage. Unshaded areas represent pores, and shaded areas represent phases other than monosulfate or ettringite (mainly C-S-H). Schematic diagrams, not to scale; the smallest ettringite crystals are <100 nm long and the lengths of the sections are ~5 μm.

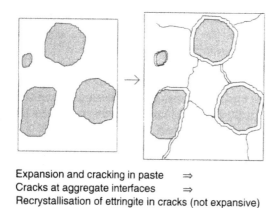

Expansion and cracking in paste ⇒
Cracks at aggregate interfaces ⇒
Recrystallisation of ettringite in cracks (not expansive)

FIGURE 10.4 Suggested expansion mechanism (schematic).

10.9.2 Factors controlling the rate and extent of expansion

The factors influencing expansion will be considered under three headings, namely:

  (i) Chemistry,
 (ii) Paste microstructure,
(iii) Mortar or concrete microstructure.

10.9.2.1 Chemistry

This is only one factor influencing expansion, and there is no simple or general relation between the content of ettringite in a material that has undergone DEF and expansion [16,20]. There is not even a general relation between expansion and the amount of ettringite formed during storage. In a general way, however, these latter amounts could explain the existence of pessimum contents of SO_3 and Al_2O_3. Low contents of either SO_3 or Al_2O_3 in the cement restrict the amount that can be formed. If SO_3 is sufficiently high, ettringite is present at the elevated temperature and less can therefore be formed during storage. If Al_2O_3 is high, monosulfate rather than ettringite will be present after storage. The pessimum content of each variable probably depends on the content of the other. A majority of cements probably have Al_2O_3 contents higher than the pessimum for the given SO_3 content. For such cements, anything that decreases the effective content of Al_2O_3 will increase expansion, unless the decrease is so great as to move the effective content to the other side of the pessimum. This may explain the effects of some other variables. Thus, observed increases in expansion with MgO content could be partly or wholly due to ones in the amount of hydrotalcite-type phase formed. Small amounts of CO_2 can bind larger amounts of Al_2O_3 in $C_4A\overline{C}H_{11}$ or $C_4A\overline{C}_{0.5}H_{12}$. This could increase or decrease expansion, or have no effect, depending on amount; additions of ground limestone are reported to have little effect on expansion [18,27]. Hydrogarnet formation also binds Al_2O_3. Granulated blastfurnace slag and most pozzolanas, such as flyash, act as sources of Al_2O_3. This may partly or wholly explain why they lower or eliminate expansion, though they have additional effects that may also be relevant.

Glasser *et al.* [42] described a method for predicting the quantities of each solid phase existing at equilibrium at 25°C, 60°C and 85°C for given contents of SO_3, Al_2O_3 and equivalent Na_2O. It was based on equilibria in the CaO–Al_2O_3–SO_3–Na_2O–H_2O system. From the phase compositions at elevated and ambient temperatures, together with the densities of the phases, they predicted the increases in solid volume during subsequent storage. The authors assumed that this gave an indication of the likelihood of expansion, but recognized that some ettringite could be accommodated in available spaces without contributing to expansion. This procedure accounted for the pessimum in SO_3 content and predicted that expansion would increase with cement alkali content, but the silicates were considered to play no significant part in the reactions; this, and some other assumptions were unrealistic.

Addition of alkali hydroxides to the initial mix in some cases increases subsequent expansion, and in others has no significant effect. Addition of KOH increased expansion with cements containing 4.1–5.3 per cent of SO_3; for mixes of the same total SO_3 content, K_2SO_4 has a similar effect to KOH [39]. With a cement containing 2.9 per cent SO_3, additions of KOH, or of additional gypsum to raise the SO_3 content to 3.9 per cent, had no significant effect, but addition of K_2SO_4 to give the same SO_3 content markedly increased ettringite formation and expansion [20].

The mechanism by which alkali present in the initial mix can increase expansion is uncertain.

Expansion, when it occurs, does so most rapidly during storage in water. Under this condition, alkali is lost by leaching [20], and any effect of the alkali content of the initial mix is, therefore, on processes occurring before the end of the heat treatment. The increases in expansion have been attributed to ones in the sulfate concentration of the pore solution [28,42] but appear too large to be explained purely by this or other chemical effects and microstructural effects may be at least as important. These are discussed in the next section.

10.9.2.2 Paste microstructure

Expansion is determined by the formation of ettringite crystals in places where their growth can exert pressure. If the ettringite is deposited in places where the pores are relatively large and highly connected, little or no pressure is exerted, whereas if it is deposited in ones where they are smaller and less highly connected, significant pressures can arise. This latter situation is favoured by high degrees of hydration. This may explain the relation between expansion and early strength, since this is related to the rate of hydration. Variations in paste microstructure may contribute to the effects on expansion of the duration and temperature of heat treatment. They could also contribute to the effect of the alkali content of the initial mix, since high alkali contents can markedly accelerate alite hydration.

10.9.2.3 Mortar or concrete microstructure

This determines the ability of the material to resist the stresses associated with ettringite formation. In the interior of an initially undamaged paste, these stresses seem insufficient to cause disruption, so that expansion and cracking spread only slowly from external surfaces. In materials made with exclusively limestone aggregates, the paste–aggregate bond is relatively strong and a similar situation exists. Similarities in thermal expansion coefficients may contribute to the robustness of the material. With quartz aggregates, in contrast, the paste–aggregate bond is intrinsically weaker and may be further weakened at elevated temperatures through superficial reaction with the alkaline pore solution [1]. This allows the reaction and resulting expansion to spread more quickly. Pre-existing weaknesses, provided, for example, by ASR, freezing and thawing, thermal stresses or mechanical damage will have a similar effect.

It has long been known that elevated temperature curing can result in damage through microcracking due to thermal stresses if proper precautions are not taken. The stresses can be local, due to the differing thermal expansion coefficients of the constituent materials, or on a larger scale because of thermal gradients. The damage can be minimized by adequate precuring to ensure sufficient strength development before heat is applied, sufficiently low rates of heating and cooling and avoidance of unduly high temperatures. These aspects of the curing regime are in varying degrees relevant to DEF, but microcracking also increases susceptibility to damage from other causes.

10.9.3 Effects of processing and post-curing conditions

Variations in pre-curing time up to ~4 h have given divergent results, perhaps because of differences between the cements used [2,15,27]. Extension of the pre-curing time to 7–28 days lowers or eliminates expansion [20,31]. The explanation of this is not yet established.

Variations in the temperature or duration of heat treatment probably have several effects. With relatively low temperatures or short times, some ettringite may be present at the end of the heat treatment. Increase in time or temperature will counteract this, and thus increase expansion. It will also increase the degree of hydration by the end of the heat treatment, thus producing a finer and less highly-connected pore structure, which will render a given quantity of ettringite more effective in producing expansion. The effect of temperature may be further enhanced by an increasing tendency for the hydration products to be concentrated in the immediate surroundings of the clinker grains ([70]; Figure 10.3). Hydrogarnet formation

increases with temperature and probably also with time. If not too extensive, this will increase ettringite formation and expansion by increasing the effective SO_3/Al_2O_3 ratio, but if the amount of Al_2O_3 bound in hydrogarnet exceeds a certain level, it would have the opposite effect. This may explain observations that extension of the curing time to 7–14 days at 80–110°C greatly decreases or eliminates expansion [19,20,31,32]. XRD examinations have given varying results. One study [49] showed major formation of hydrogarnet after prolonged heat treatment at 80°C or 110°C or after subsequent storage, but another study [20] showed greatly decreased ettringite formation but only weak indications of hydrogarnet. The explanation of the inhibition of expansion through prolonged heat treatment remains uncertain.

Famy [20] and Famy *et al.* [21] confirmed earlier observations that expansion after heat treatment is slower in air at 90–100 per cent RH than in water, and showed that it was even slower or barely significant in alkali hydroxide solutions. The decreased expansion was attributed to the increase in ettringite solubility with pH, which impedes ettringite precipitation. In water, substantial leaching of alkali cations occurs and is mainly balanced by leaching of OH^-, which counters this effect. These observations are consistent with the view [44] that the observed promotion of ettringite formation by ASR is due to removal of alkali from the pore solution through formation of alkali-silica gel. This would supplement any physical effect.

10.10 Conclusions

1. Concrete that has experienced a temperature $\geqslant 70°C$ can under some circumstances deteriorate through delayed ettringite formation (DEF). Evidence for such damage in materials that have not experienced such temperatures is unconvincing.

2. Ettringite is a normal hydration product of many cements and tends to recrystallize in any available cracks or voids. In such locations, it does not cause damage, though its presence

there may result from damage, which can arise from DEF or other causes.

3. In DEF, experimental studies indicate that the seat of expansion lies in the outer product of the cement paste, where ettringite is formed by reaction mainly between monosulfate, C-S-H and the pore solution.

4. The danger of expansion from DEF cannot be reliably avoided by placing special limits on cement composition, but can be eliminated by limiting the internal temperature of the concrete. In the case of elevated-temperature curing, the pre-curing time must also be adequate and the rates of heating and cooling not too great. Equivalent conditions apply to mass concretes.

10.11 Acknowledgements

We thank Professor G. W. Scherer for helpful discussions and Professor F. P. Glasser and Drs D. W. Hobbs, S. Kelham, and C. D. Lawrence for making material available in advance of publication.

10.12 References

1. Taylor, H. F. W., in *Adv. Cem. Concr.*, M. W. Grutzeck, and S. L. Sarkar (ed.). 122, American Society of Civil Engineers, New York, (1994).
2. Sylla, H.-M., *Beton*, 38(11), 449 (1988).
3. Shayan, A. and Quick, G. W., *ACI Mater. J.*, 89(4), 348 (1992).
4. Oberholster, R. E., Maree, H. and Brand, J. H. B., *9th Int. Conf. Alkali-Aggregate Reaction Concr.*, London, 1992, The Concrete Society, Slough, UK, 739 (1992).
5. Scrivener, K. L., *Proc. 18th Int. Conf. Cem. Micros.*, International Cement Microscopy Association, Duncansville, Texas, 375 (1996).
6. Hobbs, D. W., in *Ettringite – the sometimes host of destruction*, Spec. Publ. 177, B. Erlin (ed.), American Concrete Institute International, Farmington Hills, Michigan, 159 (1999).
7. Shayan, A. and Quick, G. W., *Proc. 16th Int. Conf. Cem. Micros.*, International Cement Microscopy Association, Duncansville, Texas, 68 (1994).
8. Mielenz, R. C., Marusin, S. L., Hime, W. G. and Jugovic, Z. T., *Concr. Int.*, 17(12), 62 (1995).
9. Diamond, S., *Cem. Concr. Comp.*, 18(3), 205 (1996).
10. Collepardi, M., *Concr. Int.*, 21(1), 69 (1999).
11. Seligmann, P. and Greening, N. R., *Proc. 5th Int. Congr Chem. Cem.*, Tokyo 1968, Vol. 2, Cement Association of Japan, Tokyo, 179 (1969).
12. Herfort, D., Soerensen, J. and Coulthard, E., *World Cem. Res. Dev.*, 28(5), 77 (1997).
13. Michaud, V. and Suderman, R., *Proc. 10th Int. Congr Chem. Cem.*, Gothenburg, Amarkai AB and Congrex Göteborg AB, Sweden, Vol 2. paper 2(ii)011 (10 pp.), (1997).
14. Miller, F. M. and Tang, F. J., *Cem. Concr. Res.*, 26(12), 1821 (1996).
15. Odler, I. and Chen, Y., *Cem. Concr. Res.*, 25(4), 853 (1995).
16. Lewis, M. C., Scrivener, K. L. and Kelham, S., *Mater. Res. Soc. Symp. Proc.*, 370, 67 (1995).
17. Yang, R., Lawrence, C. D. and Sharp, J. H., *Cem. Concr. Res.*, 26(11), 1649 (1996).
18. Kelham, S., in *Ettringite – the sometimes host of destruction*, Spec. Publ. 177, B. Erlin (ed.), American Concrete Institute International, Farmington Hills, Michigan, 27 (1999).
19. Lawrence, C. D., in *Ettringite – the sometimes host of destruction*, Spec. Publ. 177, B. Erlin (ed.), American Concrete Institute International, Farmington Hills, Michigan, 105 (1999).
20. Famy, C., Ph.D. Thesis, Imperial College, London. (1999).
21. Famy, C., Scrivener, K. L., Atkinson, A. and Brough, A. R., *Cem. Concr. Res.*, 31(5), 795 (2001).
22. Yang, R., Lawrence, C. D., Lynsdale, C. J. and Sharp, J. H., *Cem. Concr. Res.*, 29(1), 17 (1999).
23. Erlin, B., *Proc. 18th Int. Conf. Cem. Micros.*, International Cement Microscopy Association, Duncansville, Texas, 380 (1996).
24. Stark, J., Bollmann, K. and Seyfarth, K., *ZKG Int.*, 51(5), 280 (1998).
25. Ghorab, H. Y., Heinz, D., Ludwig, U., Meskendahl, T. and Wolter, A., *7th. Int. Cong. Chem. Cem.*, Paris 1980, Vol. 4, Editions Septima, Paris, 496 (1981).
26. Heinz, D. and Ludwig, U., *Concr. Durability, Katharine and Bryant Mather Int. Conf. (SP-100)*, Vol. 2, J. M. Scanlon (ed.), American Concrete Institute, Detroit, 2059 (1987).
27. Heinz, D., Ludwig, U. and Rüdiger, I. *Concr. Precast. Plant Technol.*, (11), 56 (1989).
28. Wieker, W. and Herr, R., *Z. Chem.*, 29(9), 321 (1989).
29. Ludwig, U., *11 Internationale Baustoff und Silikattagung [11th Int. Building Mater. Silicate Conf.]*, Vol. 1, TIZ, Weimar, 164 (1991).
30. Diamond, S. and Ong, S., *Ceram. Trans.*, 40, 79 (1994).
31. Lawrence, C. D., *Cem. Concr. Res.*, 25(4), 903 (1995).

32. Lawrence, C. D., in *Materials Science of Concrete IV*, J. Skalny and S. Mindess (ed.), American Ceramic Society, Westerville, Ohio, 113 (1995).

33. Yang, R., Lawrence, C. D. and Sharp, J. H., *Adv. Cem. Res.*, **11**(3), 119 (1999).

34. Grattan-Bellew, P. E., Beaudoin, J. J. and Vallée, V.-G., *Cem. Concr. Res.*, **28**(8), 1147 (1998).

35. Fu, Y., Xie, P., Gu, P. and Beaudoin, J. J., *Cem. Concr. Res.*, **24**(6), 1015 (1994).

36. Shao, Y., Lynsdale, C. J., Lawrence, C. D. and Sharp, J. H., *Cem. Concr. Res.*, **27**(11), 1761 (1997).

37. Shayan, A. and Quick, G. W., *Adv. Cem. Res.*, **4**(16), 149 (1992).

38. Grattan-Bellew, P. E., Beaudoin, J. J. and Vallée, V.-G., in *Materials Science of Concrete Special Volume, The Sidney Diamond Symposium* M. Cohen, S. Mindess and J. Skalny (ed.), American Ceramic Society, Westerville, Ohio, 295 (1998).

39. Kelham, S., *Cem. Concr. Comp.*, **18**(3), 171 (1996).

40. Kalousek, G. L. and Adams, M., *J. Am. Concr. Inst. Proc.*, **48**(1), 77 (1951).

41. Kalousek, G. L., *Mater. Res. Stand*, **5**(6), 292 (1965).

42. Glasser, F. P., Damidot, D. and Atkins, M., *Adv. Cem. Res.*, **7**(26), 57 (1995).

43. Lieber, W., *Zem.-Kalk-Gips*, **16**(9), 364 (1963).

44. Brown, P. W. and Bothe, J. V., *Adv. Cem. Res.*, **5**(18), 47 (1993).

45. Scrivener, K. L. and Taylor, H. F. W., *Adv. Cem. Res.*, **5**(20), 139 (1993).

46. Skibsted, J., Jensen, O. M. and Jakobsen, H. J., *Proc. 10th Int. Cong. Chem. Cem.*, Gothenburg, Amarkai AB and Congrex Göteborg AB, Sweden, Vol. 2, paper 2(ii)056 (8 pp.), (1997).

47. Li, G., Le Bescop, P. and Moranville-Regourd, M., *Cem. Concr. Res.*, **27**(1), 7 (1999).

48. Clark, B.A. and Brown, P.W., *J. Am. Ceram. Soc.*, **82**(10), 2900 (1999).

49. Crumbie, A. K., Pratt, P. L. and Taylor, H. F. W., *9th Int. Cong. Chem. Cem.*, New Delhi, India, National Council for Cement and Building Materials, New Delhi, Vol. 4, 131 (1992).

50. Paul, M. and Glasser, F. P., *Cem. Concr. Res.*, **30**(2), 1869 (2000).

51. Rodger, S. A. and Groves, G. W., *J. Am. Ceram. Soc.*, **72**(6), 1037 (1989).

52. Escalante-Garcia, J. I. and Sharp, J. H., *J. Am. Ceram. Soc.*, **82**(11), 3237 (1999).

53. Richardson, I. G. and Groves, G. W., *J. Mater. Sci.*, **28**(1), 265 (1993).

54. Copeland, L. E., Bodor, E., Change, T. N. and Weise, C. H., *J. Portland Cem. Assoc. Res. Dev. Lab.*, **9**(1), 61 (1967).

55. Odler, I., *7th Int. Cong. Chem. Cem.*, Paris 1980, Editions Septima, Paris, Vol. 4, 493 (1981).

56. Fu, Y., Xie, P., Gu, P. and Beaudoin, J. J., *Cem. Concr. Res.*, **24**(8), 1428 (1994).

57. Divet, L. and Randriambololona, R., *Cem. Concr. Res.*, **28**(3), 357 (1998).

58. Glasser, F. P., *Cem. Concr. Comp.*, **18**(3), 187 (1996).

59. Ong, S., Ph.D. Thesis, Purdue University, quoted in Ref. 9 (1993).

60. Patel, H. H., Bland, C. H. and Poole, A. B., *Adv. Cem. Res.*, **8**(29), 11 (1996).

61. Flint, E. P. and Wells, L. S., *J. Res. Nat. Bur. Stds.*, **27**, 171 (1941).

62. Marchese, B. and Sersale, R., *Proc. 5th Int. Cong. Chem. Cem.*, Tokyo 1968, Cement Association of Japan, Tokyo, Vol. 2, 133 (1969).

63. Gollop, R. S. and Taylor, H. F. W., *Cem. Concr. Res.*, **26**(7), 1029 (1996).

64. Scrivener, K. L., *Cem. Concr. Res.*, **22**(6), 1224 (1992).

65. Jie, Y., Warner, D. A., Clark, B. A., Thaulow, N. and Skalny, J., *Proc. 15th Int. Conf. Cem. Micros.*, International Cement Microscopy Association, Duncansville, Texas, 250 (1993).

66. Diamond, S., Olek, J. and Wang, Y., *Cem. Concr. Res.*, **28**(9), 1237 (1998).

67. Johansen, V., Thaulow, N. and Skalny, J., *Adv. Cem. Res.*, **5**(17), 23 (1993).

68. Scherer, G. W., *Cem. Concr. Res.*, **29**(8), 1347 and private communication, (1999).

69. Mehta, P. K., *Cem. Concr. Res.*, **3**(1) 1 (1973).

70. Kjellsen, K. O., Detwiler, R. J. and Gjørv, O. E., *Cem. Concr. Res.*, **20**(2), 308 (1990).

Chapter eleven

Chloride corrosion in cementitious system

Wiesław Kurdowski

11.1 Introduction

Chlorides are among the most detrimental agents that cause quick concrete corrosion. The natural concrete medium is very basic due to the pore solution of the paste, which is rich in alkalis, and the pH value is around 13. It is the reason why the chloride solutions will always be more acidic and very corrodible. The intensity of the corrosive action will certainly depend on the chloride solution concentration and of the kind of cation linked to the chloride ion.

The process of corrosion demands the penetration of ions within the microstructure of concrete and simultaneously, in the case of chlorides, the transport of ions in the opposite direction from the leached hydrated cement phases. In this situation, the porosity and permeability of concrete will play a decisive role.

The cement paste is the concrete component which will first of all be the subject of corrosion. In this context, the influence of chloride ions on cement paste will be primarily discussed. However, some remarks about the influence of aggregate on chloride ion diffusion are indispensable for the complete interpretation of the situation in concrete.

In the last decade, the interest for better understanding of chloride corrosion has been constantly growing because of increasing maintenance and repair costs, principally of highway networks and bridges. In particular, it is the influence of the deicer salts which can cause scaling and deterioration of concrete surfaces as well as induce the corrosion of reinforcing bars in concrete. A better knowledge of the mechanism of chloride corrosion is the only way of finding the methods to solve these problems.

11.2 Chloride ions diffusion in cement paste

Chloride ions are classified as being most mobile and it is one of the reasons for relatively rapid concrete corrosion when this material is in contact with chloride containing water.

For the durability of concrete, the velocity of the diffusion has great importance, and hence, the diffusion phenomena are studied very frequently. For measurement of chloride ion diffusion two methods are most commonly used:

- steady-state diffusion tests,
- non-steady-state diffusion experiments.

For steady-state diffusion tests, the two-compartment cell is used between which a sample in the form of a disc is mounted. One compartment is filled with the ionic solution – in the majority of cases NaCl or CaCl$_2$ – and the other with deionized water or saturated Ca(OH)$_2$ solution. Marchand et al. [1] suggest that the solutions should be selected in order to meet the following requirements:

- pore structure alterations by leaching should be minimized during the duration of the test
- solution ionic strength should not change along the path of diffusion; that is, the chemical composition of the test solutions in both compartments should be as close as possible to that of the pore solution.

These precautions should minimize the chemical activity gradient from the upstream compartment and thus reduce the pore structure alterations. However, in the published papers, these requirements are seldom met and in the majority of cases, pure chloride and calcium hydroxide solutions are used in upstream and downstream compartments, respectively, or in the first mixture of these solutions (e.g. [2] and [3]).

For calculation of the 'effective' diffusion coefficient one can use the equation derived from Fick's first law by Page et al. [4]:

$$J = \frac{V}{A} \cdot \frac{dC_2}{dt} = \frac{D}{e} \cdot (C_1 - C_2) \tag{1}$$

where:

J = ionic flux through the disc (moles $\times$ cm^{-2} s^{-1}),
V = volume of NaCl solution in upstream compartment (cm^3),
A = disc area (cm^2),
e = disc thickness (cm),
D = effective diffusion coefficient (cm$^2 \times$ s^{-1}),
C_1 = chloride concentration of solution in upstream compartment (moles $\times$ cm^{-3}),
C_2 = chloride concentration of solution in downstream compartment (moles $\times$ cm^{-3})

Integration of the above equation results in:

$$\ln\left(1 + \frac{C_2}{C_1 - C_2}\right) = \frac{D \cdot A}{V \cdot e} \cdot (t - t_0) \tag{2}$$

where t_0 is the time required to establish the diffusion rate.

If C_1 is and remains always much higher than C_2, equation (2) can be approximated by equation (3) for $t > t_0$:

$$C_2 = \frac{D, A, C_1}{V} \cdot e \cdot (t - t_0) \tag{3}$$

The value of D can thus be calculated using equation (4):

$$D = \frac{V \cdot e}{A} \cdot \frac{C_2}{C_1} \cdot \frac{1}{(t - t_0)} \tag{4}$$

The calculation of D using equation (4) implies the following assumptions:

- a quasi-steady diffusion,
- a constant flux through the entire section of diffusion,
- an unchanging concentration of chlorides at all points of the disc.

In the calculation of the diffusion coefficient, the correction should be made for the Nernst-Planck retardations of ion mobility.

In the non-steady-state, diffusion tests long cylinders or prismatic samples are immersed in a solution containing chloride ions. All sample surfaces, except one, are sealed in order to prevent ions from penetration. The remaining surface is thus exposed to a unidirectional penetration of ions.

The apparent diffusion coefficient is usually approximated using Fick's second law for unidirectional flow. The coefficient derived from this approximation is an apparent coefficient that provides very little reliable information on penetration mechanisms [1]. Additionally, such calculations are not truly accurate, since the solution of Fick's second law rests on the assumption that the apparent diffusion coefficient is constant throughout the sample depth. It is, however, not true (Figure 11.1).

In comparing both the methods, one must remember that the chloride-binding capacity will influence to a different degree the results obtained by each method since, during a non-steady-state experiment, interactions of chloride ions with the cement paste phases take place and slow down the

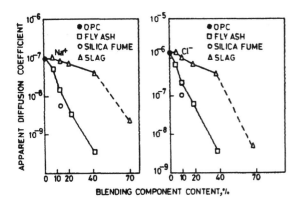

FIGURE 11.1   Apparent diffusion coefficient [1].

[1]. The application of the electric field will also create a phenomenon called electro-osmosis which will appear as a flow of ions, forming a double layer on the gel surfaces, towards one of the poles of the current source [1]. According to some authors, this phenomenon could accelerate the transport of chlorides through microcracked cement paste [1].

Many theoretical and experimental problems concerning the diffusion coefficient measurements in the cement paste are discussed in detail in Marchand *et al.* paper [1].

11.3 Binding capacity of chloride ions in cement paste

Many factors affect diffusivity of chloride ions in the cement paste and have a pronounced influence on the experimental results. First of all, chlorine can react with calcium aluminate forming Friedel's salt: $C_3A \times CaCl_2 \times 10H_2O$. When the chloride ion concentration is higher, also probably a compound containing more chloride can be formed, namely $C_3A \times 3CaCl_2 \times 30H_2O$ [7].

So the quantity of bound chloride ions will be strongly influenced by the content of C_3A in the cement. The hydrated cement pore structure seems to be another factor influencing the chloroaluminate formation. As stated by Marchand *et al.* [1] following Delagrave *et al.* [6], the chloroaluminate formation was never observed in high-performance cement paste (water-to-cement ratio w/c 0.25), even when chloride ions were present everywhere in the paste.

Several authors stated recently that the binding capacity of cement paste can be attributed not only to the C_3A content, but to the total aluminate content, that is, to the sum $C_3A + C_4AF$ [8,9]. This could be foreseen because there is an analogous compound to chloroaluminate, $C_3F \times CaCl_2 \times 10H_2O$ [10]. Also the solid solutions between aluminates and ferrite phases are well known and the formula closer to the real situation of these phases in cement paste should really be $Ca_3(Al,Fe) \times CaCl_2 \times 10H_2O$ [10,11]. However, the brownmillerite is hydrated much slower than C_3A so that its binding capacity will be

ion penetration, and the measured diffusion coefficient will be lower. During a steady-state experiment most of the interactions are completed and their effect on the chloride ions diffusion will be significantly reduced and the measured value of the diffusion coefficient will be higher.

For both described methods of diffusion coefficient measurements, the accelerated tests are frequently used by applying electrical potential to the system. They are sometimes called migration tests. The interpretation of these results essentially rests on the assumption that the mechanism of ionic migration in cement paste is very similar to that of ionic diffusion. However, there are only a few experiments comparing the results obtained by both methods. Tang [5] found a good correlation between the non-steady-state test and non-steady-state migration experiments. The opposite results were obtained by Delagrave [6], who found out that the results of steady-state migration experiments consistently underestimate the true values of the effective diffusion coefficient.

Among others, it is also assumed that the application of the electrical potential contributes only to acceleration of the transport of ions through the material and has no effect on the hydrated cement paste microstructure. Nevertheless, experimental results do not prove this assumption. On the contrary, long-term application of the electrical potential significantly modifies the paste microstructure much more time-dependent and will take place after a longer time of hydration.

The C-S-H gel is also known to chemisorb and adsorb the chloride ions. Beaudoin [13] and Ramachandran [12] studied this process, distinguishing three states of chlorides ions: free ions extracted easily with alcohol, chemisorbed on the gel surface and in the interlayer spaces which are not alcohol leachable and, finally, tightly-held chloride, probably in the solid solution in C-S-H phase, which cannot be removed by water leaching. The binding capacity of chloride ions by C-S-H phase depends on the H/S and C/S ratios, increasing with them [13].

These findings are very important for the consideration of binding capacity of the paste made of cements containing different mineral additions. For example, Diamond [15] and Page and Vennesland [14] found that the addition of silica fume significantly increases the free chloride concentration in the pore solution. In the light of experiments by Beaudoin et al. [13], it can be explained by the lowering of C/S ratio in C-S-H in the presence of silica fume.

The slag and fly ash additions will have a similar effect on the binding capacity of cement paste because they will lower the C/S ratio of C-S-H gel. For example, Uchikawa and Okamura [16] give the following value of C/S ratio: 1.7 for OPC, 1.6 with the addition of 40 per cent of slag and 1.2 for the addition of 40 per cent of fly ash. Simultaneously, these additions will lower the aluminate content to the benefit of the calcium silicate hydrate.

As it is evident from the works of several authors [1,17], the composition of the pore solution will also change the binding capacity of cement paste. Byfors [17] studied the influence of the concentration of OH^- and SO_4^{2-} in the mixing water on chloride binding capacity. He found out that the influence of concentration of both anions, that is, of OH^- and SO_4^{2-}, decrease the chloride-binding capacity. Tenoutasse [18] showed that aluminate and ferrite react first with gypsum (SO_4^{2-} ions) and only after the exhaustion of gypsum do chlorides start to react with the remaining C_3A or C_4AF. Additionally, Richartz [19] found that

chloroaluminates dissolve if exposed to sulphate solutions, increasing as a result the chloride concentration in the solution. However, it is dependent on chloride ion concentrations in the pore solution and, as a matter of fact, on pH of solution, that is, chloride-binding capacity will be dependent on chloride ion concentrations in the pore solution. These problems will be discussed in the part concerning the mechanism of cement paste destruction under the influence of chlorides.

In the context of a strong influence of the composition of pore solution on the chloride-binding capacity it is evident that the cement composition will also have a pronounced effect by influencing directly the pore solution composition.

The cation associated with the chloride ion has an important influence on the chloride-binding capacity. It was studied by Arya et al. [22]; Blunk et al. [21]; Byfors [24]; Tritthart [23] and Theissing et al. [20]. Marchand et al. [1] summarized their results stating that the sequence of ions with increasing chloride-binding capacity will be as follows: NaCl, KCl, $CaCl_2$ and $MgCl_2$. It is also in accordance with Byfors' [17] statement that the increase of OH^- ions concentration in pore solution decrease the chloride-binding capacity. Tritthart [23] pointed out also that one of the factors influencing the chloride-binding capacity will be the solubility of the hydroxide formed by a cation associating with the chloride ion.

11.4 Factors influencing the diffusion of chloride ions in the cement paste

Porosity and permeability of concrete will be the main factors influencing diffusion. Several reports indicate that the diffusion of chloride ions is directly influenced by the microstructure of the hydrated cement paste. For example, the diffusion rate has been found to decrease with a reduction of the water-cement ratio [4,23,25].

The pore structure will play a very important role in diffusion and at first the share of gel pores, smaller than 2 nm and capillary pores, up to 100 nm. Many experiments showed that the increase in the content of gel pores decrease the

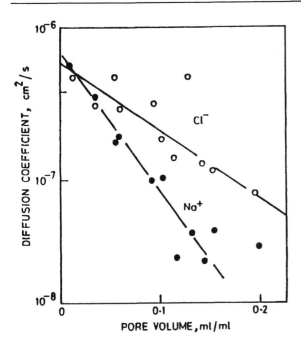

FIGURE 11.2 Diffusion coefficient as a function of the volume content of pores below 2 nm [26].

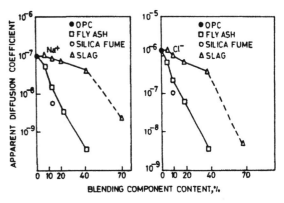

FIGURE 11.3 Diffusion coefficient of Na^+ and Cl^- ions in the cement paste with addition (45°C, w/c = 0.4) [27].

diffusion coefficient significantly. Figure 11.2, obtained by Goto and Daimon [26], can be presented as an example.

All the conditions increasing the content of gel pores will decrease the diffusion coefficient. In other words, the increase of C-S-H phase in the cement paste will influence the decreasing of the diffusion rate. For the increase of gel, two factors are most important:

- hydration time,
- addition of hydraulic and pozzolanic materials out of which blastfurnace slag and fly ash are the most popular.

The influence of additions on the diffusion coefficient [27] is presented in Figure 11.3. The chloride ion diffusivities increase in the sequence: slag cement < pfa cement < ordinary Portland cement < sulphate-resisting Portland cement [10]. Regourd [28] showed that the addition of slag and fly ash changes the pore size distribution and pore structure. When mineral additives react, there is a trend to form the finest pores and

discontinuous pores substitute for continuous capillary pores in relation to the increase in C-S-H. Consequently, it will diminish significantly the rate of diffusion.

When comparing the situation in cement paste and in concrete one must remember the differences between these materials. Note particularly the interfacial transition zone ITZ, formed at the vicinity of aggregate grains in concrete. It is a very thin layer, 20–30 mm thick, of hydrated cement paste where the microstructure differs significantly from that of the bulk volume. The ITZ is essentially characterized by a higher porosity and a gradient of the anhydrous and hydrated phases.

The aggregate present in concrete will have two opposite effects on its transport properties [29]. First, the addition of solid non-porous particles should increase the tortuosity of the pore structure. Second, the presence of porous and connected ITZ may contribute to facilitating the flow of ions or fluids through the sample. The importance of the ITZ connectivity on the transport properties has been clearly demonstrated by several authors [29–31]. Brenton *et al.* [32] and Bourdette [33] found out that the effective diffusion coefficient of chloride ions is 6–12 times larger in ITZ than in the bulk cement paste.

Among the most popular methods of eliminating ITZ one can mention the lowering of water/binder ratio and addition of silica fume. It is a widespread

299

opinion that the high-performance concrete is characterized by much lower penetration of chloride and increased durability in corrosive media.

11.5 Mechanism of cement paste destruction in the chloride medium

Three mechanisms of chloride corrosion are mentioned in the literature as a cause of concrete destruction:

- 'acid' corrosion with dissolution of $Ca(OH)_2$ and leaching of calcium from the cement paste phases,
- formation of expansive compounds, namely basic calcium and magnesium chlorides,
- osmotic pressure.

It is well known that the cement paste phases are stable in a strongly basic pore solution which has a high concentration of potassium and OH^- ions. Additionally, the pore solution is supersaturated towards all the hydrates which are stable in this condition. Nevertheless, the access of chloride ions to cement paste will disturb the thermodynamic equilibrium and, lowering pH, will cause the dissolution of these hydrates which during dissociation, will liberate OH^- ions. Simultaneously, it will be a 'defending' action of the cement paste against an external influence. As can be seen from Figure 11.4 where calculated solubilities are presented, portlandite will be the least stable component and will dissolve at pH 11 together with C_4AH_{19}. Ettringite will be the most stable. Yet, the stability of ettringite needs the presence of gypsum. When gypsum is absent, ettringite will decompose at pH 10.7.

The low concentration of lime will cause the decomposition of the C-S-H phase and in this condition the decalcification of this phase will start and the Ca/Si ratio will be lowered (Figure 11.5). As can be seen from Figure 11.5, the destruction of this phase took place when the concentration of lime in solution drops under 2 mM/l and this phase will convert into a porous, hydrous form of silica. For the calcium concentration between 20 and 2 mM/l of Ca^{2+} it is the C-S-H phase which will deliver these ions to the pore solution. Vernet [34], however, presents the opinion that the fall in pH below 11 in concrete can be attained only with great difficulty because the hydrates present will give a buffer effect. From the analysis of the solubility products, it is clear that in the case of the penetration of chloride ions the precipitation of Friedel's salt $C_3ACaCl_2 10H_2O$ will take place first, but the ettringite will be stable even at high-chloride concentration.

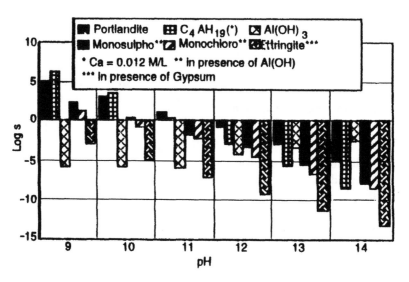

FIGURE 11.4 Solubilities of hydrates as a function of the pH level in the lime solution [34].

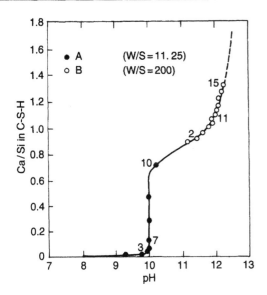

FIGURE 11.5 CaO/SiO₂ ratio in C-S-H as a function of calcium concentration in solution.

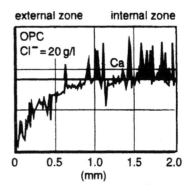

FIGURE 11.6 Distribution of the amounts of calcium and chlorides throughout the thickness of the sample (3 years, pH 11.5) [35]

As was mentioned earlier, the rate of diffusion of chloride ions is high, being higher than the diffusion of the cations. So, for retaining electrical neutrality, the diffusion of OH^- ions in the opposite direction must take place. It is the reason for the fall in pH and the dissolution of the cement paste hydrates.

There are discussions about the increase of $Ca(OH)_2$ solubility in the presence of NaCl or KCl [1]. But, in fact, the following exchange reaction always appears:

$$Ca(OH)_2 + 2NaCl \rightarrow CaCl_2 + 2NaOH$$

The solubility of $CaCl_2$ is obtained immediately. However, $CaCl_2$ will react with $Ca(OH)_2$, giving $Ca(OH)_2CaCl_2H_2O$, and the solubility will drop.

The decalcification of superficial layers of cement paste is well documented in the paper of Gegout et al. [35] which is shown in Figure 11.6, after 3 years of exposure to chloride solution ($Cl^- = 20$ g/l). There is no portlandite in the disc and even the C-S-H is partly decalcified.

Formation of expansive compounds the basic magnesium chloride $Mg_3Cl_2(OH)_42H_2O$ and the basic calcium chloride $Ca(OH)_2CaCl_2H_2O$, was first observed by Smolczyk [36] in the case of very high chloride ions concentration. These compounds are expansive and lead to the formation of microcracks and destruction of the paste.

There are contradictory opinions about the destructive action of Friedel's salt. Riedel [37] claims that there is no negative influence of this phase on cement paste strength. With saturated NaCl solution $C_3ACaCl_210H_2O$ is formed, and there is some loss in strength, but the effect is much lower than with $CaCl_2$ and $MgCl_2$ [37]. But, on the other hand, in the opinion of Conjeaud [38], formation of this phase participates in the destruction of the mortar due to chloride corrosion. Kurdowski et al. [39] studied the influence of the strong chloride solution on high-alumina cement (HAC) paste and found chloroaluminate formation in the external layer but without any traces of destruction or even microcrack formation. The bars of this paste were in an excellent condition after 11 years of immersion in strong chloride solution (164 g/l $CaCl_2 + 272$ g/l of $MgCl_2$).

In Table 11.1 the calculated molar volumes of the different alumina phases is shown. Consider the following reaction:

$$C_3ACa(OH)_2 \cdot 12H_2O + CaCl_2$$
$$\rightarrow C_3ACaCl_2 \cdot 10H_2O + Ca(OH)_2 + 2H_2O$$

There will be no volume change accompaning this process. However when $Ca(OH)_2$ is liberated in this reaction, it will further react with $CaCl_2$,

301

and the expansive basic calcium chloride will form with a real expansion.

The second possible reaction:

$$C_3ACaSO_412H_2O + CaCl_2$$
$$\rightarrow C_3ACaCl_210H_2O + CaSO_42H_2O$$

There will be a negative change of volume, but if we take into account the crystallization of gypsum, there will be a volume increase of 18 cm³/mol of AFm. However, in the strong chloride solution, anhydrite will crystallize instead of gypsum and the volume change will be negative.

This consideration seems to indicate that the chloroaluminate influence on paste destruction will be negligible.

A considerable increase of volume will take place when gypsum liberated in the last reaction reacts with monosulphate or C_4AH_{19} with the formation of secondary ettringite. In this case, the volume increase will be 406 and 462 cm³/mol, respectively.

TABLE 11.1   Molar volume of the different calcium aluminate hydrates

Compound	Molar volume (cm³/mol)
$C_3A3CaSO_4 \cdot 32H_2O$	725
$C_3ACaSO_4 \cdot 12H_2O$	319
$C_3ACa(OH)_2 \cdot 12H_2O$	263
$C_3ACaCl_2 \cdot 10H_2O$	263
$CaSO_4 \cdot 2H_2O$	74
$CaSO_4$	46
$Ca(OH)_2$	33

There are only a few papers treating the problem of osmotic pressure as the possible mechanism of cement paste destruction in chloride solution (39–41). Here, there are contradictions between these authors and Chatterji [42] rejects this mechanism. The results obtained by the author [43–45] give experimental proof for the larger influence of osmotic pressure. It was documented for the strong chloride solution and it will probably be much more difficult to find this phenomenon in the case of low concentrated solutions.

The experiments of chloride corrosion of cement paste covered a large number of samples prepared of different cements and also single phase pastes [43–47]. The most important samples were as follows:

- pastes of Portland cement with different C_3A content,
- high-alumina cement paste,
- slag activated cement paste,
- C-S-H paste,
- tobermorite paste.

The corrosion process always starts with shrinkage of the bars (Figure 11.7). This shrinkage, which is very quick in the first days of corrosion is caused by densification of the cement gel, resulting from osmotic pressure. This shrinkage results in the formation of microcracks which facilitate the diffusion of chloride ions and even the migration of the corrosive solution into the interior of the paste.

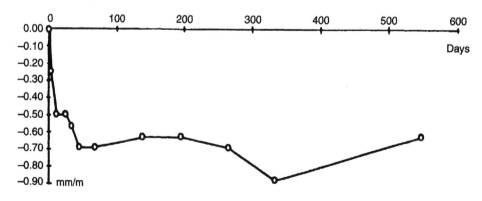

FIGURE 11.7 Shrinkage curve of clinker-free slag cement paste.

In these microcracks, basic magnesium chlorides are precipitated [43,44], but the swelling of the paste seems to be rather limited, especially at the beginning of corrosion, because there is free volume in which these new formations can crystallize (Figure 11.8). Thereby, in the instance of a

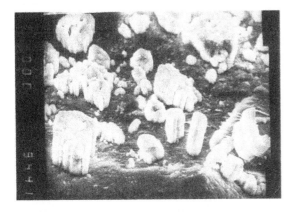

FIGURE 11.8 Crystals of basic magnesium chloride in microcracks formed during shrinkage of cement paste immersed in the strong chloride solution for 23 months.

dense C-S-H gel, the crystallization of basic magnesium chloride causes a diminution of shrinkage (period between 50 and 550 d in Figure 11.7) [45]. Subsequently the shrinkage of the paste prevails which is the result of gel densification. The external layer of the paste becomes very compact and porosity becomes very low (Figure 11.9) [46].

The process of C-S-H gel densification occurs before the chemical reactions of corrosion arise and can be seen under SEM (Figure 11.10) [47]. The densification of gel hinders the diffusion of ions which, in these conditions, is very slow. As an example, the chemical composition of the core of slag clinker-free paste can be presented (Table 11.2) [45].

The composition of the samples $40 \times 40 \times 160$ mm after 23 months of immersion in the solution of composition given in Table 11.2 shows that the changes of magnesium content in the core are imperceptible. Namely, the recalculation on the ignition free basis gives 10.2 and 6.6 per cent in external layer and in the core, respectively. The process of diffusion of chloride ions is much

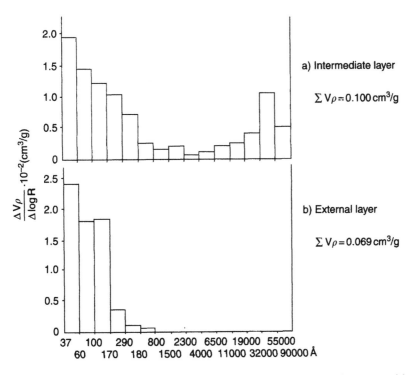

FIGURE 11.9 Porosity of slag cement paste after two years of immersion in the strong chloride solution.

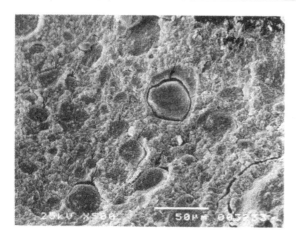

FIGURE 11.10 Dense grains (arrows) in C-S-H gel, sample after 1200 d of immersion in the strong chloride solution.

TABLE 11.2 Chemical composition of original slag, external layer and the core of the paste immersed in chloride solution for 23 months (195 g/l NaCl, 98 g/l KCl, 77 g/l $MgCl_2$)

Component	Content in mass %		
	Slag	External layer	Core
Moisture (105°C)	—	8.9	10.1
LOI	—	12.8	9.5
SiO_2	39·3	28·4	29·5
Al_2O_3	6·5	6·1	6·1
Fe_2O_3	2·6	2·1	2·2
CaO	43·7	30.8	33·0
MgO	6·4	8·4	5·5
K_2O	n.d.	1·2	0·5
Na_2O	n.d.	1·8	2·4
Cl	0·0	3·6	3·6

quicker but the new formations in the core of the sample are undetectable by X-ray analysis [45].

With the C-S-H paste, obtained by hydration of the mixture of tricalcium silicate with amorphous silica (C/S ratio equal to 1), because of the high water/binder ratio, the porosity of the paste was much larger, namely 0.59 cm^3/g for this paste and 0.12 for a slag paste. The diffusion of magnesium ions was also quicker and the decalcifying led to the destruction of C-S-H gel and destroyed the integrity of the paste [46]. The X-ray pattern shows the lines of the siliceous acid $H_2Si_3O_7$ and of magnesium silicate hydrate.

In the case of the samples which practically do not contain amorphous gel or only in very small quantities, namely, samples composed of tobermorite, no shrinkage at all is observed, even in the initial period [47]. During the time of immersion in the strong chloride solution an expansion of the samples takes place, although the chemical reactions which are involved here are the same as for the C-S-H gel. A fairly quick decalcifying of tobermorite occurs with the amorphization of this phase. Without doubt, the high porosity of the starting material (0.73 cm^3/g) accelerates the decalcifying process with the formation of basic magnesium chlorides.

The initial period of the shrinkage is also finished quickly when the Portland cement paste is immersed in the strong chloride solution, because the crystals of portlandite are quickly dissolved, which causes an increase in the porosity of cement paste. The diffusion of magnesium ions is thus accelerated, and the formation of expansive basic magnesium or calcium chlorides takes place. The shrinkage is thus transformed into expansion which leads to a quick paste destruction (Figure 11.11).

In the paste of Portland cement, by the reactions with magnesium or calcium chlorides secondary ettringite is also formed, which is the product of monosulphate decomposition by chlorides. This secondary ettringite seems to have a rather limited influence on the paste corrosion because the influence of C_3A content on mortar destruction in strong $CaCl_2$ solution is rather insignificant [48]. However, the pastes of the same cements are much more durable and the paste of Portland cement low in aluminates and with a high silica ratio is more stable. Much quicker mortar destruction is probably the result of the ITZ effect.

Because of much higher durability of slag cement, this paste becomes destroyed by shrinkage [46]. Growing shrinkage of the external part of the samples gives an increasing stress which finally becomes higher than the bending strength, and leads to the cracking of the bars perpendicularly to their longer axis.

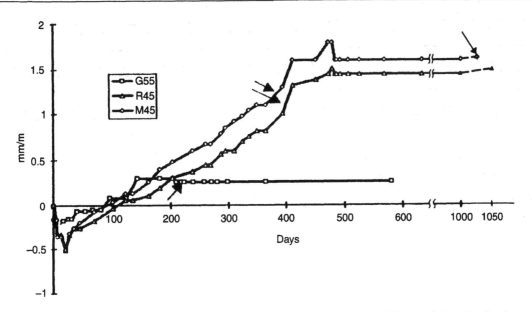

FIGURE 11.11 Expansion of the bars of the mortar and of the paste (dashed lines) of three Portland cements with different C_3A content: R 45 = 4 per cent, M 45 = 9 per cent and G 55 = 7 per cent.

In our opinion, for the paste of slag cement, a very slow process of slag hydration is also very important which makes the content of unhydrated slag high. The process of slag hydration which proceeds simultaneously with the shrinkage of cement gel under osmotic pressure gives the C-S-H phase which can fill the microcracks formed by the densification of the gel.

The unhydrated slag is very abundant in the samples of clinker-free cement paste, even after 23 months of immersion in the strong chloride solution. An example is given in Figure 11.12, in which a well-developed C-S-H phase (reticular network) and a cracked grain of unhydrated slag is visible.

There are cement pastes in which the diffusion of magnesium ions is much slower in comparison with chloride ion migration. The slag clinker-free paste and the HAC paste are examples of these types. In these conditions, the chloride ions diffusion must be balanced by OH^- ions diffusion towards the solution and on the surface of the sample a skin is formed, which is a real barrier (Figure 11.13). The formation of the barrier can be explained as a quick mechanism of precipitation which occurs when two concentration gradients of contercurrent

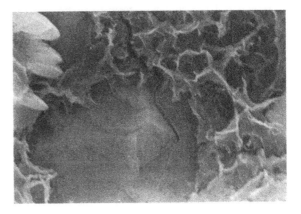

FIGURE 11.12 Clinker-free slag paste after 23 months of immersion in the strong chloride solution (Table 11.2). Grain of unhydrated slag with microcracks.

diffusion meet [34]. The precipitation must be quick in comparison with the velocity of diffusion so that the ions should be consumed on the spot without any possibility of easily crossing the crystallization frontier. Under these conditions the layer is formed, the density of which increases constantly and makes the diffusion of ions more and more difficult, representing a true barrier.

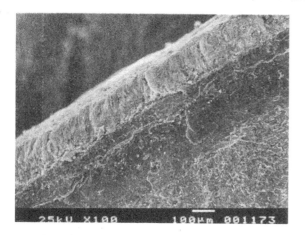

FIGURE 11.13 Skin composed of basic magnesium chloride on clinker-free slag paste after 23 months of immersion in the strong chloride solution (Table 11.2)

FIGURE 11.14 Microcrack between the skin and the surface of the sample of clinker-free slag paste after 23 months of immersion in the strong chloride solution (Table 11.2).

The X-ray patterns prove that in the magnesium chloride solution this skin is composed of brucite and basic magnesium chloride [39,44]. A particularly dense skin was found on the surface of HAC paste [39].

The role of skin itself is not clear. One thing is evident – a dense and continuous skin is formed only on the surface of a very durable cement paste. With other pastes the skin only partially covers the surface of the samples and is detached.

The stability of the linkage of the skin with the surface of the paste may be changeable because the properties of both materials are very different. Then, in some cases, between the skin and the paste surface microcracks appear (Figure 11.14), and the skin becomes easily detachable from the surface.

In turn, in some microregions under the skin the crystallization of unexpected compounds occurs, which shows that in these places the skin forms an impermeable layer for ions and water molecules. The examples of these microregions are the places in which the nests of halite and hydrargillite crystals are formed (Figures 11.15a and b).

Summing up, one can conclude that the role of the skin is variable. In some microregions it has an important significance, hindering the diffusion of the ions or, at least, making it much slower. In other microregions, because of the differences in properties, the stresses between the skin and the paste surface arise and microcracks are formed. In these cases, the skin does not play the role of a protecting barrier.

The formation of bands consisting of the corrosion products can be also formed in the interior of the paste during its corrosion. In Figure 11.16, the layers of basic magnesium chloride which lead to the cracking of the paste into parallel layers can be seen.

In case of the chloride corrosion, heterogeneities in composition of pore solution resulting from the paste microstructure are typical. Some examples, such as the band of crystallization or formation of the nests, containing crystals of one phase, were given before. Further examples can be easily found during the examination of the corroded slag cement paste. As one of them, the anhydrite crystals in Portland cement paste immersed in strong chloride solution (164 g/l $CaCl_2$ and 272 g/l $MgCl_2$),[1] can be seen (Figure 11.17).

11.6 Chloride-induced corrosion of reinforcement in concrete

The steel reinforcement in concrete is passivated in the strongly basic environment as the pore solution

[1]Some strong chloride solutions used by the author were the salt mine water.

(a)

(b)

FIGURE 11.15 HAC paste after four years of immersion in the strong chloride solution (164 g/l CaCl$_2$ and 272 g/l MgCl$_2$) (a) nest of halite crystals under the skin (b) nest of hydrargillite crystals under the skin.

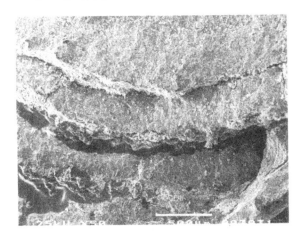

FIGURE 11.16 External part of Portland cement paste in which band crystallization of basic magnesium chloride caused parallel microcracks.

FIGURE 11.17 Anhydrite crystals in the paste of slag cement after 154 d of immersion in strong chloride solution.

has a pH around 13. In this condition, the surface of the steel is covered by a film of solid solution of Fe$_3$O$_4^-$ γ = Fe$_2$O$_3$, whose thickness is in the range 10^{-3}–10^{-1} μm [48]. The chloride ions which are not bound in the cement paste can migrate through the concrete and depassivate the reinforcement. The corrosion process can start when sufficient chloride ions reach the reinforcement. This concentration is dependent on pH of pore solution, that is, of OH$^-$ ions quantity [49]. According to Hausmann [50] the threshold value of depassivation at which the corrosion starts correspond to the ratio Cl$^-$/OH$^-$ equal to 0.6.

The action of chlorides in the corrosion process is complex and not fully understood. First of all, the chlorides decrease the resistivity of the concrete solution and in this condition the corrosion is intensified. Second, the depassivation of the

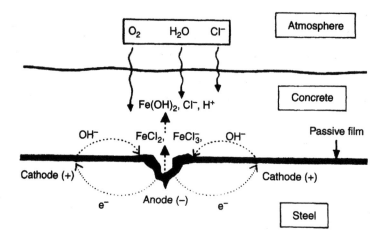

FIGURE 11.18 Mechanism of electrochemical corrosion in the presence of chlorides [49].

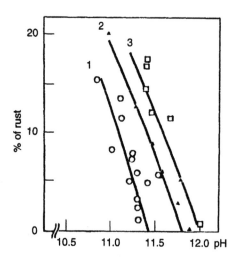

FIGURE 11.19 Influence of pH on the corrosion of reinforcement in carbonated mortars with and without sodium chloride [49] (1) without chloride; (2) with 0.12 per cent and (3) with 0.18 per cent of NaCl, respectively.

$$FeCl_3^- + 2OH^- \rightarrow Fe(OH)_2 + 3Cl^-$$

According to these reactions (Figure 11.18), the process gives decrease of pH and the recycling of chloride ions.

The formation of intermediary corrosion products containing chloride can diminish temporarily the corrosion rate, but in the case of a constant supply of Cl^- ions, their concentration increases in the anodes regions and then is distributed onto the whole reinforcement. The variations of humidity and the gradients of chlorides create new anodes which tend to couple in large corroded zones.

In the presence of chlorides the corrosion takes place even when the pH is very basic, in the neighbourhood of 12 (Figure 11.19) [49].

11.7 References

1. Marchand, J., Gerard, B. and Delgrave, A. *Mater. Sci. Concr.* V, J. Skalny and S. Mindess, (ed.), Chapter 7, pp. 307–99, American Ceramic Society, Westerville, Ohio 43081 (1998).
2. Andrade, C. *Cem. Concr. Res.*, 23, 724–742, 1993
3. Hornain, H., Marchand, J., Duhot, V. and Moranville-Regourd, M. *Cem. Concr. Res.*, 25(8), pp. 1667–78 (1995).
4. Page, C. L., Short, N. R. and El Tarras, A. *Cem. Concr. Res.*, 11, pp. 395–406 (1981).

surface layer takes place by dissolution or by diffusion of chloride ions through the oxide layer. A weak concentration of chloride ions of about 0.01 per cent changes the morphology of the passive layer with the formation of FeOOH and after the unstable $FeCl_3$ which consumes OH^- ions according to schematic reactions [49]:

$$Fe + 3Cl^- \rightarrow FeCl_3 + 2e^-$$

5. Tang, L. and Nilsson, L. O. *Cem. Concr. Res.*, 25(5), 1133–37 (1995).

6. Delagrave, A., Pigeon, M. and Revertegat, E. *Cem. Concr. Res.*, 24(8), pp. 1433 (1994).

7. Schwiete, H. E., Ludwig, U. and Albeck, *J. Zem.-Kalk-Gips*, 22(5), pp. 225–34 (1969).

8. Theissing, E. M., Mebius-van De Laar, T. and De Wind, G. *7th ICCC Paris*, Septima, Paris, Vol. IV, pp. 823–8 (1980).

9. Glasser, F. P., Luke, K. and Angus, M. J. *Cem. Concr. Res.*, 18, pp. 165–78 (1988).

10. Taylor, H. F. W. *Cem. Chem.*, Academic Press, London (1990).

11. Lea, F. M. *The Chem. Cem. and Concr.*, 3rd edn, Chemical Publishing Company, New York (1971).

12. Ramachandran, V. S. *Mater. Struct.*, 4(19), p. 312 (1971).

13. Beaudoin, J. J., Ramachandran, V. S. and Feldman, R. F. *Cem. Concr. Res.*, 20(6), pp. 875–83 (1990).

14. Page C. L. and Vennesland, O. *Mater. Constr. Struc.*, 16, pp. 19–25 (1983).

15. Diamond, S. *Cem. Concr. Aggregates*, 18(20), pp. 97–102 (1986).

16. Uchikawa, H. and Okamura, T. 'Binary and ternary components blended cement', in *Progress in Cement and Concrete*, Vol. 4, pp. 1–83, Mineral Admixtures in Cement and Concrete, L. Sarkar and S. N. Ghosh, (eds) ABI Books, New Delhi (1993).

17. Byfors, K., *Nordic Concr. Res.*, 5, pp. 27–38 (1986).

18. Tenoutasse, N., *Zem.-Kalk-Gips*, 20, p. 459 (1967).

19. Richartz, W., *Zem.-Kalk-Gips*, 22(10), pp. 447–56 (1969).

20. Theissing, E. M., Hest-Wardernier, P. V. and De Wind, G. *Cem. Concr. Res.*, 8, pp. 683–92 (1978).

21. Blunk, G., Gunkel, P. and Smolczyk, H. G. *8th ICCC*, Rio de Janeiro, Vol. 1, pp. 85–90 (1986).

22. Arya, C., Buenfeld, N. R. and Newman, J. B. *Cem. Concr. Res.*, 20, pp. 291–300 (1990).

23. Tritthart, J. Bundesministerium für Wirtschaftige Angelegenheiten Strassenforschung, Herft, Vienna, Austria (1988).

24. Byfors, K. CBI Report No 190, Swedish Cement and Concrete Institute, Stockholm (1990).

25. Midgley, H. G. and Illston, J. M. *Cem. Concr. Res.*, 14, pp. 546–58 (1984).

26. Goto, S. and Daimon, M. *8th ICCC*, Rio de Janeiro, Vol. VI, p. 405 (1986).

27. Uchikawa, H., Uchida, S. and Ogawa, K. *8th ICCC*, Rio de Janeiro, Vol. IV, pp. 251–56 (1986).

28. Regourd, M. *Mater. Res. Soc. Symp. Proc.*, Boston, 86, pp. 185–97 (1987).

29. Garboczi, E. J., Schwartz, L. M. and Bentz, D. P. *Advn. Cem. Bas, Mat.*, 2(5), pp. 169–81 (1995).

30. Winslow, D. N., Cohen, M., Bentz, D. P., Snyder, K. A. and Garboczi, E. J. *Cem. Concr. Res.*, 24(1), pp. 25–37 (1994).

31. Bourdette, B., Ringot, E. and Ollivier, J. P. *Cem. Concr. Res.*, 25(4), pp. 741–51 (1995).

32. Brenton, D., Ollivier, J. P. and Ballivy, G. 'Diffusivite des Ions Chlores dans la Zone de Transition entre Pâte de Ciment et Roche Granitique' pp. 278–88 in *Interfaces in Cementitious Composites*, J. C. Maso, (ed.) E. F. & Spon, London (1992).

33. Bourdette, B. 'Durability of Mortars' (in Fr.); Ph.D. Thesis, INSA-Toulouse, France, (1994).

34. Vernet, C. 'Stabilite chimique des hydrates – Mecanismes de defence du beton face aux agressions chimiques', pp. 129–69, in La Durabilite des Betons, Presses Ponts et Chaussees, Paris (1992).

35. Gegout, P., Revertegat, E. and Moine, G. *Cem. Concr. Res.*, 22, pp. 451–7 (1992).

36. Smolczyk, H. G. *5th ISCC*, Tokyo, Vol. III, pp. 274–80 (1969).

37. Riedel, W. *Zem.-Kalk-Gips*, 26, p. 286 (1973).

38. Conjeaud, M. *Int. Seminar on Calcium Aluminates*, pp. 171–81, M. Murat (ed.) Torino (1982).

39. Kurdowski, W., Taczuk, L. and Trybalska, B. *Calcium Aluminate Cem.*, pp. 222–9, R. J. Mangabhai (ed.), E. & F.N. Spon, London (1990).

40. Boies, D. B. and Bortz, S. *National Cooperative Highway Research Program*, 19, p.19 (1965).

41. Van Aardt, J. H. P. *4th ISCC*, Washington, Vol. II, p. 835 (1960).

42. Chatterji, S., *Cem. Concr. Res.*, 8, p. 461 (1978).

43. Kurdowski, W., Trybalska, B. and Duszak, S. *Proc. 16th Int. Conf. on Cement Microscopy*, Richmond, Virginia, p. 80, International Cement Microscopy, Association, Richardson, Texas, 19 (1994).

44. Kurdowski, W. Ceramika, 'Advances Ceramics, Glass and Mineral Binding Materials', PAN, L. Stoch, (ed.) 42, p. 35, *Proc. of 5th Polish-German Seminar*, Zakopane (in Polish) (1992).

45. Kurdowski, W., Duszak, S. and Trybalska, B. *1st Int. Conf.*, Kiev, 'Alkaline Cements and Concretes', P. V. Krivienko (ed.) Vol II, p. 961, Kiev (1994).

46. Kurdowski, W., Taczuk, L. and Trybalska, B. 'Durability of Concrete', *2nd Int. Seminar*, June 1989, p. 85, Swedish Council for Building Research, Stockholm (1989).

47. Kurdowski, W. and Duszak, S. *Advn. Cem. Res.*, (28), vol. 7, pp. 143–9 (1995).

48. Sagoe-Crentsil, K. K. and Glasser, F. P. 'Corrosion of Reinforcement in Concrete', C. L. Page, K. W. I. Treadaway, P. B. Bamforth (ed.) pp. 74–86, Elsevier Science Publishers Ltd, London (1990).

49. Duval, R., 'La durabilité des armatures et du béton d'enrobage' in La Durabilité des Bétons, pp. 173–226, Presses Ponts et Chaussées, Paris (1992).

50. Hausmann, D. A., *Mater Protection*, 4(11), pp. 19–22 (1967).

Chapter twelve

Blastfurnace cements

E. Lang

12.1 History

The manufacture and use of slag from ironworks, for a great variety of applications, goes back a long way in history. The first recorded description of its use was made by Aristotle. It was not until the eighteenth century that the use of slag became more widespread in the construction industry, which is a field that has remained a major outlet to this day.

The Englishman, John Payne, was granted a patent in 1728 for a process involving the casting of molten blastfurnace slag to basalt-like bricks that was used in a variety of ways, for example, to line cellars and build chimneys. In 1737, Bélidor recommended the use of broken blastfurnace slag in mortar and concrete. In 1862, Emil Langen discovered the latent hydraulic properties of ground granulated blastfurnace slag. 'Latent hydraulic' means, that, once activated, the slag reacts with water to give a cementitious material.

The water granulation of molten blastfurnace slag started as long ago as 1853 in a German blastfurnace plant. Initially, the slag was activated by lime and six years later, Germany introduced a slag-lime cement. This binder was used mostly for bricks, but in 1862, Emil Langen announced that the use of this material in the construction of waterways was invaluable [1].

After long years of experimentation and testing under state control, the first German standard for Portland blastfurnace slag cement (Eisenportlandzement) with a slag content of 30 per cent was introduced in 1909. The first standard for blastfurnace cement was in 1917. There was a remark on the first page that the cement could be used enjoying the same rights for public works as Portland cement or Portland blastfurnace cement.

The use of blastfurnace cements or the use of ground granulated blastfurnace slag as an addition for concrete, mortar and grout is well known in all countries with a pig-iron industry.

A great deal of experimentation and practical experience resulted in a wide field of different slag-containing cements as shown in Table 12.1. This table is based on the European standard EN 197-1, where the different categories of common cements are defined. Their field of use has expanded with the optimization of the quality of blastfurnace cements. As the demand for a high and equal pig-iron quality has grown to its present level, slag composition has also been optimized, and the differences in slag quality have decreased correspondingly. Moreover, special types of binders were developed, for example, slurry wall, inject cement and immobilization of wastes, although this subject is not dealt with as it is outside the scope of this chapter.

12.2 Slag composition and reactivity

The composition of blastfurnace slags varies widely depending upon the raw materials. Table 12.2 shows the composition of blastfurnace slags from

TABLE 12.1 Portland-slag and blastfurnace cement types and composition according to EN 197-1 'Common Cements', (2000), Proportion by mass[a]

Cement type	Designation	Notation	Main constituents		Minor additional constituents[b]
			Clinker	Blastfurnace slag	
CEM II	Portland-slag cement	CEM II/A-S	80–94	6–20	0–5
		CEM II/B-S	65–79	21–35	0–5
CEM III	Blastfurnace cement	CEM III/A	35–64	36–65	0–5
		CEM III/B	20–34	66–80	0–5
		CEM III/C	5–19	81–95	0–5

[a] The values of the table refer to the sum of the main and minor additional constituents (excluding Calcium sulphate and any additives, e.g. grinding aids).
[b] Minor additional constituents may be filler or may be one or more of the main constituents unless these are included as main constituents in the cement.

TABLE 12.2 Typical ranges of slag composition for use in blastfurnace cements or in concrete as proportion by mass

	Belgium	France	Germany	Italy	Netherlands	United Kingdom	United States
CaO	39.3–41.2	37.1–43.5	38.0–41.8	41.0–44.4	36.5	39.9–40.5	29–42
MgO	8.1–10.9	4.4–7.7	7.6–11.5	6.3–9.9	11.0	8.3–8.8	8–19
Al_2O_3	11.1–12.2	8.6–16.2	10.0–11.1	11.6–12.9	16.6	11.0–13.1	7–17
SiO_2	35.0–36.7	34.8–43.8	35.1–37.9	35.2–36.1	32.7	35.2–37.0	32–40
S	1.0–1.2	0.7–1.2	1.2–2.1	1.0	34.0	0.9–1.1	0.7–2.2
Na_2O+K_2O	0.6–0.7	0.7–1.2	0.8–1.3	0.57	1.0	0.5–0.7	
Mn	0.21–0.33	0.12–0.35	0.18–0.45	0.38–0.46	0.43	0.28–0.35	0.2–1.0
$CaO+MgO+Al_2O_3/SiO_2$	1.63–1.80	1.18–1.78	1.56–1.71	1.74–1.78	1.96	1.65–1.74	

the most important slag producers in Europe and the United States.

The glass in granulated slag is the active part. The nozzle units employed today permit a slag entry temperature of some 1500°C and water pressures of 2.0–3.5 bar for a slag-to-water ratio of between 1:6 and 1:10. These can produce granulated slag with a degree of vitrification greater than 95 per cent by mass. The free Gibb's energy is in a quenched liquid slag higher than in an air-cooled slag which is crystallized. The activation of the glassy slag in contact with water with a high alkalinity leads to a dissolution process. As the end of the interaction between the glassy slag, water, alkalis and clinker particles, calcium silicate and calcium aluminate hydrate phases are formed in slag cement.

Different tests have shown that a total glass content of 100 per cent does not give the highest reactivity. Crystal nuclei promotes the dissolution [2].

Throughout the history of the use of blastfurnace slag, attempts have been made to find a quick reproducible value for hydraulicity.

For a long time, researchers have tried to find indices resulting from chemical analysis, to estimate the hydraulicity. In some cases, such indices are added to current standards. It has been shown, that they are a reliable method to separate 'good' and 'bad' qualities, but their selectivity is not exact enough to control the quality. The indices are sometimes sufficient to describe the quality of a slag from one blastfurnace, but they are insufficient to compare slags from different plants.

Some of the most important steps are reproduced below. The basicity is well known in the following formulae [3]:

$$p_1 = \frac{CaO}{SiO_2}, \quad p_2 = \frac{CaO + MgO}{SiO_2},$$

$$p_3 = \frac{CaO + MgO}{SiO_2 + Al_2O_3}$$

For a long time in Germany, and now in Japan, cement standards have had to conform to the following formula:

$$\frac{CaO + MgO + Al_2O_3}{SiO_2} \geqslant 1$$

The European standard EN 197-1 uses the p_2-basicity ratio, where p_2 has to be greater than 1.0. Germany and other European countries had previously introduced this limit to their national standards. The F-value is widely used. The parameters in the following equation are based only on the glassy components which are soluble in hydrochloric acid [3].

$$F = \frac{CaO + CaS + 0.5MgO + Al_2O_3}{SiO_2 + MnO}$$

Estimation: $F > 1.9$ very good hydraulicity
$F = 1.5–1.9$ good hydraulicicty
$F > 1.5$ poor hydraulicity

Sopora [4] considered additionally in this equation the disadvantageous effect of FeO and MnO in an expanded F-value when the MnO-content is greater than 1 per cent.

$$F_{exp} = \frac{CaO + 0.5MgO + Al_2O_3}{SiO_2 + FeO + (MnO)^2}$$

These, and other equations of hydraulicity which are not described in this book, allow only an approximation of the hydraulicity because important factors are disregarded, for instance, the type of activator, the method and duration of storage, the grindability, the quality of the clinker and the kind of setting retarder used for the cement production.

Other different methods are only valid under defined conditions. Examples are the activation

with NaOH [5], measuring the thermal reaction with NaOH in a calorimeter [6], and measuring the glass transition [7]. These methods do not give a sufficiently accurate result for general use.

For over 20 years, the use of blastfurnace slag as a cementitious material has been investigated with the same test procedure [8]. Blastfurnace cements are produced by separately grinding blastfurnace slag, Portland cement clinker (specific surface of both 3500 and 4200 cm^2/g) and a defined mixture of anhydrite and gypsum. The blastfurnace/clinker ratios are 60/40 and 75/25. The four types of cements were tested according to the German standard DIN 1164, which has now been replaced by EN 196. All the results have been analysed using statistical methods.

More than 370 types of blastfurnace cements were produced and tested using seventy-two different blastfurnace slags and sixteen different Portland cement clinkers. The range of compressive strength of the cements with 60 per cent slag is shown in Figure 12.1. The statistical analysis of the data permits a quantitative valuation of the influence of change of the chemical composition of the strength of the blastfurnace cements. The statistical analysis of the data allowed a quantitative valuation of the strength development when the chemical composition of the blastfurnace slag changed. The results show too, that it is impossible

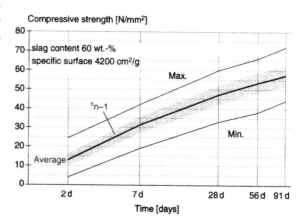

FIGURE 12.1 The range of compressive strength of 107 blastfurnace cements produced from seventy-two different blastfurnace slags and sixteen different clinkers.

to calculate the strength with sufficient precision, based only on the chemical composition and the glass content of the blastfurnace slag. The hydraulicity and the attainable strength of cements containing blastfurnace slag are the result of the interaction of one specific blastfurnace slag, one specific clinker and one specific calcium sulphate.

The following facts were established from research into the fineness and the slag/clinker ratio in respect of the range of chemical composition of the tested blastfurnace slags and clinkers:

- If the content of CaO is increasased, there is a relatively uniform rise in strength at all ages.
- The MgO-content has nearly the same influence as CaO, but higher amounts (>10–12 per cent) of the MgO-content do not increase the hydraulicity.
- An increase of the Al_2O_3-content increases the early strength. An alumina content up to about 14 per cent increases the strength after 28 or 91 days. An alumina content in excess of 14 per cent does not increase the strength further.
- Increasing the MnO-content of blastfurnace slag leads to a lower strength, if all other conditions are comparable.
- When the TiO_2-content of blastfurnace slag is less than approximately 1 per cent it does not have an influence on the compressive strength of blastfurnace cements. Higher contents ($\approx$ >1.5 per cent) reduce the strengths and the setting times will be longer.
- As the alkali content increases so does the strength, but the rate of increase is greatly influenced by the kind of clinker.

Finally, it must be considered that the kind of clinker generally has a great influence on the strength development of blastfurnace cements. It has to be taken into account that a hydrating blastfurnace cement is a complicated multi-material system. Therefore, it cannot be expected that the resulting strength can be predetermined by the use of simple hydraulic factors or the content of some chemical parameters. If slags come from one blastfurnace only, the correlation between hydraulicity and chemical composition is higher than for slags that come from different plants.

12.3 Grindability

Generally, the grinding resistance of blastfurnace slags is higher than Portland cement clinker or other main constituents, such as limestone or fly ash. The scale of the grindability is the specific energy demand. This is the ratio of the energy consumption of the cement mill to the finished material, which is usually expressed in kWh/t. The specific energy demand increases with the fineness and with components with a higher grinding resistance.

Table 12.3 shows the specific energy demand for grinding blastfurnace cements with different slag content in ball mills [9].

Keil [3] subdivided water granulated blastfurnace slags into slags with light and dark colours with different properties. Lighter slags have a foamed structure with a lower bulk density and a lighter grindability than darker ones, which have a grainy structure and a higher bulk density. Investigations by Mußgnug [10] showed a very good correlation between bulk density and grindability. When comparisons were made between work undertaken with technical plant on site and laboratory tests, it was found that this correlation arose only when slags from the same blastfurnace were used [11].

Figure 12.2 shows results of the grindability test [12] of different clinkers and blastfurnace slags. In the range of lower specific surfaces the energy demand of clinker is smaller than that of slags, but in the range of higher specific surfaces the differences will be less [13].

TABLE 12.3 Specific energy demand for grinding blastfurnace cements

Slag content (proportion of mass)	Specific surface (Blaine) (cm^2/g)	Specific energy demand (kWh/t)
<35	2700–3800	30–55
40–60	3200–4500	40–70
60–80	3200–4500	45–80

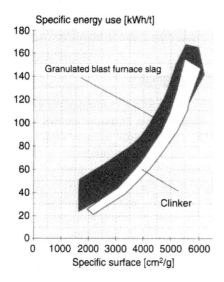

FIGURE 12.2   Range of grindability of blastfurnace slag and Portland cement clinker.

TABLE 12.4   Setting time of blastfurnace cements depends upon the slag content

Type of cement	Slag content (mass%)	Setting time	
		Initial set (h : min)	Final set (h : min)
CEM I 32.5	0	2 : 10	2 : 30
CEM III/A 42.5	45	3 : 15	3 : 50
CEM III/A 32.5	62	3 : 45	4 : 30
CEM III/B 32.5-NW/HS	73	4 : 30	5 : 20

12.4 Blastfurnace cement characterization

12.4.1 Water demand

The water demand of concrete depends upon separate components and mix proportions. Normally, blastfurnace cements show a higher fineness than Portland cement with the same strength class. Owing to this circumstance, a higher water demand can be expected. On the other hand, the spontaneous water-binding of cements with a high slag content is lower than with other cements. Both effects are contradictory and a higher water demand in the standard test does not lead to a higher water demand in the concrete. Generally, the tendency to bleed is not significantly different from Portland cement.

12.4.2 Setting time

As the slag content increases, so do the setting times of blastfurnace cements. This is an advantage, especially for the ready-mix concrete, by permitting longer transportation or waiting-time, as well as being able to concrete on hotter days or 'fresh on fresh'. Portland-slag cement (CEM II/A-S and CEM II/B-S) with a low slag content is therefore an excellent binder for concrete roads. Table 12.4 shows the typical setting times of blastfurnace cements with the same clinker and different slag contents.

As well as retarders and accelerators, super-plasticizers may also influence the setting time with important changes to the heat of hydration.

12.4.3 Curing

The strength development and durability of any concrete depends largely upon the degree of hydration of the cement. Depending upon the slag content, blastfurnace slag cement will hydrate at a slower rate than Portland cement. With lower temperatures, these differences will be higher and longer curing periods will be necessary. During periods of low ambient temperature, it may be useful to leave the formwork in place for a longer time. The main requirement to obtain the full potential of strength and durability is a sufficient curing time.

Many durability problems, for example, freeze-thaw resistance, carbonation and an unacceptably low resistance against chemical attack, result from an insufficient curing.

The maturity calculation based on the temperature measurements taken at a maximum depth of 10 mm below the concrete surface is an exact way to determine the curing period in days.

12.4.4 Strength development

The strength development of blastfurnace cement proceeds slower than Portland cement of the same strength class, as shown in Figure 12.3. However, the long-term strength development of concrete that incorporates blastfurnace cement is excellent. A thermal treatment accelerates the strength development of blastfurnace cements more than with Portland cements. The harmful effects of high concrete temperatures are significant reductions in long-term strength, and increases in coarse porosity or delayed ettringite formation. These effects are much lower in concretes incorporating blastfurnace cements than in Portland cement concretes.

On the basis of laboratory-scale experiments and tests carried out with cores taken from old buildings, it is evident that the strength of concretes made with blastfurnace cements can increase by up to 100 per cent after a period of about 25 years, compared to a 28-day compressive strength (Table 12.5) [14,15].

Generally, blastfurnace cements not only have a higher activation energy, but the rate of hydration is more dependent upon temperature than Portland cements. In practice, it leads to lower strengths for temperatures below 20°C (temperature of the standard test) and higher strengths at temperatures above 20°C.

12.4.5 Porosity

Concrete durability is closely related to its density. The higher the density of the cement paste, the higher its resistance, for example, against penetration of harmful substances into the concrete.

Besides other factors, the pore structure, and hence the density of the cement paste, depends upon the type of cement used. It has long been recognized that capillary porosity, that is, the quantity of pores with a radius of more than 30 nm (a decisive factor for concrete durability) decreases as the portion of granulated blastfurnace slag increases, under otherwise identical conditions [14,16] (Figure 12.4).

During the course of a long-term project, concrete blocks were produced with different types of cement and with water/cement ratios 0.50 and 0.70, respectively. After the blocks had been moist-cured for seven days, they were subsequently stored outdoors, and protected against rain. Figure 12.5 shows the capillary porosity of the concrete blocks after 12 years of outdoor storage. The concrete made with the cement having the highest level of granulated blastfurnace slag showed the lowest capillary porosity [14].

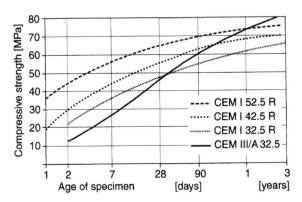

FIGURE 12.3 Compressive strength development of Portland and blastfurnace cements.

TABLE 12.5 Compressive strength development of blastfurnace cement concrete in practice

Building	Age in days	Compressive strength (MPa)	Age in years	Compressive strength (MPa)	Strength increase (%)
Sewage clarification plant	28	41	18	59	44
Cooling tower	28	40	19	61	53
Weir	28	40	23	80	100
Lock	28	29	25	54	86

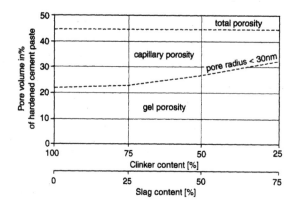

FIGURE 12.4 Pore size distribution (schematic) depends upon the slag content.

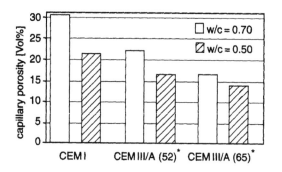

FIGURE 12.5 Capillary porosity in 12-year-old concretes; *Slag content (%).

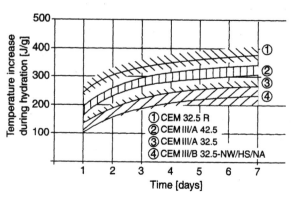

FIGURE 12.6 Heat of hydration under adiabatic conditions of Portland cement and blastfurnace cements with different slag contents.

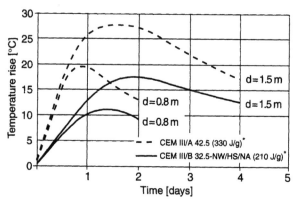

FIGURE 12.7 Increase of temperature in hardening concretes with a thickness of 0.8 m and 1.5 m; *Vadiabatic.

12.4.6 Heat of hydration

A major difference between Portland and blastfurnace cements is the development of the heat of hydration. Because of their low heat of hydration and the fact that the heat development takes place more slowly, blastfurnace cements with a high content of granulated blastfurnace slag are particularly suited for mass concrete elements, such as foundations, dams or bridge abutments. Under adiabatic conditions, the blastfurnace cement has a slower heat development than a Portland cement of the same strength class (Figure 12.6). Therefore, high-slag blastfurnace cements generally fulfil the standard demand for low heat cement.

The temperatures generated in the cores of such structures as a result of heat of hydration can be minimized by using blastfurnace cement. The differences in temperature between the core and surface can thus be reduced. The risk of cracks can consequently be minimized. Figure 12.7 shows the increase of temperature in concrete blocks with a thickness of 0.8 m and 1.5 m respectively, using CEM III/A 42.5 with a heat of hydration of 330 J/g and a CEM III/B 32.5 low heat with a heat of hydration of 210 J/g (the solution method after seven days).

The low heat of hydration of blastfurnace cements is not only very useful for mass concrete, but also for high-performance concrete with high

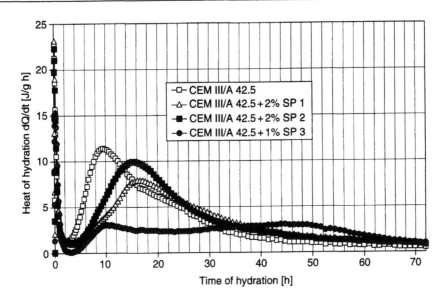

FIGURE 12.8 The influence of different superplastizicers on the heat of hydration of a CEM III/A 42.5 (SP 1: Naphthalene sulphonate; SP 2: Melamine sulphonate; SP 3: Ca-Lignosulphonate).

cement contents. Tests have shown that high-performance concrete with 550 kg blastfurnace cement CEM III/A 42.5 per cubic metre has a 20°C less maximum temperature than the same concrete with CEM I 42.5 [39].

Admixtures may have an important influence on the heat of hydration. As an example, Figure 12.8 shows the influence of three different superplasticizers on the heat of hydration of a CEM III/A 42.5 [17]. This example shows that by using a superplasticizer, the heat of hydration is reduced and the time taken to reach the second maximum is longer.

12.4.7 Colour

A specific characteristic of blastfurnace cement concrete is its blue-green appearance. Blastfurnace slag contains a small amount of sulphide which during the hydration process transmutes into complex polysulphide products that are blue-green in colour. Upon contact with air, these products oxidize to sulphate. The exact colours and intensities of the initial colours depend upon the following:

- The amounts of iron, manganese and titanium in the blastfurnace cement. Also, complexed oxides such as sulphur sesquioxide (S_2O_3) may

be formed during the oxidation process. In particular, this oxide has a blue-green colour too, but is only stable below 15°C. Above this temperature, it can undergo oxidation to other chemical states. It decomposes between 40°C and 80°C [18].

- Higher proportions of blastfurnace slag in the cement will give a deeper colour. If the slag content is very low, for example in CEM II/A-S or /B-S, the blue-green colour may not become visible.
- Concretes having the same mix proportions but different degrees of compaction will give differing colour intensities. A spotted concrete surface may denote a different degree of compaction. Well-compacted concretes show a deeper colour.
- Concretes with a low water/cement ratio will show a deeper coloration than concretes with a high water/cement ratio.
- The less porous the formwork, the greater the colour intensity. Therefore, controlled permeability formwork (CPF) liner will give a light and equable concrete surface.

Most suppliers of concrete know that the blue-green colour is a sign of a high-quality durable

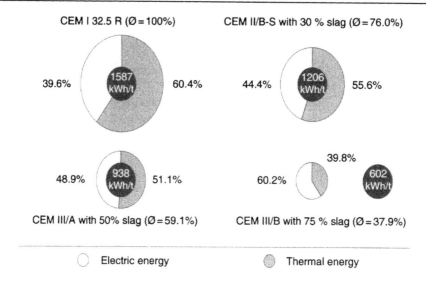

CEM I 32.5 R (Ø = 100%)

CEM II/B-S with 30 % slag (Ø = 76.0%)

CEM III/A with 50% slag (Ø = 59.1%)

CEM III/B with 75 % slag (Ø = 37.9%)

○ Electric energy   ◉ Thermal energy

FIGURE 12.9 Energy consumption for cement production.

concrete because of the dense cement pore structure. The exposure to air leads to oxidation processes of the polysulphide compounds on the concrete surface. The result is that the surface gets its normal light-grey colour. With blastfurnace cements, the colour is lighter than with the most other cements, and therefore they lend themelves to production of face concrete and of concretes with pigments.

Very dense pastes take longer to oxidise because of the fine pore structure of the cement. Generally, the surface loses its blue-green colour within a few days or weeks. The core of a high-quality concrete can keep its blue-green colour permanently.

12.4.8 Environmental aspects

The use of granulated blastfurnace slag as major constituent in cements or as ground granulated blastfurnace slag (ggbs) concrete gives rise to a number of different environmental aspects. In 1996, slag production worldwide was nearly 167 million tons. More then 60 per cent of it, almost 100 million tons, was granulated and used for a variety of purposes including cement, concrete, mortar and road construction. The use of such vast quantities of granulated slag requires every

effort to be made to minimize the effect on the environment. This is just as important as the need to husband resources and save energy.

The most effective way to minimize the impact on the environment is the substitution of Portland clinker. The reduction of energy consumption in the production of blastfurnace cements with three different slag contents in relation to Portland cement is shown in Figure 12.9 [19]. The dates are based on the statistical values of the German cement industry. In respect of the concrete production and the service life of civil constructions, the differences between the kind of cement are much lower as shown in Figure 12.9. Figure 12.10 shows the reduction of CO_2-output for the same cements [19].

12.5 Durability

12.5.1 Carbonation

In the technical literature, carbonation is often mentioned as a disadvantage of blastfurnace cement concrete. In fact, in laboratory tests at a constant temperature of 20°C and a relative humidity of 65 per cent, the carbonation rate of blastfurnace cement concrete is higher than Portland cement concrete with the same strength class.

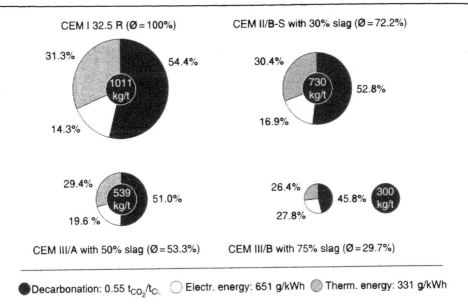

CEM I 32.5 R (Ø = 100%) CEM II/B-S with 30% slag (Ø = 72.2%)

31.3% 54.4%
1011 kg/t
14.3%

30.4% 52.8%
730 kg/t
16.9%

29.4% 51.0%
539 kg/t
19.6 %

26.4% 45.8%
300 kg/t
27.8%

CEM III/A with 50% slag (Ø = 53.3%) CEM III/B with 75% slag (Ø = 29.7%)

● Decarbonation: 0.55 t_{CO_2}/t_{Cl}. ○ Electr. energy: 651 g/kWh ◐ Therm. energy: 331 g/kWh

FIGURE 12.10 CO_2 emissions for cement production.

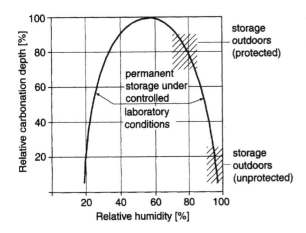

FIGURE 12.11 Influence of the relative humidity content on the carbonation depth.

This laboratory climate favoured the carbonation, see Figure 12.11 [20]. Cements with a lower degree of hydration in the first weeks of hydration are more sensitive under such conditions.

Consequently, the resistance to carbonation for cements with a low rate of hydration is lower than for cements with a high rate of hydration. Another reason for the higher rate of carbonation is the lower lime content of the hydrated cement.

The lime content decreases with a higher slag content. The higher lime content in hydrated Portland cement leads to more voluminous calcium carbonate, which partially fills up the pores. The porosity in the carbonated surface layer will decrease while in blastfurnace cement it does not. The carbonated layer of Portland cement inhibits the further carbonation process more than the blastfurnace cement.

Figure 12.12 shows the differences of the rate of carbonation over a period of 16 years, depending upon the curing conditions (laboratory climate; out of doors, under a roof and out of doors, flat). The results represent the average of twenty-seven different concretes [21]. In general, the laboratory tests are reproducible but they are not helpful in representing practical situations. In practice, the differences between blastfurnace cement and other cement concrete are much smaller or insignificant.

Carbonation is not a problem that endangers the passivation of steel reinforcement because the rate of carbonation is lower than in laboratory tests. Studies have shown that even if the carbonation frontier has been passed, reinforcement corrosion does not take place if the cover on the

319

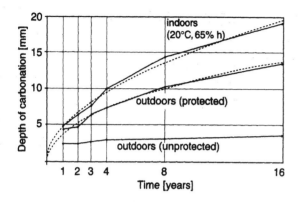

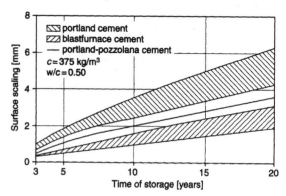

FIGURE 12.12 Rate of carbonation depending upon (micro) climatic conditions.

FIGURE 12.14 Attack of aggressive CO_2 in water dissolved concrete of the surface depending upon the kind of cement and the time.

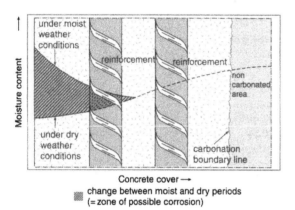

FIGURE 12.13 Area of changing moisture content, carbonation frontier and field of risk for corrosion of reinforcement.

reinforcement has been made in accordance with the regulations concrete standards. In this case, the reinforcement is beyond the zone with the fluctuating moisture content (Figure 12.13) [22]. Blastfurnace cement concrete has a finer pore structure than Portland cement concrete, and for that reason the average moisture content is higher and the moisture fluctuation smaller.

12.5.2 Resistance against chemical attacks

12.5.2.1 General

The resistance of blastfurnace cement against chemical attacks depends primarily upon the composition of the clinker and the slag content, and to a minor extent by the composition of the slag. By comparison with Portland cement, the resistance of blastfurnace cements is higher in all cases in which a lower lime content is important for resistance against chemical attack, for example, the attack of aggressive CO_2 in water or soda (Na_2CO_3). Figure 12.14 shows long-term experiences with CEM I- and CEM III-concretes stored in water with aggressive CO_2 [23].

A lower lime content may lead to hydrolytic dissociation of calcium silicate hydrates or calcium aluminate hydrates. Cements with a lower lime content do not have the same buffer capacity than cements with a high lime content. The influence of concentrated solutions, which are binding the OH-ions of the calcium hydroxide, may impair blastfurnace cements faster than Portland cement, for example, aluminium sulphate solution.

Most ammonium salts (except ammonium-carbonate, -oxalate and-fluoride) are pernicious to concrete. Therefore, the limiting values for exposure classes for chemical attack in the draft of the European concrete standard prEN 206 are limited with only $\leqslant 100\,mg/l$ for the class XA3 'Highly aggressive chemical environment'. The most dangerous ammonium salt is ammonium nitrate (NH_4NO_3), which attacks all concretes regardless of the kind of cement used. The dissolved mass of blastfurnace cement concrete, which is similar to

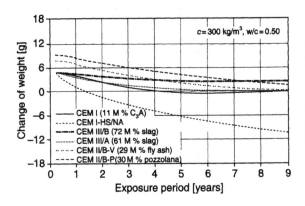

FIGURE 12.15 Loss of mass of concrete in ammonium chloride solution depending upon the kind of cement and the time.

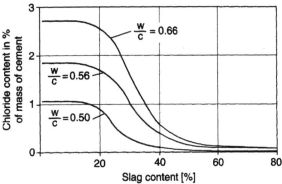

FIGURE 12.16 Chloride content in concrete depending upon the slag content and the water/cement ratio.

fly ash and pozzolana cement, is less than Portland cement concrete which has been attacked by different ammonium salts. An example of this is illustrated in Figure 12.15, which shows an attack of ammonium chloride [24].

12.5.2.2 Chloride diffusion and chloride binding

In practice, chlorides can penetrate concrete depending upon the condition of the environment. Offshore constructions, parking decks, and road bridges can be given as examples.

Figure 12.16 shows that the resistance of concrete against chloride penetration is markedly greater with a higher content of granulated blastfurnace slag in the cement and with a lower water/cement ratio [25]. The concrete cubes have been stored for one year in the 3 molar NaCl solution. The chloride content has been analysed in a layer 20–40 mm under the concrete surface.

The chloride-binding capacity of the cement paste of Portland and blastfurnace cement is a function of the total amount of chloride. Compared to Portland cement, the higher chloride-binding capacity of blastfurnace cement that contains a total amount of chloride of more than 1 per cent, related to the cement mass, is due to the fact that relatively large amounts of chloride are bound to the calcium silicate hydrates [26]. This reduces the danger of chloride attack still further.

Blastfurnace slag contains a small amount of sulphide. The supposition that this sulphide represents a risk of corrosion to prestressed concrete is unfounded [27,28].

12.5.2.3 Seawater resistance

Concrete used in offshore constructions must be resistant against seawater attack and in particular against chloride, alkali, frost and abrasion caused by the impact of waves.

RILEM advise taking a number of precautionary measures, including the use of blastfurnace slag and ensuring marine concrete constructions. The results are introduced in recommendations [29].

The successful use of blastfurnace cements in offshore constructions can be illustrated by a number of practical examples, and the results of a series of long-term trials, see Figure 12.17 [30]. The storm-surge barrier in the eastern Schelde in the Netherlands was built with blastfurnace cement. The expected service life of 200 years is an impressive sign of confidence in durable concrete. One of the world's largest sea locks at Wilhelmshaven in Germany, built with high-slag blastfurnace cement, is another European example of a construction designed for high seawater resistance [15,31].

12.5.2.4 Electrolytic resistance

Corrosion of reinforcement in concrete is an electrochemical process. Among other factors, the rate of corrosion is influenced by the electrolytic resistance of the concrete. High electrolytic resistance leads to a low corrosion current. Consequently, the rate of corrosion will be low. Research carried out on concrete corrosion cells have shown that the higher the content of granulated blastfurnace slag there is in the cement, the higher the electrolytic resistance of concrete becomes (Figure 12.18) [32]. The main reason for this efficient retardation of the corrosion process is the high density of the cement paste of blastfurnace cements.

12.5.2.5 Sulphate resistance

Sulphate solutions have been shown to have a deleterious effect on concrete by attacking the

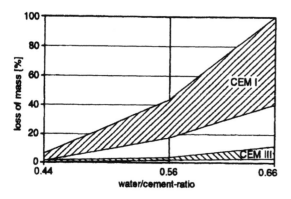

FIGURE 12.17 Loss in weight of concrete cubes with Portland and blastfurnace cements stored for 19 years in seawater, depending upon the water/cement ratio.

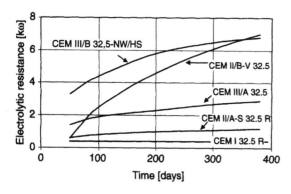

FIGURE 12.18 Electrolytic resistance of concrete with different kinds of cement.

hardened cement paste. According to different cement standards in Europe, for example, in Germany, Belgium and the Netherlands, CEM III/B, with a slag content ≥ 66 proportion by mass, is specified as a cement with a high sulphate resistance. Such cement bears the addition HS. The high sulphate resistance of blastfurnace cement is a result of three main factors: low C_3A-content, low quantity of calcium hydroxide produced during hydration and a low permeability reducing the risk of sulphate attack by penetration.

There is no consensus about the effect of the alumina content in blastfurnace slags. The influence of sulphate resistance in the use of slags with an Al_2O_3-content up to 17 per cent has, both in practice and in research in Germany and the Netherlands, been the opposite to the results that were obtained in tests carried out in the UK [33,34]. This has, unsurprisingly, led to controversial discussions between the scientists of those countries.

12.5.3 Alkali silica reaction

Aggregates containing alkali-reactive silica are found in some regions of the world. If such aggregates are used for concrete production in Germany, precautionary measures must be taken according to guidelines to prevent harmful alkali-aggregate reaction [35]. These measures include the use of cements with a low alkali content. The alkali content is limited with ≤ 0.60 per cent of Na_2O equivalent for all types of cement, with special regulations applying to blastfurnace cements (Table 12.6) [36].

TABLE 12.6 Regulations for blastfurnace cements with a low alkali content

Cement	Proportion by mass	
	Slag content	Na₂O equivalent
CEM II/B-S	20–35	≤ 0.70
CEM III/A	36–49	≤ 0.95
CEM III/A	≥ 50	≤ 1.10
CEM III/B	≥ 65	≤ 2.00

Another way to prevent an alkali-aggregate reaction is to calculate the total water-soluble alkali content in concrete [37]. This way has particular importance for the use of ggbs as an addition. In this case, use one of the following formulae and calculate either the overall alkali content of the cement, using the first formula (1), where a = the water-soluble content of ggbs (per cent Na_2O equivalent) and b = the alkali content of OPC (per cent Na_2O equivalent), or calculate the total alkali content in the concrete, using the second formula (2):

$$a \times \frac{\% \text{ ggbs}}{100} + \frac{b \times \% \text{ OPC}}{100} \qquad (1)$$

Overall the Na_2O equivalent must not exceed 0.6 per cent. The calculated alkali must not exceed $3\,kg/m^3$.

$$a \times \frac{\text{ggbs in concrete } (kg/m^3)}{100}$$

$$+ b \times \frac{\text{OPC in concrete } (kg/m^3)}{100}$$

$$+ \text{alkali from other sources} \qquad (2)$$

12.5.4 Freeze–thaw and de-icing salt resistance

The ultimate test of the durability of all structures is to measure their performance in practice. Regardless of how good laboratory tests are, and the dedication of the scientists conducting them, it is impossible to replicate real-life conditions in a laboratory. This is especially true with freeze–thaw and de-icing salt-resistance, because the stress levels that occur in practice and under laboratory conditions are too different.

Initially, concrete with high slag blastfurnace cement scales faster than Portland cement concrete. After some freeze–thaw cycles the scaling rate is lower than for Portland cement concrete without air-entraining agents and similarly for concrete with artificial air voids. The reason for the initially higher scaling rate of blastfurnace

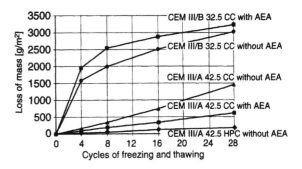

FIGURE 12.19 Loss of mass in freeze–thaw de-icing salt cycles. CC: common concrete; HPC: High Performance Concrete.

cement is mainly a thin carbonated layer of tenths of a millimetre. Figure 12.19 show the loss of mass for concrete with a different slag content, both with and without an air-entraining agent [38].

By comparison with Portland cement concrete there are no problems with air entrainment and air-void distribution in cement paste. However, the effect in high slag blastfurnace cement is not the same as in Portland cement or in cements with a low slag content. The increase in freeze–thaw resistance of concrete with high slag blastfurnace cement made more effective by reducing the water/cement ratio using a superplasticizer as an air-entraining agent. Most concretes with high slag blastfurnace cement that have not an air-entraining agent, that have been monitored a high freeze–thaw resistance [15].

12.6 References

1. Guttmann, A., Die Verwendung der Hochofenschlacke. Verlag Stahleisen m.b.H., Düsseldorf (1934).
2. Klaveren, W. van., Clinker/Slag cement. Cementfabriek Ijmuiden (CEMIJ) B.V. Netherlands.
3. Keil, F., Hochofenschlacke. Verlag Stahleisen m.b.H., Düsseldorf (1963).
4. Sopora, H., Bewertung von Hochofenschlacken für die Zementherstellung im Betriebs-laboratorium. Silikattechnik 10, pp. 361–3 (1959).
5. Lieber, W., 'Neue Schnellmethode zur Ermittlung hydraulischer Eigenschaften eines Schlackensandes', *Zem.-Kalk-Gips*, 19(3), pp. 124–7 (1966).

6. Chassevent, L., *17. Cong. Chem. Ind.*, Paris, paper 147 (1937).

7. Keil, F., *3rd Symp. Chem. Cem.*, London, pp. 547 (1952).

8. Report of Forschungsgemeinschaft Eisenhütten-schlacken (FEhS): Hüttensandkartei Duisburg, 2 (1997).

9. Kuhlmann, K., 'Verbesserung der Energieaus-nutzung beim Mahlen von Zement', *Schriftenreihe der Zementindustrie*, 44, Beton-Verlag GmbH., Düsseldorf (1985).

10. Mußgnug, G., 'Beitrag zur Frage der Mahlbarkeit von Hochofenschlacken (Zementschlacken) und Klinkern', *Zement*, 31, pp. 183–93 (1942).

11. Börner, H., 'Noch einmal: Sichter- oder Verbund-mühle?' *Zem.-Kalk-Gips*, 9(9), pp. 153–70 (1956).

12. Zeisel, H. G., 'Entwicklung eines Verfahrens zur Bestimmung der Mahlbarkeit', *Schriftenreihe der Zementindustrie*, 14, Verein Deutscher Zemen-twerke e.V. Düsseldorf (1953).

13. Forschungsgemeinschaft Eisenhüttenschlacken (FEhS) Duisburg-Rheinhausen, Germany: Statisti-cal analysis of grindability, Internal Paper (1997).

14. Smolzcyk, H.-G., 'Durabilty and pore structure on very old concretes' (Dauerhaftigkeit und Poren-struktur von sehr alten Betonen), *Beton-Informa-tionen*, 26(1), pp. 3–10 (1986).

15. Geiseler, J. and Lang, E., 'Long-term durability of non-air-entrained concrete structures exposed to marine environments and freezing and thawing cycles', *3rd CANMET/ACI Int. Conf. Durability of Concr.*, Nice, France, Supplementary Papers, pp. 715–37 (1994).

16. Romberg, H., 'Cement paste pores and concrete properties' (Zementsteinporen und Betoneigen-schaften), *Beton-Informationen*, 18(5), pp. 50–5 (1978).

17. Lang, E., 'Einfluß von Nebenbestandteilen und Betonzusatzmitteln auf die Hydratationswärme-entwicklung von Zement', *Beton-Informationen*, 37(2), pp. 22–5 (1997).

18. Koch Minerals Company, 'Greening, questions and answers'. Technical Bulletin # 10015. March (1993).

19. Ehrenberg, A. and Geiseler, J., 'Ökologische Eigen-schaften von Hochofenzement', *Beton-Informatio-nen*, 37(4), pp. 51–63 (1997).

20. Wierig, H.-J., 'Die Carbonatisierung des Betons. Die Naturstein-Industrie', 5, pp. 26–35 (1986).

21. Wierig, H.-J., Longtime studies on the carbonation of concrete under normal outdoor exposure. Proc. RILEM Seminar on the durability of concrete struc-tures under normal outdoor exposure, Hannover (Germany), 26–29 March, pp. 239–49 (1984).

22. Bakker, R. F. and Roessink, G., 'Zum Einfluß der Karbonatisierung und der Feuchte auf die Korrosion der Bewehrung im Beton', *Beton-Infor-mationen*, 31(3/4), pp. 32–5 (1991).

23. Locher, F. W., Rechenberg, W. and Sprung, S., 'Beton nach 20jähriger Einwirkung von kalklösender Kohlensäure', *Beton*, 34(5), pp. 193–8 (1984).

24. Rechenberg, W. and Sylla, H.-M., 'Die Wirkung von Ammonium auf Beton', *Beton*, 43(1), pp. 26–31 (1993).

25. Brodersen, H. A., 'Zur Abhängigkeit der Trans-portvorgänge verschiedener Ionen im Beton von Struktur und Zusammensetzung des Zementsteins', Dissertation RWTH Aachen (1982).

26. Gunkel, P., 'Die Bindung des Chlorids im Zementstein und die Zusammensetzung chlorid-haltiger Porenlösungen', Dissertation Universität Dortmund (1992).

27. RILEM CRC Committee (Corrosion of Reinforce-ment in Concrete): Steel in concrete – Corrosion of reinforcement and prestressing tendons. State of the art report (1975).

28. Lang, E., 'Hochofenzemente der Festigkeitsklasse 32,5 für Spannbeton mit sofortigem Verbund', *Beton-Informationen,* 38(2), pp. 10 (1998).

29. RILEM Technical Committee 32-RCA: Resistance of Concrete to Chemical Attacks. Subcommittee 'Long-Time Studies', Sea Water Attack on Con-crete and Precautionary Measures. *Matériaux et Constructions*, 18(105), pp. 223–6 (1985).

30. Eckhardt, A. and Kronsbein, W., 'Versuche über das Verhalten von Beton und Zement im See-wasser', Deutscher Ausschuß für Stahlbeton, 102, Verlag Ernst & Sohn, Berlin (1950).

31. de Graaf, F. F. M. and van Schaik, H. H., 'Stromvloedkering Oosterschelde, Ontwerpcriteria en berekening beton-constructie', *Cement (Gouda)*, pp. 570–6 (1979).

32. Schießl, P. and Raupach, M., 'Influence of the type of cement on the corrosion behaviour of steel in concrete', *9th Int. Cong. Chem. Cem.*, New Delhi, India, Communication Papers, Vol. 5, pp. 296–301 (1992).

33. VDZ-Kommission 'Sulfatwiderstand', Hochofenze-ment mit hohem Sulfatwiderstand. *Beton* 30(12), pp. 459–62 (1980).

34. Osborne, G. J., 'The sulphate resistance of Port-land and blastfurnace slag cement concretes', *Durability of Concr., Proc. 2nd Conf.*, Montreal, American Concrete Institut, Vol. II, SP 126, pp. 1047–71 (1991).

35. DAfStb-Richtlinie Vorbeugende Maßnahmen gegen schädigende Alkalireaktion im Beton. Deutscher Ausschuß für Stahlbeton – DAfStb, Ausgabe Dezember (1997).

36. DIN 1164-1, Zement, Zusammensetzung und Anforderungen. Beuth-Verlag Berlin 1994 und Ergänzung A 1 (1999).

37. The Concrete Society, Alkali Silica Reaction – Minimising the risk of damage to concrete. *Technical Report* No. 30 (1987).

38. Lang, E. and Geiseler, J., 'Use of blastfurnace slag cement with high slag content for high-performance concrete', *Proc. of the 4th Int. Symp. Utilization of High Strength/High Performance Concr.*, Paris, France 29 May 31, Vol. 2, S. 213–22 (1996).

39. Lang, E., Hochleistungsbeton mit Hochofenzement. *Beton-Informationen*, 38(1), pp. 11–23 (1998).

Chapter thirteen

Properties and applications of natural pozzolanas

F. Massazza

13.1 Introduction

Within the building industry, the term 'pozzolana' covers all the materials which react with lime and water giving calcium silicate and aluminate hydrates possessing cementing properties. As a consequence, all pozzolana have to be rich in reactive silica or alumina plus silica. The list of pozzolanic materials includes many natural and artificial materials of different origin and composition.

The use of both natural and artificial pozzolanas is very old and the good condition of many ancient structures still in service nowadays are the best evidence of the excellent performance of pozzolanic binders. Pozzolanic binders met with a large success in the past owing to some enhanced properties they had as compared to other customary binders, such as lime and plaster. As a matter of fact they:

- harden under water,
- have higher strength,
- have higher resistance to aggressive environments.

Pozzolana–lime mixes have been used until earlier in the last century, but nowadays they are generally replaced by pozzolanic cements which have higher early and ultimate strength, while they keep the typical property of high chemical resistance of ancient mortar and concrete, unaltered. However, pozzolana–lime mixes can be successfully used also today for casting durable mortars and concrete where high strength is not necessary, or supply of cement is difficult.

A number of papers have been published on the properties and applications of natural pozzolanas. Most of them are quoted in some reviews which should be consulted for more detailed information [1–6].

13.2 Classification of natural pozzolanas

Natural pozzolanas can be classified according to their origin and their essential active constituent; obviously any main phase can be associated with other reactive constituents. On the basis of these criteria, natural pozzolanas are usually classified as follows [6]:

- volcanic, incoherent, rich in unaltered or partially-altered glass;
- tuffs, where volcanic glass has been transformed, entirely or partially, into zeolitic compounds;
- sedimentary, rich in opaline diatoms;
- diagenetic, rich in amorphous silica, resulting from the weathering process of siliceous rocks.

All pozzolanas contain variable amounts of non-reactive or low reactive constituents (quartz, andesite, horneblende, leucite and feldspars). The active phases are conventionally assimilated to the components of pozzolana which are dissolved by an acid attack followed by a basic one [7]. The method, modified, is a part of the European Standard [8].

Pozzolanas are often polluted by clay minerals coming from the weathering of the siliceous constituents of the original deposit or, as in the case with diatomaceous earth, from the concomitant precipitation with diatoms in waters.

13.2.1 Volcanic incoherent pozzolanas

The glass of these pozzolanas has a microporous structure [9] and is associated with different crystalline minerals. Glass of Naples pozzolana appears to be unaltered [9] whereas that of Rome is altered and little transparent [10]. A pumiceous appearance characterizes the ryolitic pumicites found in USA [11], the Santorin Earth in Greece [12] and the Volvic pozzolana in France [13].

Figures 13.1 and 13.2 show the appearance of natural pozzolanas observed under a scanning electron microscope (SEM) at different magnification.

(a) (b) (c)

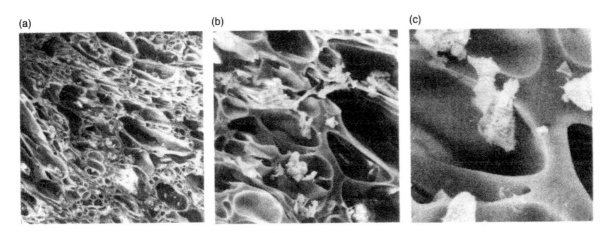

FIGURE 13.1   Bacoli pozzolana (Naples, Italy). Magnification (a) 100X; (b) 300X; (c) 1000X.

(a) (b) (c)

FIGURE 13.2   Colleferro pozzolana (Rome, Italy). Magnification (a) 100X; (b) 300X; (c) 1000X.

Reactive glass originates from explosive volcanic eruption. The gases contained in the fused magma particles are liberated during the explosion and the melted particles are quenched and transformed in a microporous glass [14]. As can be seen from Table 13.1, the main oxide is silica, accounting for more than 50 per cent, followed by alumina, Fe_2O_3 and lime; excluding some Spanish pozzolanas having MgO content as high as 9–11 per cent [2], magnesia percentage ranges between 0.2 per cent and 4 per cent. A characteristic feature of these pozzolanas is the high alkali content which, in Bacoli samples, can exceed 10 per cent. Loss on ignition (l.o.i) generally ranges between three per cent and six per cent but it can be higher in some altered materials.

Volcanic pozzolanas can undergo endogenous or exogenous actions which slowly transform them. The former have a positive effect because, through processes of zeolitization and cementation, loose pozzolanas are transformed into tuffs having marked pozzolanic properties. On the contrary, exogenous actions have negative consequences since they give rise to the formation of clay minerals [28].

13.2.2 Volcanic tuffs

The formation of compact tuffs is due to an autometamorphism process [29] which transform the glass of the incoherent pyroclastic pozzolanas in zeolitic minerals. Deposits of this type are found in many countries, that is, Italy [30], Spain [2], Germany [19], Turkey [31], Romania [22].

The transformation of volcanic glass into zeolitic compounds has been obtained also in the laboratory [30].

The range of the chemical composition is not very different from that of volcanic pozzolanas (see Table 13.1), apart from a higher loss on ignition. In these pozzolanas, the l.o.i. value has been considered to be an index of the intensity of the transformation that the original volcanic material underwent owing to weathering [30].

13.2.3 Sedimentary pozzolanas

Diatomite is a sedimentary rock formed by the skeletons of microscopic organisms (diatoms) mainly composed of opal. Large deposits are found in America, Africa and Europe: a typical example is offered by Danish moler, which has successfully been used in the past. Owing to the sedimentary origin, diatomite deposits are often polluted with clay. The chemical composition shows a high silica content and low percentages of minor oxides such as ferric oxide, lime, magnesia and alkali (see Table 13.1). A high alumina percentage means that the main component of pozzolana, opal, is associated with clay minerals.

The presence of clay creates some problems of use.

13.2.4 Diagenized materials

In some countries, pozzolanic materials with very high silica content occur. This can be due to the transformation of the original minerals into silica gel by the action of hot springs [26], or the transformation of heterogeneous sediments, partially of volcanic origin, in silica gel owing to an acid environment [32].

Silica prevails with percentages reaching 90 per cent. The amount of other oxides is generally low and depends on the mineral composition of the rocks where these pozzolanas come from (see Table 13.1).

13.3 Pozzolana–lime mixes

13.3.1 The pozzolanic reaction

The expression 'pozzolanic reaction' includes all the reactions occurring among the reactive phases of pozzolana, lime and water. According to the chemical thermodynamic rules, any blend of silica, or silicates, with lime and water should give rise to calcium silicate hydrates, however, the reaction becomes effective only if the system is activated or contains already activated phases. An example of the first case is given by the reaction between quartz and lime which becomes effective only when the mix is autoclaved at a suitable temperature [33]. In the second case, the pozzolanic reaction is possible at ordinary temperatures since

TABLE 13.1 Percentage composition of natural pozzolanas

	Country (%)	LOI (%)	SiO$_2$ (%)	Al$_2$O$_3$ (%)	Fe$_2$O$_3$ (%)	CaO (%)	MgO (%)	Na$_2$O (%)	K$_2$O (%)	SO$_3$ (%)	References
Volcanic incoherent pozzolanas											
Ciudad Real	Spain	4.60	42.20	14.90	12.10	12.00	11.00	1.70	1.60		2
Bacoli	Italy	3.05	53.08	18.20	4.29	9.05	1.23	3.08	7.61	0.65	15
Segni	Italy	4.03	45.47	19.59	9.91	9.27	4.5	0.85	6.35	0.16	15
Santorin Earth	Greece	4.74	67.98	19.61	4.34	1.25					16
Auvergne	France	0.24	46.60	17.60	11.80	9.84	5.58	3.14	1.76	0.02	17
Rhyolite pumicite	USA	3.38	70.76	12.85	1.38	1.08	1.08	0.43			11
Rhyolite pumicite	USA	3.43	65.74	15.89	2.54	3.35	1.23	4.97	1.92		11
Volcanic tuffs											
Toba Canaria	Spain	6.00	57.50	18.00	4.50	1.50	1.00	6.50	5.00		2
Trass K	Bulgaria		71.63	10.03	4.01	1.93	1.22		2.35	3.05	18
Rhenish Trass	Germany	11.10	52.12	18.29	5.81	4.94	1.20	1.48	5.06		19
Bavarian Trass	Germany	7.41	62.45	16.47	4.41	3.39	0.94	1.91	2.06		19
Selyp Trass	Hungary	16.33	55.69	15.18	6.43	2.83	1.01			0.26	20
Ratka Trass	Hungary	6.34	73.01	12.28	2.71	2.76	0.41			0.10	20
Yellow Tuff	Italy	9.11	54.68	17.70	3.82	3.66	0.95	3.43	6.38		21
Dacite Tuff	Rumania	7.27	67.70	11.32	2.66	3.73	1.64			0.18	22
Gujarat Tuff	India	12.60	40.90	12.00	14.00	14.60	1.45				23
Pella Tuff	Greece	2.23	62.22	19.78	3.99	4.57	2.70	1.58	2.25	1.6	24
Sedimentary pozzolanas											
Diatomaceous Earth	Czechoslovakia	7.62	65.22	19.23	2.76	0.80	0.25	0.43	1.42	0.31	22
Moler	Denmark	2.15	75.60	8.62	6.72	1.10	1.34	0.21	0.21	1.38	25
Diatomite	USA	8.29	85.97	2.30	1.84	9.00	0.61				11
Diatomite	USA	11.93	60.04	16.30	5.80	1.92	2.29				11
Diagenized pozzolanas											
Sacrofano	Italy	4.67	89.22	3.05	0.77	2.28	0.23				16
Beppu White Clay	Japan	4.09	87.75	2.44	0.41	0.19		0.11	0.11		26
Gaize	France	5.90	79.55	7.10	3.20	2.40	1.04			0.86	27

pozzolanas contain unstable phases which are:

- glasses, in volcanic materials and fly ashes;
- zeolitic compounds, in tuffs;
- amorphous silica, in diatomite and in diagenized natural materials;
- amorphous phases, in burnt clay minerals.

The progress in the pozzolanic reaction is displayed by the reduction of $Ca(OH)_2$ in the paste and the increase of silica and alumina dissolved by the Florentin attack in the hardened paste. Figure 13.3 [7] shows that the increase in combined lime is clearly associated with an increase in $SiO_2 + R_2O_3$ solubilized by Florentin attack.

The high percentage of R_2O_3 released by plain Roman pozzolanas before the lime attack (~30 per cent) is due to the presence in the material of minerals, such as leucite, which are soluble in HCl solution.

The pozzolanic activity includes two parameters, that is, the maximal amount of lime that a pozzolana can combine, and the rate at which such combination occurs. In practice, these two parameters cannot be separated; however, understanding of the pozzolanic reaction suggests analysing the two parameters separately. The

progress of the pozzolanic reaction is coupled with the formation of new hydrated compounds.

13.3.2 The hydrated phases

The main hydrated phases resulting from the reaction of volcanic pozzolanas with lime are [9,21,26,32,35–38].

- C-S-H,
- $4CaO \cdot Al_2O_3 \cdot 13H_2O$ (more or less carbonated),
- C_2ASH_8 (hydrated gehlenite).

Hydrated gehlenite forms in the absence of portlandite. Thus, in paste, alumina is contained only as calcium aluminate hydrates [39]. The occurrence of hydrogarnet seems to be associated with the presence of clay minerals and a long curing time (70–150 d) [26,40].

In diagenized high silica material and diatomite with low percentages of other oxides, only C-S-H forms [26].

The C/S values of C-S-H are dispersed owing to the small size of the analysed phases, which increases the error in the analytical determination and the variability in composition of calcium silicate hydrate (see Table 13.2) [26]. It seems

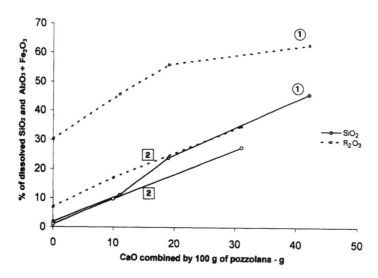

FIGURE 13.3 Silica, alumina and ferric oxide of lime–pozzolana mixes solubilized by Florentin attack versus lime combined by pozzolana. SiO_2 and $Al_2O_3 + Fe_2O_3$ expressed as percentage of the original contents. Pastes cured for different time. 1: Roman pozzolana; 2: Flegrean pozzolana [7].

that the C/S ratio depends on the presence or the absence of portlandite in the hydrated mix and, in any case, on the possibility that Ca^{2+} ion can migrate from Portlandite crystals to still unreacted pozzolana grains. As a matter of fact, as in the case of amorphous silica pozzolana which has entirely consumed the calcium hydroxide of the mix after 3 months, the C/S ratio is about 0.80, while in glassy pozzolana, which was unable to combine all the available lime, C/S ranges between 1.4 and 1.7 [26]. The former value corresponds to that (0.85) found in hardened mixes made with an opal-rich pozzolana [12].

13.3.3 Kinetics of the pozzolanic reaction

The percentage of lime combined by pozzolanas ranges between wide limits, and depends on factors related to the nature of pozzolana and to the mix characteristics. The expression '*nature of pozzolana*' includes the chemical and mineral composition as well as the specific surface of pozzolana while '*mix characteristics*' involve the conditions within which the pozzolanic reaction occurs.

The main factors affecting the pozzolanic reaction are:

- nature and composition of the active phases;
- their content in pozzolana;
- their specific surface;
- the lime/pozzolana ratio of the mix;

- the water/mix ratio;
- the curing time;
- the curing temperature.

The amount of combined lime mainly depends on the silica content or, taking into account that silica can be present also in non-reactive forms, on the active silica content of pozzolana. Moreover, since also the alumina of the active phases reacts with lime, a very good relationship occurs between combined lime and active silica plus alumina (see Figure 13.4) [41].

Table 13.3 shows that combined lime decreases from Rhine trass to obsidian glass and this can be attributed to differences in composition and specific

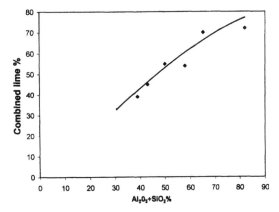

FIGURE 13.4 Combined lime versus reactive silica + alumina content of pozzolanas. Hydrated lime/pozzolana ratio = 0.8, water/binder ratio = 2, reaction time = 180 days [41].

TABLE 13.3 Lime-binding capabilities of the Principal trass minerals

Mineral component	Lime combined (mg CaO/g)
Glass phases of	
Rhenish trass	364
Bavarian trass	272
Obsidian	176
Crystal phases of Rhenish trass	
Quartz	43
Feldspar	1173
Leucite	90
Analcite	190
Kaolinite	34

TABLE 13.2 CaO/SiO_2 of C-S-H in the $Ca(OH)_2$–pozzolana mixes estimated by FESEM with EDX. Accelerating voltage 20 kV [26]

Pozzolana	CH/P ratio	Curing Temperature (°C)	Age (M)	Ca/Si molar ratio
Beppu white clay (V)	25/75	20	3	0.87–0.77
		40		0.85–0.80
		60		0.81–0.75
Higashi-Matsuyama tuff (G)	25/75	20	3	1.49–1.35
		40		1.75–1.60

surface. The role played by the chemical composition is supported by the fact that Rhine glass releases more alkalis than Bavarian and obsidian glasses [42].

Analcite, one of the components of pozzolanic tuffs, combines less lime than the glass contained in the two types of trass, but other zeolitic minerals, such as erschelite, are considered to be more active than volcanic glasses [43].

Table 13.3 shows that some finely ground crystalline minerals, such as alkali feldspar, can also bind substantial amounts of lime. This property appears to be common to many silicates and it has been evidenced that granites combine more lime than siliceous sand but less than quartz glass [44].

In the case of alkali feldspars, the formation of C_4A-hydrate occurs, whereas in the case of anorthite or kaolinite calcium aluminate hydrate or hydrogarnet form [45].

However, these reactions are not associated with any hardening comparable to that of pozzolana.

Volcanic pozzolanas from Bacoli and Segni (Italy) or Santorin Island (Greece) are considered to be very active since they are mainly glassy and contain only limited amounts of non-reactive crystalline minerals. On the contrary, the Volvic pozzolana (France) is considered to be a weak pozzolana since the glass content is relatively little and is associated with considerable amounts of crystalline minerals, mainly andesite [13]. Gaize, another French pozzolana, can contain up to 50 per cent insoluble silica after attack with hydrochloric acid and potassium hydroxide as well as considerable amounts of clay minerals [46].

Diatomaceous earth and other high-silica pozzolanas feature a very high lime-binding capacity since they are mainly formed of fine amorphous silica (see Table 13.1). Clay contamination occurring in many pozzolanas does not necessarily reduce the lime-binding capacity since clay minerals can combine remarkable percentages of lime, but it negatively affects the practical use of the material since it gives mortar a muddy consistency [47], increases the water requirement, and decreases the strength of the mixes. A typical case is offered by diatomaceous earth which, while having a remarkable pozzolanic activity, cannot

be used without previous expensive calcining when it is polluted by clay minerals [11]. Clay contamination can affect also volcanic pozzolanas [48] or high-silica diagenized pozzolanas [49].

High specific surface of pozzolana makes lime combination easier, but looking at Figure 13.5 [50] it is apparent that it definitely affects only the short-term activity of pozzolana whereas at longer ages, the pozzolanic reaction seems to be controlled by other factors such as the silica and alumina active contents.

Lime combined by pozzolana in the paste depends on the calcium hydroxide/pozzolana ratio. This dependence is observed both in dispersion and in paste [51,52]. Table 13.4 shows that high water/mix ratios significantly promote the pozzolanic reaction. Pozzolana–lime mortar requires a prolonged curing since the pozzolanic reaction is slow. The rate of combination increases also with temperature. In Figure 13.6, combined lime clearly appears to depend on the length of curing, the nature of pozzolana, the composition of the mix and the curing temperature [26].

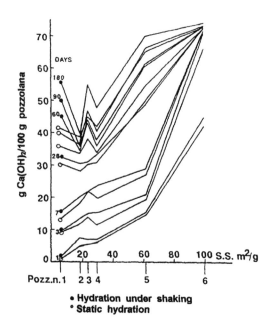

FIGURE 13.5 Calcium hydroxide combined versus specific surface of pozzolana. Hydrated lime/pozzolana ratio = 0.8, water/solid ratio = 2 [50].

TABLE 13.4 Calcium hydroxide combined by 100 g of Pozzolana after 90 days reaction [49,51].

Pozzolana (N°)	Paste[a]	Dispersion[b]
1	17.9	51
2	30.5	36
3	24.4	54
4	31.9	47
5	27.5	65
6	35.5	75
7	24.7	
8	24.9	

[a] Lime/pozzolana ratio = 0.40; water/binder ratio = 0.6;
[b] Lime/pozzolana ratio = 0.80; water/binder ratio = 2.0, continuous shacking.

13.3.4 Activation of the pozzolanic reaction

The rate of the pozzolanic reaction and hardening of pozzolana–lime mortars can be improved by resorting to different means: grinding and firing pozzolanas, compacting mixes and adding some chemicals affect the binder performance.

- Firing carried out at 600–800°C increases the performance of natural pozzolanas containing clay, but it is ineffective on glassy or zeolitic materials [11,48,53–55]. As a matter of fact, the thermal treatment destroys the crystal structure of the clay minerals which is replaced by an amorphous very reactive structure.

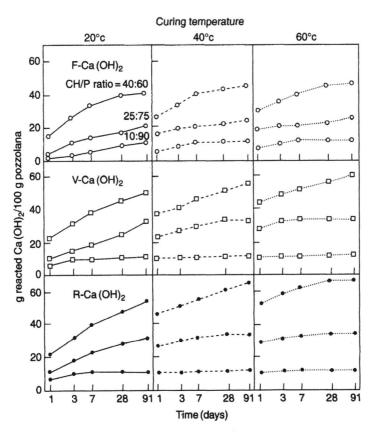

FIGURE 13.6 Ca(OH)$_2$ combined by pozzolanas F, V and R estimated by X-ray diffraction analysis. Water/binder ratio = 0.56. Hydrated lime/pozzolana ratios and curing temperatures indicated in Figure [26].

In this case, firing has two positive effects:

– it reduces the high water demand associated with the presence of clay minerals,
– it increases the active phase content.

Higher temperatures are dangerous for all the types of pozzolanas since they promote devitrification and crystallization of the active phases and decrease the specific surface of the material.

- The rate of lime combination increases in the presence of gypsum [44,56] or other chemicals such as Na_2SO_4 and $CaCl_2$. Acceleration of the pozzolanic reaction positively affects the strength development (see 13.4.5).
- Steam curing increases lime combination. When mixes of calcium hydroxide and pozzolana compacted under a pressure of about 130 MPa are cured at 70°C most of the lime reacts within 24 h, Above this temperature, reaction seems to stop or to decrease [57].

13.3.5 Mechanism of the pozzolanic reaction

The mechanism of the pozzolanic reaction is complex and not well known, however examination and discussion of the results collected so far suggest the following simplified model [26].

After mixing, the liquid quickly becomes saturated in $Ca(OH)_2$ and the pH rises to over 12.7. The active phases of pozzolana are attacked by the strong alkaline solution which dissociates the surface SiOH groups into SiO_4^{4-} and H^+ leaving the grains negatively charged. Alkalis from the pozzolana dissolve in the liquid phase and Ca^{2+} is adsorbed on the surface of the grains probably by electrostatic forces. Leaching of alkalis leaves a thin, amorphous Si and Al rich layer on the glass surface. This is unstable so that and SiO_4^{4-} and AlO_2^- gradually dissolve and combine with Ca^{2+}.

The pozzolanic reaction is likely to be the sum of topochemical reactions with dissolution and precipitation ones. The former type of reaction has been evidenced by means of SEM and optical microscope observations which have shown that C-S-H forms on the surface of zeolitic grains

dispersed with lime in excess water [58]. The optical microscope has revealed that a layer of calcium silicate hydrate replaces the external zone of pozzolana covering the unaltered grain core [59].

Apart from the Al_2O_3 combined in C-S-H, alumina seems to follow the latter process since it preferably crystallizes far from the pozzolana grains, that is, in the small holes and microcracks of the paste [59].

The calcium aluminate hydrate precipitates far from pozzolana grains since AlO_2^- is considered to diffuse more quickly than SiO_4^{4-} owing to its smaller electric charge and lesser oxygen content [26].

13.3.6 Strength

Strength of pozzolana–lime mixes increases as the amount of combined lime increases but, as shown in Figure 13.7, there is no general relationship between the two parameters even if a good correlation appears to be within each pozzolana [52].

The strength development and the ultimate strength depend on pozzolana/lime ratio and the water/binder ratio [60]. Figures 13.8 and 13.9 evidence that the rate of hardening is slow, but that the long-term strength can reach about 20 MPa and this value is high enough for many applications [61]. The maxima vary also in connection with

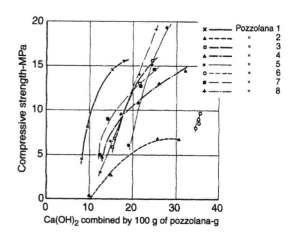

FIGURE 13.7 Compressive strength of lime/pozzolana mixes versus lime combined by pozzolanas. $Ca(OH)_2$/pozzolana ratio = 0.40; water/binder ratio = 0.6 [52].

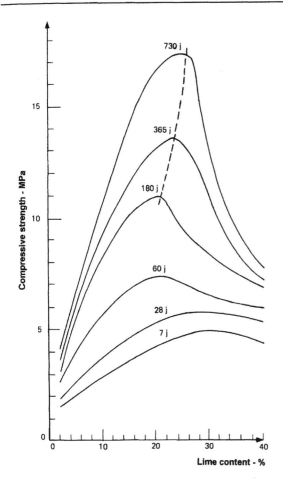

FIGURE 13.8 Compressive strength of hydrated lime-pozzolana pastes versus lime content in the mix. Water/binder ratio = 0.56; curing length indicated in figure [61].

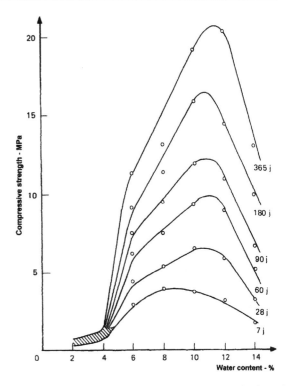

FIGURE 13.9 Compressive strength of hydrated lime-pozzolana pastes versus water content of the mix. Curing length indicated in figure [61].

other factors so that the best mixes must be established on a case-by-case basis.

A rise in temperature accelerates early age strength development of pozzolana–lime mixes, but a prolonged curing negatively affects the mechanical properties (see Figure 13.10) [62]. Strength as high as 45–55 MPa can be obtained after 17 h heating at 70°C mixes compacted under high pressure [57].

Gypsum addition increases strength of pozzolana–lime mixes (see Figure 13.11) [65]. However, the calcium sulphate content must be less than about 7 per cent since higher levels cause

strength loss, expansion, and possibly, cracking of mortar [63–65].

Addition of some chemicals such as Na_2SO_4 and $CaCl_2$ improves strength of pozzolana mortars, whereas NaCl appears to be not effective [66].

13.4 Pozzolana-containing cements

13.4.1 Classification

Pozzolana added to Portland cement reacts with the calcium hydroxide released by the calcium silicate of clinker during the hydration. Depending on the replacement level, portlandite can be partially or entirely combined giving calcium silicate and aluminate hydrates.

'Pozzolanic cements' are, by definition, blends of Portland cement and pozzolana which, cured

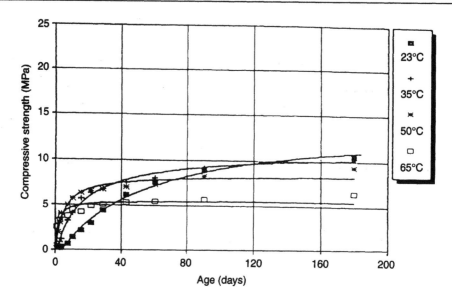

FIGURE 13.10 Effect of curing temperature on compressive strength development of pastes. Hydrated lime/pozzolana ratio = 0.25, water/binder ratio = 0.5 [62].

under certain conditions, give pastes practically without free portlandite. This requirement is ascertained by the so-called 'pozzolanicity test'. This is carried out by dispersing 20 g of cement in 100 cc of distilled water contained in a flask which is immediately sealed and kept at 40°C. After eight days, the flask is opened, the suspension filtered and the solution analysed for $[Ca^{2+}]$ and $[OH^-]$ [67,68]. The two values determine a point in the diagram $[Ca^{2+}] - [OH^-]$: the cement is pozzolanic if the point is located under the saturation curve of $Ca(OH)_2$ whereas it is not if it falls above the curve (see Figure 13.12). If the result is uncertain, the test is repeated prolonging the storage of the sample to 15 days [67].

The amount of pozzolana necessary for complying with the pozzolanicity test depends on the chemical composition of both clinker and pozzolana, that is, on the amount of calcium hydroxide coming from the hydration of silicates, and the content of the reactive phases in pozzolana.

In practice, a good active pozzolana can be blended with a higher percentage of Portland cement, and this increases early and ultimate strength of the blended cement while keeping positive the pozzolanicity test.

'Pozzolana-cements' are blends which do not comply with the pozzolanicity test inasmuch as their pozzolana content is insufficient, in terms of either quantity or quality, to combine all the calcium hydroxide formed during the hydration of the clinker fraction.

Replacement of pozzolana for Portland cement causes some changes in the microstructure, in the phase composition as well as in the engineering properties of the plain Portland cement. The changes are gradual according to the replacement level.

13.4.2 The hydrated compounds

The reaction products found in hardened pozzolana- and pozzolanic-cements are the same as those occurring in Portland cement paste, that is:

- ettringite,
- tetracalcium aluminate hydrate,
- calcium silicate hydrate,
- portlandite.

However, since the overall chemical composition is different from that of the parent Portland cement, these compounds occur in different

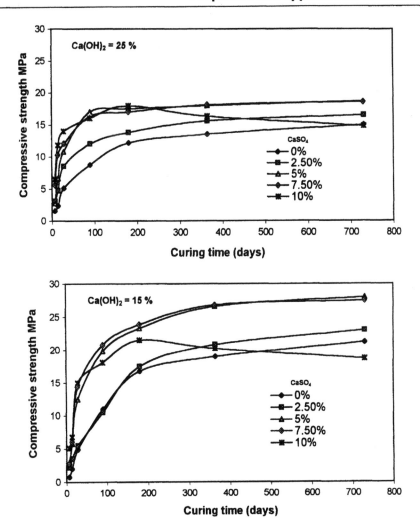

FIGURE 13.11 Compressive strength of lime–Segni pozzolana mixes containing different amounts of hydrated lime and gypsum [65].

percentages. The most important variations concern portlandite and C-S-H contents which are lesser and greater respectively, as a function of the progress of the pozzolanic reaction.

Aluminates are generally contaminated by carbon dioxide and portlandite is partially transformed into calcite.

In pozzolana-containing cements, primary C-S-H, originating from the hydration of the clinker silicates, is associated with secondary C-S-H formed by the pozzolanic reaction. C-S-H has a

lower C/S ratio and a greater alumina content as compared to those found in plain Portland cement pastes. The calcium silicate hydrate filling the space between still unreacted C_3S and pozzolana grains shows a clear zonal structure with layers having different C/S ratios [69]. The zonal structure of C-S-H covering the still unreacted pozzolana grains has been disclosed also by differences in the refractive index [70].

Portlandite persists for a long time in hardened cement paste also when the pozzolana/Portland

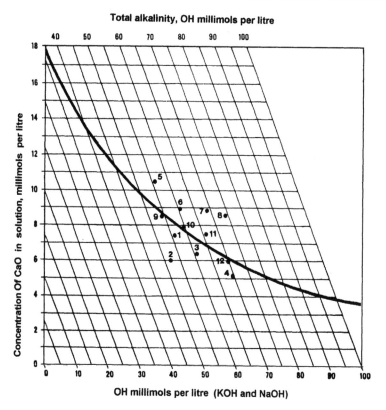

FIGURE 13.12 Solubility isotherm (40°C) for Ca(OH)$_2$ in the presence of alkalies. Points 1, 2, 3 and 4 represent pozzolanic cements; points 5, 6, 7 and 8 not pozzolanic cements [67]; results concerning points 9, 10, 11 and 12 are uncertain and the test is repeated prolonging the storage till 15 days.

cement ratio should be sufficient to combine all the calcium hydroxide formed during the hydration of Portland cement [71,72]. As expected, uncombined lime in paste decreases by increasing the pozzolana content (see Table 13.5) [71].

13.4.3 Microstructure

The main characteristic of a hardened pozzolanic cement paste is the lack of large portlandite crystals and the presence of layers of hydrated compounds covering the residual, not-yet-hydrated pozzolana particles. The ground mass of the paste does not appreciably differ from that of Portland cement since the smallest pozzolana particles are no longer perceivable because they have been eaten away by the lime attack.

Total porosity and pore size distribution of pozzolanic cement pastes are not very different from those displayed by Portland cement pastes. However, total porosity is higher, whereas the mesopores are shifted towards the smaller diameters [26,73]. The pozzolanic cement paste has a lower permeability in spite of a higher porosity because the pore structure is different. As a matter of fact, pozzolanic cement paste has a high content of relatively large pores (and this increases porosity) connected with thinner and more segmented pores (and this decreases permeability) [74]. Porosity and permeability decrease as hydration progresses.

Obviously the pozzolana-cement microstructure is more similar to that of a Portland cement as the pozzolana content decreases.

TABLE 13.5 Free lime content of cement pastes cured six months. Cements made with three different pozzolanas and three different clinkers [71]

Sample	Free CaO
Clinker A	
100%	14.90
90% + 10% Pozz. 22	14.50
80% + 20% Pozz. 22	10.35
70% + 30% Pozz. 22	8.44
60% + 40% Pozz. 22	6.30
50% + 50% Pozz. 22	3.54
Clinker B	
100%	14.00
90% + 10% Pozz. 15	14.00
80% + 20% Pozz. 15	10.85
70% + 30% Pozz. 15	8.32
60% + 40% Pozz. 15	3.30
50% + 50% Pozz. 15	4.90
Clinker C	
100%	15.65
90% + 10% Pozz. 21	14.45
80% + 20% Pozz. 21	11.80
70% + 30% Pozz. 21	9.76
60% + 40% Pozz. 21	6.97
50% + 50% Pozz. 21	5.65

TABLE 13.6 pH of pore solution squeezed from hardened cement pastes w/c = 0.60, curing length = 90 days; curing temperature $\cong 18°C$. Blended cements with 15% replacement for portland cement [75]

	OPC		HAPC	
	pH	$Na_2O_{eq.}$	pH	$Na_2O_{eq.}$
Portland cement (Plain)	13.43	0.50	13.77	0.95
Blended cement with:				
Degussa SiO_2	13.10	0.56	13.48	0.94
Silica fume			13.18	0.91
Eggborough PFA	13.50	0.96	13.64	1.35
Longannet PFA	13.43	0.60	13.62	0.99
Slag	13.54	0.54	13.68	0.93
Natural pozzolana	13.61	0.78	13.77	1.16

[a] Calculated from the composition of the blending materials.

levels greater than 10 per cent [76]. Owing to the chemical similarity with silica fume, opal-rich natural pozzolanas are expected to decrease pH by about one point for a replacement level of 20 per cent [77].

The increase of alkalis in the pore solution is associated with the decrease in the Ca^{2+} concentration [78].

13.4.4 Pore solution

Alkalinity of the pore solution in the cement paste can change in the presence of pozzolana depending on the amount of soluble alkalis contained in both pozzolana and Portland cement. That is the reason why some results appears to be contradictory. A 15 per cent replacement of volcanic pozzolanas for Portland cement with high alkali content does not change the pH of the pore solution, but it does increase it if the Portland cement has a moderate alkali level (see Table 13.6) [75]. According to other results, alkali and hydroxyl concentration in the pore solution decreases as the replacement level of pozzolana increases. However the pH decrease is small and it is about 0.1 for a replacement percentage of 30 per cent.

Silica fume causes a marked decrease of (OH^-) already after 5 h hydration [76] possibly as a consequence of a take-up of alkalis in the hydrated solid phases. Figure 13.13 shows that the pH decrease becomes dramatic with replacement

13.4.5 Kinetics of hydration

The addition of fine material to Portland cement accelerates hydration: the effect is produced either by inert powders such as rutile [79] or by active materials such as natural and artificial pozzolanas and slags [80]. Acceleration is disclosed by the increase in the initial heat evolution rate and by the increase in the calcium hydroxide content in comparison with the value calculated on the basis of the dilution factor.

The pozzolanic reaction is definitively slower than the hydration of clinker, so that the expected reduction of the calcium hydroxide content becomes perceivable only after a certain time. Practically, in the first month, the amount of portlandite in the paste is more or less the same as that formed in the plain Portland cement multiplied by the dilution factor [6].

Later on, the pozzolanic reaction gradually prevails on the $Ca(OH)_2$ formation owing to the consumption of the calcium silicates. As a

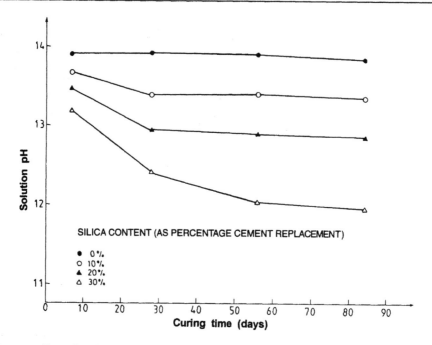

FIGURE 13.13 Effect of replacement level on the pH of the pore solution expressed from hardened cement pastes containing 0–30 per cent silica fume. Water/binder ratio = 0.5 [76].

consequence, portlandite would be expected to disappear but, on the contrary, it persists for a long time as shown in Table 13.5 [71].

Completion of pozzolanic reaction requires that Ca^{2+} can freely reach the still unreacted grains of pozzolana, but diffusion is prevented by the following hindrances:

- the progress of hydration and pozzolanic reaction sharply reduces porosity and permeability of the paste,
- portlandite crystals and pozzolana grains are covered with little permeable layers of C-S-H.

The importance of the free movement of ions in completion of the pozzolanic reaction is evidenced by the remarkable decrease in portlandite content, going from paste to mortar and concrete, as shown in Figure 13.14 [81]. The effect is mainly due to differences in the water–cement (w/c) ratio but also to the permeability of the transition zone around the sand and aggregate grains which is higher than that of the bulk paste.

13.4.6 Strength

Replacement of pozzolana for Portland cement causes a decrease in early strength, but the decrease is less than that foreseen on the basis of the dilution factor since the fine addition promotes the hydration of the Portland cement fraction [6]. Seven to fourteen days after gauging, the pozzolanic reaction starts and gives an increasing contribution to the strength development so that the ultimate strength of pozzolana-containing cements can be higher than that of the parent Portland cement. A display of this occurrence is shown in Figure 13.15 [82].

The time required to recover the initial strength loss caused by the replacement of pozzolana depends on the nature of pozzolana, including the content of the active phases and the specific surface, the replacement level, the composition and the strength class of Portland cement.

Owing to the relative slowness of the pozzolanic reaction, pozzolanic-cement mortars and concrete require longer wet-curing than Portland

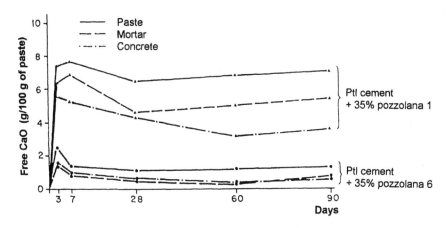

FIGURE 13.14   Free lime content of paste, mortar (1:3) and concrete (1:6) samples cured at 40°C. W/c ratio=0.5. 35 per cent of Portland cement replaced by two different pozzolanas [81].

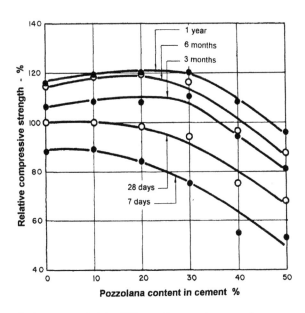

FIGURE 13.15   Effect of pozzolana replacement for Portland cement and compressive strength of standard mortar. Values expressed as percentage of the 28-day strength of the control Portland cement [82].

ones, but this handicap is associated with some advantages, namely, a lower heat of hydration and a lower rate of heat evolution.

Figure 13.15 shows that the ultimate strength of pozzolanic mortar is higher than that of a Portland one having the same 28-day strength, because the pozzolanic reaction goes on still for many years.

All other conditions being equal, tensile strength of pozzolanic cement is generally higher than that of Portland cement [3].

The water demand of a natural pozzolana-containing cement is higher than that of the parent Portland cement depending on the high specific surface of pozzolanas. As a consequence, pozzolanic-cement concrete must be gauged with a slightly higher w/c ratio and this decreases strength. A borderline event is given by diatomite or other high-silica pozzolanas with very high specific surface for which the replacement level for Portland cement cannot exceed 15–20 per cent and this is purely for workability reasons. Generally, this level is high enough to comply with the pozzolanicity test requirements, since these types of pozzolana possess a remarkable lime-binding capacity [82]. On the other hand, this replacement is low enough to keep unaltered early strength. In any case, the use of appropriate water-reducing admixtures reduces any slump loss of concrete. A positive feature of pozzolanic cement concrete is its lower tendency to segregation.

13.4.7 Shrinkage

The higher water demand of pozzolanic cements does not affect mortar shrinkage since mortars

341

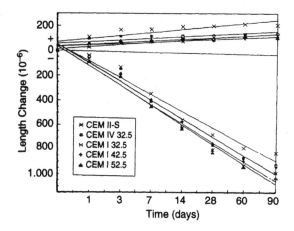

FIGURE 13.16 Length changes in standard mortar stored in water (expansion) or in 50 per cent R.H. air (shrinkage) (82).

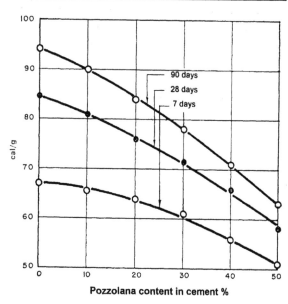

FIGURE 13.17 Effect of pozzolana replacement for Portland cement on the isothermal heat of hydration [81].

are gauged with the same w/c ratio (see Figure 13.16) [82], but it does affect the shrinkage of concrete which are generally compared on the basis of the same workability. However, the effect of the pozzolana presence is limited and contained within the wide range of the shrinkage values displayed by different Portland cements [83] and, above all, by the type of aggregate [84]. In any case, slight changes in the mix composition and the use of appropriate water-reducing admixtures can prevent any drawback.

13.4.8 Creep

Creep is strictly related to the strength of concrete at the time of loading, so no significant difference appears when concretes having the same strength are compared. Problems could arise when loading is applied too early: in this case, creep of pozzolana concrete is higher than that of a Portland one since strength is definitely lower [85]. In turn, creep of long-cured pozzolanic concrete must be lower than that of the control Portland concrete.

High creep is not necessarily a negative property, and it can be useful since it relieves stresses coming from ground sinking or from thermal or shrinkage differences occurring in structures.

13.4.9 Heat of hydration

The second reason which has contributed to the success and diffusion of pozzolanic cements is their lower heat of hydration. This property has two components, that is:

- the total heat of hydration and
- the rate of heat evolution.

The effect of the pozzolana replacement for Portland cement on the heat of hydration is shown in Figure 13.17 [81], and the effect on the heat development rate is shown in Figure 13.18 [26].

The two properties play an important role in mass concrete or concreting in summer since they reduce the peak of the temperature in the core of structure [85].

13.4.10 Durability

The ability of concrete to keep its performance unchanged over time is referred to as 'durability'. This property is of primary importance in determining the service life of a structure which, as a rule, is exposed to a more or less aggressive

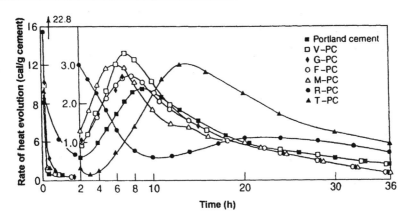

FIGURE 13.18 Effect of replacing five natural pozzolanas and one fly-ash (T) for Portland cement on the heat evolution rate. Pozzolana content = 40 per cent.

environment and sometimes can be made of detrimentally reactive materials.

Environmental attacks can have a chemical and physical origin and they produce a decline of concrete performance. Pozzolanas are not very effective in opposing physical attacks, but they significantly increase the chemical resistance of cement. It is noteworthy that pozzolanic cements began to be used just for improving the resistance of concrete to chemical attacks. The chemical resistance of a cement increases with the replacement level, so the best balance between the durability and ultimate strength is given by pozzolanic cements.

The reasons for the high chemical resistance of pozzolanic cements are many, but these can be reduced to two:

- lower content of portlandite in the paste, and
- lower permeability of the paste.

The following short comments will explain the behaviour of pozzolanic cements with regard to durability problems.

13.4.10.1 Carbon dioxide

Carbon dioxide decomposes any hydrated compounds present in the hardened cement forming calcium carbonate and leaving silica and alumina gel. The attack is severe for porous concrete, but is beneficial for the dense and compact one since calcite precipitating in capillary pores reduces porosity and permeability [85,87].

Carbonation is a risk for reinforced concrete since it decreases the pH in the pore solution and this may lead to steel corrosion.

Prevention of carbonation requires the adoption of two measures; namely, reduction of concrete permeability and increase in the cover thickness.

This is the reason why durable concrete must be compact, that is, it must be made with a low w/c ratio and a cement content not lower than 300 kg per cubic metre. In practice, a concrete having strength as high as 30–35 MPa has a low permeability which, associated with a cover of not less than 30 mm in thickness, gives concrete a good protection against the penetration of gases [88].

Under these conditions, the type of cement generally plays a secondary role.

13.4.10.2 Lime leaching

Concrete is attacked by water because calcium hydroxide is soluble and calcium silicate hydrate hydrolyses in water. However, the extent of damage is not severe and is limited to the surface, since leaching out of lime leaves a protective layer made up of silica and alumina gels which hinders the penetration of water.

Appreciable damage may occur only when concrete is permeable, the aggressive water is under pressure as in dams, or the sections of the attacked structures are thin (concrete pipes).

343

Water can be acidic owing to the presence of dissolved carbon dioxide, but the effectiveness of the attack is linked to the hardness of water [89,90].

In any case, when leaching of lime endangers the concrete structure security, pozzolanic cement contributes to slow down the attack for three reasons:

- portlandite is lacking, or any residual amount thereof is covered with little permeable layers of secondary C-S-H produced by the pozzolanic reaction
- C-S-H is more resistant to leaching owing to a lower C/S ratio and a higher alumina content,
- the C-S-H content is higher.

Figure 13.19 shows that leached lime decreases by increasing the pozzolana content in the blend [91].

13.4.10.3 Sulphate

Structures in contact with surface or ground water bearing sulphates can suffer strong attack.

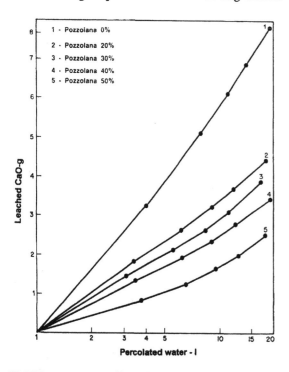

FIGURE 13.19 Effect of pozzolana content on lime leached by water percolating mortar. Cement: sand: water = 40 : 160 : 12.5. Sand grading between 0.5 and 1.0 mm. Curing length before testing 21 days [91].

Deterioration is due to the formation of expansive compounds, that is, compounds which have a definitely higher volume than the compound they replace in the concrete. This gives rise to internal stresses which make concrete swell, and crack if the tensile strength of the structure is exceeded. The effect of the attack depends on the cation associated with the sulphate ion, being the weakest with calcium sulphate and the strongest with magnesium or ammonium sulphate.

1. Deterioration caused by $CaSO_4$ is due to the formation of ettringite by reaction with the calcium aluminate hydrate according to the equation:

$$4CaO \cdot Al_2O_3 \cdot 12H_2O + 3CaSO_4 + 21H_2O$$
$$\rightarrow 3CaO \cdot Al_2O_3 \cdot 3CaSO_4 \cdot 32H_2O + Ca(OH)_2 \quad (1)$$

It is evident from equation (1) how the susceptibility of concrete to sulphate attack depends on the aluminate content of cement, so the use of low C_3A cements is suggested for preventing sulphate expansion of concrete. Replacement of pozzolana for Portland cement improves the sulphate resistance of ordinary Portland cement under moderate attack such as that given by $CaSO_4$ [18].

2. The attack carried out by Na_2SO_4 is more severe than the previous one and is initially due to the formation of $CaSO_4 \cdot 2H_2O$ according to the reaction:

$$Ca(OH)_2 + Na_2SO_4 + 2H_2O$$
$$\rightarrow CaSO_4 \cdot 2H_2O + 2NaOH \quad (2)$$

At a later stage, gypsum can react with calcium aluminate hydrate giving ettringite according to reaction (1).

Under common conditions, sodium sulphate does not attack the calcium silicate hydrates since these are less soluble than gypsum [90]. Also in this case, pozzolana proves to be able to improve the resistance of Portland cement to sulphate.

The reasons why pozzolanic cement increases the resistance of concrete to the calcium and sodium sulphate attack are still subjects of discussion. Protection against the expansive formation of gypsum (equation (2)) can be attributed to the lack of portlandite or to the occurrence of small calcium hydroxide crystals covered and shielded by C-S-H, but the prevention of the expansion due to the ettringite formation cannot be easily explained. As a matter of fact, replacement of pozzolana for Portland cement decreases C_3A, but increases the total Al_2O_3 content of cement (see Table 13.1). Thus the protective action exerted by pozzolanas cannot be interpreted on the basis of a decrease in the potential amount of calcium sulphoaluminate hydrate.

The effect of pozzolanas in preventing expansion caused by the sulphate attack has been tentatively explained by resorting to some theories based on the very low concentration of Ca^{2+} existing in the pore solution of pozzolanic cements, namely:

- formation of ettringite does not occur [92],
- ettringite formation should occur without expansion [92],
- the sulphate concentration required for ettringite formation should increase as the concentration of calcium hydroxide in solution decreases [93].

The higher C-S-H content of a pozzolanic cement paste enhances the protective action of the silicate gel on the aluminate hydrates. This supposition is supported by the higher resistance to sulphate attack given by pozzolanas with high SiO_2/Al_2O_3 ratio such as Italian 'white earth' [94].

3. Prevention of detrimental expansion due to $MgSO_4$ attack is hard since the salt attacks all the components in the hardened paste. The first reaction occurring is possibly the following:

$$Ca(OH)_2 + MgSO_4 + 2H_2O \rightarrow Mg(OH)_2 + CaSO_4 \cdot 2H_2O \quad (3)$$

Then gypsum can react with the calcium aluminate hydrate giving ettringite according to reaction (1). Calcium sulphoaluminate is however, unstable in the presence of magnesium sulphate so that it is decomposed to form gypsum, hydrated alumina and magnesium hydroxide again.

Magnesium sulphate also attacks calcium silicate hydrates according to the reaction:

$$CaO \cdot SiO_2 \cdot xH_2O + MgSO_4 + yH_2O \rightarrow Mg(OH)_2 + CaSO_4 \cdot 2H_2O + SiO_2 \cdot zH_2O \quad (4)$$

Pozzolanic cements made with volcanic pozzolanas improve the resistance of cement to magnesium sulphate, but can prevent deleterious expansion only when the concentration of solution is below 2 per cent (see Figure 13.20) [81]. Better resistance is achieved with the use of highly siliceous pozzolanas having a high silica/alumina ratio [94].

13.4.10.4 Chloride

Chloride solutions, which form in cold seasons when de-icing sodium and calcium chlorides are

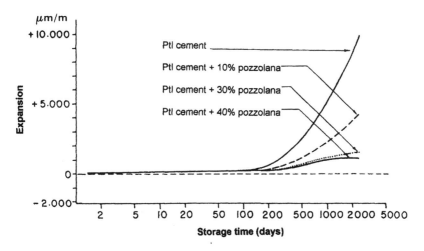

FIGURE 13.20 Effect of *a volcanic* Pozzolana on the expansion of 1:3 mortars stored in one per cent $MgSO_4$ solution [81].

345

spread on roads and bridges, are detrimental to the durability of both unreinforced and reinforced concrete.

Chloride is harmful for concrete since it promotes leaching of portlandite [95–97], and this improves porosity and reduces strength. High concentration causes swelling and crumbling of concrete [97]. Similar results occur when salts crystallize and grow into the pores of concrete, thus giving rise to stresses that cause cracks when their values become higher than the tensile strength of the cementitious material.

Cl^- promotes corrosion of the reinforcement of structures since it destroys the passivation of steel caused by the high alkalinity of the pore solution. The resulting rust has a greater volume than the corroded steel so that concrete first swells and then spalls.

The very first measure to be taken to minimize chloride penetration is to make compact concrete as well as sufficiently thick cover. Pozzolanic cement does not prevent the penetration of chloride into concrete but increases the level of protection since it reduces the depth of penetration of the aggressive ion as shown in Figure 13.21 [95].

The reasons why pozzolanic cement has a higher resistance to chloride attack than Portland cement are:

- a smaller leachable portlandite content in the paste;
- a lower effective coefficient of diffusion of Cl^- in the paste and in the concrete [98,99];
- lower permeability of paste and concrete;
- stronger interaction between Cl^- and the pore walls of C-S-H.

The last reason is based on the fact that diffusion of Na^+ and Cl^- ions through the cement paste is lower than through a thin quartzite plate [99], and this is interpreted as a consequence of an interaction between the ions and the pore walls of the cement paste [100].

Since the micropores of the pozzolanic cement paste are more segmented than those of the Portland cement one, the smaller penetration of the Cl^- ion into the hardened paste could be due to this obstacle. Moreover, it is possible that the delay in the penetration of Cl^- observed in pozzolanic cement paste is also due to an enhanced chemical interaction between Cl^- and the pore walls owing to the different composition of C-S-H [98,99].

13.4.10.5 Sea water

In spite of the high salt content in the sea water, the attack of concrete reveals itself to be less intense than that foreseen on the basis of seawater composition. Nonetheless, sea water is definitely aggressive. In the past, the problem was solved by resorting to pozzolana–lime mixes which, after thousand of years of experience, proved to resist better than other binders. These mixes are today replaced by pozzolanic cements and even if experience is limited to a hundred years, all are agreed upon their positive effect on durability of concrete [100–106], even if compactness of concrete is a prejudicial requirement for long service life of structures [106].

The positive effect concerns both permanently submerged concrete and concrete located in splash zones where the attack is due to the salt crystallization, but the reasons are different. In the former case, pozzolana-containing concrete

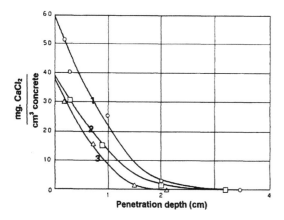

FIGURE 13.21 Depth of penetration of $CaCl_2$ into concrete after 340-day exposure: (1) OPC and solution containing $CaCl_2$ 80 g/l; (2) OPC and solution containing $CaCl_2$ 30 g/l; (3) pozzolanic cement (high chemical resistance) and solution containing $CaCl_2$ 30 g/l [95].

has a lower permeability to ionic penetration and unfavourable chemical conditions to the formation of expansive compounds. In the latter, in addition to the lower penetrability of Cl^-, a certain role could be played by the higher tensile strength/compressive strength ratio featured by pozzolanic cements. Figure 13.22 shows the different depths of chloride penetration into concretes with and without pozzolana [106].

13.4.10.6 Alkali–silica reaction

Prevention of dangerous expansion due to the reaction between the alkalis coming from the cement and some siliceous components of the aggregate is another success of pozzolanic cements containing volcanic or artificial pozzolanas such as fly ashes.

There is a general agreement on the fact that Portland cements containing less than 0.6 per cent Na_2O alkali equivalent can prevent expansion due to the alkali–silica reaction, but it is also known that modern technology, as a result of pollution prevention regulations in force in many countries, can make this target difficult to achieve.

Soon after the discovery of the alkali–silica reaction, it was found that the addition of finely-ground active materials, such as natural or artificial pozzolanas, could prevent the deleterious expansion caused by the presence of alkali-sensitive minerals in the aggregate [107].

However, in some cases, natural and artificial pozzolanas appeared to be not effective in preventing expansion and cracking, but in those cases the replacement level of pozzolana was small. As a matter of fact, small percentages of pozzolana replacement for Portland cement can increase expansion [108]. Only 30–40 per cent replacement of volcanic pozzolanas prevents this in spite of the high alkali content of pozzolanas (see Figure 13.23) [109]. This effect has been attributed to the pozzolanic reaction which, removing lime from the pore solution, changes the chemical conditions that favour the formation of the alkali–lime silicate hydrate responsible for swelling into unfavourable ones.

13.4.10.7 Thermal variations

Temperature differences occurring between surface and bulk concrete induce stresses which cause cracking whenever they become higher than tensile strength of concrete. Thermal differences occur as a consequence of the heat of hydration

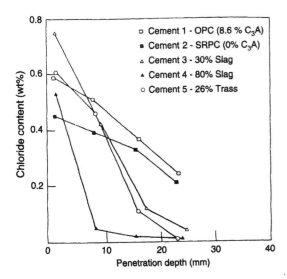

FIGURE 13.22 Effect of cement type on chloride penetration depth after six months of exposure [106].

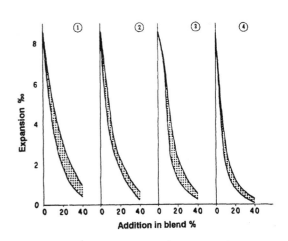

FIGURE 13.23 Variation of expansion as a function of per cent addition. Mortar bars manufactured with blended cements containing: (1) pozzolanas; (2) activated pozzolanas; (3) tuffs; (4) activated tuffs [109].

of cement, and changes in the temperature of the environment and are enhanced by the small thermal conductivity of concrete [110]. When concrete undergoes thermal gradients, pozzolanic cements behave better than Portland cements having the same compression for two reasons:

- they generally have a smaller heat of hydration and a lower heat evolution rate,
- they have a higher tensile strength.

Thermal variations which give rise to freezing and thawing of the free water contained in the concrete can cause severe deterioration. Frost resistance of concrete sharply increases with decreasing w/c ratios. However, the relationship appears to be valid only for w/c ratios being less than 0.45–0.50, that is, for high-strength concrete [90]. Frost resistance depends also on other factors, but the most important individual one is the entrainment of microscopic air bubbles into the mix. For each concrete, there is an optimum air content, but the highest resistance to frost is achieved with excellent quality concretes [111,112].

Pozzolana-containing cement does not negatively affect frost resistance, provided all other conditions are equal, and behaves worse than the counterpart Portland cement only if exposed to frost too early.

13.5 Conclusions

After thousands of years, pozzolana-containing concretes are still used today owing to their good mechanical and durability properties.

The ancient pozzolana-lime mixes are not as largely used as in the past since their early and ultimate strengths are lower than that of Portland cements. Nevertheless, the strength level that they can achieve is high enough to satisfy many structural requirements.

Pozzolanas blended with Portland cement give rise to cements whose strength is comparable to that of Portland cement. The differences with the parent Portland cement increase with the replacement level. Pozzolana-containing cements can replace the Portland cement in any current application provided they comply with the standard specifications. Moreover, pozzolanic cements have some peculiar properties on account of which they are preferred to Portland ones in special cases. Typically, the use of pozzolanic cements is suggested when a high chemical resistance or a low heat of hydration is a primary need. Other conditions being equal, pozzolana-containing cements behave like the other ones in terms of resistance to physical actions.

13.6 References

1. Lea, F. M., 'The chemistry of pozzolanas', *Proc. Symp. Chem. Cem.*, Stockholm, 460–93. INGENIORSVETENSKAPSAKADEMIEN – Stockholm (1938).
2. Calleja, J., Las Puzolanas, *Ion* 29(340), 623–38; (341) 700–713; 1970, 30(343), 81–90; (344), 154–60 (1969).
3. Malquori, G., 'Portland-pozzolan cement', *Proc. 4th Int. Symp. Chem. Cem.*, Washington, 1960, 43(2), 983–1006, National Bureau of Standard (US) Monograph (1962).
4. Lea, F. M., *The Chemistry of Cement and Concrete*, Edward Arnold Publishers Ltd, London (1970).
5. Massazza, F., 'Chemistry of pozzolanic additions and mixed cements', Principal Paper, *6th Int. Cong. Chem. Cem.*, Moscow (1974).
6. Massazza, F., 'Pozzolana and pozzolanic cements', Chapter 10, *Lea's Chemistry of Cement and Concrete*, 4th edn, P. C. Hewlett (ed.) (1998).
7. Malquori, G. and Sasso, F., Comportamento delle pozzolane flegree ad attacchi chimici con acidi ed alcali *La Ricerca Scientifica*, 6(II), 237–42 (1935).
8. CEN EN 196.2, Chemical analysis of cement: 10 Determination of residue insoluble in Hydrochloric Acid and Potassium Hydroxide – 1994.
9. Tavasci, B., 'Struttura della malta di calce e pozzolana di Bacoli', *Il Cemento*, 45(8–9), 114–204 (1948).
10. Tavasci, B., 'Struttura della pozzolana di Segni', *Il Cemento*, 42(1), 4–7; (2), 25–9 (1946).
11. Mielenz, R. C., Witte, L. P. and Glantz, O. J., 'Effect of calcination on natural pozzolanas', *Symp. Use of Pozzolanic Materials in Mortars and Concr.*, ASTM STP 99, 43–92 (1950).
12. Mehta, P. K., Studies on blended Portland cements containing Santorin earth, *Cem. Concr. Res.*, 11, 507–18 (1981).
13. Mortureux, B., Hornain, H., Gautier, E. and Regourd, M., 'Comparaison de la Reactivité de différentes pouzzolanes', *7th Int. Cong. Chemi. Cem.*, Paris, Vol. III, IV 110–15 (1980).

14. Penta, F., Sulle pozzolane del Lazio, *Annali di Chimica*, 44, 572–83 (1954).
15. Italcementi, Laboratorio Chimico Centrale – *unpublished data*
16. Battaglino, G. and Schippa, G., Su un nuovo methodo di valutazione dei materiali pozzolanici - L'industria Italiana del Cemento 38(3), 175–8 (1968).
17. Forest, J. and Demoulian, E., Appreciation de l'activité des cendres volantes et des pouzzolanes, *Revue des Matériaux de Construction* no 577, 312–17 (1963).
18. Babatchev, G. and Penchev, P. St., Résistance à la corrosion des ciments mixtes, *RILEM Int. Symp. Durability of Concr.*, Prague 1961, Final Report, 79-92, Publ. House Czech. Acad. Sci. Prague (1962).
19. Ludwig, U. and Schwiete, H. E., Untersuchungen an Deutschen Trassen – Silicates Ind. 28(10), 439–47 (1963).
20. Wessely, I., Der Trass als hydraulischer Zusatzstoff, *Zem. Kalk Gips*, 14(7), 284–93 (1961).
21. Sersale, R. and Giordano Orsini, P., 'Hydrated phases after reaction of lime with "pozzolanic" materials or blast furnace slags', *Proc. 5th Int. Symp. Chem. Cem. Tokyo*, Vol. IV 114–21 (1968).
22. Jambor, J., 'Relation between phase composition, over-all porosity and strength of hardened lime-pozzolana pastes', *Mag. Concr. Res.*, 15(45), 131–42 (1963).
23. Central Board Of Irrigation And Power Research, 'Scheme applied to river valley projects: survey of work done on pozzolana in India', Status Report, (2), pp. 177 (1971).
24. Georgiadou, G. N., 'Evaluation of pozzolanic earth occurrence in Pella, Macedonia, for the production of pozzolanic cements', *Teknika Kronika*, 93–107 (in Greek) (1971).
25. Johansson, S. and Andersen P. J., 'Pozzolanic activity of calcined moler clay', *Cem. Concr. Res.*, 20, 447–52 (1990).
26. Takemoto, K. and Uchikawa, H., 'Hydratation des ciments pouzzolaniques', *7th Int. Cong. Chem. Cem. Paris*, Vol. I, IV 2/1–2/21 (1980).
27. Baire, G., Le ciment-gaize pour travaux à la mer, *Revue des Matériaux de Construction et de Travaux Publics* 1930, no 248, 168–70 (1930).
28. Parravano, N. and Caglioti, V., 'Ricerche sulle pozzolane', *La Ricerca Scientifica*, 8(I), 271–92 (1937).
29. Scherillo, A., Petrografia chimica dei tufi flegrei – Nota II – Rendiconti dell' Accademia di Scienze Fisiche e Matematiche della Società Nazionale di Scienze, lettere ed Arti in Napoli 1955, Serie 4, 22, 345–71 (1955).
30. Sersale, R., Ricerche sulla zeolitizzazione dei vetri vulcanici per trattamento idrotermale - Rendiconti dell' Accademia di Scienze Fisiche e Matematiche della Società Nazionale di Scienze, lettere ed Arti in Napoli – Nota I, Serie 4, 26, 154–66 (1959).
31. Akman, M.S., Mazlum, F. and Esenli, F., 'A comparative study of natural pozzolans used in blended cement production', *Proc. 4th Int. Conf. 'Fly Ash, Silica Fume, Slag, and Natural Pozzolans in Concr.'*, Istanbul, Maj, ACI SP 132, Vol. I, 471–94 (1992).
32. Turriziani, R. and Corradini, G., 'Materiali pozzolanici ad alto contenuto in silice', *L'Industria Italiana del Cemento*, 29(6), 146–50 (1959).
33. Menzel C. A., 'Strength and volume change of steam-cured Portland cement mortar and concrete', *Proc. Am. Concr. Inst.* 31, 125–48 (1934).
34. Malquori, G. and Cirilli, V., 'Azione della calce sul caolino disidratato e sulle pozzolane naturali', *'La Ricerca Scientifica ed il Progresso Tecnico'*, 14(2–3), 85–93 (1943).
35. Tavasci, B., 'Struttura della malta di calce e pozzolana di Segni', *Il Cemento*, 44, 106–18 (1947).
36. Turriziani, R., Prodotti di reazione dell'idrato di calcio con la pozzolana: I, *La Ricerca Scientifica*, 24(8), 1709–17 (1954).
37. Ludwig, U. and Schwiete, H. E., Lime combination and new formations in the trass-lime reactions, *Zem. Kalk Gips*, 52, 421–31 (1963).
38. Sabatelli, V., Sersale, R. and Amicarelli, V., Ricerche sulla costituzione delle paste di calce-pozzolana lungamente stagionate in acqua dolce, 'Rendiconti dell'Accademia di Scienze Fisiche e Matematiche della Società Nazionale di Scienze, lettere ed Arti in Napoli' Serie 4, 34, 243–52 (1967).
39. Dron, R., 'Experimental and theoretical study of the CaO–Al₂O₃–SiO₂–H₂O system', *6th Int. Cong. Chem. Cem.*, Moskow, Suplementary paper Section II, 3–16 (1974).
40. Amicarelli, V., Sersale, R. and Sabatelli, V., 'Attivita' "pozzolanica" dei prodotti piroclastici "argillificati"', Rendiconti dell'Accademia delle Scienze Fisiche e Matematiche, Napoli, Serie 4, 33, 257–82 (1966).
41. Costa, U. and Massazza, F., 'Factors affecting the reaction with lime of Italian pozzolanas', *6th Int. Cong. Chem. Cem.*, Moscow, September (1974).
42. Ludwig, U. and Schwiete, H. E., "Researches on the Hydration of Trass Cements", *Proc. 4th Int. Symp. Chem. Cem.*, Washington 1960, National Bureau of Standard (USA) Monograph 43(2), 1083–1098 (1962).
43. Sersale, R. and Sabatelli, V., Sull'attivita' "pozzolanica" delle zeoliti: I – attività "pozzolanica" dell'herschelite, 'Rendiconti dell'Accademia delle Scienze Fisiche e Matematiche di Napoli', Serie 4, 27, 263–86 (1960).

44. Malquori, G. and Spadano, A., 'Azione combinata del gesso e della calce sui materiali pozzolanici', *La Ricerca Scientifica*, 7(3–4) 185–91 (1936).

45. Van Aartl, J. H. P. and Visser, S., 'Calcium hydroxide attack on feldspars and clays: possible relevance to cement-aggregate reactions', *Cem. Concr. Res.*, 7, 643–8 (1977).

46. Baire, G., 'Sur le dosage de la silice soluble dans les pouzzolanes et les ciments pouzzolaniques', Revue des Materiaux de Construction e de Travaux Publics, no 262, 268–71 (1931).

47. Feret, R., 'Additions de matières pulvérulentes aux liants hydrauliques : XX', Examen séparé des principales sortes de pouzzolanes – Revue des Materiaux de Construction, no 201, 163–8 (1926).

48. Costa, U. and Massazza, F., 'Influenza del trattamento termico sulla reattività con la calce di alcune pozzolane naturali', *Il Cemento* 74, 105–22 (1977).

49. Turriziani, R. and Corradini, G., 'Materiali pozzolanici ad alto contenuto in silice', L'Industria Italiana del Cemento 31(10), 493–8 (1961).

50. Costa, U. and Massazza, F., 'Factors affecting the reaction with lime of italian pozzolanas', *6th Int. Cong. Chem. Cem.*, Moskow, September (1974).

51. Celani, A., Collepardi, M. and Rio, A., 'Les différents mécanismes d'action de la chaux sur les matieres pouzzolaniques et sur les laitiers', Revue des Materiaux de Construction no 614, 433–9 (1966).

52. Massazza, F. and Costa, U., 'Factors determining the development of mechanical strength in lime-pozzolana pastes – XXII, *Conf. Silicate Ind. Silicate Sci.*, (Budapest, June 6–11, 1977), Vol. I, 537–52 (1977).

53. Sestini, Q., 'Pozzolane e cementi pozzolanici', *La Chimica e l'Industria*, 29, 66–9 (1937).

54. Turriziani R. and Schippa, G., 'Influenza dei trattamenti termici sulle proprieta' delle pozzolane laziali' *La Ricerca Scientifica*, 24, 600–06 (1954).

55. Rossi, G. and Forchielli, L., 'Struttura porosa ed assorbimento di calce di alcune pozzolane naturali italiane', *Il Cemento*, 73, 215–21 (1976).

56. Chapelle, J., 'Attaque sulfo-calcique des laitiers et des pouzzolanes', La Revue des Materiaux de Construction no 511, 512, 513, 514–15 et 516, pp. 63 (1958).

57. Collepardi, M., Marcialis, A., Massidda, L. and Sanna, U., 'Low pressure steam curing of compacted lime-pozzolana mixtures', *Cem. Concr. Res.* 6, 497–506 (1976).

58. Držaj, B., Hočevar, S., Slokan, M. and Zajc, A., 'Kinetics and mechanism of reaction in the zeolitic Tuff–CaO–H_2O Systems at increased temperature', *Cem. Concr. Res.*, 8, 711–20 (1978).

59. Tavasci, B., 'Struttura della malta di calce e pozzolana di Segni', *Il Cemento*, 43, 106–18 (1947).

60. Cereseto, A. and Parissi, F., Il controllo delle pozzolane, *Il Cemento Armato*, 41, 45–8 (1944).

61. Fournier, M. and Geoffray, J.-M., 'Le Liant Pouzzolanes-chaux', *Bulletin des Liaison des Laboratoires des Ponts et Chaussées*, no 93, 70–8 (1978).

62. Shi, C. and Day, R. L., 'Acceleration of strength gain of lime-pozzolan cements by thermal activation, *Cem. Concr. Res.*, 23, 824–32 (1993).

63. Parissi, F., Miscele pozzolana-calce-gesso, *Le Industrie del Cemento*, 29, 32 (1932).

64. Ferrari, F., 'Leganti Pozzolanici', *L'Industria Italiana del Cemento*, 17, 75–7 (1947).

65. Turriziani, R. and Schippa, G., 'Malte di pozzolana, calce e solfato di calcio', *La Ricerca Scientifica*, 24(9) 1895–903 (1954).

66. Shi, C. and Day, R. L., 'Chemical activation of blended cements made with lime and natural pozzolans', *Cem. Concr. Res.*, 23, 1389–96 (1993).

67. Fratini, N., 'Ricerche sulla calce di idrolisi nelle paste di cemento: Nota II – Proposta di un saggio per la valutazione chimica di cementi pozzolanici', *Annali di Chimica*, 40, 461–9 (1950).

68. CEN EN 196.V, Methods of testing cement, *Pozzolanicity test for pozzolanic cements* (1994).

69. Ogawa, K., Uchikawa, H., Takemoto, K. and Yasui, I., 'The mechanism of the hydration in the system C_3S pozzolanas', *Cem. Concr. Res.*, 10, 683–96 (1980).

70. Tavasci, B, Cereseto, A., 'Struttura del cemento pozzolanico idratato', *La Chimica e l'Industria*, 31, 392–8 (1949).

71. Sestini , Q. and Santarelli, L., 'Ricerche sulle pozzolane: IV – Analisi delle pozzolane e malte pozzolaniche', *Annali di Chimica Applicata*, 26, 534–57 (1936).

72. Turriziani, R. and Rio, A., Osservazioni su alcuni criteri di valutazione dei cementi pozzolanici, *Annali di Chimica Applicata*, 44, 787–96 (1954).

73. Costa, U and Massazza, F., 'Permeabilità di paste di cemento portalnd e pozzolanico: relazione con la tessitura porosa, Convegno AITEC" La durabilità delle opere in calcestruzzo" Padova 8–9 settembre, Vol. I, 67–73 (1987).

74. Massazza, F., 'Pozzolanic cements' *Cem. Concr. Composites*, 15, 185–214 (1993).

75. Glasser, F. P. and Marr, J., 'The effect of mineral additives on the composition of cement pore fluids', British Ceramic Proceedings 35, 419–29 (1984).

76. Page, C. L. and Vennesland, O., Pore solution composition and chloride binding capacity of silica-fume cement pastes – *Matériaux et Constructions*, 16, no 91, 19–25 (1983).

77. Andersson, K., Allard B., Bengtsson, M. and Magnusson, B., 'Chemical Composition of cement pore solutions', *Cem. Concr. Res.*, 19, 327–32 (1989).

78. Larbi, J. A., Fraay, A. L. A. and Bijen J. M., 'The chemistry of the pore fluid of silica fume-blended systems', *Cem. Concr. Res*, 20, 506–16 (1990).

79. Gutteridge, W. A. and Dalziel, J. A., 'Filler cement: the effect of the secondary component on the hydration of Portland cement – *Part 1: A fine non-hydraulic filler*', *Cem. Concr. Res.*, 20, 778–82 (1990).

80. Gutteridge, W. A. and Dalziel, J. A, 'Filler cement: the effect of the secondary component on the hydration of portland cement – *Part 2: Fine hydraulic binders*, *Cem. Concr. Res.*, 20, 853–61 (1990).

81. Massazza, F. and Costa, U., 'Aspetti dell'attività pozzolanica e proprieta' dei cementi pozzolanici', *Il Cemento* 76, 3–18 (1979).

82. Massazza, F. 'Blended cements', Special Lecture – *4th NCB Int. Sem. Cem. and Building Mater.*, New Delhi (1994).

83. Blaine, R. L. and Arni, H. T., Section 10 'Shrinkage and expansion of concrete – *Interrelations Between Cement and Concrete Properties*', Part 4, *Science Series* 15, U.S. Department of Commerce, NBS (1969).

84. Troxell, G. E., Raphael, J. M. and Davis, R. E., 'Long-time creep concrete and shrinkage tests of plain and reinforced', *Proc. ASTM*, 58, 1101–20 (1958).

85. ACI Committee 225, 'Guide to selection and use of hydraulic cements', *ACI J.*, 82, 901–29 (1985).

86. Houst, Y. F., 'Influence of microstructure and water on the diffusion of CO_2 and O_2 through cement paste', *2nd Int. Conf. Durability of Concr.*, Montreal 1991, ACI SP 126, Supplementary Papers, V. M. Malhotra (ed.) 141–59 (1991).

87. Kropp, J. and Hilsdorf, H. K., Influence of carbonation on the structure of hardened cement paste and water transport – Berichtsband des internatiolanen Kolloquiums – Werkstoffwissenschaften und Bausanierung, Esslingen, 153–57 (1983).

88. Costa, U., Facoetti, M. and Massazza, F., 'Permeability and diffusion of gases in concrete', *9th Int. Cong. Chem. Cem.*, New Delhi, 5, 107–14 (1992).

89. Tavasci, B. and Rio, A., 'La corrosione delle paste cementizie da parte dell'acqua distillatà, *La Chimica e l'Industria*, 34, 404–12 (1952).

90. Soroka, I., 'Portland cement paste and concrete', The Macmillan Press Ltd., (1979).

91. Goggi, G., 'Ulteriori progressi nel controllo del dilavamento dei cementi da parte di acque pure', *L'Industria Italiana del Cemento*, 12, 394–400 (1960).

92. Lafuma, H., 'Theorie de l'expansion des liants hydrauliques', *La Revue des Matériaux de Construction et de Travau Publics*, no 243, 441–4 and 1930, no 244, 4–8 (1929).

93. Turriziani, R. and Rio, A., 'Osservazioni su alcuni criteri di valutazione dei cementi pozzolanici', *Annali di Chimica Applicata*, 44, 787–96 (1954).

94. Turriziani, R., 'High chemical resistance pozzolanic cements', *Proc. 4th Int. Symp. Chem. Cem.*, Washington, Vol. II 1067–73 (1960).

95. Collepardi, M., Marcialis, A. and Turriziani, R., 'La cinetica di penetrazione degli ioni cloruro nel calcestruzo, *Il Cemento*, 67, 157–64 (1970).

96. Chatterji, S., 'Mechanism of the $CaCl_2$ attack on Portland Cement Concrete', *Cem. Concr. Res.*, 8, 461–8 (1978).

97. Maultzsch, M., 'Vorgange beim Angriff von Chloridlosungen auf Zementstein und Beton', *Material und Technik*, 12, 83–90 (1984).

98. Collepardi, M., Marcialis, A. and Turriziani, R., 'Penetration of chloride ions into cement pastes and concretes', *J. Am. Ceram. Soc.*, 55, 534–5 (1972).

99. Page, C. L., Short, N. R. and El Tarras, A., 'Diffusion of chloride ions in hardened cement pastes', *Cem. Concr. Res.*, 11, 395–406 (1981).

100. Regourd, M., 'Physico-chemical studies of cement pastes, mortars and concretes exposed to seawater', *Proc. Int. Cong. Performance of Concr. Mar. Environ.*, M. Malhotra, (ed.) St Andrews by the Sea, ACI SP 65, 63–82, (1980).

101. Peltier, R., 'Résultats des essais de longue durée de la résistance des ciments à la mer au laboratoire maritime de La Rochelle', *Revue des Matériaux de Construction* no 680, 31–44 (1973).

102. Mehta, P. K., 'Durability of concrete in marine environment – A review', *Int. Conf. Performance Concr. Mar. Environ.*, V. M. Malhotra (ed.) St. Andrews by the Sea, ACI SP 65, 1–20 (1980).

103. Gjorv, O. E., Vennesland, O., 'Diffusion of chloride ions from sea-water into concrete', *Cem. Concr. Res.*, 9, 229–38 (1979).

104. Rio, A., Celani A., 'L'evoluzione dei leganti pozzolanici alla luce di recenti applicazioni in opere esposte all'acqua di mare', *RILEM Int. Symp. Behaviour Concr. exposed to sea-water*, Palermo, 43–51 (1965).

105. Mehta, P. K., Haynes, H. H., Durability of concrete in seawater, *Proc. ASCE: J. Struct. Div.*, 101, 1679–86 (1975).

106. Giorv, O. E., Long-time durability of concrete in seawater, *ACI J.*, 68, 60–7 (1971).

107. Stanton, T. E., Studies of the use of pozzolans for counteracting excessive concrete expansion resulting from reaction between aggregates and the alkali in cement, *Symp. Use of Pozzolanic Mater. Mortars and Concr.*, ASTM STP 99, San Francisco (1949).

108. Porter, L. C., Small proportions of pozzolan may produce detrimental reactive expansion in mortar,

Department of the Interior, Bureau of Reclamation, Report n. C-113, Denver, US, p. 22 (1964).

109. Sersale, R., Frigione, G., Portland – zeolite cement for minimizing alkali-aggregate reaction, *Cem. Concr. Res.*, 17, 404–10 (1987).

110. Neville, A. M., Properties of concrete, Pitman Publishing, London (1972).

111. US Bureau of Reclamation, Investigation into the effect of water/cement ratio on freezing and thawing resistance of non-air and air-entrained concrete, Concr. Lab. Rep. No. C-810, Denver, Colorado (1955).

112. Blanks, R. F., Cordon, W. A., Practices, experiences and tests with air entraining agents in making durable concrete, *ACI J.*, 20, 469–87 (1949).

Chapter fourteen

Pulverized fuel ash as a cement extender

Karen Luke

14.1 Introduction

Pulverized fuel ash (PFA), or fly ash as it is also termed, has been used as a cement extender for more than half a century. Some of the first studies were performed in the late 1930s [1] and the first major application reported was in the construction of the Hungry Horse Dam, USA in 1948 [2].

Currently, it is estimated that around 500 million tons of coal ash, of which 80 per cent is PFA, are generated per annum world wide with approximately 35 per cent utilization, principally in production of lightweight aggregate and brick, as a filler in road construction and in cement and concrete manufacture. An approximate indication of percentages used in cement and concrete manufacture is given in Table 14.1. Differences in percentage utilization between countries is notable and clearly the overall trend is for increasing utilization. This growth is expected to continue not only from the need to dispose of the PFA, but also from an increase in the use of cement and demand on the cement manufacturing industry.

National standards and specifications have been established in a number of countries regulating the use of PFA, such as AS 3582-Australia,

TABLE 14.1 PFA production and utilization in cement and concrete for various countries during 1977–1995 [3]

	Production $10^3 t$			Cement and concrete manufacture % utilization		
	1977	1992	1995	1977	1992	1995
Australia	5445	8175	8773	5.33	8.45	8.67
Canada	2626	8464	—	5.02	11.02	—
Europe	42687	64330	—	1.02	15.39	—
India	9000	38889	—	—	36.79	60.3
Japan	2000	4131	4725	17.5		
USA	61495	60000	65541	—	5.60	8.43
Russia	—	62000	51809	3.70	12.00	12.34
China	—	79820	91140	—	14.00	15.10

CAN/ CSA-A23.5-M86-Canada, IS 3812-1981-India, JIS A6201-Japan, BS 3892 Part 1-UK and ASTM C 618-92a-USA. In some cases, standards have been adapted from those for natural and calcined pozzolans, in others, new standards have been established. These standards are a means used to determine the potential of a PFA for a given usage, but there remain many unanswered questions as far as the reaction mechanism of PFA as a cement extender and factors affecting performance are concerned.

14.2 PFA formation

PFA is a waste product recovered from the exhaust gas produced during the process of burning pulverized coal in thermal power plants for electricity generation (Figure 14.1). Coal is transported to the power plant and then passed to the pulverizer where it is pulverized to give fines of particle size around 70–80 per cent passing through a 200 mesh. The fines formed are transported in an air stream to the boiler where they

are burned, and the heat produced is used to generate steam in the steam-generating unit.

During the combustion process, the volatile matter is vaporized and carbon is burned off. As the particles enter the burning zone, temperatures increase rapidly and typically reach values around 1000–1500°C. Mineral components present in the coal gangue, the inorganic part of the coal, such as clays and feldspars melt and form fused droplets that on rapid cooling solidify as spherical glassy particles that comprise the coal ash. Some mineral components also remain in the crystalline phase. The coarser ash collects at the bottom of the furnace, and is known as bottom ash. The finer ash is carried in the flue gases, removed by electrical, mechanical or cyclone precipitators and collected in a series of hoppers. It is this ash that is termed PFA.

Environmental legislation, over the last few decades, has forced power plants to consider pollution control measures to reduce both SO_x and NO_x emissions into the atmosphere. This has had a detrimental impact on the standard of the PFA

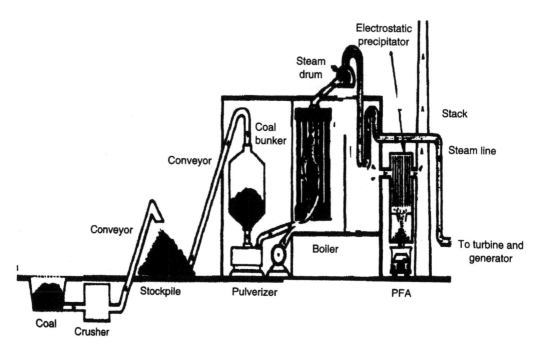

FIGURE 14.1 Schematic of coal-fired generation plant emphasising PFA generation.

produced. Two primary technologies used in SO_x reduction are flue gas desulphurization (primary desulphurization), and the use of suitable scrubbers (secondary desulphurization). In primary desulphurization, lime is added to the boiler where decarbonation occurs to give CaO and CO_2. The CaO then reacts with SO_2 to produce $CaSO_4$. The PFA is modified both by the lower temperature required for desulphurization, around 1100°C, and in that it is also mixed with the $CaSO_4$. Secondary desulphurization, on the other hand, has no effect on the PFA as it occurs after the PFA has been removed. In reducing NO_x emissions, the primary technology has been the use of low-NO_x combustion systems. This process generally gives rise to higher and more variable unburnt carbon content in the PFA [4] which can have a significant effect on concrete performance [5]. Secondary ammonia-based technologies are also being implemented to further reduce the NO_x emissions. Most of the ammonia in these processes reacts with NO_x to produce nitrogen and water. Inefficiency of the conversion process in fully utilizing the ammonia can result in ammonia slip and deposition onto the PFA particles [6,7].

The power-plant operator's primary concern is with the calorific value per ton of coal, the smooth functioning of the combustion and recovery systems, and the regulatory requirements. In the past, there was little concern with the quality and end-use of the PFA formed, though this is changing with environmental demands and costs of disposal. Beneficiation or classification technologies are being developed and implemented to improve the quality of the PFA. Methods to date include froth flotation, carbon burn-out [8], triboelectrostatic separation [9] or fluidized jet mill, and classification processes [10] to remove unburnt carbon, fluidized bed gravity separator to separate coarser and low-density particles, [11] and magnetic separator for the magnetic components.

Factors differ from one power plant to another, so essentially no two are the same, and as such the PFA produced from each is unique. There can be considerable variability between PFAs and this will be discussed in more detail in the next section. It is sufficient to say that the major factor contributing to variability relates to the inorganic chemical composition of the coal, that is, source and type of coal. However, the state of the coal, whether it contains moisture or not, the degree of pulverization, combustion conditions and flame temperature, means of collection and handling, type of emission control, form of beneficiation if any, all contribute to the variability.

14.3 Characteristics of PFA

An indication of the variability in terms of the chemical composition for PFA is presented in Table 14.2. Traditionally, it is considered that PFA derived from bituminous and anthracitic coals are low-calcium and approximate to the ASTM C 618 designation for Class F grouping. Those from lignitic or subituminous coals tend to be considered high-calcium and equivalent to the Class C grouping according to ASTM C 618. ASTM C 618 classification for the two categories is based in terms of the combined percentages of $SiO_2 + Al_2O_3 + Fe_2O_3$ and not on the CaO content directly, and there has been some question on the logic of this. In reality, there is practical need for a third category of PFAs since those with intermediate calcium content in the range of 8 to

TABLE 14.2 Typical ranges in the chemical composition (wt%) of PFAs obtained from different countries

	SiO_2	Al_2O_3	Fe_2O_3	CaO	MgO	SO_3	Na_2O	K_2O	LOI
France	29.9–54.0	10.8–33.4	5.8–15.3	1.5–38.8	1.10–4.45	0.1–7.0	0.75–0.85	0.7–6.0	0.3–15.2
Germany	34.1–49.5	21.0–29.4	8.4–28.9	2.2–11.8	0.75–4.26	0.1–2.1			1.7–20.1
UK	48.1–53.1	29.6–38.5	5.3–11.9	1.6–2.8	1.31–1.61	0.5–0.7	0.24–1.57	1.3–4.3	
Japan	51.0–62.5	21.8–30.7	3.2–7.3	0.8–5.9	0.53–1.71	0.17–0.54	0.30–1.81	0.46–1.34	1.3–3.7
USA	36.9–52.5	17.6–22.8	6.2–7.5	4.9–25.2	1.3–5.1	0.6–2.9	1.0–1.7	0.6–1.3	0.4–2.6

(a)

(b)

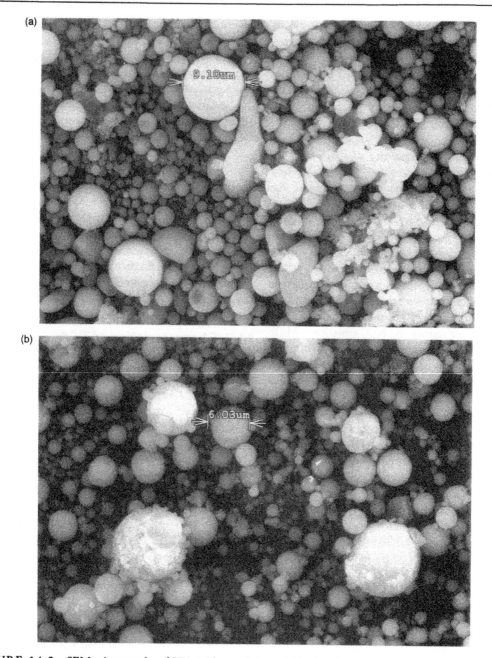

FIGURE 14.2   SEM micrographs of PFA (a) low-calcium; (b) high-calcium. (Courtesy of Professors M. Grutzeck and M. Silsbee, Pennsylvania State University.)

around 15–20 per cent behave somewhat differently from either the 'true' low-calcium or 'true' high-calcium PFAs. In the context of this chapter, PFA will be classified on the basis of the CaO content as low-calcium (<10 per cent CaO), intermediate-calcium (10–20 per cent CaO) and high-calcium (>20 per cent CaO). Variations in chemical and physical composition even within the three categories are still large. PFA also contains variable concentrations of minor components such as P_2O_5, TiO_2, MnO, BaO and SrO at less than 3 wt per cent.

The specific gravity of PFA is generally less than that of cement and ranges from 1.6 to 2.8. The lower specific gravity is associated with either a high content of unburnt carbon or hollow spherical particles or both. High-calcium PFAs tend to have specific gravities at the higher end, around 2.4–2.8, due to a higher content of small spheres that are typically solid and have a low level of unburnt carbon. High iron contents also tend to give a higher specific gravity. Particle sizes range from <1 to >150 μm with average particle size around 7–12 μm and specific surface areas are high and variable depending on the proportion of fine particles present and range from 2000 to 10,000 cm^2/g (Blaine).

Although the chemical composition gives some indication of the PFA properties, it should be noted that the PFA is a heterogenous and potentially variable material even at the micrometre and nanometre scale. PFA particles are typically spherical with surfaces that are often smooth and glassy in character (Figure 14.2). Figure 14.3 shows other features often observed in PFA samples. Occasionally, incompletely melted particles are found that are rounded to some extent and can be of variable size. Formation of these particles can result from insufficiently high temperature in the burner, too short a time of exposure, or their chemical composition. In some cases, the spheres tend to have a powdery deposit on the surface that is usually attributed to alkali sulphate deposited from the vapour phase after the spheres have solidified. Other spheres may have surfaces that appear rough at the micro-scale level. Entrapment of gas bubbles by the molten phases subsequent to cooling and solidification give rise to

hollow spheres that can be empty (cenospheres) or can contain smaller spheres packed inside them (plerospheres). The smaller spheres tend to be solid, but on rare occasions can be hollow and again either empty or packed with even smaller spheres. Conditions required for formation of hollow spheres are not evident.

The spherical glass particles tend to be amorphous though some particles cool slowly enough to undergo partial recrystallization. Structural differences in the glass occur with increase in CaO content. IR, NMR and TMS (trimethylsilylation) investigations show a decrease in the degree of silicate ion condensation with increase in CaO content [12], whereas XRD shows a change in the position of the background maxima [13]. Low-calcium PFA tends to have a XRD maxima around 23°2θ characteristic of vitreous silica whereas high-calcium PFA gives maxima around 32°2θ indicative of a calcium alumininosilicate structure (Figure 14.4). Intermediate-calcium PFA tends to have maxima in the range from 23 to 27°2θ depending on CaO content.

The glass phase in PFA constitutes from 50 to 90 per cent [14], the remainder being composed essentially of crystalline phases (Figure 14.4). These tend not to be the same as the minerals found in the original coal, with the exception of, perhaps, quartz, but secondary products formed in the burning zone. Only a few crystalline phases are observed in the low-calcium PFA, whereas in high-calcium PFA they can be numerous (Table 14.3). Some of the crystalline phases in the high-calcium PFA have a hydraulic character. C$_3$A, in particular, can be present in fairly substantial amounts while traces of C$_2$S and C$_3$S have occasionally been detected. The crystalline phases are rarely observed as individual phases in the PFA, but are predominantly embedded within the glass spheres. Selective dissolution studies [15,16] indicate that the crystalline phases are situated just below the glass surface.

14.4 Role of PFA on performance of extended systems

PFAs are reported in the literature to improve workability, reduce segregation, reduce bleedwater,

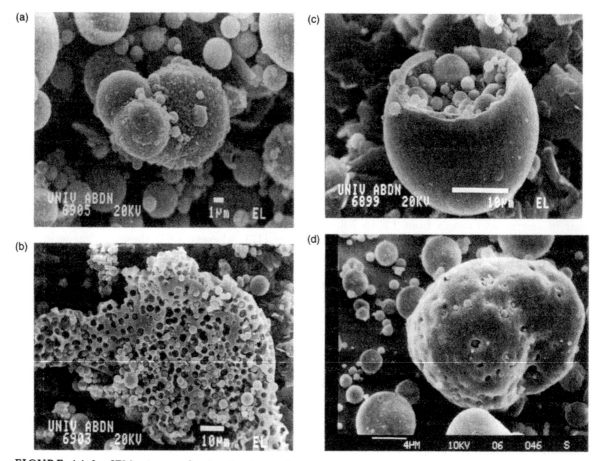

FIGURE 14.3   SEM micrographs showing (a) surface deposits of alkali sulphate; (b) unburnt coal particle; (c) plerosphere; (d) incompletely melted particle. (Courtesy of Dr. Lachowski, University of Aberdeen.)

increase pumpability and reduce equipment wear while in the plastic state, and modify setting time, increase strength, reduce heat of hydration, reduce permeability, provide resistance to sulphate attack, lessen alkali aggregate activity and improve freeze–thaw resistance in the hardened state. Much of the benefits in the plastic state are related to the spherical nature of the PFA in that it produces better plasticity and allows for a reduction in the amount of water needed in a mix. The effects on the hardened state are attributed more to chemical processes that occur between the PFA and cement. Variations in the PFA coupled with different mixing systems preclude direct comparisons being made between different studies, but there are some notable trends that relate to the role played by the PFA. Currently, most of the literature is based on low-calcium PFAs with significantly less on the high-calcium PFAs, and almost none on those of inter-mediate-calcium PFAs.

14.4.1 Early hydration reactions of PFA

PFA is reported to both accelerate and retard the setting time of cement. The general consensus seems to indicate that low-calcium PFA tends to retard [17–19], whereas high-calcium PFA is less predictable and has been reported to retard [20]

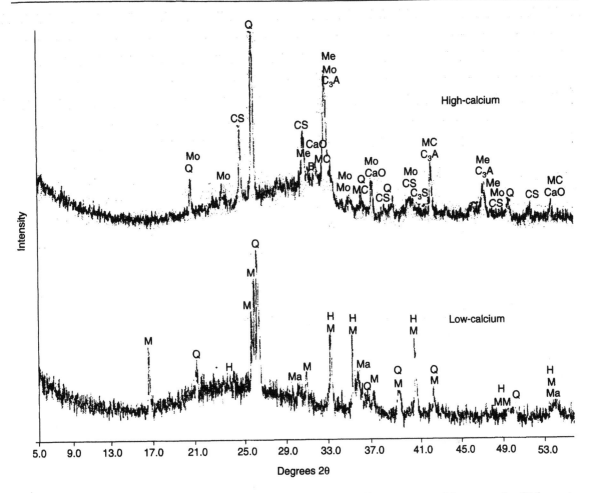

FIGURE 14.4 XRD of low- and high-calcium PFAs where M: mullite; Q: quartz; Ma: magnetite; H: hematite; Mo: monticellite; MC: $MgCO_3$; Me: merwinite; CS: anhydrite; B: C_2S. (Courtesy of Professor M. Grutzeck, Pennsylvania State University.)

or accelerate [19,21] setting time (Figure 14.5). The effect of low-calcium PFA on retardation seems to be a physical effect due simply to reduction on the cement factor and increase in the water/cement (w/c) ratio, whereas acceleration with a high-calcium PFA is due to chemical reaction. Calorimetric studies also demonstrate the retardation [22,23] and acceleration [24–26] effect of PFA blends in the early stages of hydration or the so called 'dormant' period. On setting, which correlates with the middle stage of hydration, PFA tends to accelerate the reaction though again, this is not necessarily always the case [27].

14.4.1.1 Low-calcium PFA

Reaction of pure phases with PFA gives some insight into the mechanisms involved though they do not explain all the phenomena observed. In the case of C_3S-PFA pastes, retardation in the dormant period is attributed primarily to delay in nucleation of $Ca(OH)_2$ as a result of chemisorption of some Ca^{2+} on the PFA particles [28]. Addition of NaOH, to increase the alkalinity of the paste to be more consistent with that of cement, is believed to cause the dissolution of soluble silicate and aluminate species from the PFA

359

TABLE 14.3 Common crystalline phases found in low and high-calcium PFAs

Phase	Composition
Low-calcium PFA	
Mullite	$Al_6Si_2O_{13}$
Quartz	SiO_2
Magnetite-ferrite	Fe_3O_4-$(Mg,Fe)(Fe,Mg)_2O_4$
Haematite	Fe_2O_3
Anhydrite	$CaSO_4$
High-calcium PFA	
Melilite	$(Ca,Na)(Mg,Al,Fe)(Al,Si,Fe)_2O_7$
Ferrite spinel	$(Mg,Fe)(Fe,Mg)_2O_4$
Merwinite	$Ca_3Mg(SiO_4)_2$
Bredigite, Larnite	Ca_2SiO_4
Lime	CaO
Periclase	MgO
Nepheline,	
Carnegieite	$Na,AlSiO_4$
Plagioclase	$(Ca,Na)(Al,Si)_4O_8$
Feldspar	Na,Ca,Al, silicate
Pyroxene	$(Ca,Na)(Mg,Fe,Al)(Si,Al)_2O_6$
Crisobalite, Quartz	SiO_2
Brucite	$Mg(OH)_2$
Calcite	$CaCO_3$
Gypsum	$CaSO_4 \cdot 2H_2O$
Thernadite	Na_2SO_4
C_3S	Ca_3SiO_5
C_3A	$Ca_3Al_2O_6$
Anhydrite	$CaSO_4$
Al-hauyne	$C_4A_3\bar{S}$
$C_{12}A_7$	$12CaO \cdot 7Al_2O_3$

which are considered to poison the nucleation and growth of $Ca(OH)_2$ and C-S-H. Hydration of C_3S may also be retarded by the unburnt carbon that is present in variable amounts in the PFA. PFA also retards the hydration of C_3A and C_3A/C_4AF pastes, and acts by delaying the onset of hydration as well as by slowing the reaction due to formation of a protective layer consisting of a lamellar hexagonal C_4AH_x stabilized by incorporation of sulphate from the PFA [29]. Ettringite is also observed to occur in the first few minutes of hydration. The effectiveness of the PFA in retarding the hydration of the aluminate phases is attributed essentially to the highly soluble calcium and sulphate ions present usually as surface deposits on the PFA particles. The situation in cement is more complex. Cement hydration with

PFA 'as received' and leached where the soluble fraction, the leachate, had been removed, lengthened the dormant period of both the silicate and C_3A components [30]. The leached PFA was less retarding than the 'as-received', while the leachate had no effect. This would suggest that, in the case of cement, retardation is due more to a surface phenomenon where chemisorption of Ca^{2+} ions on the PFA surface delay $Ca(OH)_2$ and C-S-H nucleation and recrystallization. However, in reality, the reaction mechanisms are more complex and depend not only on the PFA but the type of cement used. In a cement containing a high C_3A content (13.8 wt per cent), the PFA retarded the silicates but significantly accelerated the C_3A [23].

Clearly, there is still much to learn on the factors associated with PFA that affect the dormant period and hence, setting characteristics of PFA extended cements. On the other hand, there is more of a consensus that the acceleration effect of PFA on the setting or middle stage of hydration is, most likely, provision of additional nucleation sites on the PFA for C-S-H and CH formation [31,32]. This effect appears to depend on the chemical and physical properties of the surface and fineness [33]. Microstructural studies of PFA cement pastes at early ages shows deposits of a thin 'duplex film' of a single layer C-S-H and CH on the surface of the PFA grains before these have started to react [34]. Such hydration products are observed on PFA spheres within 1 h of hydration though definite signs of surface pitting on the PFA spheres associated with PFA reaction occur only after 4 h [35] or longer . This film becomes more dense with time as more hydration product is deposited from the hydrating cement. It appears that the film does not serve to protect the underlying PFA sphere from reacting if conditions for reaction occur. Hydration products from the PFA are generally not seen building up on the surface of the spheres but precipitate elsewhere after interaction with $Ca(OH)_2$ [36].

Dissolution kinetics for reaction of PFA in NaOH solution suggests that the pH and temperature are more important factors than the PFA composition [37]. There is an indication, though,

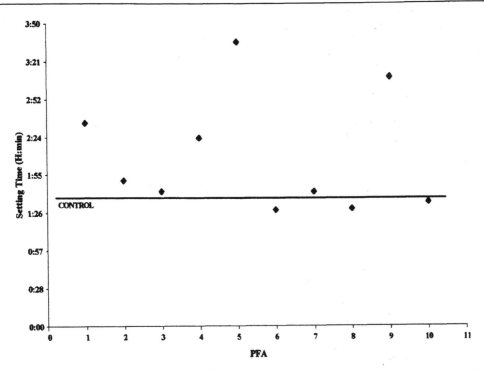

FIGURE 14.5 Retardation and acceleration effects of PFA on PFA/OPC pastes at 60°C. PFAs 1–5 are low-calcium, 6–10 are intermediate-calcium.

that glass with a low-alumina content and possibly a high-calcium content may react faster. Glass dissolution is reported as congruent or slightly incongruent and can be explained on the basis of the bulk dissolution theory. A pH > 13.2–13.3 in NaOH and above in the presence of lime is considered necessary for substantial dissolution to occur [38]. On the other hand, zeta-potential studies [39] indicate that changes occur at the PFA surface interface at much lower pH. Effect on zeta-potential on increasing pH differs depending on whether NaOH or Ca(OH)$_2$ is used. With NaOH, a minimum of −40 mV is obtained at pH 8, above this, the zeta-potential increases rapidly, whereas with Ca(OH)$_2$, a gradual increase is observed giving a positive value at around pH 9.5.

14.4.1.2 High-calcium PFA

There has been little research reported on the early hydration reactions of high-calcium PFA. It is considered that much of the early reaction is attributed to hydration of the crystalline C$_3$A with the alkali sulphates, in particular calcium sulphate, to form ettringite. This is particularly evident where additional calcium sulphate has been added. The self-cementitious properties and rapid set in some high-calcium PFA has been attributed to bridging of ettringite needles between PFA particles [40]. No indication of pitting or surface reaction on the PFA particles has been observed in the initial stages of reaction. It is suggested that the extensive chemical reactions occur around the PFA particle, but do not involve the particle itself [41].

14.4.2 PFA and strength development

The primary factors in determining strength of concrete are the amount of cement used and the w/c ratio. PFA as an extender generally requires less water to obtain the same workability as a normal concrete, and by suitable means of mix-proportioning, the desired rate of strength development can usually be achieved. Numerous methods have been reported for determining PFA proportioning [42–45]. Typically a 'PFA cementing efficiency factor', k, is used, which attempts to define a strength to w/c ratio relation considering an 'effective w/c ratio' as given by $w/(c+kf)$ [42]. This cementing efficiency factor is defined such that the mass of fly ash (f) is equivalent to the mass of cement (kf). Values of k vary according to country and percentage replacement, but typically are a single value, for example in the UK, the recommended value for k is 0.3 with PFA replacement percentages up to 50 per cent. Recently an 'overall efficiency factor' k, has been proposed that recommends more than one value of k depending on the curing age that gives a better correlation with strength values and is applicable for PFA replacement percentages from 15–75 per cent [45]. Values of k vary from 0.95–0.13, 1.15–0.33 and 1.25–0.43 at curing ages of 7, 28 and 90 days respectively. Given the same mix design, the rate of strength development can differ markedly depending on the PFA used [46] and is associated with the difference in PFA characteristics both physical and chemical. Replacing cement by a low-calcium PFA, in general, gives lower early strength with a higher long-term strength gain whereas high-calcium PFA notably leads to a more rapid strength gain [47–49]. Intermediate-calcium PFA tends to behave more like a low-calcium PFA where setting time is retarded and strength increases over longer curing periods [50]. Low-calcium PFA at 15–25 per cent replacement has also been reported to improve tensile strength whereas >25 per cent replacement causes a reduction in tensile strength but improves fracture properties [51].

It is generally considered that the fineness of PFA has a significant effect on strength development. For low-calcium fly ashes, the reactivity was found to be directly proportional to the amount of particles $<10\,\mu m$ and inversely proportional to particles $>45\,\mu m$ [47]. The effect of PFA fineness on strength can be determined if

sieved fractions of a PFA are used, and where there is little variation in the chemical composition of the fractions. Intermediate- and low-calcium PFAs sieved in a series of size fractions from $>125\,\mu m$ to $<45\,\mu m$ indicate a maximum of a 1.5-fold increase in strength at 2 days and ~2-fold increase at 90 days [52]. Direct correlation between fineness and strength is usually obscured by differences in the chemical parameters that appear to play a more significant role. Typical data on fineness parameters and strength for different PFAs are presented in Table 14.4.

PFA is considered to be pozzolanic in nature, in that it reacts with lime and water to form insoluble products, typically C-S-H or C-A-S-H gels. The significant quantities of CaO found in some high-calcium PFAs when released in solution can react with the glass phase rendering these PFAs as self-pozzolanic. This, and the hydration of the C_3A, C_2S and C_3S components give the high-calcium PFAs cementitious properties. Although PFAs clearly exhibit pozzolanic reaction, there is considerable variation as to when reaction starts, the rate of reaction, and degree of reaction over time. It is generally agreed that the pozzolanic reaction becomes apparent from 3–14 days after commencement of hydration. This delay or

TABLE 14.4 Comparison of PFA fineness (sedigraph) with compressive strength of PFA/OPC pastes at 50 per cent replacement

PFA	Fineness (cm^2/g)	Compressive strength (MPa)	
		24 h	7 days
Low-calcium	2156	20.9	20.1
	2241	21.9	24.2
	3822	22.5	25.1
	2472	26.2	33.3
	3462	24.2	24.1
	3379	19.8	23.4
	2559	17.2	17.4
	4042	19.8	29.1
	2336	18.8	23.8
Intermediate-calcium	4644	20.6	22.2
	3951	17.2	17.4
	1708	11.6	12.2
	2065	14.4	18.6

incubation period in the pozzolanic reaction may be explained by the dependency of dissolution of the glassy spheres on the pH of the aqueous phase [53].

14.4.2.1 Ageing studies on low-calcium PFA

Selective dissolution techniques have been used to determine quantitatively the fraction of PFA that remains unreacted, and hence, degree of reaction in cementitious systems. Methods include picric acid–methanol–water [54], HCl [55], salicylic acid–methanol followed by HCl [56] and a salicylic acid–methanol and KOH–sugar–water dissolution followed by QXRD determination of crystalline phases [57]. The accuracy of the data obtained by selective dissolution is questionable in that not necessarily all the hydration products or cement components are dissolved, some unreacted PFA may be dissolved or reaction products can be formed by precipitation from dissolved species [58,59]. Correction factors based on experimental data are frequently used, but they can be large. It is generally accepted, based on data from the different techniques that the PFA reacts gradually with time. Values of around 50 per cent remaining unreacted at 1 year for both 30:70 PFA–C_3S or PFA–cement blends have been reported.

In the pozzolanic reaction, CH formed during cement hydration is consumed in reaction with the dissolution products of the PFA glassy component. Comparison of the CH content in a PFA–cement blend and the corresponding control can give some useful information concerning

effect of the PFA reaction, Table 14.5. In PFA extended pastes, the consumption of $Ca(OH)_2$ appears to be gradual over time though at a rate which differs according to the PFA used. Quantitative determination on the degree of PFA reaction is, however, considered dubious as calculations based on the CH content are dependent not only on the pozzolanic reaction but also the effect of the PFA on the degree of cement hydration, and on the composition of the products.

Non-evaporable and bound water contents have also been used to give an indication of PFA reactivity in hydrated PFA–cement pastes. It would appear that for PFA–cement pastes the content on non-evaporable water is greater than those calculated on the cement fraction [26,60]. Increases in non-evaporable water were detected even at 1 day and may possibly relate to acceleration of the hydration of the cement phases attributed to the PFA. The increase at later ages would correspond to the pozzolanic reaction. There is also an indication that the nature of the hydration products of PFA–cement pastes may be different, and that they may contain a higher bound water content [60]. As in the previous cases, determination of non-evaporable water can give only an indication as to the degree of hydration as the chemical composition, water-binding capacity and quantity of the different hydration products formed are not known. Non-evaporable water is taken as the weight loss between 105°C and 1000°C and includes contributions from all hydration products. Water bound in CH is determined and subtracted from the total non-evaporable water to give the content of water bound in the silicate and aluminate hydrates.

Models based on microstructural studies have been proposed to explain the mechanism of the PFA pozzolanic reaction in C_3S and cement pastes [15,25,61]. Initial reaction is considered as diffusion of the aqueous phase through the cement paste to active sites on the PFA grains or in the macropores. Precipitation of CH to form circular masses followed by reaction with the aluminous vitreous phase [15] or dissolution of alkalies producing a negatively-charged surface layer rich in Al and Si that adsorbs Ca^{2+} from the aqueous

TABLE 14.5 Changes in CH content as a function of time for low-calcium PFA/OPC and control OPC pastes

Curing time (months)	$Ca(OH)_2$ (wt %)			
	Control	[a]Control	PFA 1	PFA 2
1	24.6	17.2	19.5	20.3
3	27.4	19.2	15.3	17.8
6	30.7	21.5	13.1	12.1
24	34.3	24.0	15.3	10.2
72	32.1	22.5	11.3	11.7

[a]Values relate to dilution expected if 30 per cent inert material were present.

phase [25] results in formation of a C-S-H gel containing aluminium. In the latter case, it is considered that continued dissolution of alkalies accelerates the attack on the PFA surface generating more SiO_4^{4-} and AlO^{2-} ions that react with Ca^{2+} and increase the surface layer. An osmotic pressure between the interior and exterior of the surface layer causes swelling, and the layer finally ruptures with release of SiO_4^{4-} and AlO^{2-} ions which diffuse to the exterior and react with Ca^{2+} ions forming precipitates away from the PFA particle. Faster diffusion of AlO^{2-} and the need for a higher Ca^{2+} concentration to form C-A-H tends to cause the C-A-H to precipitate further away from the PFA sphere surface than the C-S-H.

Although the models have some merit, they do not fully explain all the phenomena observed. As discussed previously, it would appear that the PFA is not reactive in the early stages of hydration, but behaves more as a nucleation site for precipitation of hydration products from the C_3S or cement phases. In some cases, a duplex film of hydration product coating the PFA particle surface is taken as indicative of commencement of pozzolanic reaction, although other parameters suggest this is not the case. The situation is further complicated by the fact that this coating covers the PFA spheres, and densifies due to what is assumed to be deposition of hydration products from the cement phases. A variety of different morphologies in the coating have been reported to occur on different parts of the surface of the PFA sphere [35]. It is considered that pozzolanic reaction is limited at this stage since rupture of the coating over time usually reveals unreacted PFA surfaces. Reason for rupture and time scales are not well defined at the present time. It is common to see CH crystals in intimate contact with unreacted PFA particles in the mature paste [62] with little or no reaction product between the two. Pozzolanic reaction of the PFA is considered to occur only when the aqueous phase pH > 13.2 and is evidenced by pitting and erosion of the PFA glassy spheres. However, there are considerable differences in the time scale at which the pozzolanic reaction commences, suggesting other

controlling factors are involved. Ghose and Pratt [35], for example, observed definite signs of surface pitting with growth of granular hydration products in the pits as early as 4h in some PFA particles. Diamond, Sheng and Olek [60] observed that the PFA grains are only starting to react at 1 year whereas the microstructure of the cement paste was in a mature state. A study on two British PFAs [84], Eggborough and Longannet, showed a significant difference in the degree of pozzolanic reaction between the two after 6 years of hydration. With the Eggborough PFA, there was a high percentage of almost totally unreacted particles whereas the Longannet PFA showed more particles having reacted, although in many cases not completely. On the basis of $Ca(OH)_2$ it was suggested that the pozzolanic reaction for the Longannet PFA had ceased at 2 years, whereas the Eggborough PFA showed a considerable degree of on-going reaction. The intrinsic heterogeneity within and between particles is another factor influencing the degree of pozzolanic reaction, and in mature pastes, it is not uncommon to see PFA particles in various stages of reaction ranging from completely unreacted to totally reacted, (Figure 14.6). Mineral phases encapsulated in the PFA glassy spheres tend not to react and remain in situ as the glassy phase is slowly

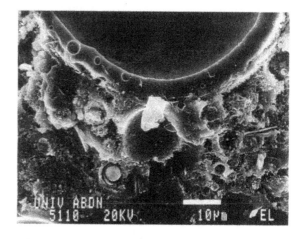

FIGURE 14.6 SEM micrograph showing a large unreacted and smaller completely reacted sphere in a 6-year-old PFA/OPC paste.

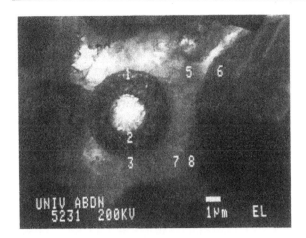

FIGURE 14.7 STEM micrograph of *in situ* mullite crystals in a PFA relic in a PFA/OPC paste. Elemental analysis indicate 1, 2, and 6 are mullite, and 3, 5, 7, 8 are C-S-H [84].

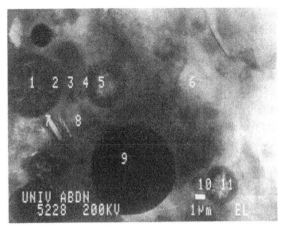

FIGURE 14.8 STEM micrograph of reaction rims around PFA particles in a hydrated PFA/OPC paste. Elemental analysis of 2, 3, 4 and 10 give mean Si/Ca 0.34 and Al/Ca 0.3:1, 9 and 11 are consistent with unreacted PFA and 6 and 8 are C-S-H.

etched away. In relics of fully reacted PFA particles, an outer layer of mullite and/or hematite crystals is frequently observed (Figure 14.7). Defined reaction rims of PFA hydration product have also been observed around PFA particles (Figure 14.8).

The presence of PFA tends to change the composition of the C-S-H. Typically, Ca/Si ratio decreases with age and percentage of PFA used. Actual ratios reported in the literature, however, are variable. The Ca/Si ratio of the inner product around C_3S in a PFA-C_3S paste was reported as 1.56 at 2 weeks and 1.45 at 1 year [63], whilst in another study the average Ca/Si ratio was determined as 1.43 for a PFA-C_3S paste compared to 1.51 for the neat C_3S paste and the Ca/Si did not vary significantly from 1 to 397 days [64]. PFA was also reported to decrease the C/S of the inner hydrate around alite in an 8-day-old PFA-cement paste in comparison to a neat cement paste from 1.71 to 1.55 [66]. A 4-year-old PFA–cement paste give C-S-H with Ca/Si ratio of 1.01 [66] while Ca/Si ratio for product close to PFA particles in a 10-year-old mortar was 1.1–1.2 [67]. An indication of the composition for 6-year-old PFA–cement pastes containing two different British PFAs, Longannet and Eggborough, is given in terms of Si/Ca

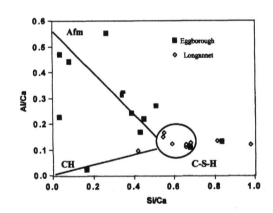

FIGURE 14.9 Al/Ca versus Si/Ca ratios of individual microanalysis of two 6-year-old PFA/OPC pastes at 30 per cent replacement. (Modified from Lachowski *et al.* [70].)

versus Al/Ca ratios of individual spot analysis (Figure 14.9). The Si/Ca ratios of the C-S-H from the hydrated cement lie between 0.56 and 0.67 (Ca/Si 1.49–1.78) and Al/Ca is approximately 0.13. The C-S-H observed in the proximity of the Eggborough PFA has Si/Ca 0.83–1.0 (Ca/Si 1.0–1.2), whereas for the Longannet PFA Si/Ca lie between 0.3–0.5 (Ca/Si 2.0–3.3). The Ca/Si values seem high for the Longannet PFA and may be due to inclusion of CH. Al/Ca is reported at around 0.3. Results for a 91-day-old paste with 28 per cent PFA and water/solid (w/s) ratio 0.5 give a mean Ca/Si ratio of 1.63 and Al/Ca ratio of 0.10 for the C-S-H from the inner product of the alite or belite, whereas near the PFA particles, Ca/Si was around 1.61 with Al/Ca approximately 0.18 [68]. The Ca/Si ratio also appears to differ in composition depending on the distance from the PFA particle [25]. It is likely that the Ca/Si ratio of the C-S-H will vary until a steady state condition is achieved between the solid and aqueous phase which for PFA, can take from 2–6 years, assuming no environmental exchanges occur [84].

14.4.2.2 Hydration mechanisms of high-calcium PFA

Hydration in high-calcium PFA appears to begin at the surface of the glass sphere, which is gradually consumed and replaced by radiating clusters or bundles of fibrous C-S-H. As the C-S-H fibre bundles grow inwards, they gradually become denser and the spheres reduce in size until they eventually disappear on complete hydration (Figure 14.10). The fibrous C-S-H has a honeycomb appearance in cross-section, and in earlier

stages, the openness of the honeycomb structure is considered to allow the free movement of aqueous phase to and from the PFA surface sphere. The Ca/Si ratio of the high-calcium PFA is consistent with formation of C-S-H, and it is suggested that the excess of aluminium ions transport to the cement matrix where they are available for formation of phases such as C_2AH_8, C_2ASH_8, monosulphate, ettringite or C_4AH_{13} [41]. Formation of ettringite predominant in high-calcium cement pastes (Figure 14.11), appears to be associated to some extent with higher Na content [69–71]. Strätlingite formation is favoured by PFA with lower total calcium and sulphur content and higher aluminium content whereas ettringite

(a)

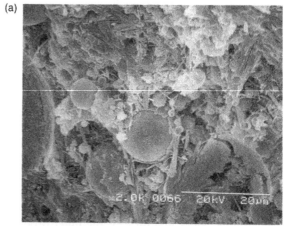

(b)

FIGURE 14.11 SEM migrographs of high-calcium PFA/OPC paste hydrated 3 months showing ettringite as binder (a) 100 per cent PFA; (b) 80/20, PFA/OPC.

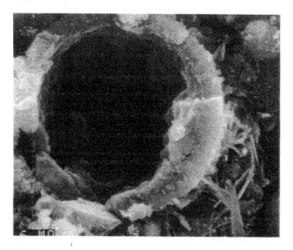

FIGURE 14.10 SEM micrograph of high-lime PFA/OPC paste showing a hydration rim with radiating bundles of C-S-H.

is associated with PFA having a higher SO_3/C_3A ratio [72]. Monosulphate tends to form over time as sulphur becomes limiting. It has been noted that PFA pastes dominated by strätlingite as hydration product give consistently higher compressive strengths in comparison to pastes containing ettringite with values at 28 days reported as 5270 psi and 950 psi respectively [69]. Not all high-calcium PFAs are reactive at early ages, Diamond *et al.* [60] observed little sign of reaction at six months with two high-calcium PFAs having CaO content of 37.1 and 26.8 per cent in PFA/cement pastes at 25 per cent PFA replacement of 20 per cent cement.

14.4.3 Porosity and permeability in extended blends

The pore structure in concrete is one of the most important factors that governs the durability in terms of attack by aggressive agents such as chlorides, sulphates, carbonates and acids. Porosity, pore size distribution, permeability and diffusivity are all means used to determine the effect of durability and numerous studies have been instigated to determine the most accurate means of determining these variables and in finding relevant correlations. Given the differences in the methods used, there is, however, a general consensus that PFA extended cements give a more impervious cement paste over time due to the pozzolanic reaction if proper curing is ensured. Initial porosity and permeabilities are greater at the time when little pozzolanic activity would have occurred. The mean pore size is much lower in PFA extended cements resulting in lower ionic diffusion rates and reduced permeabilities. PFA is reported to give a larger portion of fine pores [66,73] and more refined pore structure or smaller volume of large pores [74]. The principal *effect of adding* PFA is a considerable reduction in pores >36.8 nm and increase for pores <4 nm [75]. This also corresponds to a decrease in the relative volumes of gel + PFA and of calcium hydroxide. The volume fraction of porosity >36.8 nm tends to dominate the compressive strength. Chloride diffusion studies indicates that higher replacement levels increase the resistance of the blend to charge passage and that high-calcium PFA is more effective than low-calcium PFA.

14.4.4 Sulphate resistance effect of PFA

Reduced permeability is one factor that improves the sulphate resistance in PFA extended systems. However, some PFAs have been reported to be ineffective, while others actually cause an increase in sulphate deterioration. Attack occurs when the sulphates are able to react with $Ca(OH)_2$ to form $CaSO_4 \cdot 2H_2O$ or with aluminates to form ettringite. Formation of $CaSO_4 \cdot 2H_2O$ is considered secondary in expansion by comparison to ettringite, though no single theory exists for predicting the sulphate-resisting potential. CaO and Fe_2O_3 in the PFA have been suggested as the main contributors to resistance or susceptibility in sulphate attack and a resistance factor, $R = (C - 5)/F$, where C is the percentage CaO and F is the percentage Fe_2O_3 in the PFA was postulated [76]. An R value of <1.5 gives improved resistance whereas >3.0 produces reduced resistance and values between 1.5–3.0 indicates no significant change. The R-factor does not always hold and is considered questionable as it does not take into consideration the effect of the aluminate chemistry. Manz *et al.* [77] proposed two parameters for determining the effectiveness of PFA to sulphate resistance which take into consideration the reactive alumina and the SO_3, and they are the Calcium Aluminate Potential ($\overline{CAP}$) and the Calcium Sulphate Equivalent ($\overline{CSE}$) where $\overline{CAP} = (CaO + Al_2O_3 + Fe_2O_3)/SiO_2$ and $\overline{CSE} = $ anhydrite $+ 1.7\overline{S}$. A mathematical model has more recently been proposed by Djuric *et al.* [78] which gives a Corrosion Resistance Factor = (Flexural Strength of Corroded Sample)/(Flexural Strength of Sample Immersed in Water). In general, results seem to indicate that low-calcium PFAs tend to improve the resistance to sulphate attack or at worst have minimal effect. High-calcium PFA, on the other hand, is more controversial where, in some cases it is reported as detrimental, decreasing the resistance to sulphate [79], and in others, it improves sulphate resistance [78,80,81]. Formation of ettringite as a hydration product prior to sulphate exposure is considered to give improved sulphate resistance, whereas, formation of monosulphate, $C_3A \cdot C\bar{S}H_{18}$ and $C_3A \cdot CH.H_{18}$ are more likely to be detrimental to sulphate resistance [81].

14.4.5 Alkali aggregate reaction

PFA extended cements are well known to have beneficial properties in relation to alkali aggregate reaction (AAR) where reaction occurs between alkali and active siliceous aggregate as alkali–silica (ASR) or alkali–silicate reaction [82]. In the case of alkali–carbonate reaction (ACR), where reaction is between alkali and certain argillaceous dolomitic aggregates, the PFA has little beneficial effect. A number of theories have been proposed to explain the action of PFA in ASR or alkali silicate reactions where the PFA: (a) is considered to act as a diluent to the alkalinity of the pore solution in proportion to the quantity of cement replaced, (b) reduces the alkalinity of the pore solution due to the higher alkali binding power of the PFA cement hydration products, (c) prevents formation of a deleterious expansive reaction product through their pozzolanic reaction with $Ca(OH)_2$ and (d) lowers permeability due to pore-size and grain-size refinement which reduces the migration of alkalis towards the reactive aggregate.

It has been widely considered that it is the available alkalies that are the active part of the total alkali content in causing expansion. Although replacement with PFA can increase the total alkali content, it will reduce both the water-soluble and available alkali content. Pore solution studies on PFA cement pastes containing low-calcium PFA have been investigated by a number of researchers [84,83–90]. Two Danish PFAs having 2.4 and 3.3 per cent Na_2O_e, some of which was 'available' alkali, at 30 per cent replacement hydrated for 6 months showed the 2.4 per cent Na_2O_e PFA to behave in an entirely inert manner with respect to alkali, neither removing or contributing alkali to the pore solution, whereas, the

3.3 per cent Na_2O_e PFA, although having a higher alkali content, showed a slight removal of a Na^+ and K^+ contributed by the cement between 10 and 30 days [84]. Similar results were reported for two British PFAs, one low alkali, Na_2O 0.28, K_2O 1.39, the other high alkali, Na_2O 1.14, K_2O 3.73, at 30 per cent replacement and after 2 years of curing [84]. At 6 years, the situation was significantly different where the pore solution alkali concentrations had increased significantly in the OPC paste by 85 per cent in Na^+ and 193 per cent in K^+ but only by 52 per cent in Na^+ and 13–45 per cent in K^+ in the PFA blends. Interestingly, the Na^+ content in one of the PFA blends was higher than that of the neat OPC, though total alkali was substantially lower (Table 14.6). It is suggested that the PFA causes a decrease in the C/S ratio of inner hydrates, and these are found to incorporate more K^+ [65]. Typically, the Ca/Si ratio of C-S-H in Portland cements is about 1.5–1.8, and is lower in the presence of PFA, 1.0–1.2, as discussed previously. When the Ca/Si ratio is high, the surface charge of the C-S-H is +ve and anions (other than Na^+ and K^+ which remain in the pore solution) are adsorbed on the C-S-H fibres. When the Ca/Si ratio is lower than about 1.2–1.3, the surface charge of the C-S-H becomes −ve and incorporates alkali cations.

An investigation into the correlation between SiO_2, CaO, SO_3, available alkalies as Na_2O and the sum of $SiO_2 + Al_2O_3 + Fe_2O_3$ with reduction in mortar bar expansion give the least correlation with available alkalies and the best with SO_3 content [91]. The reason behind the correlation with

TABLE 14.6 Pore fluid data for blended cements aged 6 years at 25°C and 98 per cent relative humidity: equivalent 2 year data given in (*italics*) [84].

System	Ion concentration (mM)	
	Na^+	K^+
OPC	87.0	512
	(*47.0*)	(*175*)
Low alkali PFA/OPC	45.7	120
	(*31.0*)	(*106*)
High alkali PFA/OPC	120.0	233
	(*77.0*)	(*175*)

SO_3 is not clear. An expansion-PFA replacement relationship indicated that the higher the CaO content, the greater the level of replacement required to retard alkali silica expansion [92]. Not all alkali–silica gels are equally expansive and, it appears, the higher the $CaO/(Na_2O)_e$ ratio in the alkali–silica gel, the lower the gel's capacity for expansion [93]. Microstructural studies have shown that the presence of $Ca(OH)_2$ appears to be critical with deterioration due to expansion occuring at sites of formation of a calcium-alkali-silica rim at the cement–aggregate interface. Such rims are absent in the PFA extended systems due to a reduction in $Ca(OH)_2$ attributable to the pozzolanic reaction [94,95].

There appears to be a pessimum value in PFA replacement below which the PFA is ineffective in preventing expansion and in fact, in many cases, has been reported to increase expansion, the limit of which is inherent to each particular PFA [93,96]. Thomas *et al.* [96] report that for moderately reactive aggregates such as flint, 25 per cent PFA was found to be effective regardless of Ordinary Portland Cement (OPC) alkali content. For more reactive aggregates, higher PFA replacement levels are required to prevent deleterious expansion. Limited research has been performed on the high-calcium PFA, but what has been done indicates that higher PFA replacement levels are required for it to be effective in reducing expansion.

14.5 References

1. Davis, R. E., Carlson, R. W., Kelly, J. W. and Davis, H. E., *J. Am. Concr. Inst.*, 33, p. 577 (1949).
2. Meissner, H. S., *ASTM STP 99*, p. 16 (1949).
3. Manz, O. E., *Proc. 13th Int. Symp. on Use and Management of Coal Combustion Products*, pp. 64-1–64-6 (1999).
4. Hower, C. J., Thomas, G. A. and Trimble, A. S., *Proc. 1999 Int. Ash Utilization Symp.*, October 18–20, p. 195 (1999).
5. Hwang, J. Y., Huang, X., Gillis, J. M., Hein, A. M., Popko, D. C., Tieder, R. E. and McKimpson, *Proc. 13th Int. Symp. on Use and Management of Coal Combustion Products*, pp. 19-1–19-4 (1999).
6. Van den Berg, J. W. and Cornelissen, H. A. W., *Proc. 13th Int. Symp. on Use and Management of Coal Combustion Products*, pp. 29–1 (1999).
7. Larrimore, L., Dodgen, D. and Monroe, L., *Proc. 13th Int. Symp. on Use and Management of Coal Combustion Products*, pp. 16–1 (1999).
8. Frady, T. and Hay, P., *Proc. 13th Int. Symp. on Use and Management of Coal Combustion Products*, p. 26–1 (1999).
9. Stencel, J. M., Ban, H., Li, T. -X. and Neathery, J. K., *Proc. 13th Int. Symp. on Use and Management of Coal Combustion Products*, pp. 22–1 (1999).
10. Koshinski, C. J. and Freidrich, A., *Proc. 1999 Int. Ash Utilization Symp.*, October 18–20, p. 560 (1999).
11. Levy, E., Herrera, C., Coates, M. and Afonso, R., *Proc. 13th Int. Symp. on Use and Management of Coal Combustion Products*, pp. 17–1 (1999).
12. Uchikawa, H., Uchida, S. and Hanehara, S., *Proc. 8th ICCC*, Vol. IV, p. 245 (1986).
13. Diamond, S., *Cem. Concr. Res.*, 13, p. 459 (1983).
14. Arjunan, P., Silsbee, M. R. and Roy, D. M., *Proc. 10th ICCC*, Vol. 3, 3v020 4 pp. (1997).
15. Tenoutasse, N. and Marion, A., 'Blended Cements', *ASTM STP 897*, p. 65 (1986).
16. Scheetz, B. E., Strickler, D. W., Grutzeck, M. W. and Roy, D. M., *Mat. Res. Soc. Symp. N*, p. 24 (1981).
17. Mailvaganam, N. P., Bhagrath, R. S. and Shaw, K. L., ACI Special Publication SP-79, p. 519 (1983).
18. Lane, R. O. and Best, J. F., *Concr. Int.*, 4(7), p. 81 (1982).
19. Dodson, V. H., *Mat. Res. Soc. Symp. N*, p. 166 (1981).
20. Ramaskishan, V., Coyle, W. V., Brown, J., Tlustus, P. A. and Venkataramanujam, P., *Mat. Res. Soc. Proc. Symp. N*, p. 233 (1981).
21. Naik, T. R. and Ramme, B. W., *ACI Mater. J.*, 87, p. 619 (1990).
22. Grutzeck, M. W., Fajun, W. and Roy, D. M., *Mat. Res. Soc. Symp. Proc.*, 43, p. 65 (1985).
23. Halse, Y., Pratt, P., Dalziel, J. A. and Gutteridge, W. A., *Cem. Concr. Res.*, 14, p. 491 (1984).
24. Kovacs, R., *Cem. Concr. Res.*, 5, p. 73 (1975).
25. Takemoto, K. and Uchikawa, H., *7th ICCC*, p. 1V-2/1 (1980).
26. Abdul-Maula, S. and Odler, I., *Mat. Res. Soc. Symp. N*, p. 102 (1981).
27. Asaga, K., Kuga, H., Takahashi, S., Sakai, E. and Daimon, M., *10th ICCC*, Vol 3, p. 3ii107 (1997).
28. Jawed, I. and Skalny, J., *Mat. Res. Soc. Symp. N*, p. 60 (1981).
29. Plowman, C. and Cabrera, J., *Mat. Res. Soc. Symp. N*, p. 71 (1981).
30. Wei, F.-J., Grutzeck, M. W. and Roy, D. M., *Cem. Concr. Res.*, 15, p. 174 (1985).
31. Kawada, N. and Nemoto, A., *22nd Gen. Mtg. Cem. Assoc.*, Japan, p. 124 (1968).

32. Montgomery, D. G., Hughes, D. C. and Williams, R. I. T., *Cem. Concr. Res.*, 11, p. 591 (1981).

33. Larbi, J. A. and Bijen, J. M., *Cem. Concr. Res.*, 20, p. 783 (1990).

34. Diamond, S., Ravina, D. and Lovell, J., *Cem. Concr. Res.*, 10, p. 297 (1980).

35. Ghose, A. and Pratt, P., *Mat. Res. Soc. Symp. N*, p. 82 (1981).

36. Daimond, S., *Mat. Res. Soc. Symp. N*, p. 12 (1981).

37. Pietersen, H. S., Fraay, A. L. A. and Bijen, J. M., *Mat. Res. Soc. Symp. Proc.*, Vol 178, p. 139 (1990).

38. Fraay, A. L. A., Bijen, J. M. and de Haan, Y. M., *Cem. Concr. Res.*, 19, p. 235 (1989).

39. Nägele, E. and Schneider, U., *Cem. Concr. Res.*, 19, p. 811 (1989).

40. Diamond, S., *7th ICCC*, III, p. 1V–19 (1980).

41. Grutzeck, M. W., Roy, D. M. and Scheetz, B. E., *Mat. Res. Soc. Symp. N*, p. 92 (1981).

42. Smith, I. A., *Proc. Inst. of Civil Engrs.* 36, p. 769 (1967).

43. Gopalan, M. K. and Haque, M. N., *Symp. on Concr.*, Inst. Engrs., Australia, p. 12 (1983).

44. Ghosh, R. S., *Canadian J. Civil Engg.*, 3, p. 68 (1976).

45. Babu, K. G. and Rao, G. S. N., *Cem. Concr. Res.*, 26, p. 465 (1996).

46. Wesche, K. (ed.), RILEM TC-67-FAB (1990).

47. Mehta, P. K., *Cem. Concr. Res.*, 15, p. 669 (1985).

48. Tikalsky, P. J., Carrasquillo, P. M. and Carrasquillo, R. L., *ACI Mater. J.*, 85, p. 505 (1988).

49. Shi, C. and Day, R. L., *Proc. 7th ICCM*, p. 150 (1995).

50. Dietz, S., Miskiewicz, K. and Schmidt, M. *10th ICCC*, Vol 3, p. 3ii087 (1997).

51. Lam, L., Wong, Y. L. and Poon, C. S., *Cem. Concr. Res.*, 28, p. 271 (1998).

52. Erdoğdu, K. and Türker, P., *Cem. Concr. Res.*, 28, p. 1217 (1998).

53. Fraay, A. L. A. and Bijen, J. M., *Cem. Concr. Res.*, 19, p. 235 (1989).

54. Li, S., Roy, D. M. and Kumar A., *Cem. Concr. Res.*, 15, p. 1079 (1985).

55. Coole, M. J., *Br. Ceram. Proc.*, 35, p. 385 (1984).

56. Taylor, H. F. W., Mohan, K. and Moir, G. K. J., *Am. Ceram. Soc.*, 68, p. 685 (1985).

57. Dalziel, J. A. and Gutteridge, W. A., *Cem. Concr. Association Technical Report 560*, 28 pp. (1986).

58. Ohsawa, S., Asaga, K., Goto, S. and Daimon, M., *Cem. Concr. Res.*, 15, p. 357 (1985).

59. Luke, K. and Glasser, F. P., *Cem. Concr. Res.*, 17, p. 273 (1987).

60. Diamond, S., Sheng, Q. and Olek, J., *Mat. Res. Soc. Symp. Proc.*, Vol. 136, p. 281 (1989).

61. Tenoutasse, N. and Marion, A. M., *8th ICCC*, 4. 1, p. 1 (1986).

62. Larbi, J. A and Bijen, J. M., *Mat. Res. Soc. Symp. Proc.*, Vol. 178, p. 127 (1990).

63. Rodger, S. A. and Groves, G. W., *Adv. Cem. Res.*, 1, p. 84 (1988).

64. Mohan, K. and Taylor, H. F. W., *Mat. Res. Soc. Symp. N*, p. 54 (1981).

65. Rayment, P. L., *Cem. Concr. Res.*, 2, p. 133 (1982).

66. Uchikawa, H., *Proc. 8th ICCC*, Vol. 1, 3.2, p. 249 (1986).

67. Sato, T. and Furuhashi, I., *Review of the 36th General Meeting, Cement Association of Japan*, p. 42 (1982).

68. Harrison, A. M., Winter, N. B. and Taylor, H. F. W., *8th ICCC*, Vol 4, p. 170 (1986).

69. Schlorhiltz, S., Bergeson, K. and Demirel, T., *Mat. Res. Soc. Symp. Proc.*, Vol 113, p. 107 (1988).

70. Diamond, S., *11th ICCM*, New Orleans, p. 263 (1989).

71. McCarthy, G. J. and Solem-Tishmack, J. K., *Adv. Cem. Concr.*, Grutzeck and Sarkar (eds.), p. 103 (1994).

72. Tishmack, J., Olek, J., Diamond, S. and Sahu, S., *Proc. 1999 Int. Ash Utilization Symp.*, October 18–20, p. 624 (1999).

73. Malek, R. I. A., Roy, D. M. and Fang, Y., *Mat. Res. Soc. Symp. Proc.*, Vol. 137, p. 403 (1989).

74. Thomas, M. D. A., The Microstructure and Chemistry of Cement and Concrete, Institute of Ceramics, p. 23 (1989).

75. Patel, H. H., Pratt, P. L. and Parrott, L. J., *Mat. Res. Soc. Symp. Proc.*, Vol. 136, p. 233 (1989).

76. Dunstan, E. R., *Cem. Concr. and Aggr.*, 2, p. 22 (1980).

77. Manz, O. E., McCarthy, G. J., Docketer, B. A., Johansen, D. M., Swanson, K. D. and Steinwand, S. J., Katherine and Bryant Mather Conference on Concrete Durability, Atlanta., Paper Presentation. (1987).

78. Djuric, M., Ranogajec, J., Omorjan, R. and Miletic, S., *Cem. Concr. Res.*, 26, p. 1295 (1996).

79. Day, R. L. and Konecny, L., *Mat. Res. Soc. Symp. Proc.*, Vol. 136, p. 243 (1989).

80. Križan, D. and Živanović, B., *10th* ICCC, Vol. 4, p. 41v020 (1997).

81. Mehta, P. K., *J. ACI*, 83(6), p. 994 (1986).

82. Hobbs, D. W., Alkali Silica Reaction in Concrete, Thomas Telford Ltd., London (1998).

83. Diamond, S., *Cem. Concr. Res.*, 11, p. 383 (1981).

84. Lachowski, E. E., Kindness, A., Glasser, F. P. and Luke, K., *10th ICCC*, Sweden, p3i0091 (1997).

85. Glasser, F. P. and Marr, J., *Il Cemento*, 2, p. 85 (1985).

86. Glasser, F. P., Luke, K. and Angus, M. J., *Cem. Concr. Res.*, 18, p. 165 (1988).

87. Fraay, A. L. A., Bijen, J. M., de Haan, Y. M. and Larbi, J. M., *3rd CANMET/ ACI Int. Conf. on Fly Ash, Silica Fume & Natural Pozzolans in Concr.*, Supplementary Papers, p. 33 (1989).
88. Marr, J. and Glasser, F. P., *Proc. 6th Int. Conf. Alkalies in Concr. Res. Practice*, p. 239 (1983).
89. Luke, K. and Glasser, F. P., *Il Cemento*, 3, p. 179 (1988).
90. Luke, K. and Glasser, F. P., *Cem. Concr. Res.*, 18, p. 495 (1988).
91. Smith, R. L., *Mat. Res. Soc. Symp. Proc.*, Vol. 113, p. 249 (1980).
92. Dunstan, E. R., *Cem., Concr. and Aggr*, 3, p. 101 (1981).
93. Monteiro, P. M. J., Wang, K., Sposito, G., dos Santos, M. C. and de Andrade, W. P., *Cem. Concr. Res.*, 27, p. 1899 (1997).
94. Thomas, M. D. A., *17th ICMA*, p. 242 (1995).
95. Bleszynski, R. F. and Thomas, M. D. A., *Adv. Cem. Based Mater.*, 7(2) p. 67 (1988).
96. Thomas, M. D. A., Blackwell, B. Q. and Nixon, P. J., *Mag. Concr. Res.*, 48(177), p. 251 (1996).

Chapter fifteen

Metakaolin as a pozzolanic addition to concrete

Tom R. Jones

15.1 Introduction

Primitive cements were first discovered in the eastern Mediterranean region as long ago as 7000 BC (Bonen *et al.* 1995; Davey 1953). When man began to live in permanent settlements, he found that fire could convert soils containing gypsum, calcium carbonate and clay into low quality hydraulic materials. These were probably first used for making floor areas. Further progress was slow: it was not until about 500 BC that the benefits of blending natural pozzolanic materials with lime (calcium hydroxide) was discovered by the Greeks, and then developed into an art by the Romans. The Romans were the first to systematically use synthetic pozzolanas; they added crushed pottery tiles (i.e. amorphous silica and calcined clay) to their lime-based cements.

Many large Roman structures were made from lime mixed with natural pozzolana and crushed burned clay, and some are still in use today. The word for this wonderful material, which could be shaped and then allowed to solidify into a synthetic sculptured stone, was derived from the Roman term for hewn stone, 'caementum', which in turn was derived from 'caedere', to cut.

There was no further progress in cement technology until the eighteenth century. By 1759, Smeaton found he could make a strong hydraulic cement by blending suitable limestone-rich and clay-rich soils and then burning the mixture. This quickly led to the development of an early form of Portland cement (PC) by Aspidin in 1826, and to the extensive use of blended pozzolana – Portland cements. Slag from blast furnaces and coal ashes were used as pozzolanic additions to cement as early as the 1870s (Encyclopaedia Britannica 1876).

In the twentieth century, pozzolanas were originally used to reduce the cost of PC. In 1940, it was found that pozzolana – PC concrete could prevent the occurrence of alkali-silica reaction, ASR, and hence improve the durability of concrete. This was confirmed when the Friant Dam was successfully built in California in the early 1940s, using PC – pumicite pozzolana cement together with an active aggregate (Tuthill 1982). An early review of the chemistry of pozzolanic materials, including metakaolin (mk), was given by Malquori (1960).

In Brazil, over the period 1962–79, about 260,000 tonnes of mk were used in the construction of four dams in the Amazon basin. The local aggregate was highly alkali-active, so mk was blended with PC in order to prevent expansion due to ASR, with great success. Then, in France from 1980, Murat, Ambroise and Pera extensively examined the pozzalanic activity of mk (for example, Ambroise *et al.* 1985, 1986, 1992, 1994; Murat 1983a,b,c).

Metakaolin can be activated by alkali metal hydroxides, by alkali metal silicates, or by calcium hydroxide, which is one of the hydration products of PC. Metakaolin can be made in pure or impure form, depending on the mineralogy of the clay which is fed to the calciner. It is important to note that certain fine minerals can have a seriously deleterious effect on the strength of concrete. Therefore, mk cannot be used to make high-strength concrete if it contains more than 10–20 mass per cent of relatively inert ancillary minerals such as quartz, mica or felspar (Ambroise *et al.* 1985, 1986; Walters and Jones 1991). However, there has been much interest in using impure mk in the form of 'calcined tropical soils' to make low cost concrete for infrastructure projects, especially in developing countries, see for example Cabrera and Nwaubani (1993). Other sources of impure mk include red mud, which is a waste product of bauxite refining (Shi *et al.* 1999) and ash from the low-temperature burning of sludge from the paper recycling industry (Pera *et al.* 1998).

One of the first commercial applications of pure mk as an additive to PC was by the CemFil division of Saint Gobain to make durable glass fibre-reinforced concrete (GRC). Metakaolin inhibits portlandite from chemically attacking the (alkali resistant) glass fibres (Sohm 1988). Since the early 1990s, the commercial use of pure mk in other forms of concrete has expanded rapidly. It improves the strength and water absorption properties of concrete, and controls efflorescence; early applications include renders, mortars and paving slabs. Following long-term trials, it is now also being used in concrete where improved durability in aggressive environments is a high priority. Examples include sea walls, abrasion resistant floor slabs and roads, pre-cast bridge beams and bridge repairs. Metakaolin has been used in conjunction with other pozzolanas such as fly ash,

ground granulated slag and silica fume. This review will concentrate on on those applications where 'pure' mk is blended with PC.

15.2 Structure of metakaolin (mk)

Kaolin is the most ubiquitous of the clay minerals, but is often associated with other minerals. Deposits of sufficient size and concentration to enable commercial exploitation of the pure mineral are relatively rare. Kaolin is formed by the alteration of a variety of crystalline and amorphous rocks such as felspar and volcanic ash. Kaolin is a phyllosilicate, consisting of alternate layers of silica in tetrahedral coordination, and alumina in octahedral coordination. Kaolin is chemically quite stable under normal environmental conditions, but because of a slight mismatch in the lattice parameters of the silica and alumina layers, the crystals are inherently strained and rarely grow beyond a few micrometres in diameter. The N_2 BET surface area of pure kaolin is normally 2000–20,000 $m^2 kg^{-1}$.

The structures of clay minerals are known in detail (van Olphen 1977). Figure 15.1 shows that each silica layer is bonded to its paired alumina layer via shared oxygen atoms. For every four silicon atoms plus four aluminium atoms, there are eight hydroxyl groups. Each silica/alumina paired layer is bonded to the next pair via relatively weak hydrogen bonds.

When heated to 700–900°C kaolin loses 14 mass per cent as hydroxyl water, to form mk. The alumina and silica layers become puckered; they lose their long-range order and the powder becomes amorphous with respect to X-ray diffraction. Dehydroxylation causes the clay to become chemically reactive; in particular it is readily attacked by dilute acids and alkalis (but not water) at ambient temperatures. Another important structural change is that much of the aluminium in mk becomes tetrahedrally coordinated.[1] This is most evident from Al NMR spectroscopy (Justnes *et al.* 1990; Rocha and Klinowski

---

[1]There is also evidence that 5-coordinated Al is also present in mk, in small amounts.

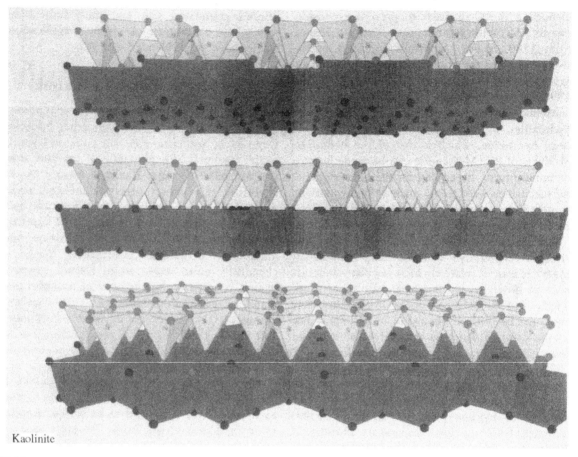

Kaolinite

FIGURE 15.1   Crystal structure of kaolin showing silica tetrahedra ▨ and alumina octahedra ■, reproduced by permission of Shape Software Inc.

1990) and soft X-ray absorption spectroscopy (Roberts *et al.* 1992); indeed the existence of a characteristic tetrahedral AlO_4 resonance is often taken as evidence that mk has been formed. Tetrahedral Al is thought to play an important role in the pozzolanic reactions of mk.

The kinetics of dehydroxylation have been studied (Meinhold *et al.* 1994). Metakaolin can be made by flash calcining at about 1000°C or by prolonged soak calcining at lower temperatures (Dunham 1992). Salvador (1995) reports that the rate of calcining does not influence the pozzolanic properties, provided that dehydroxylation is complete and that the kaolin has not been over-calcined. Above about 800°C, kaolin begins to

convert to relatively inert ceramic materials such as spinel, silica and mullite.

15.3 Pozzolanic reactions of mk

Metakaolin reacts with aqueous alkali to give a variety of products, depending on the presence of other anions and cations, the concentration of reactants, temperature, etc. Zeolites are commonly formed; they may have a well-defined crystal structure, or may be X-ray amorphous if the primary particles are of nanometre dimensions. Metakaolin is used commercially to make certain zeolites for use as cracking catalysts and ion exchange.

Ultramarine blue, a synthetic form of the natural zeolite mineral, lapis lazuli, is a commercial pigment made from mk. However, in this case, the reaction with alkali is normally carried out in the solid state at high temperature.

When mk is used for its pozzolanic properties (i.e. to make cementitious materials), the hydration products are normally nanosize crystals. They can have layered or three-dimensional structures. It is convenient to distinguish three types of pozzolanic reactions of mk: (a) with alkali metal hydroxides and soluble alkali metal silicates (b) with calcium hydroxide and (c) with PC, which generates alkalis, mainly $Ca(OH)_2$ and KOH, among its products of hydration.

15.3.1 Reaction with alkali metal hydroxides and alkali metal silicates

The reaction of mk with sodium or potassium hydroxide and sodium or potassium silicate has been extensively studied. Cements made from alkali metal hydroxides (and/or silicates) with mk are known variously as geo-polymers, earth ceramics or inorganic polymers, and have been used in many specialist applications. Bauweraerts, Wastiels *et al.* (1994) and Palomo *et al.* (1999) have examined the properties of mortars and concrete, and Davidovits (1994) has described numerous applications. Examples include rapid-setting repair mortars, heat-resistant concrete, and cement for encapsulating hazardous waste. There is general agreement that the initial reaction products are aluminosilicate zeolites, which are X-ray amorphous because of the small crystallite size. Over time, and especially at elevated temperatures, X-ray detectable faujasite may be formed.

Palomo *et al.* (1999) have investigated the chemical stability of these cements and find that they do not lose strength when immersed for 270 days in various aggressive reagents such as water, sodium sulphate solution, seawater or 0.001 m sulphuric acid. This high degree of durability, compared with the hydrated calcium silicate (CSH) phases in PC concrete, is possibly because of the three-dimensional zeolitic structure of mk/alkali silicate cements.

TABLE 15.1 Compressive strength of mk mortar, activated with KOH plus K silicate

Time of reaction @ 80°C (h)	Density (kgm^{-3})	Compressive strength (MPa)
1	1740	45.1
3	1720	60.2
24	1720	69.5
72	1720	69.7
168	1710	71.2

TABLE 15.2 Exposure to high temperature: effect on compressive strength of activated mk mortar

Exposure temperature (1 h at °C)	Density (kgm^{-3})	Compressive strength (MPa)
80	1720	69.7
250	1590	54.0
500	1620	45.0
1000	1590	26.1

One example of a simple but interesting inorganic polymer is described by Waters (2000). It was made from mk, potassium silicate, potassium hydroxide and fine sand in the dry-mass ratios of $23:15:5:57$, respectively (K_2O/SiO_2 mass ratio $= 0.26$). The ingredients were blended at a water/binder ratio of 0.49, moulded into cubes, out-gassed in vacuum, and cured for various times at 80°C. Tables 15.1 and 15.2 show that the mortar rapidly gained compressive strength to 70 MPa, and tolerated 1 h exposure to high temperatures.

Mercury intrusion porosimetry (MIP) showed that the inorganic polymer mortars described above have very fine pore structures. After curing at 80°C, the threshold pore diameter[2] was 0.03 μm and total mercury intrusion volume was 30 volume per cent. After exposure to 1,000°C for 1 h, the threshold pore diameter increased to 20 μm, and total mercury intrusion volume at 200 MPa was 28 volume per cent.

[2]The threshold diameter is a somewhat subjective measurement of the largest pore diameter at which significant pore volume is detected; it corresponds with a rapid increase in slope of the log-normal mercury intrusion curve.

15.3.2 Reaction with calcium hydroxide

People have used the reaction between mk and calcium hydroxide (CH) to make cement for several millennia. But it is only in recent years, using advanced instrumental techniques (such as solid state NMR, analytical electron microscopy and thermal analysis), together with trimethyl silylation experiments, that we have begun to understand the detailed chemistry.

There is a rapid initial reaction, corresponding to the setting of the mk/CH mixture, followed by a slow reaction, which proceeds over a period of about 30 days. A range of products is formed, including CSH, C_2ASH_8 (gehlenite hydrate) and hydrated calcium aluminates, C_4AH_{13} and C_3AH_6 (de Silva and Glasser 1990, 1992, 1993). The composition and structure of CSH is variable, as discussed in the next section. Gehlenite hydrate, also called strätlingite, is a well-defined crystalline phase, but can be of variable composition. It can incorporate relatively large amounts of Na^+ and K^+ which displace Ca^{2+} in the lattice, and this can affect the pore solution chemistry of cement paste which contains mk.

Dunster *et al.* (1993) studied the reactions of mk with CH in two systems; mk/lime/water, and mk/PC/water, respectively. They used the trimethyl silylation technique to follow the kinetics of formation of the reaction products. They concluded that over a period of 30 days, mk reacts with CH at a similar rate in both mk/lime and mk/PC pastes. Pure mk is estimated to react with up to 1.6 times its mass of CH. The stoichiometry of the pozzolanic reaction was found to be:

$$AS_2 + 5CH + 5H \rightarrow C_5AS_2H_5$$

where $C_5AS_2H_5$ is an average composition representing a mixture of CSH, C_4AH_{13}, C_3AH_6 and C_2ASH_8. This is similar to the stoichiometry proposed earlier by Murat (1983a).

McCarter and Tran (1996) showed that electrical conductivity could be used to monitor the early hydration of pozzolana/lime mixtures. The electrical conductivity of a mk:CH:water mixture (in the ratio 8:2:12) fell to about 0.2 of its initial value after 40 h.

Commercially, blends of mk and CH are used to make synthetic hydraulic lime mortars. Both natural and synthetic hydraulic lime mortars are finding increased use in new constructions, as well as for repairing old buildings (Ashall *et al.* 1996). In new constructions, cured lime-based mortar is more flexible, more permeable to water vapour, and has a lower adhesive strength than a comparable PC mortar. Therefore, mortar is less likely to crack, less likely to trap dampness, and bricks can be more easily recycled when a structure is demolished. When used to repair old buildings, lime-based mortar is compatible with the types of mortars and concrete used before 1900. Some naturally hydraulic lime mortars show variable setting rates and strengths; mk can be used to improve these properties. It is added to both hydraulic and non-hydraulic limes at about the 10 mass per cent level to give an initial set within a convenient and predictable time, typically 24 h. The mortar develops a 28-day strength of 1–4 MPa, and then slowly continues to gain strength through natural carbonation, see Figure 15.2.

15.3.3 Reaction with PC

The chemistry and microstructure of hardened PC paste is complex (Glasser 1990). When PC hydrates, a considerable amount of CH is formed, typically 20 per cent of the paste mass; the generally accepted idealized reactions are as follows:

$$2C_3S + 6H \rightarrow C_3S_2H_3 + 3CH$$
$$2C_2S + 4H \rightarrow C_3S_2H_3 + CH$$

If mk is present, it reacts rapidly with the nascent CH to give CSH phases and 'CASH' phases as discussed in Section 3.2. Therefore, mk significantly changes the chemistry of the cement paste. Dunster *et al.* (1993) concluded that over a period of 30 days, mk reacts with CH at a similar rate in both mk/lime and mk/PC pastes. However, in the mk/PC mixture, mk accelerates the hydration and polymerization of the low molecular weight silicate ions. During the first 7 days, monomeric SiO_4^{4-} is consumed more rapidly, and

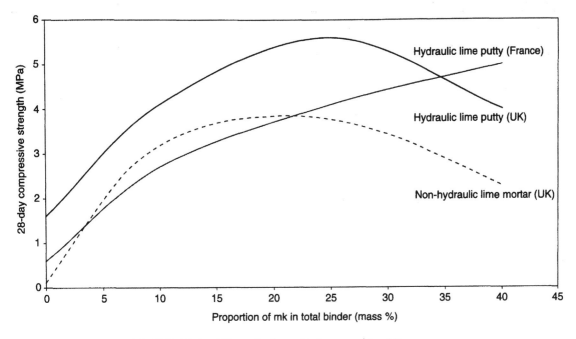

FIGURE 15.2   Effect of mk on 28-day strength of line mortars.

more silicate polymers of high molecular weight are formed, compared with PC pastes. Richardson (1999) has reviewed work on the nature of CSH in hardened cement phases, including mk/PC systems. He used trimethyl silylation, ^{29}Si NMR and TEM as complementary techniques for studying the atomic, molecular and nano size structure of the hardened paste.

Richardson concluded that, in the presence of mk, AlO_4^{5-} tetrahedra are incorporated into the CSH structure, but act solely to bridge the silicate dimers and do not form monomers or end groups.

$$O–Si_2O_5–O–AlO_2–O–Si_2O_5–O–AlO_2–O–Si_2O_5–O$$

The C/S ratio in hardened PC paste is thought to influence properties such as chemical durability. The solubility of CSH increases rapidly at C/S ratios above about 1.0 (Rahman *et al.* 1999). Microanalysis of TEM sections of CSH from PC shows that the C/S ratio typically ranges from 1.2 to 2.3. When an Al-containing pozzolana (e.g. bfs) is present, the C/(S + A) ratio ranges from 0.7 to 2.4 (Richardson 1999). This indicates a higher proportion of high molecular weight aluminosilicate polymers.

NMR studies (Pietersen and Bijen 1992b; Richardson 1999); confirm that mk increases the proportion of high molecular weight aluminosilicate polymers in cured paste; ^{29}Si spectra indicate a much higher $(Q^1 + Q^2)/Q^0$ ratio[3] compared with PC cement paste.

As well as having a major effect on the chemical structure of CSH, mk also modifies the nanostructure. Using high-resolution electron microscopy, Richardson (1999) has suggested that the CSH phase of mk/PC has a two-dimensional 'foil-like' structure, rather than the linear 'needle-like' structure of normal CSH. He suggested that the foil structure is more efficient at filling space without leaving large inter-connected capillary pores. This might account for the lower permeability and improved durability of concrete containing mk and other pozzolanas. Recent developments in high-resolution X-ray

---

[3]In ^{29}Si NMR spectroscopy: the Q^0 resonance corresponds to monomeric silicate (ortho-silicate, SiO_4^{4-}); the Q^1 resonance corresponds to dimeric silicate species ($Si_2O_7^{6-}$; the Q^2, Q^3 and Q^4 resonances correspond to Si nuclei which are connected to two, three or four other Si nuclei respectively, via oxygen bridges.

microscopy (Clark *et al.* 1999) should improve our understanding of the nanostructure of cement pastes in the near future.

15.4 Effect of mk on the basic properties of PC concrete

Cured PC paste is composed, by volume, of approximately 20 per cent CH, 30 per cent pores (filled with air or liquid) and 50 per cent of cement hydrate phases. As discussed earlier, metakaolin reacts with all or part of the CH to produce new hydrate phases; it changes the pore structure (this is particularly important in the region of the interfacial zone), and modifies the composition and structure of the gel phases. Therefore, it is not surprising that mk, like other pozzolanas, can significantly change the chemical and mechanical properties of cement paste and concrete.

15.4.1 CH content

Calcium hydroxide is a soluble and chemically reactive component of hardened cement paste. The crystals (portlandite) may be relatively large, of the order of 1–5 μm diameter, and are often concentrated within the interfacial zone, which is a layer of paste a few tens of micrometres thick adjacent to the aggregate surface. According to the stoichiometry of Dunster *et al.* (1993), pure mk reacts with 1.6 times its weight of CH. It is difficult to measure accurately low concentrations of CH in concrete, and errors can arise if the concrete is exposed to atmospheric carbon dioxide, either while curing or while preparing the sample for analysis. Differential thermal analysis appears to be the most precise method of following the concentrations of CH in cement/mk blends as they cure. Results indicate (Ambroise *et al.* 1994; Curcio *et al.* 1998; Min-Chul 1996); that mk reacts with approximately its own mass of CH, and this is supported (Kostuch *et al.* 1993) by the Chapelle test (Largent 1978). Curcio *et al.* (1998) found that, after 90 days' curing, 15 mass per cent replacement of PC by mk reduced the portlandite content of the cured paste to between 6 and

24 mass per cent of its value for the control paste, depending on the quality and particle size of the mk. To remove substantially *all* the CH from cured PC paste, approximately 20–25 mass per cent of the cement should be replaced by mk.

Klimesch *et al.* (1998a,b) measured residual mk in autoclaved blends of PC, mk and quartz. No residual mk was observed when it replaced less than 18 mass per cent of the PC. Residual mk was observed when it replaced more than 18 mass per cent. In these experiments, it is possible that some of the CH had reacted with the quartz, because the mortar, after pre-curing for 24 h, was autoclaved at 180°C for 8 h.

As discussed in later sections, removal of all or part of the CH plays an important role in:

- preventing ASR,
- improving the durability of GRC,
- modifying the structure of the interfacial zone,
- improving resistance to sulphate attack,
- controlling efflorescence.

15.4.2 Pore size distribution and interfacial zone

Mercury intrusion porosimetry (MIP) is commonly used to measure pore size distribution in cement pastes and mortars. Special corrections are needed to obtain meaningful data for gel pores less than 10 nm (0.01 μm) in diameter (Cook and Hover 1993). However, the permeability of cement paste correlates well with the threshold diameter, over a wide range of water/cement ratios (Zheng and Winslow 1995), and threshold diameters are normally within the range 0.05–0.2 μm for fully cured cement paste. Therefore, MIP is generally considered to be a valid technique to investigate the effect of pozzolanas on the transport properties of concrete. Various techniques, such as water absorption/desorption and helium pycnometry, are also available to characterize the smaller pores present in cement paste. In water desorption (Chadbourn 1997) fully saturated samples are stored in a vacuum desiccator whose internal relative humidity is controlled by a saturated salt solution. By using saturated $BaCl_2$ solution, for example, when equilibrium is attained all pores greater than 30 nm in diameter are empty; therefore, the total pore volume and the proportion of pores below 30 nm in diameter can be calculated.

The measured pore size distribution is sensitive to the exact technique employed to cast and compact the wet mixture. At low water/binder ratios, it can be difficult to achieve a flowing mixture, especially if insufficient plasticizer is used. This might result in poor mixing of the ingredients, incomplete dispersion of powder aggregates and poor compaction. Pastes at low water/binder ratios are particularly prone to errors, but mortar and concrete are generally less so because of the grinding effect of the aggregate. Sometimes, it is difficult to compare sets of pore size distribution measurements from different authors. Several reports (for example Ambroise *et al.* 1994; Chadbourn 1997; Frias and Cabrera 2000b; Khatib and Wild 1996; Larbi 1991) indicate that with cement *pastes*, addition of mk decreases the proportion of pores with diameters above about 0.1 μm. There is a corresponding increase in the proportion of pores with diameters below about 0.05 μm – that is, the

average pore diameter is reduced (refined). Using water desorption, Chadbourn (1997) found that the total porosity of mk pastes was slightly greater than that of pure pastes, indicating that mk increases the volume of fine capillary and gel pores, while decreasing the volume of coarser pores.

Asbridge (2000) has measured the pore diameter distributions of a series of fifteen cement pastes and mortars by MIP under standard conditions of preparation. The water/binder ratio, mk content, aggregate content (glass ballotini, 1.0–1.2 mm diameter) and curing temperature were systematically varied. Some typical results comparing mortars and pastes, with and without 10 mass per cent substitution of PC by mk, are shown in Figures 15.3 (mortars) and 15.4 (pastes). They illustrate the following general trends:

- The paste component of *mortar* has more coarse pores (above 1.0 μm diameter) than does pure paste
- Metakaolin does not significantly change the pore diameter distribution of paste (either pure paste, or the paste component of mortar) for pores above 0.2 μm diameter

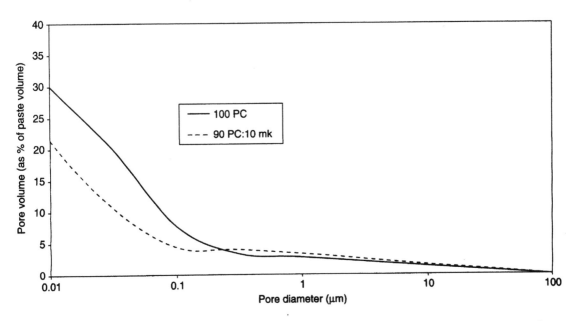

FIGURE 15.3 Pore diameter distribution of the paste component of mortar 35 vol % aggregate, 0.6 w/b ratio (Asbridge 2000).

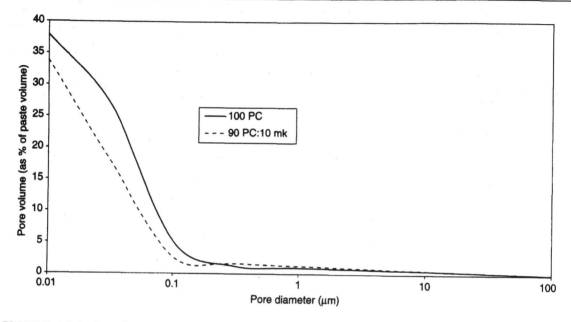

FIGURE 15.4   Pore diameter distribution of cement paste, 0.6 w/b ratio (Asbridge 2000).

- Metakaolin refines the pore structure in the region 0.01–0.2 μm, for paste and the paste component of mortar. Since mk does not reduce the total pore volume (Chadbourn 1997), this implies that mk increases the volume of pores in the region below 0.01 μm diameter.
- Metakaolin consistently reduces the threshold pore diameter, in both pure pastes and the paste component of mortars.

To summarize: addition of aggregate increases the volume of pores with diameters > 1.0 μm, and addition of mk reduces the volume of pores in the range 0.01–0.2 μm.

Interpreting the effect of mk (and other pozzolanas) on the pore structure of mortars is made difficult by the presence of an interfacial zone between the paste and the aggregate. The interfacial zone is a region of low concentration of CSH, high concentrations of CH and ettringite, and high porosity. It is typically 25–100 μm in thickness, but pozzolanas can change the structure of the interfacial zone by reacting with the CH and generating new cementitious phases. There is a very extensive literature (Maso 1996). Larbi and Bijen (1992) found that mk reduced both the amount and crystal orientation of CH in the interfacial zone. They measured the thickness of the interfacial zone to be less than 10 μm in mk mortar, compared with 30 μm in the corresponding PC mortar. This resulted in increased adhesive strength between the paste and aggregate, as measured directly by pull-off tests (Larbi 1991).

Figures 15.3 and 15.4 show that the paste component of mortar contains an extra 3.5 per cent pore volume in the region above 1 μm pore diameter, compared with pure paste. If these additional coarse pores are assumed to be concentrated in the interfacial zone, they are equivalent in volume to a void layer approximately 12 μm thick round each aggregate particle. This is not incompatible with the generally accepted thickness of the interfacial zone, viz. 25–100 μm.

Reducing the average pore size and the threshold value of cement paste is thought to have a beneficial effect on all degradation processes involving transport of gases and liquids through the paste. Examples include sulphate attack, acid attack, chloride ingress (and re-bar corrosion), ASR, and carbonation.

Larbi (1991) has reported that mk has no beneficial effect on the compressive strength of cement paste. However, as discussed above, mk replaces CH in the interfacial zone by new cementitious phases, and this improves the adhesive strength between the paste and aggregate. In concrete, this can significantly improve the compressive and tensile strengths (Larbi 1991), and related properties such as the abrasion resistance (Dhir 1999) and the adhesive bond between cement paste and steel reinforcing fibres (Banthia and Yan 2000), see later discussion.

15.5 Effect of mk on the properties of uncured concrete

Certain early curing properties of concrete are important because what happens during the first few hours, and then, over the next (say) seven days, will determine the long-term mechanical properties of the fully cured concrete. For example, freshly placed fluid concrete should set relatively quickly to prevent bleeding, and settling of aggregate. Tensile strength should increase rapidly in order to prevent any internal stresses (caused by shrinkage or drying) from causing cracks in the young concrete.

15.5.1 Dispersion and rheology

Metakaolin, manufactured for use in concrete, is milled to a fine powder, before and/or after calcining. A typical particle size distribution is 20–80 mass per cent less than $2\,\mu m$ equivalent spherical diameter, as calculated by the settling rate in water and assuming the density is $2600\,kg\,m^{-2}$. Most of the particles of mk are, in fact, aggregates formed by sintering the primary kaolin crystals. The latter are of the order of $0.5\,\mu m$ in diameter and $0.1\,\mu m$ thick, see Figure 15.5. BET nitrogen surface area is typically $10,000$–$25,000\,m^2kg^{-1}$. Therefore, compared to silica fume, mk has similar particle density and surface area, but different morphology and surface chemistry. Because of its hydrophilic surface, mk is easier to disperse into wet concrete. Metakaolin can be incorporated at any stage of concrete

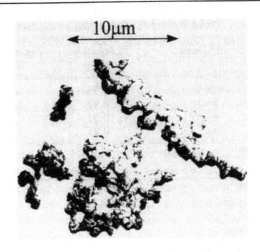

FIGURE 15.5 Typical morphology of metakaolin (scanning electron miscroscope).

production; it should be mixed in thoroughly to achieve even distribution; intensive mixing is not necessary. However, with semi-dry mixes it may be preferable to pre-mix the mk and the PC powders.

At low replacement levels (i.e. 5 mass per cent or less), mk increases the thixotropy of concrete without significantly affecting its flow properties under shear. This can reduce bleeding, improve levelling and provide a good surface finish. If appropriate, mk can be used at up to 20 or 25 mass per cent replacement of PC. However, at high mk doses and low water/binder ratios, the concrete can be difficult to work without superplasticizer (Sabir 1999); Dhir (1999) has showed that mk concrete normally requires a smaller superplasticizer dose than does the equivalent silica fume concrete. With no superplasticizer, it may be necessary to increase the water/binder ratio in order to maintain the required compaction factor, Table 15.3. This is partly due to the fact that mk has a lower density than PC so that replacing, say, 10 mass per cent PC by mk decreases the water/binder volume ratio, and slurry rheology is determined by the liquid/solid volume ratio. Using the example in Table 15.3, extra water is needed to maintain the water/binder volume ratio; because of the lower density of mk, we would need to increase the water/binder mass ratio from 0.67 to 0.68_5 when 10 mass per cent PC is replaced by mk. It is seen

TABLE 15.3 Effect of mk on concrete workability and strength

(PC:mk)	Compaction factor	Water/binder ratio	Density (kgm⁻³)	Compressive strength at 28 days (MPa)
100.0	0.90	0.67	2413	39.5
95.5	0.92	0.69	2396	41.5
90.10	0.91	0.71	2406	45.0
85.15	0.91	0.74	2400	45.0

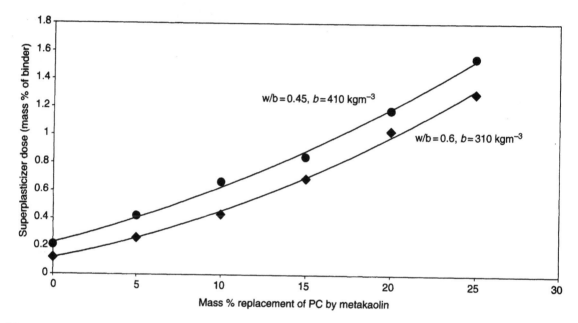

FIGURE 15.6 Superplasticizer dose required for 75 mm slump (2 concrete formulations) (Dhir *et al.* 1999).

that extra water is required on top of this, however, because of the high surface area and irregular particle shape of mk.

The viscosifying effect of mk can also be overcome by using extra superplasticizer (Dhir *et al.* 1999; Wild *et al.* 1996), reducing the sand content, or choosing coarser grades of sand (Ryle 1999). Figure 15.6 shows the superplasticizer dose required to achieve 75 mm slump, for a range of mk replacement levels (Dhir *et al.* 1999).

There have been no reports that mk reacts adversely with common additions such as (super) plasticizers, air entraining agents or waterproofing aids. If waterproofing aids are used they may need to be added after the mk has been fully incorpo-

rated into the mix. This will avoid the possibility of the waterproofing aid adsorbing on the high surface area of the mk, before the latter is properly wetted. Metakaolin is also compatible with fibrous additions such as polypropylene, glass and cellulose. The bond between the fibre and the cement matrix can be strengthened because CH in the interfacial zone is replaced by CSH and CASH phases. There is less likelihood of CH chemically attacking some fibres, such as glass (Glasser 1990) and cellulose.

15.5.2 Setting time

Metakaolin slightly reduces the initial and final setting times of cement paste (Ambroise 1994;

TABLE 15.4 Effect of mk on the initial and final setting times of cement paste

Reference	Paste composition (PC:mk)	Water/ binder ratio	Initial set (min)	Final set (min)
Moseley	100	0.275	145	200
	85:15	0.292	95	180
	90:10	0.316	125	185
Ambroise	100	0.25	115	170
	90:10	0.28	74	140
	80:20	0.34	77	154
	70:30	0.39	84	160
	60:40	0.44	106	220

Moseley 1999). Results are shown in Table 15.4. Both investigators adjusted the water/binder ratio to give constant consistency according to BS-EN 196-3: 1995.

Similar trends have been found in mortar and concrete. Zhang and Malhotra (1995) reported that a 10 per cent mk concrete showed a final setting time of 264 min, compared with 312 min for the control concrete. The water/binder ratio was 0.40.

15.5.3 Heat evolution, and effect of curing at high temperatures

The effect of mk on heat evolution in concrete has been reported by Frias et al. (2000a). They used a semi-adiabatic method, curing mortar in a Dewar flask. The water/cement ratio was 0.5, the sand/cement ratio was 3.0, and mk was used at replacement ratios of 10 and 30 mass per cent. Both mk mortars showed a maximum temperature rise of 38°C, compared with a rise of 33°C for the control. Other semi-adiabatic studies have given similar results (for example Wild 2001). Bajracharya et al. (2000) report true adiabatic temperature rise experiments in which 300 mm cubes of concrete were cured under water which was maintained at the same temperature as the centre of the cube. The concrete contained $325 \, kg \, m^{-3}$ cement, and mk was used at replacement levels of 0, 10, 15 and 20 mass per cent. All samples showed similar maximum temperature rises of 35°C.

We can conclude that, as it cures, mk concrete evolves approximately the same total quantity of heat as normal (i.e. PC) concrete with the same binder content. Therefore, under adiabatic conditions, such as the centre of a massive structure, mk would have no effect on temperature rise for a given binder content. However, there is evidence that the heat of curing may be evolved more rapidly in mk concrete, see Section 3.3. Therefore, in smaller structures, where the rate of heat dissipation significantly influences the observed temperature rise, mk concrete might show an increased temperature rise, by a few degrees Celcius, compared with the equivalent PC concrete. This corresponds to the semi-adiabatic conditions discussed above.

The effect of curing concrete at these relatively high temperatures on the mechanical properties of mk concrete has been investigated. It is of obvious importance in large-scale engineering applications of mk concrete. Asbridge (2000) showed that curing at 60°C has no deleterious effect on pore size distribution or compressive strength. In his experiments described earlier, Bajracharya initially cured the concrete for 72 h under adiabatic conditions, in which the temperature rose to about 55°C, but then continued to cure at 20°C. Compressive strength development, porosity, oxygen permeability and water permeability were measured, and compared with values obtained by normal fog curing at 20°C. Results showed that adiabatic curing was no different from fog curing, and mk was found to have a beneficial effect on all these properties.

15.5.4 Early strength development

Hannant (1994) has developed methods of measuring the tensile strength of concrete prisms as young as 1 h. The prism is supported in a split mould on a frictionless cushion of air, and the stress/strain relationship determined. Results (Hobbs 1996) show that mk significantly increases the rate at which concrete gains tensile strength during the first six hours after mixing, (Figure 15.7). (In these experiments, the binder content was $345 \, kg \, m^{-3}$, the water/binder ratio was 0.67 and

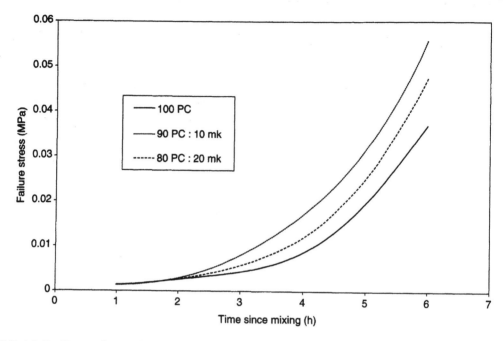

FIGURE 15.7 Very early tensile strengths of concretes – effect of mk (Hobbs 1996).

no superplasticizer was used.) The same author also measured the early development of compressive strength over the period 1–7 days and concluded that mk has no deleterious effect on the early development of compressive strength.

15.5.5 Shrinkage

Fresh concrete can shrink or expand as it cures, depending on its formulation and on environmental conditions. Several mechanisms are thought to be involved as outlined below. Once the concrete loses its plasticity, further shrinkage can cause cracking provided the shrinkage forces are greater than the developing tensile strength.

1. Chemical shrinkage is due to the fact that the volume of hydrates produced by cement reacting with water is less than the volume of reactants.
2. Autogenous shrinkage is the measured reduction in length of a concrete bar during curing, without any migration of water into or out of the specimen. The driving forces are due to

chemical shrinkage as well as surface tension forces due to internal self-desiccation.
3. Expansion can occur if the concrete is cured by immersing in water. Water is sucked into the the concrete to counteract self-desiccation and to hydrate and swell the hydrate phases. This phenomenon is mainly observed with small (i.e. 100 mm) specimens.
4. Drying shrinkage is due to surface tension forces in the near-surface capillaries as water evaporates from them. If evaporated water is not replaced by water bleeding towards the surface from the bulk concrete, shallow surface cracking can occur.

Studies on chemical shrinkage, autogenous shrinkage and swelling in mk/PC pastes have been reported (Wild *et al.* 1998). Results indicate that mk slightly elevates chemical shrinkage, typically from $0.045\,\mathrm{ml\,g^{-1}}$ for the control, to $0.05\,\mathrm{ml\,g^{-1}}$. Results for autogenous shrinkage were much lower at $0.0005\,\mathrm{ml\,g^{-1}}$ and all pastes were similar. However, there was a sudden increase in autogenous shrinkage at higher mk levels, for example

0.0012 ml g^{-1} for the 85PC:15mk blend. The expansion observed when small paste bars are cured in water is very low, typically 0.0005 ml g^{-1}, but again expansion is higher (0.0011 ml g^{-1}) at the 15 per cent mk level. It should be noted that all these shrinkage and expansion values would be considerably reduced when the binder is blended with aggregate, as in concrete.

Drying shrinkage of mk concrete has been measured (Dhir *et al.* 1999). Within the limits of experimental error, mk has no deleterious effect on drying shrinkage.

15.6 Effect of mk on the properties of hardened concrete

15.6.1 Strength

Numerous studies have shown that mk increases the compressive strength of concrete (for example Dhir *et al.* 1999; Ryle 1999; Sabir 1998 Wild *et al.* 1996). The magnitude of the effect is sensitive to water/binder ratio, being most pronounced if this ratio is less than 0.5. Best results are

obtained by adjusting the superplasticizer dose to give suitable rheological properties. Examples of strength development results are shown in Figure 15.8. It is found that mk can replace approximately twice its mass of PC while maintaining the compressive strength of the concrete. Since mk does not increase the compressive strength of PC paste (Larbi 1991), it is concluded that strength improvement in concrete is due to the changes in the structure of the interfacial zone and the increased paste-aggregate bond strength. For the same reasons, mk has a beneficial effect on flexural strength, although the magnitude of the effect is less than that observed for compressive strength.

It has been reported that the addition of certain cationic quaternary ammonium polymers (Sales *et al.* 1999) and anionic polyacrylic acids (Lota *et al.* 2000) can significantly increase the strength of mk concrete. The reasons are not clear, but the authors postulated that polyacrylic acid increases the rate of dissolution of the cement grains; subsequently, the dissolved Ca^{2+}, SiO_4^{4-} and OH^- ions readily combine with the mk, to give cementitious phases with a modified morphology.

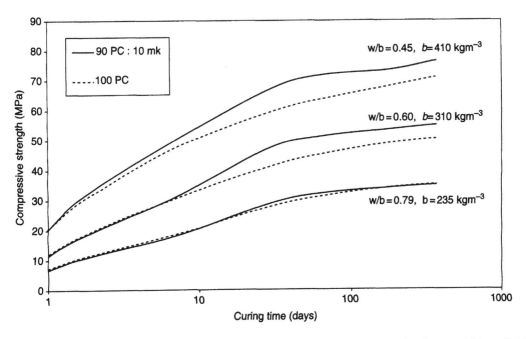

FIGURE 15.8 Compressive strength of concrete – effect of mk, binder content and w/b ratio (Dhir *et al.* 1999).

15.6.2 Creep strain

Creep strain has been measured (Dhir *et al.* 1999) for a range of mk substitution levels, and results (average of two measurements) are shown in Table 15.5. The cylinders were cured in water for 28 d and then loaded to 40 per cent of their 28-day strength for 90 days. In these experiments, there was a slight upward trend of creep strain with mk content, but it should be noted that superplasticizer dose was also varied, in order to maintain constant 75 mm slump.

TABLE 15.5   Creep strain of PC : mk concrete

Binder formulation (PC:mk)	Mass % superplasticizer on binder	28-day strength (MPa)	Creep strain ($\times 10^{-6}$)
100:0	0.13	41.0	790
95:5	0.26	44.5	800
90:10	0.43	47.0	825
85:15	0.69	49.0	840
80:20	1.04	50.5	810
75:25	1.30	50.5	835

15.6.3 Pore water chemistry

Using two samples of commercial mk, Coleman and Page (1997) found that mk reduces the pH of pore water in cured cement paste. However, the reduction is relatively small, even when PC is replaced by 20 mass per cent of mk. These results confirm studies at the BRE in the UK (Halliwell 1993) using high and low alkali cements, containing 1.38 and 0.41 mass per cent equivalent of K_2O respectively, see Table 15.6.

Coleman and Page also showed that mk-PC pastes can bind considerable quantities of dissolved chloride ion present in the pore water due to the formation of stable chloroaluminates such as Friedels salt. A selection of their data is shown in Table 15.7. Because mk 'removes' chloride ions, as well as hydroxide ions from solution, mk maintains a low $[Cl^-]/[OH^-]$ ratio; this is important with regard to rebar corrosion, as discussed in Section 7.2 (Alonso *et al.* 2000; Lambert *et al.* 1991; Page *et al.* 1991). Chloride binding also has the effect of reducing the rate of chloride ingress

TABLE 15.6   Pore water pH in PC:mk pastes

Paste formulation (PC : mk)	Pore water pH after 100 days	Pore water pH after 365 days	Reference
100:0	13.7		Coleman and Page (1997)
80:20	13.2		
100:0 (low alkali)		13.4	Halliwell (1993)
80:20		13.0	
100:0 (high alkali)		13.9	
80:20		13.4	

TABLE 15.7   Concentraton of Cl^- in the pore water of concrete containing added NaCl

Paste formulation w/b 0.5 (PC:mk)	Added Cl^- mass %	$[Cl^-]$ in pore water after 100 days (mol dm^{-1})	$[OH^-]$ in pore water after 100 days (mol dm^{-1})	$[Cl^-]/[OH^-]$
100:0	0.0	0.0	0.50	0.0
	0.1	0.01	0.53	0.02
	1.0	0.29	0.70	0.41
90:10	0.0	0.0	0.27	0.0
	0.1	0.01	0.29	0.03
	1.0	0.19	0.44	0.43

TABLE 15.8   Effect of mk on the diffusion of Cl$^-$ in cement paste and concrete

	Water/binder ratio	Cl$^-$ diffusion coefficient ($m^2 sec^{-1} \times 10^{-12}$)		Method and reference
		PC	90 PC : 10 mk	
Paste	0.4	3.37	1.06	Non-steady state (Coleman 1996)
		1.94	0.55	Steady state (Chadbourn 1997)
	0.5	9.95	1.48	Non-steady state (Coleman 1996)
		5.08	1.31	Steady state (Chadbourn 1997)
	0.6	13.5	2.62	Non-steady state (Coleman 1996)
	0.65	8.85	1.93	Steady state (Chadbourn 1997)
Concrete	0.55	7.73	2.37	Non-steady state (Chadbourn 1997)

in concrete, especially at low water/binder ratios (Jensen *et al.* 1999; Martín-Pérez *et al.* 2000).

15.6.4 Permeability

15.6.4.1 Chloride ion diffusion

Direct measurements of chloride diffusion in cement paste and concrete have been made by steady state and non-steady state methods.[4] Results show that the diffusion rate is reduced by a factor of 3 or more when 10 mass per cent of the PC is replaced by mk (Table 15.8).

Larbi (1991) found that the diffusion coefficients of Cl$^-$, Na$^+$ and K$^+$ were all reduced by factors ranging from 40 to 60 when 20 mass per cent PC in mortar was replaced by mk.

Ponding experiments, in which large prisms of concrete are exposed on their upper surfaces to saturated salt solution, have been reported (Chadbourn 1997; Gilleece *et al.* 1996). Analysis of the chloride penetration profile after 1 year typically shows that mk reduces the concentration of chloride at a depth of 20 mm by factors of 2

and 17, for 10 mass per cent and 20 mass per cent mk, respectively. Gilleece compared the effectiveness of mk (and blends of mk with fly ash and slag) with other pozzolanas and with silane treatment. Over the time scale of these experiments, the best methods of reducing chloride penetration were supplied by silane, 20 mass per cent mk, and 50 mass per cent slag.

Resistance to chloride penetration can be measured more rapidly using electrochemical methods, for example ASTM C1202. Published data shows that substitution of 10 mass per cent PC by mk reduces the Cl$^-$ diffusion coefficient in concrete by a factors of 6 (Zhang and Malhotra 1995), and 3 (Dhir 1999). This is comparable to the effect of the same concentration of silica fume on the diffusion of Cl$^-$ in concrete. Mackechnie and Alexander (2000) have compared three different rapid chloride tests, and 200-day salt ponding experiments, on concretes containing a variety of pozzolanas over a range of water/cement ratios. A selection of their results is given in Table 15.9, and confirm the trends discussed above. It is seen that, at 10 mass per cent substitution, mk and silica fume give similar results. Ground granulated blast furnace slag (gbs) at 50 mass per cent substitution gave slightly inferior results. All samples were cured for 28 days, except those used for ponding which were cured for 42 days.

15.6.4.2 Transport of water

The movement of water through concrete plays an important role in degradation processes. Moving

---

[4]In the steady state method, a diaphragm of concrete less than 5 mm thick is exposed to molar Cl$^-$ on one face and negligible Cl$^-$ on the opposite face (MacDonald and Northwood 1995). Under steady state conditions, the cement paste is fully saturated with bound Cl$^-$.

In the non-steady state metod, one face of a concrete prism is exposed to concentrated NaCl solution while other faces are sealed so that penetration is uni-directional (Bamforth 1996). After the required time, the chloride concentration profile is determined by sectioning the concrete. The concentration profile is very sensitive to the degree of chloride binding (Jensen *et al.* 1999).

TABLE 15.9 Diffusion of Cl⁻ into PC/pozzolana concretes, comparing different measurement methods

Binder	w/b	AASHTO RCPT[a] (coulombs)	Nordic[b] (D × 10⁻¹² m²s⁻¹)	Cl⁻ conductivity[c] (mScm⁻¹)	Ponding (D_C × 10⁻¹² m²s⁻¹)
100 PC	0.6	2036	20.8	1.4	7.1
	0.4	1549	12.9	0.88	2.0
90 PC : 10 sf	0.6	590	7.3	0.57	1.6
	0.4	258	1.4	0.21	0.6
90 PC : 10 mk	0.6	588	6.8	0.56	1.4
	0.4	260	1.9	0.16	0.7
50 PC : 50 gbs	0.6	467	5.3	0.33	1.6
	0.4	321	2.9	0.18	0.8

[a] AASHTO RCPT in accordance with ASTM C 1202.
[b] Nordic chloride migration test.
[c] Concrete chloride conductivity test.

water can carry aggressive agents such as acids and salts, and can also enhance the magnitude of corrosion currents by increasing electrical conductance. McCarter and Watson (1997) have measured the rate of drying, the rate of water absorption, and electrical conductivity profiles as a function of depth of concrete. Variables studied were curing regime, water/binder ratio (0.4 and 0.6), pozzolana replacement level (10.5 and 6.7 mass per cent) and type of pozzolana. Results indicated that there was little difference between the PC concrete, the silica fume concrete and the mk concrete at the higher water/binder ratio. However, at a water/binder ratio of 0.4, the mk concrete showed the lowest rates of water migration. Metakaolin decreased the equilibrium electrical conductivity, at depths of 50 mm or more, by a factor of about three. Larbi (1991) obtained comparable results for the rate of drying of wet mortars, and the rate of absorption of water into dry mortars, using 20 mass per cent replacement by mk, and other pozzolanas.

Initial surface absorption tests (ISAT) have been measured (Dhir *et al.* 1999), using the method described in BS1881:Part 5. Results are given for the water absorption after 10 min, and the N-value, which is the slope of the log (ISAT) versus log (time) plot. Table 15.10 shows a selection of results from this study: the concrete has a cement content of 410 kg m⁻³ and a water/cement

TABLE 15.10 Initial surface absorption tests for mk and silica fume concrete

Binder formulation	28-day ISAT-10 (1 m⁻² sec⁻¹)	28-day N-value
100 PC	365	0.42
90 PC : 10 mk	275	0.46
85 PC : 15 mk	225	0.47
90 PC : 10 sf	200	0.48
85 PC : 15 sf	135	0.50

ratio of 0.45. It is seen that mk reduces the rate of absorption of water, but is slightly less effective in this respect than silica fume. However, it should be noted that silica fume cannot normally be used at substitution levels of 10 or 15 mass per cent.

15.6.4.3 Transport of gases

The rate transport of oxygen or air through concrete is an important factor in re-bar corrosion. The permeability of dry concrete to air, under a pressure differential, has been reported (Basheer *et al.* 1999; Mackechnie and Alexander 1998b). Mackechnie *et al.* showed that mk significantly reduced air permeability, but Basheer found no difference between the mk and control concretes. The diffusion of dissolved oxygen through water-saturated paste has also been reported (Bajracharya 2000; Chadbourn 1997). These results indicate that mk reduces the oxygen diffusion rate, but the effect is sometimes small.

15.7 Durability of mk concrete

15.7.1 General

As discussed above, mk can have several major effects on the physical and chemical properties of the bulk paste and of the interfacial zone. Metakaolin confers a more stable chemical structure to the paste, with increased molecular weight of the silicate chains, smaller average pore size and the ability to bind chloride and alkali metal ions. Portlandite crystal size is reduced, and some or all of the portlandite is replaced by CSH and CASH phases. This has a particularly significant effect on the structure of the interfacial zone, and results in increased bond strength between paste and aggregate. Several published papers discuss the durability of mk concrete (for example Basheer 1999; Collin-Fèvre 1992; Curcio et al. 1998; Ryle 1999; Zhang and Malhotra 1995). The effect that mk has on specific aspects of concrete durability is described below.

15.7.2 Re-bar corrosion

Re-bars in concrete are normally protected against corrosion by a passive layer of iron oxide, a few nanometers thick. The oxide layer can become chemically unstable if chloride ions are present in the pore water or if the pH falls below about 12.5. The $[Cl^-]/[OH^-]$ ratio is known to be an important parameter: dissolution of the oxide layer is rapid if this ratio rises above a value which is in the range 0.5–3, depending on the formulation of the concrete and its exposure conditions (Alonso et al. 2000; Lambert et al. 1991; Page et al. 1991). Corrosion then occurs, provided that sufficient dissolved oxygen is available near the iron surface. The rate of corrosion also depends on the electrical conductivity of the pore solution and on the pore structure (which controls the diffusion of ions and oxygen).

As shown in Sections 6.3 and 6.4, mk reduces the rates of chloride diffusion and water penetration, while maintaining the pH of the pore water above 13. McCarter and Watson (1997) showed that mk also reduces the electrical conductivity of concrete by a factor ranging from 2 to 3. Coleman and Page (1997) showed that, unlike some other pozzolanas, mk does not significantly increase the Cl^-/OH^- ratio of dissolved ions in the pore water. All these observations indicate that mk should improve the corrosion resistance of steel-reinforced concrete. Chadbourn (1997) has directly confirmed this by measuring the corrosion rates (i.e. corrosion currents and electropotentials) of re-bars within concrete blocks subjected to salt ponding and drying cycles. Examples of his results are shown in Figures 15.9 and 15.10. Corrosion commenced at about 200 days for the PC concrete, but had not commenced after 500 days for the 10 mass per cent mk concrete (the experiment was terminated after 500 days).

15.7.3 Carbonation

Carbonation is sometimes a problem in reinforced concrete, because it reduces the pH of the pore water and this can result in rapid re-bar corrosion. By depleting the reservoir of CH in the cement paste, it might be expected that mk would increase the rate of carbonation. Studies at the BRE (Dunster 1996) and the University of Dundee (Dhir et al. 2000) have been carried out using a wide variety of exposure conditions. Results show that the finer pore size and reduced permeability of mk concrete effectively counteract the effect of loss of CH (Table 15.11).

15.7.4 Alkali–silica reaction

The world-wide cost of alkali–silica reaction (ASR) is extremely high taking into account repair costs and charges for importing cement and aggregates where local supplies are potentially active. ASR is caused by alkali (e.g. KOH) in the pore water, which chemically attacks certain poorly-crystalline forms of silica present in the aggregate. Calcium ions then precipitate the dissolved silica species to give an expansive gel. If sufficient gel is formed, the resulting expansive

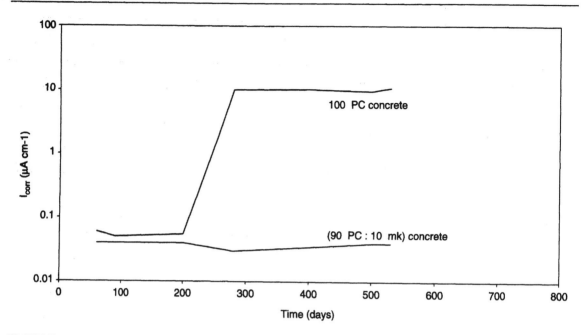

FIGURE 15.9 Ponding experiments: change in corrosion current with time for PC concrete and 10 percent mk concrete. Re-bars positioned 25 mm below surface (Chadbourn 1997).

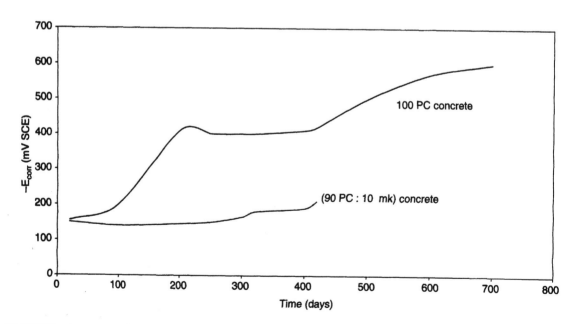

FIGURE 15.10 Ponding experiments: change in corrosion poential with time for PC concrete and 10 percent mk concrete. Re-bars positioned 25 mm below surface (Chadbourn 1997).

TABLE 15.11   Effect of mk on carbonation rate in concrete

Binder formulation	Binder (kg m⁻³)	Water/binder ratio	Mean depth of carbonation[a] after 2 years (mm) for different exposure conditions			Reference
			CEN class 1[b]	CEN class 2[b]	CEN class 3[b]	Dhir et al. (2000)
100 PC	300	0.59	3	2.5	2	
90PC:10mk	315	0.61	3	2	2	
85PC:15mk	285	0.63	7	5	3	
90PC:10csf	280	0.65	6.5	3.5	3.5	
70PC:30pfa	375	0.46	5.5	3.5	2	
60PC:40gbs	325	0.52	4.5	3.5	2	
			Indoor[c] (65% RH)	Outdoor[c] (sheltered)	Outdoor[c] (exposed)	Dunster (1996)
100 PC	305	0.66	9.3	6.8	4.1	
90PC:10mk	305	0.66	11	7.9	2.2	

[a] Mean of forty readings, twenty readings from each of two prisms.
[b] Class 1: 350 ppm CO_2, 20°C, 65 per cent RH; Class 2: as Class 1, but specimens immersed in water for 6 h every 28 days; Class 3: as Class 2, but specimens immersed every 7 days.
[c] Specimens moist-cured for 1 day, then exposed to natural carbonation.

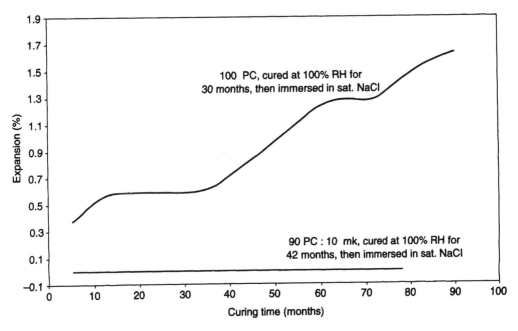

FIGURE 15.11   Expansion of concrete prisms due to ASR – effect of mk (Jones et al. 1992).

forces can crack the concrete. The phenomenon is greatly accelerated if the concrete is bathed in sodium chloride solution, and this often occurs in marine environments and where de-icing salts are used (Figure 15.11). The reason why a large excess of sodium chloride enhances expansion due to ASR is not known. One possibility is that it promotes the release of hydroxide ions from the portlandite. On the other hand geologists recognize that dissolved sodium chloride reduces the surface tension between quartz and water, enhancing the permeation of aqueous fluids through quartzite rocks (Holness 1992). Icenhower and Dove (2000) found that the rate of dissolution of amorphous silica in water at pH7 is increased by a factor of 21 in 0.05 Molar Na^+. Pozzolanas such as mk can prevent ASR by reducing the pH of pore water and removing most of the CH, which otherwise acts as a large reservoir of soluble calcium and hydroxide ions. Interestingly, geo-polymer concrete made from alkali metal hydroxides and silicates, plus mk (Section 3.1) show no expansion due to ASR, even when highly active aggregates are used (Davidovits 1994).

Several laboratory[5] studies (Jones et al. 1992; Ramlochan et al. 2000; Saad et al. 1982; Sibbick and Nixon 2000; Verein Deutscher Zement 1996; Walters et al. 1991) have shown that ASR can be prevented if 10–15 mass per cent of the PC is substituted by mk. With all but the most severely reactive aggregates, 10 mass per cent is sufficient to prevent expansion, even when the concrete is immersed in 2-molar NaCl solution. Figure 5.11 shows the typical effect of mk on ASR-susceptible concrete prisms immersed in 2-molar NaCl at 38°C.

Over the period 1962–79, four dams were constructed in the Amazon basin using highly reactive local aggregates. Approximately 260,000 tonnes of calcined kaolinitic clay was used, blended with PC, for the concrete (Saad et al. 1982). It was reported later (Andriolo and Sgaraboza 1985) that the mk had inhibited the expected ASR, and there were no signs of damage to the dams due to swelling.

15.7.5 Sulphate attack

The presence of CH and calcium aluminates in hydrated PC paste makes it vulnerable to attack

by sulphates. Sulphate ions are present in sea water and in many soils, especially on reclaimed industrial land. On entering the concrete, they form ettringite and gypsum, causing the concrete to swell and ultimately to disintegrate. Magnesium sulphate has a particularly damaging effect because Mg^{2+} ions precipitate $Mg(OH)_2$ which has the effect of reducing pore water pH and dissolving some of the calcium silicate phases. The various chemical reactions thought to be involved in sulphate attack have been discussed (Hime and Mather 1999) and there is an extensive literature (e.g., Marchand et al. 1999).

It has been shown (Khatib and Wild 1998; Osborne and Singh 1997) that mk can reduce or eliminate the deleterious effects of sulphate attack, but the mechanism is not clear. Possibly the reduction in the amount and crystal size of the portlandite in the cured paste is a factor. Also, it has been postulated that the (alumino) silicate phases of cured mk/alkali paste are thermodynamically more stable and resistant to chemical attack (Palomo et al. 1999; Richardson 1999).

Relatively high proportions of mk are required to prevent sulphate attack. For example, Khatib and Wild (1998) showed that 10 mass per cent mk prevents expansion due to sulphate attack when the PC contains 7.8 mass per cent C_3A; however, 20 mass per cent is required when the PC contains 11.7 mass per cent C_3A.

15.7.6 Other aspects relating to durability

A limited number of reports describe the effect of mk on concrete subject to attack by organic and inorganic acids (Collin-Fèvre 1992; Martin 1997) and freeze–thaw cycles (Dhir et al. 1999; Zhang and Malhotra 1995). Metakaolin improves the resistance of concrete to acids. Freeze–thaw resistance is determined by methods described in ASTM C666, Procedure A, and in CEN/TC51, Part 1: scaling, reference method A. All the published data indicates that mk does improve freeze–thaw resistance, but the effect is relatively small. Zhang et al. (1995) also tested de-icing salt scaling, according to ASTM C 672. Their results showed that the mk

---

[5]Various accelerated tests for assessing the potential reactivity of aggregates with respect to ASR have been adopted; typically these involve heating the mortar bar at 80°C in 1M NaOH solution. These tests were *not* originally designed to assess the effectiveness of candidate preventative methods. For example, it is known that mk rapidly reacts at high pH and high temperature to give well-defined crystalline zeolites, and tests showed (Sims 1995) that the apparent effectiveness of mk in suppressing ASR is reduced by a factor of about two in an accelerated test. There is further discussion in Ramlochan (2000).

concrete is marginally inferior to that of the control concrete, but the reason is not known.

Efflorescence sometimes appears as an unsightly whitish stain on the surface of concrete. There are several types of efflorescence caused, for example, by deposits of calcium carbonate or calcium sulphate, and the phenomenon is not well understood. As well as the type of cement, other ingredients and various environmental factors such as wind, changes in temperature and relative humidity, can determine whether efflorescence occurs or not. However, mk is commonly used commercially to reduce or prevent the occurrence of efflorescence on decorative concrete products such as paving slabs, flooring screeds and renders. The use of brightly coloured concrete to make sculptures, monuments and decorative public walled and paved areas is becoming popular. Metakaolin can be used to enhance the appearance of such concrete and prevent efflorescence over long time periods (Vincent 2000). Dow and Glasser (1999) have developed a semi-quantitative model to explain the most common form of efflorescence, caused by calcium carbonate deposition. A thin static film of water on the surface of the concrete is required. This film dissolves alkali from the concrete and the resulting alkaline solution rapidly absorbs carbon dioxide from the atmosphere to give carbonate and bicarbonate ions. These ions diffuse through the liquid layer to the surface of the concrete where they interact with calcium ions, which mostly derive from sparingly soluble portlandite. Metakaolin concrete is less susceptible to efflorescence, probably because it reduces the pH of the pore water, removes some of the portlandite, and the finer pore structure leads to reduced diffusion of water and ions within the concrete. However it should be noted that high levels of mk (e.g. 25 mass per cent substitution of PC) are required to ensure that most of the portlandite is consumed.

Glass fibre reinforced concrete (GRC) is an example of a specialist application where mk is used to improve durability in outdoor environments, Marikunte et al. (1997). Even alkali-resistant glass (AR-glass) fibres are rapidly weakened by the alkaline pore water, and this leads to reduced tensile strength and increased brittleness (i.e. loss of elasticity). Accelerated tests predicted that mk should maintain the strength and elasticity of GRC for at least 25 years if AR-glass fibres are used.

Steel fibres are sometimes used to make very high-strength concrete. Banthia and Yan (1996) have carried out pull-out tests on steel fibres[6] embedded in cement matrices. The effect of replacing 10 mass per cent PC with mk was to increase the maximum pull-out load from 170 to 200 N, and increase the energy absorbed at 9.75 mm slip from 680 to 1180 N mm. The pull-out load should not exceed the tensile strength of the fibre, otherwise the concrete would be undesirably brittle.

Metakaolin can be used with PC in conjunction with other pozzolanas such as silica fume, fly ash and ground granulated blast-furnace slag, ggbs. By varying the relative amounts of two pozzolanas, it may be possible to optimize the cement formulation with respect to cost and performance. Several papers have been published on this large topic (see for example, Asbridge et al. 1994; Bai et al. 2000; Gilleece et al. 1996; Sabir et al. 1999), but further discussion is outside the scope of this review.

15.8 Metakaolin in engineering concrete

Metakaolin in its 'pure' form (i.e. manufactured from kaolin greater than about 90 per cent purity) conforms to the ASTM C 618, Class N definition of a natural pozzolana suitable for use in concrete. The proposed European Standards for cement and concrete will also provide for pure mk. The British Board of Agrément (1998) has approved certain grades of mk for applications in engineering concrete. The production of commercial mk, suitable for use in high quality concrete, is normally controlled according to mineral purity, particle size distribution, pozzolanic reactivity, and if necessary colour. Table 15.12 shows some typical values for a range of commercial products from Europe and from Georgia, USA.

[6]Fibres were 30 mm long and 0.5 mm diameter with hooked ends.

TABLE 15.12 Properties of mk supplied to the concrete market (all measurements made in the laboratories of Imerys Minerals Ltd, UK)

	Mk (France)	Mk 1 (UK)	Mk 2 (UK)	Mk (USA)
Composition (mass %)				
SiO_2	55.0	55.4	51.6	51.3–52.6
Al_2O_3	37.5	40.5	41.0	43–45
Fe_2O_3	1.45	0.65	4.8	0.5–1.0
TiO_2	1.45	0.02	0.83	1.75
CaO	0.07	0.01	0.06	0.01–0.1
MgO	0.20	0.12	0.19	0.2–0.25
K_2O	1.17	2.17	0.62	0.1–0.2
Na_2O	0.2	0.13	0.24	0.2–0.3
Loss on ignition	3.0	1.0	1.0	0.7
Particle size (water sedimentation)(mass %)				
>5 μm		13		2–15
<2 μm		58		60–90
Surface area ($m^2 g^{-1}$)		15		14–22
Density (kg m^{-3})		2600	2600	
Brightness (ISO)		84	(red)	78–86
Yellowness		6		6.5–9.5
Pozzolanic reactivity[a] (mg CaO g^{-1} mk)		800		840

[a] Chapelle test, Kostuch et al. (1993).

As discussed earlier, mk is widely used in concrete for decorative applications such as paving slabs, floor screeds, renders and sculptures. Other trials, where mk concrete has been used for its engineering durability properties, have been reported. Martin (1997) described the development of a Portland/ggbs/mk concrete for acid-resistant silage clamps. After 2 years of use, the surface of the pozzolanic concrete was significantly less corroded than the control concrete. Mk concrete has been used to repair bridges (May 1999) and to construct foundations, a marine slipway, and a river diversion scheme (Asbridge et al. 1996). Large pre-stressed bridge beams are manufactured from mk concrete, and certain advantages of using mk concrete in this application have been described (Anon 2000). The beams are made from 75 MPa concrete and can span longer distances and can be used at wider spacing than conventional concrete beams. When used for motorway or rail bridges, they can be used to increase headroom clearances, and overall durability is improved.

Trade journals that serve the construction industry also report that mk concrete has been used to make roads and bridge decks (Anon 1997; Balogh 1995). Further examples of the use of mk/PC concrete and mk/alkali silicate polymers (geo-polymers) are described on various Internet web sites, and the patent literature is extensive.

Compared with other pozzolanic additions, such as silica fume, slag and fly ash, mk does not yet have a long or extensive track record. Published reports indicate that mk shows all the well-known advantages ascribed to other pozzolanic additions. To the authors knowledge, no serious disadvantages have been highlighted. The future of mk concrete would seem to depend on the balance between desirable properties (improved durability, strength etc.) and other economic factors such as the cost of raw material, plus logistical problems associated with changing the formulation and handling properties of mass concrete. The attitudes of industrial economies towards the whole-life costs of infrastructure projects is constantly evolving. The future of mk concrete promises to be an interesting subject of study over the next few decades; it will be determined by the investment decisions of the producers of mk and concrete, by concrete customers and, of course, government and regulatory agencies.

15.9 Acknowledgements

The author thanks the Directors of Imerys Minerals Ltd for permission to publish this review and to quote data obtained in the company's laboratories. Thanks are also due to many colleagues within Imerys, especially Tony Asbridge, Darren Holford and Jacek Kostuch for their help.

15.10 References

Alonso, C., Andrade, C., Castellote. M. and Castro, P. (2000). 'Chloride threshold values to depassivate reinforcing bars embedded in a standardised OPC mortar', *Cem. Concr. Res.*, 30, 1047–55.

Ambroise, J., Murat, M. and Pera, J. (1985). 'Hydration reaction and hardening of calcined clays and related minerals: V. Extension of the research and general conclusions', *Cem. Concr. Res.*, 15, 261–8.

Ambroise, J., Murat, M. and Pera, J. (1986). 'Investigations on synthetic binders obtained by middle-temperature thermal dissociation of clay minerals', *Silicates Industries*, 7–8, 99–107.

Ambroise, J., Martin-Calle, S. and Pera, J., (1992) 'Pozzolanic behaviour of thermally activated kaolin', *Proc. of 4th Int Conf. on Fly Ash, SF, Slag and Natural Pozzolans in Concrete, Turkey*, V. M. Malhotra, (ed.), 1, 731–41.

Ambroise, J., Maximilien J. and Pera, J. (1994). 'Properties of MK blended cements', *Adv. Cem. Bas. Mat.*, 1, 161–8.

Andriolo, F. R. and Sgaraboza, B. C. (1985). 'Use of pozzolan from calcined clay in preventing ASR in Brazil, *Proc. 7th Intl. Conf. on ASR*, 66–70, Noyes Publications, New Jersey.

Anon., (1997). 'N.Y. producers promote performance mixes', *Concr. Products*, October, 29.

Anon., (2000). 'Rising to the challenge for spans, depth and load bearing capacity', *Precast Update, Newsletter of the Irish Precast Concrete Association*, Autumn 2000.

Asbridge, A. H., Jones, T. R. and Osborne, G. J. (1996). 'High performance MK concrete: results of large scale trials in aggressive environments', *Proc. Conf. Concr. in the Service of Mankind, Dundee, Radical Concr. Technol.*, 13–24, E & F N Spon.

Asbridge, A. H., Walters, G. V. and Jones, T. R. (1994). 'Ternary blended concretes – OPC/ggbfs/Metakaolin', *Proc. Conf. 'Concr. across Borders'*, Odense, 547–57.

Asbridge, A. H. (2000). Private communication.

Ashall, G., Butlin, R. N., Teutonico. J. M. and Martin, W. (1996). 'Development of lime mortar formulations for use in historic buildings', *Durability of Building Materials and Components*, 7(1), C. Sjostrom (ed.), E & F N Spon.

Bai, J., Sabir, B. B., Wild, S. and Kinuthia, J. M. (2000). 'Strength Development in concrete incorporating PFA and metakaolin', *Mag. Concr. Res.*, 52(3), 153–62.

Bajracharya, Y. M., Page, C. L., Richardson, I. G. and Hassan, K. E. (2000). The engineering properties of metakaolin-OPC concrete under semiadiabatic curing, *Poster*, University of Leeds, June.

Balogh, A. (1995). High reactivity metakaolin, *Concr. Construction*, July, 604–09.

Bamforth, P. B. (1996). Definition of exposure classes and concrete mix requirements for chloride contaminated environments, *4th Int. Symp. on Corrosion of reinforcement in Concr. Construction*, Cambridge, UK.

Banthia, N. and Yan, C. (1996). Bond-slip characteristics of steel fibers in high reactivity metakaolin modified cement-based matrices, *Cem. Concr. Res.*, 26(5) 657–62.

Basheer, P. A. M., McCabe, C. C. and Long, A. E. (1999). 'The influence of metakaolin on the properties of fresh and hardened concrete', *Proc. Int. Conf. 'Infrastructure Regeneration'*, Sheffield, R. N. Swami, (ed.), Sheffield Academic Press, 199–211.

Bauweraerts, P., Wastiels, J., Faignet, S. and Wu, X. (1994). 'Rapid setting inorganic repair systems', *Proc. 5th Int CANMET/ACI Conf. on Fly Ash in Concr. (Suppl Papers)* 233–47.

Bonen, D., Tasdemir, M. A. and Sarkar, S. L. (1995). 'The evolution of cementitious materials through history', *Mat. Res. Soc, Symp. Proc.*, 370, 159–68.

Bredy, P., Chabannet, M. and Pera, J. (1989). 'Microstructure and porosity of MK blended cements', *Proc. Mat. Res. Soc. Symp.*, 136, 431–36.

British Board of Agrément. (1998). *Agrément Certificate* No. 98/3540.

Cabrera, J., Nwaubani, S. O. (1993). 'Strengths and chloride permeability of concrete containing tropical red soils', *Mag. Concr. Res.*, 45(164), 169–78.

Cabrera, J., Nwaubani, S. O. (1998). 'Microstructure and CI diffusion in cements containing metakaolin and PFA', *Proc. 6th CANMET/ACI Int. Conf. on Fly Ash, Silica Fume, Slag, and Natural Pozzolans in Concr.*, 385–400.

Chadbourn, G. A. (1997). *Chloride resistance and durability of cement paste and concrete containing metakaolin*, Ph. D. Thesis, University of Aston.

Clark, S. M., Morrison, G. R. and Shi, W. D. (1999). 'The use of scanning transmission X-ray microscopy for real-time study of cement hydration', *Cem. Concr. Res.*, 29, 1099–102.

Coleman, N. J. (1996). *Metakaolin as a cement extender*, Ph. D. Thesis, University of Aston.

Coleman, N. J. and Page, C. L. (1997). 'Aspects of the pore solution chemistry of hydrated cement pastes containing metakaolin', *Cem. Concr. Res.*, 27(1), 147–54.

Collin-Fevre, I. (1992). 'Use of metakaolin in the manufacture of concrete products', *Poster No. 479, Montreal Conference*.

Cook, R. A., Hover, K. C. (1993). 'Mercury porosimetry of cement-based materials and associated correction factors', *ACI. Mater. J.*, 90(2), 152–61.

Curcio, F., DeAngelis, B. A. and Pagliolico, S. (1998). 'Metakaolin as a pozzolanic microfiller for HP mortars', *Cem. Concr. Res.*, 28(6), 803–09.

Davey, N. (1953). *A History of Building Materials*, Phoenix House, London.

Davidovits, J. (1994). 'High alkali cements for 21st century concretes', *Proc. Symp. 'Concr. technol., past, present and future'*, P. K. Mehta (ed.), ACI, Vol V, 383–97.

Dhir, *et al.* (1999). Report: *Use of the Unfamiliar Cement to ENV 197-1 in Concrete*, DETR Research Contract 39/3/238.

Dhir, *et al.* (2000). Report: *Development of a Performance Specification for Carbonation Resistance*, DETR Research Contract 39/3/384 (submitted for publication in Mater Struct).

Dow, C. and Glasser, F. P. (1999). *Prelim. Abs., 6th Conf. Eur. Ceram. Soc.*, Brighton, 24.

Dunham, A. C. (1992). 'Developments in industrial mineralogy: I. The mineralogy of brick-making', *Proc. Yorkshire Geol. Soc.*, 49(2), 95–104.

Dunster, A. M., Parsonage, J. R. and Thomas, M. J. K. (1993). The pozzolanic reaction of MK and its effects on PC hydration, *J. Mat. Sci.*, 28, 1345.

Dunster, A. M., (1996). BRE Client Report N162/96.

Encyclopaedia Britannica. (1876). 9th (edn), Vol 4, 458–9.

Frias, M., Sanchez de Rojas, M. I. and Cabrera, J. (2000a). 'The effect that the pozzolanic reaction of metakaolin has on the heat evolution in metakaolin-cement mortars', *Cem. Concr. Res.*, 30, 209–16.

Frias, M., Cabrera, J. (2000b). 'Pore size distribution and degree of hydration of metakaolin-cement pastes', *Cem. Concr. Res.*, 30, 561–69.

Gilleece, P. R. V., Basheer, P. A. M., Long, A. E. and O'Sullivan, J. (1996). 'Time dependant changes of high performance concretes in chloride exposure tests', *Proc. 3rd CANMET/ACI Int. Conf. on Concr. in Marine Environment*, Supplementary Vol. 1–22.

Girodet, C., Chabannet, M., Bosc, J. L., Pera, J. (1997). 'Influence of the type of cement on the freeze-thaw resistance of the mortar phase of concrete', *Proc. Int. RILEM Workshop on the Frost Resistance of Concrete*, M. J. Setzer, and R. Auberg, (eds), E & FN Spon, 31–40.

Glasser, F. P. (1990). 'Cements from micro to macrostructures', *Br. Ceram. Trans. J.*, 89(6), 195–202.

Goto, S., and Roy, D. M. (1981). 'Diffusion of ions through hardened cement paste, *Cem. Concr. Res.*, 11, 751–57.

Gruber, K. A. and Sarkar, S. L. (1996). 'Exploring the pozzolanic activity of high reactivity metakaolin, *World Cem.*, 27(2), 78–80.

Halliwell, M. A. (1993). BRE Client Report TCR 24/93.

Hannant, D. J. (1994). 'Fibres in industrial ground floor slabs', *Concrete*, Jan/Feb, 16–19.

He, C., Makovicky, E. and Osbaeck, B. (1994). 'Thermal stability and pozzolanic activity of calcined kaolin, *Appl. Clay. Sci.*, 9(3), 165–87.

Hime, W. G. and Mather, B. (1999). '"Sulfate attack" or is it?', *Cem. Concr. Res.*, 29, 789–91.

Hobbs, M. (1966). 'Effect of metakaolin on the fresh and hardened properties of concrete', *MEng Project Report*, University of Surrey.

Holness, M. (1992). 'The irresistible fluid and the immovable rock', *New Scientist*, No. 1813 (21 March), 41–5.

Icenhower, J. P. and Dove, P. M. (2000). 'The dissolution kinetics of amorphous silica into sodium chloride solution: effects of temperature and ionic strength', *Geochim. et Cosmochim. Acta.*, 64(24), 4193–203.

Jensen, O. M., Hansen, P. F., Coats, A. M. and Glasser, F. P. (1999). 'Chloride ingress in cement paste and mortar', *Cem. Concr. Res.*, 29, 1497–504.

Jones, T. R., Walters, G. V., Kostuch, J. A. (1992). 'Role of MK in suppressing ASR in concrete exposed to NaCl solution', *Proc. 9th Int. Conf. on AAR in Concr.*, Vol 1, 485–96.

Justnes, H., Meland, I., Bjoergum, J. O., Krane, J. and Skjetne, T. (1990). 'NMR – a powerful tool in cement and concrete research', *Adv. in Cem. Res.*, 3(11), 105–10.

Khatib, J. M. and Wild, S. (1993). 'Sulphate resistance of metakaolin mortar', *Cem. Concr. Res.*, 28(1), 83–92.

Khatib, J. M. and Wild, S. (1996). 'Pore size distribution in MK paste', *Cem. Concr. Res.*, 26(10), 1545–53.

Klimesh, R. and Ray, A. (1998a). 'Autoclaved cement-quartz pastes with metakaolin additions', *Adv. Cem. Based Mat.*, 7, 109–18.

Klimesh, R., Lee, G., Ray, A. and Wilson, M. A. (1998b) 'Metakaolin additions in autoclaved cement-quartz pastes – a ^{29}Si and ^{27}Al MAS NMR investigation', *Adv. Cem. Bas. Mat.*, 10(3), 93–9.

Kostuch, J. A., Walters, G. V. and Jones, T. R. (1993). 'High performance concretes incorporating metakaolin – a review', in *Concrete 2000, Proc. Int. Conf., Dundee*, Vol. 2, R. K. Dhir and M. R. Jones (eds), E & F N Spon, 1799–1811.

Lambert, P. *et al.* (1991). 'Investigations of reinforcement corrosion. 2. Electrochemical monitoring of steel in chloride- contaminated concrete.' *Materials and Structures*, 24, 351–8.

Larbi, J. A. (1991). *'The Cement Paste-aggregate Interfacial zone in Concrete'*, Ph. D. Thesis, University of Delft.

Larbi, J. A. and Bijen, J. M. (1992). 'Effect of mineral admixtures on the cement paste-aggregate interface', *Proc 4th Int. Conf. on Fly Ash, Silica Fume, Slag and Natural Pozzolans in Concr.*, Istanbul, 655–69.

Largent, R. (1978). 'Bull Liasons Lab Pont Chausees', 93, 63.

Lota, J. S., Kendall, K. and Benstead, J. (2000). 'Mechanism for the modification of Portland cement hydration using polyacrylic acid', *Adv. Cem. Res.*, 12(2), 45–54.

McCarter, W. J. and Tran, D. (1996). 'Monitoring pozzolanic activity by direct activation with calcium hydroxide', *Const. Building Mats*, 10(3), 179–84.

McCarter, W. J. and Watson, D. (1997). 'Wetting and drying of cover-zone concrete', *Proc. Instn. Civ. Engrs. Structs. & Bldgs.*, 112, 227–36.

MacDonald, K. A. and Northwood, D. O. (1995). 'Experimental measurement of chloride ion diffusion rates using a 2-compartment diffusion cell: effects of material and test variables', *Cem. Concr. Res.*, 25(7), 1407–16.

Mackechnie, J. R. and Alexander, M. G. (1998). 'HPC trials with South African materials', *Proc. Int. Symp. on High Perf. and Reactive Powder Concr.*, Sherbrooke, Vol. 2, 157–67.

Mackechnie, J. R. and Alexander, M. G. (2000). 'Rapid chloride test comparisons', *Concr. Intl.*, 5 (May), 40–5.

Malisch, W. R. (1998). 'Cement containing calcined shale and clay reduces concrete's chloride permeability', *The Concr. Producer*, October.

Malquori, G. (1960). 'Portland-pozzolan cement', *Proc. 4th Int. Symp. on the Chem. Cem.*, Washington, 983–99.

Marchand, J. and Skalny, J. P. (eds) (1999). *'Materials science of concrete: sulfate attack mechanisms* (proc. seminar, Quebec 1998)', Am. Ceram. Soc.

Marikunte, S., Aldea, C. and Shah, S. P. (1997). 'Durability of glass fibre reinforced cement composites – effect of silica fume and metakaolin', *Adv. Cem. Bas. Mat.*, 5, 100–08.

Martin, S. J. (1997). 'Metakaolin and its contribution to the acid resistance of concrete', *Proc. Intl. Symp. 'Concrete for a Sustainable Agriculture' Stavanger*, May.

Martin-Perez, B., Zibara, H., Hooton, R. D. and Thomas, M. D. A. (2000). 'A study of the effect of

chloride binding on service life predictions', *Cem. Concr. Res.*, 30 1215–23.

Maso, (ed.), (1996). 'Interfacial transition zone in concrete', *Rilem Report II*, E & F N Spon.

May, M. J. (1999). 'Metakaolin concrete for bridge repair', *Concrete*, Feb., 22–3.

Meinhold, R. H., Salvador, S., Davies, T. W. and Slade, R. C. T. (1994). 'A comparison of the kinetics of flash calcination of kaolinite in different calciners', *Trans. I. Chem. E.*, 72(A), 105–13.

Min-Chul, C. (1996). 'Studies on the properties of high performance and high strength cement mortar using metakaolin and silica fume', *J. Korean. Ceram. Soc.*, 33(5), 519–23.

Morsy, M. S. *et al.* (1997). 'Microstructure and hydration of cement containing burnt kaolinite clay', *Cem. Concr. Res.*, 27(9), 1307.

Moseley. (1999). Unpublished data, Imerys Minerals Ltd.

Murat, M., (1983a) 'Hydration reaction and hardening of calcined clays: 1-Preliminary investigations on metakaolin', *Cem. Concr. Res.*, 13, 259–66.

Murat, M. (1983b). 'Hydration reaction and hardening of calcined clays: 2-Influence of mineral properties on the reactivity of metakaolinite', *Cem. Concr. Res.*, 13, 511–18.

Murat, M. and Comel, C. (1983c). 'Hydration reaction and hardening of calcined clays: 3-Influence of calcination process on mechanical strengths of hardened metakaolinite', *Cem. Concr. Res.*, 13, 631–37.

Nyame, B. K. and Illston, J. M. (1981), 'Relationship between permeability and pore structure of hardened cement paste', *Mag. Concr. Res.*, 33(116), 139–46.

Van Olphen, H. (1977). *An Introduction to clay colloid chemistry*, John Wiley & Sons.

Oriol, M. and Pera, J. (1995). Pozzolanic activity of metakaolin under microwave treatment, *Cem. Concr. Res.*, 25(2), 265–70.

Osborne, G. J. and Singh, B. (1997). *BRE Client Report: CR 46/97*.

Page, C. L. *et al.* (1991). 'Investigations of reinforcement corrosion. 1. The pore electrolyte phase in chloride-contaminated concrete.' *Materials and Structures*, 24, 243–52.

Palomo, A., Blanco-Varela, M. T., Granozo, M. L., Puertas, F., Vazquez, T. and Grutzeck, M. W. (1999). 'Chemical stability of cementitious materials based on metakaolin', *Cem. Concr. Res.*, 29, 997–1004.

Pera, J., Ambroise, J. and Oriol, M. (1997). 'Microwave processing of GR composites – modification of the microstructure', *Adv. Cem. Bas. Mat.* 6, 116.

Pera, J. and Amrouz, A. (1998). 'Development of highly reactive metakaolin from paper sludge', *Adv. Cem. Bas. Mat.*, 7, 49–56.

Pietersen, H. S. and Bijen, J. M. (1992a). 'The Hydration Chemistry of some Blended Cements', *Proc 9th Int. Congr. Cem. Chem.*, 281.

Pietersen, H. S., Kentgens, A. P. M., Nachtegaal, G. H., Veeman, W. S. and Bijen, J. M. (1992b). 'The reaction mechanism of blended cements – a ^{29}Si NMR study', *Proc. Int. Conf. on Fly Ash, Silica Fume, Slag and Natural Pozzolans in Concr.*, Ankara, 795–812.

Rahman, M. M., Nagasaki, S. and Tanaka, S. (1999). 'A model for dissolution of $CaO-SiO_2-H_2O$ gel at Ca/Si > 1', *Cem. Concr. Res.*, 29, 1091–97.

Ramlochan, T., Thomas, M. and Gruber, K. A. (2000). 'The effect of metakaolin on alkali-silica reaction in concrete', *Cem. Concr. Res.*, 30, 339–44.

Roberts, K. J., Robinson, J., Davies, T. W. and Hooper, R. M. (1992). 'Using soft X-ray absorption spectroscopy to examine the structural changes taking place around Si and Al atoms in kaolinite following flash calcination', *Proc. 7th Int. Conf. on X-ray Absorption Fine Structure*, Kobe, August.

Rocha, J. and Klinowski, J. (1990). '^{29}Si and ^{27}Al magic-angle-spinning NMR studies of the thermal transformation of kaolinite', *Phys. Chem. Minerals*, 17, 179–86.

Richardson, I. G. (1999). 'The nature of CSH in hardened cements', *Cem. Concr. Res.*, 29, 1131–47.

Ryle, R. (1999). 'Metakaolin, a highly reactive pozzolan for concrete', *Quarry Management*, Dec, 27–31.

Saad, M. N. A., de Andrade, W. P. and Paulon, V. A. (1982). 'Properties of mass concrete containing an active pozzolan made from clay', *Concr. Intl.*, July, 59–65.

Sabir, B. B. (1998). 'The effects of curing temperature and water/binder ratio on the strength of metakaolin concrete', *6th CANMET/ACI Int. Conf. on Fly Ash, Silica Fume, Slag and Natural Pozzolans in Concr.*, Bangkok, V. M. Malhotra (ed.), 1057–74.

Sabir, B. B., Wild, S. and Khatib, J. (1999). 'Strength development and workability of concrete blended with metakaolin and fly ash', *2nd Int. Conf. on HighPerformance Concr., and Performance and Quality of Concr. Structures*, Brazil, Supplementary Papers, CD ROM, SP144.doc, pp. 1–13.

Sales, G. W., Kendall, K. and Lota, J. S. (1999). 'Strength improvements in cement-based boards', *Cem. Concr. Res.*, 29, 1693–95.

Salvador, S. (1995). 'Pozzolanic properties of flash-calcined kaolinite: a comparative study with soak-calcined products', *Cem. Concr. Res.*, 25(1), 102–12.

Shi, C., Grattan-Bellew, P. E. and Stegemann, J. A. (1999). 'Conversion of a waste mud into a pozzolanic material', *Const. and Build Mat.*, 13, 279–84.

Sibbick, R. G. and Nixon, P. J. (2000). 'Investigation into the effects of MK as a cement replacement material in ASR reactive concrete', *Proc. 11th Int. Conf. on AAR in Concr.*, Quebec, 763–72.

de Silva, P. S. and Glasser, F. P. (1990) Hydration of cements based on metakaolin: thermochemistry, *Adv. Cem. Res.* 3(12), 167.

de Silva, P. S. and Glasser, F. P. (1992). 'Pozzolanic activation of metakaolin', *Adv. in Cem. Res.*, 4(16), 167–77.

de Silva, P. S. and Glasser, F. P. (1993). 'Phase relations in the system $CaO-Al_2O_3-SiO_2-H_2O$ relevant to $MK-Ca(OH)_2$ hydration', *Cem. Concr. Res.*, 23, 627–39.

Sims, I. (1995). Sandberg private client report.

Sohm, J.-M. (1988). 'Glass reinforced compositions', *U. S. Patent 4, 793, 861*.

Tuthill, L. H. (1982). 'Alkali-silica reaction – 40 years later', *Concr. Intl.* (April), 32–6.

Verein Deutscher Zementwerke. (1996). *Avoidance of ASR*, Activity report 1993–96, 76.

Vincent, C. A. Private communication.

Walters, G. V. and Jones, T. R. (1991). 'Effect of metakaolin on ASR in concrete manufactured with reactive aggregates', *Proc. 2nd Int. Conf. on Durability of Concr.*, Vol 2, V. M. Malhotra (ed.), Am. Concr. Inst., 941–47.

Waters, B. R. (2000). Unpublished results, Imerys Minerals Ltd.

Wild, S., Khatib, J. M. and Jones, A. (1996). 'Relative strengths, pozzolanic activity and hydration in superplasticised metakaolin concrete', *Cem. Concr. Res.*, 26(10), 1537–44.

Wild, S., Khatib, J. M. and Roose, L. J. (1998). 'Chemical shrinkage and autogenous shrinkage of Portland cement-metakaolin pastes', *Adv. Cem. Res.*, 10(3), 109–19.

Wild, S. (2001). Paper accepted for publication by *Cem. Concr. Composites*.

Zhang, M. H. and Malhotra, V. M. (1995). 'Characterisation of a thermally activated aluminosilicate and its use in concrete', *Proc. 2nd CANMET/AC1 Int. Sym. on Adv. in Concr. Tech.*, Las Vegas, June; and in: *Cem. Concr. Res.*, 25(8), 1713.

Zheng, L. and Winslow, D. (1995). 'Sub distribution of pore size: a new approach to correlate pore structure with permeability', *Cem. Concr. Res.*, 25(4), 769–78.

Chapter sixteen

Condensed silica fume as a cement extender

H. Justnes

16.1 Introduction

The object of the present chapter is not to give a complete review of the application of condensed silica fume (CSF) in concrete, but to explain its physical (i.e. filler) and chemical (i.e. pozzolanic) effect on the microstructure of Portland cement that in turn leads to the macro-properties utilized in concrete technology. Important reviews on condensed silica fume and concrete technology are, for instance, Fidjestø and Lewis (1998), Helland *et al.* (1988), Malhotra *et al.* (1992) and Sellevold *et al.* (1994).

CSF is, as the name indicates, the condensation of oxidized silicon monoxide coming from the gases of silicon or ferro-silicon smelter ovens;

$$2SiO(g) + O_2 = 2SiO_2(s)$$

in addition to small dust particles. Thus, the condensation and dust trapped in the filter bags are of very high surface, typically $20,000\,m^2/kg$ as measured by N_2-adsorption with a spread from 15,000 to $25,000\,m^2/kg$. The individual particles are basically spherical with an average diameter of $0.15\,\mu m$, although usually consisting of primary aggregates of fused particles. As comparison, the irregular grains of a Portland cement have typical average diameter of $10\,\mu m$. The chemical composition of CSF is rather simple as it usually consists of in excess of 90 per cent SiO_2 with a spread from 85 to 99 per cent. The major second oxide is Fe_2O_3, while the other oxides are normally below 1 per cent each.

Note that CSF in practice, and in many studies is used as a cement replacement and not added to a given mix, and it is important to differentiate between the two possibilities when evaluating the effect of CSF on properties of paste, mortar or concrete.

16.2 Physical effects

16.2.1 General

Due to it small particle size (average $0.15\,\mu m$), CSF will pack in the cavities formed by the irregular shaped cement grains (average particle size $10\,\mu m$) until a certain dosage when the CSF will disperse cement grains or its agglomerates in the fresh state. The result is a considerable void size refinement from the start. The spherical nature of the CSF particles may also lead to a ball bearing effect improving rheology. Most of all, the addition of CSF to cement mixes vastly increases the solid to liquid interface leading to increased cohesion. For instance, in a typical mix with $350\,kg$ cement of Blaine 320 and 5 per cent CSF per m^3, the cement surface is $350\,kg\cdot320\,m^2/kg = 112,000\,m^2$ while the CSF surface is $0.05\cdot350\,kg\cdot20,000\,m^2/kg = 350,000\,m^2$ or about three times that of cement.

16.2.2 Influence on rheology

The rheological properties of fresh concrete are considerably changed even for moderate additions (<10 per cent) of CSF by cement weight. The cohesion increases, the mass becomes more stable and the tendency to water separation is greatly reduced. The increased cohesion leads to more water needed to maintain the workability of the mix, but this is, in practice, counteracted by increased use of dispersants like lignosulphonate. In this way, the positive effect of CSF can be utilized in normal concrete. For high-performance concrete (HPC) that is, with low water/cement ratio (w/c), CSF is directly improving the workability, and is thereby of vital importance for high-performance practice. CSF also facilitates the production of concrete with extremely low w/c. It is very difficult to achieve lower w/c than 0.20–0.25 by normal cement, even with good dispersion, due to particle geometry and size distribution. However, addition of CSF will replace the water in cavities formed by cement grains and render the mix workable.

The effect of CSF on concrete rheology is discussed by Wallevik (1998) based on results from the two-point (Tattersall) tester. As shown in Figure 16.1(a), CSF as cement replacement will up to a threshold value reduce the plastic viscosity (μ) by up to 50 per cent. This threshold value depends on the cement content of concrete. The yield value (τ_0) is nearly constant until the threshold value of CSF is reached, while further substitutions of cement by CSF will increase the yield value (τ_0). Additions of CSF (Figure. 16.1(b)) will cause the yield value (τ_0) to increase dramatically. The plastic viscosity (μ) decreases for CSF additions up to 5 per cent and then increases for further additions.

16.2.3 Influence on permeability during setting

Another field utilizing the improvement of particle packing by CSF is oil-well cementing. CSF is used to reduce the risk of gas intrusion when the cement sets. When the oil-well cement slurry is pumped in place between the casing and the formation, the differential gas pressure is overcome by the hydrostatic pressure governed by the density of the slurry according to $P = \rho\cdot g\cdot h$. At the critical time of setting, the density is reduced from the one of the fresh slurries to the one of the percolating water phases. Then it is of vital importance that the permeability of the slurry is low to avoid gas intrusion and channel formation.

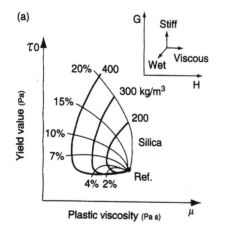 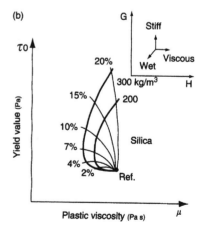

FIGURE 16.1 Effect of (a) cement replacement by CSF and (b) CSF addition on the rheology of concrete expressed by plastic viscosity (μ) and yield value τ_0).

Figure 16.2 shows the pore size distribution of oil-well cement slurries cured at 150°C and quenched in liquid nitrogen at 140 min (i.e. immediately after setting). The composition of the basic slurry per 100 kg API Class G cement was 7.5 kg CSF against gas intrusion, 35 kg quartz flour against strength retrogression, 38 kg fine manganese oxide (Mn_3O_4) as weight material, 45.7 kg fresh water and organic admixtures (retarder, defoamer, fluid loss controller and dispersant).

The pore size distributions in Figure 16.2 show that the introduction of 7.5 per cent CSF reduces the average pore opening size from 250 nm (slurry 4) to 40 nm (slurry 3).

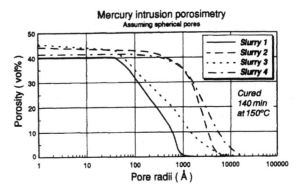

FIGURE 16.2 Pore size distribution of oil-well cement slurries immediately after set at 150°C: (1) full composition; (2) lacking CSF; (3) lacking weight material; (4) lacking both CSF and weight material.

16.3 Chemical effects

16.3.1 Pozzolanic reaction

Using the cement chemist's shorthand notation (notation C = CaO, H = H_2O, K = K_2O, N = Na_2O and S = SiO_2), the non-balanced pozzolanic reaction can be written:

$$S + CH = CSH \tag{1}$$

The presence of alkalis seems to be catalytically necessary for a rapid pozzolanic reaction of silica fume (aq = aqueous, s = solid);

$$S(s) + (N,K)H(aq) \rightarrow (N,K)SH\ (aq)$$
$$\uparrow \qquad + \tag{2}$$
$$CSH(s) + (N,K)H(aq) \leftarrow CH(aq\ or\ s)$$

since Sellevold *et al.* (1981) showed that a mix of silica fume and pure lime needed months to harden. The acceleration of the pozzolanic reaction by alkalis creating high pH in the pore water was confirmed by Justnes (1995) following the strength development for mortars with lime/silica fume binders.

The nature of the CSH-gel from the pozzolanic reaction is different due to its different origin. As sketched in Figure 16.3, two different gels are interwoven, coming from the cement hydration and the pozzolanic reaction respectively.

The calcium silicate hydrate amorphous gel can adapt to a wide composition. The general differ-

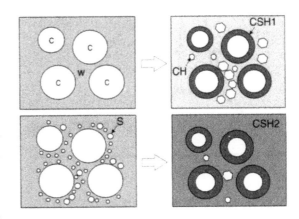

FIGURE 16.3 A sketch showing the formation of two different CSH-gels by the cement hydration (CSH1) and pozzolanic reaction (CSH2), respectively.

ences between CSH2 and CSH1 in Figure 16.3 are that CSH2 has longer linear polysilicate anions and lower C/S-ratio than CSH1. Figure 16.4 shows the ^{29}Si Magic Angle Spinning (MAS) Nuclear Magnetic Resonance (NMR) spectra of alite and a lime/SF mix after 84 days' curing (Justnes 1992b), both with water-to-solid ratio 0.70 and water of pH = 13. The two major peaks from the left is the end group and the middle group of the linear polysilicate anion, resulting in

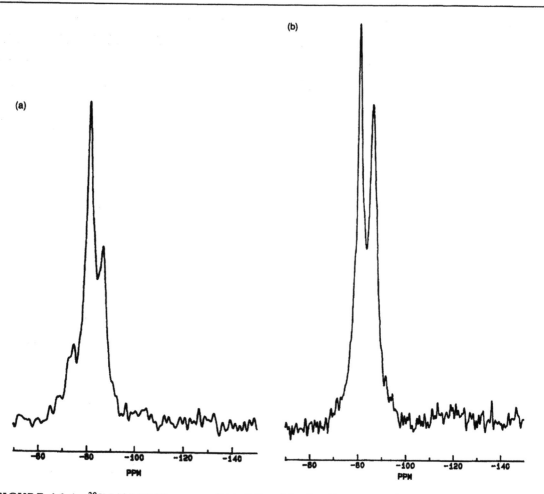

FIGURE 16.4 ^{29}Si MAS NMR spectra of (a) C$_3$S and (b) lime/CSF mix with C/S = 1.11, both with water-to-solid = 0.70 after 84 days' sealed curing (Justnes 1992b).

an average polysilicate length of 2.84 for the CSH1 of alite and of 4.09 of the CSH2 from the lime/CSF mix. The C/S-ratios of CSH1 was 1.75, while the CSH2 was 1.11 (as in the original mix).

In Figure 16.5 (from Justnes, 1998), the relative compressive strength of mortar with lime/CSF binder (corresponding to C/S = 1.11, water-to-solid = 0.70 and alkalis of K/Na = 2 to pH = 13) is plotted versus curing time. In the same figure, the degree of reaction of CSF as measured by ^{29}Si MAS NMR versus curing time is plotted as well,

with a nearly identical trend indicating a linearity between CSF conversion and strength.

16.3.2 Effect of CSF on cement hydration

Justnes *et al.* (1990a,b) showed how ^{29}Si MAS NMR can be utilized to find the degree of hydration for the silicate minerals of the cement, the degree of conversion for CSF and the average length of the linear polysilicate anion of the CSH gel in cementitious mixes containing CSF. In fact,

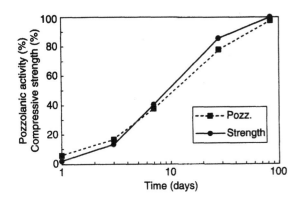

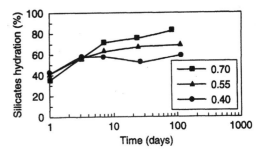

FIGURE 16.5 Comparison of compressive strength (relative to 63.9 MPa at 84 days = 100 per cent) development of mortar with reactivity of silica fume in the binder lime /CSF with C/S = 1.11 and water-to-solid ratio 0.70 (after Justnes 1998).

FIGURE 16.7 Hydration development for silicate minerals (C₃S and C₂S) in OPC with 15 per cent silica fume replacement at moderate to high w/(c+s) ratios according to ²⁹Si MAS NMR.

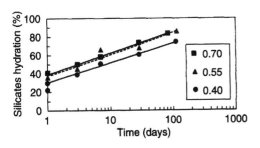

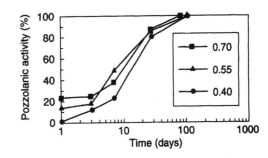

FIGURE 16.6 Hydration development for silicate minerals (C₃S and C₂S) in OPC without silica fume at moderate to high w/c ratios according to ²⁹Si MAS NMR.

FIGURE 16.8 Pozzolanic activity progress of silica fume in OPC at moderate to high w/(c+s) ratios according to ²⁹Si MAS NMR.

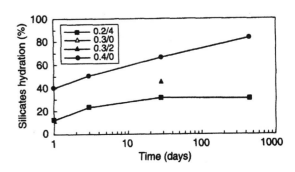

the method is the only direct method of monitoring the reactivity of amorphous silica.

Justnes *et al.* (1994) studied ordinary Portland cement (OPC), and OPC replaced with 15 per cent CSF, with water-to-cement+silica ratio (w/(c+s)) ranging from moderate to high (0.40, 0.55 and 0.70) with ²⁹Si MAS NMR. A similar study for high-strength Portland cement (HSC) with 0, 8 and 16 per cent CSF was performed by Justnes *et al.* (1992*a*) for w/(c+s) ranging from moderate to low (0.40, 0.30 and 0.20), where some of the mixes contained super-plasticizer (SP). The hydration versus time development for silicate minerals (C₃S + C₂S) without CSF, with CSF and pozzolanic activity for CSF are shown in Figures 16.6, 16.7

FIGURE 16.9 Hydration development for silicate minerals (C₃S and C₂S) in HSC without silica fume at moderate to low w/c ratios according to ²⁹Si MAS NMR.

and 16.8, respectively, for OPC and Figures 16.9, 16.10 and 16.11, respectively, for HSC.

If the degree of hydration for C₃S (see Justnes *et al.* 1990*b*) is plotted against the logarithm of curing

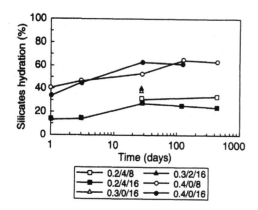

FIGURE 16.10 Hydration development for silicate minerals (C_3S and C_2S) in HSC with 8 and 16 per cent silica fume replacement at moderate to low w/(c+s) ratios according to ^{29}Si MAS NMR.

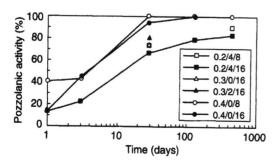

FIGURE 16.11 Pozzolanic activity progress of silica fume in HSC at moderate to low w/(c+s) ratio according to ^{29}Si MAS NMR.

time, the relation is fairly linear. This holds also for the OPC (see trend lines in Figure 16.6) which is high in alite (61 per cent C_3S and 14 per cent C_2S), indicating that the C_3S and C_2S hydrate in a consecutive manner, or with the same rate from the very start of hydration. The HSC has a higher C_2S/C_3S-ratio (44 per cent C_3S and 33 per cent C_2S) than the OPC, and the plot for w/c=0.40 in Figure 16.9 is slightly curved. The curvature may be explained by two lines (C_3S and C_2S) with different slopes displaced in time. At 28 days, the degree of hydration is already greater (66.4 per cent) than a complete hydration of C_3S only (51.9 per cent).

Figures 16.6 and 16.9 show further that the degree of hydration decreases with decreasing

water/cement ratio (w/c) at a given age, and that the hydration rate as a function of log (time) is particularly curved for HSC when w/c is 0.20. However, the first point on the curve may be unusually low due to retardation of the cement by the high dosage (4 per cent) of SP. The reason why virtually nothing has happened between 28 and 442 days' sealed-curing, may be self-desiccation at such a low w/c, or that the hydration products are particularly dense, leading to a strongly reduced diffusion-controlled hydration. Self-desiccation may be a result of increased ionic strength in the remaining pore water as well as of a pure capillary tension effect. Both reduce the relative humidity (RH). Measurement of RH during sealed-curing by Sellevold and Justnes (1992) show that RH falls below 80 per cent at 28 days for the low w/c mixes (0.20).

The effect of SP on the HSC paste with w/(c+s)=0.30 and 16 per cent SF was, increased degree of hydration (Figure 16.10) and increased pozzolanic activity (Figure 16.11), probably due to better dispersion of the paste. The effect on the paste (w/c=0.30) without CSF was only marginal (Figure 16.9). Comparing Figures 16.6 and 16.7 for OPC and Figures 16.9 and 16.10 for HSC, it may be seen that the hydration rate from 28 to 112/126/ 442 days for paste with w/(c+s)=0.40 is strongly reduced for the paste with CSF relative to the paste without CSF. This may be caused by a densification of the CSH-gel around the cement grains by CSF, or by the fact that increased CSF dosage produces more gel which adsorbs water, and thus, cement hydration stops due to lack of 'free' water. In spite of the halt of cement hydration, CSF consumption is complete within 100 days for w/(c+s)⩾0.40 and more than 80 per cent for lower w/(c+s) before 400 days.

16.3.3 Effect of CSF on porosity, strength and permeability

The effect of CSF on strength and permeability is related to the refinement of porosity, including the interface between binder and aggregate. The general refinement of the binder porosity is caused by the pozzolanic reaction resulting in more CSH with gel-porosity. The initial packing of irregular cement grains against the much larger aggregate surface will also lead to irregular voids, just as for the packing of cement grains in paste, and these voids fill be filled by the smaller CSF particles refining the interface already before set. The pozzolanic reaction will further prevent large platelets of calcium hydroxide from enriching on the aggregate surface. Since the fresh binder is more cohesive when CSF is included, this will prohibit segregation and internal bleeding leading to pockets or a film of water under larger aggregates, also indirectly leading to a better interface. The impact of CSF on interface is probably greater for moderate to high w/c. Kjellsen et al. (1999) presented results on strength of paste and corresponding concrete, indicating that there was no effect of interface since the strength gain due to CSF in paste was equal to concrete. However, they worked with rather low w/c (0.25–0.45), which could explain the contradictory result compared with, for instance, Bentur and Goldman (1989) and Scrivener et al. (1988).

The first application of CSF in concrete was in fact to improve strength or to make normal strength concrete of higher w/c. However, the weakness of such an utilization with increased permeability combined with reduced alkaline reserve was realized. A later application was HSC, and term used now is high performance concrete (HPC) to include the inherent reduced permeability and the indirectly increased durability versus external aggressive like chloride (e.g. Gautefall 1986).

16.3.4 Effect of CSF on alkalinity, carbonation and alkali aggregate reactivity

CSF reacts quickly with the alkalis inherent in cement to alkali silicate and later with calcium hydroxide (CH) to CSH in accordance with equation (2). Havdahl and Justnes (1993) and Justnes and Havdahl (1991) investigated the CH reserve in cement paste cured at ordinary and elevated

temperatures. The CH content in cementitious paste with $w/(c+s)=0.40$ according to thermo-gravimetry versus CSF dosage and sealed curing time at 20°C is plotted in Figure 16.12 together with the pH of pore water expressed from the same samples at 90 days. Even though 15 per cent CSF at 420 days' and 20 per cent CSF at 90 days' curing is sufficient to deplete CH below the detection limit, the pH of the pore water of paste with 15 per cent CSF after 420 days was 12.63. This may be explained by minor contents of alkali silicate ($0.1M$ $Na_2SiO_3\rightarrow$ pH $=12.6$) or carbonate ($0.1M$ $Na_2CO_3\rightarrow$ pH $=11.6$). Nevertheless, the alkaline reserve is considerably reduced due to the pozzolanic reaction when adding CSF. As can be seen by comparing Figures 16.6 and 16.7 and Figures 16.9 and 16.10, the inclusion of CSF actually decreases the degree of hydration at longer ages. Thus, less CH is actually formed that should be consumed by CSF. The effect is more pronounced when the initial curing temperature is elevated. Havdahl and Justnes (1993) showed that OPC with 16 per cent CSF and $w/(c+s)=0.40$ cured initially (72 h) at 50°C and the rest at 20°C had no CH detectable by TG at 7 days (first terminus), and did not show any until 360 days. The reduction of cement hydration by CSF is probably due to the encapsulation of the residual cement grain in a denser CSH than for the reference. In other words, the hydration becomes limited by diffusion at an earlier stage.

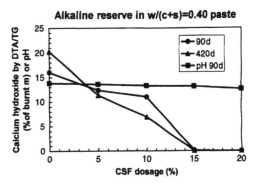

FIGURE 16.12 The effect of CSF dosage and curing time on CH content and pH of pore water (after Havdahl and Justnes 1991).

To avoid rapid carbonation, it is important to use CSF at such a low w/c that the permeability is not increased. Johansen (1981) investigated carbonation depth versus CSF dosage and w/c, and found that CSF does not increase carbonation rate significantly when w/c < 0.60.

The immediate conversion of alkali hydroxides to silicates by CSF will result in reduced alkali aggregate reactions (AAR). However, the implication of equation (2) is that this may only delay AAR unless CH is depleted. Bérubé and Duchesne (1992) showed that CSF merely postpones expansion due to AAR. Nevertheless, CSF as a remedy against AAR, together with other improvements in construction procedures, has found its application in Iceland where all cement has been interground with 7–8 per cent CSF to combat the problem (Asgeirsson 1986).

16.3.5 Effect of CSF on delayed ettringite formation (DEF)

Rønne et al. (1995) published the effect of CSF on expansion due to delayed ettringite formation (DEF), and the measurements have continued since then until 2 years' age as plotted in Figure 16.13 for OPC with 0 and 4 per cent CSF and w/(c + s) = 0.40. The concrete was resting for 6 h,

the heated to the set temperature of 20°C, 70°C, or 85°C with a rate of 12°C/h in a water bath that thereafter cooled down slowly to the ambient temperature of 20°C. The rest of the curing time was at 20°C and the volume of the specimens were monitored by weighing in water and air according to the principle of Archimedes. SEM confirmed ettringite formation in interfaces and cracks for the expanding specimens.

The reason why CSF seems to prevent DEF may be the depletion of CH as discussed in Section 16.3.4 and that the aluminates at early age then may form strätlingite, C_2ASH_8, with soluble silicates from alkalis and CSF. Alternatively, aluminates may be incorporated in CSH with lower C/S than in mixes without CSF. Finally, it is possible that the addition of CSF may form silicon-rich aluminate phases in the hydroglossular solid solution series C_3AH_6-C_3AS_3. However, it is unknown whether such phases will render the aluminates stable against reformation with sulphates to ettringite or not.

16.3.6 Effect of CSF on chemical shrinkage

With the more widespread use of HPC in Norway, the problem of early cracking (often < 12 h) has

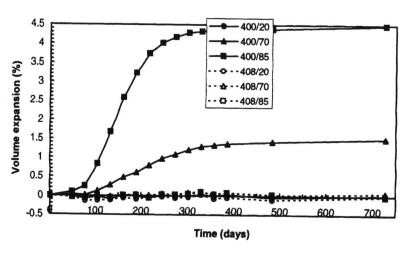

FIGURE 16.13   Volume changes of OPC with w/(c + s) = 0.40 with 0 per cent (legend 400) and 8 per cent (legend 408) CSF initial cured at 20°C, 70°C and 85°C followed by 20°C water curing.

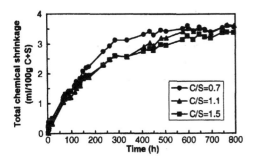

FIGURE 16.14 The total chemical shrinkage of lime–CSF slurry mixes with water of pH = 13 and different C/S-ratios as a function of curing time at 20°C.

been more frequent, especially, on horizontal bridge decks. Plastic shrinkage cracking is usually associated with evaporation of water, but the phenomenon occurs even when curing membranes are applied. The inclusion of microparticles like CSF refines the pore structure from the very start and may set up stronger capillary forces after less drying than reference concrete. However, the phenomenon is believed to be associated with the low w/(c + s) used as such. Early cracking and late age surface-crazing of HPC has led to an interest in the chemical shrinkage of the pozzolanic reaction. Justnes *et al.* (1998) studied the chemical shrinkage of pozzolanic reactions. Their plot of total chemical shrinkage from dilatometry (i.e. pipette method) versus time for different C/S ratios of lime–CSF mixes is reproduced in Figure 16.14. Based on this and NMR data from Justnes (1992b), they estimated the chemical shrinkage to be 8.8 ml/100 g reacted CSF for C/S = 1.11 as compared with the usual 6.25 ml/100 g reacted cement. On a volume basis, the shrinkage of the pozzolanic reaction is less than for cement hydration.

16.4 Conclusions

CSF is spherical particles of much smaller average size (about 0.15 μm) than OPC grains (about 10 μm). Thus, they will easily pack in the cavities formed between irregular grains or between cement grain and aggregate interface. This initial physical effect is beneficial to obtain concrete with low w/c and with an improved rheology

when plasticizers are used. The initial pore refinement also leads to less permeability immediately after set, as utilized in oil-well cementing to avoid gas intrusion from surrounding formation.

CSF is an active pozzolana that leads to strength increase. When CSF is not used to obtain equal strength at higher w/c than reference, but to obtain HPC with lower w/(c + s) including plasticizers, no durability problems are known. On the contrary, the application of CSF reduces the permeability to waterborne aggressives (e.g. chlorides), prevents DEF, and postpones (may even prevent) AAR. Carbonation rate is about equal to reference if w/(c + s) < 0.6.

16.5 References

Asgeirsson, H. (1986). 'Silica Fume in cement and silane for counteracting of alkali-silica reactions in Iceland', *Cem. Concr. Res.*, **16**(3).

Bentur, A. and Goldman, A. (1989). 'Bond effects in high-strength silica fume concrete', *ACI Mater. J.*, **86**(5), 440–47.

Bérubé, M. -A. and Duchesne, J. (1992). 'Does silica fume merely postpone expansion due to alkali-aggregate reactivity?', *Proc. 9th Int. Conf. Alkali-Aggregate Reaction in Concr.*, London, 27–31 July. 71–80.

Fidjestøl, P. and Lewis, R. (1998). 'Microsilica as an addition', Chapter 12 in *Lea's Chemistry of Cement and Concrete*, 4th edn, P. C. Hewlett (ed.), pp. 675–708, Arnold Publishers, London (ISBN 0 340 56589 6).

Gautefall, O. (1986). 'Effect of condensed silica fume on the diffusion of chlorides through hardened cement paste', *Proc. 2nd CANMET/ACI Int. Conf. on the Use of Fly Ash, Silica Fume, Slag and Natural Pozzolana in Concr.*, Madrid, paper SP91-48.

Havdahl, J. and Justnes, H. (1993) 'The alkaline reserve in cementitious pastes with microsilica cured at ambient and elevated temperatures' *Nordic Concr. Res., Publication*, no. 12, pp. 42–56.

Helland, S., Acker, P., Gram, H.E. and Sellevold, E.J. (1998). 'Condensed silica fume in concrete', *FIP State of the Art Report*, Thomas Telford, London, (ISBN 0 7277 1373 6), 37 pp.

Johansen, R. (1981). 'Silica in Concrete, Report 6: Long term effects', *SINTEF Report STF65 A81031*.

Justnes, H., Meland, I., Bjørgum, J. O. and Krane, J. (1990a). 'Nuclear magnetic resonance (NMR) -A powerful tool in cement and concrete research', *Adv. Cem. Res.*, **3**(11) 105–10.

Justnes, H., Meland, I., Bjørgum, J. O. and Krane, J. (1990b). 'A ^{29}Si MAS NMR study of the pozzolanic activity of condensed silica fume and the hydration of di- and tricalcium silicate', *Adv. Cem. Res.*, 3(11) 111–16.

Justnes, H. and Havdahl, J. (1991). 'The effect of curing temperature on the microstructure of cementitious paste for light LWA concrete', *Proc. 'Blended Cem. Construction'*, Sheffield, England, September 9–12, pp. 138–51.

Justnes, H., Sellevold, E. J. and Lundevall, G. (1992a). 'High strength concrete binders. Part A: reactivity and composition of cement pastes with and without condensed silica fume', *4th Int. Conf. Fly Ash, Silica Fume, Slag and Natural Pozzolana in Conc.*, Istanbul, May 3–8, CANMET/ACI SP 132–47 (Vol. 2) pp. 873–89.

Justnes, H.(1992b). 'Hydraulic binders based on condensed silica fume and slaked lime', *9th Int. Cong. Chem. Cem.*, New Delhi, India, November 23–28, Vol. III, pp. 284–90.

Justnes, H., Meland, I., Bjørgum, J. O. and Krane, J. (1994). 'The mechanism of silica fume action in concrete studied by solid state ^{29}Si MAS NMR', in *'Applications of NMR to Cement Science'*, P. Colombet and A. R. Grimmer (eds), pp. 213–28, A Gordon and Breach book, ISBN: 2-88124-965-5, April 1994.

Justnes, H. (1995). 'Accelerated hardening of mortars with hydraulic binders of silica fume/lime', *Nordic Concr. Res.*, Publication No. 17, 2/1995, pp. 30–41.

Justnes, H. (1998). 'Kinetics of reaction in cementitious pastes containing silica fume as studied by ^{29}Si MAS NMR' in *'Nuclear Magnetic Resonance Spectroscopy of Cement-based Materials'* P. Colombet, A.-R. Grimmer, H. Zanni and P. Sozzani (eds), pp. 245–268. Springer Verlag, Berlin.

Justnes, H., Ardouille, B., Hendrix, E., Sellevold, E. J. and Van Gemert, D. (1998). 'The chemical shrinkage of pozzolanic reaction products', *6th Int. Conf. Fly Ash, Silica Fume, Slag and Natural Pozzolana in Concr.*, Bangkok, Thailand, CANMET/ACI SP-178, pp. 191–205.

Kjellsen, K. O., Wallevik, O. H. and Hallgren, M. (1999). 'On the compressive strength development of high-performance concrete and paste – effect of silica fume', *Mater. and Struct.*, 32, January-February, pp. 63–9.

Malhotra, V. M., Caretta, G. G. and Sivasundaram, V. (1992). 'Role of silica fume in concrete: a review', *Adv. Concr. Technol.*, 2nd edn, by V. M. Malhotra (ed.), CANMET, Ottawa, Canada, pp. 915–90.

Rønne, M., Hammer, T. A., Justnes, H., Meland, I. S. and Jensen, V. (1995). 'Chemical Stability of LWAC Exposed to High Hydration Generated Temperatures', *Proc. Int. Symp. on Structural Lightweight Aggregate Concr.*, Sandefjord, Norway, June 20–24, pp. 505–16.

Scrivener, K. L., Bentur, A. and Pratt, P. (1998). 'Quantitative characterization of the transition zone in high strength concrete', *Adv. Cem. Res.*, 1(4) 230–7.

Sellevold, E. J., Bager, D. H., Klitgaard Jensen, E. and Knudsen, T. (1981). 'Silica fume cement pastes: hydration and pore structure', *Proc. Nordic Res. Sem. 'Condensed Silica Fume in Concr.'*, December 10, Report BML 82.610, NTH, 1982-02-15, pp. 19.

Sellevold, E. J. and Justnes, H. (1992). 'High-strength concrete binders. Part B: non-evaporable water, self-desiccation and porosity of cement pastes with and without condensed silica fume', *4th Int. Conf. Fly Ash, Silica Fume, Slag and Natural Pozzolana in Concr.*, Istanbul, May 3–8, CANMET/ACI SP 132–48 (Vol. 2), pp. 891–902.

Sellevold, E. J., Justnes, H., Smeplass, S. and Aasved Hansen, E. (1994). 'Selected properties of high performance concrete', *Proc. Engg. Foundation Conf.' Advances in Cem. and Concr.'*, M. W. Grutzeck and L. S. Shondeep (eds), American Society of Civil Engineers, New York, July 24–29, pp. 562–609.

Wallevik, O. (1998). 'Practical description of the rheology of fresh concrete', *Proc. Borregaard Symp. Workability and Workability Retention*, Hankø, Norway, September, 20 pp.

Chapter seventeen

Cement-based composite micro-structures

Stephen P. Bailey, David O'Connor, Sally L. Colston, Paul Barnes, Herbert Freimuth and Wolfgang Ehrfeld

17.1 Introduction

This chapter describes how cement-based composites have been utilized in designed micro-structures. In recent years, ceramic micro-structures (e.g. see Figure 17.1) have been mainly fabricated out of materials like alumina and zirconia for engineering ceramics (Stadel *et al*. 1996) and lead zirconate titanate (PZT) for piezoelectric ceramics (Wang *et al*. 1999). Well-known technologies like slurry casting (Bauer *et al*. 1998), hot embossing and injection moulding (Freimuth *et al*. 1997) have been modified in order to produce ceramic micro-structures. All these processes require sophisticated production techniques and are prone to limited quality: warping and cracking during drying or sintering stages may occur as well as non-linear shrinkage. The work described here, however, covers the first (to our knowledge)

FIGURE 17.1 A ceramic flow-sensor in which the turbine wheel and its housing were separately fabricated, followed by removal of the wheel base plate by grinding (Stadel *et al*. 1996).

design and fabrication of micro-structured cement composite devices.

The authors are aware of only the work by themselves in the production of designed micro-structures from cement-based materials. This follows from the original prototype study by Colston *et al*. (1996) which demonstrated that, as a medium for micro-structures, cements compare very favourably with the previous ceramic micro-structures, but with many superior attributes such

409

FIGURE 17.2 SEM micrograph of an OPC/organo-clay micro-structure which displays excellent micro-feature reproduction and, in testing, showed excellent strength and surface-hardness properties. This micro-structure has been formed inside a polymer micro-mould designed for opto-electronic applications (Colston et al. 2000; reproduced with permission of Kluwer Academic Publishers).

as ease of formation and excellent replication of micro-features, as can be seen in Figure 17.2. The ability of cement to replicate micro-structures at the micron-level was, initially unexpected, given that the average grain size of even well-ground Ordinary Portland Cement (OPC) is typically 20–100 μm, which is considerably larger than the smallest micro-features under consideration (~10 μm). However, cement hydrates (principally C-S-H and Calcium hydroxide (CH)) appear to grow in such a manner as to fill the available space in the micro-features, making cement materials quite unique in the field of micro-structures.

The chief drawbacks of the early cement micro-structures were the unpredictable filling in the third dimension (perpendicular to the plane of the micro-structure) over the range of vertical heights (5–120 μm) of typical features and the tendency to crack during the pyrolysis stage used to separate the cement micro-structure from the mould. These deficiencies have been largely overcome by the use of OPC/latex/organo-clay mixes (~6 per cent clay, 20 per cent latex, by weight of OPC), such that current cement micro-structures appear to show superior

reproduction of micro-detail and smoother surfaces than micro-structures fabricated out of conventional powder ceramics. More recent studies (Colston *et al.* 2000) have sought to improve replication and strength of the cement micro-structure through the adjustment of several production parameters. These parameters include a lengthened heating cycle to reduce mechanical stress and the incidence of bulk substrate fracture; an increased water content from 30–50 per cent is found to improve the vertical filling of the micro-moulds; an increased latex content from 20–40 per cent combined with a small quantity of organo-clay (6 per cent) notably increases the cement 'handling strength'. Excellent feature reproduction can be obtained with cements, and this is among the best achieved for any ceramic material up to the present day.

A second aim has been exploitation of the cement as a binder of 'functional' materials including zeolites/catalysts and magnetic phases. The use of zeolites, as functional agents within the cement, for applications involving gaseous separation is the basis underpinning the more recent studies. Using electron microscopy and synchrotron X-ray imaging (TEDDI: Tomographic Energy-dispersive Diffraction Imaging), the functional phases are found to be distributed throughout the cement bulk, substrate and micro-features. This observation is pivotal since the effective functioning of such composites requires that the hosted material is present at the micro-feature level.

Current research emphasis is to form a micro-structural cement composite device (MSCCD) uisng polyurethane micro-moulds fabricated by either the Advanced Silicon Etching (ASE) process (a dry etching process using fluorocarbon-based chemistry; Franssila *et al.* 2000) or the related Lithographie-Galvanoformung-Abformung (LIGA) process (Ehrfeld and Lehr 1995). Such moulds can be designed to contain structural features with depths up to several hundred μm and aspect ratios as high as 30, and to cover a diverse range of potential applications involving chemical sensors, ion-exchange, X-ray beam focusing, chemical catalysis, and gaseous separation.

In the production of an MSCCD, OPC binds the various components within the hardened paste, but there is also the potential for the internal matrix of the cement-composite to be 'engineered' to possess a desired average pore size. This requires adjustment of several production parameters, including the cement:water ratio and admixture (silica fume, super-plasticizer, styrene butadiene rubber-latex) type/percentage. Engineering of the cement micro-structure to produce a desirable porosity might, for example, prove valuable with MSCCD applications involving gaseous separation. Silica fume and silicon carbide can be additionally incorporated into the cement paste producing a final composite with increased density and strength and effective reinforcement of the micro-features, the smallest of which (~10 μm) are very prone to fracture during the de-moulding process.

In order to meet the physical requirements of a MSCCD, a number of cement-composite formulations have been devised from which an optimum selection is made on the basis of measured parameters such as porosity, tensile strength, and surface-hardness. The first MSCCDs produced were formed inside micro-moulds possessing a simple design geometry and examples of these are shown in Figure 17.3. In Figure 17.3a, striations are seen parallel to the channels indicating the detailed replication of the plastic mould surfaces; in Figure 17.3b, the spherical features at the end of the channels are the result of air bubbles trapped during placement. Clearly, the production of a complete MSCCD is a challenging task and, at time of writing, the production of several MSCCDs has been accomplished with promising results. However, the final test remains the production of a fully functional MSCCD.

17.2 Cement-composite development

Cement pastes for micro-structural use have been designed with specific properties in mind. These properties (tensile strength, hardness, porosity, and elasticity) are important for the manufacture of any device and its subsequent application. Table 17.1 gives the range of compositions that have been used, with some typical values. The best

(a)

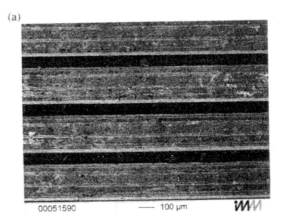

(b)

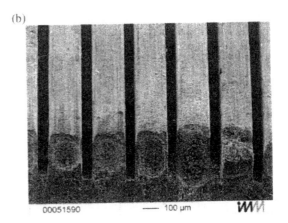

(c)

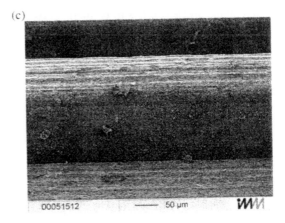

FIGURE 17.3 SEM micrograph of an OPC/water micro-structure (w/c ratio = 0.4) consisting of simple parallel rectangular channels seen in (a) plan view; (b) end view; and (c) side view.

TABLE 17.1 Example of optimum formulations for zeolite–cement-composites (expressed as percentage weight of total cement solids) against tensile strength and surface-hardness

A (%)	B (%)	C (%)	D (%)	E (%)	F (%)	G (%)	H (%)	Tensile strength (MPa)	Hardness (kg mm^{-2})
65.8	7.8	20.3	—	—	4.7	1.4	—	35	107
50.0	19.2	23.1	7.7	—	—	—	—	25	75
41.8	31.0	21.7	—	—	4.6	0.9	—	23	107
22.3	37.0	22.2	3.0	3.7	4.4	—	7.4	14	40
30.4	42.8	21.5	—	—	4.6	0.8	—	14	107
14.3	43.2	24.5	3.6	3.6	3.6	—	7.2	11	42
7.2	51.5	22.8	7.4	3.7	—	—	7.4	7	7
6.9	55.6	30.6	6.9	—	—	—	—	5	7

Key	Phase	Function	Typical composition range %
A	Portland cement	binder	07–65
B	Zeolite	catalysis/separation etc.	07–56
C	water	hydration	11–32
D	Silica spheres (2.8–4.5 μm)	marker/pozzolanic	2.8–20
E	Silicon carbide (1.5–10 μm)	reinforcement	3.5–25
F	Silica fume (~size 1 μm)	marker/pozzolanic	2.0–5
G	Super-plasticizer	rheology agent	0.3–1.5
H	SBR latex	rheology agent	07–16
I	Organo-clay	rheology and strength	0–6

combinations of tensile strength and surface-hardness are, not surprisingly, obtained with cements containing lower levels of zeolite content (see Table 17.1); higher surface-hardness values (for 40–55 per cent zeolite levels) can still, however, be obtained by the addition of silica fume and super-plasticizers, though in practice, an eventual compromise still has to be made between functionality and replication/strength. Rheology agents improve the flow of mixes into the microstructure features, and then continued hydration further completes the process of in-filling and complete replication.

17.3 Micro-mould design

Micro-mould designs start with a computer-aided drawing of a micro-structural MSCCD (e.g. using a package such as TurboCAD). The micro-mould geometry and dimensions need to conform to the tolerances associated with the production of the mould by the ASE process; typically, these are that the micro-features should have aspect ratios below 30 and minimum sizes of 5 μm or more. The

moulds must also ensure compatibility of the resultant MSCCD with an eventual device housing. Micro-mould designs collectively possess a range of feature dimensions that extend in width from 500 μm down to a minimum of 10 μm or less. A typical micro-mould design is shown in Figure 17.4. Small feature widths are often desirable; for example, a thin cement wall might be required in order to function as a selectively permeable membrane in a process of gaseous separation.

17.4 Mould fabrication

Moulds can be prepared by a combination of the ASE process and two successive polymer moulding steps. The full process (consisting of six stages) are given below and depicted in Figure 17.5 with a schematic of a prototype functional device as might be used for gas separation.

1: Planning and drawing of micro-mould designs.
2: Processing of micro-mould design files into a mask software file.
3: Production of a chromium photomask.
4: Ultraviolet (UV) lithography of a SiO$_2$ wafer.

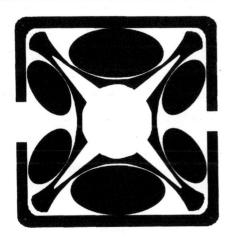

FIGURE 17.4 Example of a micro-mould design drawn using computer-aided design software; the black areas will become recessed areas of the micro-mould. The micro-mould is aimed at a MSCCD for applications involving gas-separation.

5: Micro-structuring of the wafer using reactive-ion ASE.
6: Production of a polyurethane micro-mould.

17.4.1 UV lithography

In this procedure (performed in a high-level clean room), an SiO_2 wafer is coated with a photo-sensitive chemical forming the resist layer. A chromium mask (with the micro-mould designs) is aligned with the (underlying) wafer. The mask is then exposed to broad-band UV radiation, which transmits through the non-chromium parts of the mask exposing the resist on the wafer surface to chemical degradation. Following irradiation, the wafer is chemically developed to dissolve away the exposed areas on the resist leaving behind a two-dimensional form of the micro-mould design within the structured resist. Subsequent etching transfers the mask pattern into the resist layer which acts as a mask in the ASE process.

17.4.2 Advanced silicon etching

The SiO_2 wafer is structured by reactive-ion etching using fluorocarbon-based chemistry. Fluoro-carbon species are unable to attack the SiO_2 on the wafer surface. However, those parts of the wafer not shielded by the resist are attacked by ion bombardment and silicon erosion takes place resulting in the micro-mould design outlines being etched into the wafer surface to a depth of a few hundred μm.

17.4.3 Micro-moulding

The etched wafer becomes the 'master'. In the first stage, semi-viscous silicone rubber is poured into a container engulfing the suspended wafer, and entering its micro-features. When solid, the wafer is carefully peeled away to reveal the micro-structural pattern in the silicone rubber. The silicone rubber now becomes the working mould. In the final stage, liquid polyurethane resin is poured into the silicone rubber mould, and is cured through polymerization. When solid, the polyurethane form is carefully peeled away from the silicone rubber. Examples of polyurethane micro-moulds are shown in Figure 17.6.

17.5 MSCCD applications

To date, no regularly working MSCCDs have been made. The long-term goal, however, is to make cheap functional MSCCDs benefiting from the ceramic properties of cement-composites (see above: mouldability, strength, porosity). Some potential applications are:

- *Chemical sensors*: by attachment of electrodes to a cement-composite surface thereby maximizing accessibility to the sample fluid;
- *Gaseous purification/separation*: the MSCCD would contain large quantities of a selectively permeable zeolite supported on a cement micro-channel gas flow network;
- *Electromagnetic field sensor/actuator*: the cement-composite would contain magnetic (e.g. magnetite) or piezoelectric (e.g. PZT powder) material;
- *Drying, hydrocarbon fractionation/cracking, ion-exchange, detoxification, chemical cataly-sis*: again, the MSCCD would contain zeolites,

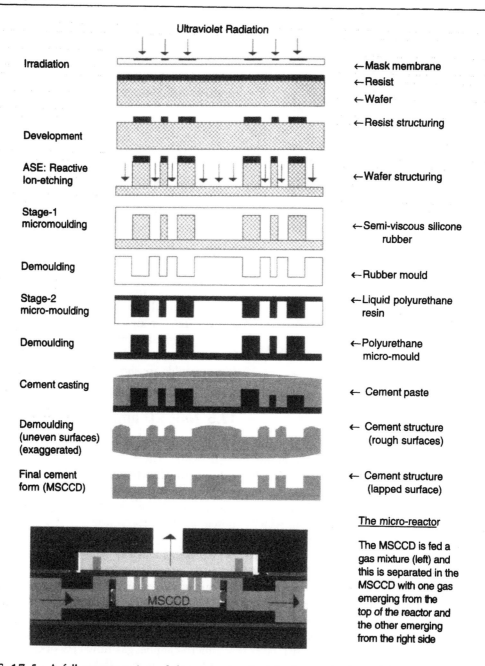

FIGURE 17.5 A full representation of the stages involved in the production of the cement micro-structure (MSCCD). The ASE/LIGA process covers all stages up to the production of the polyurethane micro-mould.

chosen for catalytic functions, supported on the cement micro-structure which also provides inflow/outflow routes for the ingredients and products;

- *X-ray compound refractive lens:* concave lenses, used for X-ray beam conditioning at synchrotron beam lines, requiring low atomic number elemental composition (Figure 17.7).

(a)

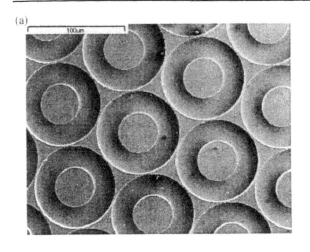

(b)

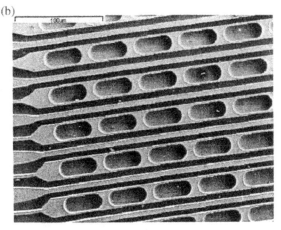

(c)

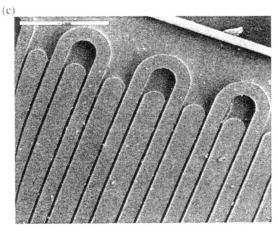

FIGURE 17.6 SEM micrograph of a polyurethane micro-mould (a) consisting of a two-dimensional array of recessed circular features; (b) consisting of a uni-directional array of recessed linear features and used to produce the cement micro-structure shown in Figure 17.10b; (c) the scale bar represents ~1 mm.

17.6 MSCCD production

MSCCD production involves several (following) stages, which are illustrated by appropriate SEM micrographs. The solid components of the cement-composite are ground into a fine powder and formed into a paste. The wet paste is evacuated using a vacuum pump to extract most of the trapped air-bubbles. The paste is placed into a micro-mould and agitated to release any remaining trapped air, and to encourage the paste to settle into the micro-features of the mould. The mould is then stored/cured in a sealed container

at 100 per cent RH. After curing (for typically days to months) the MSCCD is extracted from the polyurethane mould by heating in water at 70°C. At this temperature, polyurethane softens and can be (very carefully) peeled away from the MSCCD. The cement micro-features are examined and, if intact, the MSCCD is processed further. This may involve cutting the MSCCD to the required dimensions using a wafer saw (see Figure 17.8). It is also necessary for the micro-features to have a flat surface so that they can form a gas-tight seal against a gasket for MSCCDs used in

415

FIGURE 17.7   SEM micrograph image of an X-ray lens geometry formed in cement. The minimum feature width along the optical axis is 15 μm (a close-up view of another cement lens is shown in Figure 17.10d).

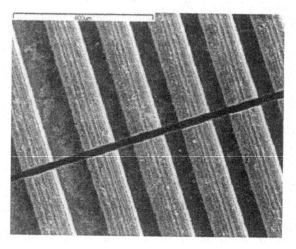

FIGURE 17.8   SEM micrograph showing a narrow (20 μm wide) wafer saw cut perpendicular to the cement micro-features. In this case, a clean cut has been produced with no damage inflicted around the area of incision.

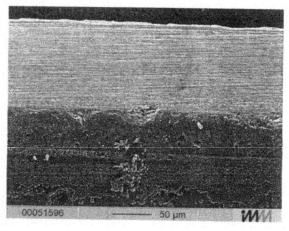

FIGURE 17.9   The SEM micrograph image shows a cement micro-feature whose upper surface has been lapped producing a profile with a flatness variation of less than 2 μm. Computer-controlled lapping is performed by erosion of the micro-feature surface at ~1 μm per lapping-pass.

gaseous separation. This is achieved through a lapping process which smoothens the cement surface (see Figure 17.9) without noticeable debris, whereas unlapped surfaces display surface debris and a partially formed skin of C-S-H (Colston *et al.* 2000).

17.7 Quality of MSCCD features

Figure 17.10 shows SEM micrographs of MSC-CDs (formulated from OPC, superplasticizer, silica fume and zeolite combinations) extracted from a mould after just 4 days of curing. As a result of the investigation into various cement-composite formulations (see Table 17.1), considerable progress has been made in reducing the hydration time required for good replication and in avoidance of damage during de-moulding; the inner fine channels visible inside the main feature of both Figures 17.10a,b have fairly high aspect ratios (~8) and serve as a spectacular illustration of the durability of current micro-features during de-moulding. Diagnostic techniques for assessing the quality of MSCCD micro-features includes

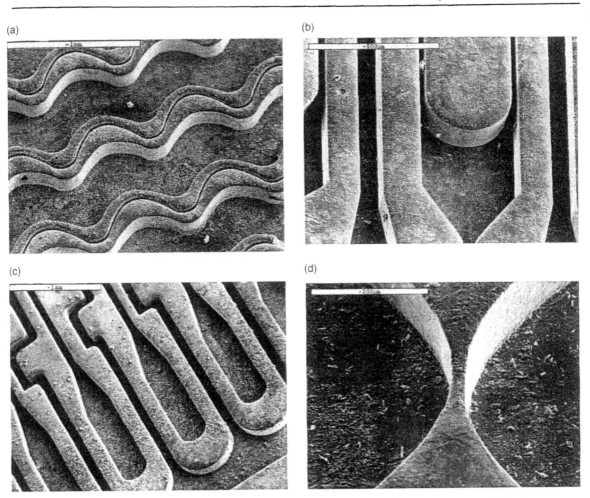

(a) (b) (c) (d)

FIGURE 17.10 SEM micrographs of four MSCCDs formed from polyurethane micro-moulds: the narrowest channel is (a) 17μm (b) 25μm or (c) 100 μm wide by ~150 μm high, and the overall MSCCD has been designed for gas-separation process; (d) close-up of a cement X-ray lens of which the narrowest part is 30 μm wide by ~150 μm high. This MSCCD has been designed for focussing high-energy X-rays.

measurement of tensile strength and surface-hardness in addition to scanning electron microscopy (with energy-dispersive X-ray (EDAX) microprobe analysis) which provides information on the quality of the micro-feature replication and composition of the cement phases present. During early hydration (first weeks), CH forms smooth tile-like coverings over large surface regions with angular discontinuities whereas C-S-H forms a continuous sheet-like surface covering with curved discontinuities (e.g. Figure 17.9). In the latter stages of curing, the surface covering becomes predominantly C-S-H with fewer discontinuities. An analysis of the cement surface using EDAX fluorescence gives information on elemental composition and can be used, for example, to identify the 'sodium signature' of a zeolite such as clinoptilolite and mordenite. TEDDI has been used to map the distribution of a zeolite within the cement-composite. The most important attributes required are:

- overall integrity and reproducibility of the micro-features without micro-cracking;

417

(a)

(b)

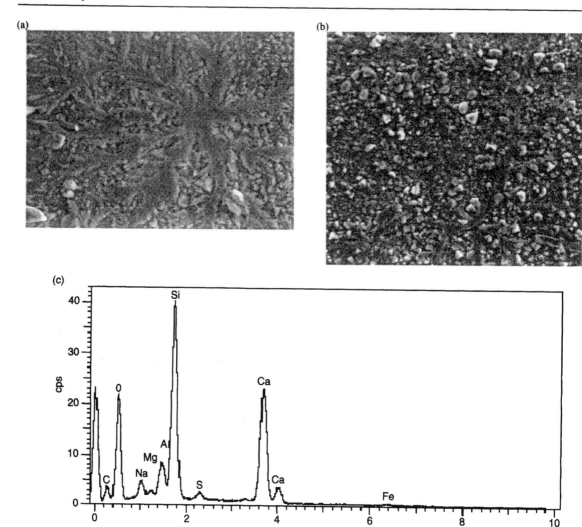

(c)

(d)

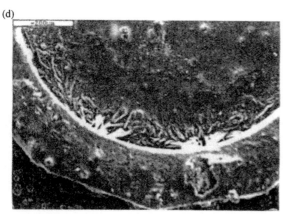

FIGURE 17.11 (a) SEM micrograph of an MSCCD. This is an image of the base of the cement wall. The dendritic outcrops are zeolitic Na-clinoptilolite crystallites; (b) SEM micrograph of the same MSCCD as (a) but viewed at the top of the cement wall. The crystalline outcrops are again visible; (c) EDAX *spectrum* of the dendritic crystallite region on the cement surface showing a significant sodium peak which identifies the clinoptilolite; (d) SEM micrograph of a different MSCCD showing crystalline outcrops at both the base and top of the cement wall.

- true replication of the micro-features;
- dispersion of the functional material (e.g. zeolite) into the host cement matrix.

Clearly the limits of the replication process, that is the reproducibility in forming small micro-features with cement is a vital aspect, with values of 5–10 μm being desired for many applications. Severe inhomogeneity of the composites in micro-features would be disadvantageous so it is of obvious interest to examine whether the hosted material (zeolites, etc.) can be truly carried into the smallest micro-features. These considerations are illustrated in Figure 17.11 with examples taken from the authors' work in devising MSCCDs for gas-separation applications. Figure 17.11 also shows the CH crystals as light tablet-shaped crystallites scattered over the surface. With continued hydration, these crystallites continue to grow but, in the highly confined space of the cement/mould interface, assume a two-dimensional tile-like form and integrate with the C-S-H skin and zeolite outcrops.

17.8 Conclusions

It has been demonstrated that cement-based materials can be successfully used in the production of functional cement-composites. The following steps have been prime goals in this pursuit:

- Design and fabrication of micro-moulds,
- Design and fabrication of a micro-reactor for MSCCD functional testing,
- Tailoring the cement-composite formulation to the MSCCD function,
- Development of a cement casting method,
- Replication accuracy and MSCCD strength,
- Extraction of MSCCD intact from the micro-mould.

The ultimate aim is to produce high quality MSCCDs with the same range of applications open to conventional metal and ceramic materials, but benefiting from the cheapness and simplicity of production.

17.9 Acknowledgements

We wish to thank Mr. P. Livesey (Castle Cement Co) for providing the Portland cement used in this work. We also wish to thank Dr. A. Beard at University College London (UCL) for providing the SEM facilities; Mr. S. Boone and fellow UCL technicians for demonstrating items of mechanical test apparatus, and Mr. P. Stukas (Birkbeck College) for manufacturing various items of apparatus. We are indebted to the many staff at IMM involved in the manufacture of the micro-moulds and micro-reactor. We also wish to thank the EPSRC, Daresbury Laboratory/ESRF, the European Commission Large Scale Facility (LSF) at IMM. Finally, we wish to acknowledge those not mentioned here who in some way have helped to advance this research.

17.10 References

Bauer, W., Ritzhaupt-Kleissl, H. J. and Hausselt, J. H. (1998). *Microsystem Techn.*, 4, 125–7.

Colston, S. L., Barnes, P., Freimuth, H., Lacher, M. and Ehrfeld, W. (1996). *J. Mater. Sci. Lett.*, 15, 1660–3.

Colston, S.L., O'Connor, D., Barnes, P., Mayes, E. L., Mann, S., Freimuth, H. and Ehrfeld, W. (2000). *J. Mater. Sci. Lett.*, 19, 1085–8.

Ehrfeld, W. and Lehr, H. (1995). *Radiat. Phys. Chem.*, 45, No. 3, 349–65.

Franssila, S., Kiihamaeki, J. and Karttunen, J. (2000). *Microsystem Technol.*, 6, 141–4.

Freimuth, H., Stadel, M., Hessel, V., Lacher, M. and Ehrfeld, W. (1997). In *Proc. Euromat 97* L. A. J. L. Sarton and H. B. Zeedijk (eds), Maastricht, pp. 285–8.

Stadel, M., Freimuth, H., Hessel, V. and Lacher, M. (1996). *Keram. Z.*, 48, 1112.

Wang, S., Li, J., Watabe, R. and Esahi, M. (1999). *J. Amer. Ceram. Soc.*, 82, 213–5.

Chapter eighteen

X-ray powder diffraction analysis of cements

J. C. Taylor, L. P. Aldridge, C. E. Matulis and I. Hinczak

18.1 Introduction

One of the most important techniques for characterizing cementitious materials is X-ray diffraction (XRD). Provided a powdered material contains crystals, XRD can be used to identify the crystalline component. The XRD pattern is like a finger print of a particular crystal and provided the correct procedures are followed, it is generally possible to find and quantify crystalline components which have over 1 per cent abundance in a sample. Furthermore if the correct experimental techniques are used XRD can give a semi-quantitative or quantitative analysis of the components of the crystalline fraction. However,

amorphous components such as the calcium silicate paste (C-S-H) give few XRD lines and features commonly called 'amorphous humps' that cannot be used to give useful quantitative analysis and are very difficult to identify. Thus, XRD gives most information from the crystalline components of cementitious materials.

It is important to stress that XRD must not be used as a 'black box' to characterize cementitious materials. Ordinary Portland cement (OPC) is not made from pure minerals but from quarries that generally contain substantial amounts of impurities. Thus, the alite component of Portland cement is not pure C_3S but is generally a solid solution containing Mg, Al and Si. These impurities can, and mostly do, have a substantial effect on the XRD patterns and should be always considered before interpreting XRD patterns of cements.

Bragg–Brentano diffractometers are almost universally used in industrial laboratories, including cement laboratories. They produce patterns with high diffracted intensity and resolution. In fact, well-collimated modern machines ($\sim 0.05°2\theta$ halfwidth) have resolutions approaching that of a synchrotron.

XRD patterns are now much easier to interpret because of the development of well-designed software and the enormous power of modern personal computers. We use the Australian software Traces Version 5 (sold by Diffraction Technology, Australia) to examine XRD patterns and Siroquant™ Version 2.5 (sold by Sietronics, Australia) for quantitative analysis. Other software is available and is doubtless just as effective as the named products.

Both cement producers and cement researchers have made extensive use of XRD to identify and quantify the phases in cement clinkers, ground cements and the cement hydration products. An example of one common application used in cement works is the use of XRD to check the form of sulphate in ground cement. Cement clinker and gypsum are inter-ground together and, in some cases, the heat generated by grinding dehydrates the gypsum so that bassanite (hemihydrate) or anhydrite is formed. Some technically acute cement manufacturers deliberately add gypsum in different forms to obtain 'optimum sulphate additions' to control the rate and extent of the cement hydration. XRD is ideal for this task as cement can be quickly loaded onto a diffractometer, and after scanning between 5–40°2θ in 30 min, a definitive identification of the sulphate types can be made. In addition, by semi-quantitative analysis an estimate of the amounts of the sulphates can be made.

18.2 X-ray diffraction from cement

The quality of any XRD pattern will depend critically on:

- the alignment of the diffractometer;
- the choice of the diffraction conditions (time, slits, monochromator, and filter);
- the sample preparation.

Such factors are more than adequately treated in standard books on XRD (see Jenkins and Snyder (1996) or the classic of Klug and Alexander (1974)) and need not be addressed here. For identification of cement phases, we would run diffraction patterns between 5° and 80°2θ, either with CoK$_\alpha$ or CuK$_\alpha$, at a stepping rate of 1°2θ a minute. For quantitative analysis, we prefer to use CoK$_\alpha$ radiation, with the same 2θ limits and a step size of at least 0.02° while counting each step for at least 10 s. In order to get accurate analyses, cement samples must be ground either by hand for about 5 min using an agate mortar and pestle

using moderate pressure until the cement feels smooth, or wet ground (we use anhydrous methanol) by a mill such as the McCrone micronizer. It is common to drive off the grinding fluid by heating with an infrared lamp. We have found that this can dehydrate gypsum, so we now heat the fluid in an oven at 60°C under vacuum.

X-ray diffraction patterns from cement are not simple to interpret. Computers have made interpretation simpler by search-match, graphical displays, and the visualization by programs such as 'Powder Cell' of diffraction patterns taken under different conditions. The following factors should be taken into account when interpreting XRD patterns.

Phases found in cement have diffraction peaks many of which overlap, as shown in Figure 18.1. In earlier days, XRD patterns were compared

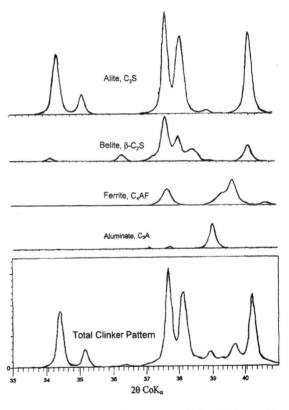

FIGURE 18.1 Breakdown of a clinker XRD profile showing the alite, belite, ferrite and aluminate components, and the total pattern.

using a light box with the patterns from standard cements to highlight differences. Today, using good computer software, the same effect can be obtained. In addition, the phases in cement are not the pure phases suggested by the formulae C_3S, C_2S, C_3A or C_4AF, but are, in fact, solid solutions. Furthermore, these solid solutions will differ between works depending on the raw materials used. Thus, one works may have substantial amounts of Mn in its raw feed, and Mn can end up in the C_4AF phase (Kennerley 1977). The effect of such solid solution replacement on the XRD patterns is significant and the extent of the change will depend on differences between the substituting and the substituted atom in scattering factors, sizes, and chemistry. Sometimes the change is so great that the symmetry of the crystalline phase changes, which can have substantial effects on the diffraction peaks. The substitution of Na into C_3A will change the symmetry from cubic to orthorhombic or monoclinic. It has been shown that when Na is substituted into C_3A that a solid solution forms with the general formula $Na_{2x}Ca_{3-x}Al_2O_6$ (with x varying between 0 and 0.25). The crystalline chemistry of the cement components is reviewed in H. W. F. Taylor's book 'Cement Chemistry' (Taylor (1997), Regourd and Guinier (1976), Regourd (1983), Taylor *et al.* (2000) and Regourd (1986)). Such detailed reviewing of the subject need not be redone here.

Another problem is diffraction from amorphous materials. Crystalline materials diffract X-rays and the amount of diffraction depends on the degree of order in a crystal. Amorphous materials gives rise to very diffuse and weak diffraction peaks. The hydrated cement paste major component is nearly amorphous C-S-H with a broad diffraction peak which gives little useful information and complicates the interpretation of the diffraction peaks of the crystalline components. Similar problems are found when trying to interpret the XRD patterns of blends of cement containing nearly amorphous supplementary cementitious materials. Ground granulated blast furnace slag, some fuel ashes, and silica fume are commonly found in blended cements because they improve cement performance, and generally, the improvements in performance are related to the amorphous materials they contain. XRD can only give limited information on these amorphous materials.

OPC is generally believed to be completely crystalline (Taylor 1997, p. 78). However, Regourd and Guinier (1976) did find a few clinkers that gave a strong diffuse XRD peak between $30–35°2\theta$ (CuK_α) indicating amorphous material. The belief that there were insignificant amounts of amorphous material in OPC was recently challenged by Suherman *et al.* (2000).

For quantitative analysis, physical problems must be considered; for example, preferred orientation, absorption, and particle statistics. These effects on the intensities of the peaks are well covered by standard texts, but it is surprising how often they are ignored.

18.3 Qualitative analysis of cements

The use of XRD to determine which phases are present in a mixture has been used for many different applications and is well described in many different reference books. Chapter 12 of Jenkins and Snyder (1996) gives a good review of the methods of identifying which crystals are present in a particular diffraction pattern. They also review the techniques used in commercial search-match programs. While in principle search-match programs should give unique solutions, factors such as poor peak resolution, solid solutions, and peak overlap do give ambiguities that only practice and patience can resolve. Although the authors have used a number of search-match programs (such as μPDSM, Socabim and Traces), they believe that great care must be taken when identifying cement minerals because of the peak overlap and solid solution problems already mentioned.

The best advice a novice can be given is to:

1. read and understand the sections of Taylor's (1997) book 'Cement Chemistry' that relate to the problem being studied;
2. read and understand Jenkins and Snyder's book (1996) 'Introduction to X-ray Powder Diffractometry' and

3. practise identification on samples such as NIST reference cements that have been previously studied by other workers.

18.4 The method of X-ray diffraction

18.4.1 Conventional (non-Rietveld) methods

The basic principles of XRD have been already well set out in a number of standard texts (Bish and Post 1989; Cullity 1978; Jenkins and Snyder 1996; Klug and Alexander 1974), and it is recommended that new practitioners consult these. A recent volume edited by Smith and Chung (2000), 'Industrial Applications of X-ray Diffraction', is well worth consulting, and shows the growing use of Rietveld analysis, both for crystal structure determination-refinement and quantitative analysis. Chapters 1 and 2 on principles and practice of XRD analysis are good starting points. A simplified overview of XRD principles relevant to cement analysis is now given below.

The intensity of a line (h,k,l) from a phase depends on the phase crystal structure, and is:

$$I(h,k,l)_\alpha \propto F(h,k,l)_\alpha^2 \cdot M(h,k,l)_\alpha \cdot (1/V_\alpha^2) \cdot$$
$$Lp(h,k,l)_\alpha \cdot E(h,k,l)_\alpha \cdot P(h,k,l)_\alpha \cdot \tau_\alpha \quad (1)$$

where:

$I(h,k,l)_\alpha$ is the integrated intensity of the line (h,k,l);

$F(h,k,l)_\alpha$ is the 'structure factor' of phase α;

$M(h,k,l)_\alpha$ is the multiplicity of the family of equivalent planes;

$Lp(h,k,l)$ is an angular Lorentz-polarization factor term;

V_α is the volume of the unit cell of phase α;

τ_α is an absorption contrast (microabsorption) factor when the matrix is more than one phase;

$P(h,k,l)_\alpha$ is the preferred orientation factor for non-random crystal orientations;

$E(h,k,l)_\alpha$ is a primary extinction factor.

Some of these intensity-correction factors can be severe, especially P and τ, and may be a factor of two. Other corrections can arise, for example, an absorption correction to the experimental pattern. The correct aberrations must be identified because of possible correlation of parameters with systematic errors.

18.4.2 Conventional XRD phase quantification

Because of the factors identified earlier, conventional XRD quantification using one or two lines from the phases is often unreliable.

If we take a ratio of two nearby lines for phases α and β in a two-phase mixture, equation (1) gives:

$$wt\%(\alpha) = 100(1 + I_\beta/I_\alpha \cdot \rho_\beta/\rho_\alpha \cdot V_\beta^2/V_\alpha^2 \cdot$$
$$M(h,k,l)_\alpha/M(h,k,l)_\beta \cdot F(h,k,l)_\alpha^2/$$
$$F(h,k,l)_\beta^2 \cdot Lp(h,k,l)_\alpha/Lp(h,k,l)_\beta \cdot$$
$$\tau_\alpha/\tau_\beta \cdot E(h,k,l)_\alpha/E(h,k,l)_\beta \cdot P(h,k,l)_\alpha/$$
$$P(h,k,l)_\beta)^{-1} \quad (2)$$

where ρ_α is the density and V_α is the unit cell volume of α.

Such calculations become onerous when there are many phases and lines. It is also not known which lines are most subject to the above aberrations. When the line density increases with more phases, a stage is reached when line superposition makes conventional XRD quantitative analysis impossible.

Smith and Chung (2000) discuss other integrated intensity XRD quantitative methods such as internal standard addition, absorption–diffraction, spiking and use of reference intensity ratios (Hubbard and Snyder (1988)). These tend to be semi-quantitative due to peak overlap, difficult physical corrections, and incomplete use of all the information in the diffraction pattern, but can be quantitative with very careful work on the less complex systems.

18.4.3 The Rietveld method

We are indebted to H. M. Rietveld (1967, 1969), who solved the superposition problem, and with this advance also allowed better corrections for the physical effects. He took a data point as one-step-scan intensity y_i along the pattern, instead of integrated intensities of phase lines. This gave several thousand intensities y_i along the XRD profile as the data. The Rietveld formula was originally used for refinement of the positions (x, y, z) of the atoms in the unit cell of one phase, that is, crystal structure refinement, in a non-linear least-squares refinement fitting a calculated XRD profile to the observed one. We now show the Rietveld profile formula is a simple extension of the integrated intensity formula (1) above. In the notation of (1) the Rietveld formula is for one phase α

$$
\begin{aligned}
(y_i)_{cal} = {} & SCALE_\alpha \cdot \tau_\alpha \cdot ASYM_\alpha \cdot \Sigma_{h,k,l}(F(h, k, l)_\alpha^2 \cdot \\
& M(h, k, l)_\alpha \cdot Lp_i \cdot P(h, k, l)_\alpha \\
& E(h, k, l)_\alpha \cdot SHAPE_\alpha / H_\alpha)
\end{aligned} \tag{3}
$$

The only difference between (1) and (3) is that all lines (h, k, l) are considered, the data points are at the step angles i, and the phase has a Rietveld scale $SCALE_\alpha$, the lines can be corrected for an asymmetric profile (ASYM), and there is a lineshape function $SHAPE_\alpha$. H_α is the value of the line halfwidth at the point i. The various lineshape functions are given (Young 1993) in Chapter 1 of the text 'The Rietveld Method'. In the Siroquant program (Taylor J. C. 1990; Taylor and Clapp 1992) the $F(h, k, l)_\alpha$ is precalculated from a crystal structure, or is an observed $F(h, k, l)_\alpha$ which fits a measured pattern.

The Rietveld equation is thus similar to the integrated intensity formula (2), except that the (h, k, l) line profiles are replaced by a line profile. The whole XRD profile is then the sum of all such line profiles.

18.4.4 Rietveld phase quantification

It was not until the late 1980s that it was realized that the Rietveld phase scales contained quantification information (Bish and Howard 1988; Hill and Howard 1987; O'Connor and Raven 1988). Chapter 5 by Snyder and Bish in Bish and Post (1989) only briefly mentions Rietveld quantification. A summary of Rietveld analysis to 1993 is given in 'The Rietveld Method' (Young 1993), but is mainly concerned with specialist basic physics, structure analysis and instrumentation, while there is no chapter on phase quantification.

For j phases, the multiphase Rietveld formula is from (3):

$$
(y_i)_{cal} = \sum_j \left(SCALE_j \cdot \tau_j \cdot \sum_{h,k,l}^j (F(h, k, l)_j^2 \cdot \right.
$$

$$
M(h, k, l)_j \cdot Lp_i \cdot P(h, k, l)_j \cdot E(h, k, l)_j \cdot
$$

$$
SHAPE_j / H_j \cdot ASYM_j)) \tag{4}
$$

The least-squares refinement is based on a few thousand differences $((y_i)_{cal} - (y_i)_{obs})$, and $\Sigma((y_i)_{cal} - (y_i)_{obs})^2$ is minimized. The Rietveld scales are part of the refineable parameters.

The weight fraction of any phase j is:

$$
wt_j = \frac{SCALE_j \cdot MASS_j \cdot VOL_j / \tau_j}{\sum_{(i=1, n)}(SCALE_i \cdot MASS_i \cdot VOL_i / \tau_i)} \tag{5}
$$

where $MASS_j \cdot VOL_j$ are the mass and volume of the unit cell of phase j, and there are n phases.

It should be realized that running a Rietveld program does not guarantee a correct answer. Like any other analytical method, the quality of the results depends on the skill of the operator. Madsen *et al.* (2000) are completing an international round robin that shows that quantitative phase analysis of even simple systems by Rietveld methods cannot be blindly carried out using a 'black-box approach'.

18.4.5 Non-Rietveld (conventional) XRD quantification of cement

Generally, the cement industry relies on the Bogue (1929) method where results from chemical analysis are used to calculate the phase composition assuming that only the phases C_3S, C_2S, C_3A, C_4AF, free lime and gypsum are present in cement. Such assumptions are known to be in error. XRD has been used for many years to give quantitative information on the amount of the crystalline phases present in a mixture. Klug and Alexander's (1974) classic book details the methods and techniques. The application of these methods to cement powders has not always been successful. Quantitative XRD of cement is not easy to perform and the results are not always an improvement on the Bogue (1929) method.

It is generally recognized that major problems in cement analyses are due to:

- Overlapping peaks – in particular, the C_3S peaks swamp and overlap the C_2S peaks.

- Variability of the XRD peaks in Portland cement. Such effects are due to both compositional variation and peak broadening. Peak broadening could occur due to quenching, overgrinding, compositional zoning, or imperfect crystallinity.
- Preferred orientation effects in the XRD peaks.

Taylor (1997), pp. 106–09 gives an admirable review of the history of quantitative X-ray diffraction of cements. Struble (1991) discussed recent developments and summarized efforts within ASTM to develop a standard test method of quantitative X-ray analysis of Portland cement clinkers. The well-known Gutteridge (1984) method also used a wide range of references chosen to correspond with the clinker under investigation.

Most conventional quantitative X-ray methods for cements are based on single or multiple peaks for each phase, covering a restricted angle range in the pattern. Such methods can give acceptable accuracy providing accurate standards are available. Aldridge (1982a) reported a round-robin study where six clinkers were analysed by optical microscopy and the results compared to the cement analysis carried out by conventional XRD. One laboratory gave results that were in good agreement with the optical results. Obviously the standards that this laboratory used were carefully chosen to reflect the mineralogy of the cement phases. Aldridge studied accuracy and precision further (Aldridge 1982b). Thus, provided a high degree of skill and care is used, non-Rietveld methods can be used to give adequate analyses of Portland cement based products.

18.5 Physical factors affecting XRD quantitative analysis

18.5.1 Absorption contrast (microabsorption)

A powder mixture may have some phases which are high absorbers of the X-rays, and some which are low absorbers. If so, then there is an absorption contrast effect, which reduces the diffraction of the high absorbers relative to the low absorbers. Brindley (1945) calculated correction factors τ_i for the different phases i as a function of $(\mu_i - \mu_i')$. R_i, where μ_i and μ_i' are linear absorption coefficients of phase i and the matrix, and R_i is the effective particle radius of phase i. He stressed the importance of using samples which were not too coarse, and gave criteria for ascertaining whether the sample was in the 'fine' range. This range depends on the absorption by the phase. Ideally, the particles should be $1\,\mu$m radius or smaller.

To determine the magnitude of the effect in Rietveld quantification, Taylor and Matulis (1991) analysed nine known mixtures of $Pb(NO_3)_2$ (linear absorption coefficient $\mu = 1046\,cm^{-1}$) and LiF ($\mu = 52\,cm^{-1}$) using the Brindley algorithm. The samples were finely ground, but the effective radii R were not known. The R_i were assumed equal for the phases, and R_i was adjusted for each mixture until the program gave weight percentages which were equal to the weighed amounts. It was found that R_i was about $5\,\mu$m for all the mixtures. Without Brindley corrections, answers for the weight fractions were in error by as much as 100 per cent, but with the correction, good quantifications were found for weight percentages of $Pb(NO_3)_2$ along the whole range from 0 to 100 per cent, with a R_i value of $5\,\mu$m. This study confirmed the validity and usefulness of the Brindley (1945) absorption contrast algorithm.

Absorption contrast, sometimes called microabsorption, is an important effect, and even though the cement case has fairly similar absorbers, the correction is still significant, especially for C_4AF. The program had default Brindley R_i values of $5\,\mu$m, which work well for ground cement. It is advisable to have fine powders, and better to use the defaults rather than have no microabsorption correction at all.

18.5.2 Anomalous dispersion

Matulis and Taylor (1994) have pointed out that neglect of anomalous dispersion corrections to the atomic scattering factors can cause significant errors in XRD phase quantification. A case in point is Fe with CoK_α radiation. It was shown that neglect of one correction in this case caused the Rietveld quantification of Fe_2O_3 in a 55.0:45.0 per cent mixture with corundum to give an answer of 40.9 per cent for corundum instead of 55 per cent. There was only a small absorption contrast in this case. After including anomalous dispersion, the Fe_2O_3 result was nearly correct (53.3 per cent), and after microabsorption correction was 54.4 per cent, close to the weighed 55.0 per cent.

The effect of anomalous dispersion with Fe is not so severe with CuK_α radiation, but is still significant. Unfortunately, CuK_α has a huge microabsorption effect with Fe_2O_3, so CuK_α radiation is quite unsuitable for Fe compounds (see below). The anomalous dispersion effect can be calculated exactly, but microabsorption sometimes renders analysis of Fe compounds with CuK_α radiation impossible.

The above example shows that all relevant physical corrections should be applied before the 'last resort' interpretation of 'amorphous content' is invoked. In the above example, the apparent amorphicity would be about 25 per cent in the 55 per cent Fe_2O_3 example. The real amorphicity for the hematite in question was about zero. The anomalous dispersion effect causes the Fe_2O_3 phase to assay too low with CoK_α unless the correction is applied. To be safe, anomalous dispersion should always be switched on; then it can be forgotten. CoK_α is still a suitable radiation for Fe-containing materials as the anomalous dispersion correction can be exactly computed, while microabsorption is not a serious problem.

18.5.3 Extinction

Primary extinction reduces the powder intensities of low angle lines with large $F(h,k,l)$, and is an interference effect from multiple reflection in perfect crystal regions. The effect has been described by Sabine (1993) and Zachariasen (1945). One correction is $SQ/\tanh(SQ)$ where:

$$SQ = K \cdot D \cdot d \cdot p^{0.5} F(h,k,l)^2/V \qquad (6)$$

where K is a constant, D is the crystal thickness (refineable), d the d-spacing of the plane (h,k,l), and p the polarisation factor(1 for neutrons). Large crystal size, large d (small 2θ) and large $F(h,k,l)$ typify the lines reduced the most in intensity. The effect has been shown to be significant in XRD powder diffraction by Cline and Snyder (1987), who studied mixtures of the hard materials Si, Al_2O_3, and TiC. The effect was small below $D = 1\,\mu m$, but could reduce some intensities by 50 per cent in the size range 5–10 μm. To avoid the effect in well-crystallized substances, one should grind to less than 1 μm size.

18.5.4 Preferred orientation

The powder method assumes that the ground crystals are randomly oriented. This is not the case when the crystals are platy or needle-shaped; and the Rietveld equation then becomes invalid unless an orientation correction is applied, or the effect is reduced experimentally. Smith *et al.* (1979) solved the ever-present preferred orientation problem by spray-drying, when the coated crystals were converted to spherical particles. This is not always convenient experimentally; the effect is now commonly corrected in the Rietveld refinement by applying the March (1932) function (Dollase 1986). Preferred orientation is practically eliminated by grinding to <1 μm size, and back-filling the specimen holders.

Figure 18.2 shows how a clinker pattern can change markedly as it is packed in different ways in the Bragg–Brentano sample holder. Orientation takes place on the C_3S pseudo-hexagonal plane. In this case, the March orientation corrections were essential.

18.5.5 Particle statistics

Smith (1992) pointed out that 52,900 crystallites must be in diffraction for the standard error in a measured intensity to be less than 1 per cent. For a sample with 40 μm crystallites, there are only a surprising twelve crystals diffracting! For 10 μm size, there are 760, and for 1 μm there are 38,000. It is clear that fine grinding is essential, for even if

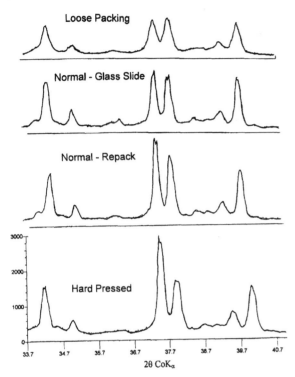

FIGURE 18.2 Changes in intensity in the XRD pattern of a cement, caused by preferred orientation with different packing methods.

all the crystals are ground to less than 1 μm diameter, the measured intensity will have greater than 1 per cent error. In fact, fine grinding helps to overcome most physical effects, but one must be sure that the grinding does not cause a phase transition or a breakdown in the crystal structure. Elton and Salt (1996) have also studied this effect.

18.6 Choice of radiation in cement XRD studies

CuK$_\alpha$ radiation is installed in most XRD machines in cement laboratories. However, a little thought shows that this radiation is quite unsuitable for all non-white cements because of the large absorption by Fe of CuK$_\alpha$ radiation and high fluorescent background. If there is more than 10 per cent C$_4$AF in the clinker, then the absorption contrast effect causes a flawed analysis. One can get any desired answer for the weight

TABLE 18.1 Effect of radiation choice on the Rietveld quantification of hematite

Brindley particle diameter (μm)	% Hematite, CuKα	% Hematite, CoKα
1	54	53
10	71	54
20	84	55

of the ferrite by varying the Brindley particle size radius, and thus changing the ferrite concentration returned by the Rietveld program. The following hypothetical example illustrates this point. For an approximately 54:46 per cent mixture of hematite and corundum we obtain, with the Siroquant^{TM} program (anomalous dispersion switched on) for the CuK$_\alpha$ and CoK$_\alpha$ cases, the result in Table 18.1.

Table 18.1 illustrates the great difficulty of quantifying iron minerals with XRD methods using CuK$_\alpha$ radiation. CoK$_\alpha$, on the other hand, is more suitable. With CoK$_\alpha$, the results are not sensitive to microabsorption and there is an added advantage of greater angular dispersion of the lines because of the longer wavelength. The microabsorption effect here works in the opposite direction to anomalous dispersion.

Microabsorption affects quantification results for real clinkers. For NIST SRM 8486 clinker and CuK$_\alpha$ radiation, the C$_4$AF result goes from 11.4 per cent for a particle diameter of 1 μm to 16.2 per cent for a diameter of 20 μm, a relative change of 40 per cent. CuK$_\alpha$ is obviously unsuitable. With CoK$_\alpha$, the C$_4$AF result goes from 12.3 per cent for a particle diameter of 1 μm to 12.9 per cent for a diameter of 20 μm, only a very slight effect.

18.7 Amorphous (non-diffracting) content

18.7.1 Determination with conventional XRD

Amorphous (non-diffracting) material may be present in a sample, with a weight fraction w_a.

427

Further, there may be peaks from a minor phase which have been lost in the background, or under the main phase peaks. Small particle sizes cause peak broadening, and if broad enough, may merge with the background. Glassy materials, such as non-crystalline quartz, give broad amorphous intensity humps, distinguishable from the background, arising from the radial distribution function of the atoms in the glass.

Bish and Chipera (1988) state that very broad amorphous humps may only become visible at $w_a = 30$ per cent. Snyder and Bish, in Bish and Post (1989), p. 126, state that the area under a hump is related to the concentration of the amorphous material and could be treated as a normal diffraction in the usual non-Rietveld XRD methods, namely absorption–diffraction, internal standard or RIR.

Yang (1996) used a modified Ruland (1961) method to determine the crystallinities of some synthesized clinker minerals. This technique does not depend on the measurement of the area under a hump, but is experimentally and computationally laborious, and should only be done by an XRD specialist. He found crystallinities (in per cent) for β-C_2S between 85 and 99, alite 88, C_4AF 88, C_3A 93, ettringite between 65 and 75, portlandite 86, gypsum 97, calcite 97, quartz 93 and corundum 92 per cent. He estimated the absolute errors in the crystallinity determination were greater than 3 per cent absolute. As computed errors in Rietveld quantification are known to be low, (probably by a factor of two, see below), by analogy some of these crystallinities may not differ significantly from 100 per cent, especially when the systematic errors and the 3σ levels are considered.

Vonk (1983) has used the modified Ruland (1961) method to determine the crystallinity of the glass ceramic $Li_2Si_2O_5$ with Zn, P and K inclusions. For the slow-cooled sample, the crystallinity was 0.47 ± 0.007, and for the fast cooled sample, the crystallinity was 0.14 ± 0.013. Here, the errors were relatively low. The Ruland (1961) method would obviously be much better for cement phases if the computed errors could be further reduced from those given by Yang (1996).

18.7.2 Determination with Rietveld XRD

Amorphous content can be determined in this method by adding a spike, say 10 per cent corundum, to the sample. If amorphous material is present, it will not register in the Rietveld refinement, but it does in the weighing. The computed Rietveld weight percentage of the spike, x, will thus be greater than the weighed amount, y. It can be shown that in the spiked sample (see Siroquant™ Manual Version 2, p. 53) that

$$w_a = 1 - y/x \qquad (7)$$

The sensitivity, $y - x$, for $w_a = 0.20$, say, has a maximum at $x = 40$–50 per cent. However, because of sample dilution by the spike, and consequent reduction of the sample pattern intensity, it is best to add about 10–15 per cent of spike.

The equation was checked by grinding some glass, and making a 1/3, 1/3, 1/3 mixture of glass, quartz and ZnO as the sample. Then a corundum spike was added. The spiked sample was 25.5 per cent quartz, 25.8 per cent ZnO, 26.3 per cent glass and 22.4 per cent corundum.

After full Rietveld refinement, the weight per cent of the corundum was returned as 30.0(7) per cent, instead of 22.4 per cent. This gave from the above formula 25.3(17) per cent glass in the spiked sample. The final results were for the unspiked sample (Table 18.2).

This Rietveld refinement picked up the glass component with the glass hump removed with the background. The program estimated standard deviation on the glass percentage is about 2 per cent absolute from the program, but this is probably low because of systematic errors. The error is about the same as found by Yang (1996) for the Ruland (1961) method, but the Rietveld analysis

TABLE 18.2 Amorphicity determination with Siroquant™

Phase	Program (wt%)	Weighed (wt%)
Quartz	33.9(4)	32.9
ZnO	33.5(4)	33.3
Glass	32.6(21)	33.9

is far easier to use. Both types of crystallinity analysis say nothing about the chemical nature of the non-diffracting material.

From the equation $w_a = 1 - y/x$, if we assume no error in y, and an absolute 0.5 per cent error (esd.) in x, w_a has an error of ± 2.5 per cent at the 0 per cent amorphicity level. This is about the same error (esd.) found in our analysis above (Table 18.2). Realistically, we need to double these esd's, so it is doubtful whether amorphicity levels below $\sim 2.5 \times 2 \times$ (two esd's) $= 10$ per cent could be detected with present methods in OPC X-ray analysis.

18.8 Background intensity

The background can be visually estimated as the trough level between peaks. This may be too high if there are many weak reflections between the main peaks, as can occur in OPC patterns. This is

illustrated in Figure 18.3. A preferable method is to use a polynomial function with refineable coefficients, which can give lower least-squares residuals. As polynomial functions are slow-changing, they will not cope with sharp bends, for example, intrusion of the main beam at low 2θ angles, so care must be taken in their use.

18.9 Rietveld refinement of OPC XRD patterns

Rietveld refinement is particularly difficult in the case of OPC, as there is extreme overlap of the stronger lines of the main phases C_3S and C_2S in the middle of the pattern 33–43°2θ for CoK$_\alpha$. Figures 18.4 and 18.5 give calculated profiles for the C_3S and C_3A polymorphs for this 2θ range.

Because the Rietveld method analyses the total XRD pattern, it can use numerical methods to get better corrections for preferred orientation than

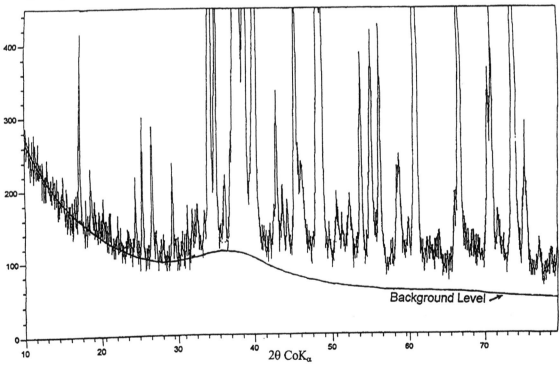

FIGURE 18.3 Background level in Australian clinker XRD pattern (CoK$_\alpha$) showing the true background is below the troughs in the high angle end of the pattern, due to the large number of weaker lines there, mainly from C_3S.

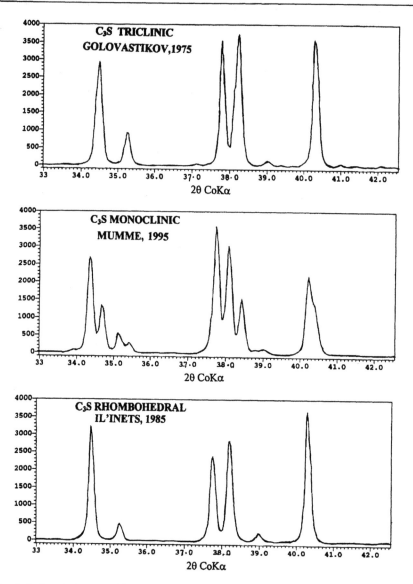

FIGURE 18.4 Calculated XRD profiles of selected forms of C₃S, free from preferred orientation effects.

would be possible with conventional methods. It allows for compositional variations in the phases to some extent by adjusting the cell parameters. However, some crystal stucture determinations on real clinker crystals are now becoming available, for example, Mumme (1995), which can be used as the Rietveld standards, and help in further addressing the solid solution problems. Madsen and Scarlett (2000) used electron probe micro-

analysis (EPMA) to determine the composition of the major phases of their cements, and thus, get better data for Bogue analysis and for reduction of the Rietveld results to oxide percentages.

There are many lines in a cement pattern, even outside the region of strong lines. For the NIST SRM 8486 clinker, and a CuK_α pattern from 10° to 85°2θ, the number of phase lines is 5862 (Table 18.3).

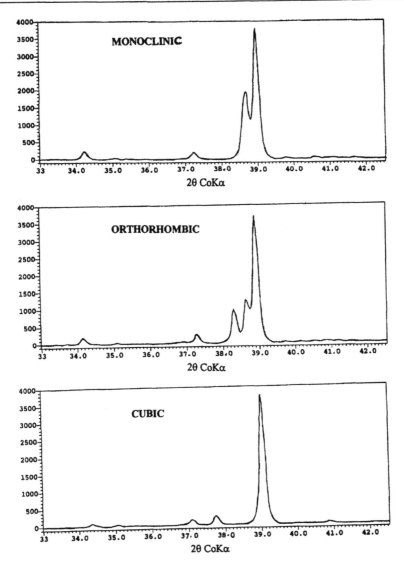

FIGURE 18.5 Calculated XRD profiles of selected forms of C_3A, free from preferred orientation effects.

Some of these lines have weak intensities, because of pseudosymmetry in the case of C_3S, and can be removed from the (h, k, l) files if necessary. One Rietveld cycle with a pattern of 3751 step scan points on a 200 MHz Pentium PC with 5862 XRD lines takes 37 s; a Rietveld quantification can thus be carried out by an experienced operator on such a computer in about 5 min.

Rietveld work on OPC started only in about 1990, even though Rietveld's first paper was written in 1967. The C_3S XRD profile can vary from one clinker type to another because of compositional changes and changes in the C_3S polymorph ratios.

To address the composition problem, some workers have used a modified Rietveld method, first proposed by Taylor and Zhu Rui (1992), in which calculated $F(h, k, l)$ for C_3S in the phase (h, k, l) file are replaced by observed $F(h, k, l)$. This allows a measured alite pattern, rather than

TABLE 18.3 Number of lines in NIST SRM 8486 clinker

Phase	Number of lines
C_3S (triclinic)	2646
C_3S (monoclinic)	250
C_3S (rhombohedral)	241
C_2S (beta)	213
C_2S (alpha)	206
C_2S (α-K_2SO_4 type)	115
C_3A (monoclinic)	2087
C_4AF	86
MgO	8
Lime	10
Total number of lines in clinker pattern	5862

a calculated one, to be used in the Rietveld refinement. Taylor and Aldridge (1993a, 1994) and Taylor, Zhu Rui and Aldridge (1993) used this approach in a number of Rietveld cement quantifications. Typically, they found r.m.s. deviations of the Rietveld method from MPC of 2.1 per cent for alite, 3.1 per cent for belite, 1.7 per cent for aluminate and 1.3 per cent for ferrite, while precision in repeated analyses was 1.3 per cent for alite, 0.8 per cent for belite, 0.5 per cent for aluminate and 0.5 per cent for ferrite. Amorphous material was not definitely found in their clinkers. Moeller (1998), Madsen and Scarlett (2000) and Scarlett *et al.* (2000) have also used the observed (h, k, l) file Rietveld approach to quantify clinkers. In this method, the observed C_3S profile used must be typical of the cement being analysed.

An alternative to the above approach is to include more than one polymorph of each cement phase in the Rietveld task. For C_3S, a representative structure for each of the C_3S crystal systems found, namely triclinic (T), monoclinic (M) and rhombohedral (R) can be used, also polymorphs of C_2S other than the β-form and two polymorphs of C_3A, cubic and monoclinic (orthorhombic C_3A is not necessary, as orthorhombic and monoclinic C_3A have nearly identical patterns).

There is only one T-form crystal structure known, namely the excellent analysis of synthetic C_3S by Golovastikov *et al.* (1975), so this must be used. The Mumme (1995) M3 structure

determination was done on a clinker crystal, so this model is a suitable standard for the M-forms. The R crystal system is suitably covered by the room temperature study of Il'Inets *et al.* (1985) on a Sr-doped crystal.

Taylor *et al.* (2000) carried out Rietveld quantifications for the three NIST standard clinkers SRM's 8486, 8487 and 8488 with one, two and three C_3S polymorphs in the refinement. Best comparisons with microscopic results were found when all three polymorphs (R, M and T) were included in the so-called RMT refinements. The precision and accuracy of the RMT refinements were then the same as for the microscopic data. This full Rietveld method is more convenient than the microscope method, as it is far less time-consuming. Observed and calculated profiles for NIST SRM 8486 clinker are shown in Figures 18.6a (CoK_α) and 18.6b (CuK_α).

The RMT Rietveld refinement method is more flexible than fixed C_3S models when quantifying different types of clinker, as it can adjust to different polymorph ratios. It would be improved if more model structures for R, and especially T forms, could be obtained from real clinker crystals.

18.10 Refinement strategy for Rietveld XRD assays of cements

Some refineable parameters are global, while the rest are phase-dependent.

Some are robust (R), while others are 'soft', (S). If soft parameters are refined first, the least-squares process will fail, with a calculation divergence, or arrival at a false value of the quantity minimized, $\Sigma(y_o - y_c)_i^2$. We assume the calculation has no structure refinement.

If a phase has a low concentration, then its lines will be weak. Approximate concentration levels are indicated, below which a certain parameter should not be refined. These levels may vary slightly from system to system. The global and phase parameters and limits are shown in Tables 18.4 and 18.5 respectively.

About six refinement stages are needed for a successful strategy. There are many cycles; this is unimportant with modern PCs. A possible strategy

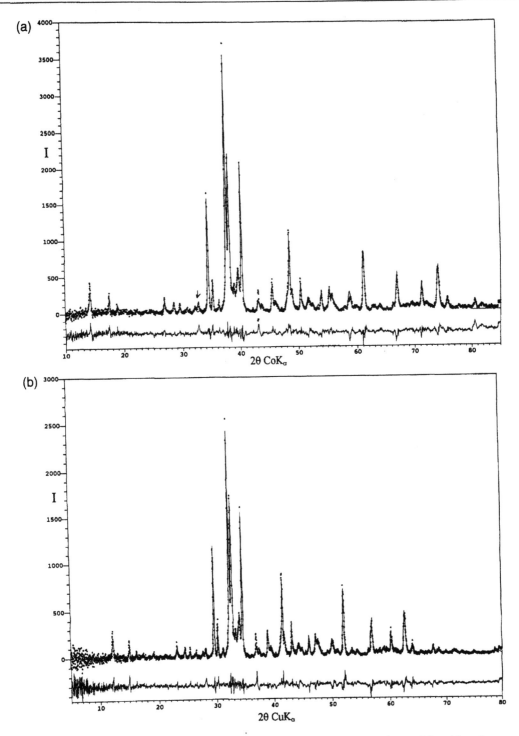

FIGURE 18.6 Observed (dots), calculated (line) and difference profiles for a Rietveld refinement of Bragg–Brentano XRD data for NIST SRM 8486 Standard clinker: (a) CoK_α and (b) CuK_α.

TABLE 18.4 Global parameters

Global parameter	Stability	Comment
Instrument zero, ZERO	R	Switched on, refine immediately
Pattern background, BACK	S	Only refine necessary coefficients. Too many coefficients may cause false minimum[a]

[a] See Figure 18.3 for an example of 'hidden background'.

TABLE 18.5 Phase parameters

Phase parameters, phase i	Stability	Concentration limit (%) for refinement
Scales, $SCALE_I$	R	0.3
Unit cell, $CELL_I$	R	2
Single lineshape, $SHAPE_i$	R	15 (program defaults usually OK)
Split shape, $SPLITSHAPE_I$	R or S	25
Linewidth, W_I	R	2
Linewidths, U_i, V_i	S	20
Preferred orientation, P_i	R or S	5
Extinction, E_i	S	15 (rarely invoked as it can correlate with systematic errors)

TABLE 18.6 Refinement strategy

Stage	Parameters On or Off	Cycles	Damp factor
1	On: ZERO, $SCALE_i$	10	0.5
2	Off: ZERO. On: $SCALE_i$, $CELL_i$, W_i	5	1
3	On: $SCALE_i$, $CELL_i$, P_i. Off: W_i	10	0.5
4	On: $SCALE_i$, $CELL_i$, P_i, W_i	10	0.5
5	On: ZERO, $SCALE_i$, $CELL_i$, P_i, W_i, (and U_i, V_i, $SHAPE_i$, $SPLITSHAPE_i$ if needed)	10	0.4

is in Table 18.6. This procedure takes about 5 min on a 400 MHz Pentium computer with an Australian cement, the pattern being stepped in 0.02° from 5–80°2θ.

18.11 Errors in Rietveld quantifications

The errors (esd's) from Rietveld programs are the result of matrix inversion. They are necessarily lower than the real errors, since all systematic errors cannot be allowed for in the model. Prince (1993) suggests the ratio of the profile R-factor $\Sigma|y_o - y_c|_i / \Sigma|y_o|_i$ to the R-factor expected on statistics alone is an estimate of the factor to be applied to the program esd's to get the real esd's. This is usually about a factor of two. It is a good general rule to double all computed esd's when assessing

the significance of any Rietveld quantification. It may be expected that some other processes, for example the Ruland (1961) method, could also need a similar factor; thus computed esd's should always be treated with caution. This factor of two was applied in the discussion of the significance of Ruland (1961) amorphous contents above. It is a good aim to reduce computed esd's to the lowest possible level, especially in amorphicity analysis.

18.12 Crystal structure analysis of Portland cement phases by the Rietveld method

Rietveld quantitative analysis is often referred to as 'standardless'. This is true in that an internal standard does not need to be added to the sample if amorphous content is not required. However, there is an in-built phase standard in the Rietveld equation, namely the term $F^2(h, k, l)$, which is the characteristic crystal structure of the phase. This is why precise crystal structure analysis of the OPC phases, especially C_3S, is so important in laying a good foundation for the Rietveld quantification of cement.

Very detailed descriptions and illustrations of the structures of the four main cement phases and their polymorphs have been given in many reviews, for example Lea (1970). The most recent are by Madsen and Scarlett (2000) and Taylor *et al.*

TABLE 18.7   Recent structural work on cement phases by XRD Rietveld Power diffraction

OPC phase	Formula	Method	Reference
β-C_2S	$(Ca_{1-x}Sr_x)_2SiO_4$	R	Fukuda et al. (1996)
β-C_2S		R	Daimon et al. (1992)
C_4AF	$C_4AF_{0.7}M_{0.3}S_{0.3}$	R	Neubauer et al. (1996a)
C_4AF	$Ca_2Fe_{2-x}Al_xO_5$ phases	R	Guirado et al. (1996)
C_4AF	$Ca_2Fe_{1.4}Mg_{0.3}Si_{0.3}O_5$ and $Ca_2Fe_{1.07}Al_{0.67}Mg_{0.13}Si_{0.13}O_5$	H	Lewis et al. (1996)
Brownmillerite	Mg-doped	R, N-R	Jupe et al. (2000/2001)
C_3A		R	Goetz-Neunhoeffer and Neubauer(1997)
C_3A	$Ca_{3-x}Sr_xAl_2O_6$, 27 samples	N-R	Prodjosantoso et al. (2000)
CA	$CA_{1-x}F_x$	R	Guirado et al. (1994)
γ-$CaSO_4$, and hydrates	γ-$CaSO_4, CaSO_4 \cdot 0.5H_2O$, and $CaSO_4 \cdot 0.6H_2O$	N-R	Bezou et al. (1995)
Thaumasite	$C_3S\bar{S}\bar{C}H_{15}$	R-H	Barnett et al. (1999)
OPC Phases	β-C_2S, $C_3A(c)$, C_4AF	N-R	Berliner et al. (1997)
Ettringite and monocarbonate	$C_3A \cdot 3CaSO_4 \cdot 32H_2O$ $C_3A \cdot CaCO_3 \cdot 11H_2O$	R	Fullmann et al. (1999a)
Ettringite	$C_3A \cdot 3CaSO_4 \cdot 32H_2O$	R	Kuzel (1994)

R: Rietveld X-ray; R-H: High resolution Rietveld XRD; R-N: Rietveld neutron.

(2000), so there is no need to repeat these descriptions here. Instead, we summarize recent structural work not mentioned in the previous reviews.

Despite the fact that over 50 Rietveld investigations of cement systems have taken place, we are surprised how little use has been made of the technique to investigate the *in situ* crystal chemistry of cement clinkers. Unlike earlier investigations where large single crystals had to be grown in order to determine the crystal structure, the Rietveld method can be applied to determine crystal chemistry in powder mixtures diffracting X-rays or neutrons *in situ* in a number of environments. The pioneering studies of Barnes' group on brownmillerite (Jupe *et al.* 2000/2001; Lewis *et al.* 1996) show how such studies can be used to carry out exciting science from which real commercial benefits may be drawn. We hope that other workers will apply these techniques to C_3A and the other silicate phases in cement to define the role of the impurities in modifying cement behaviour.

In recent XRD powder structure work, there is a tendency to refine OPC-related crystal structures, but not include new structural details (such as atomic coordinates) in the paper. Such work, in our opinion, should be published with structural details.

18.13 Quantitative analysis of hydrating phases

Substantial work has been reported using intense synchrotron X-ray sources to study synthesized pure phases of cement compounds. Unpublished work of Aldridge, Taylor and co-workers on hydrating cements have shown both the potential and the problems of using Rietveld analysis to quantitatively analyse hydrating cements. However, substantial progress has been made. It is believed that this technique will give suitable results provided that suitable crystal structures are known, a suitable internal standard is incorporated in the mix, and the diffraction patterns from amorphous material can be treated.

Many groups have used XRD analysis to quantify the amount of CH, AFt and AFm, produced in hydrating cements as a function of time. Such work has been most useful in understanding the chemistry and reactivity of hydrating cements. Odler and Abdul-Maula (1984) have shown that

TABLE 18.8   XRD powder structural investigation of hydrated cement phases since 1990

Sample	XRD Method	Reference
Synthetic ettringite, monosulphate	R	Fullmann *et al.* (1999)
Mexican OPC	C	Escalante-Garcia and Sharp (1998)
Ferrite hydrates	C	Emanuelson *et al.* (1996, 1997)
Ettringite	C	Yang *et al.* (1996)
C_3A hydrates	R	Kuzel (1996)
C_3S paste	C	Omotoso *et al.* (1996)
Ettringite	C	Talero (1996)
OPC cement (XRD, synchrotron, neutron). Time-resolved powder diffraction	Profile fitting	Clark and Barnes (1995)
Calcium aluminoferrite hydrates	C	Liang and Nanru (1994)
C_3A hydrates	XRD identification	Kuzel and Pollmann (1991)
OPC hydration	C	Parrott *et al.* (1990)
Rich C_2S cements	C	Ftikos and Philippou (1990)

R: Rietveld method; C: conventional XRD method (Gutteridge 1984).

ill-crystalline peaks make the quantitative XRD inferior to thermal methods of analysis. However, other workers have produced informative results from the technique and recently a full-scale Rietveld refinement was made of the same system (Fullmann *et al.* 1999).

It must be emphasized that good sample preparation, techniques, and a critical error analysis should be used with this application of XRD diffraction analysis. Table 18.8 lists some recent work in this area. Conventional XRD methods have mainly been used with cement hydrates because of the difficulty of Rietveld analysis with some poorly defined crystal structures.

18.14 Recent Rietveld studies of real cements

It seems certain that Rietveld analysis will eventually supersede the method of Bogue phase analysis, since it is phase-sensitive and more accurate when properly applied, and is as good as the microscope method. While microscope techniques are usually accurate, they require an experienced operator devoting a large amount of time. Sometimes, the interstitial phases of cement, C_4AF and

C_3A, as well as very small crystals, are difficult to identify with the microscope. Rietveld analysis lends itself to a rapid quantitative analysis and computer automation. Modern position-sensitive X-ray detectors allow the collection of XRD data in minutes, compared with the 10 h needed with Bragg–Brentano machines when a full angular range and very high intensities are needed.

Rietveld-based quality control clinker quantification systems have now been installed in some cement plants (Moeller 1995, 1998; Neubauer 1998; Scarlett *et al.* 2000). The long development time for automation has been due to the special difficulty of the clinker phase line overlaps, and the consequent need for basic research to obtain the best model for the complex C_3S crystal chemistry.

Accurate automation (black-box) may not fully be with us yet, as the C_3S structural problems still need further work. The C_4AF and C_3A phases are already approaching the black-box stage for Rietveld analysis (Thiesen 1999). The excellent work on C_4AF solid solutions by Lewis *et al.* (1996) has given good real models for this phase. The makers of Siroquant^{TM} are developing template task files which make it possible to run 'black box' applications providing the cement is

TABLE 18.9   Rietveld phase quantification studies of real cements

Sample type	Reference
OPC clinker (plant quality control)	Neubauer (1998)
OPC clinker	Meyer et al. (1998)
Sludge-based clinker	Goetz-Neunhoeffer and Neubauer (1998)
OPC clinker	Pollmann et al. (1997)
OPC, HAC, building products	Pollmann et al. (1998)
OPC clinker, plus oil and sludge	Neubauer and Pollmann (1997)
OPC clinker, OPC and OPC/HAC blends	Neubauer (1997)
OPC clinker	Neubauer et al. (1997)
Synthetic alite and belite mixtures	Neubauer and Sieber (1996)
Synthetic and technical clinkers	Neubauer et al. (1996b)
15 technical clinkers	Thiesen (1999)
Cement, clinker, slag	Bellotto and Signes-Frehel (1998)
NIST standard clinkers	Taylor et al. (2000)
OPC clinkers and cements (plant quality control)	Moeller (1995, 1998)
OPC clinker, cement, standard mixtures	Mansoutre and Lequeux (1996)
OPC clinker	Motzet et al. (1997)
OPC clinker, cement (pattern fitting)	Stutzman (1996)
HAC	Motzet and Poellmann (1998)
Calcium sulpho-aluminate cement	Schmidt and Pollmann (2000)
High, medium and low aluminate cements	Fullmann et al. (1999b)
HAC	Guirado et al. (2000)
OPC, rapid on-line XRD plant quality control	Scarlett et al. (2000)

OPC: ordinary Portland cement; HAC: high alumina cement.

adequately characterized. We believe that accurate automatic Rietveld analysis will in the future be used for quality control. However, we must once again reiterate the precautions already discussed which are needed to run and interpret XRD patterns of cements. Table 18.9 lists some recent Rietveld studies of real cements.

18.15 Conclusions

XRD analysis methods as described above have become an essential part of the analysis in cement research laboratories. In the last 10 years, the power of the Rietveld method in cement characterization has been realised, and much research has been done to produce accurate analyses of cements. Careful work needs to be performed to set up Rietveld analysis in new plants, but once this work has been done, routine quality control can be automated. Some installations of on-line phase quantification systems already exist in cement plants (see Table 18.9). Conventional non-Rietveld XRD techniques can be used for accurately analysing cement, but these must be expertly applied.

18.16 References

Aldridge, L. P. (1982a). 'Accuracy and precision of phase analysis in Portland cement by Bogue, Microscopic and X-ray diffraction methods', *Cem. Concr. Res.*, 12, 381–98.

Aldridge, L. P. (1982b). 'Accuracy and precision of an X-ray diffraction method for analysing Portland cements', *Cem. Concr. Res.*, 12, 437–46.

Barnett, S. J., Adam, C. D., Jackson, A. R. W. and Hywel-Evans, P. D. (1999). 'Identification and characterization of thaumasite by XRPD techniques', *Cem. Concr. Compos.*, 21(2), 123–8.

Bellotto, M. and Signes-Frehel, M. (1998). 'The role of powder X-ray diffraction in the cement industry. Recent advances and future developments', *Mater. Sci. Forum*, 278–81 (Pt. 2, *Proc. 5th European Powder Diffraction Conf.*, 1997), 846–51.

Berliner, R., Ball, C. and West, P. B. (1997). 'Neutron powder diffraction investigation of model cement compounds', *Cem. Concr. Res.*, 27(4), 551–75.

Bezou, C., Nonat, A., Mutin, J.-C., Christensen, A. N. and Lehmann, M. S. (1995). 'Investigation of the crystal structure of γ-CaSO$_4$, CaSO$_4$.0.5H$_2$O and CaSO$_4$.0.6H$_2$O by powder diffraction methods', *J. Solid State Chem.*, 117, 165–76.

Bish, D. L. and Chipera, S. J. (1988). 'Problems and solutions in quantitative analysis of complex mixtures by X-ray powder diffraction', *Adv. X-ray Anal.*, 31, 295–308.

Bish, D. L. and Howard, S. A. (1988). 'Quantitative phase analysis using the Rietveld method', *J. Appl. Cryst.*, 21, 86–91.

Bish, D. L. and Post, J. E. (eds), (1989). 'Modern Powder Diffraction. Reviews in Mineralogy', Mineralogical Society of America, Vol. 20.

Bogue, R. H. (1929). 'Calculation of the compounds in Portland cement', *Ind. Engg. Chem. (Anal.)*, 1(4), 192–7.

Brindley, G. W. (1945). 'The effect of grain or particle size on X-ray reflections from mixed powders and alloys considered in relation to the quantitative determination of crystalline substances by X-ray methods', *Phil. Mag.*, 36, 347–69.

Clark, S. M. and Barnes, P. (1995). 'A comparison of laboratory, synchrotron and neutron diffraction for the real time study of cement hydration'. *Cem. Concr. Res.*, 25(3), 639–46.

Cline, J. P. and Snyder, R. L. (1987). 'The effects of extinction on X-ray powder diffraction intensities', *Adv. X-ray Anal.*, 30, 447–56.

Cullity, B. D. (1978). *Elements of X-ray Diffraction*, Addison-Wesley, Reading, Massachusetts.

Daimon, Masaki, Hirano, Yoshinobu, Tsurumi, Takaaki and Asaga, Kiyoshi. (1992). 'Crystal structure analysis of major constituent phases in ordinary Portland cement', *9th, Int. Congr. Chem. Cem.*, 17–22.

Dollase, W. A. (1986). 'Correction of intensities for preferred orientation in powder diffractometry: application of the March model', *J. Appl. Cryst.*, 19, 267–72.

Elton, N. J. and Salt, P. D. (1996). 'Particle statistics in quantitative X-ray diffractometry', *Powder Diffr.*, 11, 218–29.

Emanuelson, A., Henderson, E. and Hansen, S. (1996). 'Hydration of ferrite Ca$_2$AlFeO$_5$ in the presence of sulphates and bases', *Cem. Concr. Res.* 26(11), 1689–94.

Emanuelson, A. and Hansen, S. (1997). 'Distribution of iron among ferrite hydrates', *Cem. Concr. Res.* 27(8), 1167–77.

Escalante-Garcia, J. I. and Sharp, J. H. (1998). 'Effect of temperature on the hydration of the main clinker phases in Portland cements. Part I, neat cements', *Cem. Concr. Res.*, 28(9), 1245–57.

Ftikos, C. and Philippou, T. (1990). 'Preparation and hydration study of rich C$_2$S cements', *Cem. Concr. Res.*, 20(6), 934–40.

Fukuda, K., Maki, I., Ito, S. and Ikeda, S. (1996). 'Structure change in strontium oxide-doped dicalcium silicates', *J. Am. Ceram. Soc.*, 79(10), 2577–81.

Fullmann, T., Neubauer, J. and Walenta, G. (1999a). 'Quantitative Rietveld phase analysis of hydrated Portland cements. I. Quantitative analysis of synthetic AFm and AFt phases', *Proc. 21st, Int. Conf. Cem. Microsc.*, 103–13.

Fullmann, T., Walenta, G., Bier, T., Espinosa, B. and Scrivener, K. L. (1999b). 'Quantitative Rietveld phase analysis of calcium aluminate cements', *World Cem.*, 30(6), 91–6.

Goetz-Neunhoeffer, F. and Neubauer, J. (1997). 'Crystal structure refinement on Na-substituted C$_3$A by rietveld analysis and quantification in OPC', *Proc. 10th Int. Congr. Chem. Cem.*, li056–8 pp.

Goetz-Neunhoeffer, F. and Neubauer, J. (1998). 'Effects of raw meal substitution by sewage sludge on OPC clinker studied by Rietveld analysis', *Proc. 20th Int. Conf. Cem. Microsc.*, 130–8.

Golovastikov, R., Matveeva, R. G. and Belov, N. V. (1975). 'Crystal structure of the tricalcium silicate 3CaO·SiO$_2$=C$_3$S', *Sov. Phys. Crystallogr.*, 20(4), 441–5.

Guirado, F., Gali, S. and Chinchon, J. S. (1994). 'The crystallography of CA$_{1-x}$F$_x$ using X-ray powder diffraction techniques', *Cem. Concr. Res.*, 24(5), 923–30.

Guirado, F., Gali, S. and Chinchon, J. S. (1996). 'X-ray profile analysis of Ca$_2$Fe$_{2-x}$Al$_x$O$_5$ solid solutions', *World Cem.*, December, 73–6.

Guirado, F., Gali, S. and Chinchon, S. (2000). 'Quantitative Rietveld analysis of aluminous cement clinker phases', *Cem. Concr. Res.*, 30(7), 1023–9.

Gutteridge, W. A. (1984). 'Quantitative X-ray powder diffraction in the study of some cementive materials', *Proc. Br. Ceram. Soc.*, 35, 11–23.

Hill, R. J. and Howard, C. J. (1987). 'Quantitative phase analysis from neutron powder diffraction data using the Rietveld method', *J. Appl. Cryst.*, 20, 467–74.

Hubbard, C. R. and Snyder, R. L. (1988). 'Reference intensity ratio-measurement and use in quantitative XRD', *Powder Diffr.*, 3, 74–8.

Il'Inets, A. M., Malinovskii, Y. and Nevskii, N. N. (1985). *Dokl. Akad. Nauk SSSR*, 281, 332–6.

Jenkins, R. and Snyder, R. L. (1996). *Introduction to Powder Diffractometry.* John Wiley and Sons, New York.

Jupe, A. C., Cockcroft, J. K., Barnes, P., Colston, S. L., Sankar, G. and Hall, C. (2001). 'The site occupation of Mg in the brownmillerite structure and its effect on hydration properties: an X-ray/neutron diffraction and EXAFS study'. *J. Appl. Cryst.*, 34, 55–61.

Kennerley, R. A. (1977). 'Composition of high manganese Portland cement', *Cem. Concr. Res.*, 7(5), 565–74.

Klug, H. P. and Alexander, L. E. (1974). *X-ray Diffraction Procedures for Polycrystalline and Amorphous Materials*, 2nd edn, Wiley-Interscience, New York.

Kuzel, H. J. (1994). 'Formation of AFm and AFt phases in hydrating Portland cements', *Proc. 16th Int. Conf. Cem. Microsc.*, 125–36.

Kuzel, H. J. (1996). 'Rietveld quantitative XRD analysis of Portland cement: Part I. Theory and application to the hydration of C_3A in the presence of gypsum', *Proc. 18th Int. Conf. Cem. Microsc.*, 87–99.

Kuzel, H. J. and Pöllmann, H. (1991). 'Hydration of C_3A in the presence of $Ca(OH)_2$, $CaSO_4 \cdot 2H_2O$ and $CaCO_3$', *Cem. Concr. Res.*, 21, 885–95.

Lea, F. M. (1970). 'The Chemistry of Cement and Concrete', 3rd edn, Edward Arnold Ltd, London.

Lewis, A. C., Cockcroft, J. K., Barnes, P., Hall, C., Cernik, R. J., Tang, C. C. and Pollmann, H. (1996). 'High resolution X-ray powder diffraction studies of some Mg- and Si-substituted brownmillerites', *Mat. Sci. Forum*, Vols 228–31, pp. 759–64.

Liang, T. and Nanru, Y. (1994). 'Hydration products of calcium aluminoferrite in the presence of gypsum', *Cem. Concr. Res.*, 24, 150–8.

Madsen, I. C. and Scarlett, N. V. Y. (2000). Cement: Quantitative Phase Analysis of Portland Cement Clinker. Chapter 16, *Industrial Applications of X-ray Diffraction*, Smith and Chung (eds), Marcel Dekker, New York.

Madsen, I. C., Scarlett, N. V. Y., Cranswick, L. M. D. and Lwin, T. (2000). 'Outcomes of the international union of crystallography commission on powder Diffraction round robin on quantitative phase analysis: samples 1A to 1H', *J. Appl. Cryst.* (submitted for publication).

Mansoutre, S. and Lequeux, N. (1996). 'Quantitative phase analysis of Portland cements from reactive powder concretes by X-ray powder diffraction', *Adv. Cem. Res.*, 8(32), 175–82.

March, A. (1932). *Z. Krist.*, 81, 285.

Matulis, C. E. and Taylor, J. C. (1994). 'The effect of anomalous dispersion on the Rietveld quantification of minerals containing iron', Siroquant™ Application Note No. 9.

Meyer, H. W., Neubauer, J. and Malovrh, S. (1998). 'New quality control with standardless clinker phase determination using the Rietveld refinement', *ZKG Int.*, 51(3), 152–6, 158–62.

Moeller, H. (1995). 'Standardless quantitative phase analysis of Portland cement clinkers', *World Cem.*, 26(9), 75–6, 78, 80, 82, 84, 109–11, 116–18.

Moeller, H. (1998). 'Automatic profile investigation by the Rietveld method for standardless quantitative phase analysis', *ZKG Int.*, 51(1), 40–2, 44–50.

Motzet, H. and Poellmann, H. (1998) 'Quantitative phase analysis of high alumina cements', *Proc. 20th Int. Conf. Cem. Microsc.*, 187–206.

Motzet, H., Poellmann, H., Koenig, U. and Neubauer, J. (1997). 'Phase quantification and microstructure of a clinker series with lime saturation factors in the range of 100', *Proc. 10th Int. Congr. Chem. Cem.*, Ii039–8 pp.

Mumme, W. G. (1995). 'Crystal structure of tricalcium silicate from a Portland cement clinker and its application to quantitative XRD analysis', *Neues Jahrb. Miner. Abh.*, 169(1), 35–68.

Neubauer, J. (1997). 'Application of Rietveld quantitative X-ray diffraction analysis on technically produced OPC clinkers, OPC's and OPC/HAC blends', *13th Int. Baustofftag.*, 1-0127-1/0136.

Neubauer, J. (1998). 'Introduction of Rietveld quantitative phase analysis in OPC clinker production', *Proc. 20th Int. Conf. Cem. Microsc.*, 103–19.

Neubauer, J., Kuzel, H.-J. and Sieber, R. (1996b). 'Rietveld quantitative XRD analysis of portland cement. Part II. Quantification of synthetic and technical Portland cement clinkers', *Proc. 18th Int. Conf. Cem. Microsc.*, 100–11.

Neubauer, J. and Mayerhofer, W. (2000). 'Solid solution series of ferrate and aluminate phases in OPC: Part I. The ferrate phase', *Proc. 22nd Int. Conf. Cem. Microsc.*, 53–64.

Neubauer, J. and Pöllmann, H. (1997). 'Rietveld calculation – A high performance instrument in automatic quality control systems of clinker and cement production', *Proc. 19th Int. Conf. Cem. Microsc.*, 295–305.

Neubauer, J., Pöllmann, H. and Meyer, H. W. (1997). 'Quantitative X-ray analysis of OPC clinker by Rietveld refinement', *Proc. 10th Int. Congr. Chem. Cem'.*, 3v007, 12pp.

Neubauer, J. and Sieber, R. (1996). 'Quantification of a mixture of synthetic alite and belite by the Rietveld method', *Mater. Sci. Forum.*, 228–31 (Pt. 2, European Powder Diffraction Conference: EPDIC IV, Pt. 2), 807–12.

Neubauer, J., Sieber, R., Kuzel, H. J. and Ecker, M. (1996a). 'Investigations on introducing Si and Mg into brownmillerite – a Rietveld refinement', *Cem. Concr. Res.*, 26(1), 77–82.

O'Connor, B. H. and Raven, M. D. (1988). 'Application of the Rietveld refinement procedure in assaying powdered mixtures', *Powder Diffr.* 3, 2–6.

Odler, I. and Abdul-Maula, S. (1984). 'Possibilities of quantitative determination of the AFt-(ettringite) and AFm-(monosulphate) phases in hydrated cement pastes', *Cem. Concr. Res.*, 14(1), 133–41.

Omotoso, O. E., Ivey, D. G. and Mikula, R. (1996). 'Quantitative X-ray diffraction analysis of chromium(III) doped tricalcium silicate pastes', *Cem. Concr. Res.*, 26(9), 1369–79.

Parrott, L. J., Geiker, M., Gutteridge, W. A. and Killoh, D. (1990). 'Monitoring Portland cement hydration: comparison of methods', *Cem. Concr. Res.*, 20(6), 919–26.

Pöllmann, H., Neubauer, J., Konig, U. and Motzet, H. (1997). 'Clinker quality control using combined methods. A study using microscopy and X-ray techniques – especially Rietveld method', *Proc. 19th Int. Conf. Cem. Microsc.*, 195–221.

Pöllmann, H., Rohleder, M., Neubauer, J., Riedmiller, A. and Giske, J. (1998). 'Quantitative phase analysis of cements, lime, gypsum, building and construction products by using a new software for quantification', *Proc. 20th Int. Conf. Cem. Microsc.*, 159–74.

Prince, E. (1993). Mathematical aspects of Rietveld refinement. Chapter 3 in 'The Rietveld Method', R. A. Young (ed.), Oxford University Press, Oxford, UK.

Prodjosantoso, A. K., Kennedy, B. J. and Hunter, B. A. (2000). 'Synthesis and structural studies of strontium-substituted tricalcium aluminate, $Ca_{3-x}Sr_xAl_2O_6$', *Aust. J. Chem.*, 53(3), 195–202.

Regourd, M. (1986). 'Application of X-ray diffraction to some problems of the cement industry', *Chem. Scr.*, 26A, 37–45.

Regourd, M. (1983). 'Crystal chemistry of Portland cement phases. Chapter 3 in 'Structure and Performance of Cements', P. Barnes (ed.), Applied Science Publishers, London, New York.

Regourd, M. and Guinier, A. (1976). *Proc. 6th Int. Congr. Chem. Cem.*, Moscow, Vol. 1, p. 25, quoted in Taylor (1997) p. 425, Ref R1, and cited on p. 75.

Rietveld, H. M. (1967). 'Line profiles of neutron powder-diffraction peaks for structure refinement', *Acta Cryst.*, 22, 151–2.

Rietveld, H. M. (1969). 'A profile refinement method for nuclear and magnetic structures', *J. Appl. Cryst.*, 2, 65–71.

Ruland, W. (1961). 'X-ray determination of crystallinity and diffuse disorder scattering', *Acta Cryst.*, 14, 1180–5.

Sabine, T. M. (1993). Chapter 3 in 'The Rietveld Method', R. A .Young (ed.), Oxford University Press.

Scarlett, N. V. Y., Madsen, I. C., Manias, C. and Retallack, D. (2001). 'On-line X-ray diffraction for quantitative phase analysis: Application in the Portland cement industry. Powder Diffraction', (submitted for publication).

Schmidt, R. and Pöllmann, H. (2000). 'Quantification of calcium sulfoaluminate cement by Rietveld analysis', *Mater. Sci. Forum.*, 321–4 (Pt. 2, EPDIC 6, Proc 6th European Powder Diffr. Conf. 1998), 1022–7.

Smith, D. K. (1992). 'Particle statistics and whole-pattern methods in quantitative X-ray powder diffraction analysis', *Adv. X-ray Anal.* 35, 1–14.

Smith, D. K. and Chung, F. H. (eds), (2000). 'Industrial Applications of X-ray Diffraction', Marcel Dekker, New York.

Smith, S. T., Snyder, R. L. and Brownell, W. E. (1979). 'Minimisation of preferred orientation in powders by spray drying', *Adv. X-ray Anal.* 22, 77–88.

Struble, L. J. (1991). 'Quantitative phase analysis of clinker using X-ray diffraction', *Cem. Concr. Aggreg.*, 13(2), 97–102.

Stutzman, Paul, E. (1996). 'Pattern fitting for quantitative X-ray powder diffraction analysis of Portland cement and clinker', *Proc. 18th Int. Conf. Cem. Microsc.*, 10–20.

Suherman, P. M., van Riessen, A. and O'Connor, B. H. (2000). 'Curtin University of Technology', Western Australia, private communication.

Talero, R. (1996). 'Comparative XRD analysis of ettringite originating from pozzolan and from Portland cement', *Cem. Concr. Res.*, 26(8), 1277–83.

Taylor, H. F. W. (1997). *Cement Chemistry*, 2nd edn, Thomas Telford, UK.

Taylor, J. C. (1990). 'Computer programs for standardless quantitative analysis of minerals using the full powder diffraction profile', *Powder Diffr.* 6(1), 2–9.

Taylor, J. C. and Zhu Rui (1992). 'Simultaneous use of observed and calculated standard profiles in quantitative analysis of minerals by the multiphase Rietveld method: the determination of pseudorutile in mineral sands products', *Powder Diffr.* 7(3), 152–61.

Taylor, J. C. and Aldridge, L. P. (1993a). 'Full-profile Rietveld quantitative XRD analysis of Portland cement: standard XRD profiles for the major phase tricalcium silicate (C_3S: $3CaO \cdot SiO_2$)', *Powder Diffr.* 8(3), 138–44.

Taylor, J. C. and Aldridge, L. P. (1993b). 'Phase analysis of Portland cement by full profile standardless quantitative X-ray diffraction-accuracy and precision', *Adv. X-ray Anal.* 36, 309–14.

Taylor, J. C. and Aldridge, L. P. (1994). 'Quantitative X-ray diffraction analysis of Portland cements', *Int. Ceram. Monogr.*, 1(1), *Proc. Int. Ceram. Conf.*, 141–6.

Taylor, J. C., Zhu Rui and Aldridge, L. P. (1993). 'Simultaneous use of observed and calculated standard patterns in quantitative XRD analysis of minerals by the multiphase Rietveld method: application to phase quantitation of mineral sands and Portland cement', *Mater. Sci. Forum.*, 133–6 (*Proc. 2nd European Powder Diffr. Conf. 1992*, Pt. 1), 329–34.

Taylor, J. C. and Matulis, C. E. (1991). 'Absorption contrast effects in the quantitative XRD analysis of powders by full multiphase profile refinement', *J. Appl. Cryst.*, 24, 14–17.

Taylor, J. C. and Clapp, R. A. (1992). 'New features and advancd applications of Siroquant™: A personal computer XRD full profile quantitative analysis software package', *Adv. X-ray Anal.*, 35, 49–55.

Taylor, J. C., Hinczak, I. and Matulis, C. E. (2000). 'Rietveld full-profile quantification of Portland cement clinker: The importance of including a full crystallography of the major phase polymorphs', *Powder Diffr.*, 15(1), 7–18.

Theisen, Kirsten. (1999). 'Phase composition of clinker measured by microscopy compared with quantitative X-ray diffraction (Rietveld) and Bogue results', *Proc. 21st Int. Conf. Cem. Microsc.*, 353–66.

Vonk, C. G. (1983). 'The determination of the crystallinity in glass ceramic materials by the method of Ruland', *J. Appl. Cryst.*, 16, 274–6.

Yang, R. (1996). 'Crystallinity determination of pure phases used as standards for QXDA in cement chemistry', *Cem. Concr. Res.*, 26(9), 1451–61.

Yang, R., Lawrence, C. D. and Sharp, J. H. (1996). 'Delayed ettringite formation in 4-year old cement pastes', *Cem. Concr. Res.*, 26(11), 1649–59.

Young, R. A. (1993). 'The Rietveld Method. IUCR Monographs on Crystallography 5', Oxford University Press, Oxford, UK.

Zachariasen, W. H. (1945). 'Theory of X-ray diffraction in crystals', Dover Publications Inc., New York.

Chapter nineteen

Electrical monitoring methods in cement science

W. J. McCarter, G. Starrs and T. M. Chrisp

19.1 Introduction

A variety of techniques have been exploited in the study of cementitious systems. These are, most notably, scanning electron microscopy, X-ray diffraction, mercury intrusion porosimetry (MIP), isothermal conduction calorimetry, magic angle spinning, nuclear magnetic resonance, small angle neutron scattering and small angle X-ray scattering. Electrical property measurements, as applied to cementitious materials, represent an additional (and still developing) investigative technique in the study of these materials both at the micro- and macro-scale. This does not imply that electrical measurements on cementitious systems are a recent development. On the contrary, fixed frequency electrical resistance measurements on Portland cement pastes date back to the 1930s [1–4]. However, the advent of variable frequency *impedance response analysers*, with the capability of obtaining electrical measurements over several decades of frequency, has renewed interest in the development of this methodology as a characterization technique [5,6].

Electrical data can be presented in a variety of formalisms,

(i) the use of complex-plane plotting techniques as an aid in developing an equivalent electrical model for the impedance response and then ascribing a physical interpretation to the model. This format of presentation is extensively used in the study of ceramics and solid state ionic materials and has been transferred to cement-based systems;
(ii) to present the electrical response in terms of the intrinsic electrical properties of the material in the frequency domain. This format of presentation is undertaken to identify the fundamental processes which give rise to electrical response.

All formats of presentation should be exploited in order to gain an insight into the nature of electrical response within cement-based systems, only then can advances be made. The combination (i) and (ii) above represent the field of immittance spectroscopy.

19.2 Immittance formalisms

19.2.1 Background

Consider a cement/water mixture contained between the electrode plates of a dielectric cell (Figure 19.1). Under the action of an alternating electrical field ions in solution are free to drift

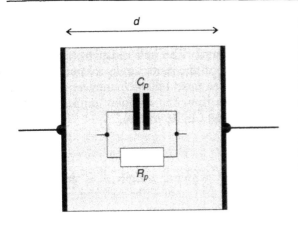

FIGURE 19.1 Idealized electrical representation of cement-paste.

through the continuous, interstitial aqueous phase and discharge at the electrodes thereby producing an ionic conduction effect. This effect is quantified by the sample resistance. Under the applied field ions are 'blocked' at pore water/crystal boundary interfaces; other charges, electrostatically held onto grain/crystal surfaces, can oscillate or diffuse (short-range) in sympathy with the applied alternating field. Such charges do not contribute to ionic conduction, but contribute to polarization processes within the system [7]. These two particular mechanisms are described as, respectively, interfacial polarization and double-layer polarization. Polarization can be regarded as an oscillation or reorientation of charges in an applied field from their zero field equilibrium position and is quantified by the sample capacitance. Double-layer effects have been shown to produce anomolously high capacitance values in particulate systems [8–10]. Likewise, thin, highly resistive sheets can induce a large capacitance effect [11,12]. Other polarization processes include dipole polarization; atomic polarization and electronic polarization, although these represent high/ultra-high frequency (high Megahertz/Gigahertz) polarization processes [7]. The electrical response at frequencies below 100 MHz (approximately) may be more informative with regard to microstructural changes and pore structure development within the paste.

Capacitance is a measure of the sum total of all polarization processes operative at a particular frequency of applied field, although a particular polarization process tends to dominate over a certain frequency range. For example, the double-layer polarization process dominates in the low kilohertz region, whereas at frequencies in the megahertz region, such a process would cease to operate and hence, its contribution to the overall polarizability of the cement paste would be negligible. Thus, as the frequency of applied field increases, the polarizability of the system reduces and the measured capacitance decreases. This manifests itself as a region of dispersion or relaxation in the frequency domain.

19.2.2 Impedance response

The electrical response of cement paste to a small-amplitude, sinusoidal, alternating electrical field can, in general terms, be represented by a parallel combination of resistor (R_p) and capacitor (C_p) as depicted in Figure 19.1. If R_p and C_p in Figure 19.1 are assumed independent of frequency, the complex admittance of the system, $Y(\omega)$, at any angular frequency, ω, is represented in rectangular form as,

$$Y(\omega) = 1/R_p - j\omega C_p \text{ Siemens (S)} \quad (1)$$

where, $j = \sqrt{-1}$ and $\omega = 2\pi f$, where f is the frequency of the applied electrical field in Hertz. The complex impedance, $Z(\omega)$, is,

$$Z(\omega) = \frac{1}{Y(\omega)} = \frac{1}{\dfrac{1}{R_p} + j\omega C_p} \text{Ohms } (\Omega) \quad (2)$$

Rationalizing equation (2) gives,

$$Z(\omega) = \frac{R_p}{1 + (\omega R_p C_p)^2} - j\frac{\omega R_p 2 C_p}{1 + (\omega C_p R_p)^2} \quad (3)$$

or,

$$Z(\omega) = Z'(\omega) - jZ''(\omega) \quad (4)$$

where,

$$Z'(\omega) = \frac{R_p}{1 + (\omega R_p C_p)^2} \text{ and}$$

$$Z''(\omega) = \frac{\omega R_p C_p}{1 + (\omega C_p R_p)^2} \quad (5)$$

$Z'(\omega)$ and $Z''(\omega)$ represent the parametric equations of a semi-circle and depend solely on ω. By plotting the real ($Z'(\omega)$) and imaginary ($Z''(\omega)$), parts of the complex impedance over a sufficiently wide frequency range will result in a semicircle as shown in Figure 19.2a with centre on the real axis and frequency increasing from right to left across the arc. If this response was produced from a cement paste, then it could be modelled by the parallel circuit shown in Figure 19.2a and the values of R_p and C_p obtained from the complex plot.

Work has shown that for cementitious systems, this ideal behaviour does not occur [5,6] and the response comprises a low frequency spur and a circular arc with centre depressed below the real axis as shown in Figure 19.2b. The spur has been identified as that part of the complex response resulting from polarization effects at the electrode cement-paste interface – so called electrode polarization – whereas the circular arc represents the response from the bulk material. The depressed semi-circular arc can be described by a parallel combination of resistance R and capacitor C. Regarding the latter, the capacitance is more correctly termed a constant phase element (CPE) [13]. The CPE is not a pure capacitor, but is introduced in order to explain the depressed centre of the arc resulting from the dispersive contribution due to the bulk dielectric response (discussed below). R is the bulk ionic resistance of

the system and is obtained from the intercept with the real axis at the low-frequency end of the arc as indicated. The bulk resistivity (ρ) or conductivity (σ) of the paste can be accurately identified and is the most reliable parameter which can be obtained from the complex impedance plot. The following can be written,

$$\rho = \frac{RA}{d} = \frac{1}{\sigma} \tag{6}$$

where, for a prismatic sample, A is the electrode surface area (m^2) and d the electrode separation. The CPE is quantified by,

$$C_{CPE} = \frac{(2\pi f_c)^{-(1-\frac{2\beta}{\pi})}}{R} \tag{7}$$

where β is the arc depression angle (in radians) and f_c is the characteristic frequency at which the bulk arc maximizes. An apparent dielectric constant, ε_a, is then given by,

$$\varepsilon_a = \frac{C_{CPE} d}{\varepsilon_0 A} \tag{8}$$

and ε_o the permittivity of free space (8.854×10^{-12} F/m).

The complex impedance plot allows accurate de-convolution of electrode and bulk processes, thereby allowing evaluation of the bulk resistance of the system, R. Complex impedance measurements

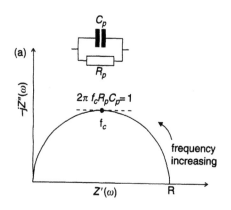

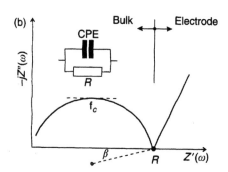

FIGURE 19.2 (a) Complex impedance response for parallel combination of resistor (R_p) and capacitor (C_p); f_c is the frequency at which $-jZ''(\omega)$ maximizes; (b) Schematic diagram of the complex impedance response for a porous ionic conductor placed between a pair of electrodes.

are thus essential in a priori optimization of the frequency at which a single frequency measurement should be taken in order to minimize errors resulting from electrode polarization effects.

19.2.3 Permittivity response

The complex impedance format of presentation is useful in developing an equivalent circuit representation for cement-paste. It must be emphasized that the bulk electrical properties of the material (i.e. its resistance and capacitance) are frequency-dependent resulting from motion of both free and bound charges within the system. If a frequency dependence for R_p and C_p in Figure 19.1 is introduced as $R_p(\omega)$ and $C_p(\omega)$, then, using equation (1) and (8) above, the complex dielectric constant of the system, $\varepsilon_r(\omega)$, is,

$$\varepsilon_r(\omega) = \frac{-jd}{\omega\varepsilon_0 A} \ Y(\omega) = \frac{d}{\varepsilon_0 A}\left\{C_p(\omega) - \frac{j}{\omega R_p(\omega)}\right\} \quad (9)$$

or simply,

$$\varepsilon_r(\omega) = \varepsilon_r'(\omega) - j\varepsilon_r'(\omega) \quad (10)$$

where $\varepsilon_r(\omega)$, is the relative permittivity of the system, comprising the dielectric constant, $\varepsilon_r'(\omega)$, and the dielectric loss $\varepsilon_r''(\omega)$; A, d and ε_0 are as previously defined. The frequency dependency of all parameters is maintained. In general terms, the dielectric constant accounts for energy storage resulting from the polarization of charges within the material, and loss accounts for the dissipation of energy incurred in the polarization process (e.g. frictional effects) together with any direct transfer of charge through the material. Loss may also be represented in terms of conductivity, $\sigma(\omega)$, which is related to $\varepsilon_r''(\omega)$ by the expression,

$$\varepsilon_r''(\omega) = \frac{\sigma(\omega)}{\omega\varepsilon_0} \quad (11)$$

Conductivity can result from both direct (ionic) conduction processes and dissipative polarization processes. If the system has a known value of d.c. (or low frequency) ionic conductivity (from equation (6)) and discussed in Section 19.2.2 above, it

may be desirable to express the effects of ionic conduction separately from purely dissipative dielectric effects, in which case,

$$\sigma(\omega) = \sigma_d(\omega) + \sigma(0) \quad (12)$$

where, $\sigma(0)$ is the low-frequency, bulk ionic conductivity obtained from equation (6), and $\sigma_d(\omega)$ represents the dielectric conductivity due to dissipative processes resulting from energy lost by motion of the polarized charges in the applied electrical field. More specifically, the relative permittivity can be written as,

$$\varepsilon_r(\omega) = \varepsilon_r'(\omega) - j\left\{\varepsilon_r''(\omega) + \frac{\sigma(0)}{\omega\varepsilon_0}\right\} \quad (13)$$

According to the classical approach of Debye [7], the dispersion of the relative permittivity resulting from a single polarization mechanism can be described by the expression,

$$\varepsilon_r(\omega) = \varepsilon_{r\infty} + \frac{\varepsilon_{rs} - \varepsilon_{r\infty}}{\left(1 + (j\omega\tau_0)^{(1-\alpha)}\right)} \quad (14)$$

where, τ_0 is the relaxation time-constant of the process ($\tau_0 = 1/\omega_0$, and $\omega_0 = 2\pi f_0$, where f_0 is the relaxation frequency), ε_{rs} is the dielectric constant of the material as ω approaches low frequencies (i.e. $\omega \ll \omega_0$), and $\varepsilon_{r\infty}$ is the dielectric constant as ω approaches high frequencies ($\omega \gg \omega_0$). The exponent α ($0 < \alpha < 1$) is a dispersion factor which is included to account for the statistical distribution of relaxation times about a mean value of τ_0 which is observed in real materials. Separation of the real and imaginary parts of equation [14], and comparison with equation [10] yields expressions for dielectric constant, $\varepsilon_r'(\omega)$, and conductivity, $\sigma(\omega)$, as a function of frequency. When $\alpha \to 0$, the resulting response has the form shown in Figure 19.3. In a material such as concrete, mortar or cement-paste, several superimposed relaxation mechanisms can be operating simultaneously. Each of these will make its own contribution to the bulk capacitive response and will relax in accordance with its own time-constant thereby increasing α from is 'ideal' value of 0. If the relaxation time of the polarization processes are sufficiently disparate for each mechanism, then the

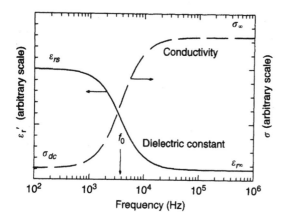

FIGURE 19.3 Showing dispersion in dielectric constant (ε_r') and conductivity (σ) with relaxation frequency, f_0 (determined at the centre of region of dispersion).

individual features of the overall response will be discernible in the frequency domain.

There is no simple or direct relationship between the value of the impedance level relaxation frequencies (f_c), and that occurring at the permittivity level (f_0). It could, therefore, be incorrect to interpret relaxation occurring in the impedance plane as intrinsic dielectric relaxation. Also, viewing the electrical response at the impedance level MAY TELL little about the intrinsic mechanisms which give rise to the response at the permittivity level. In order to develop an understanding of the underlying mechanisms responsible for the electrical response, the presentation format of the data should not be restricted to complex-plane plotting. This will only give a partial picture and immittance spectroscopy can be advantageously employed. The term 'immittance' covers several levels of response that can be obtained from electrical measurements, provided these are observed over a wide-enough frequency range. These are: relative permittivity $\varepsilon_r(\omega)$ and its reciprocal electric modulus $M(\omega)$; impedance $Z(\omega)$ and its reciprocal admittance $Y(\omega)$.

19.2.4 An overview

Immittance spectroscopy, as applied to cement-based systems, represents a developing field and it

cannot be said that all aspects have been systematically studied. This is further complicated by the fact that, as yet, there are no standardized tests procedures for immittance measurements which include, electrode configuration, test-cell geometry, sample mixing procedures, operating voltages, etc. This results in researchers developing their own testing protocols and, as a consequence, it is difficult to compare results from different laboratories. Regarding interpretation, one of the main difficulties lies in isolating the contribution from each of the factors which directly influence the electrical response of the system together with inadequate theoretical models, although advances have being made in this field [e.g., 11,14–20].

The role of both water, ionic species contained therein, and the continuously evolving pore network are critical in all aspects of the science of cement-based materials including strength, permeability, diffusion, sorptivity, creep and shrinkage. Concerning durability-related problems with concrete structures, it is the ingress of water and water-containing deleterious ions (e.g. sulphates, chlorides) via the connected pore network that is instrumental in deterioration processes. The electrical response of a cementitious system will be dependent upon the nature and topography of the pore network (e.g. pore tortuosity, pore constriction, pore connectivity), the distribution of water within the system, and ionic concentration within the pore water. In addition, reactions at interfaces play a major role in determining the subsequent chemistry and pore structure development. Immittance measurements are thus quantifying those factors which are of relevance to all the important characteristics of the cement.

19.3 Application of electrical measurements

The application of electrical measurements in the study of cement-based systems both at the micro- and macro-scale is reviewed below. While the review is by no means exhaustive, it highlights the way in which the various immittance formalisms (and variations thereof) can be exploited.

19.3.1 Electrical measurements on cementitious systems during early hydration

19.3.1.1 Monitoring early hydration

Monitoring the resistance, R, at a fixed frequency of applied electrical field has proved useful in studying the early hydration of cementitious binders [21–28] and quantifying the influence of admixtures on setting of cements [29–32]. Consider, for example, Figure 19.4a, which displays the change in conductance, G ($=1/R$), for ground granulated blast-furnace slag (GGBS), metakaolin (MK), micro-silica (MS) and pulverised fuel ash (PFA) which have been mixed with an alkaline activator [33]. The frequency at which the conductance was determined (10 kHz) had been optimized such that electrode effects were reduced to minimal proportions.

An increase in rigidity of the binder is clearly discernible by a rapid decrease in G. Another informative plot can be obtained by taking the derivative of this response and would reflect the intensity of chemical activity within the binder [27,33]. For illustrative purposes, Figure 19.4b presents the derivative of the GGBS curve in Figure 19.4a. From the dG/dt curve, several stages in the early hydration can be identified and indicated as I to V

on this plot. These periods would correspond to, respectively, the initial (dissolution), dormant, acceleration, deceleration and steady-state (diffusion controlled), and not dissimilar to those identified from conduction calorimetry studies.

Monitoring the change in the as-measured capacitance at a spot frequency, that is, $C_p(\omega)$ in equation (9) can also be informative with regard to microstructural changes within the paste. $C_p(\omega)$ is not the same as the CPE, the latter being a curve-fit quantity derived from the complex impedance plot through equation (7) and depicted in Figure 19.2b. Figures 19.5a and b present the temporal change in capacitance, $C_p(\omega)$, for the binders in Figure 19.4a and determined at the same spot frequency as the conductance. Apart from PFA (Figure 19.5b, note change in scale in from Figure 19.5a), the response is dominated by a maximum in the central portion of each response. This feature has been observed in other types of cementitious binders at frequencies extending over the range 1 kHz to 30 MHz [22,25,32,34,35]. Of particular interest, the peak in capacitance coincides with time at which dG/dt maximizes (see Figure 19.4b for GGBS paste) and is the period of rapid increase in rigidity of the system.

A thin-blocking-layer analogy has been invoked to explain the transitory peak in capacitive

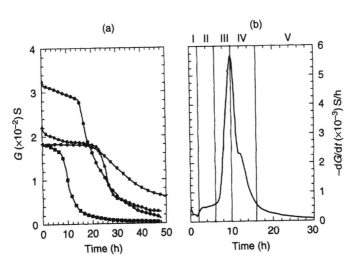

FIGURE 19.4 (a) Variation in conductance, G, for alkali-activated pozzolans: ■ GGBS; ▲ MK; ● MS; and ▼ PFA; (b) Derivative of the GGBS response curve presented in (a). (From [33] by permission Thomas Telford Ltd.)

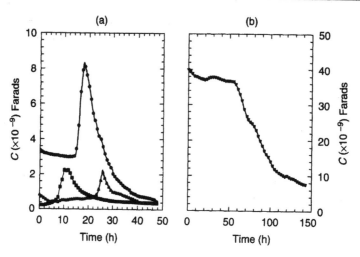

FIGURE 19.5   Variation in capacitance, *C*, for the binders in Figure 19.4 (a) ■ GGBS; ▲ MK; ● MS; (b) ▼ PFA. (From [33] by permission Thomas Telford Ltd.)

response [12,33]. The capacitance, $C_p(\omega)$, of a material contained between the plates of a dielectric cell is given by $C_p(\omega) = \varepsilon_r'(\omega)\varepsilon_0 A/d$ (see equation (9)) where, in this instance, $\varepsilon_r'(\omega)$ is the dielectric constant of the products of hydration. It is evident that by reducing *d*, a thin resistive sheet can produce a high capacitance. The schematic diagram in Figure 19.6 depicts fibrillar outgrowths from the grain surface into the aqueous phase and could be regarded as a series of thin blocking layers. Under the action of an applied alternating field, ions of opposite charge are blocked on either side of the layers and this is analogous to a sequence of small capacitors whose capacitance will be linked to the thickness and surface area of the blocking hydrate. Ions trapped in the aqueous phase between the products of hydration are unavailable for ionic conduction processes but produce a solid–liquid-interface capacitive effect. This is an interfacial polarization process, which will be frequency dependent. It would be anticipated that the prominence of the capacitance peak would decrease with increasing frequency and would not be observed at very high frequencies as in the case of ordinary Portland cement (OPC) and OPC/GGBS pastes tested at frequencies in the Gigahertz region [36–38]. The prominence of the peak would also be dependent

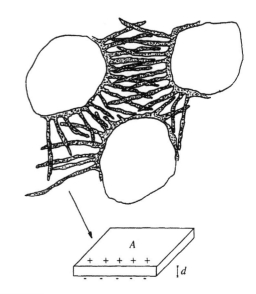

FIGURE 19.6   Schematic representation of hydrate formation to show possible origin of peak in capacitance over setting. (From [33] by permission Thomas Telford Ltd.)

upon the intensity of reaction and a sluggish reaction may not produce a capacitance peak, as in the case of the PFA in Figure 19.5b. The peak is only transitory as rapid infilling and accretion with products of hydration result in increasing thickness, and reducing surface area, thereby decreasing capacitance.

19.3.1.2 PFA/OPC mixtures

Some interesting studies have been published [39–41] on the impedance response on freshly mixed pastes, mortars and concretes (60 min after gauging), where the OPC binder has been partially replaced with PFA obtained from a bituminous coal conforming to BS3892:Part 1:1993/ASTM Class F. Figure 19.7 presents the complex impedance plots for freshly mixed mortar specimens with and without PFA replacement. The frequency range represented in this plot is 1 Hz–1 MHz. The specimens with plain OPC binder display a typical two-region response comprising an electrode polarization spur and bulk arc to the right and left, respectively, of the minimum point on the plot. The bulk arc is only partially developed due to the upper frequency limit of 1 MHz. The specimens with PFA display a plot that is now characterized by three distinct zones: the electrode response; a *plateau* region due to inclusion PFA, and the high frequency arc. In a conventional impedance analysis using the CPE concept, two completely separate arc depression parameters (β) would be assigned to both the plateau region and high frequency arc region, and

could give a false impression of the underlying effects responsible for the electrical response.

The dielectric constant and conductivity can be de-imbedded from the impedance data using equation (9) and the resulting dielectric and conductivity dispersion curves plotted. These are presented in Figures 19.8a and b for the curves in Figure 19.7.

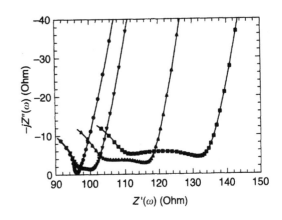

FIGURE 19.7 Complex impedance response for freshly mixed mortars taken 60 min after mixing: ● plain OPC mortar; ▼ 10 per cent PFA replacement; ▲ 25 per cent PFA replacement; ■ 40 per cent PFA replacement. (From [41] by permission Elsevier Science.)

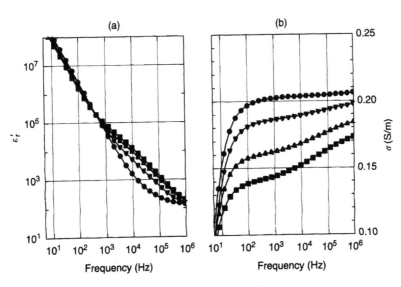

FIGURE 19.8 (a) Dispersive behaviour of dielectric constant for mixtures in Figure 19.7: ● plain OPC mortar; ▼ 10 per cent PFA replacement; ▲ 25 per cent PFA replacement; ■ 40 per cent PFA replacement. (b) Increase in bulk conductivity for mortar mixtures in (a). (From [41] by permission Elsevier Science.)

The decrease in dielectric constant and rise in conductivity with increasing frequency is indicative of a region of dispersion resulting from relaxation of a polarization process within the system. At frequencies below approximately 1 kHz, the dielectric constant rises to extremely high values, and is attributed to the increasing influence of the electrode/sample interfacial polarization effect. The inclusion of PFA in the binder results in an enhancement of the bulk polarization processes in comparison with the plain OPC mix, particularly over the range 10 kHz–1 MHz. Further supporting evidence can be obtained from the conductivity dispersion curves. The conductivity curves can be divided into two regions above and below 1 kHz (approximately). Below 1 kHz, and particularly over the region 1–100 Hz, the sharp increase in conductivity with increasing frequency is as a result of the reduction in electrode effects (as noted above from the dielectric dispersion data). At frequencies in excess of 1 kHz, there is a more gradual rise in conductivity with increasing frequency; of particular interest is that this dispersion in conductivity is more apparent in the PFA mixtures. The frequency range over which the increase in conductivity occurs is the same as that for the dielectric enhancement effect of the PFA. The increase in conductivity with increasing frequency of applied field is as a direct result of relaxation of the polarization mechanism, which, in addition to ionic conduction effects will contribute to loss processes (see equation (12)). The overall result will be an increase in measured conductivity. It is also noticeable that increasing the level of PFA replacement results in a progressive decrease in bulk conductivity.

A double-layer polarization process on the grain surface has been postulated to account for such a response which is a low-frequency polarization echanism relaxing in the kilohertz region. A double-layer polarization process has been known to produce anomalously high dielectric constants in particulate systems [8–10] and it is postulated that the enhancement in the PFA system is due to the distinctive spherical nature of the PFA particle [41]. Theoretical plots have been produced for dielectric constant versus frequency, conductivity versus frequency, and complex impedance [41]. The calcu-

lated response is based on a polarization process relaxing in the kilohertz region with a dispersion spreading factor of 0.194 (α in equation (14)); electrode effects have not been modelled in the theoretical curves. Observed and calculated response are generally very good. This approach thus allows two apparently unrelated effects in the complex impedance response to be attributed to a single polarization process, thereby simplifying interpretation of what is happening electrically within the material.

19.3.2 Electrical measurements on hardened systems

Complex impedance measurements on hardening cementitious systems have provided useful insights into their evolving microstructure. The change in diameter of the bulk arc (hence R), the arc depression angle (β), and the characteristic frequency (f_c) have all been used to characterize the underlying pore structure [see, for example, 5,12,17,20,42–44]. The bulk resistance, R, is the most reliable electrical parameter that can be determined from impedance measurements. This parameter can be converted to conductance, G ($=1/R$); conductivity, σ, through equation (6), or resistivity, ρ ($=1/\sigma$). The bulk resistance is dependent upon the connectedness of the fluid-filled capillary pores and ionic concentration within the pore water. As a consequence, such measurements can be used to study the long-term refinement of the capillary pore network.

Since the movement of water under a pressure differential or the movement of ions under a concentration gradient is analogous to the movement of current under a potential difference, there is growing interest in the application of conductivity measurements on saturated cement-based materials to assess diffusivity and permeability. Outlined below is representative work on the application of resistance measurements as a characterization parameter in hardening cements.

19.3.2.1 Bulk Conductivity and normalized conductivity

The decrease in conductivity, σ, with time for OPC mortar mixes, with and without pozzolanic additions, is presented in Figure 19.9a [45]. Although the conductivity of the mixes with pozzolanic additions is initially higher than that of the plain OPC mortar, in the longer term, these mixes achieve a considerably lower value. In order to evaluate the relative effects of changing microstructure and pore-fluid conductivity, the normalized conductivity format can be used. Figure 19.9b presents the change in the normalized conductivity, N, for the same mixes over the initial 28-days' hydration where N is defined as,

$$N = \sigma/\sigma_{pore} \qquad (15)$$

and σ_{pore} is the respective value of pore fluid conductivity at that point in time. Since N decreases with time, this implies that the bulk conductivity is decreasing at a faster rate than changes in pore-fluid conductivity. Microstructural changes (e.g. increasing pore tortuosity, pore constriction) are thus exerting a more dominant effect on the conductivity than changes in pore-fluid chemistry.

19.3.2.2 Conductivity, diffusion and permeability (at 20°C)

Provided that the solid phase can be regarded as an insulator (in comparison to that of the aqueous phase), diffusion and ionic conductivity of a saturated porous system are connected through the Nernst–Einstein relationship. This relationship has been adapted for cementitious materials [12,46–52] in the form,

$$\frac{\sigma}{\sigma_{pore}} \sim \frac{D_{eff}}{D_0} = Q \qquad (16)$$

where D_0 is the diffusivity of the species at infinite dilution; D_{eff} the effective diffusivity of the same species in the mortar, and Q is termed the diffusibility of the medium. The normalized conductivity, N, in equation (15) above and presented in Figure 19.9b, would thus represent the change in diffusibility with time for the mortar mixes. Equation (16), however, may only be valid provided a suitable value of D_0 is chosen [52], and using the value of the species at infinite dilution may be inappropriate in predicting D_{eff}. Notwithstanding this difficulty, a bulk conductivity measurement may, on its own, allow an assessment of D_{eff} [53]. Considering the difficulties associated with divided-cell, diffusion tests, conductivity techniques could thus be a viable alternative allowing a rapid, relative assessment of diffusivity.

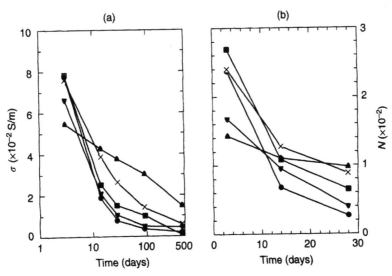

FIGURE 19.9 (a) Decrease in conductivity with time for five binders types: ▲ plain OPC; × OPC/GGBS; ▼ OPC/MK; ● OPC/MS; and ■ OPC/MK/GGBS [45]; (b) Variation in normalized conductivity, N, for mixtures in (a) over initial 28-days' hydration [45].

Regarding permeability measurements, the Katz–Thompson approach to predict the permeability of porous rock formations [54] has been applied to cement-based systems [12,55,56]. Katz and Thompson developed a relationship involving two microstructural parameters:

(i) a characteristic length scale, obtained as the threshold pore diameter, d_c, from MIP tests (at the inflection point in the dV/dP versus P curve); and,

(ii) a measure of the pore-system connectivity obtained from σ/σ_{pore} as defined above.

The Katz–Thompson equation is written as,

$$k' = \frac{1}{226} \cdot \frac{\sigma}{\sigma_{pore}} \cdot d_c \qquad (17)$$

where k' is the intrinsic permeability in m^2. This can be converted to permeability in m/s through the relationship,

$$k = k' \cdot \frac{\rho g}{\eta} \qquad (18)$$

where ρ is the density of water; g is the gravitational acceleration and η the viscosity of the permeating fluid. The applicability of this approach to cementitious systems has been criticized [55,57] due to the inherent problems in the mercury intrusion process associated with sample size and preparation techniques. Some limited studies utilizing both MIP and conductivity data on young cement-pastes (< 28 days old) indicate reasonable agreement with equation (17) above [56] although conflicting results have been published [58].

In connection with assessment of concrete permeability, electrical conductivity measurements could gain acceptance as an alternative to the ASSHTO-T277/ASTM-C1292 rapid chloride permeability test [59,60]. As with diffusion tests, a bulk conductivity measurement may thus be sufficient to assess the relative performance of cementitious systems in this respect.

19.3.2.3 Conductivity within cover-zone concrete

The performance of the cover-zone is acknowledged as a major factor governing the rate of degradation of reinforced concrete structures and provides the principal barrier to water and aggressive agents. It is clear that there exists a need to determine quantitatively those near-surface permeation properties of concrete which promote the ingress of gases or liquids containing dissolved contaminants. In addition, *in situ* monitoring of the long-term change in such properties could assist in making realistic predictions as to the in-service performance of the structure; likely deterioration rates for a particular exposure condition, or compliance with the specified design life.

Electrical measurements are being developed to study water and ionic movement within the cover-zone of reinforced concrete with direct application in durability monitoring of concrete structures [61–64]. Electrode arrays embedded within the surface region (50 mm) of concrete represent a permanent monitoring system and allow evaluation of conductivity at discrete depths from the exposed concrete surface. The advance of a waterfront into concrete can be studied and conductivity profiles through the cover-zone can be evaluated. For example, Figure 19.10a, presents the relative change in bulk resistance, denoted R_R, across embedded electrodes as water is absorbed into the cover-zone. R_R is the ratio R_t/R_0 where R_0 is the measured resistance across an electrode pair just prior to the start of the absorption test and R_t the resistance measured at time, t, after the start of the test [64]. Figure 19.10b also presents bulk conductivity profiles within the covercrete over consecutive cycles of drying and wetting namely, just before and immediately after the period of water absorption [64]. Such measurements can be used to:

1. evaluate the depth and rate of water penetration into the cover zone [62];
2. determine the conductivity of the concrete to rebar level (i.e. 50 mm). It is now recognized that once de-passivated, the electrical properties of the concrete in the vicinity of the rebar play an important role in corrosion kinetics [65]. Chloride ingress can also be studied [63,64];
3. quantify drying effects, and hence the extent of the convective zone [66];

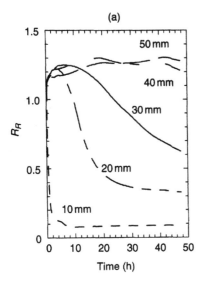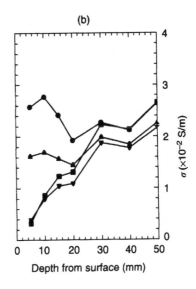

(a)   (b)

FIGURE 19.10   (a) Relative change in resistance, R_R, at discrete depths (indicated) within the cover-zone during a 48 h water absorption test. (From [64] by permission Thomas Telford Ltd.); (b) Variation in conductivity through the cover-zone before and after cycles of drying and subsequent water absorption: ■ prior to 1st absorption cycle; ● end of 1st absorption cycle; ▼ prior to 6th absorption cycle; ▲ end of 6th absorption cycle.

4. study on-going hydration and pozzolanic activity within cover-zone concrete [64];
5. assess the efficacy of surface treatments on concrete substrates [67,68].

Long-term durability monitoring using embedded electrode arrays is a technically valid and cost-effective method of deferring repairs, evaluating repair options and monitoring the structure and its environment with respect to corrosion and degradation processes. Electrical measurements can be collected at regular time intervals; such data inputs could then be used by engineers in the development of service-life prediction models.

19.4 Concluding remarks

The work outlined above has served to present the basic theory underpinning electrical measurements together with applications to hydrating cementitious materials. It is evident that in order to exploit electrical measurements in cement and concrete science, data should be considered in a range of formats. The applications section, although by no means exhaustive, has highlighted

several presentation formalisms other than complex impedance. Not only does the testing methodology represent a laboratory technique for studying cementitious systems, but applications for use in concrete structures are also being developed. Other areas where electrical methods are finding application include, for example,

(a) the use of the electrical properties of concrete incorporating carbon-fibre, with potential application in *smart-concrete* structures [69–71];
(b) in the quality control of fresh (i.e. liquid state) concrete [72];
(c) the study of the aggregate/cement-paste interface region [73].

Electrical methods also have a number of distinct advantages,

1. measurements can be taken on large samples hence bulk effects can be assessed. Furthermore, samples need not necessarily be restricted to neat cement pastes, but mortars and concretes can also be studied;
2. measurements are carried out at normal pressures thereby avoiding water removal, hence

453

the cement microstructure is not disrupted or damaged;

3. the test method is rapid, non-invasive and non-destructive and allows virtually continuous monitoring of a sample.

As noted in Section 19.2.4 above, if standardized testing procedures can be developed then the testing methodology could become a standard investigative method in cement and concrete science and represent an additional technique in the cement scientist's armoury.

19.5 Acknowledgements

Much of the authors' work cited within the text has been undertaken within a series of research grants from the Engineering and Physical Sciences Research Council (EPSRC), United Kingdom. The authors wish to acknowledge this financial support.

19.6 References

1. Petin, N. and Gajsinovitch, E., 'A study of the setting process of cement paste by electrical conductivity methods', *J. Gen. Chem.*, U.S.S.R., 2, 614–29 (1932).
2. Dorsch, K. E., 'The hardening and corrosion of cement-IV', *Cement and Cement Manufacture*, 6(4), April, 131–42 (1933).
3. Boast, W. B., 'A conductometric analysis of Portland cement pastes and mortars and some of its applications', *J. Amer. Concr. Inst.*, 33, Nov.–Dec. pp. 131–46 (1936).
4. Kind, V. A. and Zhuraler, V. F., 'Electrical conductivity of setting Portland cements', *Tsement* 5(9), 21–6 (1937).
5. McCarter, W. J., Garvin, S. and Bouzid, N., 'Impedance measurements on cement paste', *J. Mater. Sci. Lett.*, 7, 1056–7 (1988).
6. McCarter, W. J. and Brousseau, R., 'The A.C. response of hardened cement paste', *Cem. Concr. Res.*, 20(6), 891–900 (1990).
7. Hasted, J. B., *Aqueous Dielectrics*, Chapman and Hall, London (1973).
8. Schwan, P., Schwarz, G., Maczuk, J. and Pauly H. 'On the low-frequency dielectric dispersion of colloidal particles in electrolytic solutions', *J. Phys. Chem.*, 66, 2626–35 (1962).
9. Schwarz, G., 'A theory of the low-frequency frequency dielectric dispersion of colloidal particles in electrolyte solution', *J. Phys. Chem.*, 66, 2636–42 (1962).
10. Chew, W. C. and Sen P. N., 'Dielectric enhancement due to the electrochemical double layer: Thin double layer approximation', *J. Chem. Phy.*, 77(9), 4683–93 (1982).
11. Coverdale, R. T., Christensen, B. J., Mason, T. O., Jennings, H. M. and Garboczi, E. J., 'Interpretation of the impedance spectroscopy of cement paste via computer modelling: dielectric response', *J. Mater. Sci.*, 9, 4984–92 (1994).
12. Christensen, B. J., Coverdale, R. T., Olsen, R. A., Ford, S. J., Garboczi, E. J., Jennings, H. M. and Mason, T. O., 'Impedance spectroscopy of hydrating cement-based materials: measurement, interpretation, and applications', *J. Am. Ceram. Soc.*, 77(11), 2789–804 (1994).
13. Jonscher, A. K., 'Analysis of the alternating current properties of ionic conductors', *J. Mater. Sci.*, 13, 553–62 (1978).
14. Christensen, B. J., Mason, T. O. and Jennings, H. M., 'Influence of silica fume on the early hydration of Portland cements using impedance spectroscopy', *J. Am. Ceram. Soc.*, 75(4), 939–45 (1992).
15. Brantervik, K. and Niklasson, G. A., 'Circuit models for cement based materials obtained from impedance spectroscopy', *Cem. Concr. Res.*, 21(4), 496–508 (1991).
16. Gu, P., Xie, P., Beaudoin, J. J. and Brousseau, R., 'AC impedance spectroscopy (I): A new equivalent circuit model for hydrated Portland cement paste', *Cem. Concr. Res.*, 22(5), 833–40 (1992).
17. Gu, P., Xie, P., Xu, Z. and Beaudoin, J. J., 'A rationalized AC impedance model for microstructural characterisation of hydrating cement systems', *Cem. Concr. Res.*, 23(2), 359–67 (1993).
18. Coverdale, R. T., Garboczi, E. J., Jennings, H. M., Christensen, B. J. and Mason, T. O., 'Computer simulation of impedance spectroscopy in two dimensions: application to cement paste', *J. Am. Ceram. Soc.*, 76(6), 1153–60 (1993).
19. Xu, Z., Gu, P., Xie, P. and Beaudoin, J. J., 'Application of AC impedance techniques in studies of porous cementitious materials I: Influence of solid phase and pore solution on high frequency resistance', *Cem. Concr. Res.*, 23(3), 531–450 (1993).
20. Xu, Z., Gu, P., Xie, P. and Beaudoin, J. J., 'Application of AC impedance techniques in studies of porous cementitious materials II: Relationship between ACIS behaviour and the porous structure', *Cem. Concr. Res.*, 23(4), 853–62 (1993).
21. Tamas, F. D., 'Electrical conductivity of cement pastes', *Cem. Concr. Res.*, 12, 115–20 (1982).
22. McCarter, W. J. and Afshar, A. B., 'A study of the early hydration of Portland cement', *Proc. Instn. Civ. Engrs.*, 2(79), 585–604 (1985).
23. Tamas, F., Farkas, E., Voros, M. and Roy, D. M., 'Low-frequency electrical conductivity of cement,

clinker and clinker mineral phases', *Cem. Concr. Res.*, 17(2), 340–8 (1987).

24. Vernet C, 'Hydration kinetics and mechanical evolution of concrete during the first days: study of the hardening mechanism', *Proc. 10th Int. Congr. Chem. Cem.*, New Delhi, Vol. IV, 511–17 (1992).

25. Gu, P., Ramachandran, V. S. and Beaudoin, J. J., 'Electrochemical behaviour of Portland-cement/high-alumina-cement systems at early hydration times', *J. Mater. Sci. Lett.*, 12, 1771–3 (1993).

26. Tashiro, C., Ikeda, K. and Inoue, Y., 'Evaluation of pozzolanic activity by the electric resistance measurement method', *Cem. Concr. Res.*, 24, 6, 1133–9 (1994).

27. McCarter, W. J. and Tran, D., 'Monitoring pozzolanic activity by direct activation with calcium hydroxide', *Constr. Bldg. Mater.*, 10(3), 179–84 (1996).

28. McCarter, W. J., Chrisp, T. M. and Starrs, G., 'The early hydration of alkali-activated slag: developments in monitoring techniques', *Cem. Concr. Composites*, 21, 277–83 (1999).

29. McCarter, W. J. and Garvin, S., 'Admixtures in cement: a study of dosage rates on early hydration', *Mater. and Structures*, 22, 112–20 (1989).

30. Vernet, C. and Noworyta, G., 'Conductometric test for cement-admixture systems', *Proc. 9th Int. Congr. Chem. Cement*, New Delhi, Vol. IV, 627–33 (1992).

31. Abd-el-Wahed, M. G., Helmy, I. M., El-Didamony, H. and Ebied, D., 'Effect of admixtures on the electrical behaviour of Portland cement', *J. Mater. Sci. Lett.* 12, 40–2 (1993).

32. Pu, P., Ramachandran, V. S. and Beaudoin, J. J., 'Study of early hydration of high alumina cement containing phosphonic acid by impedance spectroscopy', *J. Mater. Sci. Lett.*, 14, 503–05 (1995).

33. McCarter, W. J. and Ezirim, H., 'Monitoring the early hydration of pozzolan-$Ca(OH)_2$ mixtures using electrical methods', *Adv. Cem. Res.*, 10(4), 161–8 (1998).

34. McCarter, W. J. and Ezirim, H., 'Characterising reactivity of alkali-activated pozzolans', *Proc. 10th Int. Congr. Chem. Cem.*, Gothenberg, 3(ii), 112–4 (1997).

35. Hilhorst, M. A., Van-Breugel, K., Plumigraff, D. J. M. and Stenfert-Kroese, W., 'Dielectric sensors used in environmental engineering', *Mat. Res. Soc. Symp. Proc.*, 411, 401–06 (1996).

36. Zhang, X., Yang, Y. and Ong, C. K., 'Microwave study of hydration of slag cement blends in early period', *Cem. Concr. Res.*, 25(5), 1086–94 (1995).

37. Zhang, X, Yang, Y. and Ong, C. K., 'Dielectric and electrical properties of ordinary Portland cement and slag cement in the early hydration period', *J. Mater. Sci.*, 31, 1345–52 (1996).

38. Zhang, X., Yang, Y. and Ong, C. K. 'Study of early hydration of OPC-HAC blends by microwave and calorimetry technique', *Cem. Concr. Res.*, 27(9), 1419–29 (1997).

39. McCarter, W. J., 'A parametric study of the impedance characteristics of cement-aggregate systems during early hydration', *Cem. Concr. Res.*, 24(6), 1097–110 (1994).

40. McCarter, W. J., 'The a.c. impedance response of concrete during early hydration', *J. Mater. Sci.*, 31, 6258–92 (1996).

41. McCarter, W. J., Starrs, G. and Chrisp, T. M., 'Immittance spectra for Portland cement/fly ash-based binders during early hydration', *Cem. Concr. Res.*, 29(3), 377–87 (1999).

42. Berg, A., Niklasson, G. A., Brantervik, K., Hedberg, B. and Nilsson, L. O., 'Dielectric properties of porous cement mortar: fractal surface effects', *Solid State Commun.*, 79(1), 93–6 (1991).

43. MacPhee, D. E., Sinclair, D. C. and Stubbs, S. L. 'Electrical Characterisation of pore reduced cement by impedance spectroscopy', *J. Mater. Sci. Lett.*, 15, 1566–8 (1996).

44. Cormack, S. L., MacPhee, D. E. and Sinclair, D., 'An AC impedance spectroscopy study of hydrated cement pastes', *Adv. Cem. Res.*, 10(4), 151–50 (1998).

45. McCarter, W. J., 'Pore structure characterisation of cementitious systems using electrical property measurements', *EPSRC (Swindon, U.K.), Final Report GR/K65089*, May, 17p. (1999).

46. Atkinson, A. and Nickerson, A. K., 'The diffusion of ions through water-saturated cement', *J. Mater. Sci.*, 19, 3068–78 (1984).

47. Buenfeld, N. R. and Newman, J. B., 'The resistivity of mortars immersed in sea-water', *Cem. Concr. Res.*, 16(4), 511–24 (1986).

48. Andrade, C., Alonso, C. and Goni, S., 'Possibilities for electrical resistivity to universally characterise mass transport processes in concrete', *Proc. Concr. 2000 Conf.*, Dundee, R. K. Dhir and M. R. Jones (eds), Vol. 2, 1639–52, (E & F N Spon, 1993).

49. Kyi, A. and Batchelor, B., 'An electrical conductivity method for measuring the effects of additives on effective diffusivities in Portland cement pastes', *Cem. Concr. Res.*, 24(4), 752–64 (1994).

50. Lu, X., 'Application of the Nernst-Einstein equation to concrete', *Cem. Concr. Res.*, 27(2), 293–302 (1997).

51. Tumidajski, P. J. and Schumacher, A. S., 'On the relationship between the formation factor and propan-2-ol diffusivity in mortars', *Cem. Concr. Res.*, 26(9), 1301–06 (1996).

52. Shi, J., Stegemann, A. and Caldwell, R. J., 'Effect of supplementary cementing materials on the specific conductivity of pore solution and its implications

on the rapid chloride permeability test (AASHTO T277 and ASTM C1202) results', *Am. Concr. Inst. Mater. J.*, July/August, 389–94 (1998).

53. Ampadu, K. O., Torii, K. and Kawamura, M., 'Beneficial effect of fly ash on chloride diffusivity of hardened cement paste', *Cem. Concr. Res.*, 29(4), 585–90 (1999).

54. Katz, J. and Thompson, A. H., 'Quantitative prediction of permeability in porous rock', *Phys. Rev. B: Condens. Matter*, 34(11), 8179–81 (1986).

55. El-Dieb, S. and Hooton, R. D., 'Evaluation of the Katz-Thompson model for estimating the water permeability of cement-based materials from mercury intrusion porosimetry', *Cem. Concr. Res.*, 24(3), 443–55 (1994).

56. Christensen, B. J., Mason, T. O. and Jennings, H. M., 'Comparison of measured and calculated permeabilities for hardened cement pastes', *Cem. Concr. Res.*, 26(9), 1325–34 (1996).

57. Ollivier, J. P. and Massat, M., 'Permeability and microstructure of concrete: a review of modelling', *Cem. Concr. Res.*, 22(2/3), 503–14 (1992).

58. Tumidajski, P. J. and Lin, B., 'On the validity of the Katz-Thompson equation for permeabilities in concrete', *Cem. Concr. Res.*, 28(5), 643–7 (1998).

59. Zhao, T. J., Zhou, Z. H., Zhu, J. Q. and Feng, N. Q., 'An alternating test method for concrete permeability', *Cem. Concr. Res.*, 28(1), 7–12 (1998).

60. Liu, Z. and Beaudoin, J. J., 'An assessment of the relative permeability of cement systems using A.C. impedance techniques', *Cem. Concr. Res.*, 29(7), 1085–90 (1999).

61. Schießl, P. and Raupach, M., 'New approaches for monitoring the corrosion risk for the reinforcement – installation of sensors', *Proc. Concr. across Borders Conf.*, Odense, Denmark, Vol. 1, pp. 65–77 (1994).

62. McCarter, W. J., Emerson, M. and Ezirim, H., 'Properties of concrete in the cover zone: developments in monitoring techniques', *Mag. Concr. Res.*, 47(172), 243–51 (1995).

63. McCarter, W. J., Ezirim, H. and Emerson, M., 'Properties of concrete in the cover zone: water penetration, sorptivity and ionic ingress', *Mag. Concr. Res.*, 48(176), 149–56 (1996).

64. McCarter, W. J., Chrisp, T. M. and Ezirim, H. C., 'Discretized conductivity measurements to study wetting and drying of cover zone concrete', *Adv. Cem. Res.*, 10(4), 195–202 (1998).

65. Lopez, W. and Gonzalez, J. A., 'Influence of degree of saturation on the resistivity of concrete and the corrosion rate of steel reinforcement', *Cem. Concr. Res.*, 23(2), 368–76 (1993).

66. McCarter, W. J. and Watson, D. W., 'Wetting and drying of cover-zone concrete', *Proc. Instn. Civ. Engrs., Structures and Buildings*, 122(2), 227–236 (1997).

67. Watson, D. W. and McCarter, W. J., 'The efficacy of hydrophobic surface treatments for concrete and their influence on moisture movement within the cover zone', *Proc. Int. Conf. on New Requirements for Mater. Struct.*, Prague, Sept., 23–28, (ISBN 80-01-01838-5) (1998).

68. Buenfeld, N. R. and Zhang, J-Z., 'AC impedance study of sealer- and coating-treated mortar specimens', *Adv. Cem. Res.*, 10(4), 169–78 (1998).

69. Fu, X., Ma, E., Chung, D. D. L. and Anderson, W. A., 'Self-monitoring in carbon fiber reinforced mortar by reactance measurement', *Cem. Concr. Res.*, 27(6), 845–52 (1997).

70. Fu, X. and Chung, D. D. L, 'Effect of curing age on self-monitoring behavior of carbon fiber reinforced mortar', *Cem. Concr. Res.*, 27(9), 1313–18 (1997).

71. Sihai, W. and Chung, D. D. L., 'Carbon fiber reinforced cement as a thermistor', *Cem. Concr. Res.*, 29(6), 961–5 (1999).

72. McCarter, W. J. and Puyrigaud, P., 'Water content assessment of fresh concrete', *Proc. Instn. Civil Engrs., Structures and Buildings*, 110(4), 417–25 (1995).

73. Garboczi, E. J., Schwartz, L. M. and Bentz, D. P., 'Modelling the D.C. electrical conductivity of mortar', *Maters. Res. Soc. Symp. Proc.*, 370, 429–36 (1995).

Chapter twenty

Nuclear magnetic resonance spectroscopy and magnetic resonance imaging of cements and cement-based materials

Jørgen Skibsted, Christopher Hall and Hans J Jakobsen

20.1 Introduction

Solid-state nuclear magnetic resonance (NMR) spectroscopy is firmly established as a powerful research tool for characterization and structural analysis of cement pastes and cement-based materials using several different NMR nuclei as probes (principally ^1H, ^{27}Al and ^{29}Si; and to a less extent ^{17}O, ^2D, ^{23}Na and ^{35}Cl). Generally speaking, NMR spectroscopy is most sensitive to the local ordering and structure around the spin nucleus, and this permits structural studies not only of crystalline minerals, but also of poorly crystalline and amorphous materials, such as the calcium silicate

hydration products produced in the setting of Portland cement. The high sensitivity to local ordering and geometries stands in contrast to the long-range periodicities detected in X-ray diffraction measurements. Thus, the combination of solid-state NMR and X-ray diffraction may often give a more detailed description of the structure and composition of microcomposite materials such as cements than either alone. We emphasize also that in solid-state NMR, a single element (which must, of course, have a nuclear spin) is detected at a time: a valuable feature, for example, in selective studies of the aluminate phases or of calcium silicate structures in Portland cements using respectively ^{27}Al or ^{29}Si as local probes. These features illustrate the power and versatility of solid-state NMR. That two international conferences on NMR spectroscopy of cement-based materials (Colombet and Grimmer 1994; Colombet *et al.* 1998) were held in the 1990s is a sign of the popularity and vitality of these methods. In this chapter, we give a short account of the development and application of solid-state NMR and MRI as they are applied to cement minerals, cements and cement-based materials. We place particular emphasis on state-of-the-art experiments and recent work.

Shortly after the discovery of NMR, the potential of the technique in studies of cementing minerals was demonstrated by Pake (Pake 1948) who observed the ^1H–^1H dipolar coupling in the water molecules of gypsum, and showed that information about the inter-proton distance may be derived from either ^1H single-crystal or powder NMR spectra. In the following three decades, ^1H NMR was used in a small number of studies of the hydration reactions and kinetics of cements by detecting the variations in ^1H spin-lattice and spin–spin relaxation times (Blaine 1960; Blinc *et al.* 1978; Kawachi *et al.* 1955; Seligmann 1968). When the use of high-field superconducting magnets and the magic-angle spinning (MAS) technique (Andrew *et al.* 1958, Lowe 1959) became routine in NMR in the early 1980s, the possibility of high-resolution studies of a number of important spin nuclides such as ^{29}Si and ^{27}Al in solid powders immediately became apparent. Pioneering work on ^{29}Si MAS NMR by Lippmaa *et al.* showed that the ^{29}Si isotropic *chemical shift* (that is, the variation in the resonance frequency) reveals the degree of condensation of SiO_4 tetrahedra in silicate minerals (Lippmaa *et al.* 1980) including calcium silicates (Mägi *et al.* 1984) and in the amorphous calcium silicate hydrate (C-S-H) produced by Portland cement hydration (Lippmaa *et al.* 1982). ^{29}Si MAS NMR has subsequently been employed in a large number of detailed studies of the hydration of the calcium silicates alite and belite and of these phases in Portland cement-based materials. From ^{27}Al MAS NMR experiments Müller *et al.* showed that aluminium in tetrahedral and octahedral coordination can be clearly distinguished by their different ^{27}Al isotropic chemical shifts (Müller *et al.* 1977, 1981). This result was subsequently utilized in studies of the hydration of calcium aluminate (CA), high aluminate cements, and aluminium minerals in Portland cements. Furthermore, magic-angle spinning has been combined with multi-pulse decoupling sequences to reduce the line broadening which arises from ^1H–^1H dipolar couplings in ^1H NMR spectra. Grimmer and Rosenberger (Grimmer and Rosenberger 1978, Rosenberger and Grimmer 1979) used this technique in their measurements of ^1H isotropic chemical shifts in hydrous silicates. These results have since been used to distinguish different ^1H species in the C-S-H produced in Portland cement hydration. The hydration reactions of calcium silicate minerals have also been studied by ^{17}O MAS NMR (Cong and Kirkpatrick 1993*a*) which requires the use of ^{17}O ($I = 5/2$, 0.04 per cent natural abundance) isotopically enriched starting materials. Furthermore, specially designed NMR equipment has been developed for a number of applications to cement materials. Notable are high-temperature NMR probes for investigating the clinkering reactions of model raw-mix materials and the structure and dynamics of molten silicates and calcium aluminates at elevated temperatures (Bonafous *et al.* 1995; Massiot *et al.* 1995).

20.2 Solid-state NMR methods

The NMR spectrum of a spin nucleus in a diamagnetic solid may be influenced by the chemical shift interaction, homo- and heteronuclear direct (dipolar) and indirect (J-coupling) spin-spin couplings, and the quadrupole coupling interaction for spin nuclei with $I > 1/2$. The orientational dependence of these internal interactions in a powder sample means that the NMR resonance can have a width of several kHz. To reduce or eliminate this broadening, several techniques have been invented which employ either mechanical rotation of the sample to reduce the spatially dependent part of the interactions or radio frequency (rf) pulse schemes to modify the spin-dependent part. Of these experimental techniques, MAS, ^1H heteronuclear decoupling and $^1H \rightarrow X$ cross polarization (CP) represent the most fundamental methods for improving the resolution and sensitivity in solid-state NMR spectra.

The line narrowing effect of MAS is founded in the fact that these internal interactions have a orientational dependence (to first order) which includes the factor $(3\cos^2\vartheta - 1)$, where ϑ is the angle between the static magnetic field and a specific molecule-fixed direction. Employing the heteronuclear dipolar interaction as the illustrative example, this direction is the internuclear vector (r) between the two spins (I and S). Rotation of the sample about an axis at the angle θ relative to the magnetic field (Figure 20.1) rotates the vector r in a conical path described by the angles α and β. The average dipolar interaction depends on the time-average of the factor $(3\cos^2\vartheta - 1)$, which at a rotation frequency of $\nu_R = \omega_R/2\pi$ takes the form:

$$<3\cos^2\vartheta - 1> = \frac{1}{2}(3\cos^2\theta - 1)(3\cos^2\beta - 1)$$

$$+ \frac{3}{2}[\sin2\theta\sin2\beta\cos(\omega_R t + \alpha)$$

$$+ \sin^2\theta\sin^2\beta\cos(2\omega_R t + \alpha)] \quad (1)$$

At a sufficiently high spinning speed the time-dependent terms, $\cos(\omega_R t + \alpha)$ and $\cos(2\omega_R t + \alpha)$, average to zero and thus, the time average

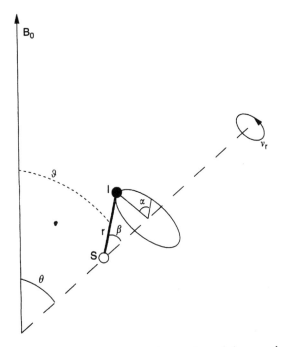

FIGURE 20.1 Macroscopic rotation of the sample at an angle θ relative to the direction of the external magnetic field (B_0). The drawing illustrates the orientation of an internuclear $I-S$ vector in a powder sample undergoing spinning around the angle θ with a rotation frequency $\nu_R = \omega_R/2\pi$ and defines the angles ϑ, α, and β which are used in equation (1) to describe to orientation of the $I-S$ vector relative to B_0.

$<3\cos^2\vartheta - 1>$ is proportional to $(3\cos^2\theta - 1)$. The angle θ is under the control of the spectroscopist and if $\theta = \theta_M = \cos^{-1}(1/\sqrt{3}) \approx 54.736°$ then $<3\cos^2\vartheta - 1> = 0$. θ_M is the so-called 'magic-angle' and spinning the sample about this angle improves the resolution of the resonances by elimination/reduction of the anisotropic interactions. However, a complete elimination of these interactions requires that the spinning speed is larger than the width of the resonance in a static powder experiment. For the dipolar and chemical shift anisotropy (CSA) interactions this situation can, in many cases, be achieved by spinning the sample at spinning speeds in the range 5–35 kHz. If this condition is not fulfilled, the terms $\cos(\omega_R t + \alpha)$ and $\cos(2\omega_R t + \alpha)$ will not average to zero and result in spinning sidebands in the MAS NMR spectrum. As an example, a series of ^{29}Si NMR spectra of belite (β–C_2S) is shown in Figure 20.2 recorded under static conditions and with increasing spinning speeds up to $\nu_R = 7.0$ kHz. The line width of the static spectrum originates mainly from the ^{29}Si CSA interaction and it is apparent that the intensity distribution of the spinning sidebands in the slow-speed ($\nu_R = 700$ Hz) spectrum reflects the profile of the static-powder NMR spectrum. This fact can be used to determine the principal elements of the CSA tensor from simulations of the spinning sideband intensities as first demonstrated by Herzfeld and Berger (Herzfeld and Berger 1980). Furthermore, it is apparent that a spinning speed of $\nu_R = 7.0$ kHz (Figure 20.2) results in an 'isotropic' spectrum, where only a single resonance from the unique ^{29}Si site in β-C_2S is observed.

For dilute spin $I = 1/2$ nuclei (for instance ^{13}C and ^{29}Si) single-pulse NMR experiments often suffer from inherently low sensitivity, and the fact that spin-lattice relaxation times can be very long means that long recycle delays are required. These disadvantages have been largely overcome by the CP technique introduced by Pines et al. (Pines et al. 1973), which transfers magnetization from an abundant spin (e.g., ^1H or ^{19}F) to the dilute spin via the heteronuclear dipolar interactions. In a {^1H} ^{29}Si CP/MAS NMR experiment (Figure 20.3), the first step is to excite the ^1H spins by a

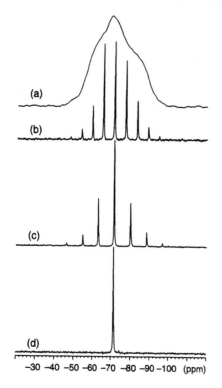

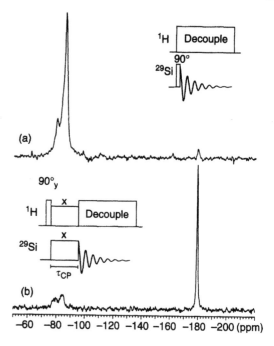

FIGURE 20.2 ^{29}Si NMR spectra of a powder sample of β-C$_2$S recorded at 14.1 T illustrating the line-narrowing effect of magic-angle spinning. The spectrum in (a) is recorded under static conditions while the spectra in (b)–(d) employ MAS with spinning frequencies of (b) $\nu_R = 700$ Hz, (c) $\nu_R = 1000$ Hz, and (d) $\nu_R = 7000$ Hz. The asymmetric lineshape observed for the static-powder spectrum and the manifolds of spinning sidebands in (b) and (d) reflect the anisotropy of the chemical shift interaction. Simulation of the spinning sidebands in (b) results in the principal elements $\delta_{33} = -87.9$ ppm, $\delta_{22} = -70.3$ ppm, and $\delta_{11} = -55.7$ ppm of the chemical shift tensor for ^{29}Si in β-C$_2$S while the isotropic chemical shift ($\delta_{iso} = -71.34$ ppm) can be obtained from the isotropic peak in either of the MAS NMR spectra.

90$^\circ_y$ -pulse, and subsequently to lock the ^{1}H magnetization by a long ^{1}H x-phase pulse. At this point the ^{29}Si rf channel is switched on and the amplitude of this field (B_{1Si}) is adjusted to the so-called Hartmann–Hahn matching condition (Hartmann and Hahn 1962), $\gamma_H B_{2H} = \gamma_{Si} B_{1Si}$. Under this condition, and during the CP contact time (τ_{CP}), magnetization is transferred from ^{1}H to ^{29}Si for spins that are dipolar coupled. This may result in a maximum enhancement of ^{29}Si magnetization

FIGURE 20.3 Comparison of ^{29}Si MAS NMR spectra of a synthetic cement paste containing 15 per cent w thaumasite recorded (a) without and (b) with {^{1}H} ^{29}Si cross polarization (CP). The single-pulse spectrum (a) employed ^{1}H high-power decoupling, a repetition delay of 60 s and 4096 scans, corresponding to a total instrument time of 68 h while the CP/MAS spectrum (b) was obtained with a 5 s repetition delay, a CP contact time of $\tau_{CP} = 600$ µs, and 2048 scans, resulting in an instrument time of less than 3 h. The insets in (a) and (b) illustrate the basic rf pulse sequences for the ^{1}H decoupled single-pulse experiment and for the ^{29}Si {^{1}H} CP experiment, respectively.

given by the ratio of the gyromagnetic ratios for the two spins, $G_{CP} = \gamma_H / \gamma_{Si} \approx 5$, although the CP gain enhancement factor G_{CP} is generally somewhat lower in practice. Perhaps the main advantage of the CP experiment is that the recycle delay is determined by the spin-lattice relaxation times of the ^{1}H spins, which are often much shorter than those for the ^{29}Si spins. A second advantage is its ability to act as a ^{1}H filter in that only silicons which are dipolar coupled to ^{1}H are observed. To illustrate these effects, Figure 20.3 compares ^{29}Si MAS NMR spectra with and without CP of thaumasite produced by hydration of a mixture of Ca(OH)$_2$, CaCO$_3$, CaSO$_4 \cdot \frac{1}{2}$H$_2$O, and silica gel at a relative humidity of 95 per cent (Skibsted *et al.* 1995). These spectra clearly show that {^{1}H} ^{29}Si CP strongly favours the observation of thaumasite ($\delta_{iso}(^{29}$Si$) = -179.4$ ppm) relative to the C-S-H products ($\delta_{iso}(^{29}$Si$) \approx -80$ to -90 ppm). This reflects the actual CP enhancement factor $G_{CP} \approx 2.5$ and the fact that the spin-lattice relaxation time (T_1) for ^{1}H in thaumasite is significantly shorter than the corresponding time for ^{29}Si ($T_1(^1$H$) = 0.83$ s and $T_1(^{29}$Si$) = 777$ s for the actual sample) (Skibsted *et al.* 1995), which results in a saturation of the ^{29}Si magnetization in the ^{29}Si MAS NMR experiment. Furthermore, the experiment times of about 68 h and 3 h for the spectra in Figure 20.3 obtained without and with {^{1}H} ^{29}Si CP, respectively, illustrate the advantages of the CP experiment. The pulse sequence shown in Figure 20.3 also includes high-power ^{1}H decoupling during acquisition, which eliminates line-broadening from ^{1}H–^{29}Si dipolar couplings. For most dilute spin systems, the major source of line-broadening is heteronuclear dipolar couplings to abundant spins (e.g., ^{1}H and ^{19}F) and thus, high-power decoupling is often important for obtaining high-resolution solid-state NMR spectra.

For half-integer spin ($I = 3/2, 5/2, \ldots$) quadrupolar nuclei (e.g., ^{17}O, ^{23}Na, ^{27}Al, ...) high resolution NMR spectra cannot be obtained by MAS alone. Although the first-order quadrupolar interaction is proportional to ($3\cos^2\vartheta - 1$), the second-order contribution includes an angle dependency proportional to ($35\cos^4\vartheta - 30\cos^2\vartheta + 3$), which is not reduced to zero at the magic angle but rather at the angles $\vartheta = 30.56°$ and $70.12°$. For strong quadrupolar interactions, this results in second-order quadrupolar lineshapes for the central transition ($m = 1/2 \leftrightarrow m = -1/2$) as shown in Figure 20.4 for the two ^{27}Al resonances observed in a synthetic sample of C$_3$A (Skibsted *et al.* 1991). This spectrum also illustrates that the width of the lineshape may be several tens of kHz, and shows that high spinning speeds are necessary to obtain lineshapes from the central transitions with no distortion or overlap from the first-order spinning sidebands (for C$_3$A at a 9.4 T magnetic field, this requires a spinning speed of $\nu_R = 18$ kHz, see Figure 20.4). A full interpretation of the spectrum

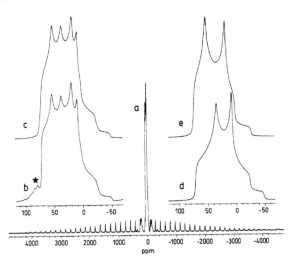

FIGURE 20.4 ^{27}Al MAS NMR spectra of C$_3$A recorded at 9.4 T employing a spinning speed of $\nu_R = 18.0$ kHz and a spectral width of 1.0 MHz. Experimental spectrum of (a) the central and satellite transitions and (b) expansion of the centerband from the central transition in (a). Part (c) illustrates the simulated spectrum of the centerband in (b) which is composed of the two second-order quadrupolar lineshapes shown in (d) and (e). These lineshapes correspond to the quadrupole coupling and isotropic chemical shift parameters $C_Q = 8.69$ MHz, $\eta_Q = 0.32$, $\delta_{iso} = 79.5$ ppm for Al(1) (e) and $C_Q = 9.30$ MHz, $\eta_Q = 0.54$, $\delta_{iso} = 78.3$ ppm for Al(2) (d) in C$_3$A [Skibsted *et al.* 1991]. The asterisk in (b) indicates the centerbands for the ($\pm 1/2, \pm 3/2$) satellite transitions. (Reproduced by permission of Academic Press, USA from Skibsted *et al.* 1991.)

in Figure 20.4a requires simulation of the second-order quadrupolar lineshapes from which we may then determine the quadrupole coupling constant C_Q (which describes the magnitude of the quadrupole coupling interaction), the associated asymmetry parameter η_Q, (which reflects the symmetry of the electric field gradients at the nuclear site) and the isotropic chemical shift. The determination of these parameters from MAS NMR spectra of either the central transition (Kundla *et al.* 1981; Samoson *et al.* 1982) or the manifold of spinning sidebands from the satellite ($m = \pm 1/2 \leftrightarrow m = \pm 3/2$ and $m = \pm 3/2 \leftrightarrow m = \pm 3/2$ for $I = 5/2$) transitions (Jakobsen *et al.* 1989, Skibsted *et al.* 1991) is of general interest since C_Q and η_Q are sensitive measures of the electronic environment around the quadrupole nucleus. In addition to a correct determination of δ_{iso}, which requires knowledge of C_Q and η_Q (or at least the product $C_Q(1+\eta_Q^2/3)^{1/2}$) (Jakobsen *et al.* 1989; Samoson *et al.* 1982), knowing these parameters is often a prerequisite for an unambiguous interpretation of MAS NMR spectra of complex cement-based materials.

Improved resolution in NMR spectra of quadrupolar nuclei can be obtained by a full averaging of the second-order quadrupolar interaction (Llor and Virlet 1988) which was firstly achieved in the double-rotation (DOR) NMR experiment (Chmelka *et al.* 1989; Samoson *et al.* 1988), in which the sample is spun continuously about two different angles, and in the dynamic-angle spinning (DAS) experiment (Chmelka *et al.* 1989; Mueller *et al.* 1990), where averaging is achieved by spinning at two sequential angles. More recently, the multiple-quantum (MQ) MAS experiment has been introduced (Frydman and Harwood 1995; Medek *et al.* 1995), which by correlating multiple-quantum and single-quantum coherences under MAS gives a two-dimensional (2D) spectrum where the second-order quadrupolar broadening is removed in one (the isotropic) dimension. Compared to DOR and DAS NMR the MQMAS experiment has the great advantage of mechanical simplicity in only requiring sample rotation around one fixed axis. The basic two-pulse sequence for the MQMAS experiment is shown in Figure 20.5. In this experiment triple-quantum coherences are created by the first rf pulse and converted into observable single-quantum coherences by the second pulse employing appropriate phase cycling of the rf pulses and the receiver. The delay (t_1) between the pulses is incremented and accumulation of free induction decay (FID) for each increment gives after a 2D Fourier transformation a correlation spectrum where triple-quantum coherences are correlated with single-quantum coherences, corresponding to the isotropic and anisotropic dimensions of the 2D spectrum. As an example, Figure 20.6 illustrates a ^{27}Al MQMAS spectrum of gibbsite (γ-AH$_3$), which clearly resolves the two resonances from the two distinct AlO$_6$ sites in γ-AH$_3$ in the isotropic dimension, in agreement with a similar spectrum reported for gibbsite (Faucon *et al.* 1998). Summation over each of the

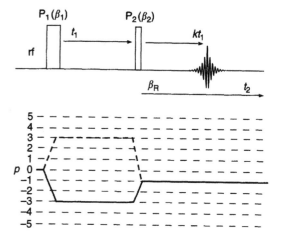

FIGURE 20.5 rf pulse scheme for the two-pulse triple-quantum MQMAS experiment selecting the coherence transfer pathway shown below, where $p = 2m$ is the order of the coherence. The evolution of single-quantum coherence will after a certain time ($t_2 = k \cdot t_1$) result in an echo, where effects from the second-order quadrupolar interaction are refocussed. For quadrupolar nuclei with $I = 3/2$ and $5/2$, k equals $7/9$ and $19/12$, respectively. The basic six-step phase cycle of the two pulses (P$_1$ and P$_2$) and of the receiver phase (β_R) is: $\beta_1 = 0°$, $60°$, $120°$, $180°$, $240°$, $300°$; $\beta_2 = 0°$, $\beta_R = 0°$, $180°$.

resonances in the anisotropic dimension may give second-order quadrupolar lineshapes which allow determination of the quadrupole coupling parameters and isotropic chemical shifts from one-dimensional (1D) simulations of these lineshapes. The triple-quantum shifts (δ_{3Q}), observed in the isotropic dimension, can be calculated from δ_{iso}, the quadrupolar product parameter $C_Q(1+\eta_Q^2/3)^{1/2}$, and the Larmor frequency (Massiot *et al.* 1996). For gibbsite the δ_{3Q} shifts observed in Figure 20.6 are in full accord with those calculated from C_Q, η_Q, and δ_{iso} parameters obtained from ^{27}Al MAS NMR spectra of the central and satellite transitions (Skibsted *et al.* 1993).

20.3 Structure and bonding in cement minerals

20.3.1 Anhydrous minerals

In ^{29}Si MAS NMR, different crystallographic structures and crystallographically distinct sites

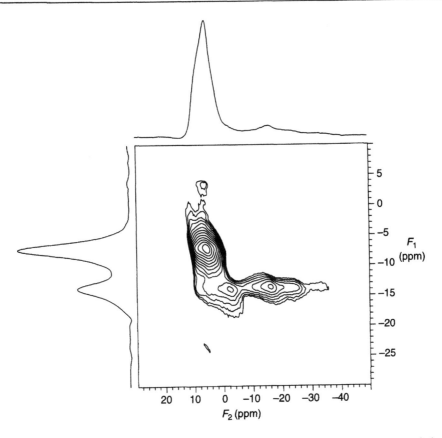

FIGURE 20.6 Contour plot of the ^{27}Al MQMAS NMR spectrum of gibbsite (γ-AH$_3$) recorded at 7.1 T using the standard two-pulse MQMAS sequence, ^{1}H decoupling, and a spinning speed of $\nu_R = 9.2$ kHz. The projections on to the isotropic (F_1) and anisotropic (F_2) dimensions correspond to summations over the 2D spectrum. The spectrum is referenced to a 1.0 M aqueous solution of AlCl$_3$·6H$_2$O in both dimensions.

within individual structures result in distinct ^{29}Si isotropic chemical shifts (δ_{iso}) in the range from -60 ppm to -125 ppm (relative to tetramethylsilane, TMS) for SiO$_4^{4-}$ tetrahedra in inorganic silicates (Lippmaa *et al.* 1980, Mägi *et al.* 1984). The ^{29}Si chemical shift mainly reflects the degree of condensation of SiO$_4^{4-}$ tetrahedra (that is, Q^n, $n = 0$, 1, 2, 3, 4) where increasing condensation displaces the chemical shift by about 5–10 ppm to lower frequency for each successive degree of condensation (Grimmer *et al.* 1983, 1985; Lippmaa *et al.* 1980; Mägi *et al.* 1984). The anhydrous calcium silicates in Portland cements all contain isolated SiO$_4^{4-}$ tetrahedra, which have ^{29}Si chemical shifts in the range from -65 to -75 ppm. The

pure triclinic form of C$_3$S includes nine distinct ^{29}Si sites (Golovastikov *et al.* 1976) in agreement with the ^{29}Si MAS NMR spectrum of this phase which exhibits distinct resonances from the nine different ^{29}Si environments (Grimmer and Zanni 1998; Mägi *et al.* 1984; Skibsted *et al.* 1990) (Figure 20.7). The resonances are observed in a rather narrow chemical shift range between -68.9 and -74.3 ppm where the variation of the individual chemical shifts can be related to small differences in mean Si–O bond lengths of the SiO$_4^{4-}$ tetrahedra (Skibsted et al. 1990). In contrast, all Si atoms in β-C$_2$S are crystallographically equivalent and the ^{29}Si MAS NMR spectrum consists of a single resonance (Figure 20.2). However,

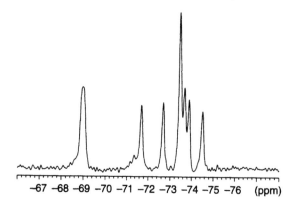

-67 -68 -69 -70 -71 -72 -73 -74 -75 -76 (ppm)

FIGURE 20.7 ^{29}Si MAS NMR spectrum (7.1 T, $\nu_R = 6.5$ kHz) of triclinic C$_3$S. The spectrum includes nine resonances when it is taken into account that the broad peak at -69 ppm contains two resonances ($\delta_{iso} = -68.93$ ppm and $\delta_{iso} = -69.04$ ppm) and that the tall narrow peak at $\delta_{iso} = -44.34$ ppm exhibits a relative intensity of two compared to the other resonances.

there are small differences in the ^{29}Si chemical shifts for the α_L-, β- and γ-polymorphs of C$_2$S (Grimmer *et al.* 1985; Skibsted *et al.* 1990), which can be ascribed to minor variations in the geometry (e.g. mean Si–O bond length) of the SiO$_4^{4-}$ tetrahedra in these polymorphs (Skibsted *et al.* 1990). For alite in production clinkers, the presence of substitutional impurities results in a stabilization of either the monoclinic M$_I$ or M$_{III}$ forms of alite. These impurity ions slightly perturb the local environment of the SiO$_4^{4-}$ tetrahedra and result in a line broadening of the individual resonances in the ^{29}Si MAS NMR spectrum. This gives a complex lineshape for the overlapping ^{29}Si resonances from monoclinic alite. Nonetheless, Skibsted *et al.* (1995) have shown that the triclinic and the two monoclinic forms of tricalcium silicate can be distinguished and that the characteristic ^{29}Si lineshapes, observed for M$_I$ and M$_{III}$ alite, can be utilized in a quantitative determination of alite and belite in production clinkers and cements by deconvolving the ^{29}Si lineshape as a sum of resonances from these components (Figure 20.8).

In anhydrous calcium aluminates, structural information can be obtained using ^{27}Al MAS NMR (Cong and Kirkpatrick 1993; Müller *et al.* 1986; Skibsted *et al.* 1991, 1993) although such

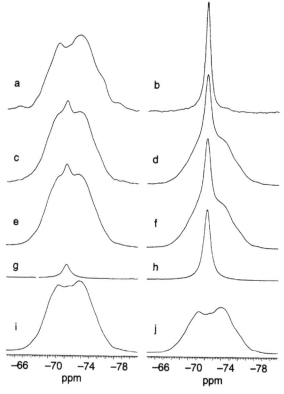

a   b

c   d

e   f

g   h

i   j

-66 -70 -74 -78   -66 -70 -74 -78
ppm   ppm

FIGURE 20.8 ^{29}Si MAS NMR spectra (7.1 T) of synthetic samples of (a) the monoclinic M$_{III}$ form of alite and (b) the monoclinic β-form of belite. Parts (c) and (d) illustrate ^{29}Si MAS NMR spectra of two ordinary Portland cements containing a relatively low (c) and high (d) content of β-C$_2$S. Optimized simulations of the lineshapes in (c) and (d) are shown in (e) and (f) which are composed of the simulated subspectra of belite and M$_{III}$ alite illustrated in parts (g)–(j), respectively. The deconvolved spectra give the compositions of 4.6 wt. per cent C$_2$S and 75.1 wt. per cent C$_3$S for the Portland cement studied in part (c) and 25.4 wt. per cent C$_2$S and 57.5 wt. per cent C$_3$S for the spectrum of the Portland cement in part (d). (Reproduced by permission of the Royal Society of Chemistry, Cambridge, UK from Skibsted *et al.* 1995.)

spectra are more complex than ^{29}Si MAS NMR spectra as a result of the ^{27}Al quadrupole coupling interaction. As shown earlier, this interaction implies that the quadrupole coupling parameters C_Q and η_Q must be determined as well as the ^{27}Al isotropic chemical shift in order to fully describe a resonance from a single aluminium site. On the other hand, this gives an improved parameterization of the Al environments since the ^{27}Al isotropic chemical shift allows a clear distinction between aluminium in tetrahedral and octahedral coordination, whereas the ^{27}Al quadrupole coupling parameters (C_Q and η_Q) are related to the distortion of these coordination polyhedra from ideal symmetry. For example, the ^{27}Al MAS NMR spectrum of the central transition for C$_3$A shown in Figure 20.4 is the superposition of quadrupolar lineshapes from two non-equivalent crystallographic sites (Skibsted *et al.* 1991). The ^{27}Al isotropic chemical shifts determined from the simulation of the overall lineshape demonstrate that the two Al sites are tetrahedrally coordinated to four oxygen atoms, whereas the strong quadrupole couplings indicate a significant distortion of the local geometry for these tetrahedra. ^{27}Al MAS NMR spectra of calcium aluminates are rather complex and require that numerical simulations of either the central transition or the complete manifold of spinning sidebands from the satellite transitions is performed. For CA$_2$, CA, C$_{12}$A$_7$ and C$_3$A, isotropic chemical shifts and quadrupole coupling parameters of high precision for all Al crystallographic sites have been determined from ^{27}Al MAS NMR spectra employing such simulation methods (Skibsted *et al.* 1993). Elsewhere, it has been confirmed (Skibsted *et al.* 1998) that the ^{27}Al resonance from C$_4$AF is almost unobservable as a result of the severe line broadening produced by nuclear-electron dipolar couplings with paramagnetic Fe^{3+} ions. ^{27}Al MAS NMR can, therefore, be used to estimate the C$_3$A content of production clinkers and cements, since the ferrite phase does not contribute to the spectrum. However, significant amounts of Al are also present as 'guest' ions in the alite and belite phases of production clinkers, and these Al species can also be observed and quantified even at these rather low levels by ^{27}Al MAS NMR (Skibsted *et al.* 1994). ^{27}Al MAS NMR spectra of synthetic samples of C$_2$S and C$_3$S doped with Al, and of Portland cements, show that Al only occupies tetrahedral sites replacing Si^{4+} in the calcium silicates; the ^{27}Al isotropic chemical shift for Al incorporated in

belite ($\delta_{iso} = 96.1$ ppm) is the most deshielded shift so far reported for any Al site tetrahedrally coordinated to four oxygen atoms (Skibsted *et al.* 1994).

20.3.2 Hydrates and hydration reactions

MAS NMR techniques for hydrated phases are broadly similar to those for anhydrous phases, but must take account of the dipolar coupling of the observed nuclei (^{27}Al or ^{29}Si) with the ^{1}H nuclei to which they are connected by chemical bonds. These heteronuclear dipolar couplings are much reduced by MAS, but can be eliminated entirely by high-power ^{1}H decoupling, where the sample is exposed during the acquisition of the NMR signal to rf irradiation at the ^{1}H resonance frequency. Alternatively, the ^{1}H dipolar couplings can be utilized in CP MAS NMR experiments to selectively observe species with ^{1}H in the near proximity to the observed NMR nuclei.

^{29}Si and ^{27}Al MAS NMR provides information on structure and chemical bonding in the hydrates formed in cement hydration processes. For crystalline minerals, NMR provides information about bonding and especially about strain in bond length and bond angle in minerals whose crystal structure is usually (but not always) known by X-ray or neutron diffraction; but NMR is especially valuable in revealing structural features of amorphous phases where diffraction methods are silent. Above all, ^{29}Si MAS NMR has been a rich source of unique information on silicate species present in C-S-H, both in simple Portland cements and in blended systems containing silica, slags and other pozzolans. ^{29}Si NMR provides the principal evidence for the condensation of silicate tretrahedra during hydration, a process in which isolated silicate tetrahedral units fuse to form dimers, which, in turn, form longer oligomers linked through so-called bridging tetrahedra. The chemical shift of ^{29}Si depends primarily on the nature of the Si–O bonds in the tetrahedral unit, so that isolated units (Q^0), units in simple dimers and end units in longer chains (Q^1; $\delta \approx -79$ ppm) and linking units within chains (Q^2; $\delta \approx -85$ ppm) give rise to distinct resonances. Q^3 and Q^4 units are not found in C-S-H produced at ambient temperatures and pressures so it appears that all oligomers are linear rather than branched. ^{29}Si MAS NMR also resolves bridging and non-bridging tetrahedra. The coexistence of these silicate oligomers was first discovered in NMR spectra by Lippmaa (Lippmaa *et al.* 1982) and has been the subject of numerous investigations since (Barnes *et al.* 1985; Brough *et al.* 1994, 1994a, 1995, 1996; Bresson and Zanni 1998; Clayden *et al.* 1984; Cong and Kirkpatrick 1995, 1996a,b; Dobson *et al.* 1988; Grimmer *et al.* 1982; Hjorth *et al.* 1988; Masse *et al.* 1993, 1995; Philippot *et al.* 1996; Rodger *et al.* 1988; Richardson *et al.* 1994; Richardson 1999; Stade *et al.* 1985; Sun *et al.* 1999; Young 1988; Zanni *et al.* 1994, 1996). Furthermore, it has been shown that the incorporation of an AlO_4 tetrahedron in the chain of silicate tetrahedra results in a separate resonance ($\delta \approx -82$ ppm) which originates from SiO_4 tetrahedra bonded to the AlO_4 tetrahedron, thereby corresponding to a $Q^2(1Al)$ site (Figure 20.9) (Richardson *et al.* 1993). Thus, from a determination of the relative intensities observed for the Q^1, $Q^2(0Al)$, and $Q^2(1Al)$ resonances, it is possible to estimate the chain length distribution in C-S-H and the average Al/Si ratio for the C-S-H. For a synthetic C-S-H produced in the absence of aluminium, Klur *et al.* (2000) have observed three resonances in the range from -81.4 to -84.9 ppm in addition to the Q^1 peak at -78.9 ppm. These resonances were ascribed to a non-bridging Q^2 site (-84.9 ppm) and two bridging Q^2 sites, where the SiO_4 tetrahedra share two of their oxygen atoms with the calcium ions in between the silicate planes that form the tobermorite-like *dreierketten* structure.

There have also been extensive ^{29}Si MAS NMR studies of the numerous crystalline silicate hydrates including the tobermorites and jennites (Cong and Kirkpatrick 1996a); and minerals which are commonly formed in hydrothermal cement hydration such as xonotlite, hillebrandite and calciochondrite (Bell *et al.* 1990); and chemical degradation processes, notably thaumasite (Skibsted *et al.* 1995).

^{27}Al MAS NMR spectra have been reported for numerous crystalline aluminate hydrates (Skibsted *et al.* 1993; Faucon *et al.* 1998; Chudek *et al.* 2000; Moulin *et al.* 1900; Mejlhede Jensen *et al.* 2000) including C_3AH_{13}, C_3AH_6, C_2AH_8 and

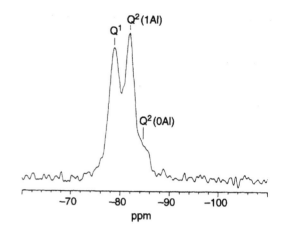

FIGURE 20.9 {^1H} ^{29}Si CP/MAS NMR spectrum (9.4 T) of a synthetic slag glass activated by a 5 M KOH solution using a solution/solid ratio of 0.4. The ^{29}Si resonances are observed at $\delta(Q^1) = -79$ ppm, $\delta(Q^2(1Al)) = -82$ ppm, and $\delta(Q^2(0Al)) = -85$ ppm. (Reproduced by permission from the American Ceramic Society, from Richardson *et al.* 1993.)

CAH_{10}; gibbsite AH_3; and complex aluminates such as ettringite, monosulphate, Friedel's salt, hydrocalumite and hydrotalcite. For most of these hydrates, ^{27}Al isotropic chemical shifts and quadrupole coupling parameters (C_Q and η_Q) have been determined from analysis of either the quadrupolar lineshape for the central transition, or the manifolds of spinning sidebands observed from the satellite transitions. These parameters have been useful for the identification and quantification of ettringite and monosulphate in hydrated Portland cements (Skibsted *et al.* 1993, 1997).

In Al mineral hydrates, the Al ion is invariably in an octahedral environment, and the isotropic chemical shifts therefore lie within the narrow range $\delta_{iso} = 5–15$ ppm. Detailed comparisons of NMR and crystallographic data show good consistency, except that the MAS NMR spectrum of ettringite indicates that all Al sites are equivalent (Skibsted *et al.* 1993) while the crystal structure shows two non-equivalent sites (Moore and Taylor 1970). For purposes of identification, the quadrupole coupling parameters are in some ways more useful than the chemical shifts. It is clear that MQMAS NMR measurements can provide the additional information to resolve overlapping resonances from ions in similar environments, as shown by Faucon *et al.* (1998) in the case of C_4AH_{13}.

The NMR spectra of anhydrous and hydrated phases provide a basis for study of the hydration process itself, both of pure phases and whole cements. For silicate systems, the main interest has been in exploring the rates of formation of the various condensed silicate species; and especially the modification of the primary cementing reactions in the presence of slags and other pozzolans. We show in Figure 20.10, the evolution of the ^{29}Si MAS NMR spectrum of a reaction mixture of C_3S, silica and KOH (Brough *et al.* 1996); and in Figure 20.11 ^{27}Al MAS NMR spectra of a hydrating ordinary Portland cement showing resonances from Al in alite/belite and in ettringite and monosulphate (Skibsted *et al.* 1997).

^{27}Al and ^{29}Si NMR yields little direct information about the location of water and hydroxide species. Unfortunately, ^1H MAS NMR spectra are generally featureless because the protons are distributed across a large number of chemically different sites between which there are only small chemical shift variations. Multi-pulse CRAMPS techniques have however been used to allow protons associated with Si–OH and Ca–OH environments to be distinguished in the hydration of C_3S (Heidemann *et al.* 1994; Zanni *et al.* 1996; Bresson *et al.* 1997; Bresson and Zanni 1998).

20.4 Proton relaxation and pore structure

We have described the kind of structural information which may be obtained at the nanoscale from MAS NMR of *solid* phases, mainly using ^{27}Al and ^{29}Si as local probes. As we have noted, solid-state proton NMR has not yet proved to be as useful at this structural level.

However, there is a long history of research into the proton resonance of the *liquid* water involved in the hydration process. During hydration, the water molecules are progressively bound into the solid matrix by incorporation into hydration products both crystalline and gel, and this

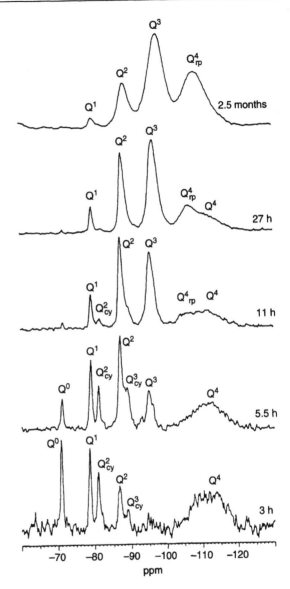

FIGURE 20.10 ^{29}Si MAS NMR spectra showing the hydration of a blend of C_3S and silica in 5 M KOH solution. Q^2_{cy} and Q^3_{cy} denote Q^2 and Q^3 species in cyclic trimers; Q^4_{rp} denotes Q^4 species in reaction products. A 10 Hz exponential linebroadening was applied to the spectra. (Reproduced by permission of Chapman and Hall from Brough *et al.* 1996, Figure 20.7.)

leads to a corresponding shortening of the proton NMR relaxation times. The measurement of transverse spin relaxation time of more or less mobile water provides therefore a means of

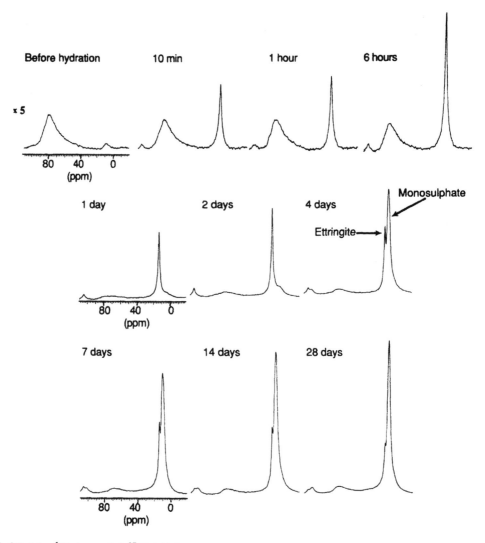

FIGURE 20.11 ^{1}H decoupled ^{27}Al MAS NMR spectra (7.1 T, $\nu_R = 7.1$ kHz) following the hydration of an OPC for 28 days. The intensities for the upper row spectra are multiplied by a factor of five compared to the other spectra. The broad asymmetric resonance around 80 ppm originates from Al incorporated in alite and belite while narrow centerbands are observed from ettringite ($\delta_{iso} = 13.1$ ppm) and monosulphate ($\delta_{iso} = 11.8$ ppm). At this magnetic field the calcium aluminate phase (C₃A) is not observed as a result of the strong [27]Al quadrupole coupling interactions which result in a severe linebroadening of the centerband resonances for this phase at 7.1 T.

exploring the development of pore structure at the microscopic rather than nanoscopic level. Interpretations of ^{1}H relaxation data are necessarily model-dependent since the relaxation parameters are for the most part statistical averages over a number of structural environments, but NMR data place constraints on model descrip-

tions and NMR-derived estimates of gel and capillary water and the surface area of gel phases can be compared with independent experimental estimates provided by adsorption and scattering methods. Pioneering work in this field was done by Blinc *et al.* (Lahajnar *et al.* 1977; Blinc *et al.* 1978). The most comprehensive recent studies have been by Halperin and coworkers (Bhattacharja *et al.* 1993; Mendelsen *et al.* 1994; Halperin *et al.* 1994; Jehng *et al.* 1996), who have derived estimates for the evolution of gel and capillary pore size distributions in Portland cements during hydration.

A large change in NMR relaxation time of water protons also accompanies the freezing/melting phase change. In the liquid, dipole–dipole interactions are greatly reduced by the averaging effects of molecular motion; in the solid they are much larger. In porous media such as cement pastes, freezing occurs over a range of temperature (perhaps 20 K) because the freezing temperature of water decreases with diminishing pore size r ($\Delta T = 50/r$ nm K). Strange *et al.* (1993) has demonstrated how data on NMR relaxation time and magnetization intensity as a function of temperature provide a means of estimating the size distributions of fine-scale porosity (pore sizes below about 100 nm). This method of NMR *cryoporometry* has now been applied to cement pastes (Jehng *et al.* 1996; Leventis *et al.* 2000; Prado *et al.* 1998a; Strange and Webber 1997.)

20.5 Other nuclei

Many elements have an isotope with nuclear spin that can be detected by NMR, although in most cases isotopes of non-zero spin are not abundant. For cement materials, isotopic enrichment allows the NMR spectra of the rare isotopes ^2D (Rakiewicz *et al.* 1998), ^{43}Ca (Klur *et al.* 1998; Nieto *et al.* 1995; Zanni *et al.* 1996) and ^{17}O (Cong and Kirkpatrick 1993a, 1996a; Mueller *et al.* 1991) to be observed. Isotopic enrichment has also been used to increase sensitivity in Si MAS NMR (Brough *et al.* 1994) where the natural abundance of ^{29}Si is only 4.7 per cent. Enrichment also provides a way to *label* a selected mineral participating in a complex reaction and to track the selected element through the transformations occurring in hydration (Brough *et al.* 1995). ^{35}Cl NMR has been the subject of recent reports (Kirkpatrick *et al.* 2001; Skibsted *et al.* 2001; Yu and Kirkpatrick 2001) related to chloride binding and migration in cementitious matrices. Alkali cation binding and migration can be observed by means of ^{23}Na and ^{133}Cs MAS NMR (Viallis *et al.* 1999). There are also several widely studied NMR-active nuclei which are of potential interest in relation to special cements and the mode of action of cement additives, notably ^2D, ^7Li, ^{11}B, ^{13}C, ^{19}F and ^{31}P: but these have so far been little exploited in cement materials research.

20.6 Magnetic resonance imaging

Spatially-resolved NMR or MRI has found spectacular application in biomedicine. Lauterbur (1973) first proposed and demonstrated the idea of using a linear magnetic field gradient superimposed on the static polarizing field to relate the NMR resonance frequency to the spatial position within an extended sample. Such frequency-encoding and the complementary phase-encoding can be extended to 2 and 3 dimensions with orthogonal gradient fields (Callaghan 1993) and is the basis of biomedical and clinical MRI, in which the distribution of mobile protons in water and other biological fluids is imaged. The fluid content of biological soft tissue is high and the other constituents exert only a weak perturbation on the liquid phase proton resonance.

MRI also has limited but valuable uses in materials science and engineering (Blumich 2000), not least in cement-based materials. Indeed, a cement-sand-lime mortar and a gypsum plaster bar were among the materials examined in one of the earliest reports of MRI in non-biological materials (Gummerson *et al.* 1979). In this study, the distribution of water in long bars of various porous materials was measured during a series of capillary absorption tests (Figure 20.12). The profile of water content with distance from the inflow surface was measured as a function of time and the data used to obtain a capillary diffusivity for each material. In this simple arrangement, MRI was used to obtain 1D moisture content profiles: the imposed magnetic gradient was parallel to the flow direction.

There has been steady development in the use of MRI as a laboratory method for obtaining moisture content distributions in partially saturated

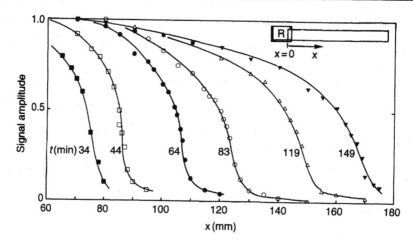

FIGURE 20.12 Water content profiles obtained by NMR during capillary absorption of water by a plaster bar. Elapsed times *t* are shown. The bar ($235 \times 33 \times 33$ mm, volume fraction porosity 0.45) was cast from a mix (1:2 by volume) of a commercial hydrated lime and a commercial retarded hemihydrate plaster. Inset sketch shows the bar with attached reservoir R to supply water to the inlow face during the experiment. (Reproduced by permission of Macmillan from Gummerson *et al.* 1979.)

porous materials since that first study, and it remains one of the most significant applications. In particular, it has been recognized that the structure and mineralogical composition of materials such as cements, ceramics and rocks perturb the proton resonance of liquid water much more than do the constituents of biological soft tissue (Osment *et al.* 1990). In fine-grained porous inorganic materials, mobile water molecules within the pores have greatly reduced relaxation times as a result of dipolar interactions with surface sites and especially with any paramagnetic species which may be present. In addition, the strong contrast of magnetic susceptibility between the solid and liquid phases creates large local magnetic field variations within the pore structure which create an effective increase in NMR linewidth. As a result, standard MRI equipment is poorly suited to imaging cementitious and similar materials. A successful approach has been to use purpose-built MRI instruments (Kopinga *et al.* 1994), typically providing strong magnetic field gradients and having electronics able to deal with the rapidly varying signals which arise from the short relaxation times. Such machines usually operate at relatively low magnetic field strengths and have small sample compartments, typically in the form of an open-

ended cylinder 50–100 mm dia. This allows specimens of unlimited length to be accommodated. With such equipment, Pel and coworkers have reported a number of studies of water migration in construction materials, including measurements of water migration in mortars (Pel *et al.* 1998) and investigations of water transfer through the brick/mortar interface (Brocken *et al.* 1998b).

In using MRI to obtain moisture distributions, we wish (ideally) to obtain a quantitative map of water concentration throughout the sample. While the primary magnetization is indeed proportional to the concentration of liquid water at each point, there may be strong variations of relaxation time through the sample (related to variations in pore size distribution or mineralogical composition from place to place) and these may introduce systematic errors in the image as a result of a relaxation-time bias. This is especially probable if water distributions are being measured through a large range of saturation: water in drier regions of the sample occupies the finer part of the pore size distribution with correspondingly shorter relaxation time. The risk of systematic errors in converting measured MRI magnetization amplitudes to water concentrations calls for careful calibration on control samples of known moisture content. We note, however, that for some purposes, a relaxation-time weighted image may provide a valuable form of image contrast if the purpose of imaging is to reveal textural and compositional features (Blumich 2000).

The problems of imaging in highly heterogeneous inorganic materials can also be overcome by the use of special rf pulse and gradient switching sequences designed to eliminate the effects of susceptibility gradient distortions. Thus Balcom, Bremner and coworkers have proposed and exploited the use of *SPRITE* (Balcom *et al.* 1996) and *turbo spin echo* (Beyea *et al.* 1998, 2000) imaging methods in a number of studies on cement and concrete materials, including the measurement of water distributions in drying (Beyea *et al.* 1998a), and in freezing (Beyea *et al.* 1998a; Choi *et al.* 2000; Prado *et al.* 1997; 1998; 1998a).

Another approach to imaging in materials in which resonances have large linewidths is to use strong magnetic field gradients. Large and stable *static* field gradients (as high as 20 T/m) are available in the vicinity of high-field superconducting magnets and *stray-field magnetic resonance imaging* (STRAFI) exploits these magnetic field gradients to image liquid-state protons and even solid-state nuclei in highly heterogeneous materials (Blümich 2000; McDonald 1998). In STRAFI, no electronic switching of field gradients is possible, so the sample must be translated bodily in the stray field during imaging; STRAFI is, however, well suited to obtain 1D moisture distribution profiles and has been applied to several cementitious materials (Bohris *et al.* 1998; Leventis *et al.* 2000; McDonald 1998; Nunes 1996).

The spatial resolution which can be obtained in MRI depends on the material and on the magnetic field gradient strengths employed, but is rarely better than about 0.2 mm. MRI does not at present image at the level of the pore structure in cement-based materials but can image aggregate particles (Figure 20.13). Furthermore, as applied today, it does not directly image solid components of the material but rather images liquids within the pore structure. It is, therefore, well suited to non-invasive mapping of patterns or dis-

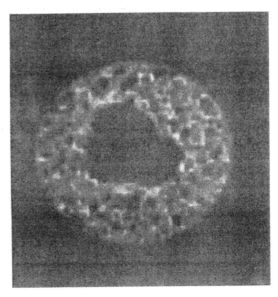

FIGURE 20.13 A 2D transverse slice from a 3D data set obtained using the T*2-weighted pure phase-encoding imaging technique SPRITE of a single large aggregate particle in a hardened mortar containing fine quartz particles. (Reproduced by permission of Academic Press from Beyea *et al.* 2000.)

tributions which are imposed on a specimen by external processes, as in the examples given of capillary water uptake, drying, chemical attack. Other larger-scale engineering features such as aggregate distributions (Beyea *et al.* 1900), cracking (Beyea *et al.* 1998), the distribution of voids (Fordham 1991, Fordham *et al.* 1994) and the dewatering of slurries and wet mortars (Brocken *et al.* 1998a) can also be well delineated.

Nearly all MRI on cementitious materials is water proton resonance imaging. However, other nuclei in the liquid phase may be imaged, although with less sensitivity. Thus Pel (Pel *et al.* 2000) has used MRI to observe the migration of sodium ions as an aqueous solution of sodium chloride is absorbed into an initially dry calcium silicate brick. The ^{23}Na resonance from the Na cation in solution is easy to observe. The Na profile lags the water profile (Figure 20.14). ^{35}Cl can also be observed by NMR and direct imaging of Cl$^-$ distributions may be a potential useful technique in future studies.

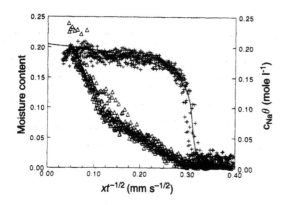

FIGURE 20.14 The moisture (+) and Na ion profiles (Δ) measured during the absorption of a 4.0 M NaCl solution determined by ^1H and ^{23}Na MRI. (Reproduced by permission of Institute of Physics from Pel *et al.* 2000.)

20.7 References

Andrew, E. R., Bradbury, A. and Eades, R. G. (1958). 'Nuclear magnetic resonance spectra from a crystal rotated at high speeds', *Nature*, **182**, 1659.

Balcom, B. J., MacGregor, R. P., Beyea, S. D., Green, D. P., Armstrong, R. L. and Bremner, T. W. (1996). 'Single-point ramped imaging with T_1 enhancement (SPRITE)', *J. Magn. Reson.*, **A123**, 131–4.

Barnes, J. R., Clague, A. D. H., Clayden, N. J., Dobson, C. M., Hayes, C. J., Groves, G. W. and Rodger, S. A. (1985). 'Hydration of Portland cement followed by Si-29 solid-state NMR spectroscopy', *J. Mater. Sci. Lett.*, **4**, 1293–5.

Bell, G. M. M., Bensted, J., Glasser, F. P., Lachowski, E. E., Roberts, D. R. and Taylor, M. J. (1990). 'Study of calcium silicate hydrates by solid state high resolution ^{29}Si nuclear magnetic resonance', *Adv. Cem. Res.*, **3**, 23–37.

Beyea, S. D., Balcom, B. J., Bremner, T. W., Prado, P. J., Cross, A. R., Armstrong, R. L. and Grattan-Bellew, P. E. (1998). 'The influence of shrinkage-cracking on the drying behaviour of white portland cement using single-point imaging (SPI)', *Solid State Nuclear Magn. Reson.*, **13**, 93–100.

Beyea, S. D., Balcom, B. J., Bremner, T. W., Prado, P. J., Green, D. P., Armstrong, R. L. and Grattan-Bellew, P. E., (1998). 'Magnetic resonance imaging and moisture content profiles of drying concrete', *Cem. Concr. Res.*, **28**, 453–63.

Beyea, S. D., Balcom, B. J., Prado, P. J., Cross, A. R., Kennedy, C. B., Armstrong, R. L. and Bremner, T. W. (1998). 'Relaxation time mapping of short T_2^* nuclei with single-point imaging (SPI) methods', *J. Magn. Reson.*, **135**, 156–64.

Beyea, S. D., Balcom, B. J., Mastikhin, I. V., Bremner, T. W., Armstrong, R. L. and Grattan-Bellew, P. E. (2000). 'Imaging of heterogeneous materials with a turbo spin echo single-point imaging technique', *J. Magn. Reson.*, **144**, 255–65.

Bhattacharja, S., Moukwa, M., Dorazio, F., Jehng, J. Y. and Halperin, W. P. (1993). 'Microstructure determination of cement pastes by NMR and conventional techniques', *Adv. Cem. Based Mater.*, **1**, 67–76.

Blaine, R. L. (1960). 'Proton magnetic resonance (N.M.R.) in hydrated portland cements', *National Bureau of Standards Monograph*, **43**, 501–11.

Blinc, R., Burgar, M., Lahajnar, G., Rozmarin, M., Rutar, V., Kocuvan, I. and Ursic, J. (1978). 'NMR relaxation study of adsorbed water in cement and C_3S pastes', *J. Am. Ceram. Soc.*, **61**, 35–7.

Blümich, B. (2000). *NMR imaging of materials*, Oxford University Press.

Bohris, A. J., Goerke, U., McDonald, P. J., Mulheron, M., Newling, B. and Le Page, B. (1998). 'A broad line NMR and MRI study of water and water transport in Portland cement pastes', *Magn. Reson. Imaging*, **16**, 455–61.

Bonafous, L., Bessada, C., Massiot, D., Coutures, J. P., Lerolland, B. and Colombet, P. (1995). '^{29}Si MAS NMR study of dicalcium silicate: The structural influence of sulfate and alumina stabilizers', *J. Am. Ceram. Soc.*, **78**, 2603–8.

Bresson, B., Zanni, H., Masse, S. and Noik, C. (1997). 'Contribution of ^1H combined rotation and multi-pulse spectroscopy nuclear magnetic resonance to the study of tricalcium silicate hydration', *J. Mater. Sci.*, **32**, 4633–9.

Bresson, B. and Zanni, H. (1998). 'Pressure and temperature influence on tricalcium silicate hydration. A ^1H and ^{29}Si NMR study', *J. Chimie Physique*, **95**, 327–31.

Brocken, H. J. P., Pel, L. and Kopinga, K. (1998a). 'Water extraction out of mortar during brick laying: A NMR study', in *Nuclear Magnetic Resonance Spectroscopy of Cement-Based Materials*, P. Colombet, A-R. Grimmer, H. Zanni and P. Sozzani (eds), Springer Verlag, Berlin and Heidelberg pp. 387–95.

Brocken, H. J. P., Pel, L. and Kopinga, K. (1998b). 'Moisture transport over the brick/mortar interface', in *Nuclear Magnetic Resonance Spectroscopy of Cement-Based Materials*, P. Colombet, A.-R. Grimmer, H. Zanni and P. Sozzani (eds), Springer Verlag, Berlin and Heidelberg, pp. 397–402.

Brocken, H. J. P., Spiekman, M. E., Pel, L., Kopinga, K. and Larbi, J. A. (1998). 'Water extraction out of mortar during brick laying: A NMR study', *Mater. Struct.*, **31**, 49–57.

Brough, A. R., Dobson, C. M., Richardson, I. G. and Groves, G. W. (1994) 'In situ solid-state NMR studies of Ca_3SiO_5: hydration at room temperature and at elevated temperatures using ^{29}Si enrichment', *J. Mater. Sci.*, 29, 3926–40.

Brough, A. R., Dobson, C. M., Richardson, I. G. and Groves. G. W. (1994a). 'Application of selective Si-29 isotopic enrichment to studies of the structure of calcium silicate hydrate (C-S-H) gels', *J. Am. Ceram. Soc.*, 77, 593–6.

Brough, A. R., Dobson, C. M., Richardson, I. G. and Groves, G. W. (1995). 'A study of the pozzolanic reaction by solid-state ^{29}Si nuclear magnetic resonance using selective isotopic enrichment', *J. Mater. Sci.*, 30, 1671–8.

Brough, A. R., Dobson, C. M., Richardson, I. G. and Groves, G. W. (1996). 'Alkali activation of reactive silicas in cements: *in situ* ^{29}Si MAS NMR studies of the kinetics of silicate polymerization', *J. Mater. Sci.*, 31, 3365–73.

Callaghan, P. T. (1993). *Principles of Magnetic Resonance Microscopy*, Oxford University Press.

Chmelka, B. F., Mueller, K. T., Pines, A., Stebbins, J. F., Wu, Y. and Zwanziger, J. W. (1989). 'O-17 NMR in solids by dynamic-angle spinning and double rotation', *Nature*, 339, 42–3.

Choi, C., Balcom, B. J., Beyea, S. D., Bremner, T. W., Grattan-Bellew, P. E. and Armstrong, R. L. (2000). 'Spatially-resolved pore-size distribution of drying concrete with magnetic resonance imaging', *J. Appl. Phys.*, 88, 3578–81.

Chudek, J. A., Hunter, G., Jones, M. R., Scrimgeour, S. N., Hewlett, P. C. and Kudryavtsev, A. B. (2000). 'Aluminium-27 solid state NMR spectroscopic studies of chloride binding in Portland cement and blends' *J. Mater. Sci.*, 2000, 35, 4275–88.

Clayden, N. J., Dobson, C. M., Hayes, C. J. and Rodger, S. A. (1984). 'Hydration of tricalcium silicate followed by solid-state Si-29 NMR spectroscopy', *J. Chemical Society: Chemical Communications*, 1396–7.

Colombet, P. and Grimmer, A.-R. (eds). (1994). *Application of NMR Spectroscopy to Cement Science*, Gordon and Breach, Amsterdam.

Colombet, P., Grimmer, A.-R., Zanni, H. and Sozzani, P. (eds). (1998). *Nuclear Magnetic Resonance Spectroscopy of Cement-Based Materials*, Springer Verlag, Berlin and Heidelberg.

Cong, X. and Kirkpatrick, R. J. (1993). 'Hydration of calcium aluminate cements: a solid-state 27-Al study', *J. Am. Ceram. Soc.*, 76, 409–16.

Cong, X. D. and Kirkpatrick, R. J. (1993a). 'O-17 and Si-29 MAS NMR study of beta-C$_2$S hydration and the structure of calcium silicate hydrates', *Cem. Concr. Res.*, 23, 1065–77.

Cong, X. and Kirkpatrick, R. J. (1995). 'A ^{1}H-^{29}Si CPMAS NMR study of the structure of calcium silicate hydrate', *Adv. Cem. Res.*, 7, 103–11.

Cong, X. D. and Kirkpatrick, R. J. (1996). 'Si-29 and O-17 NMR investigation of the structure of some crystalline calcium silicate hydrates', *Adv. Cem. Based Mater.*, 3, 133–43.

Cong, X. D. and Kirkpatrick, R. J. (1996). 'Si-29 MAS NMR study of the structure of calcium silicate hydrate', *Adv. Cem. Based Mater.*, 3, 144–56.

Dobson, C. M., Goberdhan, D. G. C., Ramsay, J. D. F. and Rodger, S. A. (1988). '^{29}Si MAS NMR study of the hydration of tricalcium silicate in the presence of finely divided silica', *J. Mater. Sci.*, 23, 4108–14.

Faucon, P., Charpentier, T., Bertrandie, D., Nonat, A., Virlet, J. and Petit, J. C. (1998). 'Characterization of calcium aluminate hydrates and related hydrates of cement pastes by ^{27}Al MQ-MAS NMR', *Inorg. Chem.*, 37, 3726–33.

Fordham, E. J., Roberts, T. P. L., Carpenter, T. A., Hall, L. D., Maitland, G. C. and Hall, C. (1991). 'Nuclear magnetic resonance imaging of simulated voids in cement slurries', *Am. Inst. Chemical Eng. J.* 37, 1895–9.

Fordham, E. J., Hall, C., Roberts, T. P. L., Hall, L. D. and Maitland, G. C. (1994). 'Imaging voids in cement slurries', *Application of NMR Spectroscopy to Cem. Sci.*, P. Colombet and A.-R. Grimmer (eds), pp. 337–52, Gordon and Breach, Yverdon.

Frydman, L. and Harwood, J. S. (1995). 'Isotropic spectra of half-integer quadrupolar spins from bidimensional magic-angle spinning NMR', *J. Am. Chem. Soc.*, 117, 5367–8,

Golovastikov, N. I., Matveeva, R. G. and Belov, N. V. (1996). 'Crystal structure of tricalcium silicate 3CaO·SiO$_2$=C$_3$S', *Sov. Phys., Crystallogr.*, 20, 441–5.

Grimmer, A.-R. and Rosenberger, H. (1978). 'Proton solid-state NMR as a contribution to the study of water-containing silicates', *Zeitschrift für Chemie*, 18, 378.

Grimmer, A.-R., Wieker, W., Lippmaa, E., Mägi, M. and Tarmak, M. (1982). 'High resolution Si-29 NMR in solid-state bodies: studies on the hydration of Ca$_3$SiO$_5$', *Zeitschrift für Chemie*, 22, 347.

Grimmer, A.-R., von Lampe, F., Mägi, M. and Lippmaa, E. (1983). 'Solid-state high-resolution Si-29 NMR and Si–O–Si bond angle and chemical shift in disilicates', *Monatshefte für Chemie*, 114, 1053–7.

Grimmer, A-R., von Lampe, F., Mägi, M. and Lippmaa, E. (1985). 'High-resolution solid-state ^{29}Si NMR of polymorphs of Ca$_2$SiO', *Cem. Concr. Res.*, 15, 467–73.

Grimmer, A.-R., Zanni, H. (1998). '^{29}Si NMR study of chemical shift tensor anisotropy of tricalcium silicate', in P. Colombet, A.-R. Grimmer, H. Zanni and P. Sozzani, (eds), *Nuclear Magnetic Resonance Spectroscopy of Cement-Based Materials*, pp. 57–68. Springer Verlag, Berlin and Heidelberg.

Gummerson, R. J., Hall, C., Hoff, W. D., Hawkes, R., Holland, G. N. and Moore, W. S. (1979). 'Unsaturated water flow within porous materials observed by NMR imaging', *Nature*, 281, 56–7.

Halperin, W. P., Jehng J. Y. and Song, Y. Q. (1994). 'Application of spin-spin relaxation to measurement of surface area and pore-size distributions in a hydrating cement paste', *Magn. Reson. Imaging*, 12, 169–73.

Hartmann, S. R. and Hahn, E. L. (1962). 'Nuclear double resonance in the rotating frame', *Phys. Rev.*, 128, 2042–53.

Heidemann, D. (1994). 'Proton high-resolution solid-state NMR spectroscopy using CRAMPS techniques for studies in silicates and cement science', in P. Colombet and A.-R. Grimmer (eds), *Applications of NMR Spectroscopy to Cement Science*, pp. 77–102, Gordon and Breach, Amsterdam.

Herzfeld, J. and Berger, A. E. (1980). 'Sideband intensities in NMR spectra of samples spinning at the magic angle', *J. Chem. Phys.*, 73, 6021–30.

Hjorth, J., Skibsted, J. and Jakobsen, H.J. (1988). '^{29}Si MAS NMR studies of Portland cement components and effects of microsilica on the hydration reaction', *Cem. Concr. Res.*, 18, 789–98.

Jakobsen, H. J., Skibsted, J., Bildsøe, H. and Nielsen, N. C. (1989). 'Magic angle spinning NMR spectra of satellite transitions for quadrupolar nuclei', *J. Magn. Reson.*, 85, 173–80.

Jehng, J.-Y., Sprague, D. T. and Halperin, W. P. (1996). 'Pore structure of hydrating cement paste by magnetic resonance relaxation analysis and freezing', *Magn. Reson. Imaging*, 14, 785–91.

Kawachi, K., Murakami, M. and Hirahara, E. (1955). 'The hydration and hardening of cement: the nuclear magnetic resonance absorption of water molecules', *Bulletin of the Faculty of Engineering, Hiroshima University*, 4, 95–101.

Kirkpatrick, R. J., Yu, P. and Kalinichev, A. (2001). 'Chloride binding to cement phases: exchange isotherm, ^{35}Cl NMR and molecular dynamics modeling studies', in J. Skalny (ed.), *The Role of Calcium Hydroxide in Cement Pastes*, American Ceramic Society.

Klur, I., Pollet, B., Virlet, J. and Nonat, A. (1998). 'C-S-H structure evolution with calcium content by multinuclear NMR', in P. Colombet, A. R. Grimmer, H. Zanni, and P. Sozzani, (eds), *Nuclear Magnetic Resonance Spectroscopy of Cement-Based Materials*, pp 119–41, Springer Verlag, Berlin and Heidelberg.

Klur, I., Jacquinot, J-F., Brunet, F., Charpentier, T., Virlet, J., Schneider, C. and Tekely, P. (2000). 'NMR cross-polarization when $T_{IS} > T_{1\rho}$; examples from silica gel and calcium silicate hydrates', *J. Phys. Chem.*, B 104, 10162–67.

Kopinga, K. and Pel, L. (1994). 'One dimensional scanning of moisture in porous materials with NMR', *Rev. Sci. Instrum.*, 65, 3673–81.

Kundla, E., Samoson, A. and Lippmaa, E. (1981). 'High-resolution NMR of quadrupolar nuclei in rotating solids', *Chem. Phys. Lett.*, 83, 229–32.

Lahajnar, G., Blinc, R., Rutar, V., Smole, V., Zupancic, I., Kocuvan, I. and Ursic, J. (1977). 'On the use of pulse NMR techniques for the study of cement hydration', *Cem. Concr. Res.*, 7, 385–94 (1977).

Lauterbur, P. C. (1973). 'Image formation by induced local interactions. Examples employing nuclear magnetic resonance', *Nature*, 242, 190–1.

Leventis, A., Verganelakis, D. A., Halse, M. R., Webber, J. B. and Strange, J. H. (2000). 'Capillary imbibition and pore characterisation in cement pastes', *Transport in Porous Media*, 39, 143–57.

Lippmaa, E., Mägi, M., Engelhardt, G. and Grimmer, A.-R. (1980). 'Structural studies of silicates by solid-state high-resolution silicon-29 NMR', *J. Am. Chem. Soc.*, 102, 4889–993.

Lippmaa, E., Mägi, M., Tarmak, M., Wieker, W. and Grimmer, A.-R. (1982). 'A high resolution Si-29 NMR study of the hydration of tricalcium silicate', *Cem. Concr. Res.*, 12, 597–602.

Llor, A. and Virlet, J. (1988). 'Towards high-resolution NMR of more nuclei in solids: sample spinning with time-dependent spinner axis angle', *Chem. Phys. Lett.*, 152, 248–53.

Lowe, I. J. (1959). 'Free induction decay of rotating solids', *Phys. Rev. Lett.*, 2, 285–7.

Mägi, M., Lippmaa, E., Samoson, A., Engelhardt, G. and Grimmer, A.-R. (1984). 'Solid-state high-resolution Si-29 chemical shifts in silicates', *J. Phys. Chem.*, 88, 1518–22.

Masse, S., Zanni, H. Lecourtier, J. Roussel, J. C. and Rivereau, A. (1993). '^{29}Si solid state NMR study of tricalcium silicate and cement hydration at high temperature', *Cem. Concr. Res.*, 23, 1169–77.

Masse, S., Zanni, H., Lecourtier, J., Roussel, J. C. and Rivereau, A. (1995). 'High temperature hydration of tricalcium silicate, the major component of Portland cement: a silicon-29 NMR contribution', *J. Chim. Phys. PCB.*, 92, 1861–6.

Massiot, D., Trumeau, D., Touzo, B., Farnan, I., Rifflet, J.-C., Douy, A. and Coutures, J.-P. (1995). 'Structure and dynamics of $CaAl_2O_4$ from liquid to glass: A high temperature ^{27}Al NMR time-resolved study', *J. Phys. Chem.*, 99, 16455–9.

Massiot, D., Touzo, B., Trumeau, D., Coutures, J. P., Virlet, J., Florian, P. and Grandinetti, P. J. (1996). 'Two-dimensional magic-angle spinning isotropic reconstruction sequences for quadrupolar nuclei', *Solid State Nucl. Magn. Reson.*, 6, 73–83.

McDonald N. (1998). 'Stray field magnetic resonance imaging', *Rep. Prog. in Phy.*, 61, 1441–93.

Medek, A., Harwood, J. S. and Frydman, L. (1995). 'Multiple-quantum magic-angle spinning NMR: A new method for the study of quadrupo-lar nuclei in solids', *J. Am. Chem. Soc.*, 117, 12779–87.

Mejlhede Jensen, O., Korzen, M. S. H., Jakobsen, H. J. and Skibsted, J. (2000). 'Influence of cement constitution and temperature on chloride binding in cement paste', *Adv. Cem. Res.*, **12**, 57–64.

Mendelson, K. S., Halperin, W. P., Jehng, J. Y. and Song, Y. Q. (1994). 'Surface magnetic-relaxation in cement pastes', *Magn. Reson. Imaging*, **12**, 207–8.

Moore, A. E. and Taylor, H. F. W. (1970). 'Crystal structure of ettringite', *Acta Crystallogr.*, **B26**, 386–93.

Moulin, I., Stone, W. E. E., Sanz, J., Bottero, J.-Y., Mosnier, F. and Haehnel, C. (2000). 'Retention of zinc and chromium ions by different phases of hydrated calcium aluminate: a solid-state ^{27}Al study', *J. Phys. Chem. B*, **104**, 9230–8.

Mueller, K. T., Sun, B. Q., Chingas, G. C., Zwanziger, J. W., Terao, T. and Pines, A. (1990). 'Dynamic angle spinning of quadrupolar nuclei', *J. Magn. Reson.*, **86**, 470–87.

Mueller, K. T., Wu, Y., Chmelka, B. F., Stebbins, J. and Pines, A. (1991). 'High-resolution oxygen-17 NMR of solid silicates', *J. Am. Chem. Soc.*, **113**, 32–8.

Müller, D., Gessner, W. and Grimmer, A.-R. (1977). 'Determination of the coordination number of aluminium in solid aluminates from the chemical shift of aluminium-27', *Zeitschrift für Chemie*, **17**, 453–4.

Müller, D., Gessner, W., Behrens, H. J. and Scheler, G. (1981). 'Determination of the aluminium coordination in alumnium-oxygen compounds by solid-state high-resolution Al-27 NMR', *Chem. Phys. Lett.*, **79**, 59–62.

Müller, D., Gessner, W., Samoson, A., Lippmaa, E. and Scheler, G. (1986). 'Solid-state ^{27}Al NMR studies on polycrystalline aluminates of the system CaO-Al$_2$O$_3$', *Polyhedron* **5**, 779–85.

Nieto, P., Dron, R., Thouvenot, R., Zanni, H. and Brivot, F. (1995). 'Study by ^{43}Ca NMR spectroscopy of the sol-gel transformation of the calcium silicate complex', *Comptes Rendues de l'Académie des Sciences Série II*, **320**, 485–8,

Nunes, T., Randall, E. W., Samoilenko, A. A., Bodart, P. and Feio, G. (1996). 'The hardening of Portland cement studied by ^{1}H NMR stray-field imaging', *J. Phys. D: Appl. Phys.*, **29**, 805–8.

Osment, P. A., Packer, K. J., Taylor, M. J., Attard, J. J., Carpenter, T. A., Hall, L. D., Herrod, N. J. and Doran, S. J. (1990). 'NMR imaging of fluids in porous solids', *Philos. Trans., R. Soc. London A*, **333**, 441–52.

Pake, G. E. (1948). 'Nuclear resonance absorption in hydrated crystals: fine structure of the proton line', *J. Chem. Phys.*, **16**, 327–36.

Pel, L. Hazrati, K., Kopinga, K. and Marchand, J. (1998). 'Water absorption in mortar determined by NMR', *Magn. Reson. Imaging*, **16**, 525–8.

Pel, L., Kopinga, K. and Kaasschieter, E. F. (2000). 'Saline absorption in calcium silicate brick observed by NMR scanning', *J. Phys. D: Appl. Phys.*, **33**, 1380–5.

Philippot, S., Masse, S., Zanni, H., Nieto, P., Maret, V. and Cheyrezy, M. (1996). 'Si-29 NMR study of hydration and pozzolanic reactions in reactive powder concrete (RPC)', *Magn. Reson. Imaging*, **14**, 891–3.

Pines, A., Gibby, M. G. and Waugh, J. S. (1973). 'Proton-enhanced NMR of dilute spins in solids', *J. Chem. Phys.*, **59**, 569–90.

Prado, P. J., Balcom, B. J., Beyea, S. D., Armstrong, R. L. and Bremner, T. W. (1997). 'Concrete thawing studied by single-point ramped imaging', *Solid State Nucl. Magn. Reson.*, **10**, 1–8,

Prado, P. J., Balcom, B. J., Beyea, S. D., Armstrong, R. L., Bremner, T. W. and Grattan-Bellew, P. E. (1998). 'Concrete/mortar water phase transition studied by single-point MRI methods', *Magn. Reson. Imaging*, **16**, 521–3.

Prado, P. J., Balcom, B. J., Beyea, S. D., Bremner, T. W., Armstrong, R. L., Pishe, R. and Grattan-Bellew, P. E. (1998). 'Spatially resolved relaxometry and pore size distribution by single-point MRI methods: porous media calorimetry', *J. Phys. D: Appl. Phys.*, **31**, 2040–50.

Prado, P. J., Balcom, B. J., Beyea, S. D., Bremner, T. W., Armstrong, R. L. and Grattan-Bellew, P. E. (1998). 'Concrete freeze/thaw as studied by magnetic resonance imaging', *Cem. and Concr. Res.*, **28**, 261–70.

Prado, P. J., Balcom, B. J., Beyea, S. D., Bremner, T. W., Armstrong, R. L., Pishe, R. and Grattan-Bellew, P. E. (1998a). 'Spatially resolved relaxometry and pore size distribution by single-point MRI methods: porous media calorimetry', *J. Phy. D: Appl. Phys.*, **31**, 2040–50.

Rakiewicz, E. F, Benesi, A. J, Grutzeck, M. W. and Kwan, S. (1998). 'Determination of the state of water in hydrated cement phases using deuterium NMR spectroscopy', *J. Am. Chem. Soc.*, **120**, 6415–6.

Richardson, I. G., Brough, A. R., Brydson, R., Groves, G. W. and Dobson, C. M. (1993). 'Location of aluminium in substituted calcium silicate hydrate (C-S-H) gels as determined by Si-29 and Al-27 NMR and EELS', *J. Am. Ceram. Soc.*, **76**, 2285–8.

Richardson, I. G., Brough, A. R., Groves, G. W. and Dobson, C. M. (1994). 'The characterization of hardened alkali-activated blast-furnace slag pastes and the nature of the calcium silicate hydrate (C-S-H) phase', *Cem. Concr. Res.*, **24**, 813–29.

Richardson, I. G. (1999). 'The nature of C-S-H in hardened cements', *Cem. Concr. Res.*, **29**, 1131–47.

Rodger, G. C. D. (1988). 'Hydration of tricalcium silicate followed by Si-29 NMR with cross-polarization', *J. Am. Ceram. Soc.*, **71**, 91–6.

Rosenberger, H. and Grimmer, A.-R. (1979). 'Studies on hydrogen-bonded compounds with high-resolution NMR spectroscopy in solids. I The relation between the chemical shifts of protons and the character of hydroxyl groups in some crystalline silicates', *Z. Anorg. Allg. Chem.*, 448, 11–22.

Samoson, A., E., Kundla, E. and Lippmaa, E. (1982). 'High-resolution MAS-NMR of quadrupolar nuclei in powders', *J. Magn. Reson.*, 49, 350–7.

Samoson, A., Lippmaa, E. and Pines, A. (1988). 'High resolution solid state NMR averaging of second order effects by means of a double-rotor', *Mol. Phys.*, 65, 1013–18.

Seligmann, P. (1968). 'Nuclear magnetic resonance studies of the water in hardened cement paste', *J. Portland Cem. Assoc. Res. Dev. Lab.*, 10, 52–65.

Skibsted, J., Nielsen, N. C., Bildsøe, H. and Jakobsen, H. J. (1991). 'Satellite transitions in MAS NMR spectra of quadrupolar nuclei', *J. Magn. Reson.*, 95, 88–117.

Skibsted, J., Bildsøe, H. and Jakobsen, H.J. (1991). 'High-speed spinning versus high magnetic field in MAS NMR of quadrupolar nuclei. ^{27}Al MAS NMR of $3CaO \cdot Al_2O_3$ (C_3A)', *J. Magn. Reson.*, 92, 669–76.

Skibsted, J., Hjorth, J. and Jakobsen, H. J. (1990). 'Correlation between ^{29}Si NMR chemical shifts and mean Si–O bond lengths for calcium silicates', *Chem. Phys. Lett.* 172, 279–83.

Skibsted, J., Henderson, E. and Jakobsen, H. J. (1993). 'Characterization of calcium aluminate phases in cements by ^{27}Al MAS NMR spectroscopy', *Inorg. Chem.* 32, 1013–27.

Skibsted, J., Hjorth, L. and Jakobsen, H. J. (1995) 'Quantification of thaumasite in cementitious materials by $^{29}Si\{^1H\}$ cross-polarization magic-angle spinning NMR spectroscopy', *Adv. Cem. Res.*, 7, 69–83.

Skibsted, J., Jakobsen, H.J. and Hall, C. (1994). 'Direct observation of aluminium guest-ions in the silicate phases of cement minerals by ^{27}Al MAS NMR spectroscopy', *J. Chem. Soc.: Faraday Trans.*, 90, 2095–8.

Skibsted, J., Jakobsen, H.J. and Hall, C. (1995). 'Quantification of calcium silicate phases in Portland cements by ^{29}Si MAS NMR spectroscopy', *J. Chem. Soc.: Faraday Trans.* 91, 4423–30.

Skibsted, J., Mejlhede Jensen, O. and Jakobsen, H. J. (1997). 'Hydration kinetics for alite, belite, and calcium aluminate phase in Portland cements from ^{27}Al and ^{29}Si MAS NMR Spectroscopy', *Proc. 10th Int. Congr. Chem. of Cem.*, Gothenburg, Sweden, June 2–6, 2, paper 2ii056.

Skibsted, J., Jakobsen, H.J. and Hall, C. (1998). 'Quantitative aspects of ^{27}Al MAS NMR of calcium aluminoferrites', *Adv. Cem. Based Mater.*, 7, 57–9.

Skibsted, J. Mejlhede Jensen, O., Jakobsen, H. J. (2001). 'High-field ^{35}Cl NMR investigation of chloride ions in inorganic solids, A variable-temperature ^{35}Cl MAS NMR study of the phase transition in Friedel's salt ($Ca_2Al(OH)_6Cl \cdot 2H_2O$)'. In preparation.

Stade, H., Grimmer, A.-R., Engelhardt, G., Mägi, M. and Lippmaa, E. (1985). 'On the structure of ill-crystallized calcium hydrogen silicates. 7. Solid-state Si-29 NMR studies on C-S-H (di, poly)', *Z. Anorg. Allg. Chem.*, 528, 147–51.

Strange, J. H., Rahman, M. and Smith, E. G. (1993). 'Characterization of porous solids by NMR', *Phys. Rev. Lett.*, 71, 3589–91.

Strange, J. H. and Webber, J. B. (1997). 'Multidimensional resolved pore size distributions', *Appl. Magn. Reson.*, 12, 231–45.

Sun, G., Brough, A. R. and Young, J. F. (1999). '^{29}Si NMR study of the hydration of Ca_3SiO_5 and beta-Ca_2SiO_4 in the presence of silica fume', *J. Am. Ceram. Soc.*, 82, 3225–30.

Viallis, H., Faucon, P., Petit, J. C. and Nonat, A. (1999). 'The interaction between salts (NaCl and CsCl) and calcium silicate hydrates', *J. Phys. Chem. B*, 103, 5212–19.

Young, J. F. (1988). 'Investigations of calcium silicate hydrate structure using Si-29 nuclear magnetic resonance spectroscopy', *J. Am. Ceram. Soc.*, 71, C118–C120.

Yu, P. and Kirkpatrick, R. J. (2001). '^{35}Cl NMR relaxation study of cement hydrate suspensions', *Cem. Concr. Res.*, 31, 1079–85.

Zanni, H., Nieto, P., Fernandez, L., Couty, R., Barret, P., Nonat, A. and Bertrandie, D. (1994). 'Sol-gel transition in silica alkaline solutions. A NMR study', *J. Chim. Phys. PCB.*, 91, 901–8.

Zanni, H., Cheyrezy, M., Maret, V., Philippot, S. and Nieto, P. (1996). 'Investigation of hydration and pozzolanic reaction in reactive powder concrete (RPC) using ^{29}Si NMR', *Cem. Concr. Res.*, 26, 93–100.

Zanni, H., Rassem-Bertolo, R., Masse, S., Fernandez, L., Nieto, P. and Bresson, B. (1996a). 'A spectroscopic NMR investigation of the calcium silicate hydrates present in cement and concrete', *Magn. Reson. Imaging*, 14, 827–31.

Chapter twenty-one

The use of synchrotron sources in the study of cement materials

Paul Barnes, Sally L. Colston, A. C. Jupe, S. D. M. Jacques,
M. Attfield, R. Pisula, S. Morgan, C. Hall, P. Livesey and S. Lunt

21.1 Introduction

It seems surprising that only twenty years ago electron synchrotrons were primarily the property of particle physicists; the electromagnetic radiation they created was then merely an inconvenient by-product. Since then, the number of dedicated synchrotron sources has increased enormously, worldwide, serving across all scientific disciplines. Also between 1994 and 1998, the first three 'third generation' super synchrotrons were completed (the European Synchrotron Radiation Facility (ESRF) in France; the APS in USA; the SPring8 in Japan); in future years, we might well look back at the synchrotron as one of the pre-eminent scientific revolutions of the late twentieth century. The X-ray flux that one can bring to bear upon an experiment has been fundamentally changed, by orders of magnitude, since the 1980s through the impact of the spectral brilliance of modern synchrotrons. This chapter will examine the benefits of this 'discontinuity' in the field of cement chemistry.

21.2 The synchrotron

It is a fundamental principle of physics, that when charged particles are accelerated they emit electromagnetic radiation; an everyday example of this effect is the radio-transmitter in which the particles being accelerated are the electrons in the transmitter mast and the radiation produced falls within the radio-frequency range. With high energy synchrotrons the radiation extends into the infrared, visible, ultra-violet and X-ray portions of the electromagnetic spectrum. This radiation is produced as the electrons are steered around the synchrotron orbit via electromagnets, usually dipole magnets, which produce a vertical magnetic field in the gap between their poles, thus imparting a Lorentz force and centripetal acceleration to the electrons to keep them in the correct stable orbit (see Figure 21.1). In practice, three different kinds of magnet devices are used, termed

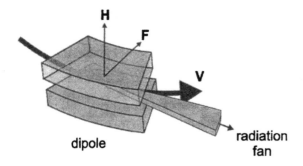

FIGURE 21.1 Schematic of dipole magnet indicating the electron beam trajectory/velocity (v), magnetic field (H) and resulting centripetal force (F) and X-radiation fan.

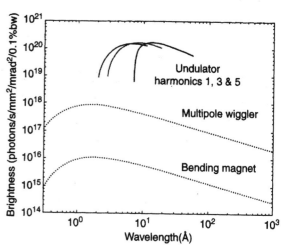

FIGURE 21.2 Typical synchrotron X-ray intensity spectra obtained from dipole (bending magnet), wiggler and undulator devices.

dipoles, wigglers and undulators, that produce synchrotron X-radiation with individual spectral characteristics tailored to the needs of the user (see Figure 21.2). The whole configuration of synchrotron magnets forms a closed loop (or 'ring') so that electrons circulate continuously, their energetic losses (principally from synchrotron radiation) being restored as they pass through high frequency radio cavities placed around the ring, and the electron losses (e.g. from collisions with stray gas molecules, in spite of the 10^{-10} Torr vacuum) being restored by injection cycles performed typically once or twice per day. In order to maintain synchronism with the RF cavities, the electrons have to travel around the ring in 'bunches'; this requirement coupled with the fact that the electrons are travelling very close to the speed of light (e.g. ~0.9999 of the speed of light at 2 GeV energy) means that the synchrotron X-ray light is in actuality 'flashing', though at a rate of around 10^9 times per second this would appear to be continuous to most observers! The relativistic nature of the electrons (they have a relativistic mass typically of many thousands of times the rest mass) also confers other unique properties to the fan of radiation emanating from the magnet devices. The attributes that are most relevant to the subject under discussion here are:

- the X-ray beam is intense. Intensity is often measured in units of 'brilliance' which

normalizes the number of X-ray photons produced to a 0.1 per cent bandwidth (to account for its wide spectral range of wavelengths), per mrad^2 (the concentration within a narrow solid angle) and per mm^2 (the small effective source size). A typical brilliance these days is such that up to 10^{12} photons per second might be incident on a sample, and this confers the possibilities of excellent counting statistics and/or short collection times during measurement;
- the radiation is horizontally polarized in the plane of the electron orbit by comparison to laboratory X-ray sources where the polarization is equal in all directions. There are several ways in which this polarization can be exploited in diffraction and spectroscopy measurements;
- the X-ray beam is highly collimated, with a typical working divergence in the range of mrads. This results in less wastage of radiation during its passage through the optical components towards the sample, and a greater angular resolution in eventual measurement;
- the radiation displays a smooth continuous spectrum thus offering the choice of conducting experiments with *white* X-ray radiation,

or alternatively allowing the free choice of wavelength, by the use of a monochromator, without being tied to metal edges such as the 1.542 Å copper Kα laboratory source.

In consideration of the background, one really needs to first appreciate the change in work-practice, and indeed in lifestyle, that is necessary when conducting experiments on a centralized facility. At present, synchrotrons that produce substantially hard X-ray beams only exist as centralized facilities at one or a few sites within a country, for reasons of geography, size and cost: for example, the Spring8 synchrotron in Japan has a ring circumference of 1436 m sited at a seismically stable region (Harima); the more modestly-sized DIAMOND synchrotron planned for the UK by around 2006 is costed at over 200 million pounds sterling. Conventional research centres, whether academic, industrial or institutional, are, with a few exceptions, almost entirely centred around in-house facilities. The idea of taking a manufacturing or testing plant away from the 'home site', and purpose-building an equivalent compatible version on a radiation beam-line, represents a distinct change in research philosophy and the social implications, of a research team working regularly at large distances away from home on a 24 h per day schedule, need to be addressed. One of the earliest example of such an approach in the field of cements was the work by Christensen *et al.* (1984, 1986, 1988), but using a neutron rather than synchrotron centralized facility, to follow the hydration of various calcium aluminate systems at temperatures up to 100°C. One of the drawbacks of neutron diffraction is the need to use deuterium oxide (heavy water) to avoid the large incoherent scattering from hydrogen, which surprisingly slows down the hydration kinetics substantially at ambient temperatures (Clark and Barnes 1995) though this difference diminishes with higher temperatures (>100°C) into the autoclave condition (Polak *et al.* 1990).

One of the first recorded uses of the synchrotron in cement research was that of X-ray diffraction-topography to analyse the defect content of calcium hydroxide crystals (Barker and Barnes 1984; Ghose and Barnes 1983), though since then the overwhelming use has been through the techniques of powder diffraction. We will find (see below) that a particularly rewarding technique has been time-resolved powder energy-dispersive diffraction, as developed for cement studies by the current authors since the late 1980s, though the first publications appeared somewhat later (Barnes *et al.* 1990; Barnes 1991) in recognized journals other than in-house reports. However, the power of synchrotron radiation has not been particularly well-exploited by the cement community at large. Its usage appears to be restricted to a small number of individual groups and researchers who have mainly benefited from the resolution and wavelength choice/range with synchrotron powder diffraction (see Section 21.3) for analysing structure/composition of cement-related phases (e.g. Mumme *et al.* 1995) though small-angle X-ray scattering (e.g. Henderson and Komanaschek 2000), Ca K-edge EXAFS (e.g. Kirkpatrick *et al.* 1997; Lequeux *et al.* 1999) and micro-probe analysis (Butler *et al.* 2000) have also found use in the study of cement systems. However, the vast majority of applications of synchrotron radiation in cement research have been undertaken by the present authors, and the remaining sections of this chapter will concentrate on these studies. The order of presentation has been chosen to highlight the particular advantages of each synchrotron technique.

21.3 High-quality/resolution powder diffraction

Powder diffraction has become a standard tool in modern materials science research laboratories, mainly for the identification of unknown samples, for quantitative composition analysis and for crystal structure refinement by Rietveld refinement techniques (Rietveld 1969). Thus, in the field of powder diffraction, the synchrotron does not offer an exclusive technique but simply enhances the laboratory version. The four synchrotron attributes listed can all enhance the functioning of powder diffraction by virtue of the following:

- The intense X-ray beam permits the collection of powder diffraction patterns with excellent counting statistics or, alternatively, with short collection times enabling time-resolved analysis of rapid reaction sequences.
- By configuring a synchrotron powder diffractometer such that the detector rotates about a horizontal axis the normal reduction, by $(1 + \cos^2 2\theta)/2$, of diffracted intensity with angle 2θ, is eliminated since the polarization (being perpendicular to the X-ray direction) remains horizontal after diffraction in the vertical plane. This means that the rapid fall-off in intensity (with 2θ) associated with laboratory diffractometers is reduced; thus, the high angle peaks are improved.
- The superior collimation of the synchrotron X-ray beam can be particularly beneficial by reducing the broadening of diffraction peaks and therefore making them sharper and better-resolved. In some cases, this improvement can be the difference between being able to solve a crystal structure with synchrotron powder diffraction data and not being able to effect a solution with the equivalent laboratory data.
- The free choice of X-ray wavelength permits collection of powder data at an X-ray absorption edge of one of the constituent atoms within the sample, thereby aiding the crystallographic phase problem in structure solution. This, however, has not been exploited in cement studies. The alternative benefit is that the synchrotron smooth continuous radiation spectrum offers the choice of conducting experiments with *white* X-ray radiation; this is exploited in energy-dispersive diffraction as detailed later.

Figure 21.3 shows powder diffraction patterns (Turrillas *et al.* 1995), not of cementitious material but of a zirconium ortho-sulphate, to illustrate the variation in pattern quality that can be obtained from different systems: the progression

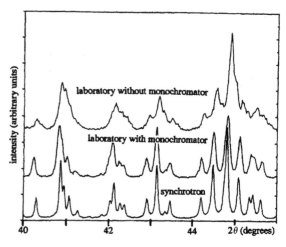

FIGURE 21.3 Powder diffraction patterns of a zirconium ortho-sulphate collected with a laboratory system without a monochromator (top pattern: full Cu-K$\alpha_1\alpha_2$ doublet with mean wavelength $\lambda = 1.5418$ Å), the same with a 100-germanium monochromator (middle: Cu-Kα_1 singlet with $\lambda = 1.5406$ Å) and on station 2.3 of the Daresbury SRS synchrotron (bottom: $\lambda = 1.51603$ Å). The progressive improvement in pattern quality is self-evident. (Reprinted from Turrillas *et al.* (1995)). 'Synchrotron-related studies on the dynamic and structural aspects of Zirconia synthesis for ceramic and catalytic applications', pp. 491–508, with permission from Elsevier Science.

in quality from a typical laboratory to a high resolution synchrotron powder diffractometer is fairly obvious from the increase in number of diffraction peaks that can be resolved. Schematics describing the essentials of such diffractometer systems (laboratory and synchrotron) are given in Figure 21.4. However, it must be immediately stated that this range of improvement can only be realized with high quality powder specimens. The sources of peak broadening in powder diffraction patterns are two-fold: specimen broadening and instrumental broadening. If the specimen contribution is high, then it will dominate and improving the diffractometer system will have little benefit. Real manufactured cements tend to be multi-phase systems with high heterogeneity and poor crystallinity with the result that their diffraction patterns are dominated by sample broadening. This point is illustrated by

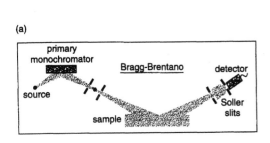

(a)

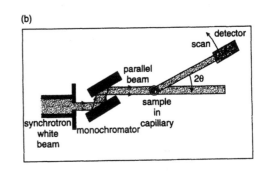

(b)

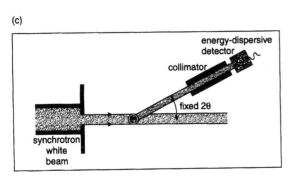

(c)

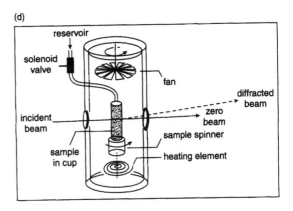

(d)

FIGURE 21.4 Schematics of a typical laboratory (Bragg–Brentano) and synchrotron powder diffractometers: (a) Laboratory Bragg–Brentano angle-scanning powder diffractometer; (b) Synchrotron (medium-high resolution) powder diffractometer in Debye–Scherrer/capillary mode; (c) Energy-dispersive (powder) diffraction mode; (d) Energy-dispersive diffraction in side view to illustrate the range of experimental (bulk) possibilities.

Figure 21.5 (Colston *et al.* 1998) which compares the powder patterns for Ordinary Portland cement (OPC) collected by a low-resolution (energy-dispersive) powder diffractometer and a high-resolution synchrotron powder diffractometer: the improvement in effective quality is minimal; there might be highly specialized situations where high resolution is essential, but such situations are not likely to be so common with commercial cements. However, with synthetic cements, the situation can be quite different since good sample preparation can be synonymous with high purity and crystallinity. Such an example concerns doped brownmillerite where carefully synthesized powders of $Ca_2Fe_{0.95}Al_{0.95}Mg_{0.05}Si_{0.05}O_5$, have been obtained (Jupe 1998; Jupe *et al.* 2001). This

powder does yield a good powder diffraction pattern (Figure 21.6) on the Daresbury SRS (medium) high-resolution powder diffractometer (station 2.3). Because of the under-determinancy of X-ray data on their own in relation to solving structures with multiple occupation of tetrahedrally- and octahedrally-coordinated sites (i.e. multi-occupation by Fe, Al, Si, Mg) the Rietveld structure refinement eventually had to be completed using, in addition, data from both neutron diffraction and X-ray absorption spectroscopy. This resulted in a definitive structure solution which is discussed further in Section 21.6. It serves as an example of the use that can be made of high-resolution synchrotron powder diffraction in the field of synthetic cement chemistry.

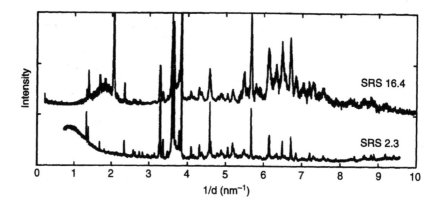

FIGURE 21.5 Comparison of powder diffraction patterns obtained from an oilwell Portland cement collected using low resolution (EDD: SRS station 16.4, merged data sets from three angles of 3.04°, 5.84° and 8.75°; collection time 300s) and high-resolution (HRPD: SRS station 2.3 at wavelength of 1.200 Å; collection time 4 h) powder diffractometers on the SRS synchrotron. The broad 'hump' in the EDD pattern around $2\,nm^{-1}$ is due to scattering from the polymeric specimen holder, and the rise at $<2\,nm^{-1}$ in the HRPD pattern results from glancing X-ray incidence at low angles.

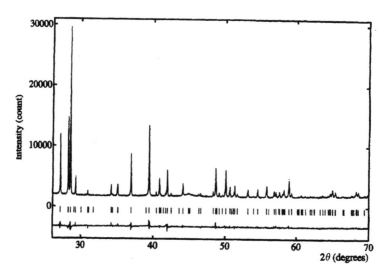

FIGURE 21.6 A Rietveld plot for doped ($Ca_2Fe_{0.95}Al_{0.95}Mg_{0.05}Si_{0.05}O_5$) brownmillerite, obtained from high-resolution powder diffraction data (station 2.3 of the SRS) using an X-ray wavelength of 1.3000(2) Å radiation over the 2θ range 6–70° in steps of 0.01°, the time steps ranging from 1 to 14 s. The points in the top curve represent the actual diffraction data after background subtraction, the continuous curve represents the equivalent prediction from the (Rietveld) refined crystallographic structure, the sticks show the predicted peak positions for the refined unit cell, and the lower curve represents the difference between the first two items (Jupe *et al.* 2001, and reproduced by permission from Munskgaard International Publishers, Denmark).

21.4 Single crystal micro-diffraction

Single crystal X-ray diffraction is normally the first method of choice for determining the crystal structures of materials, provided that suitable single crystals can be grown. The high level of sophistication of the structure solution (direct methods and Patterson methods) and refinement programs in existence can make the determination of a new crystal structure almost routine.

Often, the main problem with this approach is that of growing single crystals of the material that are large enough to collect the diffraction data. Typically, for a single crystal diffractometer, with an X-ray tube source and single-reflection detector, crystals with dimensions greater than 0.1 mm are required to collect data. This size may be reduced by using a more powerful X-ray source, such as, a rotating anode, or by using an area detector, for example, a CCD or image plate detector. The latter enables longer collection times for each diffraction spot, since many spots are collected simultaneously.

Harnessing the high intensity of a synchrotron beam has enabled the collection of useable diffraction data from crystals (micro-crystals) with dimensions as small as 5 μm, and has thereby opened up the technique of *single crystal micro-diffraction*. This technique has been used (Pisula, Attfield *et al.* 2001) on station 9.8 of the Daresbury SRS using a wavelength of 0.4852 Å with cooling of a synthesized mono-sulphate crystal (size *circa* $20 \times 100 \times 120$ μm) to yield a preliminary structure for $Ca_6Al_3(OH)_{18}(SO_4)_{1.5} \cdot 16.5H_2O$ with the following structural parameters: trigonal Space Group R3; unit cell dimensions: $a = 5.741(2)$Å, $c = 30.68(1)$Å, $Volume = 875.6(5)$ Å³; $Z = 1$, $Mw = 1068.59$, $D_c = 2.03$ g cm⁻³. The basic structure is illustrated in Figure 21.7; it consists of $[Ca_2(H_2O) Al(OH)_3]^{4+}$ layers with charge compensating SO_4^{2-} groups and water molecules in-between these layers. The SO_4^{2-} groups and water molecules between the layers are highly disordered. The success of this short study suggests that the structures of many more cement hydrate phases could be determined using single crystal micro-diffraction.

21.5 Energy-dispersive powder diffraction

Synchrotron energy-dispersive diffraction (EDD) of cements has already produced remarkable successes. Although the nature of EDD makes it less suitable for many academic studies, including crystallographic structure analysis/refinement, we will see that it is ideally suited to many industrial

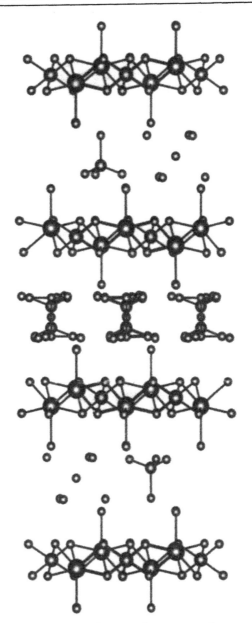

FIGURE 21.7 The crystal structure of monosulphate, $Ca_6Al_3(OH)_{18}(SO_4)_{1.5} \cdot 16.5H_2O$, obtained from a synchrotron micro-crystal diffraction experiment. The structure is viewed along the [1$\bar{1}$0] direction. Because of the disordered state of the SO_4^{2-} groups and H_2O molecules, the two orientations of the SO_4^{2-} groups and the cluster of H_2O molecules are shown separately in the top and bottom interlayer galleries, and as a combined crystallographic view in the middle gallery. The atoms, in order of decreasing size, are Ca, Al, S and O.

applications. Although conventional powder diffraction is a well-known technique to many researchers, the concept of EDD is less well-known, and therefore requires some initial explanation. Diffraction of X-rays by a crystal lattice is a well-known phenomenon which occurs under discrete conditions described by Bragg's Law:

$$\lambda = 2d \sin\theta \qquad (1)$$

where λ is the radiation wavelength, d the interplanar spacing and θ the Bragg angle. Conventional diffractometers seek these Bragg conditions using a fixed wavelength, λ, and scanning a detector through a range of 2θ angles. By contrast, EDD uses the white (continuous) X-rays to irradiate a crystal and an energy-dispersive (ED) detector is set at a fixed scattering angle. However, only those X-rays with wavelengths satisfying Bragg's law, for the fixed θ, will reach the detector. An ED-detector determines the energy, E, rather than the wavelength of X-ray photons, these two parameters being connected by a fundamental relationship, $E = hc/\lambda$ (h = Planck's constant; c = velocity of light), such that Bragg's law can be re-written as:

$$Ed \sin\theta = hc/2 \approx 6.2\,\text{keV}\cdot\text{Å} \qquad (2)$$

A schematic of the EDD-mode is included in Figures 21.4c and d. EDD was first demonstrated by Giessen and Gordon (1968) and Buras *et al.* (1968) in 1968, though it was the additional ingredient of intense synchrotron radiation (Bordas *et al.* 1977; Buras *et al.* 1976) in the mid-1970s which transformed it into a viable technique. Later, the modification of multi-angle EDD (Barnes *et al.* 1998; Colston *et al.* 1998) extended the range of accessible d-spacings. The main features of EDD are:

- It has a fixed geometry (no moving parts during collection) which makes the design of industrial environmental cells much easier due to the fixed entrance and exit beam directions; further the lack of moving parts greatly aids the collection of rapid, time-resolved, diffraction data.

- The white X-ray beam from a synchrotron is energetic (20–100 keV is typical) enabling penetration through large bulk objects, in extreme cases to 100 mm of concrete!
- The technique yields low-resolution patterns which may be disadvantageous if the diffraction peaks are congested; otherwise it is not a problem.

These attributes have been well exploited in cement research, yielding significantly new results that, frankly, were either impossible to discern or too easily confused using conventional laboratory diffraction. Since diffraction techniques link to crystalline (rather than amorphous) structure, the power of EDD is linked to its ability to discern the presence of micro-crystalline phases; for example, calcium hydroxide rather than amorphous C-S-H in hydrated OPC. There are, however, a rich variety of micro-crystalline phases to be found in silicate and aluminate cements, particularly if the field of temperature and admixtures is included. Space permits only a brief outline of the most important discoveries.

21.5.1 The rapid hydration of tricalcium aluminate

This was a later study, chronologically, but it is considered first as it has an important object lesson. It has long been known that high reactivity of C_3A can result in rapid hydration, eventually to the stable hydrate C_3AH_6. Its precise hydration sequence has now been studied (Jupe *et al.* 1996) with unprecedented time- and space-resolution using beamline ID30 on the ESRF at Grenoble, France. The speed of the reaction, even at nominally ambient temperatures, necessitated rapid time-resolved EDD (diffraction pattern collection times of either 5 s or 300 ms) and remotely-controlled addition of water to the C_3A powder. Repeating the experiment gave, consistently, the same sequence which is illustrated in Figure 21.8 showing that the growth of the final C_3AH_6 only commences after full development of the intermediate phase which, from its diffraction pattern, appears to resemble the known C_2AH_8 hydrate.

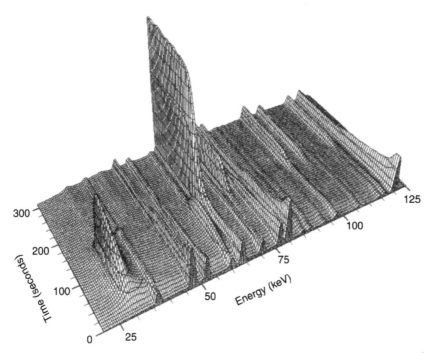

FIGURE 21.8 Rapid time-resolved EDD patterns showing the rapid hydration of C_3A. Initial C_3A peaks are clearly identified by their in-filling on the time zero section (e.g. the three largest at *circa* 48, 80, 120 keV), the intermediate phase appears between *circa* 40–160 s (two prominent peaks at *circa* 30 and 60 keV) and the final C_3AH_6 forms after *circa* 160 s (the two most prominent peaks being *circa* 65 and 75 keV). Each diffraction pattern has been collected over 5 s, and the detector angle was set at 2.2° (Jupe *et al.* 1996).

The remarkably consistent time-sequence, rather than co-existence, of $C_3A \rightarrow$ intermediate $\rightarrow C_3AH_6$ even at these rapid rates, strongly supports the notion of the intermediate providing initial nucleation sites for the growth of C_3AH_6 until it becomes self-propagating. Classical laboratory powder diffraction analysis does not reveal this clear time-separation of phases because of its inferior time- and space-resolution. Although the former deficiency might have been expected, the latter is less obvious: The time-resolved EDD experiments were performed on a small volume element of $\sim 0.1 \, mm^3$ at a fixed-depth well inside the C_3A sample; this means that the reaction sequence does not become spread out in time across the sample as with laboratory diffractometers where the effective sample volume (typically, a 10 mm diameter disc within 50 μm of the surface) results in a spread of reaction rates. This

effect can easily be simulated (Figure 21.9), showing that with just 50 s worth of effective time/space-smearing, the essential $C_3A \rightarrow$ intermediate $\rightarrow C_3AH_6$ sequence is irretrievably lost. Thus, previous reaction data on this system collected over many years by conventional laboratory systems unavoidably presented a false picture of the true nature of this important hydration sequence.

21.5.2 Conversion of calcium aluminate (high alumina) cements

This study bears similarity with that of the previous section (21.5.1). The work of Muhamad *et al.* (Barnes *et al.* 1992; Muhamad M. N. 2001), using commercial forms of calcium aluminate (high alumina) cements (CACs), showed how to optimize the EDD configuration for monitoring the course of hydration or conversion; however,

485

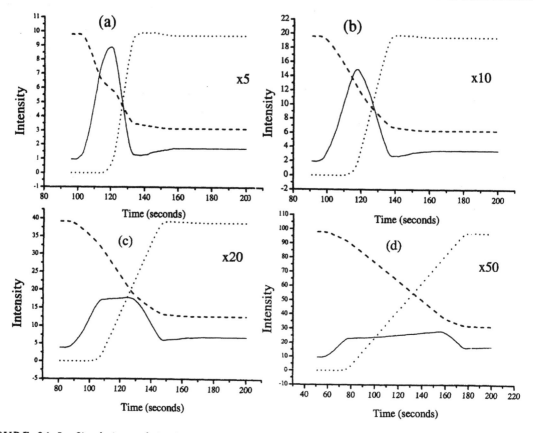

FIGURE 21.9 Simulations of time/space-smearing applied to the hydration of C_3A, using individual rapid time-resolved EDD-data (---C_3A;—intermediate; ····C_3AH_6). As time-smearing is progressively increased ((a) 5 s; (b) 10 s; (c) 20 s; (d) 50 s) the concentration plots no longer present a distinguishable $C_3A \rightarrow$intermediate$\rightarrow C_3AH_6$ pattern but rather develop into a non-sequential form of general growth/loss (Barnes *et al.* 2000, reproduced by permission from Munskgaard International Publishers, Denmark).

the complexity of manufactured CACs was such that peak-overlap in the EDD patterns made the analysis very difficult, at least initially. A breakthrough came by way of moving on to using pure synthetic phases (CA; CAH_{10}) for which the EDD peak-overlap was far less severe, enabling a clearer picture to emerge from the time-resolved data. The conversion of CAH_{10} at temperatures of up to 90°C was studied (Rashid *et al.* 1992, 1994) using time-resolved EDD. In all cases, the loss of CAH_{10} and rise in C_3AH_6 peaks is bridged by an intermediate phase (probably C_2AH_8) in an analogous manner to the previous study with C_3A hydration. This persists even up to 90°C when the

conversion starts within minutes and is completed in little over 5 min with the intermediate phase being detectable for only ~2 min. Clearly, such rapid events can only be detected with a system that is capable of collecting diffraction patterns from a bulk sample with acquisition times of ~20 s. That the same full sequence, with the intermediate phase, is found at these high reaction rates is powerful evidence in favour of the essential importance of the intermediate phase acting as a nucleating agent in the formation of C_3AH_6; but once sufficient C_3AH_6 is formed it becomes self-nucleating without further need of the intermediate phase. At slightly lower temperatures, 50°C,

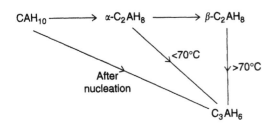

FIGURE 21.10 Schematic representation of the various pathways of conversion of CAH_{10} to C_3AH_6, according to their respective kinetics.

the sequence is considerably retarded and the intermediate phase splits into a doublet (the maxima being separated in time by about 50 min) with a characteristic d-spacing doublet that corresponds to the α-C_2AH_8 and β-C_2AH_8 hydrates. This prompted the idea that the conversion route, from CAH_{10} to C_3AH_6, has several pathways of very differing rate kinetics according to the schematic given in Figure 21.10; with sufficient time (e.g. at 50°C) the α-C_2AH_8 intermediate transforms to the denser β-C_2AH_8 phase prior to that of C_3AH_6, whereas at higher temperatures (e.g. at 90°C) the α-C_2AH_8 transforms directly to C_3AH_6.

The hydration of pure CA at high temperatures (80°C and 90°C) was also included in this study. The results were consistent with those of conversion (see above) with the condition that at these temperatures and high-rate kinetics, there is insufficient time for the CAH_{10} to form, and therefore, the hydration proceeds directly to the C_3AH_6 form after initial nucleation by the intermediate phase.

21.5.3 Ettringite mine-packing cements

This is another example of a system where high reaction rates require the fast time-resolved mode of the EDD method to fully reveal the mode of hydration. The system that has been studied (Muhamad *et al.* 1993) is based on the following reaction:

$$CA + 3C\bar{S} + 2CH + 30H \rightarrow C_3A \cdot 3C\bar{S} \cdot H_{32} \quad (3)$$

in which the reactants are pumped into the mine via two slurries, one containing the CA and the other the remaining ingredients, to yield a rapid-hardening space-filling ettringite product (Brooks and Sharp 1990) that is ready for service in 30 min. This system was studied using time-resolved EDD in which, for most cases, the diffraction patterns were collected in 20 s from one minute after initial mixing; the initial delay was necessitated by the synchrotron safety search procedures, though the sequence has also been monitored from 'time zero' by the use of remotely-controlled mixing procedures. The growth of ettringite monitored by the 110-diffraction peak displays a sigmoidal growth (Figure 21.11a) that bears some similarity with the strength-time plots (Brooks and Sharp 1990). More intriguing is the time-variation in unit cell *a*-parameter that can be determined from the position of the 100 or 110 peaks in the EDD patterns. This showed considerable variation from one experiment to another, probably reflecting sensitivity to the effectiveness of the mixing procedure. Figure 21.11b shows an example where a considerable range of variation, from 11.14 Å to the standard 11.23 Å for ettringite, is experienced. This has been interpreted (Muhamad *et al.* 1993) in terms of a contraction of the ettringite lattice due to the replacement of sulphate by hydroxy/carbonate ions during the early stages when insufficient calcium sulphate has dissolved into solution. Based on the solid solution crystal chemistry of ettringite (Pollman and Kuzel 1990), such contractions are indicative of 50–70 per cent sulphate replacement.

21.5.4 The ettringite→monosulphate transformation in OPC

Although the ettringite→monosulphate transformation in OPC has been well documented, it has remained fairly elusive to diffraction studies. However, it is readily observed (Barnes *et al.* 1996) by EDD within 180 min of OPC hydration at temperatures rising from ambient to above 70°C: a clear example of this observation, together with that of the further transformation

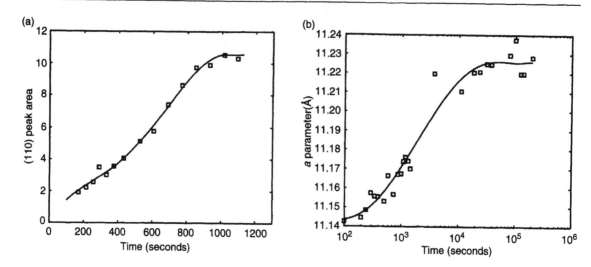

FIGURE 21.11 (a) The rapid growth of ettringite during the first 1000 s from mixing, as determined from the 110-EDD peak; (b) The variation in ettringite unit cell *a*-parameter during growth over 100,000 seconds, as determined from the 110- and 100-EDD peaks (Muhamad *et al.* 1993).

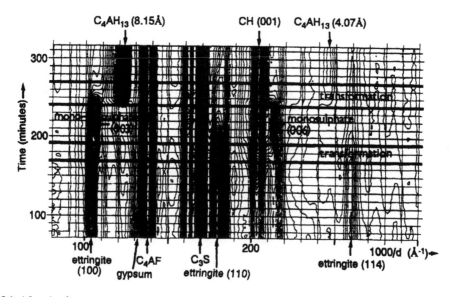

FIGURE 21.12 A classic example of the double transformation, ettringite to monosulphate to C_4AH_{13} revealed by time-resolved *in situ* EDD in a two-dimensional countoured representation of the data in which the principal peaks are indicated (Barnes *et al.* 1996; reproduced by permission of The Royal Society of Chemistry).

to C_4AH_{13} (at or above 80°C), is shown in Figure 21.12. It is well-known, and covered elsewhere in this book, that the degree of transformation has profound implications for delayed ettringite formation and therefore to the long-term performance of cement. A study was therefore made (Barnes *et al.* 1996) into the variation in kinetics of the transformation, Figure 21.12 being just one example from this study set. The full study suggests that slower the ettringite→monosulphate transformation is, the more susceptible is the cement to delayed ettringite attack as assessed by separate expansion tests in the corresponding concretes cured at 90°C. Autoclave conditions (i.e. sample confinement), while suppressing the subsequent monosulphate→C_4AH_{13} transformation, do not suppress the ettringite→monosulphate transformation; these factors might be of industrial interest in view of the degree of containment that must exist in large-scale hydrating structures.

21.5.5 Ambient temperature OPC hydration with/without different levels of admixture

For studying cement hydration under ambient conditions, it is convenient to use a multiple 'carousel' cell which permits time-resolved EDD data collection from up to five samples over long time periods. This maximizes the sample throughput for a given amount of beam time, and is ideal for comparing the relative performance from related samples (e.g. OPC control plus different levels of an admixture). Figure 21.13 shows output from such data for the case of an OPC hydrated with various dosages of a commercial calcium lignosulphonate water-reducing admixture (0, 0.25, 0.4 and 0.7 per cent, which can be described as zero (control), low, medium and high dosages). It is immediately apparent that the C_3S dissolution and calcium hydroxide (CH) formation show clearly identifiable 'take-off' points and that these are increasingly retarded by increasing the admixture doseage. The pattern of gypsum consumption markedly changes, showing a plateau with increasing dosage, whereas the

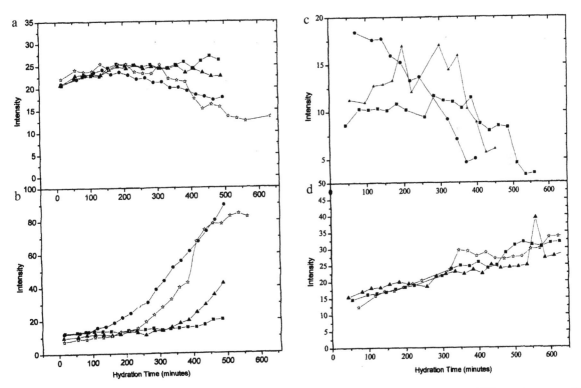

FIGURE 21.13 Resultant data obtained from time-resolved EDD on an OPC containing different doseages of a calcium lignosulphonate water-reducing admixture (● control; ☆ 0.25 per cent; ▲ 0.4 per cent; ■ 0.7 per cent). These show the changing patterns of (a) C_3S dissolution; (b) CH formation; (c) gypsum formation/dissolution and (d) ettringite growth.

TABLE 21.1 Numerical parameters from the EDD-data to indicate the effect of dosage of a retarder (a commercial calcium lignosulphonate water-reducing admixture) on the hydration sequence

	C_3S (dissolution starts) (min)	CH (end of induction period) (min)	Gypsum (complete dissolution) (min)
OPC-Control	137	44	370
+0.25%	234	128	
+0.4%	356	259	420
+0.7%	—	369	540

ettringite production/consumption is unaffected by the presence of the admixture as is also the brownmillerite phase (not shown); indicative numerical data are shown in Table 21.1. Whereas a complete description of the admixture action might not be immediately forthcoming, one can immediately perceive from these data that the retardation action pertains to the C_3S dissolution/CH formation rather than the ettringite formation.

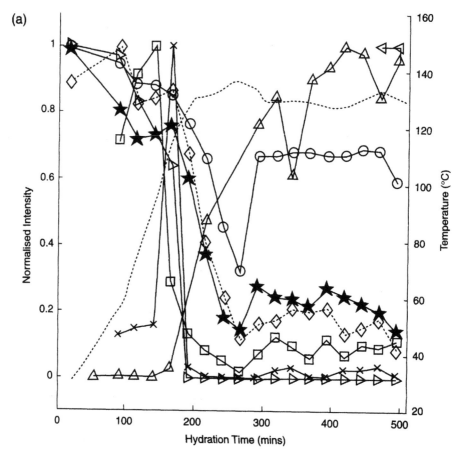

FIGURE 21.14 Resultant data obtained from time-resolved EDD on an oilwell ('Dyckerhoff') cement hydrating up to the set temperature of 130°C (temperature profiles the key to the phases in each figure; all phases are normalized to a maximum value of 1.0). (a) results obtained with negligible water/pressure loss; (b) results obtained with considerable water/pressure loss which brings about a more complex sequence including the appearance of three forms of monosulphate (14-, 10- and 12-water forms respectively).

21.5.6 Autoclave hydration of oilwell cement with/without pressure loss

Studies of cements under autoclave conditions, using *in situ* EDD, are particularly valuable since there are really no other techniques (apart from complementary neutron diffraction) that can, for example, chart the progress of oilwell cements hydrating under actual working conditions. In some cementing oilwell applications, geothermal heat and (pressure) containment produce conditions in which the temperature can reach 100–150°C, and sometimes even as high as 250°C or more, with corresponding (autogenous) pressures of tens of atmospheres (bars). For these studies,

small hydrothermal bombs, manufactured from PEEK-polymer, have been designed to perform up to 150°C while permitting transmission of X-rays through their relatively transparent walls. One of the unexpected outcomes from such studies is the wide variation in result that arises from varying the water/steam-pressure; this effect was initially discovered through accidental failures of the seal on the PEEK hydrothermal bombs, but the added value is that such variations do represent the range of pressure confinements, from atmospheric to autogenous pressure, that can exist in oilwells. Figure 21.14 gives a good illustration of this effect, comparing the hydration sequences obtained for a Dyckerhoff Class G oilwell cement

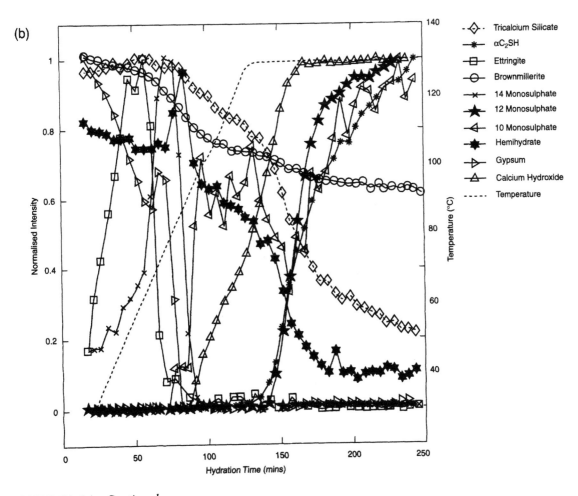

FIGURE 21.14 *Continued*

hydrating at 130°C for the two extreme cases of negligible (<0.2 per cent per hour) and significant (~1 per cent per hour) water loss over each experiment. Certain features are common to both sequences such as the growth of ettringite, subsequent conversion of ettringite to 14-water monosulphate, loss of gypsum, dissolution of C_3S and growth of CH, brownmillerite (C_4AF) consumption, and growth of 10-water monosulphate; the individual detail is remarkable, down to the increased rate in C_4AF consumption as the temperature passes 70°C, the temporary arrest in gypsum loss as the ettringite converts to 14-water monosulphate, and the interplay between the three calcium sulphates and between the three monosulphates (14-, 10- and 12-water). The

pressure/water-loss, as previously stated, itself brings about significant changes in the rates and sequence of events: for example with pressure/water-loss, the principal events (C_3S dissolution, CH growth, ettringite→monosulphate conversion) are significantly accelerated and different endpoints (additional presence of 12-water monosulphate and α-dicalcium silicate hydrate) are obtained. It is noticeable that the α-dicalcium silicate hydrate and 12-water monosulphate start and continue to grow together, suggesting a common 'trigger' such as the water-loss in the system. The data also suggest that the source material for the α-dicalcium silicate hydrate is the C-S-H matrix, which is metastable at these temperatures, since there is no decline in the CH nor any increase in

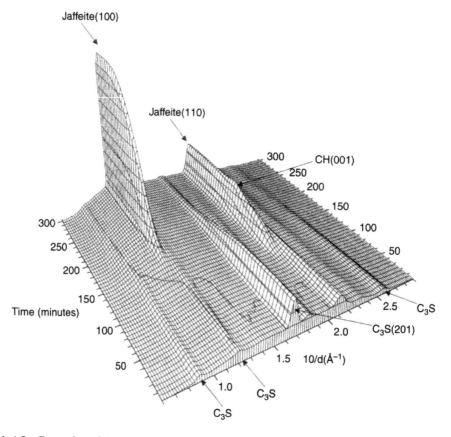

FIGURE 21.15 Examples of *in situ* EDD patterns taken during the hydration of (a) C_3S and (b) C_4A+gypsum at 150°C, the principal phases/peaks being indicated. In (a) the CH grows between 60 and 110 min followed by Jaffeite commencing at 130 min (note proximity of Jaffeite 100- and CH 001-peaks). In (b) the sequence is more complex but quite unambiguous. (BM = brownmillerite; Mono = monosulphate)

the rate of C_3S consumption. In conclusion, we can already see the quality and depth of detail that can be obtained from *in situ* EDD- data, such as in Figure 21.14, and that this is providing a fuller description of cement hydration under a regime that has been largely inaccessible by conventional analysis.

21.5.7 High-temperature hydration of model systems (C_3S to jaffeite; C_4AF)

It is often valuable to run *in situ* hydration studies on pure (model) systems in parallel to those on real cements. The diffraction patterns for synthetic materials are generally simpler and therefore it is much easier to chart, without ambiguity, the sequence of events with model systems. Such studies can then aid interpretation of the more

complex systems provided due allowance is made for the non-additive nature of a total system. We present two simple examples based on the high-temperature hydration of pure C_3S and C_4AF+gypsum. In these examples the pure materials were the laboratory synthesized compounds (C_3S; C_4AF:gypsum=1:1.4; Morgan 2001) hydrated at a water:solid ratio of 0.4 at temperatures of 22°C, 70°C, 100°C and 150°C. It is known from the literature (Aitken and Taylor 1962; Kalousek, 1969; Meducin *et al.* 2000; Sarp and Peacor 1989) that at high temperature the crystalline hydrate jaffeite ($Ca_6Si_2O_7(OH)_6$) forms though a typical prescription for its production involves hydration at 214°C for 7 days! It is now possible to chart the production with a little more precision. Figure 21.15a gives the example of C_3S hydrating at 150°C. It is seen that jaffeite formation closely follows that of CH; a similar pattern

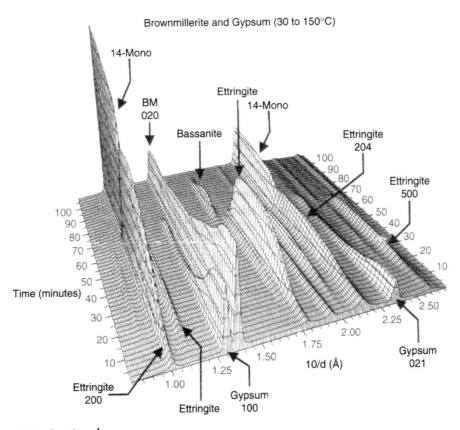

FIGURE 21.15 *Continued*

occurs at 100°C but not at 70°C. Later the CH-peak starts to decrease coincident with further 'secondary' jaffeite formation from CH and C-S-H and a significant slowing up in the consumption of C_3S. The effect of retarders are easily examined, for example (Morgan 2001) phosphonate-based retarders being more effective than gluconate-based retarders at ambient temperatures, and vice versa at higher temperatures.

In Figure 21.15b the situation is more complex but quite unambiguous, the sequence seen at 150°C being:

C_4AF + gypsum → ettringite → monosulphate + gypsum; gypsum → bassanite, the monosulphate being the 14-water variety.

The power of such *in situ* analysis is immediately apparent: Jaffeite formation occurs at lower temperatures and faster rates than previously realized and the mode of growth (i.e. two stages) and effect of retarder is easily uncovered. Discovery of such information by conventional means would have been prohibitively difficult or impossible.

21.6 Extended X-ray absorption fine structure

Extended X-ray Absorption Fine Structure (EXAFS) is a generic description covering a range of techniques that exploit the fine detail in the absorption versus energy (or wavelength) spectrum of a sample. The basic principle is illustrated in Figure 21.16: An absorption versus energy spectrum will display large upward jumps (termed K, L edges) at energies characteristic of specific electron excitations in constituent atoms of a sample. Closer examination of these edges show small oscillations in absorption within hundreds of eV above the edge which are the *fine structure* referred to in EXAFS. These oscillations result from interference effects between an outgoing electron wave from the excited atom and that fraction which is scattered back by the surrounding atoms. An appropriately weighted Fourier transform of these oscillations will produce a density plot versus distance (termed a 'radial distribution plot') of the atoms surrounding the excited atom. The great advantages of the EXAFS method is that it is

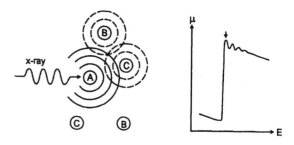

FIGURE 21.16 Schematic illustrating the basic principles of EXAFS: On excitation of an atom A by an X-ray photon, the outgoing electron wave interferes with the fraction scattered back by the surrounding atoms B and C. This results in fine oscillations (arrowed) in the absorption (μ) spectrum on the high energy (E) side of the edge.

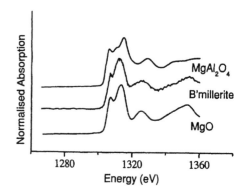

FIGURE 21.17 Comparison of EXAFS–XANES patterns for $MgAl_2O_4$-spinel, MgO and doped brownmillerite in the vicinity of the Mg K-edge. This demonstrates that the Mg-coordination in brownmillerite more closely resembles the octahedral coordination in MgO than the tetrahedral coordination in $MgAl_2O_4$ (Jupe *et al.* 2001, reproduced by permission from Munskgaard International Publishers, Denmark).

atom-specific and the sample need not be crystalline: so each EXAFS spectrum yields an individual atom-type's view of its local environment out to a distance of, typically 5 Å. Practical collection times on the synchrotron are typically minutes to hours, whereas it is days to weeks using laboratory 'brehmsstrahlung' sources and, with such a difference, the technique has now become almost exclusively practised on the synchrotron.

In Figure 21.17 an example is shown of an EXAFS pattern very close to the Mg K-edge (Jupe *et al.* 2001) from a doped ($Ca_2Fe_{0.95}Al_{0.95}$-$Mg_{0.05}Si_{0.05}O_5$) brownmillerite; such 'restricted' patterns are referred to as XANES (X-ray Absorption Near Edge Structure). In this particular case, the XANES pattern was found to be typical (Figure 21.17) of structures with octahedral, rather than tetrahedral, Mg-coordination, and this preference was also reflected in the average Mg–O bond distance ($\sim 2.06\,\text{Å}$) obtained. This feature turned out to be a vital clue in elucidating a definitive structural description of doped brownmillerite: It essentially overcame the under-determinancy associated with X-ray diffraction on its own (see Section 21.3) to yield the published structure (Jupe *et al.* 2001): this structure sees the Mg taking the octahedral sites and Si the tetrahedral sites; by contrast, the Fe/Al distribution is not uniquely split but rather displays a $2.7:1$ preference of Fe for octahedral and Al for tetrahedral sites in a similar fashion to pure brownmillerite (Smith 1962).

21.7 X-ray microscopy

X-ray microscopy, XRM (Jacobsen *et al.* 1991; Kirz *et al.* 1995), makes for an interesting and valuable alternative to optical and electron microscopy methods. The technique relies on the use of micro-zone plates which condense an appropriate X-ray beam onto the chosen sample. An image of the illuminated area is generated by using a (second) objective-zone-plate, though there is an alternative imaging system which uses mechanical translation of the specimen in the beam; the two alternatives are analogous to the TEM and SEM modes in electron microscopy. The particular advantages of the XRM are that resolutions of the order of tens of nanometres can be achieved without the need for a vacuum or dry environment, the samples can be several microns thick (e.g. $10\,\mu\text{m}$) unlike conventional TEM, and that the X-ray wavelength can be uniquely matched to the sample. One can even obtain element-specific images of a specimen by tuning the X-ray wavelength to a specific atomic absorption edge, but even coarser wavelength-tuning

brings the unique advantage of making the specimen effectively 'water transparent' due to the low X-ray absorption of water, between $2\frac{1}{2}$ and $4\frac{1}{2}\,\text{nm}$, compared to higher elements than C, N and O. These features have an obvious potential benefit for the study of cement hydration. We present one example (Pisula *et al.* 2001) using the XRM on the ASTRID synchrotron at the University of Aarhus, Denmark (Medenwaldt and Uggerhøj 1998) to observe the hydration of OPC. Figure 21.18 shows two XRM micrographs of OPC after $4\,\text{h}$ of hydration, with and without a lignosulphonate plasticizer-retarder. These micrographs are obtained while in the fully hydrated (wet) state, using an abnormally high water/cement ratio in order to disperse the hydrates across the field of view, the cement/water film being contained between two silicon wafers spaced 100–$150\,\text{nm}$ apart. However, the wavelength used, $24\,\text{Å}$, renders the water effectively 'invisible'. The ettringite needles are clearly evident in the OPC control (Figure 21.18a) but absent due to the inhibiting effect of the retarders in the case of Figure 21.18b.

21.8 Tomographic energy-dispersive diffraction imaging

The authors have been involved in the development of a new kind of synchrotron-based tomography (Barnes *et al.* 2000, 2001; Colston *et al.* 2000; Hall *et al.* 1998, 2000), termed Tomographic Energy-Dispersive Diffraction Imaging (TEDDI), which has found several applications to cementitious systems. This technique exploits the well-defined diffracting volume arising from a stationary energy-dispersive diffractometer (Figure 21.4c) operating with a configured entrance beam and collimated-diffracted beam. With TEDDI the sample is systematically translated, in one, two or three-dimensions (1D, 2D, 3D), between each data collection so that the diffracting volume is made to visit all required parts of the specimen. After all the EDD patterns have been collected, an appropriate image can then be re-constructed based on the variation in intensity

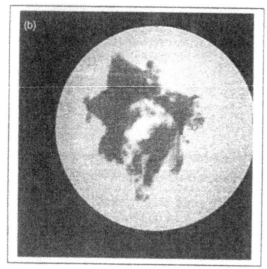

FIGURE 21.18 Two examples of X-ray micrographs of OPC after 4 h of hydration (a) OPC, and (b) OPC with a lignosulphonate plasticizer-retarder. The specimens are contained between silicon wafers (100–150 nm thick) and the circular field of view has a diameter of 10 μm. Ettringite needles are evident only in (a).

of one or more diffraction peak(s); such images are easily translated into compositional maps by using diffraction peak(s) representative of different chemical phases. In practice, collection times for 3D-maps are prohibitively slow; 2D-maps are often

feasible and 1D-profiles can be made relatively rapidly; the range of possible spatial resolutions is extremely diverse (from the micron-level in idealized cases (Hall *et al.* 1998), to millimetres). An example of a low-resolution 2D map is shown in Figure 21.19, where the concentrations in aggregate (calcite, dolomite) and cement-hydrate (portlandite, ettringite) can be compared over a 13 mm × 6 mm region just below the surface of a 77 mm thick concrete block made using a Maryland aggregate as part of a Federal Highway research project. The spatial interplay between the aggregate and cement-hydrate regions is immediately evident and the potential of the methodology is obvious considering:

- the technique is non-destructive for cores up to 80 mm thick;
- the analysis is relatively quick compared to laborious optical/electron microscopy methods;
- the same region can be re-visited indefinitely over many years to monitor progress of, for example reaction attack, compared to conventional sectioning which can be described as a 'one-shot' method.

This success has initiated a number of projects, particularly those exploiting 1D-TEDDI profiles which can be collected relatively quickly (typically 1–2 h collection time). For example, Figure 20.20 shows the 1D-profile of concrete cores that have been subjected to 24 years of exposure at a site in West Scotland (Barnes *et al.* 2001). The 1D-TEDDI-profile clearly shows the extent (8–10 mm) of carbonation from the degree of portlandite→calcite ($Ca(OH)_2 \rightarrow CaCO_3$) conversion. Similar profiles have been obtained from more rapid carbonation scenarios such as experienced with high pressure super-critical carbon dioxide diffusion (significant changes within minutes, Barnes *et al.* 2000) and with high pressure steam cured aerated autoclaved concrete (significant changes within weeks, Barnes *et al.* 2001). Already, it is clear that this tomographic technique holds particular promise in the field of cement chemistry which involves large bulk

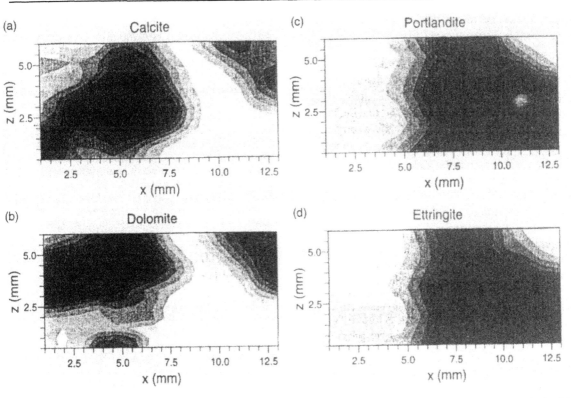

FIGURE 21.19 Four 2D-TEDD-maps of a 13 mm × 6 mm region under the concrete block surface (dark = high to white = low concentration) showing variations in (a) calcite (104-diffraction peak); (b) dolomite (104-peak); (c) portlandite (020-peak) and (d) ettringite (110-peak). (a) and (b) define the aggregate whereas (c) and (d) show the intervening cement hydrate regions. (Reprinted from Hall *et al.* (2000). 'Non-destructive Tomographic Energy-Dispersive Diffraction Imaging (TEDDI) of the interior of bulk concrete', pp. 491–495, with permission from Elsevier Science).

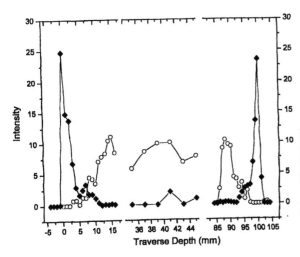

FIGURE 21.20 1D-TEDDI profiles of calcite (◆) and portlandite (O) concentration (expressed as relative diffraction intensities) over a 100 mm long concrete core that had been weathered for 24 years up to 1995. Each end displays a carbonation depth of about 8 mm after which the portlandite reaches the same levels as those near the middle (34 to 46 mm) of the core (Barnes *et al.* 2001, and reproduced with permission of Gordon & Breach Science Publishers).

497

objects, short to long time-scales and a wide range of diffracting phases.

21.9 Conclusions

It has become apparent, over a relatively short period since the 1980s, that many intractable problems in materials science can benefit from the additional opportunities provided by synchrotron radiation. This situation includes the field of cement chemistry for which much of the traditional research has been hampered by the limitations of conventional techniques (e.g. diffraction, microscopy, spectroscopy). It has been clearly demonstrated that distinctly new understanding can result from the application of synchrotron techniques (Single Crystal Micro-Diffraction; Powder Diffraction; Energy-Dispersive Diffraction; X-ray Absorption Spectroscopy; X-ray Microscopy; Tomography) and that this understanding extends from the level of atomic structure analysis to bulk characterization, and also into the time-domain by unravelling the sequence of events occurring during hydration under a range of conditions. The potential of the synchrotron for cement science has been under-valued and it is hoped that this chapter will inspire a greater exploitation in the future.

21.10 Acknowledgements

We wish to thank our various sponsors (Castle Cement Co.; Schlumberger CR; Fosroc International) including the EPSRC (Engineering and Physical Sciences Research Council) for support and beamtime provided at the SRS, ESRF and ASTRID synchrotrons. We also thank all the personnel at these sites (including Birkbeck College) for their various help.

21.11 References

Aitken, A. and Taylor, H. F. W. (1962). *4th ISSC*, 1, 285.
Barker, A. P. and Barnes, P. (1984). *Proc. British Ceram. Soc.*, 35, 25–40.
Barnes, P. and Ghose, A. (1983). 'The Microscopy of Unhydrated Portland Cement', in *Structure and Performance of Cements*, P. Barnes (ed.), pp. 139–203, Applied Science Publishers, London.
Barnes, P., Häusermann, D. and Tarling, S. E. (1990). 'Use of Dynamic Diffraction methods to study the Synthesis of Materials under Various Conditions', in *New Materials and their Applications*, D. Holland (ed.), pp. 61–6, Institute of Physics Conference Series No. 111, Bristol, Philadelphia and New York.
Barnes, P. (1991). *J. Phys. Chem. Solids*, 52, 1299–306.
Barnes, P., Clark, S. M., Häusermann, D., Henderson, E., Fentiman, C. H., Rashid, S. and Muhamad, M. N. (1992). *Phase Transitions*, 39, 117–28.
Barnes, P., Turrillas, X., Jupe, A. C., Colston, S. L., O'Connor, D., Cernik, R. J., Livesey, P., Hall, C., Bates, D. and Dennis, R. (1996). *J. Chem. Soc. Faraday Trans.*, 92, 2187–96.
Barnes, P., Jupe, A. C., Colston, S. L., Jacques, S. D. M., Grant, A., Rathbone, T., Miller, M., Clark, S. M. and Cernik, R. J. (1998). *Nucl. Instrum. Methods*, B134, 310.
Barnes, P., Colston, S. L., Craster, B., Hall, C., Jupe, A. C., Jacques, S. D. M., Cockcroft, J. K., Morgan, S., Johnson, M., O'Connor, D. and Bellotto, M. (2000). *J. Synchrotron Radiation*, 7, 117–77.
Barnes, P., Jupe, A. C., Jacques, S. D. M., Colston, S. L., Cockcroft, J. K., Hooper, D., Betson, M., Hall, C., Barè, S., Rennie, A. R., Shannahan, J., Carter, M. A., Hoff, W. D., Wilson, M. A. and Phillipson, M. C. (2001). *Nondestructive Testing and Evaluation*, in press.
Bordas, J., Glazer, A. M., Howard, C. J. and Bourdillon, A. J. (1977). *Philos. Mag.*, 35, 311.
Brooks, S. A. and Sharp, J. H. (1990). In *Calcium Aluminate Cements*, R. J. Mangabhai (ed.), pp. 335, Chapman and Hall, London.
Buras, B., Chwaszczewska, J., Szarras, S. and Szmid, Z. (1968). *Inst. Nucl. Res.* (Warsaw) Rep. No. 894/II/PS.
Buras, B., Staun Olsen, J. and Gerward, L. (1976). *Nucl. Instrum. Methods*, 135, 193.
Butler, L. G., Owens, J. W., Cartledge, F. K., Kurtz R. L., Byerly, G. R., Wales, A. J., Bryant, P. L., Emery, E. F., Dowd, B. and Xiaogang, X. (2000). *Environmental Science & Technology*, 34, 3269–75.
Christensen, A. N. and Lehmann, M. S. (1984). *J. Solid State Chem.*, 51, 196–204.
Christensen, A. N., Fjellvag, H. and Lehmann, M. S. (1986). *Acta Scandinavica*, A40, 126–41.
Christensen, A. N., Fjellvag, H. and Lehmann, M. S. (1988). *Acta Scandinavica*, A42, 117–23.
Clark, S. M. and Barnes, P. (1995). *Cem. Concr. Res.*, 25, 639–46.
Colston, S. L., Jacques, S. D. M., Barnes, P., Jupe, A. C. and Hall, C. (1998). *J. Synchrotron Radiation*, 5, 112–7.

Colston, S. L., Jupe, A. C. and Barnes, P. (2000). 'Synchrotron Radiation Tomographic Energy-Dispersive Diffraction Imaging', in *Radiation in Art and Archeometry*, D. C. Creach and D. A. Bradley (eds), pp. 129–50, Elsevier, Oxford.

Geissen, B. C. and Gordon, G. E. (1968). *Science*, 159, 973.

Hall, C., Barnes, P., Cockcroft, J. K., Colston, S. L., Häusermann, D., Jacques, S. D. M., Jupe, A. C. and Kunz, M. (1998). *Nucl. Instrum. Methods Phys. Research* B, 140, 253–7.

Hall, C., Colston, S. L., Jupe, A. C., Jacques, S. D. M., Livingston, R., Ramadan, E.-S. and Barnes, P. (2000). *Cem. Concr. Res.*, 30, 491–5.

Jacobsen, C., Williams, S., Anderson, E., Browne, M. T., Buckley, C. J., Kern, D., Kirz, J., Rivers, M. and Zhiang, X. (1991). *Opt. Commun.*, 86, 351–64.

Jupe, A. C. (1998). Ph.D. Thesis, University of London.

Jupe, A. C., Cockcroft, J. K., Barnes, P., Colston, S. L., Sankar, G. and Hall, C. (2001). *J. Appl. Crystallogr.*, 34, 55–61.

Jupe, A. C., Turrillas, X., Barnes, P., Colston, S. L., Hall, C., Häusermann, D. and Hanfland, M. (1996). *Phys. Rev. B*, 53, 14697–700.

Kalousek, G. L. (1969). *5th ISSC*, 3, 523.

Kirkpatrick, R. J., Brown, G. E., Xurick, N. and Congfer, X. (1997). *Adv. Cem. Res.*, 9, 31–6.

Kirz, J., Jacobsen, C. and Howells, M. (1995). *Q. Rev. Biophys.*, 28, 33–130.

Meducin, F. (2000). http://www.infobiosud.univ-montp1.fr/RM ... Posters/solid/fabiennemeducin 75231.html

Lequeux, N., Morau, A., Phillippot, S. and Boch, P. (1999). *J. Amer. Chem. Soc.*, 82, 1299–306.

Morgan, S. (2001). Ph.D. Thesis, University of London, in preparation.

Medenwaldt, R. and Uggerhój, E. (1998). Review of Scientific Instruments, 69, 2974–77.

Muhamad, M. N., Barnes, P., Fentiman, C. H., Hausermann, D., Pollman, H. and Rashid, S. (1993). *Cem. Concr. Res.*, 23, 267–72.

Muhamad, M. N. (2001). Ph.D. Thesis, University of London, (in preparation).

Mumme, W. G., Hill, R. J., Bushnell-Wye, G. and Segnit, E. R. (1995). *N. Jb. Miner. Abh.*, 169, 35–68.

Pisula, R., Attfield, M., Barnes, P. and Teat, S. (2001). in Preparation.

Polak, E., Munn, J., Barnes, P., Tarling, S. E. and Ritter, C. (1990). *J. Appl. Crystallogr.*, 23, 258–62.

Pöllman, H. and Kuzel, H.-J. (1990). *Cem. Concr. Res.*, 20, 941.

Rashid, S., Barnes, P. and Turrillas, X. (1992). *Advances in Cem. Res.*, 4, 61–7.

Rashid, S., Barnes, P., Bensted, J. and Turrillas, X. (1994). *J. Mater. Sci. Lett.*, 13, 1232–4.

Rietveld, H. M. (1969). *J. Appl. Crystallogr.*, 2, 65–71.

Sarp, H. and Peacor, D. R. (1989) *Amer. Miner.*, 74, 1203–6.

Shaw, S., Henderson, C. M. B. and Komonschek, B. U. (2000). *Chem. Geol.*, 167, 141–59.

Smith, D. K. (1962). *Acta Crystallographica*, 15, 1146–58.

Turrillas, X., Barnes, P., Gascoigne, D., Turner, J. Z., Jones, S. L., Norman, C. J., Pygall, C. F. and Dent, A. J. (1995). *Radiat. Phys. Chem.*, 45, 491–508.

Chapter twenty-two

Electron microscopy of cements

I. G. Richardson

22.1 General introduction

Electron microscopy techniques have now been used in a very large number of studies on cement and concrete. The purpose of this chapter is to outline the main techniques available today – mostly by the use of examples and by highlighting some key works – and to place in historical context their development and application to cement science. I have used a 'timeline' (Table 22.1) to allow reference in the available space to a reasonable number of the more significant papers. The timeline contains abbreviated author entries and keywords. The full references (given at the end) are arranged by year of publication to facilitate cross-referencing with the timeline. For example, the 1939 timeline entry with the abbreviated author R,M&E (b) corresponds to the paper Radczewski, O.E., Müller, H.O., Eitel, W. (1939b), Zur Hydratation des Trikalziumsilikats (The hydration of tricalcium

silicate), *Die Naturwissenschaften*, 27, 807. Many of the keywords in the timeline are from the lists provided by the journal *Cement and Concrete Research*, but I have found it necessary to use additional words. Some understanding is needed of the main principles of electron microscopy techniques, especially the origin of the utilized signals, so introductory points are incorporated in the text at appropriate points.

22.2 Historical context

22.2.1 Early light microscopical studies

In 1882 Henry Le Chatelier published the first results of his classic works on cement chemistry,[1] in which he identified various constituents of Portland cement clinker using microscopic examination in polarized light (Le Chatelier 1882).[2] In 1897 A. E. Törnebohm reported four constituents

---

[1] Le Chatelier published many papers on cements, including two long works in the *Annales des Mines* (1887, 1893). His ideas were soon incorporated into textbooks for example, Spalding (1900).
[2] This appears to be Le Chatelier's earliest publication concerning cements. Interestingly, a paper published the following year (Le Chatelier 1883) is cited as his first in Frederick Lea's classic textbook (see p.9 and p.99, Lea 1970). Somewhat surprisingly the error is not Lea's, but is due to Cecil H. Desch. Desch was a co-author on the first edition of Lea's book which was originally intended to be a 2nd edition of Desch's earlier text (Desch 1911). Lea, however, made no reference to Desch in the 1970 3rd edition, despite parts of the

TABLE 22.1 Timeline: Electron microscopy studies of cement. See the main text for discussions of the Key Works. Abbreviations are given at the bottom of the table

Year	Authors	EM	TEM sample type		Keywords		
1939	R,M&E (a)	TEM	Disp.	CH	Carbonation		
	R,M&E (b)	TEM	Disp.	C$_3$S	C-S-H	CH	
	R,M&E (c)	TEM	Disp.	C$_3$A			
	E,M&R	TEM	Disp.	Clay			
1940	E&R	TEM	Disp.	Clay			
1941	E	TEM	Disp.	CH	C$_3$A		
1942	E	TEM	Disp.	Cement			
1943	S,G&K	TEM	Disp.	PC	Hydr. Products	CH	
1947	MM	TEM	Disp.	C$_3$A	Aluminoferrite	PC	
1950	B	TEM	Disp.	PC			
1954	G	TEM	Disp.	C-S-H	Carbonation	SAED	
1955	C&T	TEM	Rep.	Clay			
	Gil	TEM	Rep.	Paste			
	G	TEM	Disp.	C-S-H	Microstructure	SAED	
	K	TEM	Disp.	C-S-H	Tobermorite	Hydrothermal	
	S&H	TEM	Disp.	C$_3$S	Hydr. Products	Ball mill	
	VB	TEM	Disp.	C-S-H	TA	XRD	
1956	G&R	TEM	Disp.	C-S-H			
	T	TEM	Disp.	Gypsum			
1957	B	TEM	Disp.	PC	Microstructure		
	G&R	TEM	Disp.	C-S-H	Hydr. Products		
	I&W	TEM	Disp.	PC	Hydration		
	L&L	TEM	Rep.				
	S	TEM	Disp.	CH	CaCO$_3$		
1958	C	TEM	Rep.	Paste			
	K&P	TEM	Disp.	C-S-H			
1959	B&T	TEM	Disp.	C$_2$S	C$_3$A	C-S-H	
	C	TEM	Rep.	Paste	Microstructure		
	G,H&T	TEM	Disp.	C-S-H	Tobermorite	SAED	
	G,C,D&G	TEM	Disp. & Rep.	Paste	Microstructure		
	S	TEM	Rep.	Paste	Fly Ash		
	S&B (a)	TEM	Disp. & Rep.	Paste			
	S&B (b)	TEM	Disp. & Rep.	Paste			
1960	C	TEM	Rep.	Cement			
	K&S	TEM	Disp.	C$_3$S, C$_2$S	SAED	TA	
1961	U&T	TEM	Microtome	Clinker	Hydr. Products		
1962	B&G	TEM	Disp.	C$_3$S	C$_2$S	Hydration	
	B	TEM	Disp.	C-S-H			
	C&J	TEM	Disp.	Aluminoferrite	SAED		
	C&S (a)	TEM	Disp.	Paste	C-S-H	Hydr. Products	
	C&S (b)	TEM	Disp.	Paste	C-S-H		
	C	TEM	Rep.	Paste			
	F	TEM	Disp.	C$_2$S	Hydration		
	G	TEM	Disp.	Paste	SAED	Hydr. Products	

TABLE 22.1 *Continued*

Year	Authors	EM	TEM sample type	Keywords		
	K&S	TEM	Disp.	C_3S	C_2S	SAED
	Y	TEM	Disp.	C_3A	Admixture	
1963	C&J	TEM	Disp.	Hydr. Products		
1964	C&J	TEM	Disp.	C_3A	Ettringite	Hydr. Products
	G	TEM	Disp.	Cement	SAED	
1965	G	TEM	Disp.	C-S-H	Microstructure	SAED
	Grz	TEM	Disp.	CH	Ettringite	
	M	EMPA		Clinker	PC	
	S,L&N	TEM	Rep.	Ettringite	C-S-H	GGBS
	SL&GC	TEM	Disp.	C-S-H		
1966	A,S&R	EMPA		Clinker	PC	
	C&J	SEM		Paste	Hydr. Products	
	S,L&J	TEM	Disp. & Rep.	C_3A	Gypsum	CaO
1967	C,B,C&W	TEM	Disp. & Rep.	C-S-H	Sulphate	Aluminate
	P	EMPA		Clinker	PC	
	S,N&B	TEM	Rep.	C_3S	CH	C-S-H
1968	F	EMPA		Clinker	PC	C_2S
	M	EMPA		Alite	PC	Clinker
	S	TEM	Disp.	SEM	C-S-H	
	S&M	TEM	Rep.	ITZ		
1969	C&K	TEM	Disp. & Rep.	PC		
	DK&T	TEM	Disp.	Aluminoferrite	Hydration	
	D	TEM	Rep.	C-S-H		
	F	EMPA		Clinker		
	F&K	TEM	Disp.	Hydration	C_3S	
	G&R	EMPA		Hydration	BEI	C_3S
	K&U	SEM		Cement	Kinetics	
	M&S	TEM	Rep.	Sulphate Attack	Ettringite	Hydrogarnet
	M (a)	TEM	Mech. Thin.	SEM	C-S-H	Disp. & Rep.
	M (b)	EMPA		Clinker	PC	
	R	TEM	Rep.	Paste	Microstructure	
	R&G	EMPA		C_3A	C_3S	Hydr. Products
	T	TEM	Disp.	C-S-H	SAED	
	Te	EMPA		Clinker	Hydr. Products	Fly Ash
	W	SEM		Paste		
	Y	TEM	Disp.	C_3A	Ettringite	
1970	B,C&MG	SEM		PC	CH	Fibres
	C,G,S&S	TEM	Rep.	C-S-H	Paste	Microstructure
	D	SEM		C-S-H	Paste	
	M&T	SEM		Concrete	Microstructure	
1971	C,G,S&S	TEM	Rep.	C-S-H		
	C&S (a)	TEM	Rep.	C_3S	C_3A	Hydration
	C&S (b)	TEM	Rep.	C_3S	C_3A	Hydration
	C&S (c)	TEM	Rep.	PC	Admixture	Hydration
	C&S (d)	TEM	Rep.	C_3S	C_2S	Admixture
	C&S (e)	TEM	Rep.	C_3A	Admixture	
	C	TEM	Rep.	SEM	CAC	
	D,U&K	SEM		Microstructure	C-S-H	C_3S

TABLE 22.1 *Continued*

Year	Authors	EM	TEM sample type	Keywords		
	H&B	TEM	Disp.	C-S-H	Hydr. Products	HVEM, SAED
	M	TEM	Mech. Thin.	SEM	Hydr. Products	Disp. & Rep.
	M&B	EMPA		C_2S		
	M&P	TEM	Rep.	Ettringite	GGBS	Supersulphated
	Mil	TEM	Disp.	SEM	Creep	C-S-H
1972	B	SEM	CH	C_3S		
	B,Y&L	SEM	Rep.	TEM	Admixture	C-S-H
	C&M	TEM	Disp.	C-S-H	Surface Area	
	D	SEM		EDX	Paste	C-S-H
	D&P	SEM				
	F,J,M&P	TEM	Rep.	Mortar		
	M	SEM		Microstructure	Cement	Ceramics
	S,B,G&M	TEM	Disp.	Drying	C-S-H	XRD, SEM
	W	TEM	Rep.	SEM	Hydr. Products	Microstructure
	W&T	SEM		C_3S	Hydr. Products	Microstructure
1973	D	TEM	Disp.	PC	Hydration	HVEM
	L&Y	SEM		EDX	C-S-H	C_3S
	M&W	SEM		C_3S	C-S-H	Gypsum
	M	TEM	Disp.	SEM	C-S-H	
	Mor	SEM		PC	Microstructure	C-S-H
	S	SEM		Fly Ash	Glass	
	Y,B&L	SEM		C-S-H	C_3S	Admixture
1974	D,Y&L	SEM		EDX	Cement	
	J,M&O	TEM	Ion Thin.	Mortar		
	W,O,T&M	SEM		PC	Microstructure	C-S-H
1975	C&L	TEM		Quantification		
	J,M,O&T	TEM	Ion Thin.	Mortar		
	M&T	TEM	Disp.	C-S-H		
1976	D (a)	SEM		EDX	C_3S	C-S-H
	D (b)	SEM		Paste	Microstructure	C-S-H
	D&H	TEM	Disp.	HVEM	Env. Cell	PC
	G&T	TEM	Disp.	C-S-H	XRD	EDX
	G,D,H&K	SEM		C_3S	C-S-H	EMPA
	M	SEM		Ettringite	C_3A	Expansion
	S&M	SEM		EDX	C-S-H	
1977	S&M	SEM		EDX	C-S-H	C_3S
1978	B,D&D	SEM		Paste	Hollow Shells	Microstructure
	B,H&B	SEM		PC	Hydration	
	D	TEM	Disp.	PC	HVEM	Env. Cell
	D,H&P	TEM	Disp.	Hydration	PC	
	GB,Q&S	SEM		Reliability	Instrumentation	
1979	B&C	TEM	Disp.	EDX	Microstructure	Paste
	C&B	SEM		Calorimetry	Temperature	Hydration
	G,J&T	SEM		Fatigue	PC	
	J&P (a)	TEM	Disp.	Cement	SEM	HVEM
	J&P (b)	TEM	Disp.	PC	Surface Layer	Induction period
	M,J,S&S	SEM		C_3S	C-S-H	Surface Layer
	T	TEM	Disp.	EDX		

TABLE 22.1 *Continued*

Year	Authors	EM	TEM sample type	Keywords		
1980	D,P&M	TEM	Ion Thin.	C_3S	PC	Microstructure
	D&L	SEM		C-S-H		
	D,R&L	SEM		Fly Ash		
	G,M,T&C	TEM	Disp.	EDX	C_3S	C-S-H
	G,J,P&B	TEM	Disp.	EDX	Env. Cell	
	J,M&S	SEM		C_3S	C_2S	ESCA
	J,P&B	TEM	Rep.	C_2S	C_3A	Alkalis
	J,B&M	TEM	Disp.	Aluminoferrite	SAED	
	J&P	TEM	Disp.	EDX	PC	Env. Cell
	L,M,T&M	TEM	Disp.	EDX	Paste	C-S-H
	M,MN,J&S	SEM		C_2S	Surface Layer	
	M,J&S	SEM		C_2S	Hydration	
	O,U,T&Y	SEM		EDX	Pozzolan	C_3S
	R&M	EMPA		C_3S	Hydration	C-S-H
1981	D&I	TEM	Ion Thin.	Hydr. Products	C-S-H	SEM
	D,G,J&P	TEM	Ion Thin.	SEM	Surface Layer	Hydration
	D	SEM		EDX	Cement	
	G (a)	TEM	Ion Thin.	CH	Microcrystalline	SAED
	G (b)	TEM	Ion Thin.	PC	Clinker	Microstructure
	J,D&P	TEM	Ion Thin.	C_3S	C-S-H	Morphology
	L,M,T,L&M	TEM	Disp.	EDX	Paste	C-S-H
	M&T	TEM	Disp.	EDX	C-S-H	C_2S
1982	D,P&T	SEM		PC	Hydration	Hollow Shells
	G	TEM	Ion Thin.	Clinker		
	J&S	SEM		C_3S	Surface Layer	XPS
	R&M	EMPA		Paste	C-S-H	
	R	EMPA		C-S-H	Alkalis	Fly Ash
1983	B	SEM		PC	Hydration	Setting
	G	TEM	Ion Thin.	C_3A	C_3S	Hydr. Products
	I,B&F	SEM		C_3S	Hydration.	
	L&D	TEM	Disp.	SEM	C-S-H	EDX
	P&G	SEM		PC	Hollow Shells	Microstructure
	S&B	TEM	Disp.	SEM	C_3S	C-S-H
1984	B&S	SEM		Cryo stage	Drying	
	H,G&P	SEM		Fly Ash	Silica Fume	Calorimetry
	H,P,D&G	SEM		TA	Fly Ash	Comp. Strength
	LS,D&G	SEM		TA	XRD	C_3S, $CaCl_2$
	R	SEM		Petrography	MIP, Porosity	Wood's metal
	R&L	EMPA		EDX	Paste	C-S-H
	S&R	SEM		EDX	Clinker	PC
	S&P	TEM	Disp.	SEM, Env. Cell	C_3A	Aluminoferrite
	S&G	TEM	Ion Thin.	C_3S	Alite	SAED
	T	SEM		Paste	Fly Ash	X-ray maps
	T&N (a)	SEM		Paste	C-S-H	X-ray maps
	T&N (b)	SEM		Paste	CH	C-S-H
	T&M	SEM		Fly Ash	MIP	PC
1985	H,T&W	SEM		EDX	Clinker	
	M,B&H	SEM		Cement	Hydration	Low Temp
	T,M&M (a)	TEM	Disp.	EDX	PC	

TABLE 22.1 *Continued*

Year	Authors	EM	TEM sample type	Keywords		
	T,M&M (b)	TEM	Disp.	EDX	Fly Ash	GGBS
	T&N	SEM		EDX	CH	C-S-H
	W,G&R	SEM		TA	Calorimetry	Fly Ash
1986	C&P	SEM		Fly Ash	Comp. Strength	XRD
	G,LS&S	TEM	Ion Thin.	C_3S	C-S-H	EDX
	G	SEM		C-S-H	CH	Crystal Structure
	H,I,N&H	TEM	Disp.	C-S-H	Jennite	HREM
	H,W&T	SEM		EDX	Blended Cement	Hydr. Products
	H&B	TEM	Disp.	C-S-H	EDX	Thin Films
	J&P (a)	SEM		C_3S	Porosity	Image Analysis
	J&P (b)	SEM		BEI	C-S-H	
	M,K&T	TEM	Disp.	C-S-H	XRD	TA
	R	SEM		Paste	C-S-H	EMPA
	S&P	SEM		BEI	Mortar	Microstructure
	U,U&O	SEM		Blended Cement	Chloride	BEI, EMPA
	U,U&H	SEM		Blended Cement	Glass	EMPA, NMR
1987	G	TEM	Ion Thin.	C-S-H	C_3S	EDX
	H,W&T (a)	SEM		EDX	GGBS	Microstructure
	H,W&T (b)	SEM		EDX	Microporosity	
	P	SEM		BEI	Drying	Porosity
	P,W&G	SEM		CAC	Refractory	MDF
	R,G,C&D	SEM		C-S-H, NMR	C_3S	BEI
	S	SEM		EDX	Image Analysis	BEI
	S,P,P&P	SEM		BEI	Porosity	Image Analysis
1988	D,G,R&R	TEM	Ion Thin.	Silica Fume	C-S-H, NMR	C_3S
	H&B	TEM	Thin Film	EDX	C-S-H	HREM
	K,B,R&R	TEM	Disp.	EDX	C-S-H	
	P&G	TEM	Ion Thin.	CAC		
	R&G	TEM	Ion Thin.	Fly Ash	C_3S	EDX
	S,B&P	SEM		BEI	Image Analysis	ITZ
	U	SEM		BEI	X-ray maps	ITZ
1989	B	EMPA		BEI	GGBS	X-ray maps
	B,G&R	TEM	Ion Thin.	Pozzolan	Hydr. Products	NMR
	F,L&G	TEM	Ion Thin.	EDX	GGBS	Paste
	G&R	TEM	Ion Thin.	NMR	Silica Fume	Microstructure
	I&N	TEM	Ion Thin.	EDX	C-S-H	Heavy Metals
	M&E	TEM	Disp.	C-S-H	C_3S	SAED
	R,W&D	SEM		TA	GGBS	Hydr. Products
	R,R&G	TEM	Ion Thin.	C-S-H, SEM	GGBS, MIP	Newton's Metal.
	R&G	TEM	Ion Thin.	EDX	Blended Cem.	Fly Ash
	S (a)	SEM		BEI	TEM	Microstructure
	S (b)	SEM		BEI	Image Analysis	Porosity
	S&S	SEM		BEI	Cement	Epoxy Impreg.
	Z&Gj	SEM		BEI	ITZ	
	Z&G	TEM	Ion Thin.	SEM	AAR	EDX
1990	E,G,G,J,M,...	TEM	Disp. & Ion.	Sorption	HREM	
	F&G	TEM	Ion Thin.	Microstructure	Hydration	GGBS
	G (a)	TEM	Disp.	C-S-H	Microstructure	SAED
	G (b)	TEM	Ion Thin.	Paste	Mortar	Microstructure
	G,R&R	TEM	Ion Thin.	Carbonation	EDX	Microstructure

TABLE 22.1 *Continued*

Year	Authors	EM	TEM sample type		Keywords	
	I,H,N,S,C,M&L	TEM	Ion Thin.	EDX	C-S-H	Heavy Metals
	K,D&G	SEM		Paste	BEI	Temperature
	P,K&W	SEM		Ettringite	XRD	
	R,R&G	TEM	Ion Thin.	EDX	C-S-H, GGBS	Leaching
	R,G&W	TEM	Ion Thin.	GGBS	γ-radiation	C-S-H
	Z,B&G	TEM	Ion Thin.	AAR	EDX	
	Z&G	TEM	Ion Thin.	Pozzolan	AAR	
1991	G,B,R&D	TEM	Ion Thin.	Carbonation	C_3S	NMR, EDX
	S,K&B	TEM	Disp.	EDX	C-S-H	
1992	B&J	SEM		C-S-H	Microstructure	ESEM
	D&B	SEM		Clinker	Image Analysis	BEI, EDX
	R&G (a)	TEM	Ion Thin.	EDX, EMPA	Microstructure	GGBS
	R&G (b)	TEM	Ion Thin.	EDX	C-S-H	Nanostructure
	R,G,B&D	TEM	Ion Thin.	Paste	Carbonation	NMR
	S	SEM		BEI	C-S-H	Temperature
	S&J	SEM		PC	ESEM	
	Z&D	SEM		Paste	BEI	Image Analysis
1993	B	SEM		BEI	Image Analysis	
	B,R,MC&G	TEM	Ion Thin.	Paste	C-S-H	EELS
	D&B	SEM		Paste	Microstructure	BEI
	H&B	TEM	Disp.	C-S-H	EDX	
	K,K&M	TEM	Disp.	C-S-H	Hillebrandite	
	L	SEM		ITZ	Aggregate	Concrete
	R&Grut	SEM		C_3S	C-S-H	pH
	R,B,B&D	TEM	Ion Thin.	C-S-H	EDX, EELS	NMR
	R&G	TEM	Ion Thin.	EDX	EMPA	C-S-H
	R,G,B&D	TEM	Ion Thin.	Silica Fume	Carbonation	NMR
	R,H&G	TEM	Ion Thin.	SAED, EDX	Oil Well Cement	Aluminoferrite
	Z&D	SEM		Paste	BEI	Image Analysis
1994	B&D	SEM		BEI	Hydr. Products	EDX
	B	SEM		BEI	CH	ITZ
	B,R&G	TEM	Ion Thin.	C-S-H	EELS	
	H,B,MC&R	TEM	Ion Thin.	C-S-H	EELS	Silicates
	L&J	TEM	Microtome	Fly Ash	Blended Cem.	C-S-H
	P&B	TEM	Ion Thin.	Fly Ash	GGBS	Blended Cem.
	R,B,G&D	TEM	Ion Thin.	EDX, SAED	C-S-H, NMR	AAS
1995	E	TEM	Disp.	Fly Ash	EDX	STEM
	K&K	TEM	Disp.	C-S-H	Hillebrandite	
	M,D&L	ESEM		C_3S	Hydr. Products	Induction period
1996	K&J	SEM		Drying	Microcracking	ESEM
	K,J&L	SEM		Hollow Shells	BEI	ESEM
	S&N	SEM		Wood's metal	BEI, ITZ	Porosity
	V,L,Y&X	TEM	Ion Thin.	Paste	C-S-H	HREM
	X&B	TEM	Ion Thin.	C-S-H	Hillebrandite	
	X&V	TEM		C-S-H	PC	HREM
1997	B&H	SEM		EDX	Clinker	Aluminoferrite
	C	TEM	Ion Thin.	PC	C-S-H	CH
	J,S,B&R	TEM	Disp.	Alkali Activated	C-S-H	
	K,L&J	SEM		Paste	BEI	Hollow Shells

TABLE 22.1 *Continued*

Year	Authors	EM	TEM sample type		Keywords		
	M,B&E	SEM		BEI	Image Analysis	Hydration	
	R&G	TEM	Ion Thin.	C-S-H	GGBS	NMR	
	R	TEM	Ion Thin.	C-S-H	GGBS	NMR	
	V,Y,X,C&K	TEM	Disp.	C-S-H			
	X&V	TEM	Ion Thin.	PC	C-S-H	CH	
1998	K,W&F	SEM		ITZ	BEI	HP Concrete	
	O,I&M	TEM	Ion Thin.	SEM, EDX	NMR	Chromium	
	W,A&L	SEM		Wood's Metal	MIP, BEI	Image Analysis	
1999	H&H	TEM	Ion Thin.	EDX			
	K,HA	SEM		BEI	Hollow Shells	Silica Fume	
	R	TEM	Ion Thin.	C-S-H, NMR	TMS, BEI	EDX, EMPA	
2000	B&A	SEM		BEI	ITZ	Image Analysis	
	G,I,B,S&C	TEM	Disp.	EDX	Aluminoferrite	EELS	
	G,I,B&C	TEM	Disp.	Aluminoferrite	EELS		
	H,D&D	SEM		Paste	Hydration	Hollow Shells	
	R	TEM	Ion Thin.	Hydr. Products	EMPA, NMR	C-S-H	
	R&C	TEM	Ion Thin.	C-S-H	GGBS, C_3S	EDX, NMR	
	Z,C,Z&O	TEM	Ion Thin.	C-S-H	HREM	EDX	

AAR – Alkali Aggregate Reaction; AAS – Alkali Activated Slag; BEI – Backscattered Electron Imaging; CAC – Calcium Alumi-nate Cement; Disp. – Dispersed TEM samples; EDX – Energy Dispersive X-ray Analysis; EELS – Electron Energy-Loss Spectroscopy; EMPA – Electron Microprobe Analysis; Env. Cell – Environmental Cell; ESEM – Environmental SEM; GGBS – Ground Granulated Blast-furnace Slag; HREM – High Resolution Electron Microscopy; HVEM – High Voltage Electron Microscopy; Ion Thin. – Ion-beam Thinned TEM samples; ITZ – Interfacial Transition Zone; MDF – Macro Defect-Free cement; Mech. Thin. – Mechanically Thinned TEM samples; MIP – Mercury Intrusion Porosimetry; NMR – Nuclear Magnetic Resonance Spectroscopy; PC – Portland Cement; Rep. – Replica TEM specimens; SAED – Selected Area Electron Diffraction; SEM – Scanning Electron Microscopy; STEM – Scanning TEM; TA – Thermal Analysis; TEM – Transmission Electron Microscopy; XRD – X-ray Diffraction.

and named them 'Alit,' 'Belit,' 'Celit,' and 'Felit.' (Törnebohm 1897). Whilst there was some agree-ment between the two studies, Le Chatelier and Törnebohm did not agree on the chemical com-position of the constituents. For example, Le Chatelier believed that the principal hydraulic component was C_3S,[3] whereas Törnebohm believed it (alite) to be a complex compound. Many other workers utilizing transmitted light microscopy soon contributed to the so-called 'alite controversy;' of particular note, was Clifford

Richardson's careful study of the binary and ternary mixtures supposed to constitute Port-land cement, from which he concluded that alite was a solid solution of C_3A in C_3S (Richardson 1904).[4] These early studies were made possible by the development of grinding and polishing techniques for the production of thin sections; techniques that were first used for systematic research[5] by Henry Clifton Sorby for the study

Contd.
two books – including the opening paragraphs – being identi-cal. It seems that the 1998 book in honour of Lea, 'Lea's chemistry of cement and concrete,' should perhaps have been 'Lea and Desch's ....'
[3]Oxide abbreviations are used in some formulae in this text: $C = CaO$, $S = SiO_2$, $A = Al_2O_3$, $H = H_2O$.

[4]A good example of a very early study of a practical nature which involved petrographic research was Harry Stanger and Bertram Blount's investigation into the effects of the practice of adding inert substances to Portland cement in order to produce a cheaper product (whilst still calling it PC). The paper includes micrographs of thin sections of briquettes containing different amounts of Kentish ragstone (Stanger and Blount 1897).
[5]The first published memoir reporting the use of grinding and polishing of thin sections for microscopical work was Henry

of rock minerals (Sorby 1858).[6] Sorby also pioneered metallography,[7] and reflected light microscopy was soon used to examine polished and etched samples of clinker yielding very useful information.[8] Developments in light optics led to further advances in the second decade of the twentieth century. Shepherd et al.'s 1911 (preliminary) report on the $CaO–Al_2O_3–SiO_2$ ternary phase diagram and its relevance to Portland cement – which included a thorough optical study by Wright – was especially significant: they claimed to have proved the existence of C_3S, and that it was identical to Törnebohm's alite.[9] They had used Richardson's

Contd.
Witham's 'Observations on fossil vegetables (*coal*), accompanied by representations of their internal structure, as seen through the microscope,' (Witham 1831). The sample preparation procedure – that Witham learnt from William Nicol – is set out in some detail (pp. 45–8). Witham appears to have been delighted by his results: 'The unexpected result (*the observation of internal structure in minute detail*) thus obtained, has enabled me to examine numerous varieties of structure in fossil plants; and I feel confident, that I should be rendering a service to science, by presenting to the public representations of some of these varieties, accompanied by others of those recent plants to which they seem decidedly to approximate.'
[6]Interestingly, the importance of Sorby's work was not recognized immediately: his 'first attempts were received almost with derision, some of the Members (of the Geological Society of London) present saying that he was drawing largely on their credulity. Later he was thoroughly avenged by the Geologists of all Nations who assembled to celebrate the Centenary of the Geological Society of London when Sorby on the results which were formerly derided was acknowledged and acclaimed by them to be the founder of modern Petrography' (Hadfield 1920).
[7]Sorby had a remarkable career. Although his microscopical research was mainly concerned with geological thin sections, the crowning achievement of his career was probably the discovery of the true composite nature of the 'Pearly Constituent' of steel; the discovery of pearlite (Sorby 1886, 1887). Sorby actually prepared his metallurgical specimens in the brief period between 1863 and 1865 but did not properly publish results for twenty years. Accounts of his metallurgical works are given in Hadfield (1920) and Edyvean and Hammond (1997), the latter including a catalogue of his extant metallurgical specimens.
[8]The technique gained an important role in routine quality control of the clinkering process. As examples, good micrographs have been given in most standard cement textbooks: for example, Desch (1911) plates I & II; Bogue (1947) pp. 64, 133–5, 137; Lea (1970) plate II; Taylor (1997) p. 94. Useful work has also been done on bulk, polished samples of hydrated cements, for example by Brownmiller (1943) and Terrier and Moreau (1964).
[9]Many more years passed before there was broad agreement on the answer to 'what is alite?' (see 'The alite problem,' pp.152–60, Bogue 1955).

composition limits and noted that his chief conclusion[10] was 'due to the inability of the older petrographic methods to disentangle these very fine-grained crystalline aggregates.'

Light microscopical studies of the products of cement hydration progressed in parallel with those on the constituents of clinker, the cement often being hydrated as thin layers on microscope slides, for example, Ambronn (1909), Stern (1909), Read (1910) and Klein and Phillips (1914). The last is particularly noteworthy: Klein and Phillips – building on the work of Shepherd et al. – prepared and hydrated synthetic compounds (CA, C_5A_3, C_3A, CS, γ-C_2S, β-C_2S and C_3S) and observed the changes by petrographic methods. They described the process of hydration of Portland cement itself in some detail. A good example of their results is the conclusion that the disintegrating action attributed to the crystallization of $3CaO \cdot Al_2O_3 \cdot 3CaSO_4 \cdot xH_2O$ had been greatly exaggerated since such crystals were extremely small and began to form before the cement had set. Such studies contributed much, of course, to early theories of cement hydration: to the 'crystalloids'[11] against 'colloids'[12] debate. It is of relevance to the later discussion on electron microscopy studies of dispersed particles to note that in preparing their specimens many early petrographers recognized the need to exclude CO_2; for example, Read (1910), and Klein and Phillips (1914).

Light microscopical studies – utilizing all the latest developments in instrumentation – contributed enormously to the considerable advances in the understanding of hydraulic cements that were achieved in the late nineteenth and early twentieth centuries. Whilst light microscopy would continue to be of great utility[13] it became

[10]Which by 1911 he had apparently admitted to be erroneous.
[11]Due to Le Chatelier.
[12]Initially proposed in 1893 by W. Michaëlis, Sen., and developed in further papers, for example, (1909), and modified by others, for example, Kühl (1909).
[13]Light microscopy is, of course, used very extensively today in both research and industrial laboratories (e.g. many papers in the Proceedings' of the International Conferences on Cement Microscopy, Pitts (1987), Mayfield (1990), French (1992), Elsen et al. (1995), Larbi and Heijnen (1997), Thaulow and Jakobsen (1997), St. John et al. (1998)).

clear that only limited progress would be made in some areas without greater resolving power:[14] for example, Le Chatelier, in 1919, commented that 'The microscope has never enabled us to see hydrated calcium silicate. Its crystals are certainly very minute; numerous analogies prevent us, however, from denying their existence.'[15]

22.2.2 Early development of the electron microscope

The development of a microscope utilizing electrons became conceivable after Louis de Broglie's (1924) suggestion[16] that electrons have a wavelength that is very much shorter than light – which was soon confirmed experimentally by Davisson and Germer (1927) – and Hans Busch's (1926) calculation that electrons could be focused using magnetic fields.[17] The possibility of constructing an electron microscope offered the prospect of forming images with considerably better resolution than those from the light microscope.[18] By 1931 Max Knoll and Ernst Ruska in

Berlin had experimentally verified Busch's theory and constructed a system which magnified tungsten wires and grids about as well as a good magnifying glass (Gabor 1974).[19] Two years later, Ruska had constructed the first 'supermicroscope,' capable of a magnification of ×12,000, and the following year Ladislaus ('Bill') Marton made the first observations of biological specimens (Marton 1934).[20] Whilst light microscopists had long recognized that advances in microscopy might be achieved if a suitable radiation with wavelength shorter than light could be found,[21] backing for the pioneers of the electron microscope, both moral and financial, was not very forthcoming (Ruska in Hawkes 1985). This was because there were serious reservations concerning its feasibility, not least the widely held belief that specimens would be too strongly

---

[14]It had been long known that the resolving power of the light microscope was limited by the wavelength of light (Ernst Abbe 1873: an English-language account of Abbe's work – which Abbe considered authoritative – is given in Dallinger 1891). The minimum resolvable separation in an object, d_0, is limited by diffraction to $0.6\lambda/n\sin\alpha$, where λ is the wavelength of the illuminant, n is the refractive index in the space between the object and the objective lens, and α is the half-angle in radians subtended by the objective at the object. $n\sin\alpha$ is known as the numerical aperture of the lens, NA. A laboratory microscope using white light with an objective of NA 0.7 would therefore have a resolution limit of about $0.5\,\mu$m (Watt 1997).

[15]Indeed, only forty years ago, Mielenz, in a review on petrography of concrete, noted that 'It is not possible yet to differentiate the various silicate, aluminate, or sulfoaluminate phases microscopically in sound Portland-cement pastes' (Mielenz 1962).

[16]De Broglie suggested the relation $\lambda = h/mv$, where $\lambda =$ the wavelength of the particle, $h =$ Planck's constant (6.6×10^{-34} Js), and m and v are the mass and velocity of the electron. For example, a 100 keV electron's velocity relative to light ($c = 2.998 \times 10^8$ m s^{-1}) is 0.5482 and mass relative to rest mass ($m_0 = 9.1 \times 10^{-28}$ g) is 1.1957, giving a wavelength of 0.0037 nm, which is about 100,000 times shorter than violet light!

[17]Surprisingly, Busch did not connect his work with the development of an electron microscope (Gabor 1974).

[18]The resolving power of a microscope system is principally determined by diffraction and spherical aberration in the objective lens (there are other factors). It is limited by diffraction to $d_0 = 0.6\lambda/n\sin\alpha$. Since for an electron lens $n = 1$, and α is small, $d_0 = 0.6\lambda/\alpha$, and so α should be *large* for best

resolving power. However, the resolving power is limited by spherical aberration by, $d_0 = C_s\alpha^3$, where C_s is the spherical aberration coefficient (which is a constant for a particular lens), so for best resolving power α should be *small*. Combining the two equations it follows that the optimum angular aperture, $\alpha_{opt} = (0.6\lambda/C_s)^{0.25}$, and the minimum $d_0 = 0.7\,C_s^{0.25}\lambda^{0.75}$. So, for example, assuming a value of 3.5 mm for C_s, 100 keV electrons ($\lambda = 0.0037$ nm) give $d_0 = 0.45$ nm, which is about 1000 times better than the light microscope. $\alpha_{opt} = 0.005$ radians, which is about 100 times smaller than the light microscope, and results in the electron lens giving a much greater depth of field than a light objective at the same resolution (Watt 1997).

[19]Knoll and Ruska gave the first public lecture on their results on the 4th June 1931. Unfortunately, four days earlier Reinhold Rüdenberg, the Chief Electrical Engineer of Siemens–Schuckert–Werke had filed a comprehensive patent application on all sorts of electron microscopes. So, as far as patent law is concerned, although he played no active part in its development, Rüdenberg invented the electron microscope (Gabor 1974). The beginnings of electron microscopy occurred at a time of great turmoil and inevitably its story contains some disturbing events. Fascinating personal accounts – including translations of letters between Knoll and Max Steenbeck (who, unintentionally, probably led to Rüdenberg's patents) – are given in Hawkes (1985), and references therein.

[20]Marton assembled his home-built microscope in Brussels under very difficult conditions, partly from components acquired in the 'flea market' (Gabor 1974; Hawkes 1985).

[21]For example, in 1920, at a symposium on *The microscope: its design, construction and application*, the President of the Royal Microscopical Society, Mr J. E. Barnard, commented that '... it is in the direction of using invisible radiations in the ultra-violet, or, it may be, radiations which are still shorter than the ultra-violet, that developments in microscopic work are, in my opinion, likely to occur.'

heated by the absorption of electron energy.[22] Whilst Marton soon found that image contrast for sufficiently thin specimens was due to electron scattering rather than absorption, specimens had to be *extremely* thin: about 100 times thinner than could be produced using the microtomes of the day! (Ruska in Hawkes 1985). Nevertheless, by 1937 Ruska and his brother-in-law, Bodo von Borries, had secured well-equipped laboratories – and, crucially, access to excellent manufacturing facilities – at Siemens and Halske: they marketed the first commercial instrument to exceed the optical microscope in resolving power in 1939 (von Borries and Ruska 1939).[23] The first electron microscope studies of cement were published in the same year: studies of the hydration of lime, tricalcium silicate, and tricalcium aluminate (Radczewski *et al.* 1939*a,b,c*). At about this time many other workers appeared on the electron microscopy scene (the interested reader is directed to Hawkes 1985): work at the Radio Corporation of America (RCA)[24] and by J. B. Le Poole[25] in the Netherlands being particularly noteworthy. In 1943, Sliepcevich *et al.* – no-doubt using an RCA

Type B instrument[26] – published an extensive study of the hydration products of Portland cement, calcium sulphate, lime, β-C_2S, C_3S, C_3A, C_4AF, and mixtures of the synthetic constituents. The electron microscope's importance was by 1943 sufficiently widely recognized that seven Type B instruments were shipped from the United States to Britain under the lend-lease arrangements which covered the wartime exchange of ideas and materials (Reed in Hawkes 1985).[27] Only ten years later they were commercially available in Britain, France, the Netherlands, Japan, Sweden, Switzerland, the United States, and Russia, with about 750 installed throughout the world (Swerdlow 1954). The scanning electron microscope was also pioneered in the 1930s and 1940s for the imaging of solid surfaces, principally by Manfred von Ardenne (1938*a,b*)[28] and in the laboratories of RCA (Zworykin *et al.* 1942). In 1948 Charles Oatley initiated a programme of work[29] in the Cambridge University Engineering Department that eventually led to the production of the first commercial instrument in 1965 (Oatley *et al.* 1965).[30] The first study of cement was published the following year (Chatterji and Jeffery 1966).[31]

[22]For example, Marton had impregnated his biological specimen – a sundew leaf – with osmium tetroxide because he, like everyone else, initially believed that electron contrast was due to absorption.

[23]The very first commercially produced electron microscope (the so-called EM1) had been made in the UK in 1936 by the Metropolitan Vickers Company for Imperial College, London (Martin *et al.* 1937). However, it did not exceed the light microscope in resolving power and in 1937 the company directed its resources to developing radar, and did not market a successful microscope (the EM2) until 1946 (Haine 1947). 'Metrovick' microscopes would later be used in studies of cement, for example, Buckle and Taylor (1959) used an EM3, and Chatterji and Jeffery (1962, 1964) used an EM6.

[24]Conscious of the approach of World War II Marton had emigrated to the US in 1938 and helped RCA to construct their first electron microscope, the so-called Type A (Süsskind in Hawkes 1985). However, the team that developed the first microscope marketed by them (in April 1941), the Type B – which was a great commercial success – was led by James Hillier (who had earlier done pioneering work at the University of Toronto with A. Prebus) and Arthur Vance (Zworykin *et al.* 1940).

[25]Le Poole's work – much of which occurred in very difficult wartime circumstances – led to the Philips series of electron microscopes (Le Poole in Hawkes 1985). Åke Grudemo used a Philips microscope in his classic studies of calcium silicate hydrates (1954, 1955). Le Poole developed the important technique of selected area electron diffraction (Le Poole 1947).

[26]RCA microscopes were used in other early studies, for example, van Bemst (1955) and Kalousek (1955).

[27]Only two of the seven went to the higher education sector: one to the University of Leeds and the other to The Cavendish Laboratory, Cambridge.

[28]Von Ardenne's research was ended prematurely when his microscope was destroyed in an air raid in 1944.

[29]Which generated numerous publications, for example, Thornley and Cartz (1962).

[30]The Cambridge Instrument Company's 'Stereoscan'. Its commercial development had to some extent been hindered by the emphasis that was generally placed on resolution: the SEM could not compete with the examination of surface replicas in the TEM. Once marketed, however, it soon became an enormous success. The first four production models were sold to the University of Leeds, the University of North Wales, Bangor, the University of Münster, and to the Central Electricity Research Laboratories. There were twelve in the next batch, followed by forty…(Smith 1998). More than 15,000 SEMs were sold in the following thirty years by more than a dozen manufacturers (Goldstein *et al.* 1992).

[31]Chatterji and Jeffery's micrographs were taken using the Cambridge Instrument Company's Stereoscan Mk1 demonstration instrument, whilst Berger *et al.* (1972) probably used a Mk2 for their work on C-S-H morphology that was later extended by Diamond (1976) into the well-known Types I, II, III, and IV classification.

22.3 Electron microscopy

22.3.1 Principles and generation of utilized signals

The principles, electron optics, imaging modes, and other aspects of scanning (SEM) and transmission electron microscopy (TEM) are covered extensively in many texts (e.g., Hirsch *et al.* 1977; Watt 1997; Reimer 1984, 1985; Goldstein *et al.* 1992; Goodhew *et al.* 2001). Brief details of both are given in Table 22.2. After striking the surface of a solid the primary electrons spread, gradually losing their energy by multiple collisions with specimen atoms. In bulk specimens this gradual reduction in energy results in a finite electron range (R) that is dependent on the specimen density and the initial energy of the electrons. The collisions may be either elastic or inelastic, where elastic scattering processes can be considered as electron–nucleus interactions and inelastic scattering processes as electron–electron interactions. These processes produce all the signals utilized in scanning and transmission electron microscopy. The most important are outlined in Table 22.3. Note that the information volumes are different for each signal (because they are due to different electron–specimen interactions).

22.3.2 Thermal and irradiation damage

In both bulk and thin specimens a significant fraction of the primary electron energy lost during the scattering processes is converted into heat. In many specimens the thermal conductivity of the material is sufficiently high that specimen heating is not a problem, but in materials of low melting or decomposition temperature or of low thermal conductivity, such as polymers, organic substances or hydrated cements, it can be a serious problem. Some materials also suffer from radiation damage by electron bombardment that can result in a considerable loss of mass. In some materials, including hydrated cements, surface damage may be caused by the formation of gaseous products inside the electron interaction volume. This may

be thermally or irradiation induced, or a combination of the two. So far as hydrated cements are concerned the effects of damage are most serious in the TEM for both imaging and X-ray analysis, and for the latter in an electron microprobe (an electron microscope that is optimized for X-ray analysis). Damage to TEM specimens has been well documented: selected examples of comments and references with micrographs showing damaged products are given in Table 22.4. It has particularly serious implications for attempts at high-resolution imaging (which is discussed later).

22.4 Transmission electron microscopy of cement

22.4.1 Dispersed specimens

The earliest TEM studies of cement were by O. E. Radczewski *et al.* on dispersed specimens (1939*a–c*).[32] Their micrographs show many interesting features, including spherulites – which they believed to be calcium hydroxide – small fibrillar packets of calcium silicate hydrate (1939*a,b*), and hexagonal crystals believed to be 'C₃AH₁₂' (1939*c*). Forty years after their work, John Bailey and Dawn Chescoe (1979) observed 'spherical-shaped' particles very reminiscent of Radczewski *et al.*'s (1939*a*) spherulites. A consideration of these and related works demonstrates well the care needed in the preparation of hydrated cements for electron microscopy. Like Radczewski *et al.*, Bailey and Chescoe interpreted the spherulites as calcium hydroxide. They described the particles as 'colloidal,' whereas Radczewski *et al.* (1939*a*) described theirs as 'in an early stage of crystallization.' However, in 1957, G. Schimmel had also observed similar

[32]Specimen preparation is quick, and so usually done just before the microscope session. Specimens are usually dispersed in a large amount of solvent in a mortar (e. g. acetone, ethanol, propan-2-ol). If necessary, the samples are ground using a pestle. Ultrasonic vibration has sometimes been used. A drop of the suspension is placed on a carbon film supported by a copper grid. The liquid is evaporated, preferably in a CO₂-free atmosphere, and transferred to the microscope.

TABLE 22.2 Brief summaries of the principles of the SEM and TEM. Text summarized from Hirsch *et al.* (1977), Reimer (1984, 1985), and Chescoe and Goodhew (1984). Figures modified from Reimer (1984, 1985).

(a) Scanning electron microscope

Electrons are emitted from a cathode (thermionic or field-emission) and accelerated by a voltage between the cathode and anode. A fine electron probe is formed on the specimen surface by a two- or three-stage lens system (minimum probe size of 5–10 nm if a thermionic electron gun, 0.5–2 nm if field emission gun). The probe is rastered across the area of interest and an image is displayed on a CRT rastered in synchronism. The intensity of the CRT is modulated by any of the signals resulting from the electron-specimen interactions (BSE = backscattered electrons, SE = secondary electrons, SC = specimen current, EBIC = electron-beam-induced current). X = X-ray detector. The magnification (M) is increased by scanning a smaller area on the sample whilst keeping the image size on the CRT constant, thus $M = D/d$ (see figure). So, for a 10 cm × 10 cm CRT display ($D = 10$ cm), the area sampled at a magnification of ×10 ($M = 10$) is 1 × 1 cm ($d = 1$ cm). For $M = 100$, $d = 1$ mm; $M = 1000$, $d = 100$ μm; $M = 10,000$, $d = 10$ μm; $M = 100,000$, $d = 1$ μm. The resolution depends on the probe size and the information volume that contributes to the signal, and thus on the specimen and the mode of operation.

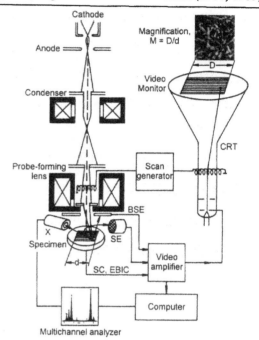

(b) Transmission electron microscope

Electrons are emitted from a cathode (thermionic or field-emission) and accelerated by a voltage between the cathode and anode. The beam of electrons is focussed down to a fairly small spot by a first condenser lens. A second condenser lens projects the beam at the specimen and enables the convergence angle and area of illumination to be controlled. An aperture controls the number of electrons allowed into the beam and hence helps control the intensity of the illumination. The electron beam illuminates the specimen (which must be very thin, typically 5 nm–0.2 μm for 100 keV electrons, depending on specimen density and elemental composition). The electrons may be undeflected (transmitted with no interaction), deflected with no loss of energy (elastically scattered), or deflected with a significant loss of energy (inelastically scattered) and the excitation of secondary electrons or X-rays. Particular groups of electrons may be selected to contribute to the final image by the use of apertures. The electron-intensity distribution behind the specimen is magnified and imaged with a three- or four-stage lens system onto a fluorescent screen. A convenient magnification for viewing hardened cement pastes is ×20,000, which might be achieved by objective lens ×25, intermediate lens ×8, and projector lens ×100.

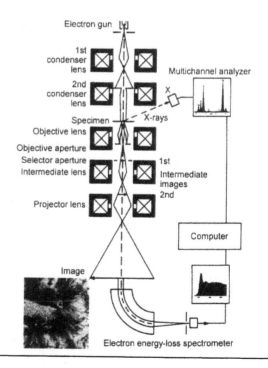

TABLE 22.3 Brief summary of the origin and information depth of useful signals generated by the interaction of the primary electron beam and a bulk or thin specimen in an electron microscope. Top diagram based on figures in Reimer (1985) and Watt (1997)

A significant fraction of the primary electrons (PE) are scattered through such large angles after multiple collisions with specimen atoms that they re-emerge from the surface. They have a broad range of energy, between the PE energy and 50 eV (by convention). Such electrons are called backscattered electrons (BSE), and they give average atomic number information (also crystallographic information on surfaces). Electrons from specimen atoms that are ejected after inelastic collisions with PE or BSE are called secondary electrons (SE). They have low energy (< 50 eV) and can only leave the specimen from a very thin surface layer, r, which is only a few nanometres thick. Consequently they give information on surface topography and surface films. Electrons transmitted through very thin specimens can give information on their internal structure (bright field and dark field imaging in the TEM), on the identity and orientation of crystals [selected area electron diffraction (SAED)], and can be used for elemental and chemical analysis [electron energy-loss spectroscopy (EELS)]. Some scattering processes result in the generation of X-rays (see diagram on the right). The two types of X-rays produced, characteristic and background (or continuum) radiation, are produced by different inelastic interactions. The interaction of an incoming electron with the Coulomb field of a nucleus normally leads to elastic scattering. However, it can sometimes lead to the emission of a background X-ray quantum of energy between zero and that of the incoming electron. The emission of this X-ray is accompanied by a corresponding deceleration of that electron. This background radiation contains no useful information for the quantification of specific elements. The interaction of an incoming electron with the inner shell electrons of an atom may result in one of these electrons being ejected from its shell. The potential energy of the atom is reduced if an electron from a higher state then fills the vacancy. The energy difference may then be emitted as an X-ray quantum or by the emission of an Auger electron (AE). The energy of the X-ray is determined by the energy levels of the ejected electron and the higher state electron that filled the vacancy. As these energy levels are sharply defined and unique to each atomic species (i.e. characteristic), the emitted X-ray provides information about the chemistry of the electron-specimen interaction volume. An illustrative spectrum from an energy dispersive X-ray (EDX) analysis is shown on the right. The Kα peaks are labeled. The small peak at around 4 keV is the Ca Kβ peak. Note that peaks can sometimes be identified incorrectly. For example, Mo sputtered on to ion-thinned TEM specimens from Mo clamping plates can be mis-identified as S. Like SE, Auger electrons can only leave the specimen from a very thin layer. They therefore give elemental and chemical information on surface layers.

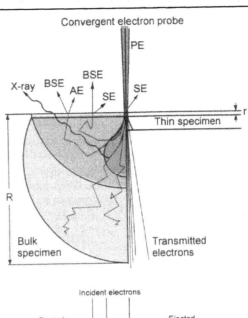

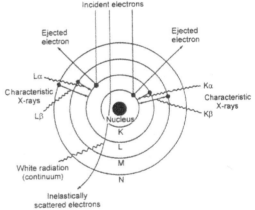

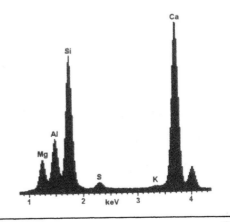

TABLE 22.4 Selected examples of comments concerning electron-beam damage in the TEM and references with micrographs showing damaged products

Radczewski *et al.* (1939c)	The crystal in their Fig. 1 appears to be electron-beam damaged
Talbot (1956)	Demonstrated the decomposition of $CaSO_4 \cdot 2H_2O$ in the electron microscope
Iwai and Watanabe (1957)	Noted the decomposition of hydration products in the EM
Kalousek and Prebus (1958)	C-S-H (Ca/Si = 1.5, prepared at room temperature or 60°C) was very susceptible to irradiation damage: their Figs. 2 and 3 show the product before and after intense electron illumination
Gaze and Robertson (1957)	Noted that '... crystals (of "tobermorite") are so thin and delicate that at ×65,000 they degrade too fast to allow good photographs to be taken'
Uchikawa and Takagi (1961)	Studied thin sections prepared using an ultra-microtome. The 'ribbons' of C-S-H in their Fig. 2 show signs of beam damage
Chatterji and Jeffery (1962)	Noted that very thin but well-formed hexagonal plate crystals (identified as C_2AH_x) formed quickly by the hydration of C_4AF were not very stable in high vacuum and under the electron beam. In the C_4AF–gypsum–water system needle shaped crystals were formed; these gave no SAED patterns but from morphology and powder XRD were identified as ettringite. They commented on the poor stability of early hydrate
Grudemo (1962)	Noted that '... under normal conditions, and when care is taken not to increase the electron beam intensity to a higher level than is strictly necessary for observing and recording the images visible on the fluorescent screen, the temperature of at least the thinner particles or particle aggregates in a sample is probably well below 100°C. Even these comparatively low temperatures, especially in combination with the vacuum maintained in an electron microscope or ED instrument, are still sufficient to cause dehydration and destruction of delicate or ill-crystallized hydrate structures. Crystal-structure transformations and changes in morphology are frequently observed in EM studies, and can sometimes be expected to have occurred even before the electron beam is switched on. Although in most cases the effects of heating can be considered negligible, one can never be quite certain that some unobservable change has not occurred.' Grudemo's 'normal conditions' would correspond to relatively modest magnifications
Grudemo (1965)	Demonstrated using SAED that calcite crystals are readily decomposed to calcium oxide (by deliberately increasing the intensity of the electron beam)
Dalgleish and Ibe (1981)	Early ion-thinned study. C-S-H in their Fig. 1 is clearly damaged. They noted that 'the effect of the electron beam was readily observed at higher magnifications'
Groves (1987)	'At high magnifications in the TEM (greater than say 30k), the pore structure of the inner gel is seen to coarsen rapidly, presumably because the hydrate decomposes to give volatile products under the electron beam'
Ivey and Neuwirth (1989)	Used a combined dimpling / ion thinning procedure. Their example micrograph of C-S-H (shown in their Fig. 2 (a)) appears beam damaged (looks like Ip with coarsened porosity)
Richardson and Groves (1993)	Discussed damage to cement hydration products. Listed many of these phases and their relative stability in the TEM. Defined the stability as being the relative dwell time that a phase can endure in the centre of the field of view before noticeable damage occurs when viewed under 'standard conditions'. The standard conditions were those established in that work as being most satisfactory for the rapid viewing of a hydrated portland cement. The scale was set from 0 to 1 with 0 corresponding to instantaneous damage on moving into the field of view – most phases this susceptible would generally already have suffered extensive damage during preparation, for example ettringite – and 1 to a dwell time of around 60 s.

TABLE 22.4   *Continued*

	The phases that extended through the full range do not damage easily. Some phases displayed marked variation. Sensitivity to beam damage correlated well with thermal stability. Micrographs were included of Ip C-S-H shortly after bringing it into the field of view and around 15 s after the first exposure. The latter was beam-damaged, displaying a coarsened structure
Viehland *et al.* (1996)	Both the Ip (in their Fig. 1(a)) and Op (1(b)) appear beam damaged. The Ip was similar to damaged examples in Richardson and Groves (1993) and Richardson (1999)

features, and identified them as calcium carbonate using selected area electron diffracton (SAED) (Schimmel 1957),[33] an interpretation that Wilhelm Eitel – a co-author of Radczewski's – later accepted (Eitel 1966: 367–8). Åke Grudemo also reported on this type of feature in his review paper at the Washington Symposium (Grudemo 1962), and emphasized the need to take rigorous precautions to exclude CO_2. Indeed, until Grudemo began his classic works in 1950 (Grudemo 1954), most workers had found it difficult to avoid contamination by CO_2 (Moore 1954). Given their method of specimen preparation,[34] it is very probable that Bailey and Chescoe's 'spherical-shaped' particles were also calcium carbonate, and that they had simply overlooked the earlier works.

[33]Schimmel (1957) observed that if limewater was allowed to dry up in air an amorphous modification of calcium carbonate was formed by the reaction of atmospheric CO_2 with $Ca(OH)_2$. When this was observed in the TEM it very rapidly crystallized into calcite under the electron beam.

[34]Bailey and Chescoe's specimens for TEM were prepared as follows. A few drops of distilled water were placed on a glass microscope slide. Two copper or nylon electron microscopy grids were then placed onto the water surface and dry cement powder sprinkled over the grids. The entire 'cell' was then sealed with a cover glass and araldite. After the required period of hydration the cell was broken open, and the grid plus cement mix allowed to partially dry for a few seconds. Excess material was removed by tapping the forceps holding the grid. Some time would then have elapsed whilst the grid was placed in the specimen stage before examination in the TEM. Thus, the specimens would have been exposed to a moist CO_2-containing atmosphere for some minutes. Indeed, it is quite conceivable that they were exposed to a moist atmosphere *enriched* in CO_2, since they may have been breathed upon during the careful handling procedure required for such small specimens.

Whilst Radczewski *et al.* (1939) published the earliest electron microscopy work on cement, Grudemo's must be the most significant on dispersed specimens (Grudemo 1954, 1955, 1962, 1965). His work on C-S-H morphology is particularly significant (see final section). Grudemo (1965) published a large work that is one of the major contributions to the study of cementitious systems by electron microscopy. It contains a very large number of well-reproduced micrographs and electron diffraction patterns (hundreds) collected largely (80 per cent) whilst he was a Guest Scientist at the Portland Cement Association (PCA) in 1957–58.[35] Of the early studies on (ground and dispersed) cement pastes, Copeland and Schulz (1962a,b) presented some excellent micrographs of C-S-H. Crystals of anomalous tobermorite (from Ewart *et al.* 1990) are shown in Figure 22.1 as an example of this type of specimen preparation.

[35]Treval C. Powers – who made some of the most significant contributions to Cement Science (most notably his and T. L. Brownyard's classic and very long paper, 'Studies of the physical properties of hardened cement paste,' which was published in the ACI Journal in nine parts between October 1946 and April 1947 (Powers and Brownyard 1946–47)) – had considered Grudemo's work to be especially important, and so invited him to go to the PCA as a Guest Scientist to work on their newly acquired electron microscope. Grudemo left them 'with a legacy of important pictures and diffraction data' (Powers 1966). Five per cent were collected by him later in Sweden, and about 15 per cent by Dr S. Chatterji during a period as a Guest Scientist at the CBI (Swedish Cement and Concrete Research Institute (Cement och Betong Institutet)) in 1959–61. All of the work was done on dispersed samples. Later work by Chatterji (at Birkbeck College) was published earlier than this in the Journal of the American Ceramic Society (Chatterji and Jeffery 1962, 1963).

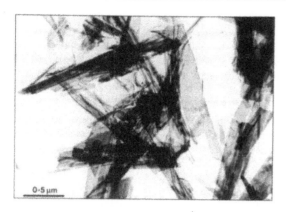

0·5 μm

FIGURE 22.1 Crystals of anomalous tobermorite (from Ewart et al. 1990).

22.4.2 Surface replica specimens

Gille (1955) appears to have been the first to use a replica method[36] for the examination of hardened cement; the paper includes an excellent micrograph showing plates of CH (his Fig. 32). Particularly significant work was conducted by Czernin et al. (1958, 1960, 1962), Hans Schwiete et al. (1965, 1966), and Ciach and co-workers (Ciach et al. 1970, 1971; Ciach and Swenson 1971a–e). Czernin's papers include good examples of replica micrographs; for instance, one would now interpret Figs. 4 and 5 of the 1958 paper as showing AFt and C-S-H respectively. Schwiete et al.'s (1965) study of the early-age hydration of Portland cement also includes some good example micrographs, especially their Figs. 7 and 8,[37]

and Fig. 20 (which is an especially nice example of ettringite from a supersulphated cement). There have been many other studies of note, for example: Saji (1959) (on fly-ash cement blends), Shpynova et al. (1967) (on the nature of CH in hydrated C_3S pastes), Dyczek (1969) (on the effect of calcium sulphate on the morphology of C-S-H), Marchese and Sersale (1969) (hydrogarnet and ettringite), Midgley and Pettifer (1971) (on supersulphated cement), and Farran et al. (1972) (on the nature of the interfacial transition zone (ITZ)).

The preparation and handling of replicas must be done with care (as should ion thinning), which probably explains the scepticism of some workers.[38] However, there is no doubt that some very striking replica micrographs exist. A great advantage (of the two-stage method) is that it provides an image of the specimen totally without contact with vacuum or electron beam, a fact noted by Groves (1990). He observed that the C-S-H morphologies present on replicas of fracture surfaces are similar to those observed by direct TEM examination of ion-thinned specimens. This is strong evidence that the morphologies observed on ion-thinned sections are not overly affected by the preparation procedure or the high vacuum of the microscope.

22.4.3 Thinned specimens

The first study of hardened cements that involved TEM of thinned rather than dispersed specimens was by Hiroshi Uchikawa and S. Takagi (1961) who examined thin-sections produced using an ultramicrotome. They claimed some success,[39]

[36]Replication procedures involve the formation of an impression of the features of a surface in a thin material that is transparent to electrons. The replica is usually made of a layer of carbon 10–20 nm thick. It is prepared either by direct evaporation onto the specimen surface, after which it is removed (this is called a single-stage replica), or by evaporation onto a preliminary plastic negative replica that was cast on and removed from the surface (a two-stage replica). Microstructural details are enhanced by 'shadowcasting,' in which a heavy metal is evaporated obliquely onto the replica. A variety of methods have been used in studies on cement: for example, many workers utilized a method from Comer and Turley (1955), who compared different methods of producing replicas of clays, and Stork and Michálková (1968) used a rapid impression method using acetyl-cellulose film and shadowing.

[37]They stated that some of their micrographs showed 'a kind of jelly-like ground mass in which a number of bar-shaped

crystals are embedded. Diameter of the cement gel ranges from 80 to 250 Å. Powers...has calculated a cement gel with a mean diameter 160 Å. The figure presented by Powers seems to correspond fairly well with our experimental results.'

[38]There was some debate on this issue during the 1960s. For example, Czernin (1962) noted: 'Contrary to Grudemo's view, however, this writer believes that replica methods when handled with skill and experience can resolve the finest details of the paste surface.'

[39]They noted that as a result of their experiments 'the process of hardening and strength development can be explained in terms of: the formation of a three-dimensional network of calcium hydroxide, the bonding of fibres and layers of the

but the technique is generally not appropriate for hydrated cements. Midgley (1969) examined pastes produced by a 'grinding-and-polishing technique' (as well as dispersed specimens and replicas). Again, his micrographs indicate some success: his Fig. 26, although described as 'a mass of small irregular plates,' appears fairly feature-less, and is quite possibly inner product (Ip) C-S-H, whilst Fig. 27, described as showing 'splines,' might be fibrillar outer product (Op) C-S-H. The earliest work with ion-beam thinned samples appears to be Javelas et al.'s (1974, 1975), which involved the examination of paste-aggregate bonds in mortars involving quartz and calcite. Their papers include some excellent micrographs: for example, Fig. 3 of their 1975 paper clearly includes fine-fibrillar Op C-S-H. Several papers were published on cement pastes during the early 1980s by workers from Peter Pratt's group at Imperial College, for example, Dalgleish et al. (1980, 1981), Dalgleish and Ibe (1981), and Jennings et al. (1981). Karen Scrivener, from the same group, later included some TEM micrographs of ion-thinned specimens of pastes less than 1 day old (that had been resin impregnated) in a review of microstructural studies on cement paste and concrete (Scrivener 1989). The most extensive body of work is due to members of Geoff Groves' group at the University of Oxford – and now continued at the University of Leeds – including studies on the components of clinker (Groves 1981b, 1982, 1983; Sinclair and Groves 1984; Richardson et al. 1993), hardened C_3S and Portland cement pastes (Groves 1981a; Groves et al. 1986; Groves 1987; Richardson and Groves 1993), high alumina cement pastes (Poon and Groves 1988), alkali-activated slags (Richardson et al. 1994; Richardson 1997), carbonation (Groves et al. 1990, 1991; Richardson et al. 1993), alkali-aggregate reaction (Zhang and Groves 1989, 1990) and on blended cements involving fly ash (Rodger and Groves 1988, 1989), silica fume (Dobson et al.

Contd.
calcium silicate hydrates by van der Waals forces and electro-static forces, and the interlocking of the calcium silicate hydrates and the calcium hydroxide by coalescence of similarly oriented lattices with cross-connections.'

1988; Groves and Rodger 1989), and slag (Richardson et al. 1989, 1990; Richardson and Groves 1992a, 1997; Richardson and Cabrera 2000). Other works are included in the timeline, Table 22.1.

Examples of micrographs of ion-thinned sections are shown in Figures 22.2a–d and in other sections of this chapter. Preparation procedures have been given by Javelas et al. (1974), Dalgleish et al. (1980), and Ivey and Neuwirth (1989). A straightforward procedure which gives good results is given in Table 22.5.

22.4.4 Selected area electron diffraction

SAED[40] in the TEM is a powerful technique for the identification and characterization of crystalline phases. It has been used extensively in studies of both clinker and hydrated phases (dispersed or ion-thinned). Åke Grudemo's work on dispersed hydrates and Geoff Groves' at the University of Oxford on ion-thinned clinker phases are especially noteworthy (see timeline). An SAED example for a cement-related mineral is given in Figure 22.3a, which is a [100]-zone axis pattern from a crystal of hillebrandite (ion-thinned) from the work of Xu and Buseck (1996). Figure 22.3b is the same pattern calculated from the atomic co-ordinates of Dai and Post (1995), and Figure 22.3c is a view of the structure along [100], calculated from the crystal structure of Dai and Post

[40]For microscopy the intermediate-projector lens system (see (b) in Table 22.2) is focused on the intermediate image plane of the objective lens, producing a magnified image of this plane on the fluorescent screen. A miniature diffraction pattern (rings or spots) is formed in the back focal plane of the objective lens. The diffraction pattern (enlarged) can form the final image if the intermediate lens is weakened to focus on this plane instead of the intermediate image plane. The electrons coming from all but a small (selected) area of the specimen can be excluded by the insertion of the selector aperture (diameter, D) into the plane of the first intermediate image. The diameter of the selected area is D/M, where M is the magnification of the objective lens (not the total magnification). So, for example, the diameter of the area selected on the sample would be 1 μm for M = ×25 and D = 25 μm. The feature of interest is selected by placing the selector aperture over it in normal (bright field) imaging mode.

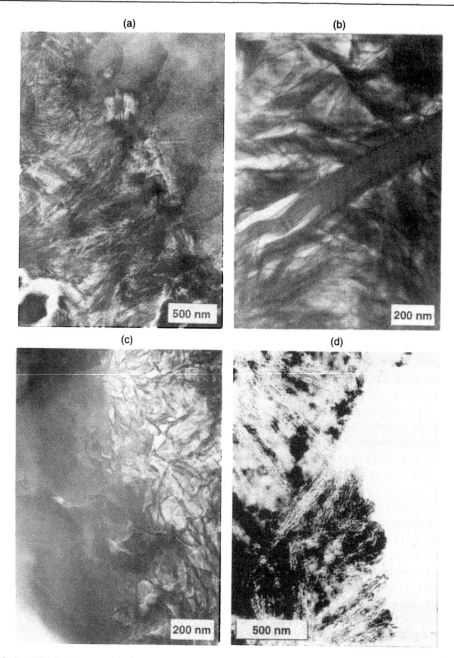

FIGURE 22.2 TEM micrographs showing (a) an inner product (Ip)/outer product (Op) interface region in a portland cement paste hydrated for three months (with fine-homogeneous Ip C-S-H (right-top), fine-fibrillar Op C-S-H (left-bottom), and an AFt relict (left-bottom corner) (modified from Richardson and Groves 1993), (b) coarse foil-like outer product C-S-H and a crystal of AFm in a KOH-activated slag paste hydrated for 8 years (the end of the crystal merges with the C-S-H) (modified from Richardson *et al.* 1994), (c) a glassy slag/coarse inner product interface in a water-activated neat slag paste hydrated for $3\frac{1}{2}$ years at 40°C, and (d) an outer product region of a C_3S paste carbonated in pure CO_2 for three days (carbonation has transformed the fibrils of C-S-H to silica (the speckled regions) and embedded them in vaterite) (modified from Groves *et al.* 1990).

(1995) (space group $Ccm2_1$, $a = 3.64$, $b = 16.31$, $c = 11.83$ Å). SiO_4 tetrahedra are black and calcium polyhedra are grey. In addition to the sharp spots (reflections) – that are characteristic of the average structure determined by Dai and Post – other patterns showed continuous streaks that led Xu and Buseck to conclude the presence of superstructure domains related by stacking faults.

SAED data can be combined very usefully with chemical information from EDX analysis. An example for a clinker phase – a calcium aluminoferrite in an oil-well cement – is given in Figure 22.4.

Figure 22.4a, is a [-313]-zone axis pattern from a crystal of ferrite from the work of Richardson *et al.* (1993). Figure 22.4b is the same pattern calculated from the atomic co-ordinates of Colville and Geller (1971). Comparison of the two shows that the experimental pattern has streaks running through the 101, 031, etc. reflections, but not through the 130, etc. The streaking was absent from the same zone axis pattern from a region with lower Fe content. An analysis of six different patterns showing streaking revealed that the ferrite was heavily faulted, the faults being caused

TABLE 22.5 Stages in TEM sample preparation. Weak specimens – for example early-age cement pastes – must be resin impregnated.

(1) Thin slices (<1 mm) cut from the bulk sample using a diamond wafering blade or microtome and vacuum-dried
(2) One side of the slice polished with fine (600/1200 grade) SiC paper
(3) Sample glued polished side down onto a glass microscope slide
(4) Sample polished down to around 30 μm thickness
(5) Glue dissolved in acetone and specimen floated off
(6) Specimen glued between two 3 mm-diameter nickel grids (using quick drying epoxy) which have central holes (2 × 1 mm) through which the sandwiched sample is exposed
(7) Specimen argon ion-beam milled until the centre of the exposed sample has thinned to a hole. Much of the region around this hole should be electron transparent. Thinning must include use of a liquid N_2-cooled stage and should be at rates <4 μm h^{-1} to minimize thermal damage
(8) Specimen carbon coated in a high-vacuum evaporation chamber

by $c/2$ displacement in the $a = 5.58$, $b = 14.60$, $c = 5.37$ Å orthorhombic structure (space group *Ibm2*). The direction of the displacement is the direction of the chains of tetrahedra in the ferrite crystal structure. These chains are illustrated in Figure 22.4c, which is a view of the ferrite structure along [110], calculated from Colville and Geller's crystal structure. (Al, Fe)–O tetrahedra are black, [Fe, Al]–O octahedra are dark grey, and calcium polyhedra are light grey. EDX data showed that the faults appeared to be associated with Fe-rich regions of ferrite.

A good example of SAED work that would undoubtedly have benefited from EDX analysis is Melzer and Eberhard's (1989). Their use of SAED in a study of ground and dispersed hardened tricalcium silicate pastes led them to postulate the existence of four previously unreported crystalline

calcium silicate hydrates, which they designated α-, β-, γ-, and δ-CSH. Groves (1990), however, demonstrated that their SAED 'evidence' was not convincing. He could not find any reflections on the diffraction patterns that could not reasonably be indexed as calcium hydroxide, calcium oxide (presumably the result of decomposition of CH in the electron beam), or calcium carbonate (presumably the result of a slight degree of carbonation). Their Fig. 8a, for example, postulated by them to be a mixture of β-, γ-, and δ-CSH, does not remotely resemble morphologically the C-S-H present in ion-thinned hardened C_3S pastes, but rather plates of CH, consistent with Groves' indexing of the corresponding diffraction pattern (their Fig. 8b). EDX analysis would presumably have shown only limited, if any, Si.

(a)
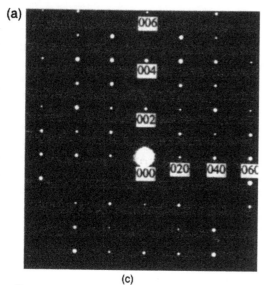

(b)
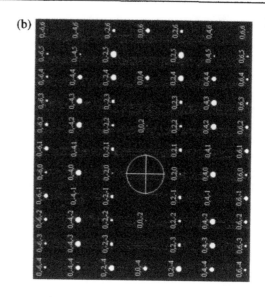

(c)
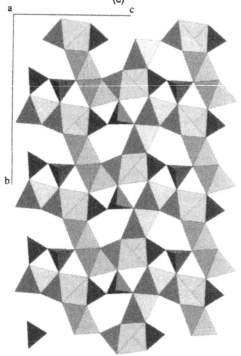

FIGURE 22.3 (a) Selected area electron diffraction pattern (100 zone axis pattern) of hillebrandite (from Xu and Buseck 1996). (b) The same pattern calculated from the atomic co-ordinates of Dai and Post (1995). (c) The hillebrandite structure viewed along [100], calculated from the crystal structure of Dai and Post (1995), space group $Ccm2_1$, $a = 3.64$, $b = 16.31$. $c = 11.83$Å. SiO_4 tetrahedra are black and calcium polyhedra are grey.

22.5 High resolution transmission electron microscopy of cement hydrate phases

Various workers have attempted high-resolution imaging of cement hydrate phases. Both dispersed crystalline calcium silicate hydrates and ion-thinned sections of hydrated cement paste have been studied. Unfortunately, these attempts have been hampered seriously by thermal and irradiation damage. Ewart *et al.* (1990), for example, undertook a TEM study of anomalous tobermorite. They found that it could be studied at moderate magnifications (say, × 50,000) for quite long periods without evident loss of crystallinity or other damage, which is not possible with C-S-H in hardened Portland cements (Richardson and Groves 1993). However, when they attempted high resolution imaging they observed lattice fringes for the tobermorite, but only as a transient phenomenon. After the initial loss of lattice fringes from the hydrate crystal as it damaged, a more stable high resolution image developed which represented a decomposition product of the initial hydrate. An example is shown in Figure 22.5. Kim *et al.* (1993) studied hillebrandite (dispersed), another crystalline calcium silicate

(a)

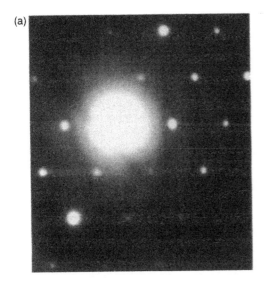

(c)

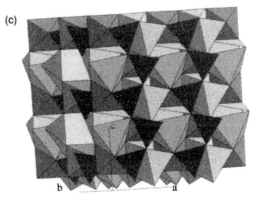

(b)

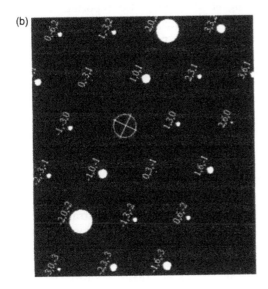

FIGURE 22.4 (a) Selected area electron diffraction pattern (-313 zone axis pattern) of ferrite in a region with Al/Fe ratio of approx. 0.75 (from Richardson *et al.* 1993). (b) The same pattern calculated from the atomic co-ordinates of Colville and Geller (1971). (c) The ferrite structure viewed along [110], calculated from the atomic co-ordinates of Colville and Geller (1971). (Al, Fe)–O tetrahedra are black, [Fe, Al]–O octahedra are dark grey, and calcium polyhedra are light grey.

hydrate, $(Ca_2(SiO_3)(OH)_2)$. They too found that the electron beam easily damaged the specimens. Xu and Buseck (1996) also studied hillebrandite (ion-beam thinned), and whilst they did obtain some high-resolution images, they noted that the main obstacle in the their TEM investigation was extremely fast electron-irradiation damage. In a study of thin films of synthetic C-S-H Henderson and Bailey (1993) noted that the practical resolution achievable was limited considerably by the low stability of the hydrates in the electron beam.

Dwight Viehland and co-workers reported results from high-resolution studies of C-S-H in hardened Portland cement pastes (1996) and of the crystalline calcium silicate hydrates 1.4 nm tobermorite and jennite (1997). The most striking aspect of their work is that they claimed to have observed lattice images of Ip C-S-H in the paste – which is known to be particularly unstable in the electron beam (Richardson and Groves 1993) – and yet were unable to obtain comparable images of the tobermorite or jennite. The authors themselves noted that their attempts at high-resolution imaging of jennite were unsuccessful because of severe radiation damage. It seems very unlikely that these crystalline calcium silicate hydrates are less stable in the electron beam than 'nanocrystalline regions' of Ip C-S-H. Indeed, their lower magnification image of Ip C-S-H (their Fig. 3, 1996) is similar to published examples of beam-damaged Ip C-S-H (Richardson

521

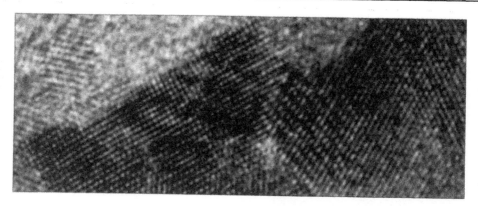

FIGURE 22.5   High resolution TEM image of tobermorite (which unfortunately had decomposed under the electron beam). The lattice spacing is about 2.4 Å (modified from Ewart *et al.* 1990).

and Groves 1993, Richardson 1999) and the region labelled as C-S-H at the bottom of their Fig. 5a (1996) appears to have experienced very severe beam damage. The authors advanced their EDX results as strong support for the presence of nanocrystalline C-S-H regions on the scale of about 5 nm, having determined variability in Ca/Si atom ratio at that scale of between 0.4 and 2.0. It is very difficult to. explain the lower value other than by beam damage,[41] which quite probably occurred before the observation at high magnification. It is certainly far too low for a dreierkette-based C-S-H with linear silicate chains.[42] The evidence from lattice spacings also seems inconclusive. The periodicities of about 0.25 and 0.28 nm, for example, that the authors suggest are consistent with CaO parts of the jennite and tobermorite structures respectively, are also consistent with the 200 and 111 planes of CaO (0.241 and 0.278 nm). CaO would result from decomposition of CH.[43]

Zhang *et al.* (2000) also reported recently a high-resolution imaging study of C-S-H in hardened cement pastes. Again, the question arises regarding the feasibility of imaging C-S-H at such high resolution with current instrumentation without it decomposing in the electron beam. It is unfortunate that the paper contains no lower magnification images containing regions imaged at high magnification, taken before and after that imaging, to help demonstrate stability. They state that they observed no change in the lattice image of the nanocrystalline regions or in chemical composition after the high-resolution observation. They claimed that this demonstrates that no beam-induced local decomposition of C-S-H occurred during the high-resolution observation, and therefore proves that the nanocrystalline regions are of C-S-H. Again, the problem is that the damage may have occurred *before* the high-resolution observation. Ewart *et al.* (1990) and Viehland *et al.* (1997), for example, observed such rapid damage to crystalline calcium silicate hydrates (tobermorite and jennite) when they attempted high-resolution imaging that they were unable to record images of the undamaged crystals. As noted already, the C-S-H in a hardened cement paste is to be expected to be much more susceptible to electron beam damage than these crystalline phases, and so it is quite conceivable that Zhang *et al.* (2000) would not have observed

[41]Or possibly by partial carbonation, which would lead to regions of silica gel surrounding nanocrystals of calcium carbonate.
[42]The chains are most unlikely to be cross-linked since ^{29}Si NMR never detects Q^3 species in such systems. Note that the minimum Ca/Si of tobermorite (with infinite dreierkette chains) is 0.67, and is higher unless the chains are completely protonated.
[43]Or calcium carbonate, if the paste were slightly carbonated.

the damage occurring. Like Viehland *et al.* (1997) they advance their EDX results as strong support for the presence of nanocrystalline C-S-H regions on the scale of about 5 nm. The reliability of those data, however, is questionable. First, the spectra display very poor resolution[44] suggesting either a very high-count rate or a problem with the detector, either of which could affect the relative intensities. Second – and assuming that the intensities are correct – there may be a problem with their quantification, one that would result in Ca/Si ratios lower than reported, possibly less than 0.6.[45] If this is correct, the same objection that applied to Viehland *et al.*'s (1996) EDX data – with extreme values of 2.0 and 0.4 – also applies to these. Again, the lower composition is far too low for a dreierkette-based C-S-H with linear silicate chains, and suggests that the C-S-H had been damaged, quite possibly before the analysis was undertaken.

22.6 Scanning electron microscopy of cement

22.6.1 Secondary electron imaging (SEI)

SEM–SEI has been applied as extensively to the study of fracture surfaces[46] of cement as to other

materials.[47] Many of the more significant studies are listed in the timeline. Of the early workers – aside from the first paper by Chatterji and Jeffrey (1966) – the most significant study was by R. Brady Williamson at the University of California. His paper includes micrographs providing evidence for Taplin's (1959) inner/outer product scheme, and showing massive crystals of CH (Williamson 1969), similar to those in Figure 22.6a. For comparison, a large region of CH as it appears in ion-thinned sections in the TEM is included, Figure 22.6b. In 1972, Williamson presented a comprehensive – and widely referenced – paper on the solidification of Portland cement and tricalcium silicate, based largely on SEI work. As the SEM became widely available[48] it soon became the principal routine tool for the study of the microstructure of cement pastes. Many studies have undoubtedly contributed much to our understanding of the microstructure of cement and concrete, and especially of three-dimensional particle-level morphologies. As examples of SEI images, three micrographs taken at increasing magnification are shown in Figures 22.7a–c. Whilst the type of information that SEI of fracture surfaces can provide is valuable, it is limited. The wide availability of the SEM and the ease with which images can be obtained has led to a large number of publications containing unnecessary, poorly interpreted or unrepresentative micrographs. Poor interpretation is often due to an inadequate appreciation of the mode of specimen preparation, that is, that the preferred fracture paths (that pass through the weaker phases) will

[44]The peaks are very broad, which is clearly demonstrated by the lack of separation of the Ca Kα and Kβ peaks.

[45]In general, quantitative analysis of thin specimens can be achieved without recourse to the complex correction procedures that are necessary for bulk specimens; if the specimen is thin enough then absorption, etc. become negligible, and the analyses need only be corrected for the efficiency of collection. This is done by the application of a so-called *k*-ratio factor that is determined by analysis of standard specimens of known composition (Cliff and Lorimer 1975). Zhang *et al.* observed Ca/Si ratios ranged between 0.87 and 3.9, and spectra for the two extreme cases are presented in their Fig. 4. Inspection of these spectra suggests that the *k*-ratio factor may have been mis-applied. One would expect the Ca/Si atom ratio calculated after the application of this factor to be less than the ratio of the raw intensities. However, crude measurement – by estimating the background and then measuring the peak heights with a ruler – of the Ca and Si intensities on their Fig. 4b indicates a significant *increase* from about 3.0–3.9. This indicates that a $k_{Ca/Si}$-factor of about × 1.3 was applied, which is confirmed by repeating the procedure with Fig. 4a (with an uncorrected Ca/Si of approx. 0.69). This seems very high. If, for the

purpose of demonstrating the serious implications of this issue, the × 1.3 factor is applied instead as $k_{Si/Ca}$ – which would not be an unreasonable figure – Ca/Si atom ratios of approx. 2.3 and 0.53 are obtained.

[46]Dried, fractured samples are mounted on stubs (often using silver paint), loose particles blown off, and carbon or gold coated. Gold sputter-coating may be preferred for imaging of rough surfaces as it provides a more uniform film than high-vacuum evaporation, with a corresponding reduction in charging effects, and hence improvement in image quality.

[47]Chatterji and Jeffery's (1966) comment in the very first SEI paper on cement has certainly been born out: '...It seems that this new technique...may open up another vista in cement research.'

[48]And as resolution improved and EDX facilities were added to help identify observed phases.

(a) (b)

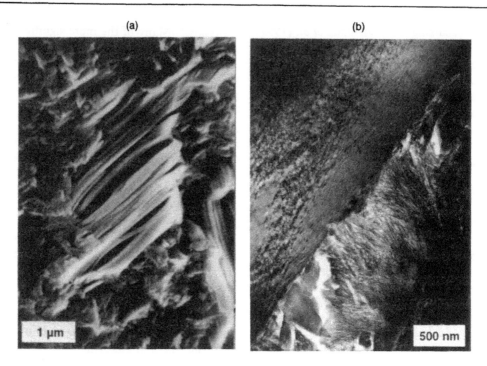

FIGURE 22.6 Crystals of CH on (a) a SEM–SEI image of a fracture surface of a Portland cement paste, and (b) a TEM image of an ion-thinned section of a Portland cement paste.

lead to over- and under-estimations of the proportions of particular features. Unrepresentative features should be avoided in publications unless there is a very good reason, however aesthetically pleasing they may be (this has been discussed previously, e.g., Grattan-Bellew *et al.* 1978). Sampling is, of course, an issue with all microscopy techniques: the micrographs in Figures 22.7a–c, for example, represent $0.056 \times 0.077\,\text{mm}$, $0.0094 \times 0.013\,\text{mm}$, and $0.0022 \times 0.003\,\text{mm}$ respectively!

22.6.2 Backscattered electron imaging

Backscattered electron (BSE) imaging provides compositional (average atomic number) contrast.[49] Attempts at quantifying various micro-structural characteristics of hardened cement, mortar, and concrete[50] – most commonly residual anhydrous phases, CH, and porosity (above $\approx 0.5\,\mu\text{m}$ in size) – using BSE images were pioneered in the 1980s by

(Reuter 1972). For homogeneous mixtures of elements at the atomic scale $\eta = \Sigma C_i \eta_i$, where i denotes each constituent, C_i is the mass concentration, and η_i the pure-element backscatter coefficient. Thus on greyscale BSE images regions of high average atomic number appear bright relative to those of lower number. η is approximately independent of beam energy in the range 10–100 keV. As an example, the formula for pure tricalcium silicate is Ca_3SiO_5, with weight fractions of Ca = 0.5266, Si = 0.1230, and O = 0.3504. Reuter's expression gives pure-element backscatter coefficients of $\eta_{Ca} = 0.2268$, $\eta_{Si} = 0.1644$, and $\eta_O = 0.0911$, and so the coefficient for tricalcium silicate is calculated as $(0.5266 \times 0.2268) + (0.1230 \times 0.1644) + (0.3504 \times 0.0911) = 0.1716$. The coefficient for pure dicalcium is calculated similarly as 0.1662, and so C_3S should appear bright relative to C_2S.

[50] A flat, polished surface is required. Specimens (dried) must usually be impregnated with resin prior to grinding and polishing, although it is sometimes possible to prepare mature pastes satisfactorily without it. Preferential polishing must be minimized. The preparation of specimens has been discussed in the literature (e.g. Struble and Stutzman 1989).

[49] The fraction of electrons that are backscattered, η (called the 'backscattering coefficient' $= n_{BSE}/n_{PE}$), increases with atomic number (Heinrich 1966) and is commonly calculated using $\eta = -0.0254 + 0.016Z - 1.86 \times 10^{-4}Z^2 + 8.3 \times 10^{-7}Z^3$

(a)

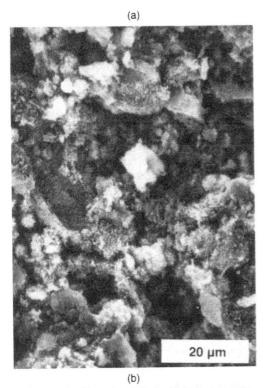

(c)

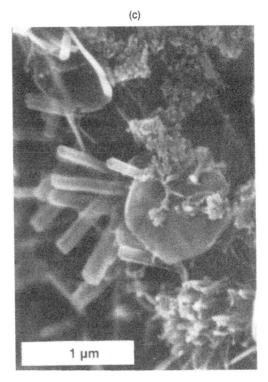

(b)

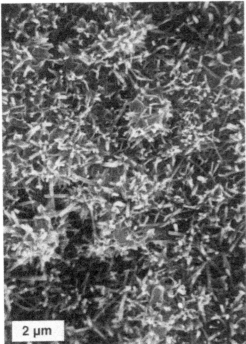

FIGURE 22.7 SEM–SEI images at three different magnifications.

Karen Scrivener and co-workers at Imperial College. Essentially, certain bins (0-255: black–white) of the greyscale histograms of a set of images are allocated to a phase and the corresponding number of pixels counted. The example, Figure 22.8, corresponds to a BSE image containing aggregate, unreacted cement phases, dense inner product (Ip) C-S-H, CH, other hydration products (mainly outer product C-S-H), and porosity. The peak centred on the 31st bin is from a single large pore. The image settings (contrast and brightness) must be carefully set. The inset in Figure 22.8 illustrates the effect of increasing the contrast. A further increase would result in the loss of information: some dark-grey pixels would become black, and some bright-grey pixels would become white. The magnification used is a compromise between the total area sampled and the smallest feature of interest: most workers have used a magnification of about ×500. This means that for a 10×10 cm CRT display (see Table 22.2), each image corresponds to

525

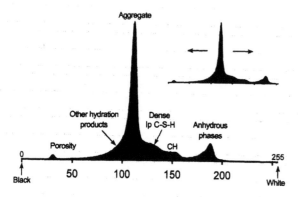

FIGURE 22.8 A grey-scale histogram from a BSE image that contained aggregate, unreacted cement phases, dense inner product (Ip) C-S-H, CH, other hydration products (mainly outer product C-S-H), and porosity.

an area of only 0.2×0.2 mm. A large number of images are required for statistical significance. The technique has been adopted widely, most notably in studies concerned with the so-called ITZ in concrete (references given in the timeline). Analysis of the images has until recently been particularly time-consuming since the aggregate-paste interface has to be identified. This has been simplified recently by the use of automated procedures utilizing information from X-ray dot maps collected at the same time as the image (Brough and Atkinson 2001). Illustrative examples are shown in Figure 22.9 and 10, one with high porosity, the other low (images from M. K. Head's Ph.D. Thesis, University of Leeds 2001).

Whilst the quantitative analysis of BSE images has produced valuable information there are some serious limitations. The main problems concern sample preparation[51] and segmentation of the image. These are well illustrated by the 'hollow shells' debate. There has been much discussion since the mid-1970s as to the nature of the formation of so-called Hadley grains or 'hollow shells' during the hydration of Portland cement based systems (e.g. Barnes et al. 1978;

[51]Drying, resin impregnation, and grinding/polishing. Since preservation of the microstructure is of paramount importance, extreme drying regimes should be avoided, for example, in an oven at 105°C (such as employed by Kjellsen and Helsing Atlassi 1999)

Dalgleish et al. 1982; Pratt and Ghose 1983; Hadley et al. 2000). Those studies utilized SEI of fracture surfaces. Figure 22.11 is an example of a Hadley grain observed on a fracture surface of a hardened Portland cement. These features have been studied recently by BSE imaging (e.g., Kjellsen et al. 1996, 1997; Kjellsen and Helsing Atlassi 1999). It is very likely that some of the regions identified as 'hollow shells' in those studies are in fact small fully hydrated particles. The much higher resolution technique of TEM (of ion-thinned sections) has shown that the hydrated remains of relatively small particles – whether of Portland cement, slag or fly ash – contain a less dense product with substantial porosity, surrounded by a zone of relatively dense C-S-H. Examples are shown in Figure 22.12. This zone, or rim, may make it difficult for the resin used during the preparation of specimens for BEI to penetrate the porous interior, which may then be removed during subsequent polishing. This is most likely for the smallest particles since they tend to have the most complete dense outer rims. Such grains would appear as isolated pores on BSE images. With regard to segmentation, the question arises as to which bins of the greyscale histogram should be assigned as porosity (there is a similar concern associated with the distinction between CH and dense Ip C-S-H)? Since low density C-S-H will appear dark on BSE images it is quite possible that many of the particles into which resin does penetrate may still be identified as isolated pores, or 'hollow-shells.' This is illustrated in Figure 22.13a, which shows a region of a Portland cement mortar (from Richardson 1999). Careful examination of the regions that are similar to those identified by other workers as 'hollow shells' reveals that they contain some material. A small fully hydrated grain (centre-left) and a large partially hydrated grain (right) are present on the enlarged part of the image in Figure 22.13b. The small particle has a well-defined rim and there is material within its interior. The Ip is darker, and so less dense, than that present in the larger grain. In the analysis of BSE images the darkest shades of grey are taken to be porosity (illustrated in Figure 22.8) so the darkest

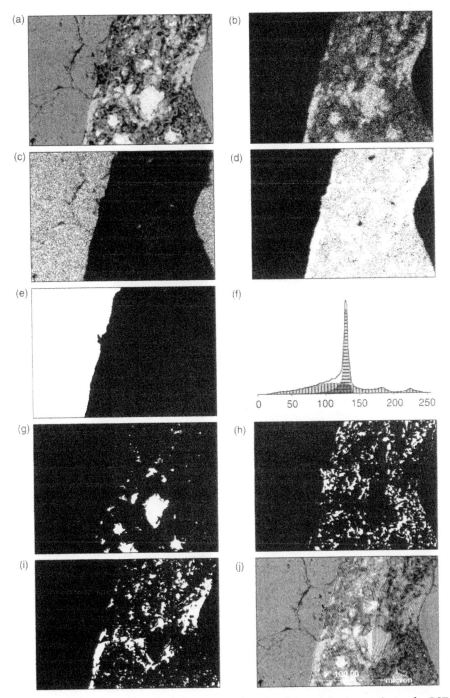

FIGURE 22.9   An illustration of an automated procedure for the quantitative analysis of a BSE image from a concrete containing much porosity. The aggregate-paste interface is identified using information from Ca and Si X-ray dot maps collected at the same time as the BSE image. The figures are (a) BSE image, (b) Ca X-ray dot map, (c) Si map, (d) Ca/Si ratio map, (e) coarse aggregate binary mask (fine aggregate mask not shown), (f) grey-scale histogram including components from aggregate (filled with horizontal lines) and cement paste (vertical lines), (g) anhydrous cement, (h) porosity, (i) CH, and (j) BSE image with contours for measuring the variation in phases with distance from the aggregate-cement paste interface.

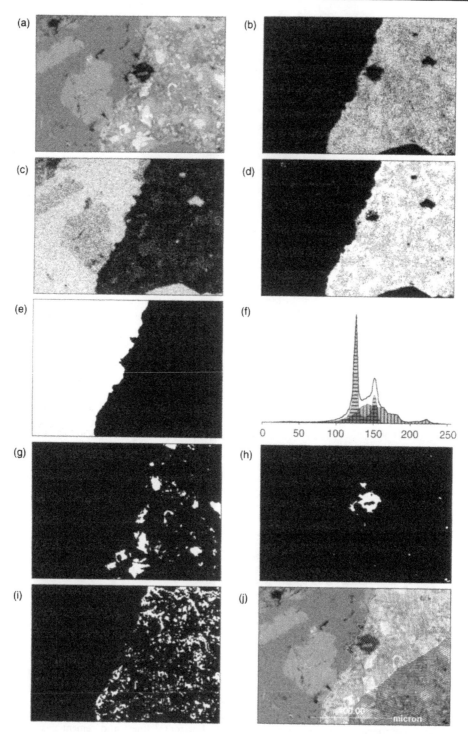

FIGURE 22.10   As Figure 22.9 but for a region containing little porosity (greater than about 0.5 μm).

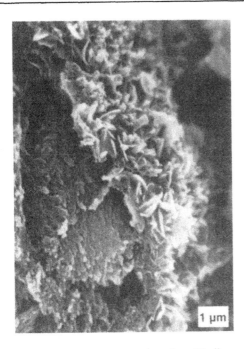

pixels correspond to the most distinct pores. For the example image, these have been assigned to the lowest thirty-two bins (arbitrarily) and are represented as white on Figure 22.13c and d. Examination of the histograms and analysis techniques used by other workers (Scrivener *et al.* 1985; Zhao and Darwin 1992; Bonen 1993) indicates that they would have assigned more bins to porosity. Figures 22.13e and f show the result of changing the threshold so that a similar number (or still possibly fewer) are included. Many of the white pixels (i.e. porosity) now fall within inner product regions. The interior of the fully hydrated particle in the enlarged area, for example, has been analysed as a pore or 'hollow shell' of about $\approx 6 \times 4 \, \mu m$. Since it is very likely that this grain contained a low density C-S-H (with foil-like morphology) it is clear that porosity and pore-size information derived from the analysis of BSE images should be interpreted with care.

Since metals appear bright on BSE images, an interesting alternative method for probing the

FIGURE 22.11 An example of a Hadley grain observed on a fracture surface of a hardened Portland cement.

(a)

(b)

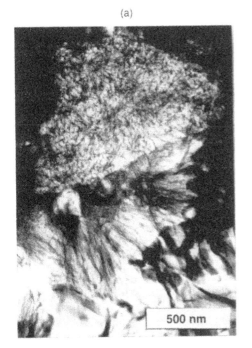

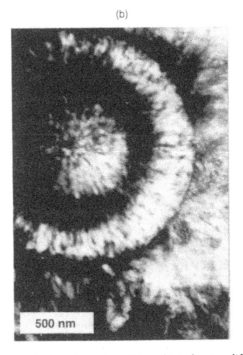

FIGURE 22.12 TEM micrographs showing the remains of fully hydrated small particles of (a) alite (modified from Richardson and Groves 1993), (b) fly ash (modified from Richardson *et al.* 1989).

(a)

(b)

(c)

(d)

(f)

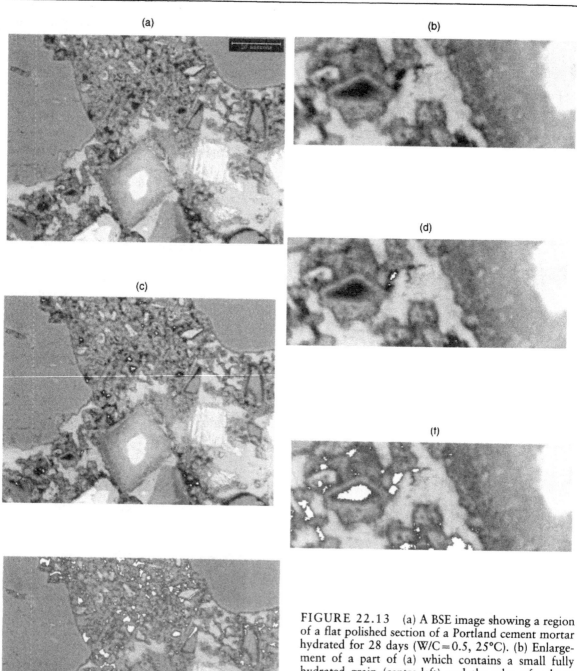

FIGURE 22.13 (a) A BSE image showing a region of a flat polished section of a Portland cement mortar hydrated for 28 days (W/C=0.5, 25°C). (b) Enlargement of a part of (a) which contains a small fully hydrated grain (centre-left) and the edge of a large partially hydrated grain (right). (c) and (d): Same as (a) and (b) but with the pixels corresponding to the darkest 32 bins of the greyscale histogram set as white. (e) and (f): Same as (c) and (d) but with more bins of the greyscale histogram assigned as porosity (see text for details). From Richardson (1999). Original micrographs courtesy of Dr Charlotte Famy.

(a)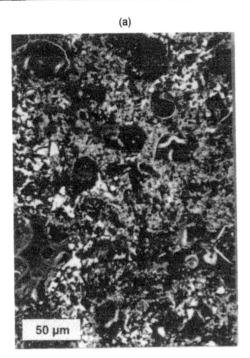

(b)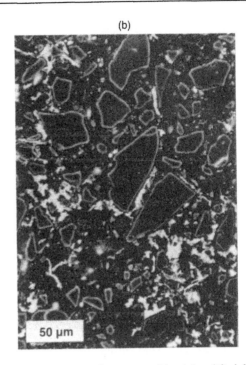

FIGURE 22.14   BSE images of (a) a neat portland cement paste and (b) a slag–cement blend (modified from Richardson *et al.* 1989), intruded with Newton's metal.

porosity of hardened cements has been to intrude them with low-melting point metals, essentially giving an insight into the physical interpretation of mercury intrusion porosimetry measurements of 'pore size distributions.' Richardson *et al.* (1989) used Newton's metal[52] and Rahman (1984), Scrivener and Nemati (1996), and Willis *et al.* (1998) used Wood's metal.[53] Figure 22.14a shows a BSE image for a neat Portland cement paste and Figure 22.14b for a slag-cement blend. The outer product region in the Portland cement paste is much better intruded. This reflects the different Op C-S-H morphologies, as observed in the TEM: the pores formed by the foil-like Op C-S-H in the blend are less well interconnected than

[52]A ternary eutectic mixture of lead, bismuth and tin, with a melting point of 95°C.
[53]Wood's metal has a slightly lower melting point than Newton's, but it also contains volatile cadmium, so the latter may be preferred for observations in an electron microscope.

FIGURE 22.15   SEM–SEI image of the Newton's metal network revealed after dissolving away the cement phases of a portland cement paste hydrated for one year.

531

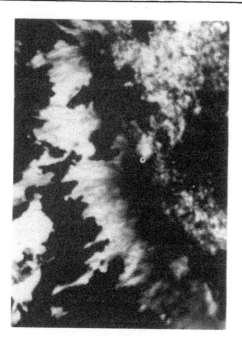

FIGURE 22.16 TEM image of an ion-thinned cement paste showing regions intruded with Newton's metal. (G. W. Groves, 1988 unpublished). Horizontal edge length = 1.5 μm.

those formed by the fibrillar C-S-H in the neat paste. The cement in the intruded sections can be dissolved out, enabling three-dimensional imaging of the pore structure by SEM–SEI: an example is shown in Figure 22.15. Alternatively, the section can be ion-thinned and observed in the TEM: an example is shown in Figure 22.16. Interestingly, some of the metal (which appears black) has penetrated the inner product region (right centre-top). This particular Ip region has the low-density morphology that is characteristic of small, fully hydrated particles rather than the dense, featureless morphology typical of larger grains.

22.7 Low-temperature and environmental electron microscopy of cement

Electron microscopes require a high vacuum so specimens are consequently observed in a dried state. Various attempts have been made to minimize the effects of drying on hydrate structures, either by the use of very low temperatures or 'environmental' microscopes or stages. Bailey and Stewart (1984), for example, examined frozen samples using a cryo-stage in a SEM, and concluded that the needle-like outgrowths (Op C-S-H) seen in vacuum-dried C_3S pastes were not a consequence of drying because they were also present in frozen samples. The environmental scanning electron microscope (ESEM) – pioneered largely by Gerry Danilatos (e.g. 1979, 1981, 1983, 1985, 1990a,b) – uses a differential pumping system that enables the sample chamber to be maintained at relatively low vacuum (around 20 torr). There is no need for a conductive coating because the gaseous layer around the specimen becomes ionized and suppresses the accumulation of charge. The gas itself is used as the detection medium. So, by using water vapour as the imaging gas, water-saturated conditions can be produced by controlling the temperature of the sample.[54] Only a few studies have been made of hydrating cement. Sujata and Jennings (1992), for example, observed the formation of a continuous layer of product on cement grains. They reported that severe drying led to damage of the layer leaving behind crumpled flakes and foils. Meredith et al. (1995) gave an introduction to the technique, and presented results from a study of pre-induction and induction period hydration of C_3S.

High voltage TEMs with environmental cells have been used to study moist hydrated cements by several workers (Double 1973, 1978; Jennings and Pratt 1979, 1980; Scrivener and Pratt 1984). Jennings and Pratt (1979, 1980) claimed to have observed C-S-H 'sheets' roll up on drying, concluding that the true (delicate) structure is not seen in dried specimens. Whilst they stated that they could not find the same morphology in dried specimens, their 'wet' C-S-H is very similar to the morphology of coarse Op C-S-H observed in

[54]The equilibrium vapour pressure of water is 6.5 torr at 5°C whilst it is 23.8 torr at 25°C.

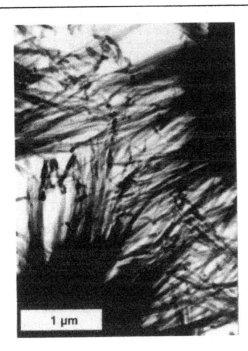

FIGURE 22.17 TEM image showing coarse Op C-S-H in a cement paste hydrated at 60°C.

22.8 Analytical electron microscopy using X-rays

22.8.1 Bulk specimens (EMPA, SEM–EDX)

Many modern SEMs are equipped with an energy dispersive X-ray (EDX) spectrometer to enable the chemical analysis of bulk specimens.[55] However, because the emphasis of SEM design is on spatial resolution, quantitative analysis is often better achieved using dedicated electron microprobes (EMPA) that are equipped additionally with one or more wavelength dispersive spectrometers (WDS).[56] The electron optics of these instruments is similar but there are a number of important differences. Some of these are summarized in Table 22.6. Since the X-ray generation volume in bulk specimens is generally 1–2 μm^3 (see top figure of Table 22.3) compositional analyses of hardened cements very often correspond to a mixture of phases. A flat, polished surface is essential for quantitative analysis as surface irregularities affect the matrix correction factors:[57] great care should be taken to minimize preferential polishing.[58] Compositional information derived from

(obviously dry) ion-thinned specimens; their 'dry' C-S-H is typical of electron-beam damaged coarse Op C-S-H. Interestingly, their specimen was hydrated at 55°C, and it is true that C-S-H formed at higher temperatures tends to be coarser. Figure 22.17 shows coarse Op C-S-H in a paste hydrated at 60°C that should be compared with Jennings and Pratt's time-sequence micrographs (1979, 1980). They claimed that this sequence of four micrographs, taken over about 30 min, illustrates the transition on drying of thin sheets to needles, the final needles being similar to 'Type I' C-S-H (C-S-H morphology is discussed in the last section of this chapter). The changes observed are typical of damage in the electron beam: many of the C-S-H fibrils in Figure 22.17 are similar to those in the first of their time-sequence, whereas the curled (beam-damaged) fibrils in the centre-left resemble those in their fourth ('dried').

[55] Commercial EDX systems were available to fit most SEMs by 1972 (Diamond 1972).

[56] The two types of spectrometer are complementary. The main advantage of the WDS is its superior sensitivity and energy resolution. This allows the separate resolution of very close lines (which would overlap using EDX), good accuracy for all elements, and the quantitative determination of trace elements. EDX, on the other hand, has the advantage that the whole spectrum can be recorded simultaneously whereas it must be recorded sequentially with WDS.

[57] In order to determine the relative proportions of the elements present, the number of characteristic X-ray counts for each element in the specimen must be compared under identical conditions with those arriving from standards of known composition (standard-less procedures do exist but should be used with caution). As the specimen differs from each standard in its density and average atomic weight, several matrix correction procedures need to be applied to obtain more accurate compositions (the so-called ZAF corrections). Since the correction factors depend on the unknown compositions they have to be applied by an iterative method. The correction procedures for bulk specimens are covered at length in the literature, for example Reed (1993), Reimer (1985), and Goldstein *et al.* (1992).

[58] After polishing, a conducting paint should be applied around the edges of the specimen followed by the evaporation of carbon onto the surface. Carbon is used because of its low atomic number, ensuring minimum absorption of the emerging X-rays.

533

TABLE 22.6   Some differences between a dedicated electron microprobe and a standard SEM

A microprobe requires a higher beam current to obtain sufficient X-ray intensity for accurate measurement and to excite the higher energy characteristic X-ray lines

An optical microscope is generally incorporated to allow accurate specimen height adjustment to satisfy the focussing condition of the WDS, and therefore enable good spectrometer performance, and to enable accurate positioning of the electron beam for standardization and analysis

The X-ray geometry must be carefully designed to enable good spectrometer performance. It is preferable to have several spectrometers fitted, each allowing a high take-off angle, and to have both energy and wavelength dispersive spectrometers to take advantage of their respective benefits

Note: These requirements involve considerable redesign of the objective lens and the specimen chamber.

fracture surfaces is at best qualitative due to the roughness of the surface. The earliest EMPA studies of cement are particularly noteworthy because they clarified the chemical composition of the constituents of clinker (Moore 1965; Amicarelli et al. 1966; Peterson 1967; Fletcher 1968, 1969; Midgley 1968, 1969b).

EDX and WD spectrometers can both be used to determine the distribution of elements in a sample (elemental mapping). This is easily achieved with EDS by scanning the beam over the area of interest but with WDS the specimen must be mechanically moved under the beam to maintain the spectrometer focusing condition. Examples of elemental maps collected using WD spectrometers are shown in Figure 22.18 for an area in a Portland cement-slag paste. The EMPA had two WD spectrometers so the maps were collected in pairs. The signal to noise ratio is excellent for the major elements and even reasonable for Mn, which is a very minor element. The brightest regions on the Ca map correspond to CH and the larger mid-grey regions to unreacted slag grains. The brighter regions on the Al and S maps which coincide (and with mid-grey on the Ca) are dominated by calcium sulpho-aluminate hydrate phases. The bright regions on the Mg, Ti and Mn maps are slag inner product.[59] Alkalis (illustrated

by the K map) are present in the Ip regions and throughout the Op, except where there are large amounts of CH or calcium sulpho-aluminate hydrates. X-ray mapping can provide quantitative spatial information, although collection of the data is time-consuming. The data can be presented usefully in several forms. An example for a cement-slag paste – which includes an elemental map (Mg atom), a contour map (Mg/Ca atom ratio), a scatter plot (Al/Ca against Si/Ca atom ratio), and data for a single line[60] – is given in Figure 22.19. The scatter plot shows clearly the nature of the phase admixing at the level of the X-ray generation volume ($1–2\ \mu m^3$), some points corresponding to essentially pure phases whilst others are due to mixtures of two or three phases.

22.9 Analytical transmission electron microscopy (TEM–EDX/EELS)

In a bulk specimen the X-ray spatial resolution is large ($1–2\ \mu m^3$) whereas in thin specimens it is of

a Mg, Al-rich hydroxide phase (with octahedral Al), and the generalized formula,

$$\left[R^{2+}_{1-k} R^{3+}_k (OH)_2 \right]^{k+} \cdot R^{r-}_{k/r} \cdot jH_2O$$

R^{2+} is principally Mg^{2+} and R^{3+} Al^{3+} (the phase also accommodates Fe, Ti, and Mn). $R^{r-} = OH^-$, SO_4^{2-}, or CO_3^{2-}. For hydrotalcite-like phases in general $\approx 0.2 \leqslant k \leqslant 0.33$ whilst in slag-containing cements k is typically ≈ 0.3.

[60]There are many line-scan and X-ray dot-map data in the literature, for example, Grutzeck and Roy (1969), Roy and Grutzeck (1969), Terrier (1969), Stucke and Majumdar (1977), Ogawa et al. (1980), Taylor and Newbury (1984), Uchikawa (1988), Richardson and Groves (1992a, 1993).

[59]Microanalysis of slag Ip in a range of systems by both EMPA and TEM has shown a linear relationship between an increase in Mg/Ca ratio and Al/Ca ratio (discussed in Richardson 1999). The Al/Ca ratio at Mg/Ca = 0 is similar to that of the Op C-S-H. This relationship is due to different levels of admixture within the analyzed volume of a C-S-H compositionally equivalent to Op C-S-H (with tetrahedral Al) and

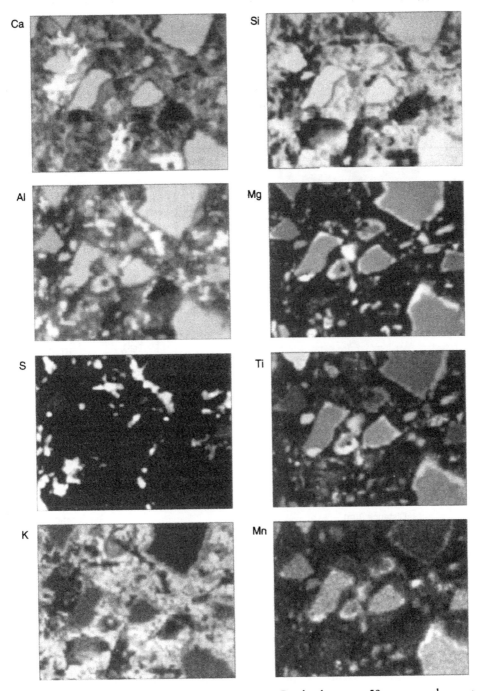

FIGURE 22.18 X-ray maps for an area in a 50 per cent Portland cement 50 per cent slag paste hydrated for $3\frac{1}{2}$ years at 20°C, W/S = 0.4. The horizontal edge length = 125 μm. The microprobe was operated at an accelerating voltage of 15 kV and probe current of 3×10^{-8} Å, and the data were collected using wavelength dispersive spectrometers.

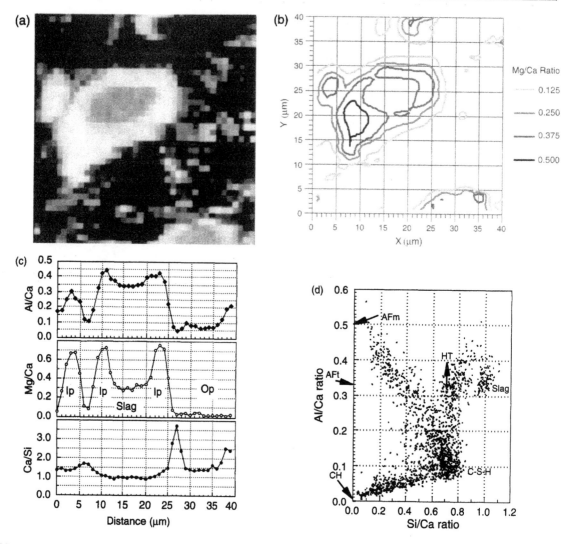

FIGURE 22.19 (a) Mg atom map of an area ($40 \times 40\,\mu m$; $1\,\mu m$ increments) in a Portland cement–slag blend containing 75 per cent slag (hydrated for 14 months at 20°C, W/S = 0.4). The map was collected simultaneously with nine other elements: Na, Al, Si, S, K, Ca, Ti, Mn, and Fe. Na and Fe were analyzed with the WD detectors, the remaining eight with the EDX detector. The data are ZAF corrected. (b) Mg/Ca contour map for the area in (a). (c) Elemental ratio plots for one of the lines from the X-ray maps in (a). (d) Al/Ca against Si/Ca atom ratio scatter plot from the X-ray mapping data of the area in (a).

the order of the probe size (see top figure in Table 22.3). In bulk specimens, therefore, analyses often correspond to a mixture of phases. Examination in the TEM, however, generally allows the unambiguous identification[61] of a phase and analysis free of admixture with other phases, although too small a probe should be not be used because of the effects of beam damage. Examples of TEM–EDX results are shown in Figures 22.20 and 20.21. Figures 22.20a–c are

[61]Although this was not possible with the earliest TEM–EDX studies which were of ground and dispersed pastes (Taylor 1979; Gard *et al.* 1980, Lachowski *et al.* 1980, 1981; Mohan and Taylor 1981).

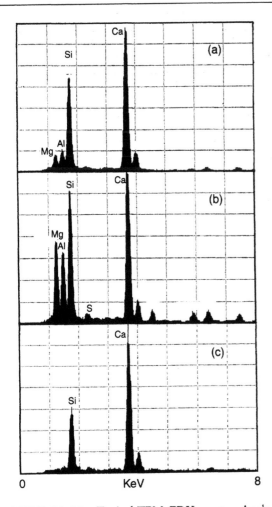

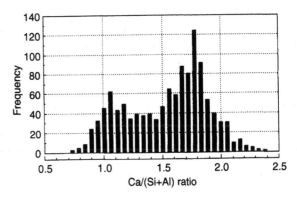

FIGURE 22.21 Ca/(Si + Al) atom ratio frequency histogram for C-S-H in a wide range of water-activated cement-slag systems (0–100 per cent slag). The plot includes 1186 individual TEM microanalyses of C-S-H free of admixture with other phases. Reproduced from Richardson (1999).

FIGURE 22.20 Typical TEM–EDX spectra. Analyses of slag ((a) and (b)) and alite (c) inner product. Be window detector. The Kα peaks are labeled. The small peak at around 4 keV is the Ca Kβ peak, and the peak at around 7.5 keV is the Ni Kα peak (sputtered from the Ni grid onto the specimen during ion-thinning).

typical EDX spectra (of slag ((a) and (b)) and alite (c) inner product). Figure 22.21 is a Ca/(Si + Al) histogram for C-S-H in a wide range of water-activated cement-slag systems (0–100 per cent slag). All the data in Figure 22.21 are from analyses of C-S-H free of admixture with other phases. It is significant that there are very few analyses <0.83 or >2.25, which are the minimum and maximum values in Taylor's model for the structure of

C-S-H (Taylor 1986), and no analyses <0.67 or >2.5, which are the minimum and maximum values in Richardson and Groves' model (1992b). Knowledge of the compositions of individual phases determined in the TEM makes the interpretation of bulk data much easier. This is illustrated in Figures 22.22 and 22.23. Figure 22.22a is an Al/Ca against Si/Ca atom ratio plot of TEM analyses of Op C-S-H in hardened pastes of a range of slag-Portland cement blends (●) and slag activated with 5 M KOH solution (■). Each point corresponds to a mean value of 20–50 analyses that are of Op C-S-H free of admixture with other phases. The regression line is Si/Ca = 0.4277 + (2.366 × Al/Ca). Interestingly, the increase in Si/Ca ratio (and Al/Ca) is accompanied by a change in Op C-S-H morphology from fibrillar to foil-like. Figures 22.23a and b are Al/Ca against Si/Ca atom ratio scatter plots for the EMPA mapping data of areas in Portland cement-slag blends containing 10 per cent and 50 per cent slag respectively (hydrated for 14 months). The full line corresponds to the regression equation from Figure 22.22a. The large numbers of data points close to the full line are thus due solely to C-S-H. The smaller number of points corresponding to mixtures of C-S-H with CH or AFm is consistent with the decrease

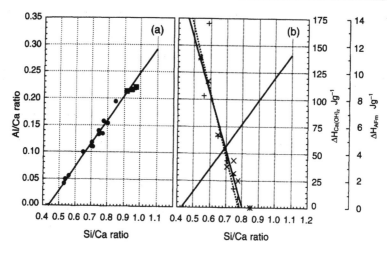

FIGURE 22.22 (a) Al/Ca against Si/Ca plot of TEM analyses of Op C-S-H in hardened pastes of a range of slag-Portland cement blends (●) and slag activated with $5\,M$ KOH solution (■). Each point corresponds to a mean value of 20 to 50 analyses that are of Op C-S-H free of admixture with other phases. The regression line is Si/Ca = $0.4277 + (2.366 \times$ Al/Ca). (b) Plots of AFm (×) and CH (+) against Si/Ca ratio of the Op C-S-H present in the cement-slag blends (integrated peak intensities from differential scanning calorimetry). From Richardson (2000).

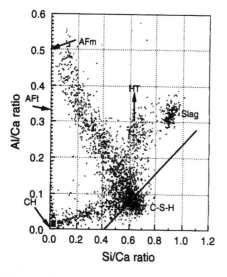
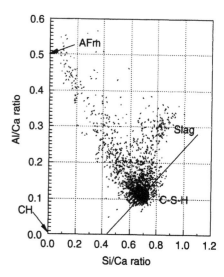

FIGURE 22.23 Al/Ca against Si/Ca atom ratio scatter plots for the EMPA mapping data of areas in portland cement-slag blends containing (a) 10 per cent and (b) 50 per cent slag respectively (hydrated for 14 months). The full line corresponds to the regression equation from Figure 22.22.

in those phases as the slag content increases, Figure 22.22b.

Additional chemical information can be obtained in the TEM from electron energy-loss spectroscopy (EELS). For example, Richardson *et al.* (1993) combined solid-state nuclear magnetic resonance spectroscopy with EELS and EDX in the TEM to locate Al in substituted C-S-H. In the inner product regions of hydrated slag particles Al in octahedral co-ordination was found to increase with an increase in Mg content (also octahedrally co-ordinated). This confirmed the conclusion from the observed linear relationship between Mg/Ca and Al/Ca that slag Ip consists of a mixture of C-S-H (with tetrahedral Al) and a Mg,Al-rich hydroxide phase (with octahedral Al). The technique has also been used to study the calcium aluminoferrite phase (Gloter and co-workers 2000).

Example of a comparison of X-ray microanalysis by EMPA, SEM–EDX, and TEM–EDX: C-S-H in hardened C_3S pastes

The different direct (EMPA, SEM–EDX, TEM–EDX) methods of determining the Ca/Si ratio of C-S-H present in hardened C_3S pastes have

all given values ranging between 1.2 and 2.0; Table 22.7 lists some typical values from the literature. A value of 1.7–1.8 is the most likely average ratio but with real local variability within a paste. This value is in good agreement with that determined by indirect methods, for example Dent Glasser *et al.* (1978) ($\approx$1.7; by quantitative XRD and TA), Clayden *et al.* (1984) (1.8 by NMR and TA), and Le Sueur *et al.* (1984) (1.7 by quantitative XRD and TA). Similar values have also been obtained for C-S-H in hardened β-C_2S pastes – for example, Taylor and Newbury (1984*b*) obtained 1.78 by EMPA and Richardson

TABLE 22.7 Typical values for the Ca/Si ratio of C-S-H in hardened C3S pastes

	Technique	Mean	Range
Grutzeck and Roy (1969)	EMPA	1.7	1.50→1.90
Stucke and Majumdar (1971)	SEM–EDX	Op 1.6 Ip 1.9	
Diamond (1976*b*)	SEM–EDX	1.9	
Gard *et al.* (1980)	TEM Disp.	1.56	1.21→1.96
Lachowski *et al.* (1981)	TEM Disp.	1.46	1.21→1.75
Rayment and Majumdar (1982)	EMPA	1.65→1.82	
Taylor and Newbury (1984*b*)	EMPA	1.72	
Lachowski and Mohan from Taylor and Newbury (1985)	TEM Disp.	1.7→1.8	
Groves *et al.* (1986)	TEM Ion Thin.	Op 1.72 Ip 1.66	1.25→2.07 1.27→1.78
Rodger (1989) specimen 26 years	TEM Ion Thin.	1.75	
Rodger (1989) specimen 3 months	TEM Ion Thin.	1.77	
Richardson and Groves (1992) specimen $3\frac{1}{2}$ years	TEM Ion Thin.	1.74	

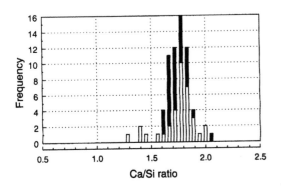
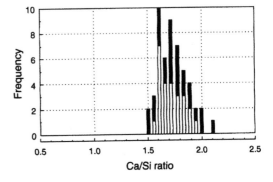

FIGURE 22.24 Ca/Si atom ratio frequency histograms for C-S-H in (a) C_3S and (b) Portland cement pastes hydrated for $3\frac{1}{2}$ years at 20°C (W/C=0.4). These TEM–EDX microanalyses are of Op (□) and Ip (■) C-S-H free of admixture with other phases. Data from Richardson and Groves (1993).

(2000) a value of 1.71 by TEM–EDX – and in neat Portland cement pastes.[62] In the latter case, small amounts of substituent ions are present most notably Al^{3+}. As examples, Ca/Si ratio frequency histograms of C-S-H are given in Figure 22.24 for C_3S (a) and Portland cement (b) pastes hydrated for $3\frac{1}{2}$ years (at 20°C, W/C = 0.4). The analyses are of Op (□) and Ip (■) C-S-H free of admixture with other phases.

22.10 Electron microscopy of cement

22.10.1 Reference microstructure

In 1959 J. H. Taplin presented a microstructural scheme in which the products of cement hydration were designated as either Op or Ip: outer products form in the originally water-filled spaces and inner products within the boundaries of the original clinker particles (Taplin 1959). Although there is not necessarily an *exact* correspondence between the positions of the outer boundaries of Ip and the original grains (Groves 1987), the scheme is straightforward and has been adopted widely. Its utility was challenged in 1993 by Diamond and Bonen (1993) who considered 'inner' and 'outer products' to be poorly suited as descriptive terms, despite their obvious physical meaning. They proposed a new classification based on the interpretation of BSE images of flat polished sections. This alternative classification has two primary morphological entities: 'phenograins' and 'groundmass'. The most important distinction is the gross porosity of the groundmass and the size of the phenograins. Phenograins are distinct grains greater than about 10 μm in size, embedded in the groundmass, which can be partially or fully hydrated clinker grains, or distinguishable crystals of CH. Smaller grains or crystals are part of the groundmass. The solid part of the groundmass, therefore, consists of the smaller hydrated clinker grains, C-S-H, CH and smaller amounts of AFm, AFt, and other phases. Whilst these phases are generally not well resolved by BEI they are readily observed by the much higher resolution technique of TEM. The utility of a classification system that is based solely on the interpretation of images with such modest resolution is limited. Its widespread adoption would be a backward step given the knowledge of spatial distributions and chemical compositions of phases from TEM studies. Taplin's scheme, on the other hand, is well supported for a whole range of cement systems by TEM evidence and by X-ray mapping of flat polished sections (using Mg as a chemical marker). (TEM: Dalgleish and Ibe 1981; Jennings *et al.* 1981, Groves *et al.* 1986; Groves 1987; Rodger and Groves 1988, 1989, Richardson *et al.* 1989, 1994; Richardson and Groves 1992a, 1993, 1997 Mg mapping: Taylor 1984; Taylor and Newbury 1984a, Richardson and Groves 1992a, 1993).

22.10.2 Morphology of C-S-H

The morphology of C-S-H is important because it largely determines the interconnectivity of the capillary pores. The change in morphology from 'foils' to 'fibres' that occurs on going from low to high Ca/Si was first described by Åke Grudemo in the earliest results of his classic TEM studies of calcium silicate hydrates (Grudemo 1954).[63] Shortly afterwards George Kalousek and Albert Prebus (1958) made a similar observation: lower

---

[62]The very high values for Op C-S-H reported by Rayment and Majumdar (1982), 2.7 and 2.6 for W/C of 0.3 and 0.6 respectively, that are often referred to in the literature (e.g. Viehland *et al.* 1996) are wrong. They were miscalculated: the values in the Ca/Si row of their Table 2 are not consistent with the Ca and Si rows; the Ca/Si should be around 2.3 and 2.1 (the values in the Ca/(Si + Al + S + Fe) row are consistent with the other rows).

[63]At low Ca/Si (1.03) 'the habit of the unit making up this aggregate is obviously that of thin, flexible foils;' SAED and X-ray diffraction show it to be CSH(I). Increasing the Ca/Si to 1.4–1.45 caused 'a more and more pronounced aggregation and denser packing of the foils.' C-S-H produced (by Hal Taylor) 'by shaking C_3S with water for several weeks until equilibrium was reached in a saturated lime solution, the final composition of the calcium silicate hydrate precipitate being approximately C_2SH_4... The morphologic appearance of these particles is substantially different from that of CSH(I). The aggregates may be described as bundles of fibres... [giving] electron diffraction diagrams which only show a diffuse line about 3.2–2.8 Å and a line at 1.82 Å.'

Ca/Si hydrates formed as crinkled foils, and higher Ca/Si (1.5) had a 'fibre habit'. Interestingly, they also observed that leaching of Ca^{2+} from a product with high initial Ca/Si ratio (1.75) led to a change from fibrillar to foil-like morphology (and ultimately to hydrated silica). A morphological classification for C-S-H formed in hydrating C_3S pastes – and so for high Ca/Si ratio – was introduced by Hamlin Jennings *et al.* based on observations of ion-thinned sections in the TEM (Jennings *et al.* 1981). It consists of Type E (or early) and Type O (or 'middle' product, formed between 4 and 24 h). Depending on the space available, Type O changes into Type 1 – slightly tapered needles radiating perpendicular from C_3S grains with an aspect ratio of about 10 (in open areas $>1\,\mu m$) – or Type 3 (where inter-particle spacing is $< 1\,\mu m$). Type 4 is a dense gelatinous inner product (Ip) that forms after 24 h, which they termed 'late' product. Groves *et al.* (1986), also using TEM of ion-thinned C_3S pastes, found it difficult to make

a sharp distinction between Jennings *et al.*'s Types 1 and 3 morphologies, and preferred instead to characterize all outer product (Op) C-S-H in hardened C_3S or Portland cement as 'fibrillar'. Since the morphology of the C-S-H is affected by the amount of available space, it seems most unlikely that there are just two distinct Op morphologies – such as Jennings *et al.*'s Types 1 and 3 – but rather fibrils possessing a range of aspect ratios dependent upon space constraints. Examples of fine fibrillar (high Ca/Si) and foil-like (lower Ca/Si) C-S-H from TEM of ion-thinned sections are shown in Figure 22.25. This change in morphology is also observable on SEI micrographs of fracture surfaces. Examples with different Ca/Si ratio are shown in Figure 22.26, but it also occurs with certain admixtures. This was first demonstrated by Berger *et al.* (1972) who reported results of TEM (replica) and SEI comparative studies (with some excellent micrographs) of the effects of 18 admixtures on the morphology of

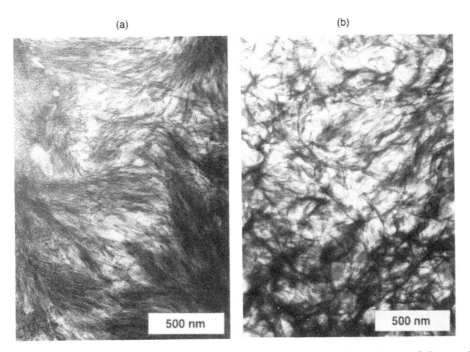

(a) (b)

FIGURE 22.25   Examples of (a) fine fibrillar (high Ca/Si) (modified from Richardson and Groves, 1992a) and (b) foil-like (lower Ca/Si) C-S-H from TEM of ion-thinned sections (modified from Richardson *et al.* 1989).

(a) (b)

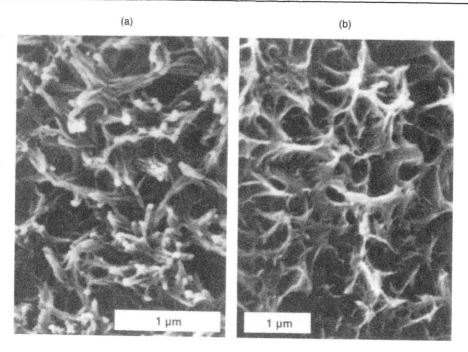

FIGURE 22.26 Examples of (a) fibrillar (Type I) and (b) foil-like (Type II) C-S-H from SEM–SEI of fracture surfaces (modified from Richardson *et al.* 1989).

C-S-H. They noted the two distinct morphologies and designated them Type I and Type II. Type I corresponds to the 'fibrillar' C-S-H as observed in TEM thin sections, and Type II to 'foil-like'. The paper includes a table outlining which admixtures produce Type I or II morphology, for example, no admixture (just water) produces Type I, and $CaCl_2$ produces Type II. Berger *et al.*'s classification was extended in 1976 by Sid Diamond (1976a)[64] to include Types III and IV, the latter being inner product.

[64]The extended classification has been referenced very widely but, despite Diamond referencing them in his paper, Berger *et al.* (1972) have received little credit, it usually being referred to as 'Diamond's,' for example, Ramachandran and Grutzeck (1993). Indeed, Diamond has himself probably contributed to this, for example, in his review paper at the 8th Congress he stated 'Approximately 10 years ago the present writer proposed a rough classification of C-S-H gel particles observable in Portland cement pastes into four morphological types' (Diamond 1986).

22.11 References

1831

Witham, H. T. M. (1831). *Observations on Fossil Vegetables, Accompanied by Representations of their Internal Structure, as seen through the Microscope.* William Blackwood, Edinburgh, and T. Cadell, London.

1858

Sorby, H. C. (1858). On the microscopical structure of crystals indicating the origin of minerals and rocks, *Q. J. Geol. Soc.*, **14**, 453.

1873

Abbe, E. (1873). Beiträge zur Theorie des Mikroskops und der mikroskopischen Wahrnehmung (A contribution to the theory of the microscope, and the nature of microscopical vision), *Archiv für mikroskopische Anatomie*, **9**, 413–68.

1882

Le Chatelier, H. (1882). Recherches expérimentales sur la constitution des ciments et la théorie de leur prise (Experimental researches on the constitution of cements and the theory of their setting), *C. R. - Acad. Sci. Paris*, **94**, 867–9.

1883

Le Chatelier, H. (1883). Application des phénomènes de sursaturation à la théorie du durcissement de quelques ciments et mastics, *C. R. - Acad. Sci. Paris*, **96**, 1056–9.

1886

Sorby, H. C. (1886). On the application of very high powers to the study of microscopical structure of steel, *J. Iron and Steel Inst.*, **1**, 140–4.

1887

Le Chatelier, H. (1887). Recherches experimentales sur la constitution des mortiers hydrauliques, *Ann. Mines, sér. 8*, **11**, 345–465.

Sorby, H. C. (1887). The microscopical structure of iron and steel, *J. Iron and Steel Inst.*, **1**, 255–88.

1891

Dallinger, W. H. (1891). Revised version of *The Microscope and its Revelations*, W. B. Carpenter, 7th Edition, J. & A. Churchill, London.

1893

Le Chatelier, H. (1893). Procédés d'essai des matériaux hydrauliques, *Ann. Mines, sér. 9*, **4**, 252–419.

1897

Stanger, W. H. and Blount, B. (1897). The adulteration of Portland cement, *J. Soc. Chem. Ind.*, **16**, 853–9.

Törnebohm, A. E. (1897). Die Petrographie des Portland-Cements, *Thonind.-Ztg.*, **21**(110), 1148–51.

1900

Spalding, F.P. (1900). *Hydraulic cement, its properties, testing, and use*. John Wiley & Sons, New York.

1904

Richardson, C. (1904). The constitution of Portland cement from a physico-chemical standpoint, *Eng. News*, **52**(6), 127–30.

1909

Ambronn, H. (1909). Ueber Umkristallisation und Gel-Bildung beim Erhärten des Zements (Crystallisation and gel formation in the hardening of cement), *Tonind.-Ztg.*, **33**(28), 270–2.

Kühl, H. (1909). The 'swelling theory' of Portland cement, *Tonind.-Ztg.*, **33**, 556–7.

Michaëlis, W. Sen. (1909). The setting of calcareous hydraulic cements, *Z. Chem. Ind. Kolloide*, **5**, 9–22.

Stern, E. (1909). Beiträge zur Kenntnis des Kleingefüges des Portlandzementes (Discussions on the microstructure of Portland cement), *Z. anorgan. Chem.* **68**, 160–7.

1910

Read, E. J. (1910). The crystalline products of the hydration of Portland cement, *J. Soc. Chem. Ind.*, **29**(12), 735–6.

1911

Desch, C. H. (1911). *The chemistry and testing of cement*. Edward Arnold, London.

Shepherd, E. S., Rankin, G. A. and Wright, F. E. (1911). Preliminary report on the ternary system $CaO-Al_2O_3-SiO_2$. A study of the constitution of Portland cement clinker, *J. Ind. Eng. Chem.*, **3**, 211–27.

1914

Klein, A. A. and Phillips, A. J. (1914). The hydration of Portland cement, *Trans. Am. Ceram. Soc.*, **16**, 313–41.

1919

Klein, A. A. (1919). The constitution and hydration of Portland cement, *Trans. Faraday Soc.*, **14**(1–2), 14–22.

Le Chatelier, H. (1919). Crystalloids against colloids in the theory of cements, *Trans. Faraday Soc.*, **14**(1–2), 8–11.

Rankin, G. A. (1919). The setting and hardening of Portland cement, *Trans. Faraday Soc.*, **14**(1–2), 23–8.

1920

Barnard, J. E. (1920). The present position and the future of the microscope – a general survey, *Trans. Faraday Soc.*, **16**(1), 37–42.

Hadfield, R. (1920). The great work of Sorby, *Trans. Faraday Soc.*, **16**(1), 114–18.

1924

De Broglie, L. (1924). A tentative theory of light quanta, *Philos. Mag.*, **47**, 446–58.

1926

Busch, H. (1926). Calculation of trajectory of cathode rays in axially symmetric electromagnetic fields, *Ann. Physik*, **81**, Ser. 4, 974–93.

1927

Davisson, C. and Germer, L. H. (1927). Diffraction of electrons by a crystal of nickel, *Phys. Rev.*, **30**, 705–40.

1934

Marton, L. L. (1934). Electron microscopy of biological objects, *Nature*, **133**, 911.

1936

Brown, L. S. and Carlson, R. W. (1936). Petrographic studies of hydrated cements, *Proc. Am. Soc. Testing Mater.*, **36**, 332–50.

1937

Martin, L. C., Whelpton, R. V. and Parnum, D. H. (1937). A new electron microscope, *J. Sci. Instr.*, **14**, 14–24.

1938

Von Ardenne, M. (1938*a*). Das Elektronen-Rastermikroskop – Theoretische Grundlagen (The scanning electron microscope – Theoretical fundamentals), *Z. Physik*, **109**, 553–72.

Von Ardenne, M. (1938*b*). Das Elektronen-Rastermikroskop – Praktische Ausführung Grundlagen (The scanning electron microscope – practical construction), *Z. technische Physik*, **19**, 407–16.

1939

Eitel, W., Müller, H. O. and Radczewski, O. E. (1939). Uebermikroskopische Untersuchungen an Tonmineralien (Examination of clay minerals with the ultramicroscope), *Berichte der Deutschen Keramischen Gesellschaft e.V.*, **20**(4), 165–78.

Radczewski, O. E., Müller, H. O. and Eitel, W. (1939*a*). Übermikroskopische Untersuchung der Hydratation des Kalkes (Supermicroscopical investigation of the hydration of lime), *Zement*, **28**, 693–8.

Radczewski, O. E., Müller, H. O. and Eitel, W. (1939*b*). Zur Hydratation des Trikalziumsilikats (The hydration of tricalcium silicate), *Die Naturwissenschaften*, **27**, 807.

Radczewski, O. E., Müller, H. O. and Eitel, W. (1939*c*). Zur Hydratation des Trikalziumaluminats (The hydration of tricalcium aluminate), *Die Naturwissenschaften*, **27**, 837–8.

Von Borries, B. and Ruska, E. (1939). Ein Übermikroskop für Forschungsinstitute (An electron microscope for research institutes), *Die Naturwissenschaften*, **27**, 577–82.

1940

Eitel, W. and Radczewski, O. E. (1940). Zur Kennzeichnung des Tonminerals Montmorillonit im Übermikroskopischen Bilde (The characteristics of the clay mineral montmorillonite as shown by the supermicroscope), *Die Naturwissenschaften*, **28**, 397–9.

O'Daniel, H. and Radczewski, O. E. (1940). Elektronen-Mikroskopie und -Beugung hochdisperser Mineralien an demselben Präparat (Electron microscopy and electron diffraction study on the same specimen of highly dispersed minerals), *Die Naturwissenschaften*, **28**, 628–30.

Zworykin, V. K., Hillier, J. and Vance, A. W. (1940). An electron microscope for practical laboratory service, *Trans. Am. Inst. Electr. Eng.*, **60**, 157–62.

1941

Eitel, W. (1941). Neuere Ergebnisse der Erforschung der Zemente (New results in the investigation of cement), *Angew. Chem.*, **54**(15–16), 185–92.

1942

Eitel, W. (1942). Elektronenmikroskopische Zementforschung, *Zement*, **31**, 489–97.

Zworykin, V. K., Hillier, J. and Snyder, R. L. (1942). A scanning electron microscope, *Bull. Am. Soc. Testing Mater.*, **117**, 15–23.

1943

Brownmiller, L. T. (1943). The microscopic structure of hydrated Portland cement, *Proc. Am. Concr. Inst.*, **39**, 193–210.

Sliepcevich, C. M., Gildart, L., Katz, D. L. (1943). Crystals from Portland cement hydration: an electron microscope study, *Ind. Eng. Chem.*, **35**(11), 1178–87.

1946

Powers, T. C. and Brownyard, T. L. (October-December 1946; January-April 1947), Studies of the physical properties of hardened cement paste, *J. Am. Concr. Inst. Proc.*, **43**, 101–32, 249–336, 469–504, 549–602, 669–712, 845–57, 865–80, 933–69, 971–92.

1947

Bogue, R. H., (1947). *The Chemistry of Portland Cement*, 1st Edition. Reinhold Publishing Corporation, New York.

Haine, M. E. (1947). The design and construction of a new electron microscope, *J. Inst. Electrical Engineers*, **94**, 447–62.

Le Poole, J. B. (1947). A new electron microscope with continuously variable magnification, *Philips Tech. Rev.*, **9**, 33–46.

McMurdie, H. F. (1947). Previously unpublished micrographs given in Bogue (1947).

1950

Boutet, D. (1950). Réflexions sur les bétons en général et sur les bétons de ciment en particulier. Etudes au microscope electronique de la prise du ciment, *Travaux (Paris)*, **183**, 1–11.

1954

3rd ISCC Proc. 3rd Int. Symp. Chem. Cem., London, 1952, Cement and Concrete Association, London.

Grudemo, Å. (1954). Discussion following the paper by J. D. Bernal on 'The structure of cement hydration compounds,' *3rd ISCC*, 247–53.

Moore, A. E. (1954). Further discussion following of Bernal's paper, *3rd ISCC*, 253–4.

Swerdlow, M. (1954). Electron microscopy, *Anal. Chem.*, **26**(1), 34–42.

1955

Bogue, R. H. (1955). *The Chemistry of Portland Cement*, 2nd Edition. Reinhold Publishing Corporation, New York.

Comer, J. J. and Turley, J. W. (1955). Replica studies of bulk clays, *J. Appl. Phys.*, 26(3), 346–50.

Gille, F. (1955). Zur Mikroskopie des Zements, *Zem.-Kalk-Gips*, 8, 128–38.

Grudemo, Å. (1955). An electronographic study of the morphology and crystallisation properties of calcium silicate hydrates, *Proc. – Swedish Cem. Concr. Res. Inst. Royal Inst. Technol., Stockholm (Handlingar – Svenska Forskningsinstitutet för Cement och Betong vid Kungliga Tekniska Högskolan i Stockholm)*, 26, 1–103.

Kalousek, G. (1955). Tobermorite and related phases in the system CaO–SiO₂–H₂O, *J. Am. Concr. Inst.*, 26(10), 989–1011.

Swerdlow, M. and Heckman, F. A. (1955). Previously unpublished micrographs given in Bogue (1955).

Van Bemst, A. (1955). Contribution à l'étude de l'hydratation des silicates de calcium purs, *Bull. Soc. Chimique de Belgique*, 64, 333–51.

1956

Gaze, R. and Robertson, R. H. S. (1956). Some observations on calcium silicate hydrate (I) – tobermorite, *Mag. Concr. Res.*, 8(22), 7–12

Talbot, J. H. (1956). Decomposition of CaSO₄·2H₂O in the electron microscope, *B. J. Appl. Phys.*, 7, 110–13.

1957

Bernard, P. (1957). Nouveau procédé pour l'étude au microscope électronique des pâtes de ciment susceptibles de durcissement, *Revue des Matér. Construction et de Travaux Publics Ciments et Betons*, 507, 351–8.

Gaze, R. and Robertson, R. H. S. (1957). Unbroken tobermorite crystals from hydrated cement, *Mag. Concr. Res.*, 9(25), 25–6.

Iwai, T. and Watanabe, K. (1957). Electronmicroscopic study on Portland cement hydration products – tricalciumaluminate, dicalciumferrite and tetracalciumaluminoferrite, *Proc. 1st Regional Conf. Electron Microscopy in Asia and Oceania*, Tokyo 1956, 288–91 and Plate LXXI.

Lukyanovich, V. M. and Leontiev, E. A. (1957). Replica techniques in electron microscope investigation of the structure of porous bodies, *Proc. 1st Regional Conf. on Electron Microscopy in Asia and Oceania*, Tokyo 1956, 331–8 and plates LXXXIII–LXXXIV.

Schimmel, G. (1957). Elektronenmikroskopische Untersuchungen an Calciumhydroxyd und Calciumcarbonat (Investigation of calcium hydroxide and calcium carbonate with the aid of the electron microscope), *Zem.-Kalk-Gips*, 10(4), 134–8.

1958

Czernin, W. (1958). Zur Elektronenmikroskopie erhärtender Zementpasten, *Zem.-Kalk-Gips*, 11(9), 381–3.

Kalousek, G. L. and Prebus, A. F. (1958). Crystal chemistry of hydrous calcium silicates: III, Morphology and other properties of tobermorite and related phases, *J. Am. Ceram. Soc.*, 41, 124–32.

1959

Buckle, E. R. and Taylor, H. F. W. (1959). The hydration of tricalcium and β-dicalcium silicates in pastes under normal and steam curing conditions, *J. Appl. Chem.*, 9, 163–72.

Czernin, W. (1959). Über die physikalische Beschaffenheit des erhärteten Zementes (About the physical conditions of hardened cement), *Betonst.-Ztg.*, 25, 376–82.

Gard, J. A., Howison, J. W. and Taylor, H. F. W. (1959). Synthetic compounds related to tobermorite: an electron-microscope, X-ray, and dehydration study, *Mag. Concr. Res.*, 11(33), 151–8.

Gille, F., Czernin, W., Danielsson, U. and Grasenick, F. (1959). Elektronenmikroskopische Untersuchungen an hydratisierten Zementen, *Zement und Beton*, 16, 21–4.

Saji, K. (1959). Zur Elektronenmikroskopie erhärtender Zementpasten, *Zem.-Kalk-Gips*, 12(9), 418–23.

Stork, J. and Bystricky, V. (1959a). Etude à l'aide du microscope électronique de la structure de la pâte de ciment durcie, *Revue des Matériaux de Construction et de Travaux Publics Ciments et Betons*, 526–7, 173.

Stork, J. and Bystricky, V. (1959b). Struktura zatvrdnutej cementovej kase zistovana pomocou elektronoveho mikroskopu, *Silikaty (Praha)*, 3(4), 321–6.

Taplin, J. H. (1959). A method for following the hydration reaction in Portland cement paste, *Austr. J. Appl. Sci.*, 10, 329–45.

1960

Czernin, W. (1960). Elektronenmikroskopische Untersuchungen an erhärtendem Zement, *Schweizer Archiv für Angew. Wissenschaft und Technik*, 26, 185–91.

Kurczyk, H.-G. and Schwiete, H. E. (1960). Electronenmikroskopische und thermochemische Untersuchungen über die Hydratation der Calciumsilikate 3CaO·SiO₂ und β-2CaO·SiO₂ und den Einfluß von Calciumchlorid und Gips auf den Hydratationsvorgang, *Tonind.-Ztg. und Keramische Rundschau*, 84, 585–98.

1961

Taylor, H. F. W. (1961). The chemistry of cement hydration, *Prog. Ceram. Sci.*, 1, 89–145.

Uchikawa, H. and Takagi, S. (1961). Elektonenmikroskopische Untersuchungen vollständig hydratisierter Klinkerkomponenten, *Zem.-Kalk-Gips*, 14(4), 153–8.

1962

4th ISCC Proc. 4th Int. Symp. Chem. Cem., Washington, 1960, US Department of Commerce, Washington DC (1962).

Brunauer, S. (1962). Tobermorite gel – The heart of concrete, *Am. Scientist*, 50, 210–29.

Brunauer, S. and Greenberg, S. A. (1962). The hydration of tricalcium silicate and β-dicalcium silicate at room temperature, *4th ISCC*, 1, 135–64.

Chatterji, S. and Jeffery, J. W. (1962). Studies of early stages of paste hydration of cement compounds, I, *J. Am. Ceram. Soc.*, 45(11), 536–43.

Copeland, L. E. and Schulz, E. G. (1962a). Discussion of 'The microstructure of hardened cement paste,' by Å. Grudemo, *4th ISCC*, 2, 648.

Copeland, L. E. and Schulz, E. G. (1962b). Electron optical investigation of the hydration products of calcium silicates and Portland cement, *J. Portland Cem. Assoc. Res. Devel. Lab.*, 4(1), 2–12.

Czernin, W. (1962). A few unsolved problems of cement hydration, *4th ISCC*, 2, 725–9.

Funk, H. (1962). Two different ways of hydration in the reaction of β-Ca_2SiO_4 with water at 25°C to 120°C, *4th ISCC*, 1, 291–5.

Grudemo, Å. (1962). The microstructure of hardened cement paste, *4th ISCC*, 2, 615–47.

Kurczyk, H. G. and Schwiete, H. E. (1962). Concerning the hydration products of C_3S and β-C_2S, *4th ISCC*, 1, 349–58.

Mielenz, R. C. (1962). Petrography applied to Portland-cement concrete, *Rev. Eng. Geol.* 1, 1–38.

Thornley, R. F. M. and Cartz, L. (1962). Direct examination of ceramic surfaces with the scanning electron microscope, *J. Am. Ceram. Soc.*, 45(9), 425–8.

Young, J. F. (1962). Hydration of tricalcium aluminate with lignosulphonate additives, *Mag. Concr. Res.*, 14(42), 137–42.

1963

Chatterji, S. and Jeffery, J. W. (1963). Studies of early stages of paste hydration of cement compounds, II, *J. Am. Ceram. Soc.* 46(4), 187–91.

1964

Chatterji, S. and Jeffery, J. W. (1964). The effect of various heat treatments of the clinker on the early hydration of cement pastes, *Mag. Concr. Res.*, 16(46), 3–10.

Gard, J. A. (1964). Electron microscopy and diffraction, *In The Chemistry of Cements*. H. F. W. Taylor (Ed.) 2, 243–70.

Terrier, P. and Moreau, M. (1964). Examens au microscope de pâtes de ciment Portland, *Rev. Matér. Construction et de Travaux Publics Ciments et Betons*, 584, 129–37.

1965

Grudemo, Å (1965). The microstructures of cement gel phases, *Trans. Roy. Inst. Technol., Stockholm, Sweden (Kungliga Tekniska Högskolans Handlingar)*, 242, 1–287.

Grzymek, J. (1965). The influence of syngenite and Friedel's salt on the hardening of cement mortars and concretes. *Proc. 7th Conf. Silic. Ind., Budapest 1963*, 265–77.

Moore, A. E. (1965). Examination of a Portland cement clinker by electron probe microanalysis, *Silicates Ind.*, 30(8), 445–50.

Oatley, C. W., Nixon, W. C. and Pease, R. F. W. (1965). Scanning electron microscopy, *Adv. Electronics and Electron Phys.*, 21, 181–247.

Schwiete, H.-E., Ludwig, U. and Niel, E. (1965). Studies on the beginning of hydration of clinker and cement, *Proc. 7th Conf. Silicate Ind., Budapest 1963*, 221–33.

Soo-Lee, T. and Gon-Chen, L. (1965). The formation of 'tobermorite-like' calcium silicate hydrates, *Proc. 7th Conf. Silic. Ind., Budapest 1963*, 293–310.

1966

Amicarelli, V., Sersale, R. and Rebuffat, P. (1966). Microanalisi del clinker di portland mediante sonda eletronica (Electron probe microanalysis of portland cement clinker), *La Ric. Sci.*, 36(7), 567–72.

Chatterji, S. and Jeffery, J. W. (1966). Three dimensional arrangement of hydration products in set cement paste, *Nature*, 209(5029), 1233–4.

Eitel, W. (1966). Silicate Science, V. *Ceramics and Hydraulic Binders*. Academic Press, New York and London.

Heinrich, K. F. J. (1966). *Proc. 4th Int. Conf. X-ray optics and microanalysis*, Ed. R. Castaing, P. Deschamps, J. Philibert, Hermann, Paris, 159.

Schwiete, H. E., Ludwig, U. and Jäger, P. (1966). Investigations in the system 3CaO · Al_2O_3–$CaSO_4$–CaO–H_2O, Symposium in Honor of Dr T. C. Powers on Structure and Performance of Portland Cement Paste and Concrete, 44th Annual Meeting of the Highway Research Board, 1965. *Highway Res. Board Special Report* Number 90, 353–67.

1967

Copeland, L. E., Bodor, E., Chang, T. N. and Weise, C. H. (1967). Reactions of tobermorite gel with aluminates, ferrites, and sulfates, *J. Portland Cem. Assoc. Res. Devel. Lab.*, 9(1), 61–74.

Peterson, O. (1967). Untersuchung von Portland-Klinker mit der Mikrosonde (Investigation of Portland cement clinker with the microprobe), *Zem.-Kalk-Gips*, 20(2), 61–4.

Shpynova, L. G., Nabitovich, I. D. and Belov, N. V. (1967). Microstructure of alite cement stone (hydrated tricalcium silicate), *Soviet Phys. Crystallogr.*, (English translation of *Krystallografiya*), **11**(6), 747–51.

1968

Fletcher, K. E. (1968). The analysis of belite in Portland cement clinker by means of an electron-probe microanalyser, *Mag. Concr. Res.*, **20**(64), 167–70.

Midgley, H. G. (1968). The composition of alite (tricalcium silicate) in a Portland cement clinker, *Mag. Concr. Res.*, **20**(62), 41–4.

Sierra, R. (1968). Étude au microscope électronique de l'hydratation des silicates calciques du ciment Portland, *J. Microscopie*, **7**, 491–508.

Stork, J. and Michálková, B. (1968). An electron-microscope study of the contact between aggregate and hardened cement paste, *Mag. Concr. Res.*, **20**(63), 67–76.

1969

5th ISCC = Proc. 5th Int. Symp. Chem. Cem., Tokyo, 1968, Cement and Concrete Association of Japan, Tokyo (1969).

Copeland, L. E. and Kantro, D. L. (1969). Hydration of Portland cement, *5th ISCC*, **2**, 387–420.

De Keyser, W. L. and Tenoutasse, N. (1969). The hydration of the ferrite phase of cements, *5th ISCC*, **2**, 379–86.

Dyczek, J. R. (1969). Written discussion on 'Crystal structures and properties of cement hydration properties (calcium silicate hydrates)' by H. F. W. Taylor, *5th ISCC*, **2**, 27–30.

Fletcher, K. E. (1969). The composition of the tricalcium aluminate and ferrite phases in Portland cement clinkers by use of an electron-probe microanalyser, *Mag. Concr. Res.*, **21**(66), 3–14.

Fujii, K. and Kondo, W. (1969). Hydration of tricalcium silicate in a very early stage, *5th ISCC*, **2**, 362–71.

Grutzeck, M. W. and Roy, D. M. (1969). Electron microprobe studies of the hydration of $3CaO \cdot SiO_2$, *Nature*, **223**, 492–4.

Kondo, R. and Ueda, S. (1969). Kinetics and mechanisms of hydration of cements, *5th ISCC*, **2**, 203–248.

Marchese, B. and Sersale, R. (1969). Stability of hydrogarnet series terms to sulphate attack, *5th ISCC*, **2**, 133–7.

Midgley, H. G. (1969a). The microstructure of hydrated Portland cement, *Proc. Br. Ceram. Soc.*, **13**, 89–102.

Midgley, H. G. (1969b). The minor elements in alite (tricalcium silicate) and belite (dicalcium silicate) from some Portland cement clinkers as determined by elec-. tron probe X-ray microanalysis, *5th ISCC*, **1**, 226–33.

Richartz, W. (1969). Electron microscopic investigations about the relations between structure and strength of hardened cement, *5th ISCC*, **3**, 119–28.

Roy, D. M. and Grutzeck, M. W. (1969). Electron microprobe studies of cement phases, *5th ISCC*, **2**, 301–10.

Taylor, H. F. W. (1969). Crystal structures and properties of cement hydration properties (calcium silicate hydrates), *5th ISCC*, **2**, 1–26.

Terrier, P. (1969). Contribution of analysis by means of an electron microprobe to the cement chemistry, *5th ISCC*, **2**, 278–87.

Williamson, R. B. (1969). Portland cement: Pseudomorphs of original cement grains observed in hardened pastes, *Science*, **164**(3879), 549–51.

Young, J. F. (1969). The influence of sugars on the hydration of tricalcium aluminate, *5th ISCC*, **2**, 256–67.

1970

Berger, R. L., Cahn, D. S. and McGregor, J. D. (1970). Calcium hydroxide as a binder in portland cement paste, *J. Am. Ceram. Soc.*, **53**(1), 57–8.

Ciach, T. D., Gillott, J. E., Swenson, E. G. and Sereda, P. J. (1970). Microstructure of hydrated Portland cement pastes, *Nature*, **227**, 1045–6.

Diamond, S. (1970). Application of scanning electron microscopy to the study of hydrated cement, *Scanning Electron Microscopy 1970; Proc. 3rd Ann. Scanning Electron Microscope Symp.*, 385–92.

Lea, F. M. (1970). *The chemistry of Cement and Concrete*, 3rd edn. Edward Arnold (Publishers) Ltd.

Marcinkowski, M. J. and Taylor, M. E. Jr. (1970). Preliminary scanning electron microscopy study of concrete fracture surfaces, *J. Appl. Phys.*, **41**, 4753–4.

1971

Ciach, T. D., Gillott, J. E., Swenson, E. G. and Sereda, P. J. (1971). Microstructure of calcium silicate hydrates, *Cem. Concr. Res.*, **1**(1), 13–25.

Ciach, T. D. and Swenson, E. G. (1971a). Morphology and microstructure of hydrating Portland cement and its constituents. I, Changes in hydration of tricalcium aluminate alone and in the presence of triethanolamine or calcium lignosulphonate, *Cem. Concr. Res.*, **1**(2), 143–58.

Ciach, T. D. and Swenson, E. G. (1971b). Morphology and microstructure of hydrating Portland cement and its constituents. II, Changes in hydration of calcium silicates alone and in the presence of triethanolamine and calcium lignosulphonate, both with and without gypsum, *Cem. Concr. Res.*, **1**(2), 159–76.

Ciach, T. D. and Swenson, E. G. (1971c). Morphology and microstructure of hydrating Portland cement and its constituents. III, Changes in the hydration of a mixture of C_3S, C_3A and gypsum with and without triethanolamine and calcium lignosulphonate present, *Cem. Concr. Res.*, **1**(3), 257–71.

Ciach, T. D. and Swenson, E. G. (1971*d*). Morphology and microstructure of hydrating Portland cement and its constituents. IV, Changes in hydration of a C_3S, C_2S, C_3A, C_4AF and gypsum paste with and without the admixtures triethanolamine and calcium lignosulphonate, *Cem. Concr. Res.*, 1(4), 367–83.

Ciach, T. D. and Swenson, E. G. (1971*e*). Morphology and microstructure of hydrating Portland cement and its constituents. V, Changes in hydration of Portland cement with and without the presence of triethanolamine and calcium lignosulphonate, *Cem. Concr. Res.*, 1(5), 515–30.

Colville, A. A. and Geller, S., The crystal structure of brownmillerite, Ca_2FeAlO_5, *Acta Crystallogr.*, B27 (1971) 2311–5.

Cottin, B. (1971). Etude au microscope electronique de pates de ciment alumineux hydratees en C_2AH_8 et en CAH_{10}, *Cem. Concr. Res.*, 1(2), 177–86.

Hale, K. F. and Henderson Brown, M. (1971). Application of high voltage electron microscopy to the study of cement, *Proc. Southampton 1969 Civil Eng. Mater. Conf.*, 1, 289–93.

Midgley, H. G. (1971). Electron microscopy of set Portland cement, *Proc. Southampton 1969 Civil Eng. Mater. Conf.*, 1, 275–87.

Midgley, H. G. and Bennett, M. (1971). A microprobe analysis of larnite and bredigite from Scawt Hill, Larne, Northern Ireland, *Cem. Concr. Res.*, 1(4), 413–8.

Midgley, H. G. and Pettifer, K. (1971). The microstructure of hydrated super sulphated cement, *Cem. Concr. Res.*, 1(1), 101–4.

Mills, R. H. (1971). Collapse of structure and creep in concrete, *Proc. Southampton 1969 Civil Eng. Mater. Conf.*, 1, 751–68.

1972

Berger, R. L. (1972). Calcium hydroxide: Its role in the fracture of tricalcium silicate paste, *Science*, 175, 626–9.

Berger, R. L., Young, J. F. and Lawrence, F. V. (1972). Discussion of the paper 'Morphology and surface properties of hydrated tricalcium silicate paste' by M. Collepardi and B. Marchese, *Cem. Concr. Res.*, 2(5), 633–6.

Collepardi, M. and Marchese, B. (1972). Morphology and surface properties of hydrated tricalcium silicate pastes, *Cem. Concr. Res.*, 2(1), 57–65.

Diamond, S. (1972). Identification of hydrated cement constituents using a scanning electron microscope – energy dispersive X-ray spectrometer combination, *Cem. Concr. Res.*, 2(5), 617–32.

Dron, R. and Petit, P. (1972). Observation au microscope électronique à balayage de l'hydratation initiale en milieu basique du laitier granulé de haut fourneau, *C. R. - Acad. Sci. Paris Série C*, 274, 1275–7.

Farran, J., Javelas, R., Maso, J. C. and Perrin, B. (1972). Existence d'une auréole de transition entre les granulats d'un mortier, ou d'un béton, et la masse de la pâte de ciment hydraté. Conséquences sur les propriétés mécaniques, *C. R. Acad. Sci. Paris Série D*, 275, 1467–8.

Majumdar, A. J. (1972). The application of scanning electron microscopy to textural studies, *Proc. Br. Ceram. Soc.*, 20, 43–69.

Reuter, W. (1972). The ionization function and its application to the electron probe analysis of thin films, *Proc. 6th Int. Conf. on X-ray optics and microanalysis*, G. Shinoda, K. Kohra, T. Ichinokawa, Ed. University of Tokyo Press, 121.

Williamson, R. B. and Tewari, R. P. (1972). Effects of microstructure on deformation and fracture of Portland cement paste, *Proc. 5th Int. Mater. Symp.*, University of California, 1971, 1223–33.

Williamson, R. B. (1972). Solidification of Portland cement, *Prog. Mater. Sci.*, 15(3), 189–286.

1973

Double, D. D. (1973). Some studies of the hydration of portland cement using high voltage (1 MeV) electron microscopy, *Mater. Sci. Eng.*, 12, 29–34.

Lawrence, F. V. Jnr. and Young, J. F. (1973). Studies on the hydration of tricalcium silicate pastes, I. Scanning electron microscopic examination of microstructural features, *Cem. Concr. Res.*, 3(2), 149–61.

Mehta, P. K. and Williamson, R. B. (1973). Hydration characteristics and properties of alite pastes containing C_3A, C_4AF and gypsum, *Proc. 11th Conf. Silic. Ind., Budapest 1973*, 1, 401–14.

Mitsuda, T. (1973). Paragenesis of 11 Å tobermorite and poorly crystalline hydrated magnesium silicate, *Cem. Concr. Res.*, 3(1), 71–80.

Morgan, D. R. (1973). Discussion of the paper 'Studies on the hydration of tricalcium silicate pastes I. Scanning electron microscope examination of microstructural features,' by F. V. Lawrence Jnr. and J. F. Young, *Cem. Concr. Res.*, 3(6), 847–9. Micrographs from D. R. Morgan, Effects of admixtures on sorption and creep properties of cement pastes, *MSc Thesis*, University of Calgary, 1969.

Sauman, Z. (1973). Significance and character of the glass phase of power station fly ashes, *Proc. 11th Conf. Silic. Ind., Budapest 1973*, 1, 461–74.

Young, J. F., Berger, R. L. and Lawrence, F. V. Jr. (1973). Studies on the hydration of tricalcium silicate pastes III. Influence of admixtures on hydration and strength development, *Cem. Concr. Res.*, 3(6), 689–700.

1974

Diamond, S., Young, J. F. and Lawrence, F. V. Jnr. (1974). Scanning electron microscopy – energy dispersive X-ray analysis of cement constituents – some cautions, *Cem. Concr. Res.*, 4(6), 899–914.

Gabor, D. (1974). The history of the electron microscope, from ideas to achievements, *Proc. 8th Int. Cong. on Electron Microscopy, Canberra 1974*, 1, 6–12.

Javelas, R., Maso, J. C. and Ollivier, J. P. (1974). Realisation de lames ultra-mince de mortier pour observation directe au microscope electronique par transmission (Realization of ultra-thin slides of mortar for direct observation in the TEM), *Cem. Concr. Res.*, 4, 167–76.

Walsh, D., Otooni, M. A., Taylor, M. E., Jnr. and Marcinkowski, M. J. (1974). Study of Portland cement fracture surfaces by scanning electron microscopy techniques, *J. Mater. Sci.*, 9, 423–9.

1975

Cliff, G. and Lorimer, G. W. (1975). The quantitative analysis of thin specimens, *J. Microsc.*, 103(2), 203–7.

Javelas, R., Maso, J. C., Ollivier, J. P. and Thenoz, B. (1975). Observation directe au microscope electronique par transmission de la liason pâte de cimentgranulat dans des mortier de calcite et de quartz (Direct observation in the TEM of the bonding between cement paste and aggregate in mortars of calcite and quartz), *Cem. Concr. Res.*, 5, 285–94.

Mitsuda, T. and Taylor, H. F. W. (1975). Influence of aluminium on the conversion of calcium silicate hydrate gels into 11Å tobermorite at 90°C and 120°C, *Cem. Concr. Res.*, 5(3), 203–10.

1976

Diamond, S. (1976a). Cement paste microstructure – an overview at several levels, *Proc. Conf. on Hydraulic Cement Pastes: their Structure and Performance, Sheffield, UK*, 2–30.

Diamond, S. (1976b). C/S mole ratio of C-S-H gel in a mature C_3S paste as determined by EDXA, *Cem. Concr. Res.*, 6(3), 413–16.

Double, D. D. and Hellawell, A. (1976). The hydration of Portland cement, *Nature*, 261, 486–8.

Gard, J. A. and Taylor, H. F. W. (1976). Calcium silicate hydrate (II), *Cem. Concr. Res.*, 6(5), 667–8.

Goto, S., Daimon, M., Hosaka, G. and Kondo, R. (1976). Composition and morphology of hydrated tricalcium silicate, *J. Am. Ceram. Soc.*, 59(7–8), 281–4.

Mehta, P. K. (1976). Scanning electron micrographic studies of ettringite formation, *Cem. Concr. Res.*, 6(2), 169–82.

Stucke, M. S. and Majumdar, A. J. (1976). The composition of the gel phase in Portland cement paste, *Proc. Conf. Hydraulic Cement Pastes: their Structure and Performance, Sheffield, UK*, 31–51.

1977

Hirsch, P. B., Howie, A., Nicholson, R. B., Pashley, D. W. and Whelan, M. J. (1977). Electron microscopy of thin crystals, Krieger Publishing Company, Malabar.

Stucke, M. S. and Majumdar, A. J. (1977). The morphology and composition of an immature C_3S paste, *Cem. Concr. Res.*, 7(6), 711–8.

1978

Barnes, B. D., Diamond, S. and Dolch, W. L. (1978). Hollow shell hydration of cement particles in bulk cement paste, *Cem. Concr. Res.*, 8(3), 263–72.

Birchall, J. D., Howard, A. J. and Bailey, J. E. (1978). On the hydration of Portland cement, *Proc. Roy. Soc. London*, A360, 445–54.

Dent-Glasser, L. S., Lachowski, E. E., Mohan, K. and Taylor, H. F. W. (1978). A multi-method study of C_3S hydration, *Cem. Concr. Res.*, 8, 733–40.

Double, D. D. (1978). An examination of the hydration of Portland cement by electron microscopy, *Silicates Ind.*, 11, 233–46.

Double, D. D., Hellawell, A. and Perry, S. J. (1978). The hydration of Portland cement, *Proc. Roy. Soc. (London)*, A 359, 435–51.

Grattan-Bellew, P. E., Quinn, E. G. and Sereda, P. J. (1978). Reliability of scanning electron microscopy information, *Cem. Concr. Res.*, 8(3), 333–42.

1979

Bailey, J. E. and Chescoe, D. (1979). Microstructure development during the hydration of Portland cement, *Proc. Br. Ceram. Soc.*, 28, 165–77.

Courtault, B. and Briand, J. P. (1979). Thermal activation of cements, *Ciments Betons Platres Chaux*, 2(717), 86–98.

Danilatos, G. D. and Robinson, V. N. E. (1979) Principles of scanning electron microscopy at high specimen pressures, *Scanning*, 2, 72–82.

Garrett, G. G., Jennings, H. M. and Tait, R. B. (1979). Fatigue hardening behaviour of cement-based materials, *J. Mater. Sci.*, 14(2), 296–306.

Jennings, H. M. and Pratt, P. L. (1979a). An experimental argument for the existence of a protective membrane surrounding Portland cement during the induction period, *Cem. Concr. Res.*, 9(4), 501–506

Jennings, H. M. and Pratt, P. L. (1979b). On the hydration of Portland cement, *Proc. Br. Ceram. Soc.*, 28, 179–93.

Ménétrier, D., Jawed, I., Sun, T. S. and Skalny, J. (1979). ESCA and SEM studies on early C_3S hydration, *Cem. Concr. Res.*, 9(4), 473–82.

Taylor, H. F. W. (1979). Mineralogy, microstructure and mechanical properties of cements, *Proc. Br. Ceram. Soc.*, 28, 147–63.

1980

7th ICCC Proceedings of the 7th International Congress on the Chemistry of Cement, Paris, 1980, Editions Septima, Paris (1980).

Dalgleish, B. J., Pratt, P. L. and Moss, R. I. (1980). Preparation techniques and the microscopical examination of portland cement paste and C_3S, *Cem. Concr. Res.*, 10, 665–75.

Diamond, S. and Lachowski, E. E. (1980). On the morphology of Type III C-S-H gel, *Cem. Concr. Res.*, 10(5), 703–5.

Diamond, S., Ravina, D. and Lovell, J. (1980). The occurrence of duplex films on flyash surfaces, *Cem. Concr. Res.*, 10(2), 297–300.

Gard, J. A., Mohan, K., Taylor, H. F. W. and Cliff, G. (1980). Analytical electron microscopy of cement pastes: I, Tricalcium silicate pastes, *J. the Am. Ceram. Soc.*, 63, 336–7.

Ghose, A., Jennings, H. M., Pratt, P.L. and Barnes, P. (1980). Fibrous growth products in the hydration of Portland cement and related systems, *7th ICCC*, 4, 599–601.

Jawed, I., Ménétrier, D. and Skalny, J. (1980). Early hydration of calcium silicates: surface phenomena, *7th ICCC*, 2, II-182 to II-187.

Jelenic, I., Bezjak, A. and Marinkovic, V. (1980). Planar interfaces in Cr_2O_3-doped calcium aluminoferrite, *J. Mater. Sci., (Lett.)*, 15, 1051–4.

Jelenic, I., Panovic, A. and Bezjak, A. (1980). Hydration and strength development in alite-C_3A-$CSbarH_2$-quartz pastes containing readily soluble alkalies, *Cem. Concr. Res.*, 10(3), 463–6.

Jennings, H. M. and Pratt, P. L. (1980). The use of a high-voltage electron microscope and gas reaction cell for the microstructural investigation of wet Portland cement, *J. Mater. Sci. (Lett.)*, 15, 250–3.

Lachowski, E. E., Mohan, K., Taylor, H. F. W. and Moore, A. E. (1980). Analytical electron microscopy of cement pastes: II, Pastes of Portland cements and clinkers, *J. Am. Ceram. Soc.* 63(7–8), 447–52.

Ménétrier, D., Jawed, I. and Skalny, J. (1980). Effect of gypsum on C_3S hydration, *Cem. Concr. Res.*, 10(5), 697–701.

Ménétrier, D., McNamara, D. K., Jawed, I. and Skalny, J. (1980). Early hydration of β-C_2S: Surface morphology, *Cem. Concr. Res.*, 10(1), 107–10.

Ogawa, K., Uchikawa, H., Takemoto, K. and Yasui, I. (1980). The mechanism of the hydration in the system C_3S-pozzolana, *Cem. Concr. Res.*, 10(5), 683–96.

Rayment, D. L. and Majumdar, A. J. (1980). The composition of the CSH phase(s) in hydrated C_3S paste, *7th ICCC*, 2, II-64 to II-70.

1981

Dalgleish, B. J., Ghose, A., Jennings, H. M. and Pratt, P. L. (1981). The early reaction between Portland cement and water, *Proc. 11th Int. Conf. Sci. Ceram.*, 2, 297–302.

Dalgleish, B. J. and Ibe, K. (1981). Thin-foil studies of hydrated Portland cement, *Cem. Concr. Res.*, 11, 729–39.

Danilatos, G. D. (1981). Design and construction of an atmospheric or environmental SEM (part 1), *Scanning*, 4, 9–20.

Diamond, S. (1981). Some illustrations of the utility of scanning electron microscopy in cement research, *Proc. 3rd Int. Conf. Cem. Microsc.*, March 16–19, 1981, Houston, USA, Ed. G. R. Gouda, 238–56.

Groves, G. W. (1981a). Microcrystalline calcium hydroxide in Portland cement pastes of low water/cement ratio, *Cem. Concr. Res.*, 11, 713–18.

Groves, G. W. (1981b). Portland cement clinker viewed by transmission electron microscopy, *J. Mater. Sci.*, 16, 1063–70.

Jennings, H. M., Dalgleish, B. J. and Pratt, P. L. (1981). Morphological development of hydrating tricalcium silicate as examined by electron microscopy techniques, *J. Am. Ceram. Soc.*, 64(10), 567–72.

Lachowski, E. E., Mohan, K., Taylor, H. F. W., Lawrence, C. D. and Moore, A. E. (1981). Analytical electron microscopy of cement pastes: III, Pastes hydrated for long times, *J. Am. Ceram. Soc.*, 64(6), 319–21.

Mohan, K. and Taylor, H. F. W. (1981). Analytical electron microscopy of cement pastes: IV, β-dicalcium silicate pastes, *J. Am. Ceram. Soc.*, 64(12), 717–19.

1982

Dalgleish, B. J., Pratt, P. L. and Toulson, E. (1982). Fractographic studies of microstructural development in hydrated Portland cement, *J. Mater. Sci.*, 17, 2199–207.

Groves, G. W. (1982). Portland cement clinker viewed by transmission electron microscopy, *Cem. Concr. Res.*, 12, 619–25.

Jawed, I. and Skalny, J. (1982). Surface phenomena during tricalcium silicate hydration, *J. Colloid Interface Sci.*, 85(1), 235–43.

Rayment, P. L. (1982). The effect of pulverised-fuel ash on the C/S molar ratio and alkali content of calcium silicate hydrates in cement, *Cem. Concr. Res.*, 12(2), 133–40.

Rayment, D. L. and Majumdar, A. J. (1982). The composition of the C-S-H phases in Portland cement pastes, *Cem. Concr. Res.*, 12(6), 753–64.

1983

Bensted, J. (1983). Chemical aspects of normal setting of Portland cement, *Eng. Found. Conf. Cem. Concr. Prediction and Characterization*, Henniker, 69–86.

Danilatos, G. D. and Postle, R. (1983). Design and construction of an atmospheric or environmental SEM (Part 2), *Micron*, 14, 41–52.

Groves, G. W. (1983). TEM studies of cement clinker compounds, their hydration and strong cement pastes, *Philos. Trans. Roy. Soc. London*, A 310, 79–83.

Ings, J. B., Brown, P. W. and Frohnsdorff, G. (1983). Early hydration of large single crystals of tricalcium silicate, *Cem. Concr. Res.*, 13(6), 843–48.

Lachowski, E. E. and Diamond, S. (1983). Investigation of the composition and morphology of individual particles of Portland cement paste: 1, C-S-H gel and calcium hydroxide, *Cem. Concr. Res.*, 13(2), 177–85.

Pratt, P. L. and Ghose, A. (1983). Electron microscope studies of Portland cement microstructures during setting and hardening, *Philos. Trans. Roy. Soc. London*, A 310, 93–102.

Stewart, H. R. and Bailey, J. E. (1983). Microstructural studies of the hydration products of three tricalcium silicate polymorphs, *J. Mater. Sci.*, 18, 3686–94.

1984

Bailey, J. E. and Stewart, H. R. (1984). C_3S hydration products viewed using a cryo stage in SEM, *J. Mater. Sci. Lett.*, 3, 411–14.

Birchall, J. D. and Thomas, N. L. (1984). The mechanism of retardation of setting of OPC by sugars, *Proc. Br. Ceram. Soc.*, 35, 305–15.

Halse, Y., Goult, D. J. and Pratt, P. L. (1984). Calorimetry and Microscopy of flyash and silica fume cements blends, *Proc. Br. Ceram. Soc.*, 35, 403–17.

Halse, Y., Pratt, P. L., Dalziel, J. A. and Gutteridge, W. A. (1984). Development of microstructure and other properties in flyash OPC systems, *Cem. Concr. Res.*, 14(4), 491–8.

Le Sueur, P. J., Double, D. D. and Groves, G. W. (1984). Chemical and morphological studies of the hydration of tricalcium silicate, *Proc. Br. Ceram. Soc.*, 35, 177–91.

Rahman, A. A. (1984). Characterization of the porosity of hydrated cement pastes, *Proc. Br. Ceram. Soc.*, 35, 249–66.

Rayment, D. L. and Lachowski, E. E. (1984). The analysis of OPC pastes: A comparison between analytical electron microscopy and electron probe microanalysis, *Cem. Concr. Res.*, 14(1), 43–8.

Reimer, L. (1984). Transmission electron microscopy, Springer-Verlag, Berlin.

Sarkar, S. L. and Roy, D. M. (1984). Important experimental prerequisites for quantitative clinker analysis by SEM-EDS, *Cem. Concr. Res.*, 14(1), 83–92.

Scrivener, K. L. and Pratt, P. L. (1984). Microstructural studies of the hydration of C_3A and C_4AF independently and in cement paste, *Proc. Br. Ceram. Soc.*, 35, 207–19.

Sinclair, W. and Groves, G. W. (1984). The structure of alites, *Proc. Br. Ceram. Soc.*, 35, 93–100.

Taylor, H. F. W. (1984). Studies on the chemistry and microstructure of cement pastes, *Proc. Br. Ceram. Soc.*, 35, 65–82.

Taylor, H. F. W. and Newbury, D. E. (1984a). An electron microprobe study of a mature cement paste, *Cem. Concr. Res.*, 14(4), 565–73.

Taylor, H. F. W. and Newbury, D. E. (1984b). Calcium hydroxide distribution and calcium silicate hydrate composition in tricalcium silicate and β-dicalcium silicate pastes, *Cem. Concr. Res.*, 14(1), 93–8.

Tenoutasse, N. and Marion, A. M. (1984). Influence of fly ash on the structure of OPC and pure calcium silicates, *Proc. Br. Ceram. Soc.*, 35, 359–74.

1985

Danilatos, G. D. (1985). Design and construction of an atmospheric or environmental SEM (part 3), *Scanning*, 7, 26–42.

Harrisson, A. M., Taylor, H. F. W. and Winter, N. B. (1985). Electron-optical analyses of the phases in a Portland cement clinker, with some observations on the calculation of quantitative phase composition, *Cem. Concr. Res.*, 15, 775–80.

Hawkes, P. W. (Ed), (1985). The beginnings of electron microscopy, *Advances in Electronics and Electron Physics, Suppl 16*, Academic Press Inc. (London) Ltd.

Monteiro, P. J. M., Bastacky, S. J. and Hayes, T. L. (1985). Low-temperature scanning electron microscope analysis of the Portland cement paste early hydration, *Cem. Concr. Res.*, 15(4), 687–93.

Reimer, L. (1985). Scanning electron microscopy, Springer-Verlag, Berlin.

Taylor, H. F. W., Mohan, K. and Moir, G. K. (1985a). Analytical study of pure and extended cement pastes: I, Pure Portland cement pastes, *J. Am. Ceram. Soc.*, 68(12), 680–5.

Taylor, H. F. W., Mohan, K. and Moir, G. K. (1985b). Analytical study of pure and extended cement pastes: II, Fly ash- and slag-cement pastes, *J. Am. Ceram. Soc.*, 68(12), 685–90.

Taylor, H. F. W. and Newbury, D. E. (1985). Reply to discussion by S. Chatterji on the paper 'Calcium hydroxide distribution and calcium silicate hydrate composition in tricalcium silicate and b-dicalcium silicate,' by H. F. W. Taylor and D. E. Newbury, *Cem. Concr. Res.*, 15(4), 741–3.

Wei, F., Grutzeck, M. W. and Roy, D. M. (1985). The retarding effects of fly ash upon the hydration of cement pastes: the first 24 hours, *Cem. Concr. Res.*, 15(1), 174–84.

1986

8th ICCC = Proc. 8th Int. Cong. on the Chem. Cem., Rio de Janeiro, 1986, Abla Gréfica e Editora Ltda, Rio de Janeiro (1986).

Cabrera, J. G. and Plowman, C. (1986). The influence of pulverised fuel ash on the early and long term strength of concrete, *8th ICCC*, 4, 84–92.

Diamond, S. (1986). The microstructures of cement paste in concrete, *8th ICCC*, 1, 122–47.

Groves, G. W., Le Sueur, P. J. and Sinclair, W. (1986). Transmission electron microscopy and microanalytical studies of ion-beam-thinned sections of tricalcium silicate paste, *J. Am. Ceram. Soc.*, 69(4), 353–6.

Grudemo, Å (1986). The crystal structures of cement hydration – a review and a new gel structure model, *CBI Report ra 1:86, Swedish Cem. Concr. Res. Inst.*, 3–16.

Hara, N., Inoue, N., Noma, H. and Hasegawa, T. (1986). Formation and characterization of jennite, *8th ICCC*, 3, 160–6.

Harrisson, A. M., Winter, N. B. and Taylor, H. F. W. (1986). An examination of some pure and composite Portland cement pastes using scanning electron microscopy with X-ray analytical capability, *8th ICCC*, 4, 170–5.

Henderson, E. and Bailey, J. E. (1986). Structure and morphological aspects of calcium silicate hydrates, *8th ICCC*, 3, 376–81.

Jennings, H. M. and Parrott, L. J. (1986a). Microstructural analysis of hardened alite pastes. Part 1: Porosity, *J. Mater. Sci.*, 21, 4048–52.

Jennings, H. M. and Parrott, L. J. (1986b). Microstructural analysis of hydrated alite pastes. Part 2: Microscopy and reaction products, *J. Mater. Sci.*, 21, 4053–9.

Mitsuda, T., Kobayakawa, S. and Toraya, H. (1986). Characterization of hydrothermally formed C-S-H, *8th ICCC*, 3, 173–8.

Rayment, D. L. (1986). The electron microprobe analysis of the C-S-H phases in a 136 year old cement paste, *Cem. Concr. Res.*, 16(3), 341–4.

Scrivener, K. L. and Pratt, P. L. (1986). A preliminary study of the microstructure of the cement/sand bond in mortars, *8th ICCC*, 3, 466–71.

Taylor, H. F. W. (1986). Proposed structure for calcium silicate hydrate gel, *J. Am. Ceram. Soc.*, 69(6), 464–7.

Uchikawa, H., Uchida, S. and Hanehara, S. (1986). Effect of character of glass phase in blending components on their reactivity in calcium hydroxide mixture, *8th ICCC*, 4, 245–50.

Uchikawa, H., Uchida, S. and Ogawa, K. (1986). Influence of character of blending component on the diffusion of Na and Cl ions in hardened blended cement paste, *8th ICCC*, 4, 251–6.

1987

Groves, G. W. (1987). TEM studies of cement hydration, *Mater. Res. Soc. Symp. Proc.*, 85, 3–12.

Harrisson, A. M., Winter, N. B. and Taylor, H. F. W. (1987a). Microstructure and microchemistry of slag cement pastes, *Mater. Res. Soc. Symp. Proc.*, 85, 213–22.

Harrisson, A. M., Winter, N. B. and Taylor, H. F. W. (1987b). X-ray microanalysis of microporous materials, *J. Mater. Sci. Lett.*, 6, 1339–40.

Parrott, L. J. (1987). Measurement and modelling of porosity in drying cement paste, *Mater. Res. Soc. Symp. Proc.*, 85, 91–104.

Pitts, J. (1987). The role of petrography in the investigation of concrete and its constituents, *Concrete (London)*, 21(7), 5–7.

Poon, C. S., Wassell, L. E. and Groves, G. W. (1987). High strength refractory aluminous cement, *Br. Ceram. Soc. Trans. J.*, 86, 58–62.

Rodger, S. A., Groves, G. W., Clayden, N. J. and Dobson, C. M. (1987). A study of tricalcium silicate hydration from very early to very late stages, *Mater. Res. Soc. Symp. Proc.*, 85, 13–20.

Scrivener, K. L. (1987). The microstructure of anhydrous cement and its effect on hydration, *Mater. Res. Soc. Symp. Proc.*, 85, 39–46.

Scrivener, K. L., Patel, H. H., Pratt, P. L. and Parrott, L. J. (1987). Analysis of phases in cement paste using backscattered electron images, methanol adsorption and thermogravimetric analysis, *Mater. Res. Soc. Symp. Proc.*, 85, 67–76.

1988

Dobson, C. M., Goberdahn, G. C. D., Ramsay, J. D. F. and Rodger, S. A. (1988). ^{29}Si MAS NMR study of the hydration of tricalcium silicate in the presence of finely divided silica, *J. Mater. Sci.*, 23, 4108–14.

Henderson, E. and Bailey, J. E. (1988). Sheet-like structure of calcium silicate hydrates, *J. Mater. Sci.*, 23, 501–8.

Komarneni, S., Breval, E., Roy, D.M. and Roy, R. (1988). Reactions of some calcium silicates with metal cations, *Cem. Concr. Res.*, 18(2), 204–20.

Poon, C. S. and Groves, G. W. (1988). TEM observations of a high alumina cement paste, *J. Mater. Sci. Lett.*, 7, 243–4.

Rodger, S. A. and Groves, G. W. (1988). The microstructure of tricalcium silicate/pulverized fuel ash blended cement pastes, *Adv. Cem. Res.*, 1(2), 84–91.

Scrivener, K. L., Bentur, A. and Pratt, P. L. (1988). Quantitative characterization of the transition zone in high strength concretes, *Adv. Cem. Res.*, 1(4), 230–7.

Uchikawa, H. (1988). Similarities and discrepancies of hardened cement paste, mortar and concrete from the standpoints of composition and structure, *Eng. Found. Conf. on 'Adv. Cem. Manufacture and Use,'* Potosi, USA.

1989

Barker, A. P. (1989). An electron optical examination of zoning in blastfurnace slag hydrates: Part I. Slag cement pastes at early ages, *Adv. Cem. Res.*, 2(8), 171–79.

Beedle, S. S., Groves, G. W. and Rodger, S. A. (1989). The effect of fine pozzolanic and other particles on the hydration of C₃S, *Adv. Cem. Res.*, 2(5), 3–8.

Feng, Q. L., Lachowski, E. E. and Glasser, F. P. (1989). Densification and migration of ions in blast furnace slag-Portland cement pastes, *Mater. Res. Soc. Symp. Proc.*, 137, 419–28.

Groves, G. W. and Rodger, S. A. (1989). The hydration of C₃S and ordinary Portland cement with relatively large additions of microsilica, *Adv. Cem. Res.*, 2(8), 135–40.

Ivey, D. G. and Neuwirth, M. (1989). A technique for preparing TEM specimens from cementitious materials, *Cem. Concr. Res.*, 19(4), 642–8.

Melzer, R. and Eberhard, E. (1989). Phase identification during early and middle hydration of tricalcium silicate (Ca₃SiO₅), *Cem. Concr. Res.*, 19(3), 411–22.

Richardson, I. G., Rodger, S. A. and Groves, G. W. (1989). The porosity and pore structure of hydrated cement pastes as revealed by electron microscopy techniques, *Mater. Res. Soc. Symp. Proc.*, 137, 313–18.

Richardson, I. G., Wilding, C. R. and Dickson, M. J. (1989). The hydration of blastfurnace slag cements, *Adv. Cem. Res.*, 2(8), 147–57.

Rodger, S. A. (1989). Unpublished results reported in Richardson and Groves (1992b).

Rodger, S. A. and Groves, G. W. (1989). Electron microscopy of ordinary Portland cement and ordinary Portland cement-pulverised fuel ash blended pastes, *J. Am. Ceram. Soc.*, 72(6), 1037–9.

Scrivener, K. L. (1989a). The microstructure of concrete, *Materials Science of Concrete I*, J. P. Skalny (Ed.), The American Ceramic Society Inc., 127–63.

Scrivener, K. L. (1989b). The use of backscattered electron microscopy and image analysis to study the porosity of cement paste, *Mater. Res. Soc. Symp. Proc.*, 137, 129–40.

Struble, L. and Stutzman, P. (1989). Epoxy impregnation of hardened cement for microstructural characterization, *J. Mater. Sci. Lett.*, 8, 632–4.

Zhang, M. H. and Gjørv, O. E., (1989). Backscattered electron imaging studies on the interfacial zone between high-strength lightweight aggregate and cement paste, *Adv. Cem. Res.*, 2(8), 141–6.

Zhang, X. and Groves, G. W. (1989). Microstructural studies of alkali-silica reaction products, *Mater. Sci. Technol.*, 5, 714–18.

1990

Danilatos, G. D. (1990a). Design and construction of an environmental SEM (part 4), *Scanning*, 12, 23–7.

Danilatos, G. D. (1990b). Mechanisms of detection and imaging in the ESEM, *J. Microsc.* 160, 9–19.

Ewart, F. T., Glasser, F., Groves, G., Jappy, T., McCrohan, R., Moseley, P. T., Rodger, S. and Richardson, I. (1990). Mechanisms of sorption in the near field, AEA Technology Harwell, AERE Report R 13800.

Feng, Q. L. and Glasser, F. P. (1990). Microstructure, mass transport and densification of slag cement pastes, *Mater. Res. Soc. Symp. Proc.*, 178, 57–65.

Groves, G. W. (1990a). A discussion of the paper 'Phase identification during early and middle hydration of tricalcium silicate', by R. Melzer and E. Eberhard, *Cem. Concr. Res.*, 20(2), 315–6.

Groves, G. W. (1990b). Transmission electron microscopy of cements and mortars, *Mater. Forum*, 14, 1–8.

Groves, G. W., Rodway, D. I. and Richardson, I. G. (1990). The carbonation of hardened cement pastes, *Adv. Cem. Res.*, 3(11), 117–25.

Ivey, D. G., Heimann, R. B., Neuwirth, M., Shumborski, S., Conrad, D., Mikula, R. J. and Lam, W. W. (1990). Electron microscopy of heavy metal waste in cement matrices, *J. Mater. Sci.*, 25, 5055–62.

Kjellsen, K. O., Detwiler, R. J. and Gjorv, O. E. (1990). Backscattered electron imaging of cement pastes hydrated at different temperatures, *Cem. Concr. Res.*, 20(2), 308–11.

Mayfield, B. (1990). The quantitative evaluation of the water/cement ratio using fluorescence microscopy, *Mag. Concr. Res.*, 42(150), 45–9.

Poellmann, H., Kuzel, H.-J. and Wenda, R. (1990). Solid solutions of ettringites. Part I: Incorporation of OH⁻ and CO₃²⁻ in 3CaO·Al₂O₃·3CaSO₄·32H₂O, *Cem. Concr. Res.*, 20(6), 941–7.

Richardson, I. G., Groves, G. W. and Wilding, C. R. (1990). Effect of γ-radiation on the microstructure of GGBFS/OPC cement blends, *Mater. Res. Soc. Symp. Proc.*, 176, 31–7.

Richardson, I. G., Rodger, S. A. and Groves, G. W. (1990). The microstructure of GGBFS/OPC hardened cement pastes and some effects of leaching, *Mater. Res. Soc. Symp. Proc.*, 176, 63–74.

Zhang, X., Blackwell, B. Q. and Groves, G. W. (1990). The microstructure of reactive aggregates, *Br. Ceram. Soc. Trans. J.*, 89, 89–92.

Zhang, X. and Groves, G. W. (1990). The beneficial effect of pozzolanas in suppressing the alkali-silica reaction, *Adv. Cem. Res.*, 3(9), 15–21.

1991

Groves, G. W., Brough, A., Richardson, I. G. and Dobson, C. M. (1991). Progressive changes in the structure of hardened C3S cement pastes due to carbonation, *J. Am. Ceram. Soc.*, 74, 2891–6.

Shrivastava, O. P., Komarneni, S. and Breval, E. (1991). Mg²⁺ uptake by synthetic tobermorite and xonotlite, *Cem Concr. Res.*, 21(1), 83–90.

1992

Bergstrom, T. B. and Jennings, H. M. (1992). On the formation of bonds in tricalcium silicate pastes as observed by scanning electron microscopy, *J. Mater. Sci. Lett.* 11, 1620–2.

Diamond, S. and Bonen, D. (1992). A chemical image analysis method for Portland cements, *Mater. Res. Soc. Symp. Proc.*, 245, 291–301.

French, W. J. (1992). Concrete petrography – a review, *Q. J. Eng. Geol.*, 24, 17–48.

Goldstein, J. I., Newbury, D. E., Echlin, P., Joy, D. C., Romig, A. D. Jnr., Lyman, C. E., Fiori, C. and Lifshin, E. (1992). *Scanning Electron Microscopy and X-ray Microanalysis*, Plenum Press, New York. 2nd Edition.

Richardson, I. G. and Groves, G. W. (1992a). Microstructure and microanalysis of hardened cement pastes involving ground granulated blast-furnace slag, *J. Mater. Sci.*, 27, 6204–12.

Richardson, I. G. and Groves, G. W. (1992b). Models for the composition and structure of calcium silicate hydrate (C-S-H) gel in hardened tricalcium silicate pastes, *Cem. Concr. Res.*, 22, 1001–10.

Scrivener, K. L. (1992). The effect of heat treatment on inner product C-S-H, *Cem. Concr. Res.*, 22(6), 1224–6.

Sujata, K. and Jennings, H. M. (1992). Formation of a protective layer during the hydration of cement, *J. Am. Ceram. Soc.*, 75(6), 1669–73.

Zhao, H. and Darwin, D. (1992). Quantitative backscattered electron analysis of cement paste, *Cem. Concr. Res.*, 22(4), 695–706.

1993

Bonen, D. (1993). A discussion of the paper 'Quantitative backscattered electron analysis of cement paste,' by H. Zhao and D. Darwin, *Cem. Concr. Res.*, 23(3), 749–53.

Brydson, R., Richardson, I. G., McComb, D. W. and Groves, G. W. (1993). Parallel electron energy loss spectroscopy study of Al-substituted calcium silicate hydrate (C-S-H) phases present in hardened cement pastes, *Solid State Commun.*, 88(2), 183–7.

Diamond, S. and Bonen, D. (1993). Microstructure of hardened cement paste – a new interpretation, *J. Am. Ceram. Soc.*, 76(12), 2993–9.

Henderson, E. and Bailey, J. E. (1993). The compositional and molecular character of the calcium silicate hydrates formed in the paste hydration of 3CaO·SiO$_2$, *J. Mater. Sci.*, 28, 3681–91.

Kim, Y. J., Kriven, W. M. and Mitsuda, T. (1993). TEM study of synthetic hillebrandite (Ca$_2$SiO$_4$·H$_2$O), *J. Mater. Res.*, 8(11), 2948–53.

Larbi, J. A., (1993). Microstructure of the interfacial transition zone around aggregate particles in concrete, *Heron*, 38(1), 5–69.

Ramachandran, A. R. and Grutzeck, M. W. (1993). Effect of pH on the hydration of tricalcium silicate, *J. Am. Ceram. Soc.*, 76(1), 72–80.

Reed, S. J. B. (1993). Electron microprobe analysis, 2nd Edition, Cambridge University Press.

Richardson, I. G., Brough, A. R., Brydson, R., Groves, G. W. and Dobson, C. M. (1993). The location of aluminium in substituted calcium silicate hydrate (C-S-H) gels as determined by ^{29}Si and ^{27}Al NMR and EELS, *J. Am. Ceram. Soc.*, 26, 2285–8.

Richardson, I. G. and Groves, G. W. (1993). Microstructure and microanalysis of hardened ordinary Portland cement pastes, *J. Mater. Sci.*, 28, 265–77.

Richardson, I. G., Groves, G. W., Brough, A. R. and Dobson, C. M. (1993). The carbonation of OPC and OPC/silica fume hardened cement pastes in air under conditions of fixed humidity, *Adv. Cem. Res.*, 5(18), 81–6.

Richardson, I. G., Hall, C. and Groves, G. W. (1993). TEM study of the composition of the interstitial phase in an oil-well cement clinker, *Adv. Cem. Res.*, 5(17), 15–21.

Zhao, H. and Darwin, D. (1993). A reply to a discussion by D. Bonen of the paper 'Quantitative backscattered electron analysis of cement paste,' by H. Zhao and D. Darwin, *Cem. Concr. Res.*, 23(3), 754–7.

1994

Bonen, D. (1994). Calcium hydroxide deposition in the near interfacial zone in plain concrete, *J. Am. Ceram. Soc.*, 77(1), 193–6.

Bonen, D. and Diamond, S. (1994). Interpretation of compositional patterns found by quantitative energy dispersive X-ray analysis for cement paste constituents, *J. Am. Ceram. Soc.*, 77(7), 1875–82.

Brydson, R., Richardson, I. G. and Groves, G. W. (1994). Determining the local coordination of aluminium in cement using electron energy loss near-edge structure, *Mikrochim. Acta*, 114/115, 221–9.

Hansen, P. L., Brydson, R., McComb, D. W. and Richardson, I. (1994). EELS fingerprint of Al-coordination in silicates, *Micros. Microanal. Microstruct.*, 5(3), 173–82.

Lu, Y. and Joy, D. C. (1994). High-resolution transmission electron microscopy studies of microstructures in blended cement pastes, *Proceedings of the Annual Meeting of the Electron Microscopy Society of America*, 52, 658–9.

Pietersen, H. S. and Bijen, J. M. (1994). Fly ash and slag reactivity in cements – TEM evidence and application of thermodynamic modelling, *Environ. Aspects Constr. Waste Mater.*, 949–60.

Richardson, I. G., Brough, A. R., Groves, G. W. and Dobson, C. M. (1994). The characterization of hardened alkali-activated blast-furnace slag pastes and the nature of the calcium silicate hydrate (C-S-H) phase, *Cem. Concr. Res.*, 24(5), 813–29.

1995

Dai, Y. and Post, J. E. (1995). Crystal structure of hillebrandite: A natural analogue of calcium silicate hydrate (CSH) phases in Portland cement, *Am. Mineralogist*, 80, 841–4.

Elsen, J., Lens, A., Aarre, T., Quenard, D. and Smolej, V. (1995). Determination of the W/C ratio of hardened cement paste and concrete samples on thin sections using automated image analysis techniques, *Cem. Concr. Res.*, 25(4), 827–34.

Enders, M. (1995). Microanalytical characterization (AEM) of glassy spheres and anhydrite from a high-calcium lignite fly ash from Germany, *Cem. Concr. Res.*, 25(6), 1369–77.

Kim, Y. J. and Kriven, W. M. (1995). A transmission electron microscopy study of the decomposition of synthetic hillebrandite ($Ca_2SiO_4 \cdot H_2O$), *J. Mater. Res.*, 10(12), 3084–95.

Meredith, P., Donald, A. M. and Luke, K. (1995). Pre-induction and induction hydration of tricalcium silicate: an environmental scanning electron microscopy study, *J. Mater. Sci.*, 30, 1921–30.

1996

Kjellsen, K. O. and Jennings, H. M. (1996). Observations of microcracking in cement paste upon drying and rewetting by environmental scanning electron microscopy, *Adv. Cem.-Bas. Mater.*, 3, 14–19.

Kjellsen, K. O., Jennings, H. M. and Lagerblad, B. (1996). Evidence of hollow shells in the microstructure of cement paste, *Cem. Concr. Res.*, 26(4), 593–9.

Scrivener, K. L. and Nemati, K. M. (1996). The percolation of pore space in the cement paste/aggregate zone of concrete, *Cem. Concr. Res.*, 26(1), 35–40.

Viehland, D., Li, J-F., Yuan, L–J. and Xu, Z. (1996). Mesostructure of calcium silicate hydrate (C-S-H) gels in Portland cement paste: short-range ordering, nanocrystallinity, and local compositional order, *J. Am. Ceram. Soc.*, 79(7), 1731–44.

Xu, H. and Buseck, P. R. (1996). TEM investigation of the domain structure and superstructure in hillebrandite, $Ca_2SiO_3(OH)_2$, *Am. Mineralogist*, 81, 1371–4.

Xu, Z. and Viehland, D. (1996). Observation of a mesostructure in calcium silicate hydrate gels of Portland cement, *Phys. Rev. Lett.*, 77(5), 952–5.

1997

Bäckström, E. and Hansen, S. (1997). X-ray mapping of interstitial phases in sulphate resisting cement clinker, *Adv. Cem. Res.*, 9(33), 17–23.

Chatterji, S. (1997). Comment on 'Mesostructure of calcium silicate hydrate (C-S-H) gels in Portland cement paste: short-range ordering, nanocrystallinity, and local compositional order,' by Viehland et al., *J. Am. Ceram. Soc.*, 80(11), 2959–60.

Edyvean, E. G. J. and Hammond, C. (1997). The metallurgical work of Henry Clifton Sorby and an annotated catalogue of his extant metallurgical specimens, *J. Histor. Met. Soc.*, 31(2) 54–85.

Jiang, W., Silsbee, M. R., Breval, E. and Roy, D. M. (1997). Alkali activated cementitious materials in chemically aggressive environments, *Proceedings of a Materials Research Society Symposium on Mechanisms of Chemical Degradation of Cement-based Systems, Boston 1995*, K. L. Scrivener, J. F. Young, (Ed.) E &FN Spon. 289–96.

Kjellsen, K. O., Lagerblad, B. and Jennings, H. M. (1997). Hollow-shell formation – an important mode in the hydration of Portland cement, *J. Mater. Sci.*, 32, 2921–7.

Larbi, J. A. and Heijnen, W. M. M. (1997). Determination of the cement content of five samples of hardened concrete by means of optical microscopy, *Heron*, 42(2), 125–38.

Mouret, M., Bascoul, A. and Escadeillas, G. (1997). Study of the degree of hydration of concrete by means of image analysis and chemically bound water, *Adv. Cem.-Bas. Mater.* 6, 109–15.

Richardson, I. G. (1997). The structure of C-S-H in hardened slag cement prastes, *Proc. 10th Int. Cong. Chem. Cem.*, 2, 2ii068 8pp.

Richardson, I. G. and Groves, G. W. (1997). The structure of the calcium silicate hydrate phases present in hardened pastes of white Portland cement/blast-furnace slag blends, *J. Mater. Sci.*, 32, 4793–802.

Smart, S. and Sims, I. (1997). Concrete petrography – A cost-effective technique, *Concrete (London)*, 31(4), 31–4.

Thaulow, N. and Jakobsen, U. H. (1997). The diagnosis of chemical deterioration of concrete by optical microscopy, *Proceedings of a Materials Research Society Symposium on Mechanisms of Chemical Degradation of Cement-based Systems*, Boston, Nov. 1995, K. L. Scrivener, J. F. Young (Ed.), 3–13. E&FN Spon.

Viehland, D., Yuan, L. J., Xu, Z., Cong, X.-D. and Kirkpatrick, R. J. (1997). Structural studies of jennite and 1.4 nm tobermorite: disordered layering along the [100] of jennite, *J. Am. Ceram. Soc.*, 80(12), 3021–8.

Watt, I. M. (1997). The principles and practice of electron microscopy, 2nd edition, Cambridge University Press.

Xu, Z. and Viehland, D. (1997). Reply to 'Comment on "Mesostructure of calcium silicate hydrate (C-S-H) gels in Portland cement paste: short-range ordering, nanocrystallinity, and local compositional order," by S. Chatterji,' *J. Am. Ceram. Soc.*, 80(11), 2961–2.

1998

Kjellsen, K. O., Wallevik, O. H. and Fjällberg, L., (1998). Microstructure and microchemistry of the paste-aggregate interfacial transition zone of high-performance concrete, *Adv. Cem. Res.*, 10(1), 33–40.

Omotoso, O. E., Ivey, D. G. and Mikula, R. (1998). Hexavalent chromium in tricalcium silicate. Part II, Effects of CrVI on the hydration of tricalcium silicate, *J. Mats Sci*, 33, 515–22.

Smith, K. C. A. (1998). Sir Charles William Oatley, O. B. E., *Biographical Memoirs of Fellows of the Roy. Soc. London*, 44, 329–47.

Willis, K. L., Abell, A. B. and Lange, D. A. (1998). Image-based characterization of cement pore structure using Wood's metal intrusion, *Cem. Concr. Res.*, 28(12), 1695–705.

1999

Hodgkinson, E. S. and Hughes, C. R. (1999). The mineralogy and geochemistry of cement/rock reactions: high-resolution studies of experimental and analogue materials, *Geol. Soc. London, Special Publication, 'Chemical Containment of Waste in the Geosphere'*, 157, 195–211.

Kjellsen, K. O. and Helsing Atlassi, E. (1999). Pore structure of cement silica fume systems; presence of hollow shell pores, *Cem. Concr. Res.*, 29, 133–42.

Richardson, I. G. (1999). The nature of C-S-H in hardened cements, *Cem. Concr. Res.*, 29, 1131–47.

2000

Brough, A. R. and Atkinson, A. (2000). Automated identification of the aggregate-paste interfacial transition zone in mortars of silica sand with Portland or alkali-activated slag cement paste, *Cem. Concr. Res.*, 30, 849–54.

Gloter, A., Ingrin, J., Bouchet, D. and Colliex, C. (2000). Composition and orientation dependence of the O K and Fe $L_{2,3}$ EELS fine structures in $Ca_2(Al_xFe_{1-x})_2O_5$, *Phys. Rev. B*, 61(4), 2587–94.

Gloter, A., Ingrin, J., Bouchet, D., Scrivener, K., and Colliex, C. (2000). TEM evidence of perovskite-brownmillerite coexistence in the $Ca(Al_xFe_{1-x})O_{2.5}$ system with minor amounts of titanium and silicon, *Phys. Chem. Miner.*, 27(7), 504–13.

Hadley, D. W., Dolch, W. L. and Diamond, S. (2000). On the occurrence of hollow-shell hydration grains in hydrated cement paste, *Cem. Concr. Res.*, 30, 1–6.

Richardson, I. G. (2000). The nature of the hydration products in hardened cement pastes, *Cem. Concr. Compos.*, 22, 97–113.

Richardson, I. G. and Cabrera, J. G. (2000). The nature of C-S-H in model slag-cements, *Cem. Concr. Compos.*, 22, 259–66.

Zhang, X., Chang, W., Zhang, T. and Ong, C. K. (2000). Nanostructure of calcium silicate hydrate gels in cement paste, *J. Am. Ceram. Soc.*, 83(10), 2600–4.

2001

Goodhew, P. J., Humphreys, J. and Beanland, R. (2001). Electron microscopy and analysis, 3rd Edition. Taylor & Francis.

Index

557

Milton Keynes UK
Ingram Content Group UK Ltd.
UKHW051856071024
449327UK00025B/1987